CULTURE
OF ANIMAL CELLS

CULTURE OF ANIMAL CELLS

A MANUAL OF BASIC TECHNIQUE
Fourth Edition

R. Ian Freshney

CRC Department of Medical Oncology
CRC Beatson Laboratories
University of Glasgow

A JOHN WILEY & SONS, INC., PUBLICATION
New York · Chichester · Weinheim · Brisbane · Singapore · Toronto

For ordering and customer service, call 1-800-CALL-WILEY.

Library of Congress Cataloging-in-Publication Data:

Freshney, R. Ian.
  Culture of animal cells : a manual of basic techniques / R. Ian Freshney.—4th ed.
    p. cm.
  Includes bibliographical references and index.
  ISBN 0-471-34889-9 (alk. paper)
  1. Tissue culture Laboratory manuals. 2. Cell culture Laboratory
manuals. I. Title.
QH585.2.F74 1994
571.6′38—dc21   99-23536

Printed in the United States of America.

10 9 8 7 6

To the late John Paul who first introduced me to the fascinating world of tissue culture through his expert guidance and excellent textbook, Cell and Tissue Culture, which was, for many years, the leading text in the field.

Contents

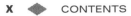

14 Cell Separation, 215

15 Characterization, 229

16 Differentiation, 259

17 Transformation, 269

Figures

Color Plates

Note: Most of these photomicrographs have been taken on an Olympus CK inverted microscope using a $10\times$ objective. Shots referred to as "low-power" in the legends were taken with a $4\times$ objective, and "high-power" shots were taken with a $40\times$ objective. Most of the exceptions to this general rule are mentioned in the appropriate legend.

 1 Human Lung Carcinoma
 2 Mouse Kidney Tubules
 3 Outgrowth from Kidney Tubules
 4 Newborn-rat Kidney
 5 Newly Subcultured Monolayer
 6 Midlog Phase Cells
 7 Early-Plateau Phase Cells
 8 Monolayer, Pretrypsin
 9 Monolayer after Trypsin Removal
10 Monolayer 1 Min after Trypsin Removal
11 Fully Disaggregated Monolayer
12 HeLa Clones
13 NRK Clone
14 Microphotograph of Breast Carcinoma, JUW,
    Cloned on Plastic
15 Microphotograph of Breast Carcinoma on
    Feeder Layer
16 Normal Human Fetal Lung Fibroblasts
17 Confluent Fibroblasts
18 A549 Cells

19 MOG-GP Astrocytoma
20 CHO-K1 Cells
21 Cos-7 Cells
22 HeLa Cells
23 Domes
24 Cytokeratin
25 GFAP in Glial Cells
26 Hemoglobin in Friend Cells
27 Normal Human Glial Cells
28 Differentiated Human Glial Cells
29 Induction of Angiogenesis
30 Mycoplasma
31 Mycoplasma, High Power
32 SCE in A2780 Cells
33 Spheroids
34 Alginate Encapsulation
35 Culture on Filter-well Inserts
36 Microcarriers
37 *In situ* Labeling of Friend Cells

Preface

In preparing this fourth edition of *Culture of Animal Cells* I have retained the emphasis of previous editions and focused on basic techniques with some examples of more specialized cultures and methods. These are presented as detailed step-by-step protocols which should give sufficient information to carry out a procedure without recourse to the prime literature. There is also introductory material to explain the background to a method and information on alternative procedures and applications. Some basic biology is provided but it is assumed that the reader will have a little knowledge of anatomy, histology, biochemistry, and cell and molecular biology. The book is targeted at those with little or no previous experience in tissue culture, including technicians in training, senior undergraduates, graduate students, postdoctoral workers, and clinicians with an interest in laboratory science.

Some chapters have been reorganized since the third edition, for example, the topics of culture vessels, serum-free media, molecular techniques, and scale-up have been revised and are presented in new, separate chapters and an entirely new chapter on problem solving has been added. The molecular techniques that are included are all seen as having direct relevance to cell culture. No attempt has been made to present basic molecular methodology as these are already available [e.g., Sambrook et al., 1989; Ausubel et al., 1996]. Similarly, the chapter on scale-up serves as an interface with biotechnology and provides some background on systems for increasing cell yield, but takes no account of full-scale biopharmaceutical production processes.

Color plates have been included for the first time, selected to supplement, rather than replace, the figures in the text. Most are microphotographs of cells, and have been taken with a 10× objective, unless stated otherwise in the legend. I have not quoted magnifications, as I feel it is more important for the reader to be able to compare the photographs to the images that they might be familiar with on their own microscope, which requires knowledge of the objective used rather than the total magnification, which is difficult to relate to a specimen without a slide micrometer.

I have increased the details of cross references to assist the reader in locating the correct sections of text and have also numbered all the protocols to make them more accessible. Some pieces of narrative text have been converted into protocol format, and there are several new protocols, some contributed by invited authors and others abridged from books in the *Culture of Specialized Cells* series (Wiley-Liss). To deal with the accumulation of more and more literature references, I have replaced some of the older references, used simply as examples, with more recent ones, while retaining some original early references to those who initiated a particular technique. Some additional text books and useful journals are listed after the references. Scientific societies, cell banks and databases are now listed in Suppliers and Other Resources in the Trade Index.

Abbreviations used in the text are listed separately after the preface. Conventions employed throughout are PBSA for Dulbecco's PBS without Ca^{2+} and Mg^{2+} and UPW for ultrapure water, regardless of how it is prepared. Concentrations are given in molarity wherever possible, and actual weights have been omitted from the media tables on the assumption that very few people will attempt to make up their own media, but will, more likely, want to compare constituents, for which molar equivalents are more useful. I am indebted to Life Technologies and Sigma Aldrich for information on the popularity of different types of media, and to the European Collection of Cell Cultures (ECACC) for information on the most frequently requested cell lines, and have ensured that the media and cell lines used most often are listed in the appropriate tables.

Protocols are identified in the text by a vertical line on the left of the text. Reagents that are specific to a particular protocol are detailed in the materials sections of the protocols, the recipes for the common reagents, such as Hanks' BSS or trypsin, are given in the Reagent Appendix at the end of the book, where you will also find details of the sources of equipment and materials in the Trade Index.

As always, I owe a great debt of gratitude to the authors who have contributed protocols, and to others who have advised me in areas where my knowledge is imperfect including Robert Auerbach, Bob Brown, Kenneth Calman, Richard Ham, Rob Hay, Stan Kaye, Nicol Keith, Wally McKeehan, Rona McKie, Stephen Merry, Jane Plumb, Peter Vaughan, Paul Workman, and the late John Paul. I am fortunate in having had the clinical collaboration of David I. Graham, David G. T. Thomas, and the late John Maxwell Anderson. In the early stages of the preparation of this book I also benefitted from discussions with Don Dougall, Peter del Vecchio, Sergey Federoff, Mike Gabridge, Dan Lundin, John Ryan, Jim Smith, and Charity Waymouth. I am eternally grateful to Paul Chapple who first persuaded me that I should write a basic techniques book on tissue culture and who, more recently, suggested the development of this text into a multimedia presentation, now published by Wiley-Liss. Many of the original illustrations were produced by Jane Gillies and Marina LaDuke. Some of the data presented were generated by those who have worked with me over the years including Sheila Brown, Ian Cunningham, Lynn Evans, Margaret Frame, Elaine Hart, Carol McCormick, Alison Mackie, John McLean, Alistair McNab, Diana Morgan, Alison Murray, Irene Osprey, and Natasha Yevdokimova. I am greatly indebted to Fiona Conway, Margaret Jenkins, and Joanne Thomson for secretarial help and to Liz Gordon for help in information retrieval.

I have been fortunate to receive excellent advice and support from the editorial staff of John Wiley. I would also like to acknowledge with sincere gratitude all those who have taken the trouble to write to me or to John Wiley with advice and constructive criticism on previous editions. It is pleasant and satisfying to hear from those who have found the book beneficial, but even more important to hear from those who have found deficiencies, some of which I can attempt to rectify. I can only hope that those of you who use this book retain the same excitement that I feel about the future prospects emerging in this field.

I would like to thank my daughter Gillian and son Norman for all the help they gave me in the preparation of the first edition, many years ago, and for their continued advice and support. Above all, I would like to thank my wife, Mary, for her hours of help in compilation, proof-reading, and many other tasks; without her help and support, the original text would never have been written and I would never have completed this revision by the assigned deadline, nor attained the necessary level of technical accuracy that is the keynote of a good tissue culture manual.

Ian Freshney

Abbreviations

ATCC	American Type Culture Collection		G_2	gap two (of the cell cycle)
bp	base pairs (in DNA)		GLP	good laboratory practice [Jacobs, 1979]
BPE	bovine pituitary extract		H&E	hemalum and eosin (stains)
BrUdR	bromodeoxyuridine		HAT	hypoxanthine, aminopterin, and thymidine
BSA	bovine serum albumin		HBS	HEPES buffered saline
BUdR	bromodeoxyuridine		HBSS	Hanks's balanced salt solution
CAM	chorioallantoic membrane		HC	hydrocortisone
CAM	cell adhesion molecule		hCG	human chorionic gonadotropin
CCD	charge-coupled device		HGPRT	hypoxanthine guanosine phosphoribosyl transferase
CCTV	closed-circuit television			
cDNA	complementary DNA		HITES	hydrocortisone, insulin, transferrin, estradiol, and selenium
CE	cloning efficiency			
CMC	carboxymethylcellulose		HPV	human papilloma virus
CMF	calcium- and magnesium-free saline		HuS	human serum
CMRL	Connaught Medical Research Laboratories		HS	horse serum
DEPC	diethyl pyrocarbonate		HSV	herpes simplex virus
DMEM	Dulbecco's modification of Eagle's medium		HT	hypoxanthine/thymidine
DNA	deoxyribonucleic acid		ITS	insulin, transferrin, selenium
DT	population doubling time		KBM	keratinocyte basal medium
EBSS	Earle's balanced salt solution		kbp	kilobase pairs (in DNA)
EBV	Epstein–Barr virus		KGM	keratinocyte growth medium
ECACC	European Collection of Animal Cell Cultures (now European Collection of Cell Cultures)		LI	labeling index
			M199	Medium 199
			MACs	mammalian artificial chromosomes
ECGF	endothelial cell growth factor		MACS	magnet-activated cell sorting
EGF	epidermal growth factor		MEM	Eagle's Minimal Essential Medium
EM	electron microscope		mRNA	messenger RNA
FBS	fetal bovine serum		MTT	3-(4,5-dimethylthiazol-2yl)2,5-diphenyltetrazolium bromide
FCS	fetal calf serum			
FGF	fibroblast growth factor		NBCS	newborn-calf serum
G_1	gap one (of the cell cycle)		NCI	National Cancer Institute

O.D.	optical density
PA	plasminogen activator
PBS	phosphate-buffered saline
PBSA	phosphate-buffered saline, solution A (Ca^{2+} and Mg^{2+} free)
PBSB	phosphate-buffered saline, solution B (Ca^{2+} and Mg^{2+})
PCA	perchloric acid
PCR	polymerase chain reaction
PDGF	platelet-derived growth factor
PE	plating efficiency
PE	PBSA/EDTA
PEG	polyethylene glycol
PHA	phytohemaglutinin
PMA	phorbol meristate acetate
PVP	polyvinylpyrrolidone
PWM	pikeweed mitogen
RNA	ribonucleic acid

RPMI	Rosewell Park Memorial Institute
RT-PCR	reverse transcriptase PCR
S	DNA synthetic phase of cell cycle
SD	saturation density
SIT	selenium, insulin, transferrin
S-MEM	MEM with low Mg^{2+} and no Ca^{2+}
SSC	sodium citrate/sodium chloride
SV40	simian virus 40
SV40LT	SV40 gene for large T-antigen
TCA	trichloracetic acid
T_D	population doubling time
TEB	Tris/EDTA buffer
TGF	transforming growth factor
TK	thymidine kinase
UPW	ultrapure water
VEGF	vascular endothelial growth factor
YACs	yeast artificial chromosomes

CHAPTER 1

Introduction

BACKGROUND

Tissue culture was first devised at the beginning of this century [Harrison, 1907; Carrel, 1912] as a method for studying the behavior of animal cells free of systemic variations that might arise in the animal both during normal homeostasis and under the stress of an experiment. As the name implies, the technique was elaborated first with undisaggregated fragments of tissue, and growth was restricted to the migration of cells from the tissue fragment, with occasional mitoses in the outgrowth. Since culturing cells from such primary explants of tissue dominated the field for more than 50 years, it is not surprising that the name "tissue culture" has remained in spite of the fact that most of the explosive expansion in this area in the second half of the 20th century utilized dispersed cell cultures.

Throughout this book, the term *tissue culture* is used as a generic term to include organ culture and cell culture. The term *organ culture* will always imply a three-dimensional culture of undisaggregated tissue retaining some or all of the histological features of the tissue *in vivo*. *Cell culture* refers to a culture derived from dispersed cells taken from original tissue, from a primary culture, or from a cell line or cell strain by enzymatic, mechanical, or chemical disaggregation. The term *histotypic culture* implies that cells have been reaggregated to re-create a three-dimensional tissue-like structure, e.g., by cultivation at high density in a filter well, perfusion and overgrowth of a monolayer in a flask or dish, reaggregation in suspension over agar or in real or simulated zero gravity, or infiltration

of a three-dimensional matrix such as collagen gel. *Organotypic* implies the same procedures but recombining cells of different lineages, e.g., epidermal keratinocytes in combined culture with dermal fibroblasts.

Harrison [1907] chose the frog as his source of tissue, presumably because it was a cold-blooded animal, and consequently, incubation was not required. Furthermore, since tissue regeneration is more common in lower vertebrates, he perhaps felt that growth was more likely to occur than with mammalian tissue. Although his technique may have sparked off a new wave of interest in the cultivation of tissue *in vitro*, few later workers were to follow his example in the selection of species. The stimulus from medical science carried future interest into warm-blooded animals, in which both normal development and pathological development are closer to that found in humans. The accessibility of different tissues, many of which grew well in culture, made the embryonated hen's egg a favorite choice; but the development of experimental animal husbandry, particularly with genetically pure strains of rodents, brought mammals to the forefront as the favorite material. While chick embryo tissue could provide a diversity of cell types in primary culture, rodent tissue had the advantage of producing continuous cell lines [Earle et al., 1943] and a considerable repertoire of transplantable tumors. The development of transgenic mouse technology [Beddington, 1992; Peat et al., 1992], together with the well-established genetic background of the mouse, has added further impetus to the selection of this animal as a favorite species.

The demonstration that human tumors could also give rise to continuous cell lines [e.g., HeLa; Gey et al., 1952] encouraged interest in human tissue, helped later by the classical studies of Leonard Hayflick on the finite life span of cells in culture [Hayflick & Moorhead, 1961] and the requirement of virologists and molecular geneticists to work with human material. The cultivation of human cells received a further stimulus when a number of different serum-free selective media were developed for specific cell types, such as epidermal keratinocytes, bronchial epithelium, and vascular endothelium. (See Selective Media in Chapter 9.) These formulations are now available commercially although the cost remains high relative to the cost of regular media.

For many years, the lower vertebrates and the invertebrates were largely ignored, although unique aspects of their development (tissue regeneration in amphibia, metamorphosis in insects) make them attractive systems for the study of the molecular basis of development. More recently, the needs of agriculture and pest control have encouraged toxicity and virological studies in insects, and developments in gene technology have suggested that insect cell lines with baculovirus and other vectors may be useful producer cell lines because of the possibility of inserting larger genomic sequences in the viral DNA and a reduced risk of propagating human pathogenic viruses. Furthermore, the economic importance of fish farming and the role of oceanic pollution have stimulated more studies of normal development and pathogenesis in fish. Procedures for handling nonmammalian cells have tended to follow those developed for mammalian cell culture, although a limited number of specialized media are now commercially available for fish and insect cells. (See Insect Cells, Fish Cells in Chapter 26.)

The types of investigation that lend themselves particularly to tissue culture are summarized in Fig. 1.1: (1) intracellular activity, e.g., the replication and transcription of deoxyribonucleic acid (DNA), protein synthesis, energy metabolism, and drug metabolism; (2) intracellular flux, e.g., RNA, the translocation of hormone receptor complexes and resultant signal transduction processes, and membrane trafficking; (3) environmental interaction, e.g., nutrition, infection, cytotoxicity, carcinogenesis, drug action, and ligand receptor interactions; (4) cell–cell interaction, e.g., morphogenesis, paracrine control, cell proliferation kinetics, metabolic cooperation, cell adhesion and motility, matrix interaction, organotypic models for medical prostheses and invasion; (5) genetics, including genome analysis in normal and pathological conditions, genetic manipulation, transformation, and immortalization; and (6) cell products and secretion, biotechnology, bioreactor design, product harvesting, and downstream processing.

The development of cell culture owed much to the needs of two major branches of medical research: the production of antiviral vaccines and the understanding of neoplasia. The standardization of conditions and cell lines for the production and assay of viruses undoubtedly provided much impetus to the development of modern tissue culture technology, particularly the production of large numbers of cells suitable for biochemical analysis. This and other technical improvements made possible by the commercial supply of reliable media and sera and by the greater control of contamination with antibiotics and clean-air equipment, have made tissue culture accessible to a wide range of interests.

An additional force of increasing weight from public opinion has been the expression of concern by many animal-rights groups over the unnecessary use

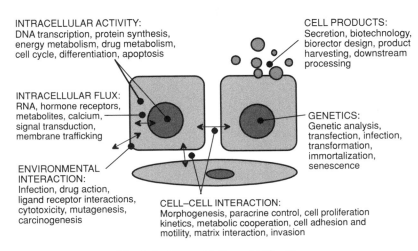

INTRACELLULAR ACTIVITY:
DNA transcription, protein synthesis, energy metabolism, drug metabolism, cell cycle, differentiation, apoptosis

CELL PRODUCTS:
Secretion, biotechnology, biorector design, product harvesting, downstream processing

INTRACELLULAR FLUX:
RNA, hormone receptors, metabolites, calcium, signal transduction, membrane trafficking

GENETICS:
Genetic analysis, transfection, infection, transformation, immortalization, senescence

ENVIRONMENTAL INTERACTION:
Infection, drug action, ligand receptor interactions, cytotoxicity, mutagenesis, carcinogenesis

CELL–CELL INTERACTION:
Morphogenesis, paracrine control, cell proliferation kinetics, metabolic cooperation, cell adhesion and motility, matrix interaction, invasion

Fig. 1.1. Tissue Culture Applications.

of experimental animals. While most accept the idea that some requirement for animals will continue for preclinical trials of new pharmaceuticals, there is widespread concern that extensive use of animals for cosmetics development and similar activities may not be morally justifiable. Hence, there is an ever-increasing lobby for more *in vitro* assays, the adoption of which, however, still requires their proper validation and general acceptance. While this seemed a distant prospect some years ago, the introduction of more sensitive and more readily performed *in vitro* assays, together with a very real prospect of assaying for inflammation *in vitro*, has promoted an unprecedented expansion in *in vitro* testing. (See Chapter 21.)

In addition to cancer research and virology, other areas of research have come to depend heavily on tissue culture techniques. The introduction of cell fusion techniques (See Somatic Cell Fusion in Chapter 27) and genetic manipulation [Maniatis et al., 1978; Sambrook et al., 1989; Ausubel et al., 1996] established somatic cell genetics as a major component in the genetic analysis of higher animals, including humans. A wide range of techniques for genetic recombination now includes DNA transfer [Ravid & Freshney, 1998], monochromsomal transfer [Newbold & Cuthbert, 1998], and nuclear transfer [Kono, 1997], which have been added to somatic hybridization as tools for genetic analysis and gene manipulation. DNA transfer itself has spawned many techniques for the transfer of DNA into cultured cells, including calcium phosphate coprecipitation, lipofection, electroporation, and retroviral infection. (See DNA Transfer in Chapter 27.)

In particular, human genetics has progressed under the stimulus of the Human Genome Project [Kornberg et al., 1995], and the data generated therefrom have recently made feasible the introduction of multigene array expression analysis [Iyer et al., 1999].

Tissue culture has contributed greatly, via the monoclonal antibody technique, to the study of immunology, already dependent on cell culture for assay techniques and the production of hematopoietic cell lines. The insight into the mechanism of action of antibodies and the reciprocal information that this provided about the structure of the epitope, derived from monoclonal antibody techniques [Kohler & Milstein, 1975], was, like the technique of cell fusion itself, a prologue to a whole new field of studies in genetic manipulation. This field has supplied much basic information on the control of gene transcription, and a vast new technology and a multibillion-dollar industry have grown up from the ability to insert exploitable genes into prokaryotic cells. Cell products such as human growth hormone, insulin, and interferon are now produced routinely by transfected prokaryotic and eukaryotic cells, although the absence of posttranscrip-

tional modifications, such as glycosylation, in bacteria suggests that mammalian cells may provide more suitable vehicles [Grampp et al., 1992], particularly in light of developments in immortalization technology. (See Immortalization in Chapter 17.)

Other areas of major interest include the study of cell interactions and intracellular control mechanisms in cell differentiation and development [Jessell and Melton, 1992; Ohmichi et al., 1998; Balkovetz & Lipschutz, 1999] and attempts to analyze nervous function [Richard et al., 1998; Dunn et al., 1998; Haynes, 1999]. Progress in neurological research has not had the benefit, however, of working with propagated cell lines from normal brain or nervous tissue, as the propagation of neurons *in vitro* has not been possible, until now, without resorting to the use of transformed cells. (See Chapter 17.) However, recent developments with human embryonal stem cell cultures [Thomson et al., 1998; Rathjen et al., 1998; Wolf et al., 1998] suggest that this approach may provide replicating cultures that will differentiate into neurons.

Tissue culture technology has also been adopted into many routine applications in medicine and industry. Chromosomal analysis of cells derived from the womb by amniocentesis (see Culture of Amniocytes in Chapter 26) can reveal genetic disorders in the unborn child, the quality of drinking water can be determined, and the toxic effects of pharmaceutical compounds and potential environmental pollutants can be measured in colony-forming and other *in vitro* assays (see Chapter 21).

Further developments in the application of tissue culture to medical problems have followed from the demonstration that cultures of epidermal cells form functionally differentiated sheets [Green et al., 1979] and endothelial cells may form capillaries [Folkman & Haudenschild, 1980], offering possibilities in homografting and reconstructive surgery using an individual's own cells [Tuszynski et al., 1996; Gustafson et al., 1998; Limat et al., 1998], particularly for severe burns [Gobet et al., 1997; Wright et al., 1998].

With the ability to transfect normal genes into genetically deficient cells, it has become possible to graft such "corrected" cells back into the patient. Transfected cultures of rat bronchial epithelium carrying the *β-gal* reporter gene have been shown to become incorporated into the rat's bronchial lining when they are introduced as an aerosol into the respiratory tract [Rosenfeld et al., 1992]. Similarly, cultured satellite cells have been shown to be incorporated into wounded rat skeletal muscle, with nuclei from grafted cells appearing in mature, syncytial myotubes [Morgan et al., 1992].

The prospects for implanting normal cells from adult or fetal tissue-matched donors or implanting genetically reconstituted cells from the same patient are now very real. The technical barriers are steadily being overcome, bringing the ethical questions to the fore. The technical feasibility of implanting normal fetal neurons into patients with Parkinson's disease has been demonstrated; society must now decide to what extent fetal material may be used for this purpose. Where a patient's own cells can be grown and subjected to genetic reconstitution by transfection of the normal gene—e.g., transfecting the normal insulin gene into β-islet cells cultured from diabetics, or even transfecting other cell types, such as skeletal muscle progenitors [Morgan et al., 1992]—it would allow the cells to be incorporated into a low-turnover compartment and, potentially, give a long-lasting physiological benefit. The ethics of this type of approach seem less contentious.

ADVANTAGES OF TISSUE CULTURE

Control of the Environment

The two major advantages of tissue culture (Table 1.1) are control of the physiochemical environment (pH, temperature, osmotic pressure, and O_2 and CO_2 tension), which may be controlled very precisely, and the physiological conditions, which may be kept relatively constant, but cannot always be defined. Most cell lines still require supplementation of the medium with serum or other poorly defined constituents. These supplements are prone to batch variation and contain undefined elements such as hormones and other regulatory substances. The identification of some of the essential components of serum (see Table 8.5), together with a better understanding of factors regulating cell proliferation (see Table 9.3), has made the replacement of serum with defined constituents feasible (see Chapter 9). As laboratories seek to express the

normal phenotypic properties of cells *in vitro*, the role of the extracellular matrix becomes increasingly important. Currently, that role is similar to the use of serum—that is, the matrix is often necessary, but not always precisely defined, yet it can be regulated and, as cloned matrix constituents become available, may still be fully defined.

Characterization and Homogeneity of Sample

Tissue samples are invariably heterogeneous. Replicates—even from one tissue—vary in their constituent cell types. After one or two passages, cultured cell lines assume a homogeneous (or at least uniform) constitution, as the cells are randomly mixed at each transfer and the selective pressure of the culture conditions tends to produce a homogeneous culture of the most vigorous cell type. Hence, at each subculture, replicate samples are identical to each other, and the characteristics of the line may be perpetuated over several generations, or even indefinitely if the cell line is stored in liquid nitrogen. Since experimental replicates are virtually identical, the need for statistical analysis of variance is reduced.

The availability of stringent tests for cell line identity (Chapter 15) and contamination (Chapter 18) means that preserved stocks may be validated for future research and commercial use.

Economy, Scale, and Mechanization

Cultures may be exposed directly to a reagent at a lower, and defined, concentration and with direct access to the cell. Consequently, less reagent is required than for injection *in vivo*, where 90% is lost by excretion and distribution to tissues other than those under study. Screening tests with many variables and replicates are cheaper, and the legal, moral, and ethical

TABLE 1.1. Advantages of Tissue Culture

Category	Advantages
Physico-chemical environment	Control of pH, temperature, osmolarity, dissolved gases
Physiological conditions	Control of hormone and nutrient concentrations
Microenvironment	Regulation of matrix, cell–cell interaction, gaseous diffusion
Cell line homogeneity	Availability of selective media, cloning
Characterization	Cytology and immunostaining are easily performed
Preservation	Can be stored in liquid nitrogen
Validation & accreditation	Origin, history, purity can be recorded
Replicates and variability	Quantitation is easy
Reagent saving	Reduced volumes, direct access, lower cost
Control of C × T	Ability to define dose, concentration, and time
Mechanization	Available with microtitration and robotics
Reduction of animal use	Cytotoxicity and screening of pharmaceutics, cosmetics, etc.

questions of animal experimentation are avoided. New developments in multiwell plates and robotics also have introduced significant economies in time and scale.

In Vivo Modeling

Perfusion techniques allow the delivery of specific experimental compounds to be regulated in concentration (C), duration of exposure (T) (see Table 1.1), and metabolic state. The development of histotypic and organotypic models also increases the accuracy of *in vivo* modeling.

LIMITATIONS

Expertise

Culture techniques must be carried out under strict aseptic conditions, because animal cells grow much less rapidly than many of the common contaminants, such as bacteria, molds, and yeasts. Furthermore, unlike microorganisms, cells from multicellular animals do not normally exist in isolation and, consequently, are not able to sustain an independent existence without the provision of a complex environment simulating blood plasma or interstitial fluid. These conditions imply a level of skill and understanding on the part of the operator in order to appreciate the requirements of the system and to diagnose problems as they arise (Table 1.2). Also, care must be taken to avoid the recurrent problem of cross contamination and to authenticate stocks. Hence, tissue culture should not be undertaken casually to run one or two experiments.

Quantity

A major limitation of cell culture is the expenditure of effort and materials that goes into the production of relatively little tissue. A realistic maximum per batch for most small laboratories (with two or three people doing tissue culture) might be 1–10 g of cells. With a little more effort and the facilities of a larger laboratory, 10–100 g is possible; above 100 g implies industrial pilot-plant scale, a level that is beyond the reach of most laboratories, but is not impossible if special facilities are provided, when kilogram quantities can be generated.

The cost of producing cells in culture is about 10 times that of using animal tissue. Consequently, if large amounts of tissue (>10 g) are required, the reasons for providing them by culture must be very compelling. For smaller amounts of tissue (∼10 g), the costs are more readily absorbed into routine expenditure, but it is always worth considering whether assays or preparative procedures can be scaled down. Semimicro- or microscale assays can often be quicker, due to reduced manipulation times, volumes, centrifuge times, etc., and are frequently more readily automated. (See Replicate Sampling in Chapter 20; Microtitration Assays in Chapter 21.)

Dedifferentiation and Selection

When the first major advances in cell line propagation were achieved in the 1950s, many workers observed the loss of the phenotypic characteristics typical of the tissue from which the cells had been isolated. This effect was blamed on *dedifferentiation*, a process assumed to be the reversal of differentiation, but was later shown to be largely due to the overgrowth of undifferentiated cells of the same or a different lineage. The development of serum-free selective media (see Selective Media in Chapter 9) has now made the isolation of specific lineages quite possible, and it can be seen that, under the right conditions, many of the differentiated properties of these cells may be restored. (See Induction of Differentiation in Chapter 16.)

Origin of Cells

If differentiated properties are lost, for whatever reason, it is difficult to relate the cultured cells to functional cells in the tissue from which they were derived. Stable markers are required for characterization of the cells (see Chapter 15); in addition, the culture may need to be modified so that these markers are ex-

TABLE 1.2. Limitations of Tissue Culture

Category	Examples
Necessary expertise	Handling
	Chemical contamination
	Microbial contamination
	Cross contamination
Environmental control	Workplace
	Incubation, pH control
	Containment and disposal of biohazards
Quantity and cost	Capital equipment
	Consumables
	Medium, serum, plastics
Genetic instability	Heterogeneity, variability
Phenotypic instability	Dedifferentiation
	Adaptation
	Selection
Identification of cell type	Expression of markers
	Histology, cytology
	Geometry and microenvironment

pressed. (See Maintenance of Differentiation in Chapter 2; Induction of Differentiation in Chapter 16.)

Instability

Instability is a major problem with many continuous cell lines, resulting from their unstable aneuploid chromosomal constitution. Even with short-term cultures of untransformed cells, heterogeneity in growth rate and the capacity to differentiate within the population can produce variability from one passage to the next. (See Genetic Instability in Chapter 17.)

MAJOR DIFFERENCES *IN VITRO*

Many of the differences in cell behavior between cultured cells and their counterparts *in vivo* stem from the dissociation of cells from a three-dimensional geometry and their propagation on a two-dimensional substrate. Specific cell interactions characteristic of the histology of the tissue are lost, and, as the cells spread out, become mobile, and, in many cases, start to proliferate, so the growth fraction of the cell population increases. When a cell line forms, it may represent only one or two cell types, and many heterotypic cell–cell interactions are lost.

The culture environment also lacks the several systemic components involved in homeostatic regulation *in vivo*, principally those of the nervous and endocrine systems. Without this control, cellular metabolism may be more constant *in vitro* than *in vivo*, but may not be truly representative of the tissue from which the cells were derived. Recognition of this fact has led to the inclusion of a number of different hormones in culture media (see Hormones, Growth Factors in Chapter 9), and it seems likely that this trend will continue.

Energy metabolism *in vitro* occurs largely by glycolysis, and although the citric acid cycle is still functional, it plays a lesser role.

It is not difficult to find many more differences between the environmental conditions of a cell *in vitro* and *in vivo* (see *In Vitro* Limitations in Chapter 21), and this disparity has often led to tissue culture being regarded in a rather skeptical light. Still, although the existence of such differences cannot be denied, many specialized functions are expressed in culture, and as long as the limits of the model are appreciated, tissue culture can become a very valuable tool.

TYPES OF TISSUE CULTURE

There are three main methods of initiating a culture [Schaeffer, 1990; see Glossary, Fig. 1.2, and Table 1.3]:

(1) *Organ culture* implies that the architecture characteristic of the tissue *in vivo* is retained, at least in part, in the culture. (See Organ Culture in Chapter 24.) Toward this end, the tissue is cultured at the liquid–gas interface (on a raft, grid, or gel), which favors the retention of a spherical or three-dimensional shape. (2) In *primary explant culture*, a fragment of tissue is placed at a glass (or plastic)–liquid interface, where, following attachment, migration is promoted in the plane of the solid substrate. (See Primary Explant in Chapter 11.) (3) *Cell culture* implies that the tissue, or outgrowth from the primary explant, is dispersed (mechanically or enzymatically) into a cell suspension, which may then be cultured as an adherent monolayer on a solid substrate or as a suspension in the culture medium. (See Primary Culture in Chapter 11; Subculture in Chapter 12.)

Because of the retention of cell interactions found in the tissue from which the culture was derived, organ cultures tend to retain the differentiated properties of that tissue. They do not grow rapidly (cell proliferation is limited to the periphery of the explant and is restricted mainly to embryonic tissue) and hence cannot be propagated; each experiment requires fresh explantations, which implies greater effort and poorer reproducibility of the sample than is achieved with cell culture. Quantitation is, therefore, more difficult, and the amount of material that may be cultured is limited by the dimensions of the explant (~ 1 mm^3) and the effort required for dissection and setting up the culture. However, organ cultures do retain specific histological interactions without which it may be difficult to reproduce the characteristics of the tissue.

Cell cultures may be derived from primary explants or dispersed cell suspensions. Because cell proliferation is often found in such cultures, the propagation of cell lines becomes feasible. A monolayer or cell suspension with a significant growth fraction (see Growth Fraction in Chapter 20) may be dispersed by enzymatic treatment or simple dilution and reseeded, or subcultured, into fresh vessels (Table 1.4; see also Subculture in Chapter 12). This constitutes a *passage*, and the daughter cultures so formed are the beginnings of a *cell line*.

The formation of a cell line from a primary culture implies (1) an increase in the total number of cells over several generations and (2) the ultimate predominance of cells or cell lineages with the capacity for high growth, resulting in (3) a degree of uniformity in the cell population (Table 1.4). The line may be characterized, and the characteristics will apply for most of its finite life span. The derivation of *continuous* (or "established," as they were once known) cell lines usually implies a phenotypic change, or *transformation*. (See Evolution of Cell Lines in Chapter 2; Chapter 17.)

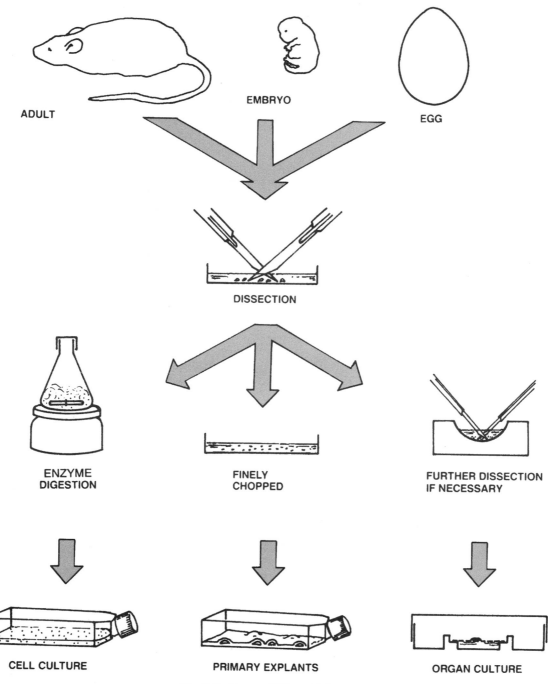

ADULT   EMBRYO   EGG

DISSECTION

ENZYME
DIGESTION

FINELY
CHOPPED

FURTHER DISSECTION
IF NECESSARY

CELL CULTURE   PRIMARY EXPLANTS   ORGAN CULTURE

Fig. 1.2. Types of Tissue Culture.

When cells are selected from a culture, by cloning or by some other method, the subline is known as a *cell strain*. A detailed characterization is then implied. Cell lines or cell strains may be propagated as an adherent monolayer or in suspension. *Monolayer* culture signifies that, given the opportunity, the cells will attach to the substrate and that normally the cells will be propagated in this mode. *Anchorage dependence* means that attachment to (and usually, some degree of spreading onto) the substrate is a prerequisite for cell proliferation. Monolayer culture is the mode of culture common to most normal cells, with the exception of hematopoietic cells. *Suspension* cultures are derived from cells that can survive and proliferate without attachment (*anchorage-independent*); this ability is restricted to hematopoietic cells, transformed cell lines, and cells from malignant tumors. It can be shown, however, that a small proportion of cells that are capable of proliferation in suspension exists in many normal tissues. (See Anchorage Independence in

TABLE 1.3. Properties of Different Types of Culture

Category	Organ Culture	Explant	Cell Culture
Source	Embryonic organs, adult tissue fragments	Tissue fragments	Disaggregated tissue, primary culture, propagated cell line
Effort	High	Moderate	Low
Characterization	Easy, histology	Cytology and markers	Biochemical, molecular, immunological, and cytological assays
Histology	Informative	Difficult	Not applicable
Biochemical differentiation	Possible	Heterogeneous	Lost, but may be reinduced
Propagation	Not possible	Possible	Standard procedure
Replicate sampling, reproducibility, homogeneity	High intersample variation	High intersample variation	Low level of variation
Quantitation	Difficult	Difficult	Easy; many techniques available

Chapter 17.) The identity of these cells remains unclear, but a relationship to the stem cell or uncommitted precursor cell compartment has been postulated. This concept implies that some cultured cells represent precursor pools within the tissue of origin. (See Origin of Cultured Cells in Chapter 2.) Cultured cell lines are more representative of precursor cell compartments *in vivo* than of fully differentiated cells, as, normally, most differentiated cells do not divide.

Because they may be propagated as a uniform cell suspension or monolayer, cell cultures have many advantages in quantitation, characterization, and replicate sampling, but lack the potential for cell–cell in-

TABLE 1.4. Subculture

Advantages	Disadvantages
Propagation	Trauma of enzymatic or mechanical disaggregation
More cells	Selection of cells adapted to culture
Possibility of cloning	Overgrowth of unspecialized or stromal cells
Increased homogeneity	Genetic instability
Characterization of replicate samples	Loss of differentiated properties (may be inducible)
Frozen storage	

teraction and cell–matrix interaction afforded by organ cultures. For this reason, many workers have attempted to reconstitute three-dimensional cellular structures using aggregates in cell suspension (see Spheroids in Chapter 24) or perfused high-density cultures on microcapillary bundles or membranes. (See Hollow Fibers in Chapter 24.) Such developments have required the introduction, or at least redefinition, of certain terms. *Histotypic* or *histiotypic* culture, or *histoculture* (I use *histotypic* culture), has come to mean the high-density, or "tissue-like," culture of one cell type, while *organotypic* culture implies the presence of more than one cell type interacting as they might in the organ of origin (or a simulation of such interaction). Organotypic culture has given new prospects for the study of cell interaction among discrete, defined populations of homogeneous and potentially genetically and phenotypically defined cells.

In many ways, some of the most exciting developments in tissue culture arise from recognizing the necessity of specific cell interaction in homogeneous or heterogeneous cell populations in culture. This recognition may mark the transition from an era of fundamental molecular biology, in which many of the regulatory processes have been worked out at the cellular level, to an era of cell or tissue biology, in which that understanding is applied to integrated populations of cells and to a more precise elaboration of the signals transmitted among cells.

CHAPTER 2

Biology of Cultured Cells

THE CULTURE ENVIRONMENT

The validity of the cultured cell as a model of physiological function *in vivo* has frequently been criticized. Often, the cell does not express the properties characteristic of the same cell type *in vivo* because the cellular environment has changed. Cells proliferate *in vitro* that would not normally *in vivo*, cell–cell and cell–matrix interactions are reduced because cell lines lack the heterogeneity and three-dimensional architecture found *in vivo*, and the hormonal and nutritional milieu is altered. This creates an environment that favors the spreading, migration, and proliferation of unspecialized cells, rather than the expression of differentiated functions. The influence of the environment on the culture is expressed via four routes: (1) the nature of the substrate or phase on or in which the cells grow, which may be solid, as in monolayer growth on plastic, semisolid, as in a gel such as collagen or agar, or liquid, as in a suspension culture; (2) the physicochemical and physiological constitution of the medium; (3) the constitution of the gas phase; and (4) the incubation temperature. The provision of the appropriate environment, including substrate adhesion, nutrients, hormones, and substrate, is fundamental to the expression of specialized functions.

Regulation of the cell phenotype in culture (see Induction of Differentiation in Chapter 16) is influenced by a number of factors—cell–cell interaction, cell–matrix interaction, and soluble factors (hormones, growth factors, nutrients, and inorganic ions), among others—and will affect the adhesion, spreading, proliferation, differentiation, shape, and migration of the cells. An introduction to some of these is presented next; more detailed accounts are available in a number of excellent textbooks [Alberts et al., 1994, 1997; Lodish et al., 1995].

CELL ADHESION

Most cells from solid tissues grow as adherent monolayers, and, unless they have transformed and become anchorage independent (see Anchorage Independence in Chapter 17), following tissue disaggregation or subculture they will need to attach and spread out on the substrate before they will start to proliferate (see also Subculture in Chapter 12). Originally, it was found that cells would attach to, and spread on, glass that had a slight net negative charge. Subsequently, it was shown that cells would attach to some plastics as well, such as polystyrene, if the plastic was appropriately treated with an electric ion discharge or high-energy ionizing radiation. We now know that cell adhesion is mediated by specific cell surface receptors for molecules in the extracellular matrix (see also Treated Surfaces in Chapter 7; Cell–Matrix Interactions in Chapter 16), so it seems likely that spreading may be preceded by the secretion of extracellular matrix proteins and proteoglycans by the cells. The matrix adheres to the charged substrate, and the cells then bind to the matrix via specific receptors. Hence, glass or plastic that has been conditioned by previous cell growth can often provide a better surface for attachment, and substrates pretreated with matrix constituents, such as fibronectin or collagen, or derivatives, such as gelatin, will help the more fastidious cells to attach and proliferate.

With fibroblast-like cells, the main requirement is for substrate attachment and spreading, because the cells migrate individually at low densities. Epithelial cells, however, appear to have to make the correct cell–cell contacts for optimum survival and growth; consequently, they tend to grow as patches.

Cell Adhesion Molecules

Three major classes of transmembrane proteins have been shown to be involved in cell–cell and cell–substrate adhesion (Fig. 2.1). Cell–cell adhesion molecules, CAMs (Ca^{2+}-independent), and *cadherins* (Ca^{2+}-dependent) are involved primarily in interactions between homologous cells. These proteins are self-interactive; that is, homologous molecules in opposing cells interact with each other [Rosenman & Gallatin, 1991; Alberts et al., 1994, 1997]. Cell substrate interactions are mediated primarily by *integrins*, receptors for matrix molecules such as fibronectin, entactin, laminin, and collagen, which bind to them via a specific motif usually containing the arginine–glycine–aspartic acid (RGD) sequence [Yamada & Geiger, 1997]. Each integrin comprises one α and one β subunit, both of which are highly polymorphic, thus generating considerable diversity among the integrins.

The third group of cell adhesion molecules is the transmembrane proteoglycans, also interacting with matrix constituents such as other proteoglycans or collagen, but not via the RGD motif. Some transmembrane and soluble proteoglycans also act as low-affinity growth factor receptors [Subramanian et al., 1997; Yevdokimova & Freshney, 1997] and may stabilize, activate, and/or translocate the growth factor

to the high-affinity receptor, participating in its dimerization [Schlessinger et al., 1995].

Disaggregation of the tissue, or an attached monolayer culture, with protease will digest some of the extracellular matrix and may even degrade some of the extracellular domains of transmembrane proteins, allowing cells to become dissociated from each other. Epithelial cells are generally more resistant to disaggregation, as they tend to have tighter junctional complexes (desmosomes, adherens junctions, and tight junctions) holding them together, while mesenchymal cells, which are more dependent on matrix interactions for intercellular bonding, are more easily dissociated. Endothelial cells may also express tight junctions in culture, especially if left at confluence for prolonged periods on a preformed matrix, and can be difficult to dissociate. In each case, the cells must resynthesize matrix proteins before they attach or must be provided with a matrix-coated substrate.

Cytoskeleton

Cell adhesion molecules are attached to elements of the cytoskeleton. The attachment of integrins to actin microfilaments via linker proteins is associated with reciprocal signaling between the cell surface and the nucleus. Cadherins can also link to the actin cytoskeleton in adherens junctions, mediating changes in cell shape. Desmosomes, which also employ cadherins, link to the intermediate filaments—in this case, cytokeratins—via an intracellular plaque, but it is not yet clear whether this linkage is a purely structural feature or also has a signaling capacity. Intermediate filaments are specific to cell lineages and can be used to characterize them. (See Lineage or Tissue Markers in Chapter 15.) The microtubules are the remaining

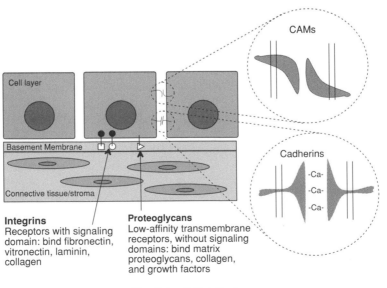

Fig. 2.1. Cell Adhesion.

component of the cytoskeleton; their role appears to be related mainly to cell motility and intracellular trafficking of microorganelles, such as the mitochondria and the chromatids at cell division.

CELL PROLIFERATION

Cell Cycle

The cell cycle is made up of four phases (Fig. 2.2). In the *M phase* (M = mitosis), the chromatin condenses into chromosomes, and the two individual chromatids, which make up the chromosome, segregate to each daughter cell. In the G_1 (Gap 1) *phase*, the cell either progresses towards DNA synthesis and another division cycle or exits the cell cycle reversibly (G_0) or irreversibly to commit to differentiation. It is during G_1 that the cell is particularly susceptible to control of cell cycle progression at a number of restriction points, which determine whether the cell will re-enter the cycle, withdraw from it, or differentiate. G_1 is followed by the *S phase* (DNA *s*ynthesis), in which the DNA replicates. S in turn is followed by the G_2 (Gap 2) *phase* in which the cell prepares for reentry into mitosis. Checkpoints at the beginning of DNA synthesis and in G_2 determine the integrity of the DNA and will halt the cell cycle to allow DNA repair or entry into apoptosis if repair is impossible. Apoptosis, or programmed cell death [al-Rubeai & Singh, 1998], is a regulated physiological process whereby a cell can be removed from a population. Marked by DNA fragmentation, nuclear blebbing, and cell shrinkage, apoptosis can also be detected by a number of marker enzymes using kits such as Apotag (Appligene Oncor) or the COMET assay [Maskell & Green, 1995].

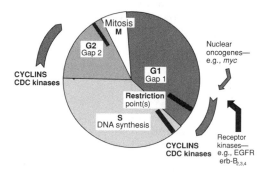

Fig. 2.2. Cell Cycle. The cell cycle is divided into four phases: G_1, S, G_2, and M. Progression round the cycle is driven by cyclins interacting with CDC kinase and stimulated by nuclear oncogenes and cytoplasmic signals initiated by receptor kinase interaction with ligand. The cell cycle is arrested at restriction points by cell cycle inhibitors such as Rb and p53.

Control of Cell Proliferation

Entry into the cell cycle is regulated by signals from the environment. Low cell density leaves cells with free edges and renders them capable of spreading, which permits their entry into the cycle in the presence of mitogenic growth factors, such as epidermal growth factor (EGF), fibroblast growth factors (FGFs), or platelet-derived growth factor (PDGF) (see Table 9.3 and Growth Factors in Chapter 9), interacting with cell surface receptors. High cell density inhibits the proliferation of normal cells (though not transformed cells). (See Contact Inhibition and Density Limitation of Growth in Chapter 17.) Inhibition of proliferation is initiated by cell contact and is accentuated by crowding and the resultant change in the shape of the cell and reduced spreading.

Intracellular control is mediated by positive-acting factors, such as the cyclins [Planas-Silva & Weinberg, 1997] (see Fig. 2.2), which are upregulated by signal transduction cascades activated by the receptor when it is bound to growth factor. Negative-acting factors such as p53 [Sager, 1992; McIlwrath et al., 1994], p16 [Russo et al., 1998], or the Rb gene product [Sager, 1992] block cell cycle progression at restriction points or checkpoints (Fig. 2.3). The link between the extracellular control elements (both positive-acting, e.g., PDGF, and negative-acting, e.g., TGF-β) and intracellular effectors is made by cell membrane receptors and signal transduction pathways, often involving protein phosphorylation and second messengers such as cyclic adenosine monophosphate (cAMP), Ca^{2+}, and diacylglycerol [Alberts et al., 1994]. Much of the evidence for the existence of these steps in the control of cell proliferation has emerged from studies of oncogene and suppressor gene expression in tumor cells, with the ultimate objective of the therapeutic regulation of uncontrolled cell proliferation in cancer. The immediate benefit, however, has been a better understanding of the factors required to regulate cell proliferation in culture [Jenkins, 1992]. These studies have had other benefits as well, including the identification of genes that enhance cell proliferation, some of which can be used to immortalize finite cell lines. (See Immortalization in Chapter 17.) The potential of this approach for cell production in biotechnology is under intense scrutiny.

DIFFERENTIATION

As stated earlier (see Limitations in Chapter 1), the expression of differentiated properties in cell culture is often limited by the promotion of cell proliferation, which is necessary for the propagation of the cell line and the expansion of stocks for use in the laboratory.

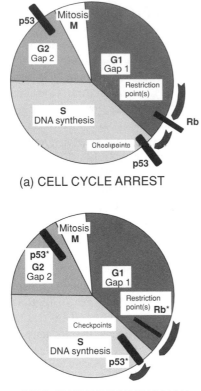

(a) CELL CYCLE ARREST

(b) CELL CYCLE PROGRESSION

Fig. 2.3. Cell Cycle Inhibition and Progression. The cell cycle is arrested at restriction points or checkpoints by the action of Rb, p53, and other cell cycle inhibitors (a). When these are inactivated, usually by phosphorylation, cells proceed round the cycle (b).

The conditions required for the induction of differentiation—a high cell density, enhanced cell–cell and cell–matrix interaction, and the presence of various differentiation factors (see Induction of Differentiation in Chapter 16)—may often be antagonistic to cell proliferation and vice versa. So if differentiation is required, it may be necessary to define two distinct sets of conditions—one to optimize cell proliferation and one to optimize cell differentiation.

Maintenance of Differentiation

It has been recognized for many years that specific functions are retained longer when the three-dimensional structure of the tissue is retained, as in organ culture. (See Organ Culture in Chapter 24.) Unfortunately, organ cultures cannot be propagated, must be prepared *de novo* for each experiment, and are more difficult to quantify than cell cultures. For these reasons, numerous researchers have attempted to re-create three-dimensional structures by perfusing monolayer cultures (see Histotypic Culture in Chapter 24; Perfusion Systems in Chapter 25) and to reproduce elements of the environment *in vivo* by culturing cells

on or in special matrices, such as collagen gel, cellulose, or gelatin sponge, or matrices from other natural tissue matrix glycoproteins, such as fibronectin, chondronectin, and laminin (see Matrix Coating and Three-dimensional Matrices in Chapter 7; Cell–Matrix Interactions in Chapter 16). A number of commercial products, the best known of which is Matrigel (Becton-Dickinson), reproduce the characteristics of extracellular matrix, but are undefined, although a growth-factor-depleted version is also available (GFR Matrigel, Becton Dickinson). These techniques present some limitations, but with their provision of homotypic cell interactions and cell–matrix interactions, and with the possibility of introducing heterotypic cell interactions, they hold considerable promise for the examination of tissue-specific functions. Expression of the differentiated phenotype will also require maintenance in the appropriate selective medium (see Selective Media in Chapter 9), with appropriate soluble inducers, such as hydrocortisone, retinoids, or planar polar compounds (see Induction of Differentiation in Chapter 16), and usually in the absence of serum.

The development of normal tissue functions in culture would facilitate the investigation of pathological behavior such as demyelination and malignant invasion. But, from a fundamental viewpoint, it is only when cells *in vitro* express their normal functions that any attempt can be made to relate them to their tissue of origin. The expression of the differentiated phenotype need not be complete, since the demonstration of a single type-specific surface antigen may be sufficient to place a cell in the correct lineage. More complete functional expression may be required, however, to place a cell in its correct *position* in the lineage and to reproduce a valid model of its function *in vivo*.

Dedifferentiation

Historically, the inability of cell lines to express the *in vivo* phenotype of the cells from which they were derived was blamed on *dedifferentiation*. According to this concept, differentiated cells lose their specialized properties *in vitro*, but it is often unclear whether (1) undifferentiated cells of the same lineage (Fig. 2.4) overgrow terminally differentiated cells of reduced proliferative capacity or (2) the absence of the appropriate inducers (hormones: cell or matrix interaction) causes deadaptation. (See Dedifferentiation in Chapter 16.) In practice, both processes may occur. Continuous proliferation may select undifferentiated precursors, which, in the absence of the correct inductive environment, do not differentiate.

An important distinction should be made between dedifferentiation, deadaptation, and selection. Dedifferentiation implies that the specialized properties of the cell are lost irreversibly. For example, a hepatocyte

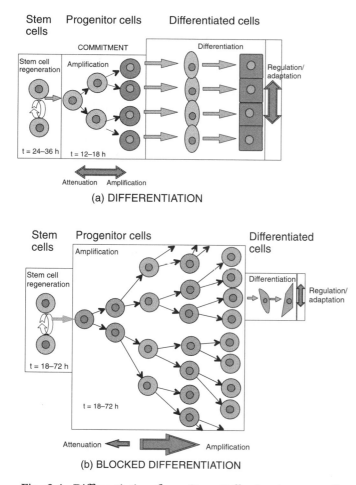

Fig. 2.4. Differentiation from Stem Cells. *In vivo,* a small stem cell pool gives rise to a proliferating progenitor compartment that produces the differentiated cell pool. *In vitro,* differentiation is limited by the need to proliferate, and the population becomes predominately progenitor cells, although stem cells may also be present. Culture conditions select for the proliferating progenitor cell compartment of the tissue or induce cells that are partially differentiated to revert to a progenitor status.

would lose its characteristic enzymes (arginase, aminotransferases, etc.) and could not store glycogen or secrete serum proteins, and these properties could not be reinduced once lost. Deadaptation, on the other hand, implies that the synthesis of specific products or other aspects of specialized function are under regulatory control by hormones, cell–cell interaction, cell–matrix interaction, etc., and can be reinduced if the correct conditions can be re-created. For instance, the presence of matrix as a floating collagen raft [Michalopoulos & Pitot, 1975] allows the induction of tyrosine aminotransferase in normal hepatocytes, and Matrigel was shown to stabilize the differentiated phenotype in hepatocytes [Bissell et al., 1987]. It is gradually becoming apparent that, given the correct culture conditions, differentiated functions can be

expressed by a number of different cell types (see Table 2.1 and Induction of Differentiation in Chapter 16), and the concept of dedifferentiation is now regarded as an unlikely explanation for the loss of specialized functions.

For induction to occur, the appropriate cells must be present. In early attempts at liver cell culture, the failure of cells to express hepatocyte properties was due partly to overgrowth of the culture by connective tissue fibroblasts or endothelium from blood vessels or sinusoids. With the correct disaggregation technique and the correct culture conditions [see Guguen-Guillouzo, 1992] (see Liver in Chapter 22), hepatocytes can be selected preferentially. Similarly, epidermal cells can be grown by using either a confluent feeder layer [Rheinwald and Green, 1975] or a selective medium [Peehl and Ham, 1980; Tsao et al., 1982]. Selective media also have been used for many other types of epithelium [Freshney, 1992]. These and other examples (e.g., selective feeder layers (see Cervix in Chapter 22; Confluent Feeder Layers in Chapter 23), D-valine for the isolation of kidney epithelium, and the use of cytotoxic antibodies (see Selective Inhibitors in Chapter 13) clearly demonstrate that the selective culture of specialized cells is not the insuperable problem that it once appeared to be. New selective media, based mainly on supplemented Ham's F12:DMEM or modifications of the MCDB series (see Selective Media in Chapter 9), are appearing all the time [Cartwright & Shah, 1994; Mather, 1998].

ENERGY METABOLISM

Most culture media contain 4 to 20 mM glucose, which is used mainly as a carbon source for glycolysis, generating lactic acid as an end product. Under normal culture conditions (atmospheric oxygen and a submerged culture), oxygen is in relatively short supply. In the absence of an appropriate carrier, such as hemoglobin, raising the O_2 tension will generate free-radical species that are toxic to the cell, so O_2 is usually maintained at atmospheric levels. This results in anaerobic conditions and the use of glycolysis for energy metabolism. However, the citric acid cycle remains active, and it has become apparent that amino acids—particularly glutamine—can be utilized as a carbon source by oxidation to glutamate by glutaminase and entry into the citric acid cycle by transamination to 2-oxoglutarate [Reitzer et al., 1979; Butler & Christie, 1994]. Deamination of the glutamine tends to produce ammonia, which is toxic and can limit cell growth, but the use of dipeptides, such as glutamyl-alanine or glutamyl-glycine, appears to minimize the production of ammonia and has the additional ad-

TABLE 2.1. Cell Lines with Differentiated Properties

Cell type	Origin	Cell line	Species	Marker	Reference
Astrocytes	Optic nerve		Rat	Glial fibrillary acidic protein	Raff et al., 1990; see Protocol 22.17
β-islet cells	Pancreas		Human	Insulin	Kinard et al., 1990
Buccal epithelium	Buccal mucosa		Human	Cytokeratin	Sundqvist et al., 1991
Cardiac myoblasts	Heart		Human	Contraction	Goldman & Wurzel, 1992
Endocrine	Adrenal cortex tumor		Rat	Steroids	Buonassisi et al., 1962
Endocrine	Pituitary tumor	GH2, GH3	Rat	Growth hormone	Buonassisi et al., 1962
Endocrine	Hypothalamus	C7	Mouse	Neurophysin; vasopressin	De Vitry et al., 1974
Endocrine	Adrenal cortex		Cow	Steroids	Simonian et al., 1987
Endothelium	Dermis	HDMEC	Human	Factor VIII, CD36	Gupta et al., 1997
Endothelium	Umbilical vein	ECV304	Human	Ulex lectin binding	Hughes, 1996
Endothelium	Pulmonary artery	CPAE	Cow	Factor VIII, angiotensin II converting enzyme	Del Vecchio & Smith, 1981
Endothelium	Hepatoma	SK/HEP-1	Human	Factor VIII	Heffelfinger et al., 1992
Epithelium	Prostate	PPEC	Human	PSA	Robertson & Robertson, 1995
Epithelium	Salivary glands		Human		Sabatini et al., 1991
Epithelium	Placenta		Human	Human chorionic gonadotrophin	Cou, 1978
Epithelium	Kidney	MDCK	Dog	Domes, transport	Gaush et al., 1966; Rindler et al., 1979
Epithelium	Kidney	LLC-PKI	Pig	Na$^+$-dependent glucose uptake	Hull et al., 1976; Saier, 1984
Epithelium	Breast	MCF-7	Human	Domes, α-lactalbumin	Soule et al., 1973
Epithelium	Gingiva		Human	Cytokeratin	Oda & Watson, 1990
Glia	Glioma	MOG-G-CCM	Human	Glutamyl synthetase	Balmforth et al., 1986
Glia	Glioma	C6	Rat	Glial fibrillary acidic protein, GPDH	Benda et al., 1968
Hepatocytes	Liver		Human	Albumin	Li et al., 1992
Hepatocytes	Hepatoma	H4-11-E-C3	Rat	Tyrosine aminotransferase	Pitot et al., 1964
Hepatocytes	Liver		Mouse	Aminotransferase	Yeoh et al., 1990
Keratinocytes	Epidermis	HaCaT	Human	Cornification	Boukamp et al., 1988
Keratinocytes	Epidermis		Mouse	Cornification	Fusenig, 1994a; Protocol 22.1
Keratinocytes	Epidermis		Human	Cornification	Rheinwald & Green, 1975
Leukemia	Spleen	Friend	Mouse	Hemoglobin	Scher et al., 1971
Melanocytes	Epidermis		Human	Melanin	Naeyaert et al., 1991; Protocol 22.18
Melanocytes	Melanoma	B16	Mouse	Melanin	Nilos & Makarski, 1978
Myeloid	Leukemia	K562	Human	Hemoglobin	Andersson et al., 1979a,b
Myeloid	Myeloma	Various	Mouse	Immunoglobulin	Horibata & Harris, 1970
Myeloid	Marrow	WEHI-3B D+	Mouse	Morphology	Nicola, 1987
Myeloid	Leukemia	HL60	Human	Phagocytosis Neotetrazolium Blue reduction	Olsson & Ologsson, 1981
Myoblasts	Skeletal muscle		Human	Myotubes, creatine kinase MM isoenzyme	Quax et al., 1992; see Protocol 22.12
Myoblasts	Skeletal muscle		Chick	Myogenesis, creatine kinase MM isoenzyme	Richler & Yaffe, 1970
Myoblasts	Skeletal muscle		Rat	Myogenesis	Richler & Yaffe, 1970
Neurons	Neuroblastoma	C1300	Rat	Neurites	Liebermann & Sachs, 1978
Osteoblasts	Bone		Rat	Mineralization	Bernier et al., 1990
Skeletal muscle		C2	Mouse	Myotubes	Morgan et al., 1992
Skeletal muscle		L6	Rat	Myotubes	Richler & Yaffe, 1970
Teratocarcinoma	Embryo	Various	Mouse	Various	Martin, 1975
Type II or Clara cell	Lung carcinoma	A549	Human	Surfactant	Giard et al., 1972
		NCI-H441	Human	Surfactant	Brower et al., 1986
Type II pneumocyte	Lung carcinoma		Mouse	Surfactant	Wilkenheiser et al., 1991

vantage of being more stable in the medium (e.g., Glutamax, Life Technologies).

INITIATION OF THE CULTURE

Primary culture techniques are described in detail in Chapter 11. Briefly, a culture is derived either by the outgrowth of migrating cells from a fragment of tissue or by enzymatic or mechanical dispersal of the tissue. Regardless of the method employed, primary culture is the first in a series of selective processes (Table 2.2) that may ultimately give rise to a relatively uniform cell line. In primary explantation (see Primary Explant in Chapter 11), selection occurs by virtue of the cells' capacity to migrate from the explant, while with dispersed cells, only those cells that both survive the disaggregation technique and adhere to the substrate or survive in suspension will form the basis of a primary culture.

If the primary culture is maintained for more than a few hours, a further selection step will occur. Cells that are capable of proliferation will increase, some cell types will survive but not increase, and yet others will be unable to survive under the particular conditions of the culture. Hence, the relative proportion of each cell type will change and will continue to do so until, in the case of monolayer cultures, all the available culture substrate is occupied.

After *confluence* is reached (i.e., all the available growth area is utilized and the cells make close contact with one another), cells whose growth is sensitive to density limitation (see Contact Inhibition and Density Limitation of Growth in Chapter 17) will stop dividing, while any transformed cells, which are insensitive to density limitation, will tend to overgrow. Keeping the cell density low (e.g., by frequent subculture) helps to preserve the normal phenotype in cul-

tures such as mouse fibroblasts, in which spontaneous transformants tend to overgrow at high cell densities [Todaro & Green, 1963].

Some aspects of specialized function are expressed more strongly in primary culture, particularly when the culture becomes confluent. At this stage, the culture will show its closest morphological resemblance to the parent tissue and retain some diversity in cell type.

EVOLUTION OF CELL LINES

After the first subculture, or passage (Fig. 2.5), the primary culture becomes known as a cell line and may be propagated and subcultured several times. With each successive subculture, the component of the population with the ability to proliferate most rapidly will gradually predominate, and nonproliferating or slowly proliferating cells will be diluted out. This is most strikingly apparent after the first subculture, in which differences in proliferative capacity are compounded with varying abilities to withstand the trauma of trypsinization and transfer. (See Subculture in Chapter 12.)

Although some selection and phenotypic drift will continue, by the third passage the culture becomes more stable and is typified by a rather hardy, rapidly proliferating cell. In the presence of serum and without specific selection conditions, mesenchymal cells derived from connective tissue fibroblasts or vascular elements frequently overgrow the culture. While this has given rise to some very useful cell lines (e.g., WI38 human embryonic lung fibroblasts [Hayflick and Moorhead, 1961], BHK21 baby hamster kidney fibroblasts [Macpherson and Stoker, 1962], COS cells [Gluzman, 1981[16]], CHO cells [Puck et al., 1958[17]] (see Table 12.2), and perhaps the most famous of all, the

TABLE 2.2. Selection in Cell Line Development

Stage	Factors influencing selection	
	Primary explant	Enzymatic disaggregation
Isolation	Mechanical damage	Enzymatic damage
Primary culture	Adhesion of explant; outgrowth (migration)	Cell adhesion and spreading
First subculture	Trypsin sensitivity; nutrient, hormone, and substrate limitations	
Propagation as a cell line	Relative growth rates of different cells; selective overgrowth of one lineage	
	Nutrient, hormone, and substrate limitations	
	Effect of cell density on predominance of normal or transformed phenotype	
Senescence; transformation	Normal cells die out; transformed cells overgrow	

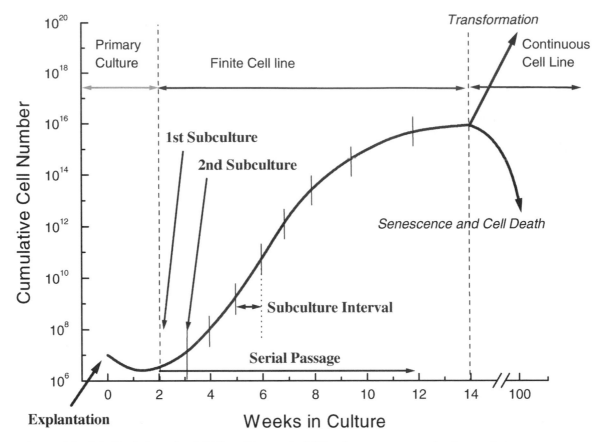

Fig. 2.5. Evolution of a Cell Line. The vertical (Y) axis represents total cell growth (assuming no reduction at passage) for a hypothetical cell culture. Total cell number (cell yield) is represented on this axis on a log scale, and the time in culture is shown on the X-axis on a linear scale. Although a continuous cell line is depicted as arising at 14 weeks, with different cells it could arise at any time. Likewise, senescence may occur at any time, but for human diploid fibroblasts it is most likely to occur between 30 and 60 cell doublings, or 10 to 20 weeks, depending on the doubling time. Terms and definitions used are as in the glossary. Transformation is explained in more detail in Chapter 15.

L-cell, a mouse subcutaneous fibroblast treated with methylcholanthrene [Earle et al., 1943; Sanford et al., 1948], this overgrowth represents one of the major challenges of tissue culture since its inception—namely, how to prevent the overgrowth of the more fragile or slower growing specialized cells such as hepatic parenchyma or epidermal keratinocytes. Inadequacy of the culture conditions is largely to blame for this problem, and considerable progress has now been made in the use of selective media and substrates for the maintenance of many specialized cell lines. (See Selective Media in Chapter 9; Chapter 22.)

Senescence

Normal cells can divide only for a limited number of times; hence, cell lines derived from normal tissue will die out after a fixed number of population doublings. This is a genetically determined event, known as *senescence*, and is thought to be determined by the ina-

bility of terminal sequences of the DNA in the telomeres to replicate at each cell division. The result is a progressive shortening of the telomeres, until, finally, the cell is unable to divide further [Bodnar et al., 1998]. Exceptions to this rule are germ cells, stem cells, and transformed cells, which often express the enzyme telomerase, which is capable of replicating the terminal sequences of DNA in the telomere and extending the life span of the cells, infinitely in the case of germ cells and some tumor cells.

THE DEVELOPMENT OF CONTINUOUS CELL LINES

Some cell lines may give rise to continuous cell lines. (See Fig. 2.5.) The ability of a cell line to grow continuously probably reflects its capacity for genetic variation, allowing subsequent selection. Genetic variation often involves the deletion or mutation of the p53 gene, which would normally arrest cell cycle progression, if DNA were to become mutated, and overexpression of the telomerase gene. Human fibroblasts remain predominantly euploid throughout their life span in culture and never give rise to continuous cell lines [Hayflick and Moorhead, 1961], while mouse fibroblasts and cell cultures from a variety of human and animal tumors often become aneuploid in culture and frequently give rise to continuous cultures. Possibly the condition that predisposes most to the development of a continuous cell line is inherent genetic variation, so it is not surprising to find genetic instability perpetuated in continuous cell lines. A common feature of many human continuous cell lines is the development of a subtetraploid chromosome number (Fig. 2.6). The alteration in a culture that gives rise to a continuous cell line is commonly called *in vitro transformation* (see Chapter 17) and may occur spontaneously or be chemically or virally induced (see Immortalization in Chapter 17). The word *transformation* is used rather loosely and can mean different things to different people. In this volume, *immortalization* means the acquisition of an infinite life span and *transformation* implies an alteration in growth characteristics (anchorage independence, loss of contact inhibition, and density limitation of growth) that will often, but not necessarily, correlate with tumorigenicity.

Continuous cell lines are usually *aneuploid* and often have a chromosome number between the diploid and tetraploid value. (See Fig. 2.6.) There is also considerable variation in chromosome number and constitution among cells in the population (*heteroploidy*). (See also Genetic Instability in Chapter 17.) It is not clear whether the cells that give rise to continuous lines are present at explantation in very small numbers or arise later as a result of the transformation of one or more cells. The second alternative would seem to be more probable on cell kinetic grounds, as continuous cell lines can appear quite late in a culture's life history, long after the time it would have taken for even one preexisting cell to overgrow. The possibility remains, however, that there is a subpopulation in such cultures with a predisposition to transform that it is not shared by the rest of the cells.

The term *transformation* has been applied to the process of formation of a continuous cell line partly because the culture undergoes morphological and kinetic alterations, but also because the formation of a continuous cell line is often accompanied by an increase in tumorigenicity. A number of the properties of continuous cell lines, such as a reduced serum requirement, reduced density limitation of growth, growth in semisolid media, aneuploidy (see also Table 17.1), and more, are associated with *malignant* trans-

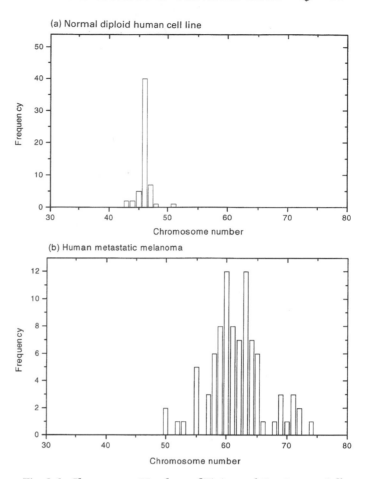

Fig. 2.6. *Chromosome Numbers of Finite and Continuous Cell Lines.* a. A normal human cell line. b. A coninuous cell line from human metastatic melanoma.

formations. (See Aberrant Growth Control in Chapter 17.) These properties are reviewed in more detail in Chapter 17. Similar morphological and behavioral changes can also be observed in cells that have undergone virally or chemically induced transformation.

Many (if not most) normal cells do not give rise to continuous cell lines. In the classic example, normal human fibroblasts remain euploid throughout their life span and at crisis (usually around 50 generations) will stop dividing, although they may remain viable for up to 18 months thereafter. Human glia [Pontén and Westermark, 1980] and chick fibroblasts [Hay and Strehler, 1967] behave similarly. Epidermal cells, on the other hand, have shown gradually increasing life spans with improvements in culture techniques [Rheinwald & Green, 1977; Green et al., 1979] and may yet be shown capable of giving rise to continuous growth. Such growth may be related to the self-renewal capacity of the tissue *in vivo*. (See next section.) Continuous culture of lymphoblastoid cells is also possible [Gjerset et al., 1990] by transformation with Epstein–Barr virus.

ORIGIN OF CULTURED CELLS

Because most people working under standard conditions do so with finite or continuous proliferating cell lines, it is important to consider the cellular composition of the culture. The capacity to express differentiated markers under the influence of inducing conditions may mean either that the cells being cultured are mature and only require induction to continue synthesizing specialized proteins or that the culture is composed of precursor or stem cells that are capable of proliferation, but remain undifferentiated until the correct inducing conditions are applied, whereupon some or all of the cells mature and become differentiated. It may be useful to think of a cell culture as being in equilibrium between multipotent stem cells, undifferentiated but committed precursor cells, and mature differentiated cells (see Fig. 2.4) and to suppose that the equilibrium may shift according to the environmental conditions. Routine serial passage at relatively low cell densities would promote cell proliferation and constrain differentiation, while high cell densities, low serum, and the appropriate hormones would promote differentiation and inhibit cell proliferation.

The source of the culture will also determine which cellular components may be present. Hence, cell lines derived from the embryo may contain more stem cells and precursor cells and be capable of greater self-renewal than cultures from adults. In addition, cultures from tissues undergoing continuous renewal *in vivo* (epidermis, intestinal epithelium, hematopoietic cells) may still contain stem cells, which, under the appropriate conditions, will have a prolonged life span, while cultures from tissues that renew solely under stress (fibroblasts, muscle, glia) may contain only committed precursor cells with a limited life span.

Thus, the identity of the cultured cell is defined not only by its lineage *in vivo* (hematopoietic, hepatocyte, glial, etc.), but also by its position in that lineage (stem cell, committed precursor cell, or mature differentiated cell). With the exception of mouse teratomas and one or two other examples from lower vertebrates, it seems unlikely that cells will change lineage (transdifferentiate), but they may well change position in the lineage, even reversibly in some cases.

When cells are cultured from a neoplasm, they need not adhere to these rules. Thus, a hepatoma from rat may proliferate *in vitro* and still express some differentiated features, but the closer they are to those of the normal phenotype, the more induction of differentiation may inhibit proliferation. Although the relationship between position in the lineage and cell proliferation may become relaxed (though not lost—B16 melanoma cells still produce more pigment at a high cell density and at a low rate of cell proliferation than at a low cell density and a high rate of cell proliferation), transfer between lineages has not been clearly established. (See also Commitment and Lineage in Chapter 16.)

CHAPTER 3

Design and Layout

PLANNING

The major requirement that distinguishes tissue culture from most other laboratory techniques is the need to maintain asepsis. While it is usually not economically viable to create large sterile areas, it is important that the tissue culture laboratory be dust free and have no through traffic. The introduction of laminar-flow hoods has greatly simplified the problem and allows the utilization of unspecialized laboratory accommodation (see also Laminar-Flow Hood in Chapter 4; Laminar Flow in Chapter 5), provided that the location satisfies the aforementioned requirements.

Several considerations need to be taken into account in planning new accommodation. Is a new building to be constructed, or will an existing one be converted? A conversion limits you to the structural confines of the building; new extracts and air-conditioning can be expensive, and structural modifications that involve load-bearing walls can be difficult to make. When a new building is contemplated, there is more scope for integrated and innovative design, and facilities may be positioned for ergonomic and energy-saving reasons, rather than structural ones. The following measures should be contemplated:

(1) Consider where air inlets and extracts must be placed. It is preferable to duct laminar-flow hoods to the exterior to improve air circulation and remove excess heat (300–500 W per hood) from the room. This also facilitates decontamination with formaldehyde, should it be required. Venting hoods to the outside will probably provide most of the air extract for the room, and it remains only to ensure that the incoming air, from a central plant or an air conditioner, does not interfere with the integrity of the airflow in the hood.

(2) Where will facilities for washing up and for sterilization be located, and will there be a difference in ceiling height between the washup and sterilization facilities and the rest of the laboratory? Will an elevator be required, or will a ramp suffice? If a ramp will do, what will be the gradient and the maximum load that you can expect to be carried up that gradient without mechanical help?

(3) Ensure that tissue culture is reasonably accessible to, but not contiguous with, the animal facility.

(4) Give your wash up, sterilization and preparation staff a reasonable visual outlook; they usually perform fairly repetitive duties, whereas the scientific and technical staff look into a laminar-flow hood and do not need a view. In fact, windows can be a disadvantage, in a tissue culture laboratory leading to heat gain, ultraviolet (UV) denaturation of the medium, and the incursion of microorganisms if they are not properly sealed.

(5) If a conversion is contemplated, then there will be significant structural limitations; choose the location carefully, to avoid space constraints and awkward projections into the room that will limit flexibility.

(6) Make sure that doorways are both wide enough and high enough and that ceilings have sufficient clearance to allow the installation of equipment such as laminar-flow hoods (which may need ad-

ditional space for ductwork), incubators, and autoclaves. Make sure that doorways and spacing between equipment provide access for maintenance.

Some questions are more general and apply to both new buildings and those that are to be converted:

(1) How many people will work in the facility, how long will they work each week, and what kinds of culture will they perform? These considerations determine how many laminar-flow hoods will be required (based on whether people can share hoods or whether they will require a hood for most of the day) and whether a large area will be needed to handle bioreactors, animal tissue dissections, or large numbers of cultures. As a rough guide, 12 laminar-flow hoods in a communal facility can accommodate 50 people with different requirements.

(2) What type of incubation will be required in terms of size, temperature, gas phase, and proximity to the work space? Will regular, non-gassed incubators or a hot room suffice, or are CO_2 and a humid atmosphere required? Generally, large numbers of flasks or large-volume flasks that are sealed are best incubated in a hot room, while open plates and dishes will require a humid CO_2 incubator.

(3) What space is required for each facility? The largest area should be given to the culture operation, which has to accommodate laminar-flow hoods, cell counters, centrifuges, incubators, microscopes, and some stocks of reagents, media, glassware, and plastics. The second largest is for washup, preparation, and sterilization, third is storage, and fourth is incubation. A reasonable estimate is 4:2:1:1, in the order just presented.

(4) The space between hoods should be approximately 500 mm (2 ft), to allow access for maintenance and to minimize interference in airflow between hoods. This space is best filled with a removable cart or trolley, which allows space for bottles, flasks, reagents, and a note book.

(5) Will people require access to the animal house for animal tissue?

(6) What proportion of the work will be cell line work, with its requirement for storage in liquid nitrogen?

(7) What quarantine facilities will be required? Newly introduced cell lines and biopsies need to be screened for mycoplasma before being handled in the same room as general stocks, and some human and primate biopsies and cell lines may carry a biohazard risk that requires containment (see Biohazards in Chapter 6).

These questions will enable you to decide what size of facility you require and what type of accommodation—one or two small rooms (Fig. 3.1, Fig. 3.2), or a suite of rooms incorporating washup, sterilization, one or more aseptic areas, an incubation room, a darkroom for fluorescence microscopy and photomicrography, a refrigeration room, and storage (Fig. 3.3).

CONSTRUCTION AND SERVICES

The rooms should be supplied with air filtered to usual industrial or office standards and be designed for easy cleaning. Furniture should fit tightly to the floor or be suspended from the bench, with a space left underneath for cleaning. Cover the floor with a vinyl or other dustproof finish, and allow a slight fall in the level toward a floor drain located toward the door side of the room (i.e., away from the sterile cabinets). This arrangement allows liberal use of water if the floor has to be washed, but, more important, it protects equipment from damaging floods if stills, autoclaves, or sinks overflow.

If the tissue culture lab can be separated from the preparation, washup, and sterilization areas, so much the better. Adequate floor drainage should still be provided in both areas, although, clearly, the washup and sterilization area will be most important. If you have a separate washup and sterilization facility, it will be convenient to have this on the same floor as, and adjacent to, the laboratory, with no steps to negotiate, so that carts or trolleys may be used. Across a corridor is probably ideal. (See Fig. 3.3; see also Washing Up in Chapter 4.)

Try to imagine the flow of traffic—people, reagents, carts or trolleys, etc.—and arrange for minimum conflict, easy and close access to stores, good access for replenishing stocks without interfering with sterile work, and easy withdrawal of soiled items.

Services that are required include power, combustible gas (domestic methane, propane, etc.), carbon dioxide, compressed air, and vacuum. Power is always underestimated, in terms of both the number of outlets and the amperage per outlet. Assess carefully the equipment that will be required, assume that both the number of appliances and their power consumption will treble within the life of the building in its present form, and try to provide sufficient power, preferably at or near the outlets, but at least at the main distribution board.

Gas is more difficult to judge, as it requires some knowledge of the local provision of power. Electricity is cleaner and generally easier to manage from a safety standpoint, but gas may be cheaper and more reliable.

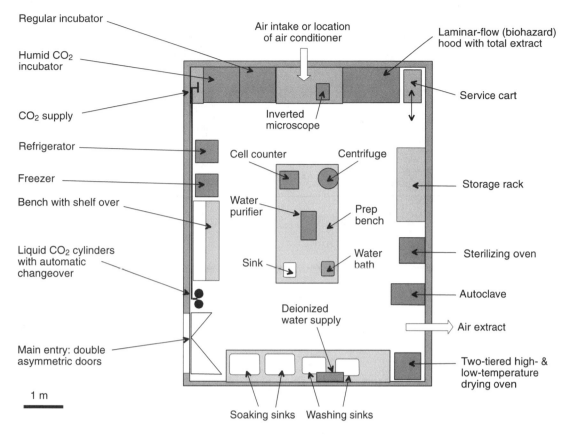

Fig. 3.1. *Small Tissue Culture Laboratory.* Suggested layout for simple, self-contained tissue culture laboratory for use by two or three persons. Dark-shaded areas represent movable equipment, lighter shaded areas fixed or movable furniture.

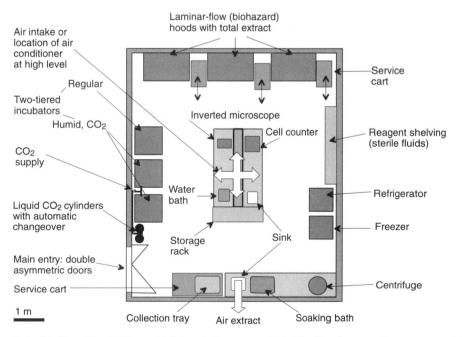

Fig. 3.2. *Medium-Sized Tissue Culture Laboratory.* Suitable for five or six persons, with washing-up and preparation facility located elsewhere. Dark-shaded areas represent movable equipment, light-shaded areas movable or fixed furniture.

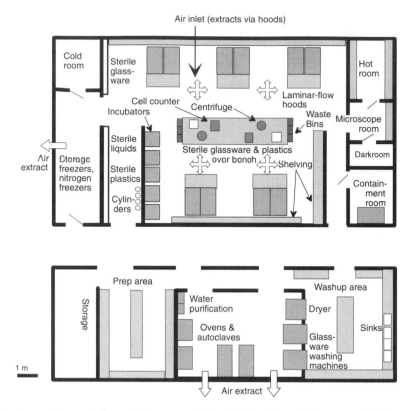

Fig. 3.3. Large Tissue Culture Laboratory. Suitable for 20 to 30 persons. Adjacent washing-up, sterilization, and preparation area. Dark-shaded areas represent equipment, light-shaded areas fixed and movable furniture.

Local conditions will usually determine the need for combustible gas.

If possible, carbon dioxide should be piped into the facility. The installation will pay for itself eventually in the cost of cylinders of mixed gases for gassing cultures, and it provides a better supply, which can be protected (see CO_2 Incubator in Chapter 4), for gassing incubators. Gas-mixing flowmeters (Gow-Mac, Platon) can be provided at workstations to provide the correct gas mixture, and gas blenders, though expensive, are now available (Hotpak, Signal) to provide different mixtures of gases at the flick of a switch or twist of a dial.

Compressed air is generally no longer required for incubators that regulate the gas mixture from pure CO_2 supplies only, but will be required if a gas mixture is provided at each workstation, via either flowmeters or a gas blender. Compressed air is also used to expel cotton plugs from glass pipettes before washing and may be required for some types of glassware washing machine (e.g., Scientek 3000).

A vacuum line can be very useful for evacuating culture flasks, but one should take several precautions to run the line successfully. A collection vessel must be present, as must a trap with a hydrophobic filter between the flasks, in order to prevent fluid, vapor, or some contaminant from entering the vacuum line and pump. Also, the vacuum pump must be protected against the line being left open inadvertently; usually this can be accomplished via a pressure-activated foot switch that closes when no longer pressed. In many respects, it is better to provide individual peristaltic pumps at each workstation (see Figs 4.8, 4.9), or one pump between two workstations.

LAYOUT

Six main functions need to be accommodated in the laboratory: sterile handling, incubation, preparation, washup, sterilization, and storage (Table 3.1). If a single room is used, create a "sterility gradient"; the clean area for sterile handling should be located at one end of the room, furthest from the door, and washup and sterilization facilities should be placed at the other end, with preparation, storage, and incubation in between. The preparation area should be adjacent to the washup and sterilization areas, and storage and incubators should be readily accessible to the sterile working area. (See Figs. 3.1, 3.2, 3.3.)

TABLE 3.1. Tissue Culture Facilities

Minimum requirements	Desirable features	Useful additions
Sterile area, clean, quiet, and with no through traffic	Filtered air (air-conditioning)	Piped CO_2 and compressed air
Separate from animal house and microbiological labs	Service bench adjacent to culture area	Storeroom for bulk plastics
Preparation area	Separate prep room	Containment room for biohazard work
Washup area (not necessarily within tissue culture laboratory, but at least adjacent to it)	Hot room with temperature recorder	Liquid N_2 storage tank ($\approx$500 l) and separate storeroom for nitrogen freezers
Space for incubator(s)	Separate sterilizing room	Microscope room
Storage areas:	Separate cylinder store	Darkroom
Liquids: ambient, 4°C, −20°C		Vacuum line
Glassware (shelving)		
Plastics (shelving)		
Small items (drawers)		
Specialized equipment (slow turnover), cupboard(s)		
Chemicals: ambient, 4°C, −20°C; share with liquids, but keep chemicals in sealed container over desiccant		
CO_2 cylinders		
Space for liquid N_2 freezer(s)		
Sink		

Sterile Handling Area

Sterile work should be located in a quiet part of the laboratory, should be restricted to tissue culture, and there should be no through traffic or other disturbance that is likely to cause dust or drafts. Use a separate room or cubicle if laminar-flow hoods are not available. The work area, in its simplest form, should be a plastic laminate-topped bench, preferably plain white or neutral gray, to facilitate the observation of cultures, dissection, etc., and to allow an accurate reading of pH when phenol red is used as an indicator, Nothing should be stored on the bench, and any shelving above should be used only in conjunction with sterile work (e.g., for holding pipette cans and instruments). The bench should be either freestanding (away from the wall) or sealed to the wall with a plastic sealing strip or mastic.

Laminar flow. The introduction of laminar-flow hoods with sterile air blown onto the work surface (see Laminar-Flow Hood in Chapter 4, and Fig. 4.1) affords greater control of sterility at a lower cost than providing a separate sterile room. Individual freestanding hoods are preferable, as they separate operators and can be moved around, but laminar-flow wall or ceiling units in batteries can be used. With individual hoods, only the operator's arms enter the sterile area, while with laminar-flow wall or ceiling units, there is no cabinet and the operator is part of the

work area. While this arrangement may give more freedom of movement, particularly with large pieces of apparatus (roller bottles, bioreactors), greater care must be taken by the operator not to disrupt the laminar flow, and it will be necessary to wear caps and gowns to avoid contamination.

Select hoods that suite your accommodation—freestanding or benchtop—and allow plenty of legroom underneath with space for pumps, aspirators, and so forth. Freestanding cabinets should be on castors so that they can be moved if necessary. Chairs should be a suitable height, with adjustable seat height and back angle, and able to be drawn up close enough to the front edge of the hood to allow comfortable working well within it. A small cart, trolley or folding flap (300–500 mm minimum) should be provided beside each hood for depositing apparatus or reagents not in immediate use.

Laminar-flow hoods should have a lateral separation of 500 mm (~2 ft), and, if hoods are opposed, there should be a minimum of 3,000 mm between the fronts of each cabinet. They should be installed as part of the construction contract as they will influence ventilation. In addition, you may wish to provide a small room or cubicle for use as a containment area (Fig. 3.3). If so, it must be separated by a door or air lock from the rest of the suite and will need its own incubators, freezer, refrigerator, centrifuge, etc. It will also require a biohazard cabinet or pathogen hood with separate extract and pathogen trap. (For a fuller description of containment facilities, see Biohazards in Chapter 6.)

Incubation

The requirement for cleanliness is not as stringent as that for sterile handling, but clean air, a low disturbance level, and minimal traffic will give your incubation area a better chance of avoiding dust, spores, and drafts that carry them.

Incubation may be carried out in separate incubators or in a thermostatically controlled hot room (Figs. 3.4, 3.5). If only one or two are required, incubators are expensive and economical in terms of space; but as soon as you require more than two, their cost is more than that of a simple hot room, and their use is less convenient. Incubators also lose more heat when they are opened and are slower to recover than a hot room. As a rough guide, you will need 0.2 m^3 (200 l, 6 ft^3) of incubation space with 0.5 m^2, 6 ft^2 shelf space per person. Extra provision may need to be made for one or more humid incubators with a controlled CO_2 level in the atmosphere. (See CO_2 Incubator in Chapter 4.)

Hot Room

If you have the space within the laboratory area or have an adjacent room or walk-in cupboard readily available and accessible, it may be possible to convert the area into a hot room (Figs. 3.4, 3.5). The area need not be specifically constructed as a hot room, but it should be insulated to prevent cold spots being generated on the walls. If insulation is required, line the area with plastic laminate-veneered board, separated from the wall by about 5 cm (2 in) of fiberglass, mineral wool, or fire-retardant plastic foam. Mark the location of the straps or studs carrying the lining panel in order to identify anchorage points for wall-mounted shelving if that is to be used. Use demountable shelving, and space shelf supports at 500–600 mm (21 in) to support the shelving without sagging. Freestanding shelving units are preferable, as they can be removed for cleaning the rack and the room. Allow 200–300 mm (9 in) between shelves, and use wider shelves (450 mm, 18 in) at the bottom and narrower (250–300 mm, 12 in) ones above eye level. Perforated shelving mounted on adjustable brackets will allow for air circulation. The shelving must be flat and perfectly horizontal, with no bumps or irregularities.

Do not underestimate the space that you will require in the lifetime of the hot room. It costs very little more to equip a large hot room than a small one. Calculate costs on the basis of the amount of shelf space you will require; if you have just started, mul-tiply by 5 or 10; if you have been working for sometime, multiply by 2 or 4.

Wooden furnishing should be avoided as much as possible, as they warp in the heat and can harbor infestations.

A small bench, preferably stainless steel or solid plastic laminate, should be provided in some part of the hot room. The bench should accommodate a microscope, the flasks that you wish to examine, and a notebook. If you contemplate doing cell synchrony experiments or having to make any sterile manipulations at 37°C, you should also allow space for a small laminar-flow unit with a 300 × 300 or 450 × 450 mm, (12–18-in) filter size, mounted either on a wall or on a stand over part of the bench. Alternatively, a small laminar-flow hood (not more than 1,000 mm (3 ft) wide) could be located in the room. The fan motor should be specified as for use in the tropics and should not run continuously. If it does run continuously, it will generate heat in the room and the motor may burn out.

Once a hot room is provided, others may wish to use the space for non-tissue-culture incubations, so the area of bench space provided should also take account of possible usage for incubation of tubes, shaker racks, etc.

Incandescent lighting is preferable to fluorescent, which can cause degradation of the medium. Furthermore, some fluorescent tubes have difficulty lighting up in a hot room.

The temperature of the hot room should be controlled within ±0.5°C at any point and at any time and depends on the sensitivity and accuracy of the control gear, the location of the thermostat sensor, the circulation of air in the room, the nature of the insulation, and the evolution of heat by other apparatus (stirrers, etc.) in the room.

Heaters. Heat is best supplied via a fan heater, domestic or industrial, depending on the size of the room. Approximately 2–3 kW per 20 m^3 (700 ft^3) will be required (or two heaters could be used, each generating 1.0–1.5 kW), depending on the insulation. The fan on the heater should run continuously, and the power to the heating element should come from a proportional controller.

Air circulation. A second fan, positioned on the opposite side of the room and with the airflow opposing that of the fan heater, will ensure maximum circulation. If the room is more than 2 × 2 m (6 × 6 ft), some form of ducting may be necessary. Blocking off the corners as in Fig. 3.5 is often easiest and most economical in terms of space in a square room. In a long, rectangular room, a false wall may be built at

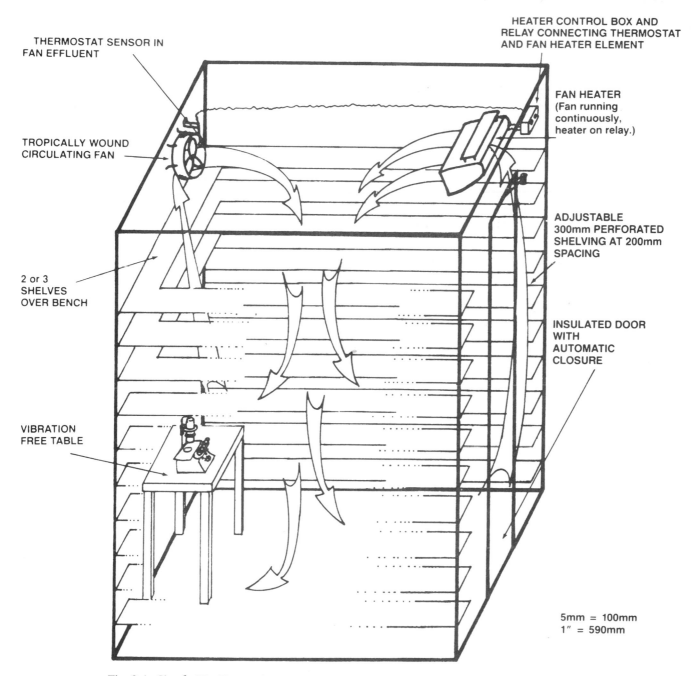

THERMOSTAT SENSOR IN
FAN EFFLUENT

TROPICALLY WOUND
CIRCULATING FAN

2 or 3
SHELVES
OVER BENCH

VIBRATION
FREE TABLE

HEATER CONTROL BOX AND
RELAY CONNECTING THERMOSTAT
AND FAN HEATER ELEMENT

FAN HEATER
(Fan running
continuously,
heater on relay.)

ADJUSTABLE
300mm PERFORATED
SHELVING AT 200mm
SPACING

INSULATED DOOR
WITH
AUTOMATIC
CLOSURE

5mm = 100mm
1″ = 590mm

Fig. 3.4. Simple Hot Room. Arrows represent air circulation. (Based on an original design by Dr. John Paul.)

either end, but be sure to insulate it from the room and make it strong enough to carry shelving.

Thermostats. Thermostats should be of the "proportional controller" type, acting via a relay to supply heat at a rate proportional to the difference between the room temperature and the set point. When the door opens and the room temperature falls, recovery will be rapid; on the other hand, the temperature will

not overshoot its mark, as the closer it approaches the set point, the less heat is supplied.

Ideally, there should be two separate heaters (H1 and H2), each with its own thermostat (HT1 and HT2). One thermostat (HT1) should be located diagonally opposite and behind the opposing fan (F2) and should be set at 37°C The other thermostat (HT2) should be located diagonally opposite and behind its opposing fan (F2) and should be set at 36°C (Fig. 3.5).

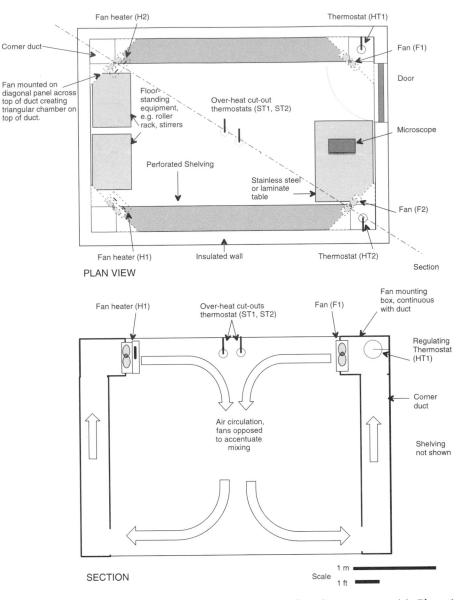

Fig. 3.5. Large Hot Room. Dual heating circuits and safety thermometers (a). Plan view. (b) Diagonal section. Arrows represent air circulation. Layout and design were developed in collaboration with Malcolm McLean of Boswell, Mitchell & Johnson (architects) and Jimmy Lindsay of Kenneth Munro & Associates (consulting engineers).

Two safety override cutout thermostats should also be installed, one in series with HT1 and set at 38°C and the other in series with the main supply to both heaters. If the first heater (H1) stays on above the set point, ST1 will cut out, and heater 2 (H2) will take over, regulating the temperature on HT2. If the second heater also overheats, ST2 will cut out all power to the heaters (Table 3.2). Warning lights should be installed to indicate when ST1 and ST2 have been activated. The thermostat sensors should be located in an area of rapid airflow, close to the effluent from the second, circulating, fan for greatest sensitivity. A rapid-response, high-thermal-conductivity sensor (thermistor or thermocouple) is preferred over a pressure-bulb type.

Overheating. Since so much care is taken to provide heat and replenish its loss rapidly, the problem of unwanted heat gain is often forgotten. It can arise because of (1) a rise in ambient temperature in the laboratory in hot weather or (2) heat produced from within the hot room by apparatus such as stirrer motors, roller racks, laminar-flow units, etc. Try to avoid heat-producing equipment in the hot room, and ar-

TABLE 3.2. Hot-Room Thermostats

Thermostat	37°C	<37°C	<<37°C	>37°C <38°C	>37°C >38°C	>37°C >39°C
HT1	O	I	I	O	O	O
ST1	I	I	I	I	O	O
HT2	O	O	I	O	O	O
ST2	I	I	I	I	I	O
Warning light ST1					◆	◆
Warning light ST2						◆

I = on; **O** = off; ◆ = pilot light illuminated.

HT1, regulating thermostat for heater H1; HT2, regulating thermostat for heater H2; ST1, safety over-ride cut-out thermostat for H1, ST2, safety over-ride cut-out thermostat for common supply to H1 and H2.

range for heat dissipation either by a thermostatically controlled fan extract (and inlet) or by an air conditioner. In either case, set the thermostat well below (>2°C) the heater thermostats so that the latter will regulate the temperature.

Access. If a proportional controller, good circulation, and adequate heating are provided, an air lock will not be required. The door should still be well insulated (with foam plastic or fiberglass), light, and easily closed—preferably, self-closing. It is also useful to have a hatch leading into the tissue culture area, with a shelf on both sides, so that cultures may be transferred easily into the room. The hatch door should have an insulated core as well. Locating the hatch above the bench will avoid any risk of creating a "cold spot" on the shelving.

Thermometer. A temperature recorder should be installed and should have a chart that is visible to the people working in the tissue culture room. The chart should be changed weekly. If possible, one high-level and one low-level warning light should be placed beside the chart or at a different, but equally obvious, location.

Service Bench

It may be convenient to position a bench for a cell counter, microscope, etc., close to the sterile handling area and either dividing the area or separating it from the other end of the lab. (See Figs. 3.1, 3.2, and 3.3.) The service bench should also provide for the storage of sterile glassware, plastics, pipettes, screw caps, syringes, etc., in drawer units below and open shelves above. The bench may also be used for other accessory equipment, such as a small centrifuge, and should make its contents readily accessible.

Preparation

The need for extensive preparation of media in small laboratories can be avoided if there is a proven source of reliable commercial culture media. While a large enterprise (approximately 50 people doing tissue culture) may still find it more economical to prepare its own media, smaller laboratories may prefer to purchase ready-made media. These laboratories would then need only to prepare reagents, such as salt solutions and ethylenediamine-tetraacetic acid (EDTA), bottle these and water, and package screw caps and other small items for sterilization. In that case, while the preparation area should still be clean and quiet, sterile handling is not necessary, as all the items will be sterilized.

If reliable commercial media are difficult to obtain, the preparation area should be large enough to accommodate a coarse and fine balance, a pH meter, and, if at all possible, an osmometer. Bench space will be required for dissolving and stirring solutions and for bottling and packaging various materials, and additional ambient and refrigerated shelf space will also be needed. If possible, an extra horizontal laminar-flow hood should be provided in the sterile area for filtering and bottling sterile liquids, and incubator space must be allocated for quality control of sterility (i.e., incubation of samples of media in broth and after plating out).

Heat-stable solutions and equipment can be autoclaved or dry-heat sterilized at the nonsterile end of the preparation area. Both streams then converge on the storage areas. (See Fig. 3.3.)

Washup

Washup and sterilization facilities are best situated outside the tissue culture lab, as the humidity and heat that they produce may be difficult to dissipate without increasing the airflow above desirable limits. Autoclaves, ovens, and distillation apparatus should be located in a separate room if possible (see Fig. 3.3), with an efficient extraction fan. The washup area should have plenty of space for soaking glassware and space for an automatic washing machine, should you require one. There should also be plenty of bench space for handling baskets of glassware, sorting pipettes, and packaging and sealing packs for sterilization. In addition, you will need space for a pipette washer and drier. If the sterilization facilities must be located in the tissue culture lab, place them nearest the air extract and farthest from the sterile handling area.

If you are designing a lab from scratch, then you can get sinks built in of the size that you want. Stainless steel or polypropylene are best, the former if you plan to use radioisotopes and the latter for hypochlorite disinfectants.

Sinks should be deep enough (450 mm, 18 in) to allow manual washing and rinsing of your largest items without having to stoop too far to reach into them. They should measure about 900 mm (3 ft) from floor to rim (Fig. 3.6). It is better to be too high than too low—a short person can always stand on a raised step to reach a high sink, but a tall person will always have to bend down if the sink is too low. A raised edge around the top of the sink will contain spillage and prevent the operator from getting wet when bending over the sink. The raised edge should go around behind the taps at the back.

Each washing sink will require four taps: a single cold-water tap, a combined hot-and-cold mixer, a cold tap for a hose connection for a rinsing device, and a nonmetallic or stainless-steel tap for deionized water from a reservoir above the sink. (See Fig. 3.6.) A centralized supply for deionized water should be avoided, as the piping can build up dirt and algae and is difficult to clean.

Trolleys or carts are often useful for collecting dirty glassware and redistributing fresh sterile stocks, but remember to allocate parking space for them.

Storage

Storage must be provided for the following items:

(1) Sterile liquids, at room temperature (salt solutions, water, etc.), at 4°C (media), and at −20°C or −70°C (serum, trypsin, glutamine, etc.)
(2) Sterile glassware, including media bottles and pipettes
(3) Sterile disposable plastics (e.g., culture flasks and Petri dishes, centrifuge tubes and vials, and syringes)
(4) Screw caps, stoppers, etc.
(5) Apparatus such as filters
(6) Gloves, disposal bags, etc.
(7) Liquid nitrogen to replenish freezers; the liquid nitrogen should be stored in two ways:
    (a) in dewars (25–50 l) under the bench, or

(b) in a large storage vessel (100–150 l) on a trolley or in storage tanks (500–1,000 l) permanently sited in a room of their own with adequate ventilation or, preferably, outdoors in secure, weatherproof housing (Fig. 3.7).

    Note: Liquid-nitrogen storage vessels can build up contamination, so they should be kept in clean areas.

△ *Safety Note.* Adequate ventilation must be provided for the room in which the nitrogen is stored and dispensed, preferably with an alarm to signify when the oxygen tension falls below safe levels. The reason for this safety measure is that filling, dispensing, and manipulating freezer stocks are accompanied by the evaporation of nitrogen, which can replace the air in the room (1 l liquid $N_2 \rightarrow$ ~700 l gaseous N_2).

(8) Cylinder storage for carbon dioxide, in separate cylinders for transferring to the laboratory as required

△ *Safety Note.* The cylinders should be tethered to the wall or bench in a rack (see Fig. 6.2).

A piped supply of CO_2 can be taken to work stations, or else the CO_2 supply piped from a pressurized tank of CO_2 that is replenished regularly (and must therefore be accessible to delivery vehicles). Which of the two means of storage you actually employ will be based on your scale of operation and unit cost, which can be obtained from suppliers. As a rough guide, 2 to 3 people will only require a few cylinders, 10 to 15 will probably benefit from a piped supply from a bank of cylinders, and for more than 15 it will pay to have a storage tank.

Storage areas 1–6 should be within easy reach of the sterile working area. Refrigerators and freezers should be located toward the nonsterile end of the lab, as the doors and compressor fans create dust and drafts and may harbor fungal spores. Also, refrigerators and freezers require maintenance and periodic defrosting, which creates a level and kind of activity best separated from your sterile working area.

The key ideal regarding storage areas is ready access for both withdrawal and replenishment of stocks. Double-sided units are useful because they may be restocked from one side and used from the other.

△ *Safety Note.* It is essential to have a lip on the edge of both sides of a shelf if the shelf is at a high level and glassware and reagents are stored on it. This prevents items being accidentally dislodged during use and when stocks are replenished.

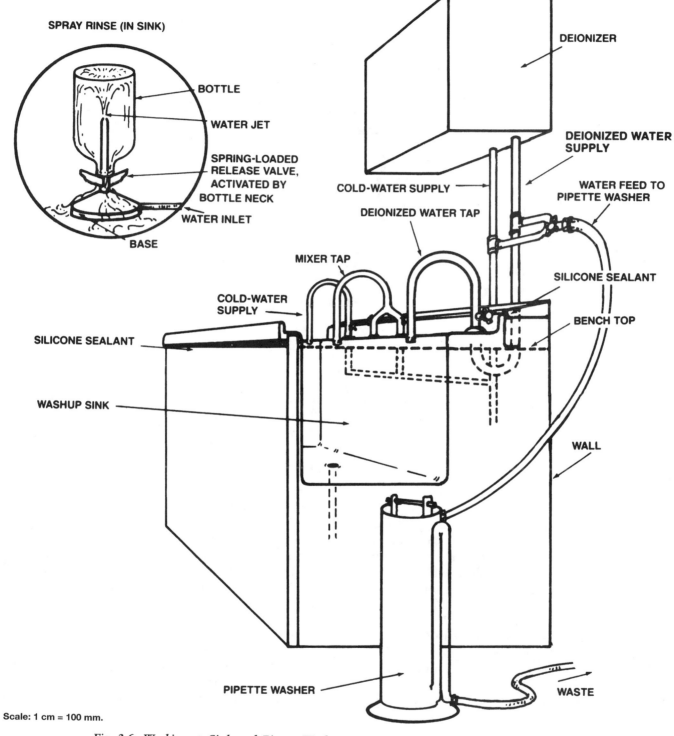

SPRAY RINSE (IN SINK)

BOTTLE

WATER JET

SPRING-LOADED RELEASE VALVE, ACTIVATED BY BOTTLE NECK

WATER INLET

BASE

DEIONIZER

DEIONIZED WATER SUPPLY

COLD-WATER SUPPLY

WATER FEED TO PIPETTE WASHER

DEIONIZED WATER TAP

MIXER TAP

SILICONE SEALANT

COLD-WATER SUPPLY

BENCH TOP

SILICONE SEALANT

WASHUP SINK

WALL

PIPETTE WASHER

WASTE

Scale: 1 cm = 100 mm.

Fig. 3.6. Washing-up Sink and Pipette Washer. Drawn to scale (bench height, 900 mm). Inset: bottle-rinsing device, located in sink.

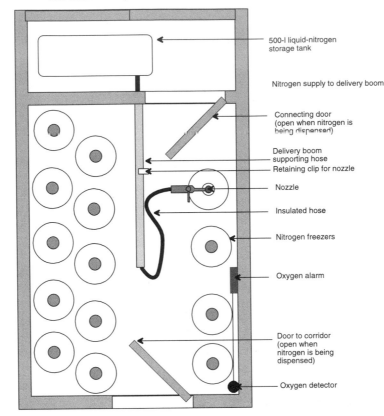

Outer slatted doors to provide ventilation to outside

500-l liquid-nitrogen storage tank

Nitrogen supply to delivery boom

Connecting door (open when nitrogen is being dispensed)

Delivery boom supporting hose

Retaining clip for nozzle

Nozzle

Insulated hose

Nitrogen freezers

Oxygen alarm

Door to corridor (open when nitrogen is being dispensed)

Oxygen detector

Fig. 3.7. Liquid-Nitrogen Store and Freezer Store. The liquid-nitrogen store is conveniently located on an outer wall with ventilation to the outside and easy access for deliveries. If the freezer store is adjacent, freezers may be filled directly from an overhead supply line and flexible hose. Doors are left open for ventilation during filling, and a wall-mounted oxygen alarm with a low-mounted detector sounds if the oxygen level falls below a safe level.

Remember to allocate sufficient space for storage, as doing so will allow you to make bulk purchases, thereby saving money, and, at the same time, reduce the risk of running out of valuable stocks at times when they cannot be replaced. As a rough guide, you will need 200 l (~8 ft³) of 4°C storage and 100 l (~4 ft³) of −20°C storage per person. The volume per person increases with fewer people. Thus, one person may need a 250-l (10-ft³) refrigerator and a 150-l (6-ft³) freezer. Of course, these figures refer to storage space only, and allowance must be made for access and working space in walk-in cold rooms and deep freezer rooms.

In general, separate −20°C freezers are better than a walk-in cold room. They are easier to clean out and maintain, and they provide better backup if one unit fails. You may also wish to consider whether a cold room has any advantage over refrigerators. No doubt, a cold room will give more storage per cubic meter, but the utilization of that space is important—how easy is it to clean and defrost, and how well can space be allocated to individual users? Several independent refrigerators will occupy more space than the equivalent volume of cold room, but may be easier to manage and maintain in the event of failure.

It is also well worth considering budgeting for additional freezer and refrigerator space to allow for routine maintenance and unpredicted breakdowns.

CHAPTER 4

Equipment

REQUIREMENTS OF A TISSUE CULTURE LABORATORY

Like those of most laboratories, the specific needs of a tissue culture laboratory can be divided into three categories: (1) essential—you cannot perform a job without them; (2) beneficial—the work would be done better, more efficiently, quicker, or with less labor; and (3) useful—they would make life easier, improve working conditions, reduce fatigue, enable more sophisticated analyses to be made, or generally make your working environment more attractive (Table 4.1).

The need for a particular piece of equipment is often very subjective—a product of personal aspirations, high-pressure salesmanship, technical innovation, and peer pressure. The real need is harder to define, but is determined objectively by the type of work, the saving in time that the equipment would produce, the greater technical efficiency in terms of asepsis, quality of data, analytical capability, and sample requirements, the saving in time or personnel, the number of people who would use the device, the available budget and potential cost benefit, and the special requirements of your own procedures. Priorities relating to equipment are given in Table 4.1.

ESSENTIAL EQUIPMENT

Laminar-Flow Hood

It is quite possible to carry out aseptic procedures without laminar flow if you have appropriate isolated, clean accommodation, with restricted access. However,

it is clear that, for most laboratories, which are typically busy and overcrowded, the simplest way to provide aseptic conditions is to use a laminar-flow hood. Usually, one hood is sufficient for two to three people. (See Laminar Flow in Chapter 5.) A horizontal flow hood is cheaper and provides the best sterile protection for your cultures, but it is really suitable only for preparing medium and other sterile reagents and for culturing nonprimate cells. It is also particularly suitable for dissecting nonprimate material prior to primary culture. For potentially hazardous materials (any primate, including human, cell lines, virus-producing cultures, radioisotopes, or carcinogenic or toxic drugs), a Class II or Class III biohazard cabinet should be used (Figs. 4.1, 6.4).

△ *Safety Note.* It is important to familiarize yourself with local and national biohazard regulations before installing equipment, as legal requirements and recommendations vary. (See Biohazards in Chapter 6.)

Choose a hood that is (1) large enough (usually, a minimum working surface of 1,200 mm [4 ft] wide × 600 mm [2 ft] deep) for one person to use at a time; (2) quiet (noisy hoods are more fatiguing); (3) easily cleaned both inside the working area and below the work surface in the event of spillage and (4) comfortable to sit at (some cabinets have awkward ducting below the work surface, which leaves no room for your knees; have lights or other accessories above that strike your head; or have screens that obscure your vision). The front screen should be able to be raised, lowered, or removed completely, to facilitate cleaning and handling bulky culture apparatus. Remember, however, that a biohazard cabinet will not give you,

TABLE 4.1. Tissue Culture Equipment

Basic requirements	Nonessential, but beneficial	Useful additions
Laminar-flow hood (biohazard if for human cells)	Cell counter	Glassware washing machine
Incubator (humid CO_2 incubator if using open plates or dishes)	Peristaltic pump	$-70°C$ freezer
5% CO_2 cylinder (for gassing cultures)	Pipettor(s)	Conductivity meter
Liquid CO_2 cylinders, without siphon (for CO_2 incubator)	pH meter	Osmometer
Balance	Sterilizing oven	Polyethylene bag sealer (for packaging sterile items for long term storage)
Sterilizer (autoclave, pressure cooker)	Hot room	Computer for freezer records and cell line database
Refrigerator	Temperature recorders on sterilizing oven and autoclave and in hot room	Colony counter
Freezer (for $-20°C$ storage)	Phase-contrast, fluorescence microscope	High-capacity centrifuge (6 × 1 l)
Inverted microscope	Pipette plugger	CCTV and printer for inverted microscope(s)
Soaking bath or sink	Pipette drier	Time-lapse video equipment
Deep washing sink	Automatic dispenser	Cell sizer (e.g., Schärfe, Coulter)
Pipette cylinder(s)	Trolleys or carts	Portable temperature recorder for checking hot room or incubators
Pipette washer	Drying oven(s), high and low temperature	Plastics shredder/sterilizer
Still or water purifier	Roller racks for roller bottle culture	Controlled-rate cooler (for cell freezing)
Bench centrifuge	Piped CO_2 supply from cylinder store	Fluorescence-activated cell sorter
Liquid N_2 freezer (~35 l, 1,500–3,000 ampules)	Automatic change-over device on CO_2 cylinders	Confocal microscope
Liquid N_2 storage dewar (~25 l)		Microtitration plate scintillation counter
Slow-cooling device for cell freezing (see Chapter 17)		Centrifugal elutriator centrifuge and rotor
Magnetic stirrer racks for suspension cultures		
Hemocytometer		

the operator, the required protection if you remove the front screen.

Insist that you be allowed to examine and sit at a hood in order to simulate using it before purchasing it. Answer the following questions:

(1) Can you get your knees under the hood while sitting comfortably and close enough to work, with your hands at least halfway into the hood?
(2) Is there a footrest in the correct place?
(3) Are you able to see what you are doing without placing strain on you neck?
(4) Is the work surface perforated, and, if so, will that give you trouble with spillage? (A solid work surface vented at the front and back is preferable.)
(5) Is the work surface easy to remove for cleaning?
(6) If the work surfaces are raised, are the edges sharp, or are they rounded so that you will not cut yourself when cleaning out the hood?
(7) Are there crevices in the work surface that might accumulate spillage and contamination? Some cabinets have sectional work surfaces that are easier to remove for cleaning, but leave capillary spaces when in place.
(8) Is the lighting convenient and adequate?

(9) Will you be able to get the hood into the tissue culture laboratory?
(10) When in place, can the hood be serviced easily? (Ask the service engineer, not the salesman!) Hoods require 500 mm (~2 ft) of lateral clearance.
(11) Will there be sufficient headroom for venting to the room or for ducting to the exterior?
(12) Will the airflow from other cabinets, the room ventilation, or independent air-conditioning units interfere with the integrity of the work space of the hood? That is, will air spill in or aerosols leak out due to turbulence? Meeting this condition will require a minimum of 1,500 mm of face-to-face separation and correct testing by an engineer with experience in microbiological safety cabinets.

Incubator

If a hot room (see Hot Room in Chapter 3) is not available, it may be necessary to buy an equivalent dry incubator. Even with a hot room, it is sometimes convenient to have another incubator close to the hood for trypsinization. The incubator should be large enough (probably 50–200 l (1.5–6 ft³) per person) and

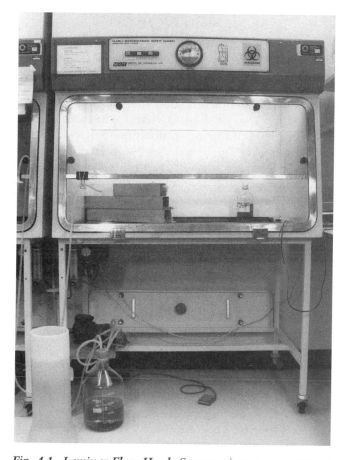

Fig. 4.1. Laminar-Flow Hood. Square pipette cans prevent rolling. An automatic pipetting device is placed at the right-hand side; the pipette hod is placed on the floor at the left. A peristaltic pump, connected to a receiver vessel, is shown on the left side below the hood, with a foot switch to activate the pump. The suction line from the pump is attached to a clip at the left of the aperture to the work area, and a delivery tube travels across, below the hood, to provide a supply of mixed CO_2 in air. The orientation would be reversed for left-handed individuals. Below the hood, on the left-hand side, is a CO_2 and air supply with flowmeters to regulate the flow rate and gas mixture. (This may be replaced with an electronic gas mixer [Signal].) The blanking screen, which may be used to seal off the hood when not in use, has been placed at the back below the hood, but can be hung alongside it.

should have forced-air circulation, temperature control to within ±0.2°C, and a safety thermostat that cuts off if the incubator overheats or, better, that regulates the incubator if the first thermostat fails. The incubator should be resistant to corrosion (e.g., it could be made of stainless steel, although anodized aluminum is acceptable for a dry incubator) and easily cleaned. A double chamber, one above the other, independently regulated (see Fig. 4.2), is preferable to one large incubator because it can accommodate more cultures with better temperature control, and if one

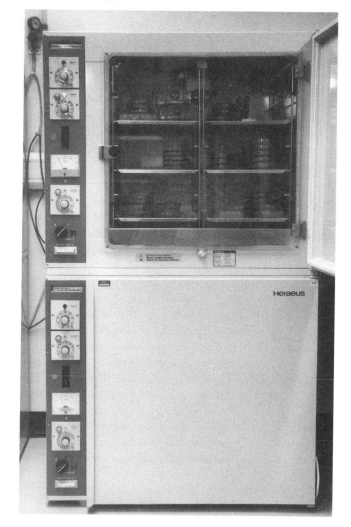

Fig. 4.2. Two-tiered CO_2 Incubator. This is a dual-chamber model (Heraeus) with the top chamber open. Dual controls for temperature regulation and the CO_2 controller are located on the side panels. (See also Fig. 4.3.)

half fails or needs to be cleaned, the other can still be used.

Many incubators have a heated water jacket to distribute heat evenly around the cabinet, thus avoiding the formation of cold spots. These incubators also hold their temperature for longer in the event of a heater failure or cut in power. However, new high-efficiency insulation and diffuse surface heater elements have all but eliminated the need for a water jacket and make moving the incubator much simpler. (A water jacket generally needs to be emptied if the incubator is to be moved.)

Incubator shelving is usually perforated to facilitate the circulation of air. However, the perforations can lead to irregularities in cell distribution in monolayer cultures, with variations in cell density following the pattern of spacing on the shelves. The variations may be due to convection currents generated over points of contact relative to holes in the shelf, or they may be related to areas that cool down more quickly when the door is opened. Although no problem may arise in routine maintenance, flasks and dishes should be placed on an insulated tile or metal tray in experiments in which uniform density is important.

CO₂ Incubator

Although cultures can be incubated in sealed flasks in a regular dry incubator or a hot room, some vessels e.g., Petri dishes or multiwell plates, require a controlled atmosphere with high humidity and elevated CO_2 tension. The cheapest way of controlling the gas phase is to place the cultures in a plastic box, or chamber (Bellco, ICN). Gas the container with the correct CO_2 mixture and then seal it. If the container is not completely filled with dishes, include an open dish of water to increase the humidity inside the chamber.

CO_2 incubators (Fig. 4.2) are more expensive, but their ease of use and superior control of CO_2 tension and temperature (anaerobic jars and desiccators take longer to warm up) justify the expenditure. A controlled atmosphere is achieved by blowing air over a humidifying tray (Fig. 4.3) and controlling the CO_2 tension with a CO_2-monitoring device, which draws air from the incubator into a sample chamber, deter-

mines the concentration of CO_2, and injects pure CO_2 into the incubator to make up any deficiency. Air is circulated around the incubator by a fan to keep both the CO_2 level and the temperature uniform. Most inexpensive CO_2 controllers need to be calibrated every few months; in contrast, most "top-of-the-line" models reset the zero of the CO_2 detector automatically.

Frequent cleaning of incubators—particularly humidified ones—is essential (see Sources of Contamination in Chapter 18), so the interior should dismantle readily without leaving inaccessible crevices or corners. Flasks or dishes, or boxes containing them, that are taken from the incubator to the laminar-flow or other sterile workstation should be swabbed with alcohol before being opened. (See also Swabbing in Chapter 5.)

Sterilizer

The simplest and cheapest sterilizer is a domestic pressure cooker that generates 100 kPa (1 atm, 15 lb/in²) above ambient. Alternatively, a bench-top autoclave (Fig. 4.4) gives automatic programming and safety locking. A larger, freestanding model with a programmable timer, a choice of pre- and poststerilization evacuation, and temperature recording (Fig. 4.5) has a greater capacity, provides more flexibility, and offers the opportunity to comply with good laboratory practice (GLP).

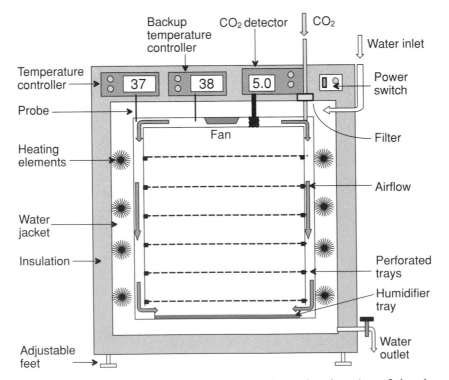

Fig. 4.3. CO₂ Incubator Design. Front view of control panel and section of chamber of a stylized humid CO_2 incubator (not representative of any particular make).

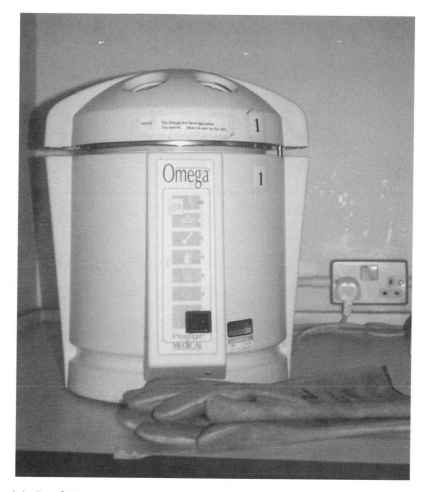

Fig. 4.4. Bench-Top Autoclave. Simple, top-loading autoclave, from Prestige Medical.

A "wet" cycle (water, salt solutions, etc.) is performed without evacuation of the chamber before or after sterilization. Dry items (instruments, swabs, screw caps, etc.) require that the chamber be evacuated before sterilization or the air replaced by downward displacement, to allow efficient access of hot steam. The chamber should also be evacuated *after* sterilization, to remove steam and promote subsequent drying; otherwise the articles will emerge wet, leaving a trace of contamination from the condensate on drying. To minimize this risk when a "postvac" cycle is not available, always use deionized or reverse-osmosis water to supply the autoclave. If you require a high sterilization capacity (>300 l, 9 ft³), buy two medium-sized autoclaves rather than one large one, so that during routine maintenance and accidental breakdowns you still have one functioning machine. Furthermore, a medium-sized machine will heat up and cool down more quickly and can be used more economically for small loads. Leave sufficient space around the sterilizer for maintenance and ventilation, provide adequate air extraction to remove heat and steam, and ensure that a suitable drain is available for condensate.

Most small autoclaves come with their own steam generator (calorifier), but larger machines may have a self-contained steam generator, a separate steam generator, or the facility to use a steam line. If high-pressure steam is available on-line, that will be the cheapest and simplest method of heating and pressurizing the autoclave; if not, it is best to purchase a sterilizer complete with its own self-contained steam generator. Such a sterilizer will be cheaper to install and easier to move. With the largest machines, you may not have a choice, as they are frequently offered only with a separate generator. In that case, you will need to allow space for the generator at the planning stage.

Refrigerators and Freezers

Usually, a domestic refrigerator or freezer is quite efficient and cheaper than special laboratory equipment. Domestic refrigerators are available without freezers ("larder refrigerators"), giving more space and elimi-

Fig. 4.5. Autoclave. Medium-sized (300 l; 10 ft³) laboratory autoclave with square chamber for maximum load. The recorder on the top console is connected to a probe in the bottle in the center of the load.

nating the need for defrosting. However, if your require 400 l, 12 ft³, or more storage (see Storage in Chapter 3), a large hospital, blood-bank, or catering refrigerator may be better.

If space is available and the number of people using tissue culture is more than three or four, it is worth considering the installation of a cold room, which is more economical in terms of space than are several separate refrigerators and is also easier to access. The walls should be smooth and easily cleaned, and the

racking should be on castors to facilitate cleaning. Cold rooms should be cleaned out regularly to eliminate old stock, and the walls and shelves should be washed to minimize fungal contamination.

Similar advice applies to freezers—several inexpensive domestic freezers will be cheaper and just as effective as a specialized laboratory freezer. Most tissue culture reagents will keep satisfactorily at −20°C, so an ultradeep freeze is not essential. Nor is a deep-freeze room recommended—they are very difficult and unpleasant to clear out, and they create severe problems in regard to relocating their contents if extensive maintenance is required.

While autodefrost freezers may be bad for some reagents (enzymes, antibiotics, etc.), they are quite useful for most tissue culture stocks, whose bulk and nature preclude severe cryogenic damage. Conceivably, serum could deteriorate during oscillations in the temperature of an autodefrost freezer, but in practice it does not. Many of the essential constituents of serum are small proteins, polypeptides, and simpler organic and inorganic compounds that are be insensitive to cryogenic damage, particularly if solutions are stored in volumes greater than or equal to 100 ml.

Inverted Microscope

It cannot be overemphasized that, in spite of considerable progress in the quantitative analysis of cultured cells and remote sensing techniques, it is still vital to look at cultures regularly. A morphological change is often the first sign of deterioration in a culture (see Cell Morphology in Chapter 12, and Fig. 12.1), and the characteristic pattern of microbiological infection (see Visible Microbial Contamination in Chapter 18, and Fig. 18.1) is easily recognized.

A simple inverted microscope is essential (Fig. 4.6). Make certain that the stage is large enough to accommodate large roller bottles between it and the condenser (see Roller Culture in Chapter 25) in case you should require such bottles. Many simple and inexpensive inverted microscopes are available on the market, but if you foresee the need for photographing living cultures, then you should invest in a microscope with high-quality optics and a long-working-distance phase-contrast condenser and objectives, with provisions for a camera (e.g., Olympus CM, Zeiss Axiovert, Leitz Diavert). The Nikon marking ring is a useful accessory to the inverted microscope. This device is inserted in the nosepiece in place of an objective and can be used to mark the underside of a dish in which an interesting colony or patch of cells is located. The colony can then be picked (see Isolation of Clones in Chapter 13, and Fig. 13.10) or the development of a particularly interesting area in a culture followed.

Washing Up

Soaking baths or sinks. Soaking baths or sinks should be deep enough so that all your glassware (except pipettes and the largest bottles) can be totally immersed in detergent during soaking, but not so deep that the weight of the glass is sufficient to break smaller items at the bottom. A sink that is 400 mm (15 in) wide × 600 mm (24 in) long × 300 mm (12 in) deep is about right.

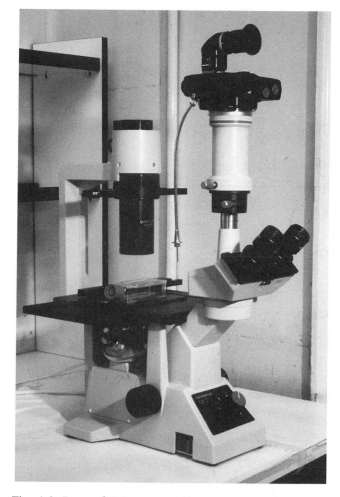

Fig. 4.6. *Inverted Microscope.* Olympus CK2 inverted microscope fitted with phase-contrast optics and trinocular head with automatic camera.

Pipette cylinders, or hods. Pipette cylinders should be made from polypropylene and should be freestanding and distributed around the lab, one per workstation, with sufficient numbers in reserve to allow full cylinders to stand for 2 h in disinfectant before washing.

Pipette washer. Reusable pipettes are easily washed in a standard siphon-type washer. (See Pipettes in Chapter 10, and Fig. 10.5.) The washer should be placed at floor rather than bench level to avoid awkward lifting of the pipettes and should be connected to the deionized water supply so that the final rinse can be done in deionized water. If possible, a simple change-over valve should be incorporated into the deionized water feed line. (See Fig. 3.6.)

Pipette drier. If a stainless-steel basket is used in the washer, pipettes may subsequently be transferred directly to an electric drier. Alternatively, pipettes can be dried on a rack or in a regular drying oven.

Sterilizing and Drying

Although all sterilizing can be done in an autoclave, it is preferable to sterilize pipettes and other glassware by dry heat, avoiding the possibility of both chemical contamination from steam condensate and corrosion of pipette cans. Such sterilization, however, will require a high-temperature (160–180°C) fan-powered oven to ensure even heating throughout the load. As with autoclaves, do not get an oven that is too big for the amount or size of glassware that you use. It is better to use two small ovens than one big one; heating is easier, more uniform, quicker, and more economical when only a little glassware is being used. You are also better protected during breakdowns.

Water Purification

Purified water is required for rinsing glassware, dissolving powdered media, and diluting concentrates. The first of these purposes is usually satisfied by deionized or reverse osmosis water, but the second and third require ultrapure water (UPW), which demands a three- or four-stage process (Fig. 4.7; see also Fig. 10.10). The important principle is that each stage be qualitatively different; reverse osmosis may be followed by charcoal filtration, deionization, and micropore filtration (e.g., via a filter from Millipore), or distillation (with a silica-sheathed element) may be substituted for the first stage. Reverse osmosis is cheaper if you pay the fuel bills, if you do not, distillation is better and more likely to give a sterile product. If reverse osmosis is used, the type of cartridge should be chosen to suit the pH of the water supply, as some membranes can become porous in extreme pH conditions.

The deionizer should have a conductivity meter monitoring the effluent, to indicate when the cartridge must be changed. Other cartridges should be dated and replaced according to the manufacturer's instructions. A total organic carbon (TOC) meter (millipore) can be used to monitor colloids.

Purified water should not be stored, but should be recycled through the apparatus continually to minimize infection with algae. Any tubing or reservoirs in

Fig. 4.7. Preparation of Ultrapure Water. Tap water first passes through a reverse-osmosis unit on the right and then goes to the storage tank on the left. It then passes through carbon filtration and deionization (center unit) before being collected via a micropore filter (Millipore MillQ).

the system should be checked regularly (every three months or so) for algae, cleaned out with hypochlorite and detergent (e.g., Chlorox or Chloros), and thoroughly rinsed in purified water before reuse.

Water is the simplest, but probably the most critical, constituent of all media and reagents, particularly serum-free media. (See Chapter 9.) A good quality-control measure is to check the plating efficiency of a sensitive cell line (see Culture Testing in Chapter 10, Protocol 20.9) at regular intervals.

Centrifuge

Periodically, cell suspensions require centrifugation to increase the concentration of cells or to wash off a reagent. A small bench-top centrifuge, preferably with proportionally controlled braking, is sufficient for most purposes. Cells sediment satisfactorily at 80–100 g; higher gravity may cause damage and promote agglutination of the pellet. A large-capacity refrigerated centrifuge, say, 4 × 1 litre or 6 × 1 litre, will be required if large-scale suspension cultures (see Suspension in Chapter 25) are contemplated.

Cryostorage Container

Procedures for cryopreservation are dealt with in detail in Chapter 19 (see Preservation), but the basic facilities are considered here. The freezing process can be carried out satisfactorily without sophisticated equipment, but storage requires a properly constructed liquid-nitrogen freezer and storage dewar. (See Fig. 19.1.) Freezers range in size from around 25 l to 500 l (i.e., 250–15,000 1-ml ampules). It is best to freeze a minimum of 5 to 10 ampules of seed stock for each cell line, plus 20 ampules of using stock for a cell line in current use, and 100 for one in continuous use. A capacity of 1,200–1,500 ampules is appropriate for most small laboratories.

The choice of freezer is determined by three factors: (1) capacity (the number of ampules the freezer can hold), (2) economy and static holding time (the time taken for all the liquid nitrogen to evaporate), and (3) convenience of access of the contents. The first two factors are governed by the evaporation rate, which in turn depends on the frequency of access. Generally speaking, number 2 is inversely proportion to each of numbers 1 and 3.

Two main types of freezer are available: narrow necked with slow evaporation, but more difficult access; and wide necked with easier access, but three times the evaporation rate. (See Fig. 19.8.) If the cost of liquid nitrogen and its supply present no problem, then a wide-necked freezer may be more convenient (e.g., a Taylor Wharton 3K or equivalent, of 3,000 ampules capacity and 3–5 l/d evaporation rate), although the holding time will be only about one week to 10 days. A narrow-necked freezer, on the other hand, will be more economical and will hold its contents up to two months (e.g., an Aire Liquide 35-1, of 1,500 ampules capacity and 0.5 l/d evaporation rate).

It is also possible to compromise by purchasing a narrow-necked freezer with a rack inventory system (e.g., from Thermolyne, Cryomed, or some other major suppliers; see Trade Index). These freezers have the advantage of storaging the ampules in rectangular drawers, rather than triangular, but still have a relatively narrow neck with consequent savings in evaporation of liquid nitrogen and a longer holding time. The total storage per unit volume is still not as great as that of a wide-necked freezer, and they can be a little awkward to use, but they do represent the best compromise between access and holding time.

If you require bulk storage (~10,000 ampules), then you will need to consider a vessel of around 300 l capacity. Wide-necked freezers are most common in this size because of the mechanical difficulties in operating narrow-necked freezers of high capacity, but the latter are available and will save a considerable amount in expenditure on liquid nitrogen. At 300 l, the evaporation rate is approximately 10 l/d in a wide-necked freezer.

The advantages of gas-phase and liquid-phase storage are discussed in Chapter 19 (see Cryofreezers), but one major implication of storing ampules in the gas phase is that the liquid phase is then necessarily reduced to the space below your ampule storage area, usually 10–20% of the full volume. Hence, the static holding time is reduced to a tenth or a fifth that of the filled freezer, filling must be carried out more regularly, and the chances of accidental thawing are increased. Where the investment is high (many ampules or rare cell strains), automatic alarm systems should be fitted, and for high-capacity freezers, an automatic filling system is recommended. However, even automatic systems can fail, and a twice-weekly check and record of liquid levels with a dipstick is essential.

An appropriate storage vessel should also be purchased to enable a backup supply of liquid nitrogen to be held. The size of the vessel depends on (1) the size of the freezer, (2) the frequency and reliability of delivery of liquid nitrogen, and (3) the rate of evaporation of the liquid nitrogen. A 40-l wide-necked freezer will require about 20–30 l/week, so a 50-l dewar flask (or two 25-l flasks, which are easier to handle) is advisable. A 35-l narrow-necked freezer, on the other hand, using 5–10 l/week, will require only a 25-l dewar. Larger freezers are best supplied on-line from a dedicated storage tank (e.g., a 160-l storage vessel linked to a 320-l freezer with automatic filling and alarm, or a 500-l tank for a larger freezer or for more smaller freezers).

Balance

Although most laboratories obtain media that are already prepared, it may be cheaper to prepare some reagents in-house. Doing so will require a balance (an electronic one with automatic tare is best) capable of weighing items from around 10 mg up to 100 g or even 1 kg, depending on the scale of the operation. If you are a service provider, it is often preferable to prepare large quantities, sometimes 10× concentrated, so the amounts to be weighed can be quite high. It may prove better to buy two balances, coarse and fine, as the outlay may be similar and the convenience and accuracy are increased.

Hemocytometer

It is essential to have some means of counting cells. A hemocytometer slide is the cheapest option and has the added benefit of allowing cell viability to be determined by dye exclusion. (See Hemocytometer in Chapter 20; Figure 20.1, and Protocol 20.1.)

BENEFICIAL EQUIPMENT

The preceding sections have described equipment that is essential for a modest tissue culture facility; but several other pieces of equipment will make your laboratory easier to use, more efficient, and better controlled. If the facility is to be used frequently for more than one or two people, then many of the items that follow will probably join the "essential" list.

Cell Counter

A cell counter (see Fig. 20.2) is a great advantage when more than two or three cell lines are carried and is essential for precise quantitative growth kinetics. Several companies now market models ranging in sophistication from simple particle counting up to automated cell counting and size analysis. For routine counting, the Schärfe Casy 1 or Coulter Z1 is more than adequate and less expensive than equipment with advanced cell-sizing capability. (See also Electronic Counting in Chapter 20.)

Aspiration Pump

A peristaltic pump may be used to remove spent medium or other reagents from a culture flask. Pumping is a rapid and efficient way to deal with several flasks at a time or with large volumes, and the effluent can be collected directly into disinfectant (see Disposal in Chapter 6) in a vented container, with minimal risk of discharging aerosol into the atmosphere, particu-

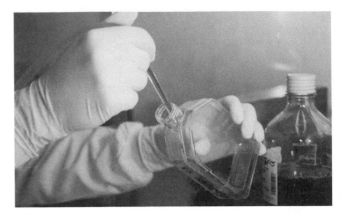

Fig. 4.8. Withdrawing Medium by Suction. Suction line connected via a peristaltic pump to a waste receiver. (See Fig. 4.9.)

larly if the vent carries a cotton plug or micropore filter. (Figs. 4.8, 4.9). The inlet line should extend further below the stopper than the outlet, by at least 5 cm (2 in), so that waste does not splash back into the vent. The pump tubing should be checked regularly for wear, and the pump should be operated by a self-canceling foot switch.

A vacuum pump, similar to that supplied for sterile filtration, may be used instead of a peristaltic pump. If necessary, the same pump could serve both purposes; however, a trap will be required to avoid the risk of waste entering the pump. The effluent should be collected in a reservoir and a disinfectant such as hypochlorite added (see Disposal in Chapter 6) when work is finished and left for at least 2 h before the reservoir is emptied. (See Fig. 4.9.) A hydrophobic micropore filter (Pall Gelman), and a second trap should be placed in the line to the pump to prevent fluid or

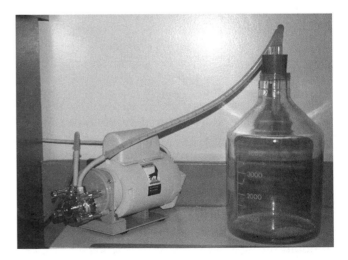

Fig. 4.9. Aspiration Pump. Peristaltic pump on a suction line from the hood to a waste receiver.

aerosol from being carried over. Do not draw air through a pump from a reservoir containing hypochlorite, as the free chlorine will corrode the pump and could be toxic. Also, avoid vacuum lines; if they become contaminated with fluids, they can be very difficult to clean out.

To keep effluent from running back into a flask, always switch on the pump before inserting a pipette in the tubing. (See Standard Procedure in Chapter 5.)

pH Meter

A simple pH meter for the preparation of media and special reagents is a useful addition to the tissue culture area. Although a phenol-red indicator is sufficient for monitoring pH in most solutions, a pH meter will be required when phenol red cannot be used (e.g., in preparing cultures for fluorescence assays and in preparing stock solutions).

Conductivity Meter

When solutions are prepared in the laboratory, it is essential to perform quality-control measures to guard against errors. (See Quality Control in Chapter 10.) A simple check of ionic concentration can be made with a conductivity meter against a known standard, such as normal saline (0.15 M).

Osmometer

One of the most important physical properties of a culture medium, and one that is often difficult to predict, is the osmolality. While the conductivity is controlled by the concentration of ionized molecules, nonionized particles can also contribute to the osmolality. An osmometer (see Fig. 8.1) is therefore a useful accessory to check solutions as they are made up, to adjust new formulations, or to compensate for the addition of reagents to the medium. Osmometers usually work by depressing the freezing point of a medium or elevating its vapor pressure. Choose one with a low sample volume ($\leq$1 ml), since you may want to measure a valuable or scarce reagent on occasion, and the accuracy ($\pm$10 mOsmol/kg) may be less important than the value or scarcity of the reagent.

Upright Microscope

An upright microscope may be required, in addition to an inverted microscope, for chromosome analysis, mycoplasma detection, and autoradiography. Select a high-grade research microscope (e.g., Leica, Zeiss, or Nikon) with regular bright-field optics up to 100× objective magnification; phase contrast up to at least 40×, and preferably 100×, objective magnification; and fluorescence optics with epi-illumination and 40× and 100× objectives for mycoplasma testing by fluorescence (see Protocol 18.2) and fluorescent

antibody observation. Leica supplies a 50× water-immersion objective, which is particularly useful for observing routine mycoplasma preparations with Hoechst stain. An automatic photographic camera or charge-coupled device (CCD) should also be fitted for photographic records of permanent preparations.

Dissecting Microscope

Dissection of small pieces of tissue (e.g., embryonic organs or tissue from smaller invertebrates) will require a dissecting microscope (Nikon, Olympus, Leica). A dissecting microscope is also useful for counting colonies—and essential for small colonies in agar—and for setting the lower size threshold for automatic counting of monolayer colonies. A dissecting microscope is essential as well for picking colonies from agar.

Temperature Recording

A recording thermometer with ranges from below −50°C to about +200°C will enable you to monitor frozen storage, the freezing of cells, incubators, and sterilizing ovens with one instrument fitted with a resistance thermometer or thermocouple with a long Teflon-coated lead.

Ovens, incubators, and hot rooms should be monitored regularly for uniformity and stability of temperature control. Recording thermometers should be permanently fixed into the hot room, sterilizing oven, and autoclave, and dated records should be kept to check regularly for abnormal behavior, particularly in the event of a problem arising.

Magnetic Stirrer

Certain specific requirements apply to magnetic stirrers. A rapid stirring action for dissolving chemicals is available with any stirrer, but for the stirrer to be used for disaggregation (see Protocol 11.5) or suspension culture (see Protocol 12.3), (1) the motor should not heat the culture (use the rotating-field type of drive or a belt drive from an external motor); (2) the speed must be controlled down to 50 rpm; (3) the torque at low rpm should still be capable of stirring up to 10 l of fluid; (4) the device should be capable of maintaining several cultures simultaneously; (5) each stirrer position should be individually controlled; and (6) a readout of rpm should appear for each position. It is preferable to have a dedicated stirrer for culture work (e.g., Techne or Bellco).

Roller Racks

Roller racks are used to scale up monolayer culture. (See Protocol 25.3.) The choice of apparatus is determined by the scale (i.e., the size and number of bottles to be rolled). The scale may be calculated from the number of cells required, the maximum attainable cell density, and the surface area of the bottles. (See Table 7.1.) A large number of small bottles gives the highest surface area, but tends to be more labor intensive in handling, so a usual compromise is bottles around 125 mm (5 in) in diameter and various lengths from 150–500 mm (6–20 in). The length of the bottle will determine the maximum yield, but is limited by the size of the rack; the height of the rack will determine the number of tiers (i.e., rows) of bottles. Although it is cheaper to buy a larger rack than several small ones, the latter alternative (1) allows you to build up your rack gradually (having confirmed that the system works), (2) can be easier to locate in a hot room, and (3) will still allow you to operate if one rack requires maintenance. Bellco or Life Sciences bench-top models may be satisfactory for smaller scale activities, and Bellco and New Brunswick Scientific make larger racks.

Fluid Handling

An increase in the caution applied to the handling of tissue cultures and reagents, combined with an increase in the scale of operation in many laboratories, has eliminated traditional mouth pipetting and transformed the use of pipettes and other devices for the handling of reagents.

Four main tasks are required for fluid-transfer devices: (1) nonsterile dispensing of water, salt solutions, and other reagents; (2) simple addition of sterile liquids to, and their removal from, small numbers of flasks, dishes, etc; (3) repeated additions to and removals from a large number of vessels, and (4) simultaneous addition of reagents to multiple replicate cultures. The choice of equipment is also determined by the volume of liquid being transferred, the frequency of the operation, the relative cost of labor or equipment, the ergonomic efficiency, and the safety of the procedure.

Removal of fluids. Fluids can be removed simply and rapidly with the use of a vacuum pump or vacuum line, with suitable reservoirs to collect the effluent and prevent contamination of the pump, or with the use of a simple peristaltic pump discharging into a reservoir with disinfectant. (See Aspiration Pump above.)

Nonsterile dispensing. This term applies to the preparation of reagents for sterilization and the dispensing of nonsterile reagents such as counting fluid. Bottle-top dispensers (Fig. 4.10) are suitable for volumes up to about 50 ml; above that level, gravity dispensing from a reservoir, which may be a graduated bottle or plastic bag, is acceptable if accuracy is not

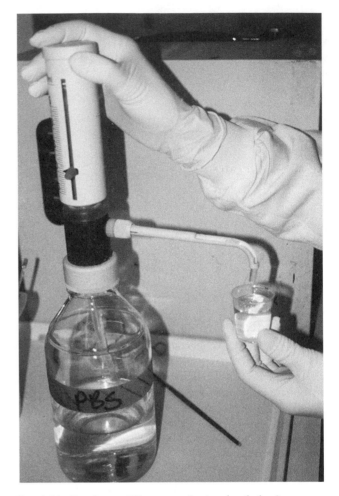

Fig. 4.10. Bottle-top Dispenser. Spring-loaded piston connected to a two-way valve, alternately drawing a preset volume from the reservoir and dispensing through a side arm. The device is useful for repetitive dispensing of nonsterile solutions such as PBSA for cell counting. Some models can be autoclaved for sterile use.

critical. If the volume dispensed is more critical, then a peristaltic pump (Fig. 4.11) is preferable. The duration of dispensing and diameter of tube determine the volume and accuracy of the dispenser. A long dispense cycle with a narrow delivery tube will be more accurate, but a wide-bore tube will be faster.

When using bottletop dispensers for cell counting, let the fluid settle for a few minutes before counting, as small bubbles are often generated by rapid dispensing, and these will be counted by the cell counter.

Simple pipetting. Simple pipetting is one of the most frequent tasks required in the routine handling of cultures. While a rubber bulb or other proprietary pipetting device (see Fig. 5.6) is cheap and simple to use, speed, accuracy, and reproducibility are greatly enhanced by a motorized pipetting aid (Fig. 4.12), which

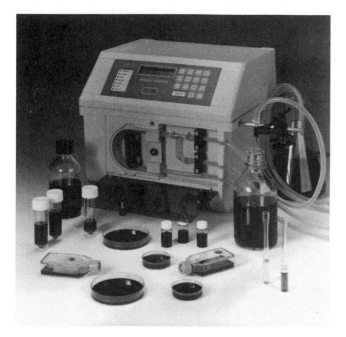

Fig. 4.11. *Automatic Dispenser.* The Perimatic Premier, suitable for repetitive dispensing and dilution in the 1–1,000-ml range. If the device is used for sterile operations, only the delivery tube needs to be autoclaved.

may be obtained with a separate or built-in pump and can be mains operated or rechargeable. (See Trade Index: Sources of Materials, Pipetting Aids.) The major determinants in choosing a pipetting aid are the weight and feel of the instrument during continuous use, and it is best to try one out before purchasing it. Pipetting aids usually have a filter at the pipette insert to minimize the transfer of contaminants. Some are disposable, and some are reusable after resterilization. (See Figure 6.1 for the proper method of inserting a pipette into a pipetting device.)

Pipettors. These devices originated from micropipettes marketed by Eppendorf and used for dispensing 10–200 μl. Since the working range now extends up to 5 ml, the term "micropipette" is not always appropriate, and the instrument is more commonly called a pipettor (Fig. 4.13). Only the tip needs to be sterile, but the length of the tip then limits the size of vessels used. If a sterile fluid is withdrawn from a container with a pipettor, the nonsterile shank must not touch the sides of the container. Reagents of 10–20 ml in volume may be sampled in 5-μl–1-ml volumes from a universal container or in 5–200-μl volumes from a bijou bottle. Eppendorf-type tubes may also be used for volumes of 100 μl–1 ml, but should be of the shrouded-cap variety and will require sterilization.

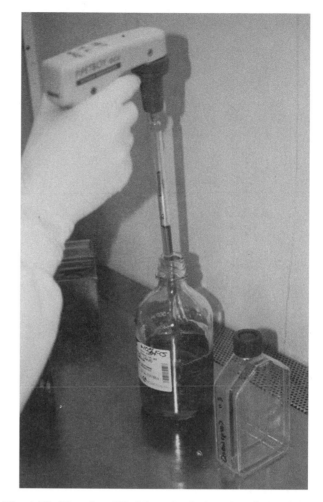

Fig. 4.12. *Pipetting Aid.* Motorized pipetting device for use with conventional graduated pipettes.

It is assumed that the inside of a pipettor is sterile or does not displace enough air for this to matter. However, it clearly does matter in certain situations. For example, if you are performing serial subculture of a stock cell line (as opposed to a short-term experiment with cells that ultimately will be sampled or discarded, but *not* propagated), the security of the line is paramount, and you must either use a plugged regular glass or a disposable pipette with a sterile length

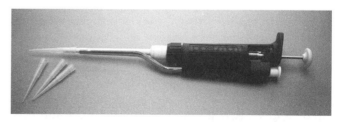

Fig. 4.13. *Pipettor.* Variable-volume pipetting device. Also available in fixed volume. The pipettor is not itself sterilized, but is used with sterilized plastic tips.

that is sufficient to reach into the vessel that you are sampling. If you are using a small enough container to preclude contact from nonsterile parts of a pipettor, then it is permissible to use a pipettor, provided that the tip is of the type that contains a filter. Otherwise, you run the risk of microbial contamination from nonsterile parts of the pipette or, more subtle and potentially more serious, cross-contamination from aerosol or fluid drawn up into the shank of the pipettor.

Routine subculture, which should be rapid and secure from microbial and cross-contamination, but need not be very accurate, is best performed with conventional glass or disposable plastic pipettes. Experimental work, which must be accurate, but should not involve stock propagation of the cells used, may benefit from using pipettors.

Pipettor tips are available with a filter near the top that prevents cross-contamination and minimizes microbial contamination, but adds considerably to the unit cost of the tip.

Tips can be bought loose and can be packaged and sterilized in the laboratory, or they can be bought already sterile and mounted in racks ready for use. Loose tips are cheaper, but more labor intensive. Prepacked tips are much more convenient, but considerably more expensive. Some racks can be refilled and resterilized, which presents a reasonable compromise.

Large-volume dispensing. When culture vessels exceed 100 ml in the volume of medium, a different approach to fluid delivery must be adopted. If only a few flasks are involved, a 100-ml pipette (Becton Dickinson) or a graduated bottle (Fig. 4.14) or bag (Sigma), may be quite adequate, but if very large volumes (>500 ml) or a large number of high-volume replicates are required, then a peristaltic pump will probably be necessary. Single fluid transfers of large volumes (10–10,000 l) are usually achieved by preparing the medium in a sealed pressure vessel and then displacing it by positive pressure into the culture vessel (Alfa Laval). It is possible to dispense large volumes by pouring, but this should be restricted to a single action with a premeasured volume. (See Pouring in Chapter 5.)

Repetitive dispensing. The traditional repetitive dispenser was the type known as a Cornwall syringe (see Fig. 10.12b), in which liquid is alternately taken into a syringe via one tube and expelled via another, using a simple two-way valve. The syringe plunger is spring loaded, so the whole procedure is semiautomatic and repetitive. There are numerous variants of this type of dispenser, many of which are still in regular use. The major problems arise from the valves sticking, but this can be minimized by avoiding the

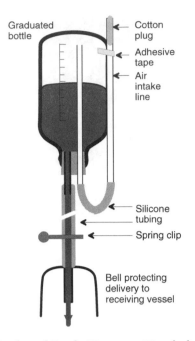

Fig. 4.14. Graduated Bottle Dispenser. Two-hole stopper inserted in the neck of a graduated bottle with a delivery line connected to a dispensing bell, a spring clip on the line, and an inlet line for balancing air. The stopper may be sterilized without the bottle and inserted into any standard bottle containing a medium as required. (From an original design by Dr. John Paul.)

drying cycle after autoclaving and flushing the syringe out with medium or a salt solution before and after use. Small-volume repetitive dispensing can also be achieved by incremental movement of the piston in a syringe (Fig. 4.15).

A peristaltic pump can also be used for repetitive serial deliveries and has the advantage that it may be activated via a foot switch, leaving the hands free. Care must be taken setting up such devices to avoid contaminating the tubing at the reservoir and delivery ends. In general, they are worthwhile only if a very large number of flasks is being handled. Automated pipetting provided by a peristaltic pump can be controlled in small increments. (See, e.g., Fig. 4.11.) In addition, only the delivery tube is autoclaved, and accuracy and reproducibility can be maintained at high

Fig. 4.15. Repette. Stepping dispenser operated by incremental movement of piston in syringe, activated by thumb button.

Fig. 4.16. Multipoint Pipettor. Pipettor with manifold to take 8 plastic tips. Also available for 4 and 12.

levels over a range from 10 ml to 100 ml. A number of delivery tubes may be sterilized and held in stock, allowing a quick change-over in the event of accidental contamination or change in cell type or reagent.

Automation. Many attempts have been made to automate changing fluids in tissue culture, but few devices or systems have the flexibility required for general use. When a standard production system is in use, automatic feeding may be useful, but the time taken in setting it up, the modifications that may be needed if the system changes, and the overriding importance of complete sterility have deterred most laboratories from investing the necessary time and funds. The introduction of microtitration trays (see Fig. 7.2) has

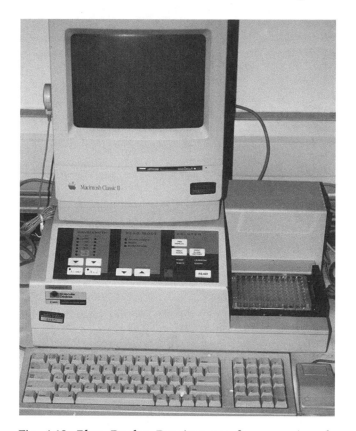

Fig. 4.18. Plate Reader. Densitometer for measuring absorbance of each well; some models also measure fluorescence.

brought with it many automated dispensers, plate readers, and other accessories (Figs. 4.16, 4.17, 4.18). Transfer devices using perforated trays or multipoint pipettes make it easier to seed from one plate to another, and plate mixers and centrifuge carriers also are available. The range of equipment is so extensive that it cannot be covered here, and the appropriate trade catalogues should be consulted. (See Sources of Materials, Microtitration in Trade Index.) Two items worthy of note are the Rainin programmable single or multitip pipettor and the Corning Costar Transtar media transfer and replica plating device (Fig. 4.19).

Robotic systems (e.g., from Packard) are now being introduced into tissue culture assays as a natural extension of the microtitration system. Robots provide totally automated procedures, but also allow for reprogramming if the analytical approach changes.

Choice of system. Whether a simple manual system or a complex automated one is chosen, the choice is governed mainly by five criteria:

(1) Ease of use and ergonomic efficiency
(2) Cost relative to time saved and increased efficiency

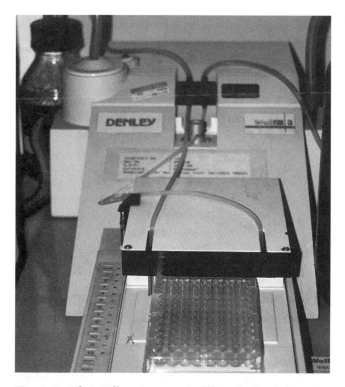

Fig. 4.17. Plate Filler. Automatic filling device for loading microtitration plates. The photo shows a nonsterile application, but the device can be used in sterile applications.

Fig. 4.19. Transfer Device. Transtar (Costar) for seeding, transferring medium, replica plating, and other similar manipulations with microtitration plates, enabling simultaneous handling of all 96 wells. (Reproduced by permission of Corning Costar.)

(3) Accuracy and reproducibility in serial or parallel delivery
(4) Ease of sterilization and effect on accuracy and reproducibility
(5) Mechanical, electrical, chemical, biological, and radiological safety

△ **Safety Note.** Most pipetting devices tend to expel fluid at a higher rate than normal manual operation does and consequently have a greater propensity to generate aerosols. This must be kept in mind when using substances that are potentially hazardous.

USEFUL ADDITIONAL EQUIPMENT

Low-Temperature Freezer
Most tissue culture reagents can be stored at 4°C or −20°C, but occasionally, some drugs, reagents, or products from cultures may require a temperature of −70°C to −90°C, at which point most, if not all, of the water is frozen and most chemical and radiolytic reactions are severely limited. A freezer that keeps materials at these low temperatures is thus a useful accessory. (See Protocol 19.1.) The chest type of freezer is more efficient at maintaining a low temperature with minimum power consumption, but vertical cabinets are much less extravagant in floor space and easier to access. If you do choose a vertical cabinet type, make sure that it has individual compartments (six to eight in a 400-l [15-ft³] freezer) with separate close-fitting doors, and expect to pay at least 20% more than for a chest type.

Low-temperature freezers generate a lot of heat, which must be dissipated for them to work efficiently (or at all). Such freezers should be located in a well-ventilated room or one with air-conditioning such that the ambient temperature does not rise above 23°C. If this is not possible, invest in a freezer designed for tropical use; otherwise you will be faced with constant maintenance problems and a shorter working life for the freezer, with all the attendant problems of relocating valuable stocks. One or two failures costing $1,000 or more in repairs and the loss of valuable material soon cancel any savings that would be realized in buying a cheap freezer.

Glassware Washing Machine
Probably the best way of producing clean glassware, is to have a reliable person do your washing up, but when the amount gets to be too great, it may be worth considering the purchase of an automatic washing machine (Fig. 4.20). Several of these are currently available that are quite satisfactory. Look for the following principles of operation:

(1) A choice of racks with individual spigots over which you can place bottles, flasks, etc. Open vessels such as Petri dishes and beakers will wash satisfactorily in a whirling-arm spray, but narrow-necked vessels need individual jets. The jets should have a cushion or mat at the base to protect the neck of the bottle from chipping.
(2) The pump that forces the water through the jets should have a high delivery pressure, requiring around 2–5 hp, depending on the size of the machine.
(3) Water for washing should be heated to a minimum of 80°C.
(4) There should be a facility for a deionized water rinse at the end of the cycle. This should be heated to 50–60°C; otherwise the glassware may crack after the hot wash and rinse. The rinse should be delivered as a continuous flush, discarded, and not

the discussion of cultures and the training of new staff or students (Fig. 4.21). Choose a high-resolution, but not high-sensitivity, camera, as the standard camera sensitivity is usually sufficient, and high sensitivity may lead to overillumination. Black and white usually gives better resolution and is quite adequate for phase-contrast observation of living cultures. Color is preferable for fixed and stained specimens. If you will be discussing cultures with an assistant or one or two colleagues, a 300- or 400-mm (12- or 15-in) monitor is adequate and gives good definition, but if you are teaching a group of 10 or more students, then go for a 500- or 550-mm (19- or 21-in) monitor.

Fig. 4.20. Glassware Washing Machine. Glassware is placed on individual jets, which ensures thorough washing and rinsing. After washing, glassware is withdrawn on the rack onto the trolley (front) and transferred to the drier (right), which is fitted with the same rails as the washing machine (Betterbuilt [Scientek]).

a

b

Fig. 4.21. Closed-circuit Television. (a) CCD camera attached to Zeiss Axiovert inverted microscope. Can be used for direct printing or for time-lapse studies when linked to a video recorder (see Time Lapse Recording in Chapter 26). Microinjection port on right. (b) Peltier temperature-controlled stage for time-lapse video or microinjection, with perfusion and sampling capabilities.

recycled. If recycling is unavoidable, a minimum of three separate deionized rinses will be required.

(5) Preferably, rinse water from the end of the previous wash cycle should be discarded and not retained for the prerinse of the next wash. Discarding the rinse water reduces the risk of chemical carryover when the machine is used for chemical and radioisotope washup.

(6) The machine should be lined with stainless steel and plumbed with stainless steel or nylon piping.

(7) If possible, a glassware drier should be chosen that will accept the same racks as the washer (see Fig. 4.20), so that they may be transferred directly, without unloading, via a suitably designed trolley.

Video Camera and Monitor

Since the advent of cheap microcircuits, television cameras and monitors have become a valuable aid to

High-resolution charge-coupled-device (CCD) video cameras can be used to record and digitize images for subsequent analysis; the addition of a video recorder will allow real-time or time-lapse recordings. (See Time Lapse Recording in Chapter 26.) Color or monochrome electronic printers are available (Polaroid, Kodak) that provide permanent video images of publishable quality.

Colony Counter

Monolayer colonies are easily counted by eye or on a dissecting microscope with a felt-tip pen to mark off the colonies, but if many plates are to be counted, then an automated counter will help. The simplest uses an electrode-tipped marker pen, which counts when you touch down on a colony. They often have a magnifying lens to help visualize the colonies. From there, a large increase in sophistication and cost takes you to a programmable electronic counter, with a preset program, which counts colonies using image analysis software. These counters are very rapid, can discriminate between colonies of different diameters, and can even cope with contiguous colonies. (See Automatic Colony Counting in Chapter 20.)

Cell Sizing

Most midrange or top-of-the-range cell counters (Schärfe, Coulter) (Fig. 20.2) will provide cell size analysis and the possibility of downloading data to a PC, directly or via a network.

Controlled-Rate Freezer

While cells may be frozen simply by placing them in an insulated box at $-70°C$, some cells may require different cooling rates or complex, programmed cooling curves. (See Cooling Rate in Chapter 19.) A programmable freezer (Planer, Cyro-med) enables the cooling rate to be varied by controlling the rate of injecting liquid nitrogen into the freezing chamber, under the control of a preset program. (See Fig. 19.7.) Cheaper alternatives for controlling the cooling rate during cell freezing are the variable-neck plug (Taylor Wharton), specialized cooling box (Nalge Nunc), a simple polystyrene foam packing container, or foam insulation for water pipes. (See Fig. 19.4–19.6.)

Centrifugal Elutriator

The centrifugal elutriator is a specially adapted centrifuge that is suitable for separating cells of different sizes. (See Centrifugal Elutriation in Chapter 14.) The device is costly, but highly effective.

Flow Cytometer

This instrument can analyze cell populations according to a wide range of parameters, including light scatter, absorbance, and fluorescence. (See Fluorescence-Activated Cell Sorting in Chapter 14; Cytometry in Chapter 20.) Multiparametric analysis can be displayed in a two- or three-dimensional format. In the analytic mode, these machines are generally referred to as *flow cytometers* (see, e.g., Becton Dickinson Cytostar), but the signals they generate can also be used in a *fluorescence-activated cell sorter* to isolate individual cell populations with a high degree of resolution (e.g., Becton Dickinson FACStar). The cost is high ($100,000–200,000), and the best results are obtained with a skilled operator.

CONSUMABLE ITEMS

This category includes general items such as pipettes and pipette canisters, culture flasks, ampules for freezing, centrifuge tubes (10–15 ml, 50 ml, 250 ml; Sterilin, Nalge Nunc, Corning), universal containers (Sterilin, Nalge Nunc), disposable syringes and needles (21–23 G for withdrawing fluid from vials, 19 G for dispensing cells), filters of various sizes (see Sterile Filtration in Chapter 10), for sterilization of fluids, surgical gloves, and paper towels.

Pipettes

Pipettes should be of the blow-out variety, wide tipped for fast delivery, and graduated to the tip, with the maximum point of the scale at the top rather than the tip. Disposable pipettes can be used, but are expensive and may need to be reserved for holidays or crises in washup or sterilization. Reusable pipettes are collected in pipette cylinders or hods, one per workstation.

Pasteur pipettes are best regarded as disposable and should be discarded not into pipette cylinders, but into secure glassware waste.

Glass pipettes are usually sterilized in aluminum or nickel-plated steel cans. Square-sectioned cans are preferable to round, as they stack more easily and will not roll about the work surface. Versions are available with silicone rubber-lined top and bottom ends to avoid chipping the pipettes during handling (Life Sciences, Bellco).

Many laboratories have adopted disposable plastic pipettes, which have the advantage of being prepacked and presterilized and do not have the safety problems associated with chipped or broken glass pipettes. Nor do they have to be washed, which is relatively difficult to do, or plugged, which is tedious. On the downside, they are very expensive and slower to use if singly packed. If, on the other hand, they are bulk packed, there is a high wastage rate unless packs are shared, which is not recommended. (See Pipetting in Chapter 5.) Plastic pipettes also add a significant burden to disposal, particularly if they have to be disinfected first.

As a rough guide, plastic pipettes cost about $2,000 per person per annum. The number of people will therefore determine whether to employ a person to wash and sterilize pipettes or whether to buy plastic pipettes. (Recall that glass pipettes must be purchased at around $200 per person per annum and require energy for washing and sterilizing.)

Culture Vessels

The choice of culture vessels is determined by (1) the yield (number of cells) required (see Table 7.1); (2) whether the cell is grown in monolayer or suspension; and (3) the sampling regime (i.e., are the samples to be collected simultaneously or at intervals over a period of time?) (see Replicate Sampling in Chapter 20). "Shopping around" will often result in a cheaper price, but do not be tempted to change products too frequently, and always test a new supplier's product before committing yourself to it. (See Quality Control in Chapter 10.)

Care should be taken to label sterile, nonsterile, tissue culture, and non-tissue-culture grades of plastics clearly and to store them separately. Glass bottles with flat sides can be used instead of plastic, provided that a suitable washup and sterilization service is available. However, the lower cost tends to be overridden by the optical superiority, sterility, quality assurance, and general convenience of plastic flasks. Nevertheless, disposable plastics can account for approximately 60% of the tissue culture budget—even more than serum.

Petri dishes are much less expensive than flasks, though more prone to contamination and spillage. Depending on the pattern of work and the sterility of the environment, they are worth considering, at least for use in experiments if not for routine propagation of cell lines. Petri dishes are particularly useful for colony-formation assays, in which colonies have to be stained and counted or isolated at the end of an experiment.

Sterile Containers

Petri dishes (9 cm) are required for dissection, 5-ml bijou bottles, 30-ml universal containers, or 50-ml sample pots for storage, 15- and 50-ml centrifuge tubes for centrifugation, and 1.2-ml plastic vials (Nalge Nunc) for freezing in liquid nitrogen. (See Protocol 19.1.)

Syringes and Needles

While it is not recommended that syringes and needles be used extensively in normal handling (for reasons of safety, sterility, and problems with shear stress in the needle when cells are handled), syringes are required for filtration in conjunction with syringe filter adapters, and needles may also be required for extraction of reagents (drugs, antibiotics, or radioisotopes) from sealed vials.

Sterilization Filters

While permanent apparatus is available for sterile filtration, most laboratories now use disposable filters. It is worthwhile to hold some of the more common sizes in stock, such as 25-mm syringe adapters (Pall Gelman, Millipore) and 47-mm bottle-top adapters or filter flasks (Falcon, Nalge Nunc). It is also wise to keep a small selection of larger sizes on hand. (See Sterile Filtration in Chapter 10.)

CHAPTER 5

Aseptic Technique

OBJECTIVES OF ASEPTIC TECHNIQUE

Contamination by microorganisms remains a major problem in tissue culture. Bacteria, mycoplasma, yeast, and fungal spores may be introduced via the operator, the atmosphere, work surfaces, solutions, and many other sources. (See Table 18.1.) Proper aseptic technique seeks to eliminate such contaminants. In this regard, it is essential to establish a strict code of practice and adhere to it, particularly if several people share the same workspace.

Contamination can be minor and confined to one or two cultures, can spread among several cultures and infect a whole experiment, or can be widespread and wipe out your (or even the whole laboratory's) entire stock. Catastrophes can be minimized if (1) cultures are checked carefully by eye and on a microscope, preferably by phase contrast, every time that they are handled; (2) cultures are maintained without antibiotics for at least part of the time to reveal cryptic contaminations (see Use of Antibiotics in Chapter 12); (3) reagents are checked for sterility (by yourself or the supplier) before use; (4) bottles of media, etc., are not shared with other people or used for different cell lines; and (5) the standard of sterile technique is kept high at all times.

Mycoplasmal infection, invisible under regular microscopy, presents one of the major threats. Undetected, it can spread to other cultures around the laboratory. It is therefore essential to back up visual checks with a mycoplasma test, particularly if cell growth appears abnormal. (For a more detailed account of contamination, see Chapter 18.)

Maintaining Sterility

Correct aseptic technique should provide a barrier between microorganisms in the environment outside the culture and the pure, uncontaminated culture within its flask or dish. Hence, all materials that will come into direct contact with the culture must be sterile and manipulations designed such that there is no direct link between the culture and its nonsterile surroundings.

It is recognized that the sterility barrier cannot be absolute without working under conditions that would severely hamper most routine manipulations. Since testing the need for individual precautions would be an extensive and lengthy controlled trial, procedures are adopted largely on the basis of common sense and experience. Aseptic technique is a combination of procedures designed to reduce the probability of infection, and the correlation between the omission of a step and subsequent contamination is not always absolute. The operator may abandon several precautions before the probability rises sufficiently that a contamination is likely to occur (Fig. 5.1). By then, the cause is often multifactorial, and consequently, no simple single solution is obvious. If, once established, all precautions are maintained consistently, breakdowns will be rarer and more easily detected.

Although laboratory conditions have improved in some respects (with air-conditioning and filtration, laminar-flow facilities, etc.), the modern laboratory is often more crowded, and facilities may have to be shared. However, with reasonable precautions, sterility is not difficult to maintain.

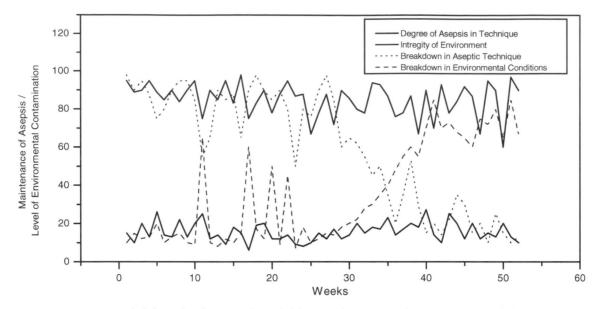

Fig. 5.1. Probability of Infection. The solid line in the top graph represents variability in technique against a scale of 100, which represents perfect aseptic technique. The solid line in the bottom graph represents fluctuations in environmental contamination, with zero being perfect asepsis. Both lines show fluctuations, the top one representing lapses in technique (forgetting to swab the work surface, handling a pipette too far down the body of the pipette, touching nonsterile surfaces with a pipette, etc.), the bottom crises in environmental contamination (a high spore count, a contaminated incubator, contaminated reagents, etc.). As long as these lapses or crises are minimal in degree and duration, the two graphs do not overlap. When particularly bad lapses in technique coincide with severe environmental crises (e.g., at the left-hand side of the chart, the dashed and dotted lines overlap briefly), the probability of infection increases. If the breakdown in technique is progressive (dotted line, center), and the deterioration in the environment is also (dashed line, center), then, when the two cross, the probability of infection is high, resulting in frequent, multispecific, and multifactorial contamination.

ELEMENTS OF ASEPTIC ENVIRONMENT

Quiet Area

In the absence of laminar flow, a separate sterile room should be used for sterile work. If this is not possible, pick a quiet corner of the laboratory with little or no traffic and no other activity. (See Sterile Handling Area in Chapter 3.) With laminar flow, an area should be selected that is free from drafts from doors, windows, etc., the area should also have no through traffic and no equipment that generates drafts (e.g., air conditioners, centrifuges, refrigerators, and freezers). Activity should be restricted to tissue culture, and animals and microbiological cultures should be excluded from the tissue culture area. The area should be kept clean and free of dust and should not contain equipment other than that connected with tissue culture. Nonsterile activities, such as sample processing, staining, or extractions, should be carried out elsewhere.

Work Surface

One of the more frequent examples of bad technique is failure to keep the work surface clean and tidy. The following rules should be observed:

(1) Start with a completely clear surface.
(2) Swab the surface liberally with 70% alcohol.
(3) Bring onto the surface only those items you require for a particular procedure.
(4) Remove everything that is not required, and swab the surface down between procedures.
(5) Arrange your work area so that you have (a) easy access to all items without having to reach over one to get at another and (b) a wide, clear space in the center of the bench (not just the front edge!) to work on. If you have too much equipment too close to you, you will inevitably brush the tip of a sterile pipette against a nonsterile surface. Furthermore, the laminar airflow will fail in

a hood that is crowded with equipment (Figs. 5.2, 5.3, 5.4).

(6) Work within your range of vision (e.g., insert a pipette in a bulb or pipetting aid with the tip of the pipette pointing away from you so that it is in your line of sight continuously and not hidden by your arm).

(7) Mop up any spillage immediately and swab the area with 70% alcohol.

(8) Remove everything when you have finished, and swab the work surface down again.

Personal Hygiene

There has been much discussion about whether hand washing encourages or reduces the bacterial count on the skin. Regardless of this debate, washing will moisten the hands and remove dry skin that would otherwise be likely to blow onto your culture. Washing will also reduce loosely adherent microorganisms, which are the greatest risk to your culture. Surgical gloves may be worn and swabbed frequently, but it may be preferable to work without them (where no hazard is involved) and retain the extra sensitivity that this allows.

Caps, gowns, and face masks are required under Good Manufacturing Practice (GMP) [Food and Drug Administration, 1992; Rules and Guidance for Pharmaceutical Manufacturers and Distributors, 1997] conditions, but are not necessary under normal conditions, particularly when working with laminar flow. If you have long hair, tie it back. When working aseptically on an open bench, do not talk. If you have a cold, wear a face mask, or, better still, do not do any tissue culture during the height of the infection. Talking is permissible when you are working in a vertical laminar flow hood, with a barrier between you and the culture, but should still be kept to a minimum.

Reagents and Media

Reagents and media obtained from commercial suppliers will already have undergone strict quality control to ensure that they are sterile, but the outside surface of the bottle they come in is not. Some manufacturers supply bottles wrapped in polyethylene, which keeps them clean and allows them to be placed in a water bath to be warmed or thawed. The wrapping should be removed outside the hood. Unwrapped bottles should be swabbed in 70% alcohol.

Fig. 5.2. Suggested Layout of Work Area. Laminar-flow hood laid out correctly. Positions may be reversed for left-handed workers.

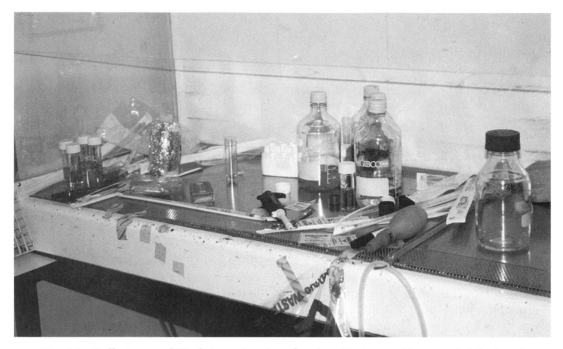

Fig. 5.3. Badly Arranged Work Area. Laminar-flow hood being used incorrectly. The hood is too full, and many items encroach on the air intake at the front, destroying the laminar airflow and compromising both containment and sterility.

Cultures

Flasks from an incubator or hot room can carry sedimentary dust into the hood and should be swabbed first. Flasks, plates, and dishes from humid incubators are particularly prone to outside contamination and should always be swabbed or else incubated within a plastic sandwich or cake box, which is itself swabbed before being taken into the hood.

Cultures imported from another laboratory also carry a high risk, because they have been contaminated either at the source or in transit. Imported cell lines should always be quarantined; i.e., they should be handled separately from the rest of your stocks and kept free of antibiotics until they are shown to be uncontaminated. They may then be incorporated into your main stock, but still maintained free of antibiotics.

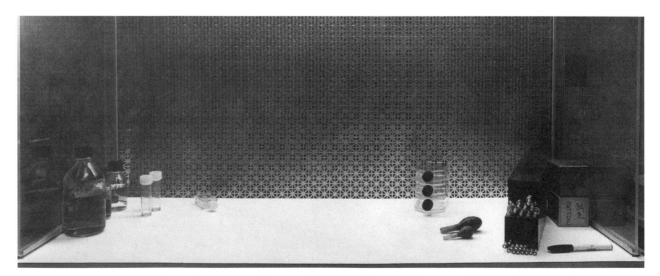

Fig. 5.4. Horizontal Laminar-Flow Hood. Layout for working properly in a horizontal laminar-flow hood. Positions may be reversed for left-handed workers.

Fig. 5.5. Layout of Work Area on Open Bench. Items are arranged in a crescent around the clear work space in the center. The Bunsen burner is located centrally, to be close by for flaming and to create an updraft over the work area.

STERILE HANDLING

Swabbing

Swab down the work surface before and during work, particularly following any spillage, and swab it down again when you have finished. Swab bottles, as well, especially those coming from the cold room, before using them for the first time each day, and also swab any flasks or boxes from the incubator. Swabbing sometimes removes labels, so try to use an alcohol-resistant marker.

Capping

Deep screw caps are preferred to stoppers, although care must be taken in washing the caps to ensure that all detergent is rinsed from behind rubber liners. Wadless polypropylene caps should be used if possible. The screw cap should be covered with aluminum foil to protect the neck of the bottle from sedimentary dust, although the introduction of deep polypropylene caps (e.g., Duran) has made foil shrouding less necessary.

Flaming

When working on an open bench (Fig. 5.5), you should flame the necks of bottles and screw caps before and after opening a bottle and before and after closing it. Pipettes should be flamed before use. Work close to the flame, where there is an updraft due to convection, and do not leave bottles open. Screw caps should be placed with the open side down on a clean surface and flamed before being replacing on the bottle. Alternatively, screw caps may be held in the hand during pipetting, avoiding the need to flame them or lay them down. (See Fig. 5.6.)

Flaming is not advisable when you are working in a laminar-flow hood, as it disrupts the laminar flow, which, in turn, compromises both the sterility of the hood and its containment of biohazardous material. An open flame can also be a fire hazard and can melt some of the plastic interior fittings.

Handling Bottles and Flasks

When working on an open bench, you should not keep bottles vertical when open; instead, keep them at an angle as shallow as possible without risking spillage. A bottle rack (ICN) can be used to keep the bottles or flasks tilted. Culture flasks should be laid down horizontally when open and, like bottles, held at an angle during manipulations. When you are working

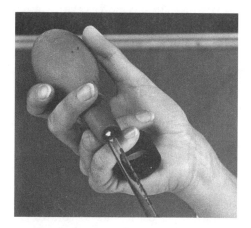

Fig. 5.6. Holding Cap. Cap may be unscrewed and held in the crook of the little finger of the hand holding the bulb or pipetting aid.

in laminar flow, bottles can be left open and vertical, but do not let your hands or any other items come between an open vessel or sterile pipette and the HEPA air filter.

Pipetting

Standard glass or disposable plastic pipettes are still the easiest way to manipulate liquids. Syringes are often used, but regular needles are too short to reach into most bottles. Syringing may also produce high shearing forces when you are dispensing cells, and the practice also increases the risk of self-inoculation. Syringes with blunt, wide-bore cannulae are preferable to pipettes, but still not as rapid to use, except when multiple-stepping dispensers (see Fig. 4.15) are used.

Pipettes of a convenient size range should be selected—1 ml, 2 ml, 5 ml, 10 ml, and 25 ml cover most requirements, although 100-ml disposable pipettes are now available (Becton Dickinson) and are useful for preparing media and fractionating them into aliquots. If you are using glass pipettes and require only a few of each, make up mixed cans for sterilization and save space. Disposable plastic pipettes should be double wrapped and removed from their outer wrapping before being placed in a hood. Unused pipettes should be stored in a dust-free container.

Mouth pipetting should be avoided, as it has been shown to be a contributory factor in mycoplasmal contamination and may introduce an element of hazard to the operator (e.g., with virus-infected cell lines and human biopsy or autopsy specimens or other potential biohazards; see Biohazards in Chapter 6). Inexpensive bulbs and electric pipetting aids are available; try a selection of these devices to find one that suits you. (See Fig. 4.12.) The instrument you choose should accept all sizes of pipette that you use without forcing them in and without the pipette falling out. Regulation of flow should be easy and rapid, but at the same time capable of fine adjustment. You should be able to draw liquid up and down repeatedly (e.g., to disperse cells), and there should be no fear of carryover. The device should fit comfortably in your hand and should be easy to operate with one hand.

Pipettors (see Fig. 4.13) (Gilson, Oxford, Eppendorf, etc.) are particularly useful for small volumes (1 ml and less), although most makes now go up to 5 ml. Because it is difficult to reach down into larger vessels without touching the side of the vessel, pipettors should only be used in conjunction with small containers, such as a universal container. Alternatively, longer tips may be used with larger volumes. Pipettors are particularly useful in dealing with microtitration assays and other multiwell dishes, but should not be used for serial propagation unless filter tips are used.

Multipoint pipettors (with 4, 8, or 12 points) are available for microtitration dishes. (See Fig. 4.16.)

It is necessary to insert a cotton plug in the top of a glass pipette before sterilization to keep the pipette sterile during use. The plug prevents contamination from the bulb or pipetting aid and reduces the risk of cross-contamination. If the plug becomes wet, discard the pipette into disinfectant for return to the washup facility. Plugging pipettes for sterile use is a very tedious job, as is the removal of plugs before washing. Automatic pipette pluggers are available and, although expensive, speed up the process and reduce the tedium. (See Fig. 10.6.)

The foregoing problems are avoided with plastic pipettes, which come already plugged. However, they are slower to use if individually wrapped, wasteful if bulk wrapped (some pipettes are always lost, as it is not advisable to share or reuse a pack once opened), and carry a slightly higher risk of contamination, because removing them from a plastic wrapper is not as clean as withdrawing them from a can. They are, however, more likely to be free of chemical and microbial contamination than are recycled glass pipettes.

Automatic pipetting devices and repeating dispensers are discussed in Chapter 4 (section titled "Fluid Handling").

Pouring

Whenever possible, do not pour from one sterile container into another, unless the bottle you are pouring from is to be used once only and, preferably, is to deliver all its contents (premeasured) in one single delivery. The major risk in pouring lies in the generation of a bridge of liquid between the outside of the bottle and the inside, which may permit contamination to enter the bottle, so bottles or flasks that are stored or incubated after pouring are at a significantly higher risk.

LAMINAR FLOW

The major advantage of working in a laminar flow hood is that the working environment is protected from dust and contamination by a constant, stable flow of filtered air passing over the work surface (Fig. 5.7; see also Fig. 4.1). There are two main types of flow: (1) *horizontal*, where the airflow blows from the side facing you, parallel to the work surface, and is not recirculated; and (2) *vertical*, where the air blows down from the top of the hood onto the work surface and is drawn through the work surface and either recirculated or vented. In most hoods, 20% is vented and made up by drawing in air at the front of the work surface. This configuration is designed to minimize

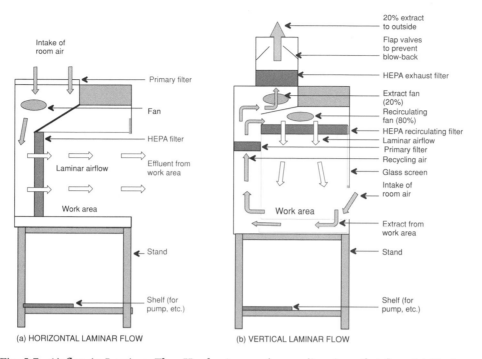

Fig. 5.7. Airflow in Laminar-Flow Hoods. Arrows denote direction of airflow. (a) Horizontal flow, (b) vertical flow.

overspill from the work area of the cabinet. Horizontal flow hoods give the most stable airflow and best sterile protection to the culture and reagents; vertical flow hoods give more protection to the operator.

A Class II vertical-flow *biohazard* hood should be used (see Fig. 6.4a and Biohazards in Chapter 6) if potentially hazardous material (human- or primate-derived cultures, virally infected cultures, etc.) is being handled. The best protection from chemical and radiochemical hazards is given by a cytotoxicity hood that is specially designed for the task and that has a carbon filter trap in the recirculating airflow or a hood with all the effluent vented to outside the building. (See Chemical Toxicity in Chapter 6.) If known human pathogens are handled, a Class III pathogen cabinet with a pathogen trap on the vent is obligatory. (See Fig. 6.4b and Biohazards in Chapter 6.)

Laminar-flow hoods depend for their efficiency on a minimum pressure drop across the filter. When resistance builds up, the pressure drop increases, and the flow rate of air in the cabinet falls. Below 0.4 m/s (80 ft/min), the stability of the laminar airflow is lost, and sterility can no longer be maintained. The pressure drop can be monitored with a manometer fitted to the cabinet, but direct measurement of the airflow with an anemometer is preferable.

Routine maintenance checks of the primary filters are required (about every three to six months). With horizontal-flow hoods, filters may be removed (after switching off the fan) and discarded or washed in soap and water, as they are usually made of polyurethane foam. The primary filters in vertical-flow and biohazard hoods are internal and may need to be replaced by an engineer. They should be incinerated or autoclaved and discarded.

Every 6 months the main high-efficiency particulate air (HEPA) filter above the work surface should be monitored for airflow and holes. (The latter are detectable by a locally increased airflow and an increased particulate count.). Monitoring is best done by professional engineers on a contract basis. Class II biohazard cabinets will have HEPA filters on the exhaust, which will also need to be changed periodically. Again, this should be done by a professional engineer, with proper precautions taken for bagging and disposing of the filters by incineration. If used for biohazardous work, cabinets should be sealed and fumigated before the filters are changed.

Regular weekly checks should be made below the work surface and any spillage mopped up, the tray washed, and the area sterilized with 5% phenolic disinfectant in 70% alcohol. Spillages should, of course, be mopped up when they occur, but occasionally they go unnoticed, so a regular check is imperative. Swabs, tissue wipes, or gloves, if dropped below the work surface during cleaning, can end up on the primary filter and restrict airflow, so take care during cleaning and check the primary filter periodically.

Laminar-flow hoods are best left running continuously, because this keeps the working area clean.

Should any spillage occur, either on the filter or below the work surface, it dries fairly rapidly in sterile air, reducing the chance that microorganisms will grow.

Ultraviolet lights are used to sterilize the air and exposed work surfaces in laminar-flow hoods between uses. The effectiveness of the lights is doubtful because they do not reach crevices, which are treated more effectively with alcohol or other sterilizing agents, which will run in by capillarity. Ultraviolet lights present a radiation hazard, particularly to the eyes, and will also lead to crazing of some clear plastic panels (e.g., Perspex) after six months to a year if used in conjunction with alcohol.

△ *Safety Note.* If ultraviolet lights are used, protective goggles must be worn and all exposed skin covered.

STANDARD PROCEDURE

The essence of good sterile technique embodies many of the principles of standard good laboratory practice. (See Table 5.1.) Keep a clean, clear space to work, and have on it only what you require at one time. Prepare as much as possible in advance, so that cultures are out of the incubator for the shortest possible time and the various manipulations can be carried out quickly, easily, and smoothly. Keep everything in direct line of sight, and develop an awareness of accidental contacts between sterile and nonsterile surfaces. Leave the area clean and tidy when you finish.

The two protocols that follow emphasize aseptic technique. Preparation of media and other manipulations are discussed in more detail under the appropriate headings. (See Reagents and Media in Chapter 10, and Protocol 12.1.)

PROTOCOL 5.1 WORKING IN LAMINAR FLOW

Outline
Clean and swab down work area, and bring bottles, pipettes, etc. Carry out preparative procedures first (preparation of media and other reagents), followed by culture work. Finally, tidy up and wipe over surface with 70% alcohol.

TABLE 5.1. Good Aseptic Technique

Subject	Do	Don't
Laminar-flow hoods	Swab down before and after use. Keep minimum amount of apparatus and materials in hood. Work in direct line of sight.	Clutter up the hood. Leave the hood in a mess.
Contamination	Work without antibiotics. Check cultures regularly, by eye and microscope. Box open plates.	Open contaminated flasks in tissue culture. Carry infected cells. Leave contaminations unclaimed; dispose of them safely.
Mycoplasma	Test cells routinely.	Carry infected cells. Try to decontaminate items.
Importing cell lines	Get from reliable source. Check for mycoplasma. Validate origin. Keep records.	Get from source far removed from originator.
Exporting cell lines	Check for mycoplasma. Validate origin. Send data sheet. Triple wrap.	
Glassware	Keep stocks separate.	Use for chemicals.
Flasks	Vent briefly if stacked.	Stack too high.
Media and reagents	Swab bottles before placing them in hood. Open only in hood.	Share among cell lines. Share with others. Pour.
Pipettes	Use plugged pipettes. Change if contaminated. Use plastic for agar.	Use the same pipette for different cell lines. Share with other people. Overfill disposal cylinders.

Materials

Sterile or aseptically prepared:

Media, stocks, etc., for immediate use

Glass pipettes, graduated, plugged, in square pipette cans, 1 ml, 5 ml, 10 ml, 25 ml, or individually wrapped plastic pipettes

Pipettes, 2 ml or Pasteur, fast flow, not plugged (for aspiration if using a pump)

Cultures for immediate attention

Culture flasks, Petri dishes, or multiwell plates

Nonsterile:

Pipetting aid or bulb (see Figs. 4.12, 5.6)

Waste beaker (Fig. 5.8) or aspiration pump and reservoir (see Figs. 4.8, 4.9)

70% alcohol in spray bottle

Lint-free swabs

Absorbent paper tissues

Pipette cylinder containing water and disinfectant (e.g., hypochlorite; see Disposal in Chapter 6)

Scissors

Marker pen

Notebook, pen, protocols, etc.

Protocol

1. Swab down work surface and all other inside surfaces of laminar-flow hood, including inside of front screen, with 70% alcohol and a lint-free swab or tissue.
2. Bring media, etc., from cold store or thawed from freezer, swab bottles with alcohol, and place those that you will need first in the hood.
3. Collect pipettes and place at the side of the work surface in an accessible position. (See Figs. 5.2, 5.4)
   (a) If glass, open pipette cans and place lids on top or alongside, with the open side down.
   (b) If plastic, remove outer packaging and stack individually wrapped pipettes, sorted by size, on a rack or in cans.
4. Collect any other glassware, plastics, instruments, etc., that you will need, and place them close by (e.g., on a trolley or an adjacent bench).
5. Slacken, but do not remove, caps of all bottles about to be used.
6. Remove the cap of your bottle(s) or flasks(s), into which you are about to pipette, and the bottle(s) that you wish to pipette from, and place the caps open side uppermost on the work surface, at the back of the hood and behind the bottle, so that your hand will not pass over them. Alternatively, if you are handling only one cap at a time, grasp the cap in the crook formed between your little finger and the heel of your hand (Fig. 5.6), and replace it when you have finished pipetting.

7. Select pipette:
   (a) If glass, take pipette from can, lifting it parallel to the other pipettes in the can and touching them as little as possible, particularly at the tops. If the pipette that you are removing touches the end of any of the pipettes still in the can, discard it.
   (b) If plastic, open the pack at the top, peel the ends back, turning them outside in, and withdraw a pipette from the wrapping without it touching any part of the outside of the wrapping; discard the wrapping into the waste bin.
8. Insert pipette in a bulb or pipette aid, pointing pipette away from you and holding it well above the graduations, so that the part of the pipette entering the bottle or flask will not be contaminated.

△ *Safety Note.* As you insert the pipette into the bulb or pipette aid, take care not to exert too much pressure; pipettes can break if forced. (See Glassware and Sharp Items in Chapter 6, and Fig. 6.1.)

9. The pipette in the bulb or pipette aid will now be at right angles to your arm. Take care that the tip of the pipette does not touch the outside of a bottle or the inner surface of the hood. Always be aware of where the pipette is. Following this procedure is not easy when you are learning aseptic technique, but it is an essential requirement for success and will come with experience.
10. Tilt the bottle towards the pipette so that your hand does not come over the open neck, withdraw the requisite amount of fluid, and transfer it to the recipient flask, also tilted. If you are pipetting into several bottles or flasks, they can be sloped by laying them on a suitable bottle rest (ICN). If you are obliged to use the recipient flasks or bottles in a vertical position, without a rack (e.g., because of lack of space), ensure that they remain well back in the hood and that your hand does not come over open necks.
11. Discard the pipette into the pipette cylinder containing disinfectant. Plastic pipettes should be discarded into double-thickness autoclavable biohazard bags.
12. Replace the cap(s) on the bottle(s) or flask(s). Bottles may be left open while you complete a particular maneuver, but should be closed if you leave the hood for more than a few minutes.

Note. In vertical laminar flow, do not work immediately above an open vessel. In horizontal laminar flow, do not work behind an open vessel.

13. On completion of the operation, tighten all caps, and remove all solutions and materials no longer required from the work surface.

Example of aseptic procedure: changing the medium (see also Protocol 12.1)

1. Check cultures, decide what they require (i.e., changing the medium or initiating a new subculture; see Protocols 12.1, 12.2), and bring them to the sterile work area.

Note. Handle one cell strain at a time with its own bottles of medium and other solutions, to prevent cross-contamination.

2. Place bottle with medium on work surface together with culture, and slacken caps.
3. Take sterile pipette and insert into bulb or pipette aid.
4. Open culture flask, withdraw medium, and discard into waste beaker. (See Fig. 5.8.) Or, preferably, aspirate medium via a suction line in the hood connected to an external pump. (See Figs. 4.8, 4.9.)
5. Discard pipette.
6. With a fresh pipette, transfer fresh medium to culture flask as before.
7. Discard pipette.
8. Tighten caps, flame necks, and replace foil.
9. Return culture flasks to incubator and all media to cold room.
10. Clear away all pipettes, glassware, etc., and swab down the work surface.

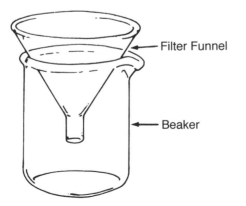

— Filter Funnel

— Beaker

Fig. 5.8. Waste Beaker. Filter funnel prevents beaker from splashing back contents.

PROTOCOL 5.2 WORKING ON THE OPEN BENCH

Outline
Clean and swab down work area, and bring bottles, pipettes, etc. (See Fig. 5.5.) Carry out preparative procedures first. Flame articles as necessary, and keep the work surface clean and clear. Finally, tidy up and wipe over surface with 70% alcohol.

Materials
Sterile or aseptically prepared:
Media, stocks, etc.
Plugged graduated pipettes
Pipettes, 2 ml or Pasteur, fast flow, not plugged (for aspiration if using a pump)
Cultures for immediate attention
Culture flasks, dishes, or plates
Nonsterile:
Pipetting aid or bulb (see Figs. 4.12, 5.6)
Waste beaker (Fig. 5.8) or aspiration pump and reservoir (see Figs. 4.8, 4.9)
70% alcohol in spray bottle
Lint-free swabs
Absorbent paper tissues
Pipette cylinder containing water and disinfectant (e.g., hypochlorite; see Disposal in Chapter 6)
Bunsen burner (or equivalent) and lighter
Scissors
Marker pen
Notebook, pen, protocols, etc.

Protocol
1. Swab down bench surface with 70% alcohol.
2. Bring media, etc., from cold store or thawed from freezer, swab bottles with alcohol, and place those that you will need first on the bench in the work area, leaving the others at the side.
3. Collect pipettes, and place at the side of the work surface in an accessible position. (See Fig. 5.5.)
   (a) If glass, open pipette cans and place lids on top or alongside, open side down.
   (b) If plastic, remove outer packaging and stack individually wrapped pipettes, sorted by size, on a rack or in cans.
4. Collect any other glassware, plastics, instruments, etc., that you will need, and place them close by.
5. Flame necks of bottles, briefly rotating neck in flame, and slacken caps.
6. Select pipette:
   (a) If glass, take pipette from can, lifting pipette parallel to the other pipettes in the can and touching them as little as possible, partic-

ularly at the tops. If the pipette that you are removing touches the end of any of the pipettes still in the can, discard it.

(b) If plastic, open the pack at the top, peel the ends back, turning them outside in, and withdraw a pipette from the wrapping without it touching any part of the outside of the wrapping; discard the wrapping into the waste bin.

7. Flame pipette by pushing it lengthwise through the flame, rotate 180°, and pull the pipette back through the flame. This should only take 2–3 s, or the pipette will get too hot. You are not attempting to sterilize the pipette; you are merely trying to fix any dust that may have settled on it. If you have touched anything or contaminated the pipette in any other way, discard it into disinfectant for return to the wash up facility; do not attempt to resterilize the pipette by flaming.

8. Insert pipette in a bulb or pipette aid, pointing pipette way from you and holding it well above the graduations, so that the part of the pipette entering the bottle or flask will not be contaminated.

△ **Safety Note.** Take care not to exert too much pressure, as pipettes can break when being forced into a bulb. (See Fig. 6.1.)

9. The pipette in the bulb or pipette aid will now be at right angles to your arm. Take care that the tip of the pipette does not touch the outside of a bottle or pipette can. Always be aware of where the pipette is. Following this procedure is not easy when you are learning aseptic technique, but it is an essential requirement for success and will come with experience.

10. Holding the pipette still pointing away from you, remove the cap of your first bottle into the crook formed between your little finger and the heel of your hand (Fig. 5.6). If you are pipetting into several bottles or flasks, they can be sloped by laying them on a suitable bottle rest (ICN). Work with the bottles tilted so that your hand does not come over the open neck. If you have difficulty holding the cap in your hand while you pipette, place the cap on the bench, open side down. If bottles are to be left open, they should be sloped as close to horizontal as possible in laying them on the bench or on a bottle rest.

11. Flame the neck of the bottle.

Note. If flaming Duran bottles, do not use with pouring ring.

12. Tilt the bottle towards the pipette so that your hand does not come over the open neck.

13. Withdraw the requisite amount of fluid and hold.

14. Flame the neck of the bottle and recap.

15. Remove cap of receiving bottle, flame neck, insert fluid, reflame neck, and replace cap.

16. When finished, tighten caps.

17. On completion of the operation, remove stock solutions from work surface, keeping only the bottles that you will require.

Example of aseptic technique on open bench: changing the medium or "feeding" a culture

1. Check cultures, decide what they require (i.e., changing the medium or initiating a new subculture; see Protocols 12.1, 12.2), and bring them to the sterile work area.

2. Swab bottles, flame necks, slacken caps, and place all on work surface together with cultures, preferably one cell strain at a time with its own bottle of medium and other solutions.

3. Take sterile pipette, flame as before, and insert in bulb or pipetting aid.

4. Flame neck of culture flask, open, withdraw medium, and discard into waste beaker. (See Fig. 5.8.) Or, preferably, aspirate medium via a suction line in the hood connected to an external pump. (See Figs. 4.8, 4.9.)

Note. If flaming plastics, pass through the flame very briefly, or plastic will melt.

5. Discard pipette into pipette cylinder.

6. With a fresh pipette, transfer fresh medium to culture flask as before.

7. Discard pipette.

8. Tighten caps.

9. Return culture flasks to incubator and medium to cold room.

10. Clear away all pipettes, glassware, etc., and swab down the work surface.

Petri Dishes and Multiwell Plates

Petri dishes and multiwell plates are particularly prone to contamination because of the following factors:

(1) The larger surface area exposed when the dish is open.

(2) The risk of touching the rim of the dish when handling an open dish.

(3) The risk of carrying contamination from the work surface to the plate via the lid if the lid is laid down.
(4) Medium filling the gap between the lid and the dish due to capillarity if the dish is tilted or shaken in transit to the incubator.
(5) The higher risk of contamination in the humid atmosphere of a CO_2 incubator.

The following practices will minimize the risk of contamination:

(1) Do not leave dishes open for an extended period, or work over an open dish or lid.
(2) When moving dishes or transporting them to or from the incubator, take care not to tilt them or shake them, to avoid the medium entering the capillary space between the lid and the base. If it does, discard the lid, blot any medium carefully from the outside of the rim with a sterile tissue dampened with 70% alcohol, and replace the lid with a fresh one. (Make sure that the labeling is on the base!)
(3) Enclose dishes and plates in a transparent plastic box for incubation, and swab the box with alcohol when it is retrieved from the incubator. (See Fig. 5.9 and "Incubators," to follow.)

The following procedure is recommended for handling Petri dishes or multiwell plates.

PROTOCOL 5.3 HANDLING DISHES OR PLATES

Materials
As for Protocol 5.1 or 5.2, as appropriate

Protocol
To remove medium, etc:
1. Stack dishes or plates on one side of work area.
2. Switch on aspiration pump.
3. Select unplugged pipette and insert in aspiration line.
4. Lift first dish or plate to center of work area.
5. Remove lid and place behind dish, open side up.
6. Grasp the dish as low down on the base as you can, taking care not to touch the rim of the dish or to let your hand come over the open area of the dish or lid. With practice, you may be able to open the lid sufficiently and tilt the dish to remove the medium without removing the lid completely. This technique is quicker and safer than the preceding one.

7. Tilt dish and remove medium.
8. Replace lid.
9. Move dish to other side of work area from the untreated dishes in the initial stack.
10. Repeat procedure with remaining dishes or plates.
11. Discard pipette and switch off pump.

To add medium or cells, etc:
1. Position necessary bottles and slacken the cap of the one you are about to use.
2. Bring dish to center of work area.
3. Remove bottle cap and fill pipette from bottle.
4. Remove lid and place behind dish.
5. Add medium to dish, directing the stream gently at the side of the base of the dish.
6. Replace lid.
7. Return dish to side where dishes were originally stacked.
8. Repeat with second dish, and so on.
9. Discard pipette.
Again, with practice, you may be able to lift the lid and add the medium without laying the lid down, as you did when you removed the medium.

APPARATUS AND EQUIPMENT

All apparatus used in the tissue culture area should be cleaned regularly to avoid the accumulation of dust and to prevent microbial growth in accidental spillages. Replacement items, such as gas cylinders, must be cleaned before being introduced to the tissue culture area, and no major movement of equipment should take place while people are working aseptically.

Incubators
Humidified incubators are a major source of contamination. (See Humid Incubators in Chapter 18.) They should be cleaned out at regular intervals (weekly or monthly, depending on the level of atmospheric contamination and frequency of access) by removing the contents, including all the racks or trays, and washing down the interior and the racks or shelves with a nontoxic detergent such as Decon or Roccall. Traces of detergent should then be removed with 70% alcohol, which should be allowed to evaporate completely before replacing the shelves and cultures.

A fungicide, such as 2% Roccall or 1% copper sulfate, may be placed in the humidifier tray at the bottom of the incubator to retard fungal growth, but the success of such fungicides is limited to the surface that they are in contact with, and there is no real substitute for regular cleaning. Some incubators have high-temperature sterilization cycles, but these are seldom able to generate sufficient heat for long enough to be effective, and the length of time that the incubator is out of use can be inconvenient. More recently, incubators have been introduced with micropore filtration and laminar airflow to inhibit the circulation of microorganisms (Forma, Jencons).

One major source of contamination in plates and petri dishes is the formation of a capillary film between the plate or dish and its lid. Most plastic dishes and plates now have raised, or "vented," lids (see Fig. 7.7) to facilitate the exchange of gases; this reduces, but does not eliminate, the risk of trapping medium. Trapped medium forms a bridge with the nonsterile outer air and may cross-contaminate wells in a multiwell plate, so it must be prevented or, having happened, eliminated. Liquid films may be removed by the careful application of a tissue moistened with 70% alcohol; the lid should be replaced with a fresh one.

Boxed Cultures

When problems with contamination recur in humidified incubators, it is advantageous to enclose dishes and plates in plastic sandwich boxes (Fig. 5.9). The box should be swabbed before use, inside and outside,

Fig. 5.9. Boxed Dishes. A transparent box, such as a sandwich box or cake box, helps to protect unsealed plates and dishes from contamination in a humid incubator. This type of container should also be used for materials that may be biohazardous, to help contain spillage in the event of an accident.

and allowed to dry in sterile air. When the box is subsequently removed from the incubator, it should be swabbed with 70% alcohol before being opened or introduced into your work area. The dishes are then carefully removed and the interior of the box swabbed prior to reuse.

Gassing with CO_2

It is common practice to place flasks, with the caps slackened, in a humid CO_2 incubator to allow for gaseous equilibration, but doing so does increase the risk of infection. It is preferable to purge the flasks from a sterile, premixed gas supply and then seal them. This avoids the need for a gassed incubator for flasks and gives the most uniform and rapid equilibration. Several manufacturers (Corning, Falcon, Nunc, Costar) now provide flasks with permeable caps to allow rapid equilibration in a CO_2 atmosphere without the risk of contamination.

TRAINING

It is vital that new staff introduced into the sterile area receive adequate training in techniques and procedures. Before they are allowed to work independently, they should be apprenticed to a skilled operator for a period of up to three months; they should not have to learn from the last recruit to come in! A brief introduction to the theory is helpful, but the major emphasis is on practical skills, so supervised operation is required for several weeks, followed by a trial period without using antibiotics. After a successful trial period all culture should be done without antibiotics.

A good starting point is the simple manipulation of fluid from one bottle to another. After that technique is mastered, the trainee should progress to changing the medium in a culture, followed by initiating a subculture. It is good practice for a novice to set up a growth curve (see Growth Cycle in Chapter 20) to gain experience in handling replicate samples, trypsinization, and cell counting. Plotting the curve will also give the novice an appreciation of how the cells grow and when they are suitable for subculture or other uses.

CHAPTER 6

Safety

In addition to the everyday safety hazards common to any workplace, the cell culture laboratory has a number of particular risks associated with culture work. In spite of the scientific background and training of most people who work in this environment, accidents still happen, as familiarity often leads to a more casual approach in dealing with regular, biological, and radiological hazards. Furthermore, individuals who service the area often do *not* have a scientific background, and the responsibility lies with those who are better informed to maintain a safe environment for all who work there. It is important to identify potential hazards, but, at the same time, not to overemphasize the risks. If a risk is not seen as realistic, then precautions will tend to be disregarded, and the whole safety code will be placed in disrepute.

RISK ASSESSMENT

Risk assessment is an important principle that has become incorporated into most modern safety legislation. Determining the nature and extent of a particular hazard is only part of the process; the way in which the material or equipment is used, who uses it, the frequency of use, training, and general environmental conditions are all equally important in determining risk (Table 6.1). Such considerations as the amount of a particular material, the degree and frequency of exposure to a hazard, the procedures for handling materials, the type of protective clothing worn, ancillary hazards like exposure to heat, frost, and electric current, and the type of training and ex-

perience of the operator all contribute to the risk of a given procedure, although the nature of the hazard itself may remain constant.

A major problem that arises constantly in establishing safe practices in a biomedical laboratory is the disproportionate concern given to the more esoteric and poorly understood hazards, such as those arising from genetic manipulation, relative to the known and proven hazards of toxic and corrosive chemicals, solvents, fire, ionizing radiation, electrical shock, and broken glass. It is important that biohazards be categorized correctly [National Research Council, 1989; US Department of Health and Human Services, 1993; Health and Safety Commission, 1994; Caputo, 1996], neither overemphasized nor underestimated, but the precautions taken should not displace the recognition of everyday safety problems (e.g., pipetting by mouth is clearly unacceptable, but the use of rubber bulbs or pipetting aids introduces a new risk of injury from inserting broken or chipped pipettes).

STANDARD OPERATING PROCEDURES

Hazardous substances, equipment, and conditions should not be thought of in isolation, but should be taken to be part of a procedure, all of whose components should be assessed. If the procedure is deemed to carry any significant risk beyond the commonplace, then a standard operating procedure (SOP) should be defined, and all who work with the material, equipment, etc., should conform to that procedure. The different stages of the procedure—procurement, storage,

TABLE 6.1. Elements of Risk Assessment

Category	Items affecting risk
Operator	
Experience	Level
	Relevance
	Background
Training	Previous
	New requirements
Protective clothing	Adequate
	Properly worn
Equipment	
Age	Condition
	Adherence to new legislation
Suitability for task	Access, sample capacity, containment
Mechanical stability	Loading
	Anchorage
	Balance
Electrical safety	Connections
	Leakage to ground (earth)
	Proximity of water
Containment	Aerosols:
	Generation
	Leakage from hood ducting
	Overspill from work area
	Toxic fumes
	Exhaust ductwork:
	Integrity
	Site of effluent and downwind risk
Heat	Generation
	Dissipation
Maintenance	Frequency
	Decontamination required?
Disposal	Route
	Decontamination required?
Physical Risks	
Intense cold	Frostbite
	Numbing
Electric shock	Loss of conciousness
	Cardiac arrest
Fire	General precautions
	Equipment wiring, installation, and maintenance
	Incursion of water near electrical wiring
	Fire drills, procedures, escape routes
	Solvent usage and storage (e.g., ether in refrigerators)
	Flammable mixtures
	Identification of stored biohazards and radiochemicals
Chemicals (including gases and volatile liquids)	
Scale	Amount used
Toxicity	Poisonous
	Carcinogenic
	Teratogenic
	Mutagenic
	Corrosive
	Irritant
	Allergenic
	Asphyxiative
Reaction with water	Heat generation
	Effervescence

TABLE 6.1. (*Continued*)

Category	Items affecting risk
Reaction with solvents	Heat generation
	Effervescence
	Generation of explosive mixture
Volatility	Intoxication
	Asphyxiation
Generation of powders and aerosols	Dissemination
	Inhalation
Import, export, and transportation	Breakage, leakage
Location and storage conditions	Access to untrained staff
	Illegal entry
	Weather, incursion of water
	Stability, compression, breakage, leakage
Biohazards	
Pathogenicity	Grade
	Infectivity
	Host specificity
	Stability
Scale	Number of cells
	Amount of DNA
Genetic manipulation	Host specificity
	Vector infectivity
	Disablement
Containment	Room
	Cabinet
	Procedures
Radioisotopes	
Emission	Type
	Energy
	Penetration, shielding
	Interaction, ionization
	Half-life
Volatility	Inhalation
	Dissemination
Localization on ingestion	DNA precursors, such as [^{3}H]thymidine
Disposal	Solid, liquid, gaseous
	Route
	Legal limits
Special Circumstances	
Pregnancy	Immunodeficiency
	Risk to fetus, teratogenicity
Illness	Immunodeficiency
Immunosuppressant drugs	Immunodeficiency
Cuts and abrasions	Increased risk of absorption
Allergy	Powders, e.g., detergents
	Aerosols
	Contact, e.g., rubber gloves
Elements of Procedures	
Scale	Amount of materials used
	Size of equipment & facilities and effect on containment
	Number of staff involved
Complexity	Number of steps or stages
	Number of options
	Interacting systems and procedures
Duration	Process time
	Incubation time
	Storage time
Number of persons involved	Increased risk?
	Diminished risk?
Location	Containment
	Security and access

operations, and disposal—should be identified, and the possibility must be taken into account that the presence of more than one hazard within a procedure will compound the risk or, at best, complicate the necessary precautions (e.g., how does one dispose of broken glass that has been in contact with a human cell line labeled with a radioisotope?).

SAFETY REGULATIONS

The following recommendations should not be interpreted as a code of practice, but rather as advice that might help in compiling safety regulations. The information is designed to provide the reader with some guidelines and suggestions to help construct a local code of practice, in conjunction with regional and national legislation and in full consultation with one's local safety committee. These recommendations have no legal standing and should not be quoted as if they do.

General safety regulations should be available from the safety office in the institution or company at which you work. In addition, they are available from the Occupational Health and Safety Administration (OSHA) in the United States. Beginning January 1, 1993, Europe, including the United Kingdom, came under new joint regulations [Management of Health and Safety at Work Regulations, 1992; Provision and Use of Work Equipment, 1992]. These regulations cover all matters of general safety. The relevant regulations and recommendations for biological safety for the United States are contained in *Biosafety in Microbiological and Biomedical Laboratories* [U.S. Department of Health and Human Services, 1993], a joint document prepared by the Centers for Disease Control in Atlanta, Georgia, and the National Institutes of Health in Bethesda, Maryland. For the United Kingdom, the Health Services Advisory Committee of the Health and Safety Commission has published two booklets: *Safe Working and the Prevention of Infection in Clinical Laboratories* [Health and Safety Commission, 1991] and *Safe Working and the Prevention of Infection in Clinical Laboratories—Model Rules for Staff and Visitors* [Health and Safety Commission, 1991]. Genetically modified cells are dealt with in *A Guide to the Genetically Modified Organisms (Contained Use) Regulations* [Health and Safety Commission, 1992] also from Her Majesty's Stationery Office. Several of the UK guidelines are under review and may be viewed at www.open.gov.uk/hse/condocs/. The advice given in this chapter is general and should not be construed as satisfying any legal requirement.

Table 6.2 provides a checklist highlighting some of the factors to be considered in elaborating safety pol-

icy in a tissue culture laboratory. In most cases, the headings and subheadings are self-explanatory, but a few need a little explanation.

GENERAL SAFETY

The next table emphasizes those aspects of general safety which are particularly important in a tissue culture laboratory (Table 6.3) and should be used in conjunction with local safety rules.

Operator

It is the responsibility of the institution to provide the correct training, or to determine that the individual is already trained, in appropriate laboratory procedures. It is the supervisor's responsibility to ensure that procedures are carried out correctly and that the correct protective clothing is worn at the appropriate times.

Equipment

A general supervisor should be appointed to be in charge of all equipment maintenance, electrical safety, and mechanical reliability, and a curator should be in charge of each specific piece of equipment to ensure that the day-to-day operation of the equipment is satisfactory and to train others in its use. Particular risks include the generation of toxic fumes or aerosols from centrifuges and homogenizers, which must be contained either by the design of the equipment or by placing them in a fume cupboard.

The electrical safety of equipment is dealt with in the United States by OHSA (www.osha.gov) and in the United Kingdom by the Provision and Use of Work Equipment Regulations [1992]. As already mentioned, the latter will shortly be incorporated into new European Community (EC) guidelines.

Glassware and Sharp Items

The most common form of injury in tissue culture results from accidental handling of broken glass and syringe needles. Particularly dangerous are broken pipettes in a washup cylinder, which result from too many pipettes—particularly Pasteur pipettes—being forced into too small a container. Needles that have been improperly disposed of together with ordinary waste or forced through the wall of a rigid container when the container is overfilled are also very dangerous.

Accidental inoculation via a discarded needle or broken glass, or due to an accident during routine handling, remains one of the more acute risks associated with handling potentially biohazardous material. It may even carry a risk of transplation when one handles human tumors [Southam, 1958; Scanlon et

TABLE 6.2. General Precautions

Category	Action
Regulatory authority	Contact national, regional inspectors
Local Safety Committee	Appoint representatives
	Arrange meetings and discussion
Guidelines	Access local and national
	Generate local guidelines if not already done
Standard operating procedures (SOPs)	Define and make available
Protective clothing	Provide, launder, and ensure that it is worn correctly
Containerization	Specify physical description (e.g., storage and packaging)
Containment levels	Specify chemical, radiological, biological levels
Training	Arrange seminars, supervision
Monitoring	Automatic smoke detectors, oxygen meter
Inspection	Arrange equipment, procedures, laboratory inspections by trained, designated staff
Record keeping	Safety officers and operatives to keep adequate records
Import and export	Regulate and record
Classified waste disposal	Define routes for sharps, radioactive waste (liquid and solid), biohazards, corrosives, solvents, toxins
Access	Limit to trained staff and visitors only
	Exclude children, except in public areas

Try to think in terms of the stages of handling: procurement, storage, operating procedure, and disposal, and be aware of how individual components of a procedure will interact and alter the level of risk. These are suggestions only and have no legal basis. Consult national legal requirements and local regulations before formulating proper guidelines.

TABLE 6.3. Safety Hazards in a Tissue Culture Laboratory

Category	Item	Risk
General	Broken glass	Injury, infection
	Pipettes	Injury, infection
	Sharp instruments	Injury, infection
	Pasteur pipettes	Injury, infection
	Syringe needles	Injury, infection
	Cables	Fire, electrocution, snagging, tripping
	Tubing	Leakage, snagging, tripping
	Cylinders	Instability, leakage
	Liquid nitrogen	Frostbite, asphyxiation, explosion
Fire	Bunsen burners; flaming, particularly in association with alcohol	Fire, melting damage
Radiological	Radioisotopes in sterile cabinet	Emission, spillage, aerosols, volatility
	Irradiation of cultures	Radiation dose
Biological	Importation of cell lines and biopsies	Infection
	Genetic manipulation	Infection, DNA transfer
	Propagation of viruses	Infection
	Position and maintenance of laminar-flow hoods	Breakdown in containment

al., 1965; Gugel and Sanders, 1986], although reports of this are largely anecdotal.

Pasteur pipettes should be discarded into a sharps bin or, if reused, handled separately and with great care. Avoid using syringes and needles, unless they are needed for loading ampules (use a blunt cannula) or withdrawing fluid from a capped vial. When disposable needles are discarded, use a rigid plastic or metal container. Do not attempt to bend, manipulate, or resheath the needle. Provide separate receptacles for the disposal of sharp items and broken glass, and do not use these receptacles for general waste.

Take care when you are fitting a bulb or pipetting device onto a pipette. Choose the correct size to guard against the risk of the pipette breaking at the neck and lacerating your hand. Check that the neck is sound, hold the pipette as near the end as possible, and apply gentle pressure with the pipette pointing away from your knuckles (Fig. 6.1).

Chemical Toxicity

Relatively few major toxic substances are used in tissue culture, but when they are, the conventional precautions should be taken, paying particular attention to the distribution of powders and aerosols by laminar-flow hoods. (See Biohazards in this chapter.) Detergents—particularly those used in automatic machines —are usually caustic and even when they are not, they

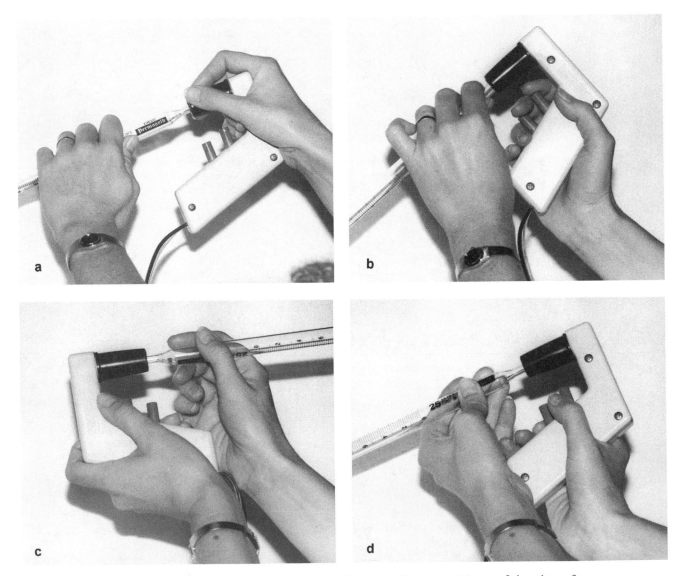

Fig. 6.1. Inserting a Pipette into a Pipetting Device. a. Wrong position. Left hand too far down pipette, risking contamination of the pipette and exerting too much leverage, which might break the pipette; right hand too far over and exposed to end of pipette or splinters, should the pipette break at the neck during insertion. b–d. Correct positions. Left hand farther up pipette, right hand clear of top of pipette.

can cause irritation to the skin, eyes, and lungs. Use liquid-dispensing devices whenever possible, wear gloves, and avoid procedures that cause powder detergent to spread as dust. Liquid-detergent concentrates are more easily handled, but often are more expensive. Chemical disinfectants such as hypochlorite should be used cautiously, either in tablet form or as a liquid dispensed from a dispenser. Hypochlorite disinfectants will bleach clothing, cause skin irritations, and even corrode welded stainless steel.

Specific chemicals used in tissue culture that require special attention are (1) dimethyl sulfoxide (DMSO), which is a powerful solvent and skin penetrant and can, therefore, carry many substances through the skin [Horita and Weber, 1964] and even through some (e.g., rubber or silicone) protective gloves, and (2) mutagens, carcinogens, and cytotoxic drugs, which should be handled in a safety cabinet. A Class II laminar-flow hood may be adequate for infrequent handling of small quantities of these substances, but it may be necessary to use a hood designed specifically for cytotoxic chemicals (see Fig. 4b). Mutagens, carcinogens, and other toxic chemicals are sometimes dissolved in DMSO, increasing the risk of uptake via the skin. Polyethylene gloves provide a reasonable barrier, but often leak and tend to give a poor grip, so a combination of two types of glove may be required.

The handling of chemicals is regulated by the Occupational Safety & Health Administration (OSHA www.osha.gov) and by the Control of Substances Hazardous to Health [Health and Safety Commission 1999] in the United Kingdom. The latter regulations are shortly to be incorporated into EC legislation and guidelines, as mentioned earlier. Information and guidelines are also available from the National Institutes of Health (www.niehs.nih.gov/odhsb/), and the National Institute for Occupational Safety and Health (NIOSH, www.cde.gov/niosh/homepage.html).

Gases

Most gases used in tissue culture (CO_2, O_2, N_2) are not harmful in small amounts, but are nevertheless dangerous if handled improperly. They should be contained in pressurized cylinders that are properly secured (Fig. 6.2). If a major leak occurs, there is a risk of asphyxiation from CO_2 and N_2 and of fire from O_2. Evacuation and maximum ventilation are necessary in each case; if there is extensive leakage of O_2, call the fire department. An oxygen monitor should be installed near floor level in rooms where N_2 and CO_2 are stored in bulk, or where there is a piped supply to the room.

If glass ampules are used, they are sealed in a gas oxygen flame. Great care must be taken both to guard the flame and to prevent inadvertent mixing of the gas and oxygen. A one-way valve should be incorpo-

Fig. 6.2. Cylinder Clamp. Clamps onto edge of bench and secures gas cylinder with fabric strap. Fits different sizes of cylinder and can be moved from one position to another if necessary. Available from most laboratory suppliers.

rated into the gas line so that oxygen cannot blow back.

Liquid Nitrogen

Three major risks are associated with liquid nitrogen: frostbite, asphyxiation, and explosion. (See Protocol 19.1.) Since the temperature of liquid nitrogen is $-196°C$, direct contact with the liquid (via splashes, etc.) or with anything—particularly something metallic—that has been submerged in it presents a serious hazard. Gloves that are thick enough to act as insulation, but flexible enough to allow the manipulation of ampules, should be worn. When liquid nitrogen boils off during routine use of the freezer, regular ventilation is sufficient to remove excess nitrogen; but when nitrogen is being dispensed or a lot of material is being inserted in the freezer, extra ventilation will be necessary. Remember, 1 l liquid nitrogen generates nearly 700 l of gas.

When an ampule is submerged in liquid nitrogen, a high pressure difference results between the outside and the inside of the ampule. If the ampule is not perfectly sealed, liquid nitrogen may be inspired, causing the ampule to explode violently when thawed. This prospect can be avoided by storing the ampules in the gas phase (see Cryofreezers in Chapter 19) or by ensuring that the ampules are perfectly sealed. Thawing material stored under liquid nitrogen should always be performed in a container with a lid, such as a plastic bucket (see Protocol 19.2), and a face shield or goggles must be worn.

FIRE

Particular fire risks associated with tissue culture stem from the use of Bunsen burners for flaming, together with alcohol for swabbing or sterilization. Keep the two separate; always ensure that alcohol for sterilizing instruments is kept in minimum volumes in a narrow-necked bottle or flask that is not easily upset (Fig. 6.3). Alcohol for swabbing should be kept in a plastic wash bottle or spray and not be used in the presence of an open flame. When instruments are sterilized in alcohol and the alcohol is subsequently burnt off, care must be taken not to return the instruments to the alcohol while they are still alight.

RADIATION

Three main types of radiation hazard are associated with tissue culture: ingestion, irradiation from labeled reagents, and irradiation from a high-energy source. Guidelines on radiological protection for the US can be obtained from the Office of Nuclear Regulatory Research, U.S. Nuclear Regulatory Commission, Washington, DC 20555, and for the UK are contained in Ionising Radiations Regulations and Approved Code of Practice [Radiological Protection Service, 1985]. These are currently being updated in line with European legislation.

Ingestion
Radiolabeled compounds can be ingested by being splashed on the hands or via aerosols generated by pipetting or the use of a syringe. Tritiated nucleotides, if accidentally ingested, will become incorporated into DNA, and, due to the short path length of the low-energy β-emission from ^3H, will cause radiolysis within the DNA. Radioactive isotopes of iodine will concentrate in the thyroid and may also cause local damage.

Labeled Reagents
The second type of risk is from irradiation from higher energy β- and γ-emitters such as ^{32}P, ^{125}I, ^{131}I, and ^{51}Cr. Protection can be obtained by working behind a 2-mm-thick lead shield and storing the concentrated isotope in a lead pot. Perspex screens (5 mm) can be used with ^{32}P at low concentrations for short periods.

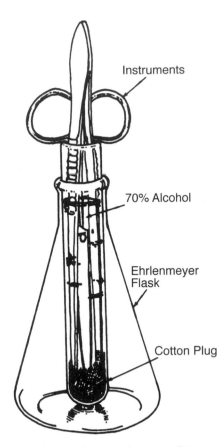

Fig. 6.3. Flask for Alcohol Sterilization of Instruments. The wide base prevents tipping and the center tube reduces the amount of alcohol required, so that spillage, if it occurs at all, is minimized. (From an original idea by M.G. Freshney.)

In both cases, one should work in a Class II hood to contain aerosols, and wear gloves. The items that you are working with should be held in a shallow tray lined with paper tissue or Benchcote to contain any accidental spillage. Whenever possible, use small pieces of equipment (e.g., a pipettor with disposable plastic tips, small sample tubes, etc.) that will generate minimum bulk when they are discarded into a radioactive waste container. Clean up carefully when you are finished, and monitor the area regularly for any spillage.

High-Energy Sources
The third type of irradiation risk is from X-ray machines, high-energy sources such as ^{60}Co, or ultraviolet (UV) sources used for sterilizing apparatus or stopping cell proliferation in feeder layers. (See Chapters 11 and 23.) Since the energy, particularly from X-rays or ^{60}Co, is high, these sources are usually located in specially designed accommodations and are subject to strict control. UV sources can cause burns to the skin and damage to the eyes; they should be carefully

screened to prevent direct irradiation of the operator, who should wear barrier filter goggles.

Consult your local radiological officer and code of practice before embarking on radioisotopic experiments. Local rules vary, but most places have strict controls on the amount of radioisotopes that can be used, stored, and discarded.

BIOHAZARDS

The need for protection against biological hazards [see Caputo, 1996] is defined both by the source of the material and by the nature of the operation being carried out. It is also governed by the conditions under which culture is performed. Using standard microbiological technique on the open bench has the advantage that the techniques in current use have been established as a result of many years of accumulated experience. Problems arise when new techniques are introduced or when the number of people sharing the same area increases. With the introduction of horizontal laminar-flow hoods, the sterility of the culture was protected more effectively, but the exposure of the operator to aerosols was more likely. This led to the development of vertical laminar-flow hoods with an air curtain at the front (see Chapters 4 and 5) to minimize overspill from within the cabinet.

Levels of Containment

We can define three levels of handling with decreasing levels of containment:

(1) A sealed pathogen cabinet with filtered air entering and leaving via a pathogen trap filter (biohazard hood or microbiological safety cabinet, Class III; Fig. 6.4c). The cabinet will generally be housed in a separate room with restricted access and with showering facilities and protection for solid and liquid waste, depending on the nature of the hazard.

(2) A vertical laminar-flow hood with front protection in the form of an air curtain and a filtered exhaust (biohazard hood or microbiological safety cabinet, Class II; Fig. 6.4a) [National Sanitation Foundation Standard 49, 1983; British Standard BS5726, 1992, European Committee for Standardisation, 1999]. If recognized pathogens are being handled, hoods such as these should be housed in separate rooms, at containment levels II, III, or IV, depending on the nature of the pathogen. If there is no reason to suppose that the material is infected, other than by adventitious agents, then hoods can be housed in the main tissue culture facility, which may be categorized as containment level I, requiring restricted access, control of waste dis-

posal, protective clothing, and no food or drink in the area.

All biohazard hoods must be subject to a strict maintenance program [Osborne et al., 1999], with the filters tested at regular intervals, proper arrangements made for fumigation of the cabinets before changing filters, and disposal of old filters made safe by extracting them into double bags for incineration.

(3) Open bench, depending on good microbiological technique. Again, this will normally be conducted in a specially defined area, which may simply be defined as the "tissue culture laboratory," but which will have level I conditions applied to it.

Table 6.4 lists common procedures with suggested levels of containment. All those using the facilities, however, should seek the advice of their local safety committees and the appropriate biohazard guidelines (see Risk Assessment in this chapter) for legal requirements.

Human Biopsy Material

Issues of biological safety are clearest when known, classified pathogens are being used, since the regulations covering such pathogens are well established both in the United States [US Department of Health and Human Services, 1993] and in the United Kingdom [Advisory Committee on Dangerous Pathogens (ACDP), 1995a,b]. However, in two main areas there is a risk that is not immediately apparent in the nature of the material. One is in the development by recombinant techniques such as transfection, retroviral infection, and interspecific cell hybridization, of new potentially pathogenic genes. Handling such cultures in facilities such as laminar-flow hoods introduces putative risks for which there are no epidemiological data available for assessment. Transforming viruses, amphitropic viruses, transformed human cell lines, human–mouse hybrids, and cell lines derived from xenografts in immunodeficient mice, for example, should be treated cautiously until there are enough data to show that they carry no risk.

The other area of risk is the inclusion of adventitious agents in human or other primate biopsy or autopsy samples or cell lines [Grizzle & Polt, 1988; Centers for Disease Control (1988); Wells et al., 1989; Tedder et al., 1995] or in animal products such as serum, particularly if those materials are derived from parts of the world with a high level of endemic infectious diseases. When infection has been confirmed, the type of organism will determine the degree of containment, but even when infection has not been confirmed, the possibility remains that the sample may yet carry hepatitis B, human immunodeficiency virus (HIV), tu-

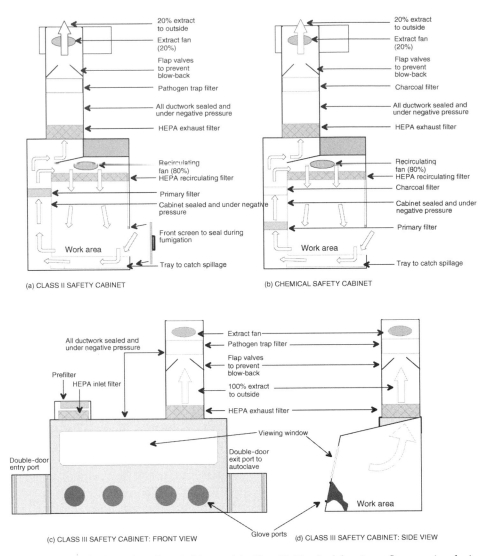

Fig. 6.4. Microbiological Safety Cabinets. (a) *Class II.* Vertical laminar flow, recirculating 80% of the air. Air (20–30%) is exhausted via a filter and ducted out of the room through an optional pathogen trap. Air is taken in at the front of the cabinet to make up the recirculating volume and prevent overspill from the work area. (b) Class II chemical safety cabinet with charcoal filters on extract and recirculating air. (c) Class III nonrecirculating, sealed cabinet with glove pockets. Works at negative pressure and with air lock for entry of equipment and direct access to autoclave, either connected or adjacent. (d) Side view of Class III cabinet.

berculosis, or other pathogens as yet undiagnosed. Confidentiality frequently prevents HIV testing without the patient's consent, and, for most of the adventitious infections, the appropriate information will not be available. If possible, biopsy material should be tested for potential adventitious infections before handling. The authority to do so should be written on the consent form that the person donating the tissue will have been asked to sign (see example, Table 11.1), but the need to get samples into culture quickly will often mean that you must proceed without this information. Such samples should be handled with caution:

(1) Transport specimens in a double-wrapped container (e.g., a universal container or screw-capped vial within a second screw-top vessel, such as a polypropylene sample jar). This in turn should be enclosed in an opaque, plastic or waterproof paper envelope, filled with absorbent tissue packing to contain any leakage, and transported to the lab by a designated carrier.

(2) Enter all specimens into a logbook on receipt, and place the specimens in a secure refrigerator marked with a biohazard label.

(3) Carry out dissection and subsequent culture work

TABLE 6.4. Biohazard Procedures and Suggested Levels of Containment*

Procedure	Level of protection
Preparation of media	Open bench with standard microbiological practice, or horizontal or vertical laminar flow
Primary cultures and cell lines other than human and other primates	Open bench with standard microbiological practice, or horizontal or vertical laminar flow
Primary cultures and cell lines, other than human and other primates, that have been infected or transfected	Class II laminar-flow hood
Primary culture and serial passage of human and other primate cells	Class II laminar-flow hood
Interspecific hybrids or other recombinants, transfected cells, human cells, and animal tumor cells	Class II laminar-flow hood
Human cells infected with retroviral constructs	Class II laminar-flow hood
	Located in a separate room with separate facilities, incubator, cell counting, centrifuge, etc., separate autoclaved or incinerated waste or chemical decontamination
Virus-producing human cell lines and cell lines infected with amphotropic virus	Class II laminar-flow hood
	Located in a separate room with separate provision for incubation, centrifugation, cell counting, etc.
	No access, except to designated personnel
	All waste, soiled glassware, etc., to be sterilized and extracted air to be filtered as it leaves the room
Tissue samples and cultures carrying known human pathogens	Class III pathogen cabinet with glove pockets, filtered air, and pathogen trap on vented air
	Located in separate room with separate provision for incubation, centrifugation, cell counting, etc. No access, except to designated personnel
	All waste, soiled glassware, etc., to be sterilized and extracted air to be filtered as it leaves the room
	Shower facilities and change of clothing on entering and leaving

*These are suggested procedures only and have no legal basis. Consult national legal requirements and local regulations before formulating proper guidelines.

in a designated Class II biohazard hood, preferably located in a separate room from that in which routine cell culture is performed. This will minimize the risk of spreading infections, such as mycoplasma, to other cultures and will also reduce the number of people associated with the specimen, should it eventually be found to be infected.

(4) Avoid the use of sharp instruments (e.g., syringes, scalpels, Pasteur pipettes) in handling specimens. Clearly, this rule may need to be compromised when a dissection is required, but that should proceed with extra caution.

(5) Put all cultures in a plastic box with tape or labels identifying the cultures as biohazardous and with the name of the person responsible and the date on them. (See Fig. 5.9)

(6) Discard all glassware, pipettes, instruments, etc., into disinfectant or an autoclave.

If appropriate clinical diagnostic tests show that the material is uninfected, and when it has been shown to be sterile and free of mycoplasma, the material may then be cultured with other stocks. However, if more than 10^9 cells are to be generated or if pure DNA is to be prepared, the advice of the local safety committee should be sought.

If a specimen is found to be infected, it should be discarded into double biohazard bags together with all reagents used with it and should then be autoclaved or incinerated. Instruments and other hardware should be placed in a container of disinfectant, soaked for at least 2 h, and then autoclaved. If it is necessary to carry on working with the material, the level of containment must increase, according to the category of the pathogen [Centers for Disease Control (1988); Advisory Committee on Dangerous Pathogens, 1995].

Genetic Manipulation

Any procedure that involves altering the genetic constitution of the cells or cell line that you are working with by transfer of nucleic acid will need to be authorized by the local biological safety committee. The current regulations may be obtained from the Environmental Protection Branch, NCI, NIH, Bldg 13, Room 2W64, Bethesda, MD., for the United States (information available on the internet at *www.nci.nih.gov/ intra/resource/biosafe.htm*) and from the Health and Safety Commission [1992] for the United Kingdom. The latter is currently under revision (publication due March, 2000) and may be reviewed at www.open.gov. uk/hse/condocs/.

Disposal

Potentially biohazardous materials must be sterilized before disposal [National Research Council, 1989; Health Services Advisory Committee (1992)]. They may be placed in unsealed autoclavable sacks and autoclaved, or they may be immersed in a sterilizing agent such as hypochlorite. Various proprietary preparations are available (e.g., Chlorox (Lab Safety Supply, Biomedical Products, Polysciences), Chloros liquid concentrate (Hays Chemical Distribution), and Precept Tablets (Johnson and Johnson)). Recommended concentrations vary according to local rules, but a rough guide can be obtained from the manufacturer's instructions. Hypochlorite is often used at 300 ppm of available chlorine, but some authorities demand 2,500 ppm (a 1:20 dilution of Chloros), as recommended in the Howie Report [1978]. Hypochlorite is effective and easily washed off those items which are to be reused, but is highly corrosive, particularly in alkaline solutions. It will bleach clothing and even corrode stainless steel, so gloves and a lab coat or apron should be worn when handling hypochlorite, and soaking baths and cylinders should be made of polypropylene.

CHAPTER 7

Culture Vessels

THE SUBSTRATE

Attachment and Growth

The majority of vertebrate cells cultured *in vitro* grow as monolayers on an artificial substrate. Hence the substrate must be correctly charged to allow cell adhesion, or at least to allow the adhesion of attachment factors, which will, in turn, allow cell adhesion and spreading. Although spontaneous growth in suspension is restricted to hemopoietic cell lines, rodent ascites tumors, and a few other selected cell lines, such as human small-cell lung cancer [Carney et al., 1981], many transformed cell lines can be made to grow in suspension and are therefore independent of the surface charge on the substrate. Indeed, these lines may actually benefit from propagation on a hydrophobic surface. However, most normal cells need to spread out on a substrate in order to proliferate [Folkman and Moscona, 1978; Ireland et al., 1989], and inadequate spreading due to poor adhesion or overcrowding will inhibit proliferation. Cells shown to require attachment for growth are said to be *anchorage dependent*; cells that have undergone transformation frequently become *anchorage independent* (see Anchorage Independence in Chapter 17) and can grow in suspension (see Propagation in Suspension in Chapter 12) when stirred or held in suspension with semisolid media such as agar.

Substrate Materials

Glass. This was the original substrate because of its optical properties and surface charge, but it has been replaced in most laboratories by synthetic plastic (usually polystyrene), which has greater consistency and superior optical properties. Glass is now rarely used, although it is cheap, is easily washed without losing its growth-supporting properties, can be sterilized readily by dry or moist heat, and is optically clear. Treatment with strong alkali (e.g., NaOH or caustic detergents) renders glass unsatisfactory for culture until it is neutralized by an acid wash. (See Glassware in Chapter 10.)

Disposable Plastic. Single-use sterile polystyrene flasks provide a simple, reproducible substrate for culture. They are usually of good optical quality, and the growth surface is flat, providing uniform and reproducible cultures. As manufactured, polystyrene is hydrophobic and does not provide a suitable surface for cell growth, so tissue culture plastics are treated by γ-irradiation, chemically, or with an electric ion discharge to produce a charged surface that is then wettable. Because the resulting product varies in quality from one manufacturer to another, samples from a number of sources should be tested by determining the plating efficiency and growth rate of cells in current use (see Protocols 20.7, 20.9) in a medium containing the optimal and half-optimal concentrations of serum. (High serum concentrations may mask imperfections in the plastic.)

While polystyrene is by far the most common and cheapest plastic substrate, cells may also be grown on polyvinylchloride (PVC), polycarbonate, polytetrafluorethylene (PTFE), melinex, thermanox (TPX), and a number of other plastics. If you need to use a different plastic, it is worth trying to grow a regular monolayer and then attempting to clone cells on it (see Chapters 11 and 19), with and without pretreating the surface. PTFE is available in a charged (hydrophilic) and uncharged (hydrophobic) form; the charged form can be used for regular monolayer cells and the uncharged for macrophages and some transformed cell lines.

CHOICE OF CULTURE VESSEL

Some typical culture vessels are listed in Table 7.1. The anticipated yield of HeLa cells is quoted for each vessel; the yield from a finite cell line (e.g., diploid fibroblasts) would be about one-fifth of the HeLa figure. Several factors govern the choice of culture vessel, including (1) the cell mass required, (2) whether the cells grow in suspension or as a monolayer, (3) whether the culture should be vented to the atmosphere or sealed, (4) the frequency of sampling, (5) the type of analysis required, and (6) the cost.

Cell Yield

For monolayer cultures, the cell yield is proportional to the available surface area of the flask (Fig. 7.1). Small volumes and multiple replicates are best performed in multiwell plates (Fig. 7.2), which can have a large number of small wells (e.g., microtitration plates 96 or 144 wells, 0.1–0.2 ml of medium and 0.25-cm² growth area or 24-well "cluster dishes" with 1–2 ml medium in each well, 1.75 cm² growth area) up to 4-well plates with each well 50 mm in diameter and using 5-ml culture medium (see Table 7.1). The middle of the size range embraces both Petri dishes (Fig. 7.3) and flasks ranging from 10 cm² to 225 cm² (Fig. 7.4). Flasks are usually designated by their surface area (e.g., No. 25 or No. 175—sometimes T25 or T175, respectively), while Petri dishes are referred to by diameter (e.g., 35 mm or 9 cm).

Glass bottles are more variable than plastic since they are usually drawn from standard pharmaceutical supplies (Fig. 7.5). Glass bottles should have (1) one reasonably flat surface, (2) a deep screw cap with a good seal and nontoxic liner, and (3) shallow-sloping shoulders to facilitate harvesting monolayer cells after trypsinization and to improve the efficiency of washing.

If you require large cell yields (e.g., $\sim 10^9$ HeLa cervical carcinoma cells or 2×10^8 MCR-5 diploid human fibroblast), then increasing the size and number of conventional bottles becomes cumbersome, and special vessels are required. Flasks with corrugated surfaces (Corning, Becton Dickinson) or multilayered flasks (Nalge Nunc) offer an intermediate step in in-

TABLE 7.1. Culture Vessel Characteristics

Culture vessel	Replicates	ml	cm²	Approximate cell yield (HeLa)
Multiwell plates				
Microtitration	96	0.1	0.3	1×10^5
Microtitration	144	0.1	0.3	1×10^5
4-well plate	4	2	2	5×10^5
6-well plate	6	2	10	2×10^6
24-well plate	24	1	2	5×10^5
Petri dishes				
35 mm diameter	1	2	8	2×10^6
50 mm diameter	1	4	17.5	4×10^6
60 mm diameter	1	5	21	5×10^6
90 mm diameter	1	10	49	1×10^7
Flasks				
#10	1	2	10	2×10^6
#25	1	5	25	5×10^6
#75	1	25	75	2×10^7
#175	1	75	175	5×10^7
#225	1	100	225	6×10^7
Roller bottle	1	200	850	2.5×10^8
Stirrer bottles				
500 ml (unsparged)	1	50		5×10^7
5,000 ml (sparged)	1	4,000		4×10^9

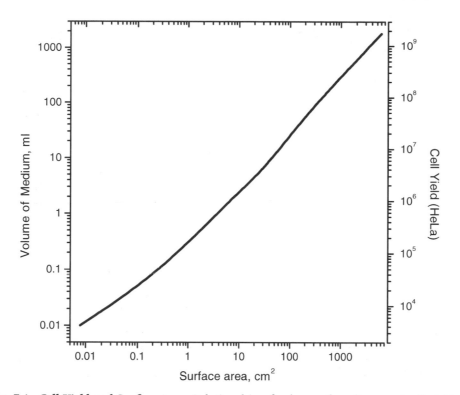

Fig. 7.1. Cell Yield and Surface Area. Relationship of volume of medium and cell yield to the surface area of a culture vessel. The graph is plotted on the basis of the volume of the medium for each size of vessel and is nonlinear, as smaller vessels tend to be used with proportionally more medium than is used with larger vessels. The cell yield is based on the volume of the medium and is approximate.

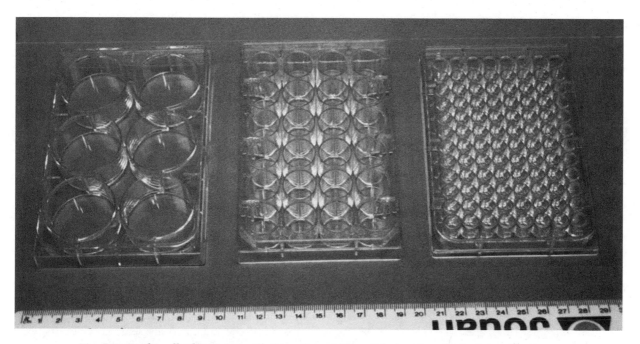

Fig. 7.2. Multiwell Plates. Six-well, 24-well, and 96-well (microtitration) plates. Plates are available with a wide range in the number of wells from 4 to 144. (See Table 7.1 for sizes and capacities.)

Fig. 7.3. Petri Dishes. Illustrated are dishes of 35 mm, 50 mm, and 90 mm diameter. Square Petri dishes are also available, with dimensions 90 × 90 mm. Larger dishes are also available but are seldom used for cell culture. A grid pattern can be provided to help in scanning the dish—for example, in counting colonies—but can interfere with automatic colony counting.

creasing the surface area. Cell yields beyond that require multisurface propagators or roller bottles on special racks. (See Monolayer in Chapter 25.) Increasing the yield of cells growing in suspension requires only that the volume of the medium be increased, as long as cells in deep culture are kept agitated and sparged with 5% CO_2 in air. (See Suspension in Chapter 25.)

Suspension Culture

Cells that grow in suspension can be grown in any type of flask, plate, or Petri dish that, although sterile, need not be treated for cell attachment. Stirrer bottles are used when agitation is required to keep the cells

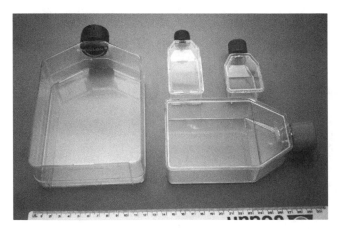

Fig. 7.4. Plastic Flasks. Sizes illustrated are 10 and 25 cm² (Falcon), 75 cm² (Corning) and 185 cm² (Nalge Nunc). (See Table 7.1 for representative sizes and capacities.)

Fig. 7.5. Glass Bottles. Standard medical flat bottle suitable for use in tissue culture. Specialized glass bottles are also available (Bellco).

in suspension. These bottles are available in a wide range of sizes, usually in glass (Bellco, Techne). Agitation is usually by a suspended paddle containing a magnet, whose rotation is driven by a magnetic stirrer (Fig. 7.6; see also Figs. 12.7 and 25.1). The rotational speed must be kept low, ~60 rpm, to avoid damage from shear stress. Generally, the pendulum design is preferable for minimizing shear. Suspension cultures can be set up as replicates or can be sampled repetitively from a side arm or the flask. They can also be used to maintain a steady-state culture by adding and removing medium continuously. (See Continuous Culture in Chapter 25.)

Venting

Multiwell dishes and petri dishes chosen for replicate sampling or cloning have loose-fitting lids to give easy access to the dish. Consequently, they are not sealed and require a humid atmosphere with the CO_2 tension controlled. (See CO_2 and Bicarbonate in Chapter 8.) Because a thin film of liquid may form around the inside of the lid, partially sealing some dishes, vented lids with molded plastic supports inside should be used (Fig. 7.7). If a perfect seal is required, some multiwell dishes can be sealed with self-adhesive film (ICN).

Fig. 7.6. Small Stirrer Flasks. Four small stirrer flasks (Techne), 250 ml capacity, with 50–100 ml medium, on four-place stirrer rack (Techne). Larger flasks, up to 10 l, are available. (See also Figs. 12.2 and 25.1.)

Flasks may be vented by slackening the caps one full turn. Flasks are vented in this way to allow CO_2 to enter (in a CO_2 incubator) or to allow excess CO_2 to escape in excessive acid-producing cell lines. Caps are also available with permeable filters that permit equilibration with the gas phase (Fig. 7.8).

Sampling and Analysis

Multiwell plates are ideal for replicate cultures if all samples are to be removed simultaneously and processed in the same way. If, on the other hand, samples need to be withdrawn at different times and processed immediately, it may be preferable to use separate vessels (flasks, test tubes, etc.) (Fig. 7.9). Individual wells

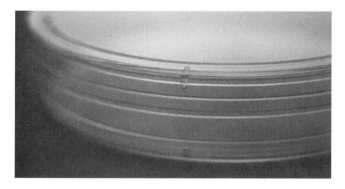

Fig. 7.7. Venting Petri Dishes. Vented dish. Small ridges, 120° apart, raise the lid from the base and prevent a thin film of liquid from sealing the lid and reducing the rate of gas exchange.

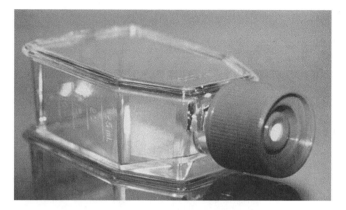

Fig. 7.8. Venting Flasks. Gas-permeable cap on 10 cm² flask (Falcon).

in microtitration plates can be sampled by cutting and removing only that part of the adhesive plate sealer overlying the wells to be sampled. Alternatively, microtitration plates are available with removable wells for individual processing, although you should ensure that the wells are treated for tissue culture if you wish to use adherent cells.

Low-power microscopic observation is performed easily on flasks, Petri dishes, and multiwell plates with the use of an inverted microscope. In using phase contrast, however, difficulties may be encountered with microtitration plates because of the size of the meniscus; even 24-well plates can only be observed satisfactorily by phase contrast in the center of the well. If microscopy will play a major part in your analysis, it may be advantageous to use a chamber slide. (See Culture Vessels for Cytology: Monolayer Cultures in Chapter 15.) Large roller bottles give problems with some microscopes; it is usually necessary to remove the condenser, in which case phase contrast will not be available.

If processing of the sample involves extraction in acetone, toluene, ethyl acetate, or certain other organic solvents, then a problem will arise with the solubility of polystyrene. Since this problem is often associated with organic solvents used in histological procedures, Lux (Bayer, ICN) supplies solvent-resistant Thermanox (TPX) plastic coverslips, suitable for histology, that fit into regular multiwell dishes (which need not be of tissue culture grade). However, these coverslips are of poor optical quality and should be mounted on slides with cells uppermost and a conventional glass coverslip on top.

Glass vessels are required for procedures such as hot perchloric acid extractions of DNA. Plain-sided test tubes or Erlenmeyer flasks (with no lip), used in conjunction with sealing tape or Oxoid caps, are quick to use and are best kept in a humid CO_2-controlled atmosphere. Regular glass scintillation vials, or "min-

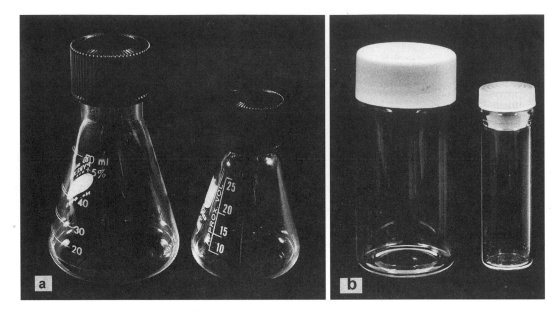

Fig. 7.9. Screw-cap Vials and Flasks. (a) Glass flasks are suitable for replicate cultures or storage of samples, particularly when plastic may not survive downstream processing. Screw caps are preferable to stoppers, as they are less likely to leak and they protect the neck of the flask from contamination. (b) Scintillation vials are particularly useful for isotope incorporation studies, but should not be reused for culture after containing scintillation fluid.

ivials," are also good culture vessels, because they are flat bottomed and have a screw closure. Once used with scintillation fluid, however, they should not be reused for culture.

Uneven Growth

Sometimes, cells can be inadvertently distributed non-uniformly across the growth surface. Vibration, caused by opening and closing the incubator, a faulty fan motor, or vibration from equipment, can perturb the medium, which can result in resonance or standing waves in the flask that, in turn, result in a wave pattern in the monolayer (Fig. 7.10), creating variations in cell density. Eliminating vibration and minimizing entry into the incubator will help reduce uneven growth. Placing a heavy weight in the tray or box with the plates and separating it from the shelf with plastic foam may also help alleviate the problem [Nielsen, 1989], but great care must be taken to wash and sterilize such foam pads, as they will tend to harbor contamination.

Cost

Cost always has to be balanced against convenience; for example, Petri dishes are cheaper than flasks with an equivalent surface area, but require humid, CO_2-controlled conditions and are more prone to infection. They are, however, easier to examine and process.

Cheap soda-glass bottles, though not always of good optical quality, are often better for culture than

higher grade Pyrex or optically clear glass, which usually contains lead. A major disadvantage of glass is that its preparation is labor intensive, because it must be carefully washed and resterilized before it can be

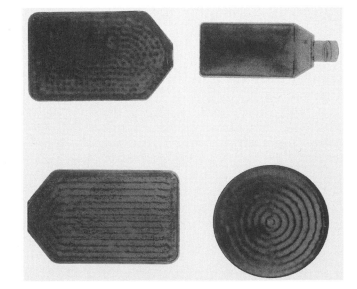

Fig. 7.10. Nonrandom Growth. Examples of ridges seen in cultured monolayers in dishes and flasks, probably due to resonance in the incubator from fan motors or to opening and closing the incubator doors. (Courtesy of Nalge Nunc.)

reused. Most laboratories now use plastic, due to its convenience, optical clarity, and quality.

SPECIALIZED SYSTEMS

Permeable Supports

Gas-permeable, but non-water-permeable, PTFE films are available as disposable Petri dishes (Petriperm, Heraeus) or as membranes to be incorporated in an autoclavable, reusable culture vessel (e.g., "Chamber/Dish," from Bionique) (Fig. 15.3). These dishes have two advantages: The substrate is permeable to O_2 and CO_2, and the plastic is thin and, therefore, well suited to histological sectioning for light or electron microscopy. Semipermeable membranes are used as gas-permeable substrates and will also allow the passage of water and small molecules (<500–1,000 da), a property that is exploited in some large-scale bioreactors. (See Membrane Perfusion Systems in Chapter 25.)

Growing cells on a water-permeable substrate contributes more than an increased diffusion of oxygen, CO_2, and nutrients. Attachment of cells to a natural substrate such as collagen may exert some biological control of phenotypic expression due to the interaction of integrin receptors on the cell surface with specific sites in the extracellular matrix. (See Cell Adhesion Molecules in Chapter 2; Cell–Matrix Interactions in Chapter 16.) Extensive use has also been made of natural gels, such as collagen. The growth of cells on floating collagen [Michalopoulos and Pitot, 1975; Lillie et al., 1980] has been used to improve the survival of epithelial cells and promote terminal differentiation. (See Cell–Matrix Interactions in Chapter 16; Filter Wells in Chapter 24.)

Filter Wells. The permeability of the surface to which the cell is anchored may induce polarity in the cell by simulating the basement membrane. Such polarity may be vital to full functional expression in secretory epithelia and many other types of cells [Gumbiner and Simons, 1986; Chambard et al. 1987; Arthursson and Magnusson, 1990; Mullin et al., 1997]. Several manufacturers now provide permeable supports in the form of disposable filter wells of many different sizes, materials, and membrane porosities (Costar, Falcon, Millipore, Nunc). Also available are supports precoated with collagen, laminin, or some other matrix material (e.g., Matrigel, Becton Dickinson). All these supports have been used extensively in studies of cell–cell interaction, cell–matrix interaction, differentiation and polarity, and transepithelial permeability. (See Filter-Well Inserts in Chapter 24).

Hollow Fibers. Knazek et al. [1972] developed a technique for growing cells on the outer surface of

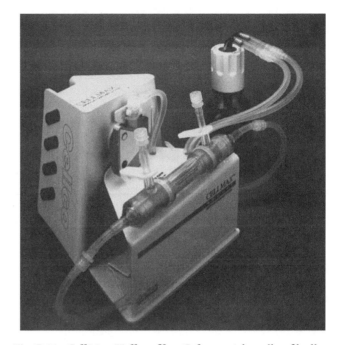

Fig. 7.11. CellMax Hollow-fiber Culture. A bundle of hollow fibers of permeable plastic is enclosed in a transparent plastic outer chamber and is accessible via either of the two side arms for seeding cells. During culture, the chamber is perfused down the center of the hollow fibers through connections attached to either end of the chamber (CellMax, Cellco; see also Fig. 25.6).

bundles of plastic microcapillaries (Fig. 7.11). (See Hollow Fibers in Chapter 24.) The plastic allows the diffusion of nutrients and dissolved gases from a medium perfused through the capillaries. Cells will grow up to several cells deep on the outside of the capillaries, and an analogy with whole tissue is suggested. Hollow fibers are also used in large-scale bioreactors. (See Perfusion Systems in Chapter 25.)

TREATED SURFACES

Matrix Coating

Cell attachment and growth can be improved by pretreating the substrate in a variety of ways [Barnes et al., 1984a]. A well-established piece of tissue culture lore has it that used glassware supports growth better than new. If that is true, it may be due to etching of the surface or minute traces of residue left after culture. The growth of cells in a flask also improves the surface for a second seeding, and this type of conditioning may be due to collagen, fibronectin or other matrix products [Crouch et al., 1987] released by the cells. The substrate can be conditioned by treating it with spent medium from another culture [Stampfer et al., 1980] or by purified fibronectin or collagen. (For the latter, see Protocol 22.9.) Treatment with denatured collagen improves the attachment of many types of cells, such as epithelial cells, and it may be necessary for the expression of differentiated functions. (See Cell–Matrix Interactions in Chapter 16; Epidermis in Chapter 22.)

Coating with denatured collagen may be achieved using rat-tail collagen or commercially supplied alternatives (Vitrogen, ICN) and simply pouring the collagen solution over the surface of the dish, draining off the excess, and allowing the residue to dry. Because this procedure sometimes leads to detachment of the collagen layer during culture, a protocol was devised by Macklis et al. [1985] to ensure that the collagen would remain firmly anchored to the substrate, by cross-linking to the plastic with carbodiimide.

Collagen may also be applied as an undenatured gel (see Cell–Matrix Interactions in Chapter 16), a type of substrate that has been shown to support neurite outgrowth from chick spinal ganglia [Ebendal, 1976] and morphological differentiation of breast [Nicosia and Ottinetti, 1990; Berdichevsky et al., 1992] and other epithelia [Sattler et al., 1978], and to promote the expression of tissue-specific functions of a number of other cells *in vitro* (e.g., see Protocol 22.2). Diluting the collagen 1:10 with culture medium and neutralizing to pH 7.4 causes the collagen to gel, so the dilution and dispensing must be rapid. It is best to add the growth medium to the gel for a further 4–24 hr to ensure that the gel equilibrates with the medium before adding cells. At this stage, fibronectin (25–50 μg/ml) or laminin (1–5 μg/ml), or both, may be added to the medium.

Evidence is gradually accumulating that specific treatment of the substrate with biologically significant compounds can induce specific alterations in the attachment or behavior of specific cell types. For example, chondronectin enhances chondrocyte adherence [Varner et al., 1984], and laminin promotes the adherence of epithelial cells [Kleinman et al., 1981]. Rojkind et al. [1980] described methods for preparing reconstituted "basement membrane rafts" from tissue extracts in order to optimize culture conditions for cell differentiation.

Commercially available matrices, such as Matrigel (Becton Dickinson) from the Engelbreth Holm Swarm (EHS) sarcoma, contain laminin, fibronectin, and proteoglycans, with laminin predominating. (See Cell–Matrix Interactions in Chapter 16.) Numerous investigations have evaluated Matrigel in studies of differentiation and malignant invasion. (See Invasiveness in Chapter 17.) Other matrix products include Pronectin F (Protein Polymer Technologies), laminin, fibronectin, entactin (UBI), heparan sulfate, EHS Natrix (Becton Dickinson), ECM (IBT), and Cell-tak (Becton Dickinson). Some of these products are purified, if not completely chemically defined; others are a mixture of matrix products that have been poorly characterized and may also contain bound growth factors. If cell adhesion for survival is the main objective, and defined substrates are inadequate, the use of these matrices is acceptable, but if mechanistic studies are being carried out, they can only be an intermediate stage on the road to a completely defined substrate.

Gelatin coating has been found to be beneficial for the culture of muscle [Richler and Yaffe, 1970] and endothelial cells [Folkman et al., 1979] (see Endothelium in Chapter 22), and it is necessary for some mouse teratomas. McKeehan and Ham [1976a] found that it was necessary to coat the surface of plastic dishes with 1 mg/ml of poly-D-lysine before cloning in the absence of serum. (See Improving Clonal Growth in Chapter 13.) At least two components of interaction with the substrate may be recognised: (1) adhesion, to allow the attachment and spreading that are necessary for cell proliferation [Folkman and Moscona, 1978], and (2) specific interactions, reminiscent of the interaction of an epithelial cell with basement membrane, with other extracellular matrix constituents, or with adjacent tissue cells, and required for the expression of some specialized functions. (See Cell Adhesion in Chapter 2; Cell–Matrix Interactions in Chapter 16.) Rojkind et al. [1980], Vlodavsky et al. [1980], and others explored the growth of cells on other natural substrates related to basement membrane. Synthetic matrices and defined-matrix macromolecules are now available for controlled studies on matrix interaction (Matrigel, Natrigel, Collagen, laminin, vitronectin [Becton Dickinson, Gibco/Life Technologies]).

While inert coating of the surface may suffice, it may yet prove necessary to use a monolayer of an appropriate cell type to provide the correct matrix for the maintenance of some specialized cells. Gospodarowicz et al. [1980] were able to grow endothelium on confluent monolayers of 3T3 cells that had been extracted with Triton X100, leaving a residue on the surface of the substrate. This so-called extracellular matrix (ECM) has also been used to promote differentiation in ovarian granulosa cells [Gospodarowicz et al., 1980] and in studying tumor cell behavior [Vlodavsky et al., 1980].

PROTOCOL 7.1. PREPARATION OF ECM

Outline

Remove a postconfluent monolayer of matrix-forming cells with detergent, wash flask or dish and seed required cells onto residual matrix.

Materials

3T3 mouse fibroblasts, MRC-5 human fibroblasts, or CPAE bovine pulmonary arterial endothelial cells (or any other cell line shown to be suitable for producing extracellular matrix)

Sterile, ultrapure water (UPW) (see Water Purification in Chapter 4)

1% Triton X100 in sterile, UPW

Protocol

1. Set up matrix-producing cultures, and grow to confluence.
2. After 3–5 days at confluence, remove the medium and add an equal volume of sterile 1% Triton X100 in UPW to the cell monolayer.
3. Incubate for 30 min at 37°C.
4. Remove Triton X solution and wash residue three times with the same volume of sterile UPW.
5. Flasks or dishes may be used directly or may be stored at 4°C for up to 3 weeks.

Feeder Layers

While matrix coating may help attachment, growth, and differentiation, some cultures of more fastidious cells, particularly at low cell densities [Puck and Marcus, 1955], require support from living cells (e.g., mouse embryo fibroblasts; see Protocol 13.3). This action is due partly to supplementation of the medium by either metabolite leakage or the secretion of growth factors from the fibroblasts, but may also be due to conditioning of the substrate by cell products. Feeder layers grown as a confluent monolayer may make the surface suitable, or even selective, for attachment for other cells. (See Selective Adhesion in Chapter 13; Cervix in Chapter 22; Confluent Feeder Layers in Chapter 23.) The survival and extension of neurites by central and peripheral neurons can be enhanced by culturing the neurons on a monolayer of glial cells, although in this case the effect is due to a diffusible factor rather than direct cell contact [Seifert and Müller, 1984].

After a monolayer culture reaches confluence, subsequent proliferation causes cells to detach from the artificial substrate and migrate over the surface of the monolayers. The morphology of the cells may change (Fig. 7.12): The cells may become less well spread, more densely staining, and more highly differentiated. Apparently, and not too surprisingly, the interaction of a cell with a cellular underlay is different from the interaction of the cell with a synthetic substrate. The former can cause a change in morphology and reduce the cell's potential to proliferate.

Three-Dimensional Matrices

It has long been realized that, while growth in two dimensions is a convenient way of preparing and ob-serving a culture and allows a high rate of cell proliferation, it lacks the cell–cell and cell–matrix interaction characteristic of whole tissue *in vivo*. The very first attempts to culture animal tissues [Harrison, 1907; Carrel, 1912] were performed with gels formed of clotted lymph or plasma on glass. In these cases, however, the cells migrated along the glass–clot interface rather than within the gel, and the tissue architecture and cell–cell interaction were gradually lost. Migration was often accompanied by proliferation of cells in the outgrowth, leading, as later studies showed, to the development of propagated cell lines. Gradually, it became apparent that many functional and morphological characteristics were lost during serial subculture, as discussed in Chapter 2 (see Dedifferentiation in Chapter 2).

These deficiencies encouraged the exploration of three-dimensional matrices, such as collagen gel [Douglas et al., 1980], cellulose sponge (either alone or coated with collagen) [Leighton et al., 1968], or Gelfoam. (See Gel and Sponge Techniques in Chapter 24.) Fibrin clots were one of the first media to be used for primary culture and are still used either as crude plasma clots (see Primary Explant in Chapter 11) or as purified fibrinogen mixed with thrombin. Both systems generate a three-dimensional gel in which cells may migrate and grow, either on the solid–gel interface or within the gel [Leighton, 1991].

Many different types of cell can be shown to penetrate such matrices and establish a tissuelike histology. Breast epithelium, seeded within collagen gel, displays a tubular morphology, while breast carcinoma shows more disorganized growth, confirming the correlation between this mode of growth and the condition *in vivo* [Berdichevsky et al., 1992]. The kidney epithelial cell line MDCK responds to paracrine stimulation from fibroblasts by producing tubular structures, but only in collagen gel [Kenworthy et al., 1992]. Neurite outgrowth from sympathetic ganglia neurons growing on collagen gels follows the orientation of the collagen fibers in the gel [Ebendal, 1976]. (See further discussion of three-dimensional cultures, Chapter 24.)

Alternative Artificial Substrates

Microcarriers. Polystyrene (Nalge Nunc, Gibco), Sephadex (ICN and Amersham Pharmacia), polyacrylamide (Biorad), and collagen (Amersham Pharmacia) or gelatin (JRH Biosciences) are available in bead form for the propagation of anchorage-dependent cells in suspension (see Chapter 25: Microcarriers).

Metallic Substrates. Cells may be grown on stainless-steel discs [Birnie and Simons, 1967] or other me-

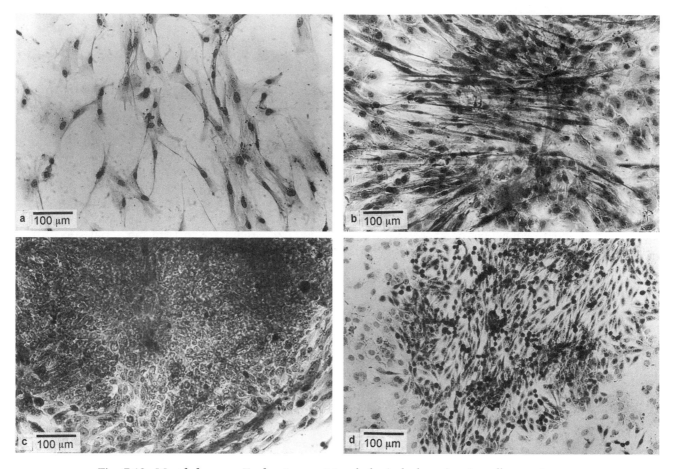

Fig. 7.12. Morphology on Feeder Layers. Morphological alteration in cells growing on feeder layers: (a) Fibroblasts from human breast carcinoma growing on plastic and (b) growing on a confluent feeder layer of fetal human intestinal cells (FHI). (c) Epithelial cells from human breast carcinoma growing on plastic and (d) on same confluent feeder layer as in (b).

tallic surfaces [Litwin, 1973]. Observation of the cells on an opaque substrate requires surface interference microscopy, unless very thin metallic films are used. Westermark [1978] developed a method for the growth of fibroblasts and glia on palladium. Using electron microscopy shadowing equipment, he produced islands of palladium on agarose, which does not allow cell attachment in fluid media. The size and shape of the islands were determined by masks made in the manner of electronic printed circuits, and the palladium was applied by "shadowing" under vacuum, as used in electron microscopy. Because the layer was very thin, it remained transparent.

Nonadhesive Substrates

Sometimes, attachment of the cells is undesirable. The selection of virally transformed colonies, which are anchorage independent, can be achieved by plating cells in agar [Macpherson and Montagnier, 1964], as the untransformed cells do not form colonies readily in

this matrix. There are two principles involved in such a system: (1) prevention of attachment at the base of the dish, where spreading and anchorage-dependent growth would occur, and (2) immobilization of the cells such that daughter cells remain associated with the colony, even if they are nonadhesive. Most commonly, agar, agarose, or Methocel (methylcellulose of viscosity 4,000 cps) is used. The first two are gels and the third is a high-viscosity sol. Because Methocel is a sol, cells will sediment slowly through it. It is, therefore, commonly used with an underlay of agar. (See Semisolid Media in Chapter 13.) Dishes that are not of tissue culture grade can be used without an agar underlay, but some attachment and spreading may occur.

Liquid–Gel or Liquid–Liquid Interfaces

While the Methocel-over-agar system usually gives rise to discrete colonies at the interface of the agar and the Methocel, some cells can migrate across the gel

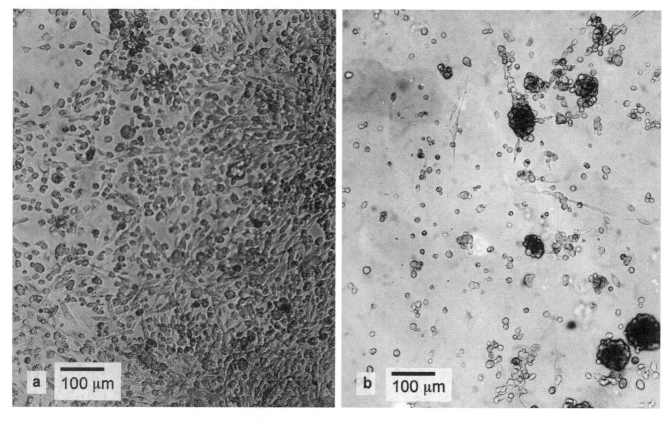

Fig. 7.13. Liquid–Gel Interface. Cell growth at interface between Methocel-containing medium and agar gel. Methocel concentration, 1.5%; agar, 1.25%. Human metastatic melanoma. In (a), 2.5×10^5 cells/ml, seeded alone. In (b), 5×10^4 cells/ml, seeded with 2×10^5 homologous feeder cells/ml.

surface and form monolayers or cords of cells (Fig. 7.13). The reason for this migration remains obscure, although the concentrations of Methocel and agar in the figure are higher than normal and may have contributed to the effect. Cell spreading and monolayer formation at the liquid–liquid interface between various fluorinated hydrocarbons (FC43, FC73) and aqueous culture media have been observed [Nagaoka et al., 1990]. The spreading and locomotion on non-rigid substrates conflict somewhat with current concepts of cell adhesion and locomotion, unless denatured serum protein or extracellular matrix forms a layer at the interface that is sufficient to permit anchorage. Methocel in particular often contains particulate debris that may help to promote such anchorage.

CHAPTER 8

Media

DEVELOPMENT OF MEDIA

Initial attempts to culture cells were performed in natural media based on tissue extracts and body fluids, such as chick embryo extract, serum, lymph, etc. With the propagation of cell lines, the demand for larger amounts of a medium of more consistent quality led to the introduction of chemically defined media based on analyses of body fluids and nutritional biochemistry. Eagle's Basal Medium [Eagle, 1955] and, subsequently, Eagle's Minimal Essential Medium [Eagle, 1959] became widely adopted, variously supplemented with calf, human, or horse serum, protein hydrolysates, and embryo extract. As more continuous cell lines became available (L929 cells, HeLa, etc.), it was apparent that these media were perfectly adequate for the majority of those lines, and most of the succeeding developments were aimed at replacing serum (see Chapter 9), optimizing media for different cell types (e.g., RPMI 1640 for lymphoblastoid cell lines), or modifying for specific conditions (e.g., Leibovitz L15 to eliminate the need for adding CO_2 and $NaHCO_3$) [Leibovitz 1963].

Currently, there is a divergence in media requirements: Some scientists wish to isolate and propagate cells of a specific lineage, while others simply require cells as substrates for the formation of products, as a host for viral propagation, or for non-cell-specific molecular studies. Production techniques make ever more use of selective media, usually serum-free, while viral and molecular work relies mainly on Eagle's MEM [Eagle, 1959], Dulbecco's modification DMEM [Dulbecco and Freeman, 1959], or, increasingly, RPMI1640

[Moore et al., 1967]. There is also a growing tendency to effect a compromise by mixing a complex medium, such as Ham's F12 [Ham, 1965], with one with higher amino acid and vitamin concentrations, such as DMEM. This alternative will horrify the purist, but it does generate a useful, all-purpose medium for primary culture as well as cell line propagation.

Serum is still widely used, as most people do not want the added complexity and cost of serum-free media. However, those culturing specialized cells and those working in the production of biopharmaceuticals, in which serum is undesirable for a number of reasons (see Disadvantages of Serum in Chapter 9), are now working without serum. This chapter concentrates on the general principles of medium composition, using the widely used serum-supplemented media as examples. Chapter 9 treats the design and use of serum-free media.

PHYSICOCHEMICAL PROPERTIES

pH

Most cell lines grow well at pH 7.4. Although the optimum pH for cell growth varies relatively little among different cell strains, some normal fibroblast lines perform best at pH 7.4–7.7, and transformed cells may do better at pH 7.0–7.4 [Eagle, 1973]. It was reported that epidermal cells could be maintained at pH 5.5 [Eisinger et al., 1979], but this level has not been universally adopted. In special cases it may prove advantageous to do a brief growth experiment (see Protocols

20.7 and 20.8) or a special function analysis (e.g., Chapter 15) to determine the optimum pH.

Phenol red is commonly used as an indicator. It is red at pH 7.4 and becomes orange at pH 7.0, yellow at pH 6.5, lemon yellow below pH 6.5, more pink at pH 7.6, and purple at pH 7.8. Since the assessment of color is highly subjective, it is useful to make up a set of standards using a sterile balanced salt solution (BSS) and phenol red at the correct concentration and in the same type of bottle, with the same headspace for air, that you normally use for preparing a medium.

PROTOCOL 8.1. PREPARATION OF pH STANDARDS

Materials

Hanks's Balanced Salt Solution (HBSS), 10× concentrate or powder, without bicarbonate or glucose, with 20 mM HEPES
Nine bottles of a size closest to your standard medium bottles or culture flasks
Ultrapure water (UPW)
Sterile 0.1 N NaOH (make up in UPW and filter sterilize; see Sterile Filtration in Chapter 10)
pH meter

Protocol

1. Make up the BSS at pH 6.5, and dispense into seven bottles of the appropriate size.
2. Allow to equilibrate with air.
3. Adjust the pH to 6.5, 6.8, 7.0, 7.2, 7.4, 7.6, and 7.8 with sterile 0.1 N NaOH, checking the pH on a pH meter.
4. Filter sterilize BSS into flasks or bottles.
5. Keep sterile and sealed.

CO₂ and Bicarbonate

Carbon dioxide in the gas phase appears in the medium as dissolved CO_2 in equilibrium with HCO_3^- and lowers the pH. Because dissolved CO_2, HCO_3^- and pH are all interrelated, it is difficult to determine the major direct effect of CO_2. The atmospheric CO_2 tension will regulate the concentration of dissolved CO_2 directly, as a function of temperature. This regulation in turn produces H_2CO_3, which dissociates according to the reaction

$$H_2O + CO_2 \Leftrightarrow H_2CO_3 \Leftrightarrow H^+ + HCO_3^-. \quad (1)$$

Since HCO_3^- has a fairly low dissociation constant with most of the available cations, it tends to reassociate, leaving the medium acid. The net result of increasing atmospheric CO_2 is to depress the pH, so the effect of elevated CO_2 tension is neutralized by increasing the bicarbonate concentration:

$$NaHCO_3 \Leftrightarrow Na^+ + HCO_3^- \quad (2)$$

The increased HCO_3^- concentration pushes equation (1) to the left until equilibrium is reached at pH 7.4. If another alkali (e.g., NaOH) is used instead, the net result is the same:

$$NaOH + H_2CO_3 \Leftrightarrow NaHCO_3 + H_2O \quad (3)$$
$$\Leftrightarrow Na^+ + HCO_3^- + H_2O$$

The equivalent $NaHCO_3$ concentrations commonly used with different CO_2 tensions are listed in Tables 8.1, 8.2, and 8.3. Intermediate values of CO_2 and HCO_3^- may be employed, provided that the concentration of both is varied proportionately. Because many media are made up in acid solution and may incorporate a buffer, it is difficult to predict how much bicarbonate to use when other alkali may also end up as bicarbonate, as in equation (3). When preparing a new medium for the first time, add the specified amount of bicarbonate and then sufficient 1 N NaOH such that the medium equilibrates to the desired pH after incubation at 37°C overnight. When dealing with a medium that is already at working strength, vary the amount of HCO_3^- to suit the gas phase (Table 8.1), and leave the medium overnight to equilibrate at 37°C. Each medium has a recommended bicarbonate concentration and CO_2 tension for achieving the correct pH and osmolality, but minor variations will occur in different methods of preparation.

With the introduction of Good's buffers (e.g., HEPES, Tricine) [Good et al., 1966] into tissue culture, there was some speculation that, since CO_2 was no longer necessary to stabilize the pH, it could be omitted. This proved to be untrue [Itagaki and Kimura, 1974], at least for a large number of cell types, particularly at low cell concentrations. Although 20 mM HEPES can control pH within the physiological range, the absence of atmospheric CO_2 allows equation (1) to move to the left, eventually eliminating dissolved CO_2, and ultimately HCO_3^-, from the medium. This chain of events appears to limit cell growth, although whether the cells require the dissolved CO_2 or the HCO_3^- (or both) is not clear. Recommended HCO_3^-, CO_2, and HEPES concentrations are given in Table 8.1.

The inclusion of pyruvate in the medium enables cells to increase their endogenous production of CO_2, making them independent of exogenous CO_2, as well as HCO_3^-. Leibovitz L15 medium [Leibovitz, 1963] contains a higher concentration of sodium pyruvate (550 mg/l), but lacks $NaHCO_3$ and does not require CO_2 in the gas phase. Buffering is achieved via the relatively high amino acid concentrations. Because it does not require CO_2, L15 is sometimes recommended for the collection of biopsy samples. Sodium β-

TABLE 8.1. Relationship between Bicarbonate, Carbon Dioxide, and HEPES

Compound	Eagle's MEM Hanks's salts	Low HCO_3^- + buffer	Eagle's MEM Earle's salts	Dulbecco's modification DMEM
$NaHCO_3$	4 mM	10 mM	26 mM	44 mM
CO_2	Atmospheric and evolved from culture	2%	5%	10%
HEPES	10 mM	20 mM	50 mM	—

glycerophosphate can also be used to buffer autoclavable media lacking CO_2 and HCO_3^- [Waymouth, 1979], and Life Technologies (Gibco) markets a CO_2-independent medium. If the elimination of CO_2 is important for cost saving, convenience, or other reasons, it might be worth considering one of these formulations, but only after appropriate testing.

In sum, cultures in open vessels need to be incubated in an atmosphere of CO_2, the concentration of which is in equilibrium with the sodium bicarbonate in the medium. (See Tables 8.1, 8.2, and 8.3.) Cells at moderately high concentrations ($\geq 1 \times 10^5$ cells/ml) and grown in sealed flasks need not have CO_2 added to the gas phase, provided that the bicarbonate concentration is kept low (~4 mM), particularly if the cells are high acid producers. At low cell concentrations, however (e.g., during cloning), and with some primary cultures, it is necessary to add CO_2 to the gas phase of sealed flasks. When venting is required, to allow either the equilibration of CO_2 or its escape in high acid producers, it is necessary to leave the cap slack or to use a CO_2-permeable cap (Corning, Falcon, Nalge Nunc).

Buffering

Culture media must be buffered under two sets of conditions: (1) open dishes, wherein the evolution of CO_2 causes the pH to rise (see CO_2 and Bicarbonate in this chapter), and (2) overproduction of CO_2 and lactic acid in transformed cell lines at high cell concentrations, when the pH will fall. A buffer may be incorporated into the medium to stabilize the pH, but in (1) exogenous CO_2 may still be required by some cell lines, particularly at low cell concentrations, to prevent the total loss of dissolved CO_2 and bicarbonate from the medium. In spite of its poor buffering capacity at physiological pH bicarbonate buffer is still used more frequently than any other buffer, because of its low toxicity, low cost, and nutritional benefit to the culture. HEPES is a much stronger buffer in the pH 7.2–7.6 range and is now used frequently at 10 or 20 mM. It has been found that, when HEPES is used with exogenous CO_2, the HEPES concentration must

be more than double that of the bicarbonate for adequate buffering. (See Table 8.1.) A variation of Ham's F12 with 20 mM HEPES, 10 mM bicarbonate, and 2% CO_2 has been used successfully in the author's laboratory for the culture of a number of different cell lines. It allows the handling of microtitration and other multiwell plates out of the incubator without an excessive rise in pH and minimizes the requirement for HEPES, which is both toxic and expensive.

Oxygen

The other major significant constituent of the gas phase is oxygen. While most cells require oxygen for respiration *in vivo*, cultured cells often rely on glycolysis, a high proportion of which, as in transformed cells, may be anaerobic. Although attempts have been made to incorporate O_2 carriers, by analogy with hemoglobin in blood, as yet this practice is not in general use, and cells rely chiefly on dissolved O_2, which can be toxic due to the elevation in the level of free radicals. Providing the correct O_2 tension is, therefore, always a compromise between fulfilling the respiratory requirement and avoiding toxicity. Strategies involving both elevated and reduced O_2 levels have been employed, as has the incorporation of free-radical scavengers, such as glutathione, 2-mercaptoethanol (β-mercaptoethanol) or dithiothreitol, into the medium. Most such strategies have been derived empirically.

Cultures vary in their oxygen requirement, the major distinction lying between organ and cell cultures. While atmospheric or lower oxygen tensions [Cooper et al., 1958; Balin et al., 1976] are preferable for most cell cultures, some organ cultures, particularly from late-stage embryos, newborns, or adults, require up to 95% O_2 in the gas phase [Trowell, 1959; De Ridder and Mareel, 1978]. This requirement for a high level of O_2 may be a problem of diffusion related to the geometry and gaseous penetration of organ cultures (see Organ Culture in Chapter 24), but may also reflect the difference between differentiated and rapidly proliferating cells. Oxygen diffusion may also become limiting in porous microcarriers [Preissman et al., 1997; see also Microcarriers in Chapter 25].

Most dispersed cell cultures prefer lower oxygen tensions, and some systems (e.g., human tumor cells in clonogenic assay [Courtenay et al., 1978] and human embryonic lung fibroblasts [Balin et al., 1976]) do better in less than the normal level of atmospheric oxygen tension. McKeehan et al. [1976] suggested that the requirement for selenium in medium is related to oxygen toxicity, as selenium is a cofactor in glutathione synthesis. Oxygen tolerance—and selenium as well—may be provided by serum, so the control of O_2 tension is likely to be more critical in serum-free media.

Because the depth of the culture medium can influence the rate of oxygen diffusion to the cells, it is advisable to keep the depth of the medium within the range 2–5 mm (0.2–0.5 ml/cm²) in static culture.

Osmolality

Most cultured cells have a fairly wide tolerance for osmotic pressure [Waymouth, 1970]. Since the osmolality of human plasma is about 290 mOsm/kg, it is reasonable to assume that this level is the optimum for human cells *in vitro*, although it may be different for other species (e.g., around 310 mOsm/kg for mice [Waymouth, 1970]). In practice, osmolalities between 260 mOsm/kg and 320 mOsm/kg are quite acceptable for most cells, but, once selected, should be kept consistent at ±10 mOsm/kg. Slightly hypotonic medium may be better for Petri-dish or open-plate culture to compensate for evaporation during incubation.

Osmolality is usually measured by depression of the freezing point (Fig. 8.1), or elevation of the vapor pressure, of the medium. The measurement of osmolality is a useful quality-control step if you are making up the medium yourself, as it helps to guard against errors in weighing, dilution, etc. It is particularly important to monitor osmolality if alterations are made in the constitution of the medium. The addition of HEPES and drugs dissolved in strong acids and bases and their subsequent neutralization can all markedly affect osmolality.

Temperature

The optimal temperature for cell culture is dependent on (1) the body temperature of the animal from which the cells were obtained, (2) any anatomical variation in temperature (e.g., the temperature of the skin and testis may be lower than that of the rest of the body), and (3) the incorporation of a safety factor to allow for minor errors in regulating the incubator. Thus, the temperature recommended for most human and warm-blooded animal cell lines is 37°C, close to body

Fig. 8.1. Osmometer. Roebling osmometer (Camlab). This model accepts samples of 50 μl.

heat, but set a little lower for safety, as overheating is a more serious problem than underheating.

Because of the higher body temperature in birds, avian cells should be maintained at 38.5°C for maximum growth, but will grow quite satisfactorily, if more slowly, at 37°C.

Cultured cells will tolerate considerable drops in temperature, can survive several days at 4°C, and can be frozen and cooled to −196°C (see Protocol 19.1), but they cannot tolerate more than about 2°C above normal (39.5°C) for more than a few hours and will die quite rapidly at 40°C and over.

Attention must be paid to the consistency of the temperature (within ±0.5°C) to ensure reproducible results. Doors of incubators or hot rooms must not be left open longer than necessary, and large items or volumes of liquid, placed in the warm room to heat, should not be put near any cultures. The spatial distribution of temperature within the incubator or hot room must also be uniform (see Incubation in Chapter 3; Incubator in Chapter 4); there should be no "cold spots," and air should circulate freely. This means that a large number of flasks should not be stacked together when first placed in the incubator or hot room; space must be allowed between them for air to circulate.

Poikilotherms (cold-blooded animals that do not regulate their blood heat within narrow limits) tolerate a wide temperature range, between 15°C and 26°C. Simulating *in vivo* conditions (e.g., for cold-water fish) may require an incubator with cooling as well as heating, to keep the incubator temperature below ambient. If necessary, poikilothermic animal cells can be maintained at room temperature, but the variability of the ambient temperature in laboratories makes this undesirable, and a cooled incubator is generally preferred.

A number of temperature-sensitive (ts) mutant cell lines have been developed that allow the expression of specific genes below a set temperature, but not above it [Su et al., 1991; Foster and Martin, 1992; Wyllie et al., 1992]. These mutants facilitate studies on cell regulation, but also emphasize the narrow range within which one can operate, as the two discriminating temperatures are usually only about 2–3°C apart. The use of ts-mutants usually requires an incubator with cooling as well as heating, to compensate for a warm ambient temperature.

Apart from its direct effect on cell growth, the temperature will also influence pH due to the increased solubility of CO_2 at lower temperatures and, possibly, due to changes in ionization and the pK_a of the buffer. The pH should be adjusted to 0.2 unit lower at room temperature than at 37°C. In preparing a medium for the first time, it is best to make up the medium complete with serum, if that is to be used, and incubate a sample overnight at 37°C under the correct gas tension, in order to check the pH.

Viscosity

The viscosity of a culture medium is influenced mainly by the serum content and in most cases will have little effect on cell growth. Viscosity becomes important, however, whenever a cell suspension is agitated (e.g., when a suspension culture is stirred) or when cells are dissociated after trypsinization. Any cell damage that occurs under these conditions may be reduced by increasing the viscosity of the medium with carboxymethylcellulose (CMC) or polyvinylpyrrolidone (PVP) [Cherry & Papoutsakis, 1990: see also Reagent Appendix]. This becomes particularly important in low-serum concentrations, in the absence of serum, and in stirred bioreactor cultures (see Mixing and Aeration in Chapter 25), in which Pluronic F68 is often used, although its effect is probably pleiotropic.

Surface Tension and Foaming

The effects of foaming have not been clearly defined, but the rate of protein denaturation may increase, as may the risk of contamination if the foam reaches the neck of the culture vessel. Foaming will also limit gaseous diffusion if a film from a foam or spillage gets into the capillary space between the cap and the bottle, or between the lid and the base of a Petri dish.

Foaming can arise in suspension cultures when 5% CO_2 in air is bubbled through medium containing serum. The addition of a silicone antifoam (Dow Chemical) or Pluronic F68 (Sigma), 0.01–0.1%, helps prevent foaming in this situation by reducing surface tension and may also protect cells against shear stress from bubbles.

BALANCED SALT SOLUTIONS

A balanced salt solution (BSS) is composed of inorganic salts and may include sodium bicarbonate and, in some cases, glucose. The compositions of some common BSS's are given in Table 8.2. HEPES buffer (5–20 mM) may be added to these solutions if necessary and the equivalent weight of NaCl omitted to maintain the correct osmolality. BSS forms the basis of many complete media, and catalogues cite, often for example, Eagle's MEM with Hanks's salts [Hanks and Wallace, 1949] or Eagle's MEM with Earle's salts [Earle et al., 1943], indicating which BSS formulation was used; Hanks's would imply the use of sealed flasks with a gas phase of air, while Earle's salts would imply a higher bicarbonate concentration compatible with growth in 5% CO_2.

BSS is also used as a diluent for concentrates of amino acids and vitamins to make complete media, as a washing or dissection medium, and for short incubations up to about 4 h (usually with glucose). These recipes are often modified—for instance, by omitting glucose or phenol red from Hanks's BSS or by leaving out Ca^{2+} or Mg^{2+} ions from Dulbecco's PBS [Dulbecco & Vogt, 1954]. PBS without Ca^{2+} and Mg^{2+} is known as PBS Solution A, and the convention PBSA will be used throughout this book to indicate the absence of these divalent cations. One should always check for modifications when purchasing BSS and should quote any modifications to the published formula in one's own reports and publications.

The choice of BSS is dependent on both the CO_2 tension (see CO_2 and Bicarbonate in this chapter and Tables 8.1 and 8.2) and the intended use of the solution for tissue disaggregation or monolayer dispersal; in these cases Ca^{2+} and Mg^{2+} are usually omitted, as in Moscona's [1952] calcium- and magnesium-free saline (CMF) or PBSA. (See Table 8.2.) The choice of BSS also is dependent on whether the solution will be used for suspension culture of adherent cells. S-MEM, based on Eagle's Spinner salt solution, is a variant of Eagle's [1959] minimum essential medium that is de-

TABLE 8.2. Balanced Salt Solutions

Component	M.W.	Earle's BSS g/l	Earle's BSS mM	Dulbecco's PBS A g/l	Dulbecco's PBS A mM	Dulbecco's PBS B g/l	Dulbecco's PBS B mM	Hanks's BSS g/l	Hanks's BSS mM	Spinner salts (as in S-MEM) g/l	Spinner salts (as in S-MEM) mM
Inorganic salts											
$CaCl_2$ (anhydrous)	111	0.02	0.18			0.2	1.80	0.14	1.3		
KCl	74.55	0.4	5.37	0.2	2.7	0.2	2.68	0.4	5.4	0.40	5.37
KH_2PO_4	136.1			0.2	1.5	0.2	1.47	0.06	0.4		
$MgCl_2 \cdot 6H_2O$	203.3					0.00		0.1	0.5		
$MgSO_4 \cdot 7H_2O$	246.5	0.2	0.81			0.98	3.98	0.1	0.4	0.20	0.81
NaCl	58.44	6.68	114.31	8	136.9	8	136.89	8	136.9	6.80	116.36
$NaHCO_3$	84.01	2.2	26.19			0.00		0.35	4.2	2.20	26.19
$Na_2HPO_4 \cdot 7H_2O$	268.1			2.2	8.1	2.16	8.06	0.09	0.3		
$NaH_2PO_4 \cdot H_2O$	138	0.14	1.01							1.40	10.14
Total salt			147.86		149.1		154.88		149.4		158.87
Other components											
D-glucose	180.2	1	5.55					1	5.5	1.00	5.55
Phenol red	354.4	0.01	0.03					0.01	0.0	0.01	0.03
Gas phase		5% CO_2				Air		Air		5% CO_2	
Buffer											
HEPES, Na salt	260.3	13.02	50.00	5.21	20.0	5.21	20	2.08	8.0	13.02	50.00

Note. If HEPES is used, the equivalent molarity of NaCl must be omitted and osmolality must be checked.

ficient in Ca^{2+} in order to reduce cell aggregation and attachment. (See Table 8.2.).

HBSS, EBSS, and PBS rely on the relatively weak buffering of phosphate, which is not at its most effective at physiological pH. Paul [1975] constructed a tris-buffered BSS that is more effective, but for which the cells sometimes require a period of adaptation. HEPES (10–20 mM) is currently the most effective buffer in the pH 7.2–7.8 range, and TRICINE in the pH 7.4–8.0 range, although both tend to be expensive if used in large quantities.

COMPLETE MEDIA

Complete media range in complexity from the relatively simple Eagle's MEM [Eagle, 1959], which contains essential amino acids, vitamins, and salts, to complex media such as Medium 199 (M199) [Morgan et al., 1950], CMRL 1066 [Parker et al., 1957], MB 752/1 [Waymouth, 1959], RPMI 1640 [Moore et al., 1967], and F12 [Ham, 1965] (Table 8.3), and a wide range of serum-free formulations (see Tables 9.1 and 9.2). The complex media contain a larger number of different amino acids, including nonessential amino acids and additional vitamins, and are often supplemented with extra metabolites (e.g., nucleosides, tricarboxylic acid cycle intermediates, and lipids) and minerals. Nutrient concentrations are, on the whole, low in F12 (which was optimized by cloning) and high in Dulbecco's modification of Eagle's MEM (DMEM) [Dulbecco and Freeman, 1959; Morton, 1970], optimized at higher cell densities for viral propagation, although the latter solution has fewer constituents. Barnes and Sato [1980] employed a 1:1 mixture of DMEM and F12 as the basis for their serum-free formulations to combine the richness of F12 and the higher nutrient concentration of DMEM. Although not always entirely rational, this combination has provided an empirical formula that is suitable as a basic medium for supplementation with special additives for many different cell types.

Amino Acids

The essential amino acids (i.e., those which are not synthesized in the body) are required by cultured cells, together with cysteine and tyrosine in addition, although individual requirements for amino acids will vary from one cell to another. Other nonessential amino acids are often added as well, to compensate either for a particular cell type's incapacity to make them or because they are made, but lost by leakage into the medium. The concentration of amino acids usually limits the maximum cell concentration attainable, and the balance may influence cell survival and growth rate. Glutamine is required by most cells, although some cell lines will utilize glutamate; evidence suggests that glutamine is also used by cultured cells as a source of energy and carbon [Butler and Christie, 1994; see also Energy Metabolism in Chapter 2].

Vitamins

Eagle's MEM contains only the water-soluble vitamins (the B-group, plus choline, folic acid, inositol, and nicotinamide, but excluding biotin; see Table 8.3); other requirements presumably are derived from the serum. Biotin is present in most of the more complex media, including the serum-free recipes, and p-aminobenzoic acid (PABA) is present in M199, CMRL 1066 (which was derived from M199), and RPMI 1640. All the fat-soluble vitamins (A, D, E, K) are present only in M199, while vitamin A is present in LHC-9 and vitamin E in MCDB110. (See Table 9.1.) Some vitamins (e.g., choline and nicotinamide) have increased concentrations in serum-free media. Vitamin limitation—for example, by precipitation of folate from concentrated stock solutions—is usually expressed in terms of reduced cell survival and growth rates, rather than maximum cell density. Like those of the amino acids, vitamin requirements have been derived empirically and often relate to the cell line originally used in their development; e.g., Fischer's medium has a high folate concentration because of the folate dependency of L5178Y, which was used in the development of the medium [Fischer and Sartorelli, 1964].

Salts

The salts are chiefly those of Na^+, K^+, Mg^{2+}, Ca^{2+}, Cl^-, SO_4^{2-}, PO_4^{3-}, and HCO_3^- and are the major components contributing to the osmolality of the medium. Most media derived their salt concentrations originally from Earle's (high bicarbonate; gas phase, 5% CO_2) or Hanks's (low bicarbonate; gas phase, air) BSS. Divalent cations, particularly Ca^{2+}, are required by some cell adhesion molecules, such as the cadherins [Yamada and Geiger, 1997]. Ca^{2+} also acts as an intermediary in signal transduction [Alberts et al., 1994], and the concentration of Ca^{2+} in the medium can influence whether cells will proliferate or differentiate. (See Soluble Inducers in Chapter 16; Epidermis in Chapter 22). Na^+, K^+, and Cl^- regulate membrane potential, while SO_4^{2-}, PO_4^{3-}, and HCO_3^- have roles as anions required by the matrix and nutritional precursors for macromolecules, as well as regulators of intracellular charge.

Calcium is reduced in suspension cultures in order to minimize cell aggregation and attachment. (See Balanced Salt Solution in this chapter.) The sodium bicarbonate concentration is determined by the concentration of CO_2 in the gas phase (see CO_2 and Bicarbonate in this chapter) and has a significant nutritional role in addition to its buffering capability.

Glucose

Glucose is included in most media as a source of energy. It is metabolized principally by glycolysis to form pyruvate, which may be converted to lactate or acetoacetate and may enter the citric acid cycle to form CO_2. The accumulation of lactic acid in the medium, particularly evident in embryonic and transformed cells, implies that the citric acid cycle may not function entirely as it does *in vivo*, and recent data have shown that much of its carbon is derived from glutamine rather than glucose. This finding may explain the exceptionally high requirement of some cultured cells for glutamine or glutamate.

Organic Supplements

A variety of other compounds, including proteins, peptides, nucleosides, citric acid cycle intermediates, pyruvate, and lipids, appear in complex media. Again, these constituents have been found to be necessary when the serum concentration is reduced, and they may help in cloning and in maintaining certain specialized cells, even in the presence of serum.

Hormones and Growth Factors

Hormones and growth factors are not specified in the formulas of most regular media although they are frequently added to serum-free media. (See Growth Factors and Hormones in Chapter 9.)

Antibiotics

Antibiotics were originally introduced into culture media to reduce the frequency of contamination. However, the use of laminar-flow hoods, coupled with strict aseptic technique, makes antibiotics unnecessary. Indeed, antibiotics have a number of significant disadvantages:

(1) They encourage the development of antibiotic-resistant organisms.
(2) They hide the presence of low-level, cryptic contaminants that can become fully operative if the antibiotics are removed, the culture conditions change, or resistant strains develop.
(3) They may hide mycoplasma infections.
(4) They have antimetabolic effects that can cross-react with mammalian cells.
(5) They encourage poor aseptic technique.

For all these reasons, it is strongly recommended that routine culture be performed in the absence of antibiotics and their use be restricted to primary-culture or large-scale labor-intensive experiments with a high cost of consumables. If conditions demand the use of antibiotics, then they should be removed as soon as possible, or, if they are used in the long term, parallel

TABLE 8.3. Frequently Used Media

Component	MEM	DMEM	F12	DMEM/F12	αMEM	CMRL 1066	RPMI 1640	M199	L15	McCoy's 5A	Fischer	MB 752/1
Amino acids												
L-alanine			1.0E-04	5.0E-05	2.8E-04	2.8E-04		2.8E-04	2.5E-03	1.5E-04		
L-arginine	6.0E-04	4.0E-04	1.0E-03	7.0E-04	6.0E-04	3.3E-04	1.1E-03	3.3E-04	2.9E-03	2.0E-04	7.1E-05	3.6E-04
L-asparagine			1.0E-04	5.0E-05	3.3E-04	3.8E-04	1.7E-03	3.0E-04	7.6E-05	1.5E-04		4.5E-04
L-aspartic acid			1.0E-04	5.0E-05	2.3E-04	2.3E-04	1.5E-04	2.3E-04		2.0E-04		5.0E-04
L-cysteine			2.0E-04	1.0E-04	5.7E-04	1.5E-03		5.6E-07	9.9E-04		9.9E-05	6.3E-05
L-cystine	1.0E-04	2.0E-04		1.0E-04	1.0E-04	8.3E-05	2.1E-04	9.9E-05		1.5E-04		1.0E-03
L-glutamic acid			1.0E-04	5.0E-05	5.1E-04	5.1E-04	1.4E-04	4.5E-04		1.5E-04	1.4E-03	2.4E-03
L-glutamine	2.0E-03	4.0E-03	1.0E-03	2.5E-03	2.0E-03	6.8E-04	2.1E-03	6.8E-04	2.1E-03	1.5E-03		6.7E-04
Glycine		4.0E-04	1.0E-04	2.5E-04	6.7E-04	6.7E-04	1.3E-04	6.7E-04	2.7E-03	1.0E-04	3.9E-04	8.3E-04
L-histidine	2.0E-04	2.0E-04	1.0E-04	1.5E-04	2.0E-04	9.5E-05	9.7E-05	1.0E-04	1.6E-03	1.5E-04		1.9E-04
L-hydroxy-proline						7.6E-05	1.5E-04	7.6E-05			5.7E-04	3.8E-04
L-isoleucine	4.0E-04	8.0E-04	3.0E-05	4.2E-04	4.0E-04	1.5E-04	3.8E-04	1.5E-04	9.5E-04	3.0E-04	2.3E-04	1.3E-03
L-leucine	4.0E-04	8.0E-04	1.0E-04	4.5E-04	4.0E-04	4.6E-04	3.8E-04	4.6E-04	9.5E-04	3.0E-04	2.7E-04	3.4E-04
L-lysine HCl	4.0E-04	8.0E-04	2.0E-04	5.0E-04	4.0E-04	3.8E-04	2.2E-04	3.8E-04	5.1E-04	2.0E-04	6.7E-04	3.0E-04
L-methionine	1.0E-04	2.0E-04	3.0E-05	1.2E-04	1.0E-04	1.0E-04	1.0E-04	1.0E-04	5.0E-04	1.0E-04	4.1E-04	4.3E-04
L-phenylalanine	2.0E-04	4.0E-04	3.0E-05	2.2E-04	1.9E-04	1.5E-04	9.1E-05	1.5E-04	7.6E-04	1.0E-04	1.4E-04	
L-proline			3.0E-04	1.5E-04	3.5E-04	3.5E-04	1.7E-04	3.5E-04		1.5E-04		
L-serine		4.0E-04	1.0E-04	2.5E-04	2.4E-04	2.4E-04	2.9E-04	2.4E-04	1.9E-03	2.5E-04	3.4E-04	6.3E-04
L-threonine	4.0E-04	8.0E-04	1.0E-04	4.5E-04	4.0E-04	2.5E-04	1.7E-04	2.5E-04	2.5E-03	1.5E-04	3.4E-04	2.0E-04
L-tryptophan	4.9E-05	7.8E-05	1.0E-05	4.4E-05	4.9E-05	4.9E-05	2.5E-05	4.9E-05	9.8E-05	1.5E-05	4.9E-05	2.2E-04
L-tyrosine	2.0E-04	4.0E-04	3.0E-05	2.1E-04	2.3E-04	2.2E-04	1.1E-04	2.2E-04	1.7E-03	1.2E-04	3.3E-04	5.6E-04
L-valine	4.0E-04	8.0E-04	1.0E-04	4.5E-04	3.9E-04	2.1E-04	1.7E-04	2.1E-04	8.5E-04	1.5E-04	6.0E-04	
Vitamins												
p-Aminobenzoic acid						3.6E-07	7.3E-06	3.6E-07		7.3E-06		
L-Ascorbic acid					2.5E-04	2.8E-04		2.8E-04				9.9E-05
Biotin			3.0E-08	1.5E-08	4.1E-07	4.1E-08	8.2E-07	4.1E-08		8.2E-07	4.1E-08	8.2E-08
Calciferol								2.5E-07				
Choline chloride	7.1E-06	2.9E-05	1.0E-04	6.4E-05	7.1E-06	3.6E-06	2.1E-05	3.6E-06	7.1E-06	3.6E-05	1.1E-05	1.8E-03
Folic acid	2.3E-06	9.1E-06	2.9E-06	6.0E-06	2.3E-06	2.3E-08	2.3E-06	2.3E-08	2.3E-06	2.3E-05	2.3E-05	9.1E-07
myo-inositol	1.1E-05	4.0E-05	1.0E-04	7.0E-05	1.1E-05	2.8E-07	1.9E-04	2.8E-07	1.1E-05	2.0E-04	8.3E-06	5.6E-06
Menadione						2.0E-07		6.9E-08				
Nicotinamide	8.2E-06	3.3E-05	3.3E-07	1.7E-05	8.2E-06	2.0E-07	8.2E-06	2.0E-07	8.2E-06	4.1E-06	4.1E-06	8.2E-06
Nicotinic acid						2.0E-07		2.0E-07		4.1E-06	2.1E-06	4.2E-06
D-Ca pantothenate	4.2E-06	1.7E-05	2.0E-06	9.4E-06	4.2E-06	4.2E-08	1.1E-06	4.2E-08	4.2E-06	8.4E-07	2.1E-06	4.2E-06
Pyridoxal HCl	4.9E-06	2.0E-05		1.0E-05	4.9E-06	1.2E-07		1.2E-07		2.5E-06	2.5E-06	4.9E-06
Pyridoxine HCl			3.0E-07	1.5E-07		1.2E-07	4.9E-06	1.2E-07		2.4E-06	1.3E-06	2.7E-06
Riboflavin	2.7E-07	1.1E-06	1.0E-07	5.8E-07	2.7E-07	2.7E-08	5.3E-07	2.7E-08	1.9E-07	5.3E-07	1.3E-06	2.7E-06
Thiamin	3.0E-06	1.2E-05	1.0E-06	6.4E-06	3.0E-06	3.0E-08	3.0E-06	3.0E-08	2.4E-06	5.9E-07	3.0E-06	3.0E-05

	1	2	3	4	5	6	7	8	9
Thiamin mono PO4									
α-tocopherol							2.3E-08		
Retinol acetate							3.5E-07		
Vitamin B12		1.0E-06	5.0E-07	1.0E-06		3.7E-09	1.5E-07	4.8E-06	1.5E-07
Antioxidants									
Glutathione					3.0E-05	3.0E-06		1.5E-06	4.5E-05
Inorganic salts									
CaCl2	1.8E-03	3.0E-04	1.1E-03	1.8E-03	1.8E-03		1.3E-03	9.0E-04	8.2E-04
KCl	5.3E-03	3.0E-03	4.2E-03	5.3E-04	5.3E-03	5.3E-03	5.3E-03	5.3E-03	2.0E-03
KH2PO4							4.4E-04		5.9E-04
MgCl2				1.2E-01					1.2E-03
MgSO4	8.1E-04	8.1E-04	4.0E-04	8.1E-04	8.1E-04	4.0E-04	8.1E-04	8.1E-04	8.1E-04
NaCl	1.2E-01	1.3E-01	1.2E-01		1.2E-01	1.0E-01	1.4E-01	1.1E-01	1.0E-01
NaHCO3	2.6E-02	1.4E-02	2.9E-02		2.6E-02	2.6E-02		2.6E-02	2.7E-02
NaH2PO4	4.4E-02		4.5E-04				4.2E-03	4.2E-03	
Na2HPO4	9.1E-04	1.0E-03	5.0E-04		1.0E-03	5.6E-03	4.0E-04	1.6E-03	2.1E-03
Trace elements									
CuSO4·5H2O	2.5E-07	1.6E-08	7.8E-09						
Fe(NO3)3·9H2O			1.2E-07						
FeSO4·7H2O		3.0E-06	1.5E-06						
ZnSO4·7H2O		3.0E-06	1.5E-06						
Bases, nucleosides, etc.									
Adenine SO4							5.4E-05		
Adenosine				3.7E-05					
AMP							5.8E-07		
ATP							1.8E-05		
Cytidine				4.1E-05					
Deoxyadenosine				4.0E-05	4.0E-05				
Deoxycytidine				4.2E-05	3.8E-05				
Deoxyguanosine				3.7E-05	3.7E-05				
2-Deoxyribose									
DPN					9.5E-06		3.7E-06		
FAD					1.2E-06				
Glucuronate, Na					1.9E-05				
Guanine							1.6E-06		
Guanosine				3.5E-05					
Hypoxanthine		3.0E-05	1.5E-05				2.2E-06		
5-Me-deoxycytidine					4.1E-07				
D-Ribose			1.5E-06	4.1E-05			3.3E-06		
Thymidine	3.0E-06	3.0E-06			4.1E-05				
Thymine							2.4E-06		
TPN					1.3E-06				
Uracil									
Uridine				4.1E-05			2.7E-06		
UTP					1.8E-06				
Xanthine							2.0E-06		

TABLE 8.3. Frequently Used Media (*Continued*)

Component	MEM	DMEM	F12	DMEM/F12	αMEM	CMRL 1066	RPMI 1640	M199	L15	McCoy's 5A	Fischer	MB 752/1
Energy metabolism												
Cocarboxylase						2.2E-06						
Coenzyme A						3.3E-06						
D-galactose									5.0E-02			
D-glucose	5.6E-03	2.5E-02	1.0E-02	1.8E-02	5.6E-03	5.6E-03	1.1E-02	5.6E-03		1.7E-02	5.6E-03	2.8E-02
Sodium acetate						6.1E-04		4.5E-04				
Sodium pyruvate		1.0E-03	1.0E-03	1.0E-03	1.0E-03				5.0E-03			
Lipids and precursors												
Cholesterol						5.2E-07		5.2E-07				
Ethanol (solvent)		3.0E-07				3.5E-04						
Linoleic acid			1.0E-06	1.5E-07								
Lipoic acid				5.1E-07	9.7E-07							8.9E-05
Tween 80						1.8E-05		1.8E-05				
Other components												
Peptone, mg/ml										0.6		
Phenol red	2.7E-05	4.0E-05	3.2E-05	3.6E-05	2.9E-05	5.3E-05	1.3E-05	4.5E-05	2.7E-05	2.9E-05	1.3E-05	2.7E-05
Putrescine			1.0E-06	5.0E-07								
Gas Phase												
CO$_2$	5%	10%	2%	7%	5%	5%	5%	5%	Air	5%	2%	5%

All concentrations are molar, and computer-style notation is used (e.g., 3.0E-2 = 3.0 × 10⁻² = 30 mM). Molecular weights are given for root compounds; although some recipes use salts or hydrated forms, molarities will, of course, remain the same. *Synonyms and abbreviations:* AMP, adenosine monophosphate; ATP, adenosine triphosphate; **b**iotin = vitamin H; calciferol = vitamin D$_2$; FAD, flavine adenine dinucleotide; lipoic acid = thioctic acid; menadione = vitamin K$_3$; *myo*-inositol = *I*-inositol; nicotinamide = niacinamide; nicotinic acid = niacin; pyridoxine HCl = vitamin B$_6$; thiamin = vitamin B$_1$; α-tocopherol = vitamin E; retinol = vitamin A$_1$; TPN, triphosphopyridine nucleotide; UTP, uridine triphosphate; vitamin B$_{12}$ = cobalamin. See text for references.

TABLE 8.4. Antibiotics Used in Tissue Culture

| Antibiotic | Concentration, $\mu g/ml$ (unless otherwise stated) | | Activity against |
	Working	Cytotoxic	
Amphotericin B (Fungizone)	2.5	30	Fungi, yeasts
Ampicillin	2.5		Bacteria, gram positive and gram negative
Ciprofloxacin	100		Mycoplasma
Erythromycin	50	300	Mycoplasma
Gentamycin	50	>300	Bacteria, gram positive and gram negative; mycoplasma
Kanamycin	100	10 mg/ml	Bacteria, gram positive and gram negative; mycoplasma
MRA (ICN)	0.5		Mycoplasma
Neomycin	50	3,000	Bacteria, gram positive and gram negative
Nystatin	50		Fungi, yeasts
Penicillin-G	100 U/ml	10,000 U/ml	Bacteria, gram positive
Polymixin B	50	1 mg/ml	Bacteria, gram negative
Streptomycin SO_4	100	20 mg/ml	Bacteria, gram positive and gram negative
Tetracyclin	10	35	Bacteria, gram positive and gram negative
Tylosin	10	300	Mycoplasma

cultures should be maintained free of antibiotics. (See Use of Antibiotics in Chapter 12.)

A number of antibiotics used in tissue culture are moderately effective in controlling bacterial infections (Table 8.4). However, a significant number of bacterial strains are resistant to antibiotics, either naturally or by selection, so the control that they provide is never absolute. Fungal and yeast contaminations are particularly hard to control with antibiotics; they may be held in check, but are seldom eliminated. (See Eradication of Contamination in Chapter 18.)

SERUM

Serum contains growth factors, which promote cell proliferation, and adhesion factors and antitrypsin activity, which promote cell attachment. Serum is also a source of minerals, lipids, and hormones, many of which may be bound to protein (Table 8.5). The sera used most in tissue culture are calf (bovine), fetal bovine, horse, and human serum. Calf (CS) and fetal bovine serum (FBS) are the most widely used, the latter particularly for more demanding cell lines and for cloning. Human serum is sometimes used in conjunction with some human cell lines, but it needs to be screened for viruses, such as HIV and hepatitis B. Horse serum is preferred to calf serum by some workers, as it can be obtained from a closed herd and is often more consistent from batch to batch. Horse se-

rum may also be less likely to metabolize polyamines, due to lower levels of polyamine oxidase; polyamines are mitogenic for some cells [Hyvonen et al., 1988; Kaminska et al., 1990].

Protein

Although proteins are a major component of serum, the functions of many proteins *in vitro* remain obscure; it may be that relatively few proteins are required other than as carriers for minerals, fatty acids, and hormones. Those proteins for which requirements have been found are albumin [Iscove and Melchers, 1978; Barnes and Sato, 1980], which may be important as a carrier of lipids or minerals and globulins [Tozer and Pirt, 1964]; fibronectin (cold-insoluble globulin), which promotes cell attachment [Yamada and Geiger, 1997; Hynes, 1992]; and α2-macroglobulin, which inhibits trypsin [de Vonne and Mouray, 1978]. Fetuin in fetal serum enhances cell attachment [Fisher et al., 1958], and transferrin [Guilbert and Iscove, 1976] binds iron, making it less toxic but bioavailable. Other proteins, as yet uncharacterized, may be essential for cell attachment and growth.

Protein also increases the viscosity of the medium, reducing shear stress during pipetting and stirring, and may add to the medium's buffering capacity.

Growth Factors

Natural clot serum stimulates cell proliferation more than serum from which the cells have been removed

TABLE 8.5. Constituents of Serum

Constituent	Range of concentration[a]
Proteins and Polypeptides	40–80 mg/ml
Albumin	20–50 mg/ml
Fetuin[b]	10–20 mg/ml
Fibronectin	1–10 μg/ml
Globulins	1–15 mg/ml
Protease inhibitors: α_1-antitrypsin, α_2-macroglobulin	0.5–2.5 mg/ml
Transferrin	2–4 mg/ml
Growth factors:	
EGF, PDGF, IGF-1 and 2, FGF, IL-1, IL-6	1–100 ng/ml
Amino acids	0.01–1.0 μM
Lipids	2–10 mg/ml
Cholesterol	10 μM
Fatty acids	0.1–1.0 μM
Linoleic acid	0.01–0.1 μM
Phospholipids	0.7–3.0 mg/ml
Carbohydrates	1.0–2.0 mg/ml
Glucose	0.6–1.2 mg/ml
Hexosamine[c]	6–1.2 mg/ml
Lactic acid[d]	0.5–2.0 mg/ml
Pyruvic acid	2–10 μg/ml
Polyamines:	
Putrescine, spermidine	0.1–1.0 μM
Urea	170–300 μg/ml
Inorganics	0.14–0.16 M
Calcium	4–7 mM
Chlorides	100 μM
Iron	10–50 μM
Potassium	5–15 mM
Phosphate	2–5 mM
Selenium	0.01 μM
Sodium	135–155 mM
Zinc	0.1–1.0 μM
Hormones	0.1–200 nM
Hydrocortisone	10–200 nM
Insulin	1–100 ng/ml
Triiodothyronine	20 nM
Thyroxine	100 nM
Vitamins	10 ng–10 μg/ml
Vitamin A	10–100 ng/ml
Folate	5–20 ng/ml

[a]The range of concentrations is very approximate and is intended to convey only the order of magnitude. Data are from Evans and Sanford [1978], Bergman et al. [1962], and Cartwright and Shah [1994].

[b]In fetal serum only.

[c]Highest in human serum.

[d]Highest in fetal serum.

physically (e.g., by centrifugation). This increased stimulation appears to be due to the release of platelet-derived growth factor (PDGF) from the platelets during clotting. PDGF [Antoniades et al., 1979; Heldin et al., 1979] is one of a family of polypeptides with mitogenic activity and is probably the major growth factor in serum. PDGF stimulates growth in fibroblasts and glia, but other platelet-derived factors, such as TGF-β, may inhibit growth or promote differentiation in epithelial cells [Lechner et al., 1981].

Other growth factors (see Table 9.3), such as fibroblast growth factors (FGFs) [Gospodarowicz, 1974], epidermal growth factor (EGF) [Cohen, 1962; Carpenter and Cohen, 1977; Gospodarowicz et al., 1978a], endothelial cell growth factors such as vascular endothelial growth factor (VEGF) and angiogenin [Hu et al., 1997; Folkman and d'Amore, 1996; Joukov et al., 1977; Folkman et al., 1979; Maciag et al., 1979], and insulinlike growth factors IGF-1 and IGF-2 [le Roith and Raizada, 1989], which have been isolated from whole tissue or released into the medium by cells in culture, have varying degrees of specificity [Hollenberg and Cuatrecasas, 1973] and are probably present in serum in small amounts [Gospodarowicz and Moran, 1974]. Many of these growth factors are available commercially (see Sources of Materials and Growth Factors in Trade Index) as recombinant proteins, some of which also are available in long-form analogues (Sigma) with increased mitogenic activity and stability.

Hormones

Insulin promotes the uptake of glucose and amino acids [Kelley et al., 1978; Stryer, 1995] and may owe its mitogenic effect to this property or to activity via the IGF-1 receptor. IGF-1 and IGF-2 bind to the insulin receptor, but also have their own specific receptors, to which insulin may bind with lower affinity. IGF-2 also stimulates glucose uptake [Sinha et al., 1990]. Growth hormone may be present in serum—particularly fetal serum—and, in conjunction with the somatomedins (IGFs), may have a mitogenic effect. Hydrocortisone is also present in serum—particularly fetal bovine serum—in varying amounts and it can promote cell attachment [Ballard and Tomkins, 1969; Fredin et al., 1979] and cell proliferation [Guner et al., 1977; McLean et al., 1986; see also Epidermis, Breast, and Cervix in Chapter 22], but under certain conditions (e.g., high cell density) may be cytostatic [Freshney et al., 1980a,b] and can induce cell differentiation [Moscona and Piddington, 1966; Ballard, 1979; McLean et al., 1986; Speirs et al., 1991; McCormick et al., 1995, 2000].

Nutrients and Metabolites

Serum may also contain amino acids, glucose, keto-acids, nucleosides, and a number of other nutrients and intermediary metabolites. These may be important in simple media but less so in complex media, particularly those with higher amino acid concentrations and other defined supplements.

Lipids

Linoleic acid, oleic acid, ethanolamine, and phospho-ethanolamine are present in serum in small amounts, usually bound to proteins such as albumin.

Minerals

Serum replacement experiments [Ham and McKeehan, 1978] have also suggested that trace elements and iron, copper, and zinc may be bound to serum protein. McKeehan et al. [1976] demonstrated a requirement for selenium which probably helps to detoxify free radicals as a cofactor for GSH synthetase.

Inhibitors

Serum may contain substances that inhibit cell proliferation [Harrington and Godman, 1980; Liu et al., 1992; Varga, Weisz, and Barnes, 1993]. Some of these may be artefacts of preparation (e.g., bacterial toxins from contamination prior to filtration, or antibodies, contained in the γ-globulin fraction, that cross-react with surface epitopes on the cultured cells), but others may be physiological negative growth regulators, such as TGF-β [Massague et al., 1992]. Heat inactivation removes complement from the serum and reduces the cytotoxic action of immunoglobulins without damaging polypeptide growth factors, but it may also remove some more labile constituents and is not always as satisfactory as untreated serum.

SELECTION OF MEDIUM AND SERUM

All 12 media described in Table 8.3 were developed to support particular cell lines or conditions. Many were developed with L929 cells or HeLa, and Ham's F12 was designed for Chinese hamster ovary (CHO) cells; all now have more general applications and have become classical formulations. Among them, data from suppliers would indicate that RPMI 1640, DMEM, and MEM are the most popular, making up about 75% of sales. Other formulations seldom account for more than 5% of the total; most constitute 2–3%, although blended DMEM/F12 comes closer, with over 4%.

Eagle's Minimal Essential Medium (MEM) was developed from Eagle's Basal Medium (BME) by increasing the range and concentration of the constituents.

For many years, Eagle's MEM had the most general use of all media. Dulbecco's modification of BME (DMEM) was developed for mouse fibroblasts for transformation and virus propagation studies. It has twice the amino acid concentrations of MEM, has four times the vitamin concentrations, and uses twice the HCO_3^- and CO_2 concentrations to achieve better buffering. αMEM [Stanners et al., 1971] has additional amino acids and vitamins, as well as nucleosides and lipoic acid; it has been used for a wide range of cell types, including hematopoietic cells. Ham's F12 was developed to clone CHO cells in low serum; it is also used widely, particularly for clonogenic assays and primary culture. Ham's F12 has also been combined with DMEM, 50:50 v/v, to produce a compromise between high concentrations and a wide range of ingredients. The combination has been used for many primary cultures, for more fastidious cell lines, and as a basis for serum-free media. (See Table 9.2.)

CMRL 1066, M199, and Waymouth's were all developed to grow L929 cells in a serum-free medium, but have been used alone or in combination with other media, such as DMEM or F12, for a variety of more demanding conditions. RPMI 1640 and Fischer's were developed for lymphoid cells—Fischer's specifically for L5178Y lymphoma, which has a high folate requirement. RPMI 1640 in particular has quite widespread use, often for attached cells, in spite of being designed for suspension culture and lacking calcium. L15 medium was developed specifically to provide buffering in the absence of HCO_3^- and CO_2. It is often used as a transport and primary culture medium for this reason, but its value was diminished by the introduction of HEPES and the demonstration that HCO_3^- and CO_2 are often essential for optimal cell growth, regardless of the requirement for buffering.

Information regarding the selection of the appropriate medium for a given type of cell is usually available in the literature in articles on the origin of the cell line or the culture of similar cells. Information may also be obtained from the source of the cells. Cell banks, such as ATCC and ECACC, provide information on media used for currently available cell lines, and data sheets can be accessed from their Web sites. (See Suppliers and Other Resources in Trade Index; see also Table 8.6 and Cell Banks in Chapter 19.) Failing this, the choice is made either empirically or by comparative testing of several media as for selection of serum (see Testing Serum in this chapter.)

Many continuous cell lines (e.g., HeLa, L929, BHK21), primary cultures of human, rodent, and avian fibroblasts, and cell lines derived from them can be maintained on a relatively simple medium such as Eagle's MEM, supplemented with calf serum. More complex media may be required when a specialized function is being expressed (see also Chapter 22) or when cells are subcultured at low seeding density (<1 × 10³/ml), as in cloning. (See Stimulation of Plating Efficiency in Chapter 13.) Frequently, the more demanding culture conditions that require complex media also require fetal bovine serum rather than calf or horse serum, unless the formulation specifically allows for the omission of serum.

Some suggestions for the choice of medium and several examples of cell types and the media used for them are given in Table 8.6. (See also Mather [1998].) If information is not available, a simple cell growth experiment with commercially available media and multiwell plates (see Protocols 20.7 and 20.8) can be carried out in about two weeks. Assaying for clonal growth (see Protocol 20.9) and measuring the expres-

sion of specialized functions may narrow the choice further. You may be surprised to find that your best conditions do not agree with those mentioned in the literature; reproducing the conditions found in another laboratory may be difficult, due to variations in preparation or supplier, the impurities present in reagents and water and differences between batches of serum. It is to be hoped that as serum requirements are reduced and the purity of reagents increases, the standardization of media will improve.

Finally, you may have to compromise in your choice of medium or serum because of cost. Autoclavable media are available from commercial suppliers (ICN, Gibco). They are simple to prepare from powder and are suitable for many continuous cell strains. They may need to be supplemented with glutamine for most cells and usually require serum. The cost of

TABLE 8.6. Selecting a Suitable Medium

Cells or cell line	Medium	Serum
3T3 cells	MEM, DMEM	CS
Chick embryo fibroblasts	Eagle's MEM	CS
Chinese hamster ovary (CHO)	Eagle's MEM, Ham's F12	CS
Chondrocytes	Ham's F12	FB
Continuous cell lines	Eagle's MEM, DMEM	CS
Endothelium	DMEM, M199, MEM	CS
Fibroblasts	Eagle's MEM	CS
Glial cells	MEM, DMEM/F12	FB
Glioma	MEM, DMEM/F12	FB
HeLa cells	Eagle's MEM	CS
Hematopoietic cells	RPMI 1640, Fischer's, αMEM	FB
Human diploid fibroblasts	Eagle's MEM	CS
Human leukemia	RPMI 1640	FB
Human tumors	L15, RPMI 1640, DMEM/F12	FB
Keratinocytes	αMEM	FB
L cells (L929, LS)	Eagle's MEM	CS
Lymphoblastoid cell lines (human)	RPMI 1640	FB
Mammary epithelium	RPMI 1640, DMEM/F12	FB
MDCK dog kidney epithelium	DMEM, DMEM/F12	FB
Melanocytes	M199	FB
Melanoma	MEM, DMEM/F12	FB
Mouse embryo fibroblasts	Eagle's MEM	CS
Mouse leukemia	Fischer's, RPMI 1640	FB, HoS
Mouse erythroleukemia	DMEM/F12, RPMI 1640	FB, HoS
Mouse myeloma	DMEM, RPMI 1640	FB
Mouse neuroblastoma	DMEM, DMEM/F12	FB
Neurons	DMEM	FB
NRK rat kidney fibroblasts	MEM, DMEM	CS
Rat minimal-deviation hepatoma (HTC, MDH)	Swim's S77, DMEM/F12	FB
Skeletal muscle	DMEM, F12	FB, HoS
Syrian hamster fibroblasts (e.g., BHK 21)	MEM, GMEM, DMEM	CS

Abbreviations: CS, calf serum; FB, fetal bovine serum; HoS, horse serum. SF12 is Ham's F12 plus Eagle's essential amino acids and nonessential amino acids as in DMEM (available as 100× stock ICN, Gibco, etc.). Further recommendations on the choice of medium can be found in McKeehan [1977], Barnes et al. [1984(a–d)], and Mather [1998].

serum should be calculated on the basis of the volume of the medium when cell yield is not important, but if the objective is to produce large quantities of cells, one should calculate serum costs on a per-cell basis. Thus, if a culture grows to 10^6/ml in serum A and 2×10^6 ml in serum B, serum B becomes the less expensive by a factor of two, given that product formation or some other specialized function is the same.

If fetal bovine serum seems essential, try mixing it with calf serum. This may allow you to reduce the concentration of the more expensive fetal serum. If you can, leave out serum altogether, or reduce the concentration, and use a serum-free formulation (see Chapter 9).

Batch Reservation

Considerable variation may be anticipated between batches of serum. Such variation results from differing methods of preparation and sterilization, different ages and storage conditions, and variations in animal stocks from which the serum was derived, including different strains and disparities in pasture, climate, and other environmental conditions. It is important to select a batch, use it for as long as possible, and replace it, eventually, with one as similar to it as possible.

Serum standardization is difficult, as batches vary considerably, and one batch will last only about six months to a year, stored at $-20°C$. Select the type of serum that is most appropriate for your purposes, and request batches to test from a number of suppliers. Most serum suppliers will normally reserve a batch until a customer can select the most suitable one (provided that this does not take longer than three weeks or so). When a suitable batch has been selected, the supplier is requested to hold the appropriate volume for up to one year, to be dispatched on demand. Other suppliers should also be informed, so that they may return the rejected batches to their stocks.

Testing Serum

The quality of a given serum is assured by the supplier, but the firm's quality control is usually performed with one of a number of continuous cell lines. If your requirements are more discriminating, then you will need to do your own testing. There are four main parameters for testing serum.

Plating Efficiency. During cloning, the cells are at a low density and hence are at their most sensitive, making this a very stringent test. Plate the cells out at 10 to 100 cells/ml, and look for colonies after 10 days to two weeks. Stain and count the colonies (see Protocol 20.9), and look for differences in plating efficiency (survival) and colony size (cell proliferation).

Each serum should be tested at a range of concentrations from 2% to 20%. This approach will reveal whether one serum is equally effective at a lower concentration, thereby saving money and prolonging the life of the batch, and will show up any toxicity at a high serum concentration.

Growth curve. A growth curve should be plotted for cell growth in each serum (see Protocols 20.7 and 20.8), so that the lag period, doubling time, and saturation density (density at "plateau") can be determined. A long lag implies that the culture is having to adapt to the serum, short doubling times are preferable if you want a lot of cells quickly, and a high saturation density will provide more cells for a given amount of serum and will be more economical.

Preservation of Cell Culture Characteristics. Clearly, the cells must do what you require of them in the new serum, whether they are acting as host to a given virus, producing a certain cell product, differentiating, or expressing a characteristic sensitivity to a given drug.

Sterility. Serum from a reputable supplier will have been tested and shown to be free of microorganisms. However, in the unlikely event that a sample of serum is contaminated, but has escaped quality control, the fact that it is contaminated should show up in mycoplasma screening. (See Mycoplasma in Chapter 18.)

Heat Inactivation

Originally, heating was designed to inactivate complement for immunoassays, but it may achieve other effects not yet documented. Often, heat-inactivated serum is used because of the adoption of a previous protocol, without any concrete evidence that it is beneficial. Claims that heat inactivation removes mycoplasma are probably unfounded, although heat treatment may reduce the titer for some mycoplasma.

Serum is heat inactivated by incubating it for 30 min at $56°C$. It may then be dispensed into aliquots and stored at $-20°C$.

OTHER SUPPLEMENTS

In addition to serum, tissue extracts and digests have traditionally been used as supplements to tissue culture media. Many such supplements are derived from microbiological culture techniques and autoclavable broths. Bactopeptone, tryptose, and lactalbumin hydrolysate (Difco) are proteolytic digests of beef heart or lactalbumin and contain mainly amino acids and small peptides. Bactopeptone and tryptose may also contain nucleosides and other heat-stable tissue constituents, such as fatty acids and carbohydrates.

Embryo Extract

Embryo extract is a crude homogenate of 10-day-old chick embryo that is clarified by centrifugation. (See Reagent Appendix.) The crude extract was fractionated by Coon and Cahn [1966] to give fractions of either high or low molecular weight. The low-molecular weight fraction promoted cell proliferation, while the high-molecular-weight fraction promoted pigment and cartilage cell differentiation. Although Coon and Cahn did not fully characterize these fractions, more recent evidence would suggest that the low-molecular-weight fraction probably contains peptide growth factors and the high-molecular-weight fraction proteoglycans and other matrix constituents.

Embryo extract was originally used as a component of plasma clots (see Initiation of the Culture in Chapter 2; Primary Explant in Chapter 11) to promote cell migration from the explant. It has been retained in some organ culture techniques (see Organ Culture in Chapter 24) and can still be used in nerve and muscle culture (see Neurons in Chapter 22), although in the latter it is gradually being replaced with defined growth factors and matrix components. It has been shown that embryo extract can be replaced by hemin in the induction of skeletal muscle differentiation [Smith and Schroedl, 1992].

Conditioned Medium

Puck and Marcus [1955] found that the survival of low-density cultures could be improved by growing the cells in the presence of feeder layers. (See Stimulation of Plating Efficiency in Chapter 13.) This effect is probably due to a combination of effects including conditioning of the substrate and conditioning of the medium by the release into it of small molecular metabolites and growth factors [Takahashi and Okada, 1970]. Hauschka and Konigsberg [1966] showed that the conditioning of culture medium that was necessary for the growth and differentiation of myoblasts was due to collagen released by the feeder cells. Using feeder layers and conditioning the medium with embryonic fibroblasts or other cell lines remains a valuable method of culturing difficult cells [e.g., Stampfer et al., 1980].

Attempts have been made to isolate active fractions from conditioned medium, and the original supposition is still probably close to the correct interpretation. Conditioned medium contains both substrate-modifying matrix constituents, like collagen, fibronectin, and proteoglycans, and growth factors, such as those of the heparin-binding group (FGF, etc.), insulinlike growth factors (IGF-1 and -2), PDGF, and several others (see Table 9.3), in addition to the intermediary metabolites previously proposed. However, conditioned medium adds undefined components to medium and should be eliminated after the active constituents are determined. Still, they often constitute an easy alternative to many months of tedious attempts to optimize the medium and may be the only economical route to successful culture, given that you wish to invest your time dealing with the specific problem of interest, rather than devloping culture conditions.

CHAPTER 9

Serum-Free Media

Although most cell lines still require that the medium in which they are growing be supplemented with serum, in many instances cultures may now be propagated in serum-free media (Tables 9.1, 9.2). The twin needs to standardize media among laboratories and to supply the more demanding and specific requirements of primary cultures and cell cloning, as well as the desire to remove uncontrolled, and often contaminated, natural products—particularly serum—from the culture and its products, led to the development of the more complex media, such as M199 of Morgan et al. [1950], CMRL 1066 of Parker et al. [1957], NCTC109 [Evans et al., 1956], Waymouth's MB 572/1 [1959], NCTC135 [Evans and Bryant, 1965], and Birch and Pirt [1971], for L929 mouse fibroblast cells, and Ham's F10 [1963] and F12 [1965] clonal growth media for Chinese hamster ovary (CHO) cells. Serum-free media were also developed for HeLa human cervical carcinoma cells [Blaker et al., 1971; Higuchi, 1977].

While a degree of selection may have been involved in the adaptation of continuous cell lines to serum-free conditions, the MCDB series of media [Ham and McKeehan, 1978; see also Table 9.1], Sato's DMEM/F12-based media [Barnes and Sato, 1980], and others based on RPMI 1640 [Carney et al., 1981; Brower et al., 1986; see also Table 9.2) demonstrated that serum could be reduced or omitted without apparent cellular adaptation if appropriate nutritional and hormonal modifications were made to the media [Barnes et al., 1984a–d; Cartwright and Shah, 1994; Mather, 1998]. These also provided selective conditions for primary culture of particular cell types. Specific formulations

(e.g., MCDB 110 [Bettger et al., 1981]) were derived to culture human fibroblasts [Ham, 1984], many normal and neoplastic murine and human cells [Barnes and Sato, 1980], lymphoblasts [Iscove and Melchers, 1978], and several different primary cultures [Mather and Sato, 1979a,b; Sundqvist et al., 1991; Gupta et al., 1997; Keen and Rapson, 1995; Vonen et al., 1992] in the absence of serum, with, in several cases, some protein added [Tsao et al., 1982; Benders et al., 1991]. This list now covers a wide range of cell types, and many of the media are available commercially. (See Sources of Materials in Trade Index.)

DISADVANTAGES OF SERUM

Using serum in a medium has a number of disadvantages:

(1) **Physiological Variability.** The major constituents of serum such as albumin and transferrin, are known, but serum also contains a wide range of minor components that may have a considerable effect on cell growth (see Table 8.5). These components include nutrients (amino acids, nucleosides, sugars, etc.), peptide growth factors, hormones, minerals, and lipids, the concentrations and actions of which have not been fully determined.

(2) **Shelf Life and Consistency.** Serum varies from batch to batch, and at best a batch will last one year, perhaps deteriorating during that time. It must then be replaced with another batch that

TABLE 9.1. Examples of Serum-Free Media Formulations

Component	Mol. wt.	MCDB 110 Human lung fibroblasts [Bettger et al., 1981]	MCDB 131 Human vascular endothelium [Knedler and Ham, 1987; Gupta et al., 1997]	MCDB 170 Mammary epithelium [Hammond et al., 1984]	MCDB 202 Chick embryo fibroblasts [McKeehan and Ham, 1976b]	MCDB 302 CHO cells [Hamilton and Ham, 1977]	MCDB 402 3T3 cells [Shipley and Ham, 1983]	WAJC 404 Prostatic epithelium [McKeehan et al., 1984; Chaproniere and McKeehan, 1986]	MCDB 153 Keratinocytes [Peehl and Ham, 1986]	Iscove's lymphoid cells [Iscove and Melchers, 1978]	LHC-9 Bronchial epithelium [Lechner and LaVeck, 1985]
Amino acids											
L-alanine	89	1.0E-04	3.0E-05	1.0E-04	1.0E-04		1.0E-04	1.0E-04	2.8E-04	1.0E-04	
L-arginine	211	1.0E-03	3.0E-04	3.0E-04	3.0E-04	1.0E-03	3.0E-04	1.0E-03	1.0E-03	4.0E-04	2.0E-03
L-asparagine	132	1.0E-04	1.0E-04	1.0E-03	1.0E-03	1.1E-04	1.0E-04	1.0E-04	1.0E-04	1.9E-04	1.0E-04
L-aspartic acid	133	1.0E-04	1.0E-04	1.0E-04	1.0E-04	1.0E-04	1.0E-05	3.0E-05	3.0E-05	2.3E-04	3.0E-05
L-cysteine	176	5.0E-05	2.0E-04	7.0E-05	2.0E-04	1.0E-04		2.4E-04	2.4E-04		2.4E-04
L-cystine	240			2.0E-04	2.0E-04		4.0E-04			2.9E-04	
L-glutamic acid	147	1.0E-04	3.0E-05	2.0E-04	1.0E-04	1.0E-04	1.0E-05	1.0E-04	1.0E-04	5.1E-04	1.0E-04
L-glutamine	146	2.5E-03	1.0E-02	2.0E-03	1.0E-03	3.0E-03	5.0E-03	6.0E-03	6.0E-03	4.0E-03	6.0E-03
Glycine	75	3.0E-04	3.0E-05	1.0E-04	1.0E-04	1.0E-04	1.0E-04	1.0E-04	1.0E-04	4.0E-04	1.0E-04
L-histidine	210	1.0E-04	2.0E-04	1.0E-04	1.0E-04	1.0E-04	2.0E-03	8.0E-05	8.0E-05	2.0E-04	1.6E-04
L-isoleucine	131	3.0E-05	5.0E-04	1.0E-04	3.0E-04	3.0E-05	1.0E-03	1.5E-05	1.5E-05	8.0E-04	3.0E-05
L-leucine	131	1.0E-04	1.0E-04	3.0E-04	2.0E-04	1.0E-04	2.0E-03	5.0E-04	5.0E-04	8.0E-04	1.0E-03
L-lysine HCl	183	2.0E-04	1.0E-03	2.0E-04	3.0E-05	2.0E-04	8.0E-04	1.0E-04	1.0E-04	8.0E-04	2.0E-04
L-methionine	149	3.0E-05	1.0E-04	3.0E-05	3.0E-05	3.0E-05	2.0E-04	3.0E-05	3.0E-05	2.0E-04	6.0E-05
L-phenylalanine	165	3.0E-05	2.0E-04	3.0E-05	5.0E-05	3.0E-05	3.0E-04	3.0E-05	3.0E-05	4.0E-04	6.0E-05
L-proline	115	3.0E-04	1.0E-04	5.0E-05	3.0E-04	3.0E-04	3.0E-04	3.0E-04	3.0E-04	3.5E-04	3.0E-04
L-serine	105	1.0E-04	3.0E-04	3.0E-04	3.0E-04	1.0E-04	1.0E-04	6.0E-04	6.0E-04	4.0E-04	1.2E-03
L-threonine	119	1.0E-04	1.0E-04	3.0E-04	3.0E-05	1.0E-04	5.0E-04	1.0E-04	1.0E-04	8.0E-04	2.0E-04
L-tryptophan	204	1.0E-05	2.0E-05	3.0E-05	5.0E-05	1.0E-05	1.0E-05	1.5E-05	1.5E-05	7.8E-05	3.0E-05
L-tyrosine	181	3.5E-05	1.0E-04	5.0E-05	5.0E-05	4.4E-05	2.0E-04	1.5E-05	1.5E-05	4.6E-04	3.0E-05
L-valine	117	1.0E-04	1.0E-03	3.0E-04	3.0E-04	1.0E-04	2.0E-03	3.0E-04	3.0E-04	8.0E-04	6.0E-04
Vitamins											
Biotin	244	3.0E-08	3.0E-08	3.0E-08	3.0E-08	3.0E-08	3.0E-08	6.0E-08	6.0E-08	5.3E-08	6.0E-08
Choline chloride	140	1.0E-04	1.0E-04	1.0E-04	1.0E-04	1.0E-04	1.0E-04	1.0E-04	1.0E-04	2.0E-05	2.0E-04
Folic acid	441	1.0E-04	1.0E-04			3.0E-06		1.8E-06	1.8E-06	9.1E-06	1.8E-06
Folinic acid	512	1.0E-09	1.0E-06	1.0E-08	1.0E-08		1.0E-06				
myo-Inositol	180	1.0E-04	4.0E-05	1.0E-04	1.0E-04	1.0E-04	4.0E-05	1.0E-04	1.0E-04	4.0E-05	1.0E-04
Nicotinamide	122	5.0E-05	5.0E-05	5.0E-05	5.0E-05	3.0E-07	5.0E-05	3.0E-07	3.0E-07	3.3E-05	3.0E-07
Pantothenate	238	1.0E-06	5.0E-05	1.0E-06	1.0E-06	1.0E-06	5.0E-05	1.0E-06	1.0E-06	1.7E-05	1.0E-06
Pyridoxal HCl	204									2.0E-05	
Pyridoxine HCl	206	3.0E-07	1.0E-05	3.0E-07	3.0E-07	3.0E-07	3.0E-07	3.0E-07	3.0E-07		3.0E-07
Riboflavin	376	3.0E-07	1.0E-08	3.0E-07	3.0E-07	1.0E-07	1.0E-06	1.0E-07	1.0E-07	1.1E-06	1.0E-07
Thiamin HCl	337	1.0E-06	1.0E-05	1.0E-06	1.0E-06	1.0E-06	1.0E-04	1.0E-06	1.0E-06	1.2E-05	1.0E-06
α-Tocopherol	430	1.4E-07									
Retinoic acid	300										3.3E-07

Retinol acetate	329	4.2E-07									
Vitamin B_{12}	1355	1.0E-07	1.0E-08	1.0E-07	1.0E-07	1.0E-07	1.0E-08	3.0E-07	3.0E-07	9.6E-09	3.0E-07
Antioxidants											
Dithiothreitol	154	6.5E-06									
Glutathione	307	6.5E-07									
Inorganic salts											
$CaCl_2$	147	1.0E-03	1.6E-03	2.0E-03	2.0E-03	6.0E-04	1.6E-03	1.3E-04	3.0E-05	1.5E-03	1.1E-04
KCl	75	5.0E-03	4.0E-03	3.0E-03	3.0E-03	3.0E-03	4.0E-03	1.5E-03	1.5E-03	4.4E-03	1.5E-03
KNO_3	160					1.6E-08				7.5E-07	
$MgCl_2$	203					6.0E-04		6.0E-04	6.0E-04		2.2E-02
$MgSO_4$	247	1.0E-03	1.0E-02	1.5E-03	1.5E-03	6.1E-10	8.0E-04			8.1E-04	
NaCl	58	1.1E-01	1.1E-01	1.2E-01	1.2E-01	1.3E-01	1.2E-01	1.2E-01	1.2E-01	7.7E-02	1.0E-01
$NaHCO_3$	84		1.4E-02			1.4E-02	1.4E-02	1.4E-02	1.4E-02	3.6E-02	1.2E-02
Na_2HPO_4	120	3.0E-03	5.9E-04	5.0E-04	5.0E-04	1.2E-03	5.0E-04	2.0E-03	2.0E-03	1.0E-03	2.0E-03
Trace elements											
$CuSO_4 \cdot 5H_2O$	160	1.0E-09	7.5E-09	1.0E-09	1.0E-C9		5.0E-09	1.0E-09	1.1E-08		1.0E-08
$FeSO_4$	278	5.0E-06	1.0E-06	5.0E-06	5.0E-06	3.0E-06	1.0E-06	5.0E-06	5.0E-06		5.4E-04
$MnSO_4 \cdot H_2O$	169	1.0E-09	1.2E-09	5.0E-10	5.0E-10		1.0E-09	1.0E-09	1.0E-09		1.0E-09
$(NH_4)_6Mo_7O_{24}$	1236	1.0E-09	3.0E-09	1.0E-09	1.0E-09	1.0E-08	3.0E-09	1.0E-09	1.0E-09		1.0E-09
$NiCl_2$	238	5.0E-10	4.2E-10	5.0E-12	5.0#-12		3.0E-10	5.0E-10	5.0E-10		5.0E-10
H_2SeO_3	129	3.0E-08		3.0E-08	3.0E-08	1.3E-08	1.0E-08	3.0E-08	3.0E-08	1.0E-07	3.0E-08
Na_2SiO_3	122	5.0E-07	2.3E-05	5.0E-07	5.0E-07		1.0E-05	5.0E-07	5.0E-07		5.0E-07
$SnCl_2$	190	5.0E-10		5.0E-12	5.0E-12			5.0E-10	5.0E-10		5.0E-10
NH_4VO_3	117	5.0E-09	5.1E-09	5.0E-09	5.0E-09	1.0E-08	5.0E-09	5.0E-09	5.0E-09		5.0E-09
$ZnSO_4 \cdot 7H_2O$	288	5.0E-07	1.0E-09	1.0E-07	1.0E-07	3.0E-06	1.0E-06	5.0E-07	5.0E-07		4.8E-07
Lipids and precursors											
Cholesterol	387	7.6E-06									
Ethanolamine	61			1.0E-04					1.0E-04		
Linoleic acid	280			2.0E-07	2.0E-07	3.0E-07	3.0E-07				
Lipoic acid	206	1.0E-08	1.0E-08	1.0E-08	1.0E-08	9.8E-07	1.0E-08	1.oE-06	1.0E-06		1.0E-06
Phosphoethanolamine	141			1.0E-04							5.0E-07
Soya lecithin μg/ml		6									
Soybean lipid μg/ml										50	
Sphingomyelin μg/ml		1									
Hormones and growth factors											
EGF (ng/ml)		30		10				25	25		5
Epinephrine	183										2.7E-06
Hydrocortisone[1]	362	5.0E-07		1.4E-07				5.0E-07	1.4E-07		2.0E-07
Insulin (μg/ml)		1		5				10	5		5
Prolactin μg/ml		1		5							
PGE_1	355	2.5E-08							2.5E-08		
Triiodothyronine	673										1.0E-08
Nucleosides, etc.											
Adenine SO_4	184	1.0E-05	1.0E-06	1.0E-06	1.0E-06		1.0E-06	1.8E-04	1.8E-04		1.8E-04
Hypoxanthine	136					3.0E-05					
Thymidine	242	3.0E-07	1.0E-07	3.0E-07	3.0E-07		1.0E-06	3.0E-06	3.0E-06		3.0E-06

TABLE 9.1. Examples of Serum-Free Media Formulations (*Continued*)

Component	Mol. wt.	MCDB 110 Human lung fibroblasts [Bettger et al., 1981]	MCDB 131 Human vascular endothelium [Knedler and Ham, 1987; Gupta et al., 1997]	MCDB 170 Mammary epithelium [Hammond et al., 1984]	MCDB 202 Chick embryo fibroblasts [McKeehan and Ham, 1976b]	MCDB 302 CHO cells [Hamilton and Ham, 1977]	MCDB 402 3T3 cells [Shipley and Ham, 1983]	WAJC 404 Prostatic epithelium [McKeehan et al., 1984; Chaproniere and McKeehan, 1986]	MCDB 153 Keratinocytes [Peehl and Ham, 1986]	Iscove's lymphoid cells [Iscove and Melchers, 1978]	LHC-9 Bronchial epithelium [Lechner and LaVeck, 1985]
Energy metabolism											
D-glucose	180	4.0E-03	5.6E-03	8.0E-03	8.0E-03	1.0E-02	5.5E-03	6.0E-03	6.0E-03	2.5E-02	6.0E-03
Phosphoenolpyruvate	190	1.0E-05									
Sodium acetate·3H_2O	136						1.0E	3.7E-03	3.7E-03		3.7E-03
Sodium pyruvate	110	1.0E-03	1.0E-03	1.0E-03	5.0E-04	1.1E+02		5.0E-04	5.0E-04	1.0E-03	5.0E-04
Other components											
Cholera toxin	~90 KDa							2.0E-10			
HEPES, Na salt	260	3.0E-02		3.0E-02	3.0E-02			2.8E-02	2.8E-02	2.5E-02	2.3E-02
Phenol red	376	3.3E-06	3.3E-05	3.3E-06	3.3E-06	3.3E-06	3.3E-05	3.3E-05	3.3E-05	4.0E-05	3.3E-06
Putrescine 2HCl	161	1.0E-09	1.2E-09	1.0E-09	1.0E-09	1.0E-06	1.0E-09	1.0E-06	1.0E-06		1.0E-06
Protein supplements											
BPE μgP/ml²								25			35
BSA, mg/ml		70									
Dialyzed FBS, μgP/ml				5					1	0.5–10	
Transferrin, Fe^{3+} saturated (μg/ml)										30–300	10
Gas phase											
CO_2	44	2%	5%	2%	2%	5%	5%	5%	5%	10%	5%

¹Soluble analogues of hydrocortisone, such as dexamethasone, can be used.

²Ovine prolactin can be substituted for BPE.

Most concentrations are molar and computer-style notation is used, e.g., 3.0E-2 = 3.0×10^{-2} = 30 mM. Molecular weights are given for root compounds; although some recipes use salts or hydrated forms, molarities will, of course, remain the same. The units are given in the component column where molarity is not used.

Abbreviations: BPE, bovine pituitary extract; BSA, bovine serum albumin; EGF, epidermal growth factor; FBS, fetal bovine serum. See text for references.

TABLE 9.2. Examples of Serum-free Media; Supplemented Basal Media

Medium	Carney et al., 1981 HITES	Brower et al., 1996	Masui et al., 1986b	Bottenstein, 1984 N3	Michler-Stucke and Bottenstein, 1982	Taub, 1984 K-1	Taub, 1984 K-2	Chopra and Xue-Hu, 1993	Robertson and Robertson, 1995	Naeyaert et al., 1991
Cell type	Human small cell lung carcinoma	Human non-small-cell lung carcinoma	Human lung adenocarcinoma	Human neuroblastoma, LA-N-1	Rat glial cells	MDCK (dog kidney)	LLC-PK$_1$ (pig kidney)	Human parotid	Human prostate	Human melanocytes
Basal medium	RPMI 1640	RPMI 1640	DMEM/F12	DMEM/F12	DMEM	DMEM/F12	DMEM/F12	MCDB 153	αMEM/F12	M199
Supplements										
Arg·VP, μU/ml										
BPE, μgP/ml		5.0					10	25	25	
BSA, mg/ml					0.3					
Cholera toxin										1.0E-09
Cholesterol							1.0E-08			
EGF (ng/ml)		10						10	10	1
Epinephrine										
Estradiol	1.0E-08									
Ethanolamine			5.0E-07						1.0E-03	
FGF-2 (basic FGF) ng/ml										10
Glucagon, μg/ml			0.2							
Glutamine (additional)		2.0E-03							2.0E-03	
Hydrocortisone	1.0E-08	5.0E-08				5.0E-08	2.0E-07	1.4E-06	1.0E-06	1.4E-06
Insulin (μg/ml)	5.0	20.0	5.0	5.0	0.5	5.0	25.0	5.0	5.0	10.0
Na pyruvate (additional)		5.0E-04								
Na_2SeO_3	3.0E-08	2.5E-08	2.5E-08	3.0E-08						
Phosphoethanolamine										
Progesterone				2.0E-08	2.0E-07				1.0E-04	
Prolactin										
Prostaglandin E$_1$						7.0E-08				
Putrescine				1.0E-04	1.0E-07					
Transferrin, Fe^{3+} saturated (μg/ml)	100	10		50	100	5	10		5	10
Triiodothyronine		1.0E-10	5.0E-10		4.9E-07	5.0E-12	1.0E-09			1.0E-09
Thyroxine					4.5E-07					

Most concentrations are molar and computer-style notation is used, e.g., 3.0E-2 = 3.0×10^{-2} = 30 mM. Molecular weights are given for root compounds; although some recipes use salts or hydrated forms, molarities will, of course, remain the same. The units are given in the component column where molarity is not used. Soluble analogues of hydrocortisone, such as dexamethasone, can be used. *Abbreviations:* Arg·VP, arginine vasopressin; BPE, bovine pituitary extract; BSA, bovine serum albumin; EGF, epidermal growth factor; FBS, fetal bovine serum; FGF, fibroblast growth factor.

may be selected as similar, but will never be identical, to the first batch.

(3) **Quality Control.** Changing serum batches requires extensive testing to ensure that the replacement is as close as possible to the previous batch. This can involve several tests (for growth, plating efficiency, and special functions; see Testing Serum in Chapter 8) and a number of different cell lines.

(4) **Specificity.** If more than one cell type is used, each type may require a different batch of serum, so that several batches must be held on reserve simultaneously. Coculturing different cell types will present an even greater problem.

(5) **Availability.** Periodically, the supply of serum is restricted due to drought in the cattle-rearing areas, the spread of disease among the cattle, or economic or political reasons. This can create problems at any time, restricting the amount of serum available and the number of batches to choose from, but can be particularly acute at times of high demand. Today, demand is increasing, and it will probably exceed supply unless the majority of commercial users are able to adopt serum-free media. While an average research laboratory may reserve 100–200 l of serum per year, a commercial biotechnology laboratory can use that amount or more in a week.

(6) **Downstream Processing.** To anyone interested in recovering cell products, the presence of serum creates a major obstacle to purification and may even limit the pharmaceutical acceptance of the product.

(7) **Contamination.** Serum is frequently contaminated with viruses, many of which may be harmless to cell culture, but represent an additional unknown factor outside the operator's control. Fortunately, improvements in serum sterilization techniques have virtually eliminated the risk of mycoplasma infection from sera from most reputable suppliers, but this cannot be guaranteed for viral infection, in spite of claims that some filters may remove viruses (Pall Gelman). Because of the risk of spreading bovine spongiform encephalitis among cattle, cell cultures and serum shipped to the United States or Australia require information on the country of origin and the batch number of the serum. Serum derived from cattle in New Zealand probably has the lowest endogenous viral contamination, as many of the viruses found in European and North American cattle are not found in New Zealand.

(8) **Cost.** Cost is often cited as a disadvantage of serum supplementation. Certainly, serum constitutes the major part of the cost of a bottle of medium (more than 10 times the cost of the chemical constituents), but if it is replaced by defined constituents, the cost of these may be as high as that of the serum. However, as the demand for such items as transferrin, selenium, insulin, etc., rises, the cost is likely to come down with increasing market size, and serum-free media will become relatively cheaper. The availability of recombinant growth factors, coupled with market demand, particularly for growth factors in the form of pharmaceuticals, may help to reduce their intrinsic cost.

(9) **Growth Inhibitors.** As well as its growth-promoting activity, serum contains growth-inhibiting activity, and although stimulation usually predominates, the net effect of the serum is an unpredictable combination of both inhibition and stimulation of growth. While substances such as PDGF may be mitogenic to fibroblasts, other constituents of serum can be cytostatic. Hydrocortisone, present at around 10^{-8} M in fetal serum, is cytostatic to many cell types, such as glia and lung epithelium, at high cell densities (though it may be mitogenic at low cell densities), and TGF-β, released from platelets, is cytostatic to many epithelial cells.

(10) **Standardization.** Standardization of experimental and production protocols is difficult, both at different times and among different laboratories, due to batch-to-batch variations in serum.

ADVANTAGES OF SERUM-FREE MEDIA

As well as eliminating the foregoing disadvantages of serum, serum-free media have two major positive benefits.

Selective Media
One of the major advantages of the control over growth-promoting activity afforded by serum-free media is the ability to make a medium selective for a particular cell type. (See Tables 9.1 and 9.2.) The long-standing problem of overgrowth by stromal fibroblasts can now be tackled effectively in breast and skin cultures by using MCDB 170 [Hammond et al., 1984] and 153 [Peehl and Ham, 1980], melanocytes can be cultivated in the absence of fibroblasts and keratinocytes [Naeyaert et al., 1991], and separate lineages and even stages of development may be selected in hematopoietic cells by choosing the correct growth factor or group of growth factors (see Hematopoietic Cells in Chapter 22).

Regulation of Proliferation and Differentiation

Add to the ability to select for a specific cell type the possibility of switching from a growth factor, after necessary amplification of the culture, to a differentiation factor or set of factors, and the amplified culture may then be made to perform one or more specialized functions.

DISADVANTAGES OF SERUM-FREE MEDIA

Serum-free media are not without disadvantages:

(1) **Multiplicity of Media.** Each cell type appears to require a different recipe, and cultures from malignant tumors may vary in requirements from tumor to tumor, even within one class of tumors. While this degree of specificity may be an advantage to those isolating specific cell types, it presents a problem for laboratories maintaining cell lines of several different origins.
(2) **Selectivity.** Unfortunately, the transition to serum-free conditions, however desirable, is not as straightforward as it seems. Some media may select a sub-lineage that is not typical of the whole population, and even in continuous cell lines, some degree of selection may still be required. Cells at different stages of development (e.g., stem cells vs. committed precursor cells) may require different formulations, particularly in the growth factor and cytokine components.
(3) **Reagent Purity.** The removal of serum also requires that the degree of purity of reagents and water and the degree of cleanliness of all apparatus be extremely high, as the removal of serum also removes the protective, detoxifying action that some serum proteins may have. While removing this action is no doubt desirable, it may not always be achievable, depending on resources.
(4) **Cell Proliferation.** Growth is often slower in serum-free media, and fewer generations are achieved with finite cell lines.
(5) **Availability.** Although improving steadily, the availability of properly controlled serum-free media is quite limited, and the products are often more expensive than conventional media.

REPLACEMENT OF SERUM

The essential factors in serum have been described (see Serum in Chapter 8) and include (1) adhesion factors such as fibronectin; (2) peptides, such as insulin, PDGF, and TGF-β, that regulate growth and differentiation; (3) essential nutrients, such as minerals, vitamins, fatty acids, and intermediary metabolites; and (4) hormones, such as insulin, hydrocortisone, estrogen, and triiodothyronine, that regulate membrane transport, phenotypic status, and the constitution of the cell surface. While some of these are catered for in the formulation of serum-free media others are not and may require an alteration in procedures.

Subculture

Adhesion Factors. When serum is removed, it may be necessary to treat the plastic growth surface with fibronectin (25–50 μg/ml) or laminin (1–5 μg/ml), added directly to the medium [Barnes et al., 1984a; see Protocol 22.9]. Pretreating the plastic with poly-L-lysine, 1 mg/ml, and washing off was shown to enhance the survival of human diploid fibroblasts [McKeehan and Ham, 1976a; see also Treated Surfaces in Chapter 7 and Barnes et al., 1984a].

Protease Inhibitors. Following trypsin-mediated subculture, the addition of serum inhibits any residual proteolytic activity. Consequently, protease inhibitors such as soya bean trypsin inhibitor or 0.1 mg/ml aprotinin (Sigma) must be added to serum-free media after subculture. Furthermore, because crude trypsin is a complex mixture of proteases, some of which may require different inhibitors, it is preferable to use pure trypsin (e.g., Sigma Gr. III) followed by a trypsin inhibitor. Alternatively, one may wash cells by centrifugation to remove trypsin. It is possible that Pronase can be inactivated by dilution without subsequent neutralization in serum-free conditions [McKeehan, personal communication].

Trypsin Temperature. Special care may be required when trypsinizing cells from serum-free media, as the cells are more fragile and may need to be chilled to reduce damage [McKeehan, 1977].

Hormones

Hormones that have been used to replace serum include growth hormone (somatotropin), 50 ng/ml, insulin at 1–10 U/ml, which enhances plating efficiency in a number of different cell types, and hydrocortisone, which improves the cloning efficiency of glia and fibroblasts (see Tables 9.1 and 9.2 and Improving Clonal Growth in Chapter 13) and has been found necessary for the maintenance of epidermal keratinocytes and some other epithelial cells (see Epidermis in Chapter 22). Barnes and Sato [1980] described 10 pM triiodothyronine (T_3) as a necessary supplement for MDCK (dog kidney) cells, and it has also been used

TABLE 9.3. Growth Factors and Mitogens

Name and synonyms	Abbreviation	Mol. wt. (KDa)	Source[a]	Function
Acidic fibroblast gf; aFGF, heparin binding gf 1, HBGF-1; endothelial cell gf (ECGF); myoblast gf (MGF)	FGF-1	13 h	Bovine brain; pituitary	Mitogen for endothelial cells
Activin; TGF-⟨beta⟩ family		g	Gonads	Morphogen; stimulates FSH secretion
Angiogenin		16	Fibroblasts, lymphocytes, colonic epithelial; cells	Angiogenic; endothelial mitogen
Astroglial growth factor-1; member of acidic FGFs	AGF-1	14	Brain	Mitogen for astroglia
Astroglial growth factor-2; member of basic FGFs	AGF-2	14	Brain	Mitogen for astroglia
Basic fibroblast gf; bFGF; HBGF-2; Prostatropin	FGF-2	13 h	Bovine brain; pituitary	Mitogen for many mesodermal and neuroectodermal cells; adipocyte and ovarian granulosa cell differentiation
Brain-derived neurotrophic factor	BDNF	28	Brain	Neuronal viability
Cachectin	TNF-α	17	Monocytes	Catabolic; cachexia; shock
Cholera toxin	CT	80–90	Cholera bacillus	Mitogen for some normal epithelia
Ciliary neurotrophic factor; member of IL-6 group	CNTF		Eye	
Endothelial cell growth factor; acidic FGF family	ECGF	h	Recombinant	Endothelial mitogenesis
Endothelial growth supplement; mixture of endothelial mitogens	ECGS		Bovine pituitary	Endothelia mitogenesis
Endotoxin			Bacteria	Stimulates TNF production
Epidermal growth factor, Urogastrone	EGF	6	Submaxillary salivary gland (mouse) human urine; guinea pig prostate	Active transport; DNA, RNA, protein, synthesis; mitogen for epithelial and fibroblastic cells; synergizes with IGF-1 and TGF-⟨beta⟩
Erythropoietin	EPO	34–39 g	Juxta-glomerular cells of kidney	Erythroid progenitor proliferation and differentiation
Eye-derived growth factor-1; member of basic FGFs	EDGF-1	14	Eye	Mitogen; morphogen; angiogenic
Eye-derived growth factor-2; member of acidic FGFs	EDGF-2	14	Eye	Mitogen; morphogen; angiogenic
Fibroblast gf-3; product of int-2 oncogene	FGF-3	14 h	Mammary tumors	Mitogen; morphogen; angiogenic
Fibroblast gf-4; product of hst/KS3 oncogene	FGF-4	14 h	Embryo; tumors	Mitogen; morphogen; angiogenic
Fibroblast gf-5	FGF-5	14 h	Fibroblasts; epithelial cells; tumors	Mitogen; morphogen; angiogenic
Fibroblast gf-6; product of hst-2 oncogene	FGF-6	14 h	Testis; heart; muscle	Mitogenic for fibroblasts; morphogen
Fibroblast gf-9	FGF-9	14 h		
Granulocyte colony-stimulating factor; pluripoietin; CFS-β	G-CSF	18–22	Recombinant	Granulocyte progenitor proliferation and differentiation

Granulocyte/macrophage colony-stimulating factor CSA; human CSFα	GM-CSF	14–35 g		Granulocyte/macrophage progenitor proliferation
Heparin-binding EGF-like factor	HB-EGF	h	Recombinant	
Hepatocyte gf, HBGF-8; Scatter factor	HGF		Fibroblasts	Epithelial morphogenesis; hepatocyte proliferation
Heregulin; erbB2 ligand	HRG	70	Breast cancer cells; recombinant	Mammary and other epithelial cell mitogen
Immune interferon; macrophage-activating factor (MAF)	IFN-γ	20–25	Activated lymphocytes	Antiviral; activates macrophages
Inhibin; TGF-β family		31 g	Ovary	Morphogen; inhibits FSH secretion
Insulin	Ins	6	⟨beta⟩ islet cells of pancreas	Glucose uptake and oxidation; amino acid uptake; glyconeogenesis
Insulinlike gf 1; somatomedin-C; NSILA-1	IGF-1			Mediates effect of growth hormone on cartilage sulphation; insulinlike activity
Insulinlike gf2; MSA in rat	IGF-2	7	BRL-3A cell-conditioned medium	Mediates effect of growth hormone on cartilage sulphation; insulinlike activity
Interferon-α1; leukocyte interferon	IFNα1	18–20	Macrophages	Antiviral; differentiation inducer; anticancer
Interferon-α2; leukocyte interferon	IFN-α2	18–20	Macrophages	Antiviral; differentiation inducer; anticancer
Interferon-β; fibroblast interferon	IFN-β1	22–27 g	Fibroblasts	Antiviral; differentiation inducer; anticancer
Interferon β2; fibroblast interferon; IL-6, BSF-2 (see also IL-6)	IFN-β2	22–27 g	Activated T-cells; fibroblasts; tumor cells	Keratinocyte differentiation; PC12 differentiation (see also IL-6)
Interferon γ; immune interferon	IFNγ		Activated lymphocytes	Antiviral, macrophage activator; antiproliferative on transformed cells
Interleukin-1; lymphocyte-activating factor (LAF); B-cell-activating factor (BAF); hematopoietin-1	IL-1	12–18	Activated macrophages	Induces IL-2 release
Interleukin-2; T-cell gf (TCGF)	IL-2	15	CD4+ve lymphocytes (NK); murine LBRM-5A4 and human Jurkat FHCRC cell lines	Supports growth of activated T-cells; stimulates LAK cells
Interleukin-3; multipotential colony-stimulating factor; mast cell growth factor	IL-3	14–28 g	Activated T-cells WEHI-3b myelomonocytic cell lines	Granulocyte/macrophage production and differentiation
Interleukin-4; B-cell gf, BCGF-1; BSF-1	IL-4	15–20	Activated CD4+ve lymphocytes	Competence factor for resulting B-cells; mast cell maturation (with IL-3)
Interleukin-5; T-cell-replacing factor (TRF); eosinophil-differentiating factor (EDF) BCGF-2	IL-5	12–18 g	T-lymphocytes	Eosinophil differentiation; progression factor for competent B-cells
Interleukin-6; Interferon β-2; B-cell-stimulating factor (BSF-2); hepatocyte-stimulating factor; hybridoma-plasmacytoma gf	IL-6	22–27 g	Activated T-cells macrophage/monocytes; fibroblasts; tumor cells	Acute phase response; B-cell differentiation; keratinocyte differentiation; PC12 differentiation
Interleukin-7; hematopoietic growth factor; lymphopoietin 1	IL-7	15–17 g	Bone marrow stroma	Pre- and pro-B-cell growth factor

TABLE 9.3. Growth Factors and Mitogens (*Continued*)

Name and synonyms	Abbreviation	Mol. wt. (KDa)	Source[a]	Function
Interleukin-8; monocyte-derived neutrophil chemotactic factor (MDNCF); T-cell chemotactic factor; neutrophil-activating protein (NAP-1)	IL-8	8–10 h	LPS monocytes PHA lymphocytes; endothelial cells; IL-1- and TNF-stimulated fibroblasts and keratinocytes	Chemotactic factor for neutrophils, basophils, and T-cells
Interleukin-9; human P-40; mouse T-helper gf; mast-cell-enhancing activity (MEA)	IL-9	30–40	CD4+ve T-cells; stimulated by anti-CD4 antibody PHA or PMA	Growth factor for T-helper, megakaryocytes, mast cells (with IL-3)
Interleukin-10; cytokine synthesis inhibitory factor (CSIF)	IL-10	20		Immune suppressor
Interleukin-11; adipogenesis inhibitory factor (AGIF)	IL-11	21		Stimulates plasmacytoma proliferation and T-cell-dependent development of Ig-producing B-cells
Interleukin-12; cytotoxic lymphocyte maturation factor	IL-12	40, 35 subunits		Activated T-cell and NK cell growth factor; induces IFN-γ
Keratinocyte gf, FGF-7	KGF	14 h	Fibroblasts	Keratinocyte proliferation and differentiation; prostate epithelial proliferation and differentiation
Leukemia inhibitory factor; HILDA; member of IL-6 group	LIF	24	SCO cells	Inhibits differentiation in embryonal stem cells
Lipopolysaccharide	LPS	10	Gram-positive bacteria	Lymphocyte activation
Lymphotoxin	TNF-β	20–25	Lymphocytes	Cytotoxic for tumor cells
Macrophage inflammatory protein-1α	MIP-1α	10	Macrophages	Hematopoietic stem cell inhibitor
Monocyte/macrophage colony-stimulating factor CSF-1	M-CSF	47–74	B- and T-cells, monocytes, mast cells, fibroblasts	Macrophage progenitor proliferation and differentiation

Name	Abbreviation	Source	MW	Biological activity
Mullerian inhibition factor	MIF	Testis		Inhibition of Mullerian duct; inhibition of ovarian carcinoma
Nerve gf, β	β NGF	Male mouse submaxillary salivary gland	27	Trophic factor; chemotactic factor; differentiation factor; neurite outgrowth in peripheral nerve
Oncostatin M; member of IL-6 group	OSM	Activated T-cells and PMA-treated monocytes	28	Differentiation inducer (with glucocorticoid); fibroblast mitogen
Phytohemagglutinin	PHA	Red kidney bean (Phaseolus vulgaris)	30	Lymphocyte activation
Platelet-derived endothelial cell growth factor; similar to gliostatin	PD-ECGF	Blood platelets, fibroblasts, smooth muscle	~70	Angiogenesis; endothelial cell mitogen; neuronal viability; glial cytostasis
Platelet-derived growth factor	PDGF	Blood platelets	30	Mitogen for mesodermal and neuroectodermal cells; wound repair; synergizes with EGF and IGF-1
Phorbol meristate acetate; TPA; phorbol ester	PMA	Croton oil	0.617	Tumor promoter; mitogen for some epithelial and melanocytes; differentiation factor for HL-60 and squamous epithelium
Pokeweed mitogen	PWM	Roots of pokeweed (Phytolacca americana)		Monocyte activation
Stem cell factor; mast cell growth factor; steel factor; c-kit ligand	SCF	Endothelial cells, fibroblasts, bone marrow, Sertoli cells	31 g	Promotes first maturation division of pluripotent hematopoietic stem cell
Transferrin	Tfn	Liver	78	Iron transport; mitogen
Transforming growth factor α	TFG-α		6	Induces anchorage-independent growth and loss of contact inhibition
Transforming growth factor β (six species)	TGF-β 1-6	Blood platelets	23-25 dimer	Epithelial cell proliferation inhibitor; squamous differentiation inducer

[a]Sources described are some of the original tissues from which the natural product was isolated. In many cases the natural product has been replaced by cloned recombinant material that is available commercially. (See Sources of Materials in Trade Index.) *Abbreviation* (other than in column 2): gf, growth factor; g, glycosylated; h, heparin binding; HBGF, heparin-binding growth factor. Information in this table was taken from Barnes et al. [1984a], Lange et al. [1991], Jenkins [1992], and Smith et al. [1997].

for lung epithelium [Lechner and LaVeck, 1985; Masui et al., 1986b]. Various combinations of estrogen, androgen, or progesterone with hydrocortisone and prolactin at around 10 nm can be shown to be necessary for the maintenance of mammary epithelium [Klevjer-Anderson and Buehring, 1980; Hammond et al., 1984; Strange et al., 1991; Lee et al., 1996].

Other hormones with activities not usually associated with the cells they were tested on were found to be effective in replacing serum (e.g., follicle-stimulating hormone (FSH) with B16 murine melanoma [Barnes and Sato, 1980]). It is possible that sequence homologies exist between some growth-stimulating polypeptides and well-established peptide hormones. Alternatively, the processing of some of the large proteins or polypeptides may release active peptide sequences with quite different functions.

Growth Factors

The family of polypeptides that has been found to be mitogenic *in vitro* is now quite extensive (Table 9.3) and includes the heparin-binding growth factors (including the FGF family), EGF, PDGF [Barnes et al., 1984a,c], IGF-1 and -2, and the interleukins [Thomson, 1991] that are active in the 1–10-ng/ml range. Growth factors and cytokines tend to have a wide-ranging specificity, except for some that are active in the hematopoietic system [Barnes et al., 1984d; see also Hematopoietic Cells in Chapter 22). Keratinocyte growth factor (KGF) [Aaronson et al., 1991], besides showing activity with epidermal keratinocytes, will also induce proliferation and differentiation in prostatic epithelium [Planz et al., 1998; Thomson et al., 1997]. Hepatocyte growth factor (HGF) [Kenworthy et al., 1992] is mitogenic for hepatocytes, but is also morphogenic for kidney tubules [Furue and Saito, 1997; Montesano et al., 1997; Balkovetz and Lipschutz, 1999]. Growth factors and cytokines acquire their specificity by virtue of the fact that their production is localized and that they have a limited range. Most act as paracrine factors (they are active on adjacent cells) and not by systemic distribution in the blood.

Growth factors may act synergistically or additively with each other or with other hormones and paracrine factors, such as prostaglandin $F_{2\alpha}$ and hydrocortisone [Westermark and Wasteson, 1975; Gospodarowicz, 1974]. The action of interleukin 6 (IL-6) and oncostatin M on A549 cells is dependent on dexamethasone, a synthetic hydrocortisone analogue [McCormick et al., 1995, 2000]. The action is due to the production of a heparan sulphate proteoglycan (HSPG) [Yevdokimova and Freshney, 1997]. The requirement for heparin or HSPG was first observed with FGF [Klagsbrun and Baird, 1991], but may be a more general phenomenon, as β-glycan has been shown to be involved in the cellular response to TGF-β [Lopez-Casillas et al., 1993]. Some growth factors are dependent on the activity of a second growth factor before they act [Phillips and Christofalo, 1988]; e.g., bombesin alone is not mitogenic in normal cells, but requires the simultaneous or prior action of insulin or one of the IGFs [Aaronson et al., 1991].

Nutrients in Serum

Iron, copper, and a number of minerals have been included in serum-free recipes, although evidence that some of the rarer minerals are required is still lacking. Selenium (Na_2SeO_3), at around 20 nM, is found in most formulas, and there appears to be some requirement for lipids or lipid precursors such as choline, linoleic acid, ethanolamine, or phosphoethanolamine.

Proteins and Polyamines

The inclusion in medium of proteins such as bovine serum albumin (BSA) or tissue extracts often increases cell growth and survival, but adds undefined constituents to the medium. BSA, fatty acid free, is used at 1–10 mg/ml. Transferrin, at around 10 ng/ml, is required as a carrier for iron and may also have a mitogenic role. Putrescine has been used at 100 nM.

Matrix

One of the properties of serum is to provide a number of proteins, such as fibronectin, that coat the plastic and make it more adhesive. (See Adhesion Factors in this chapter, and Matrix Coating in Chapter 7.) In the absence of serum, the plastic substrate may need to be coated with fibronectin (see Protocol 22.9) or polylysine [McKeehan and Ham, 1976; (see also Tables 9.1 and 9.2 and Improving Clonal Growth in Chapter 13)].

SELECTION OF SERUM-FREE MEDIUM

If the reason for using a serum-free medium is to promote the selective growth of a particular type of cell, then that reason will determine the choice of medium (e.g., MCDB 153 for epidermal keratinocytes, LHC-9 for bronchial epithelium, HITES for small-cell lung cancer, MCDB 130 for endothelium, etc.; see Tables 9.1 and 9.2). If the reason is simply to avoid using serum with continuous cell lines, such as CHO cells or hybridomas, in order to reduce the likelihood of contamination or serum proteins in the cell product, then the choice will be wider, and there will be several commercial sources to choose from (Table 9.4). When a cell line is obtained from the originator or a reputable cell bank, the supplier will recommend the appropriate medium, and the only reason to change will be if the medium is unavailable or is incompatible with other stocks. If possible, it is best to stay with the originator's recommendation, as this may be the only way to ensure that the line exhibits its specific properties. Table 9.4 summarizes the availability and selection of the serum-free media listed in Tables 9.1 and 9.2 and makes a few additional suggestions. (See also Mather [1998], Barnes et al. [1984(a–d)], and Ham and McKeehan [1979].)

Commercially Available Serum-Free Media

Several suppliers (see Table 9.4 and Sources of Materials in Trade Index) now make serum-free media. Some are defined formulations, such as MCDB 131 (Sigma) for endothelial cells and LHC-9 (Biofluids) for bronchial epithelium, while others are proprietary formulations, such as CHO-S-SFM for CHO-K1 cells and Opti-MEM for hematopoietic cells (Gibco). Many are designed primarily for culture of hybridomas culture, when the formation of a product that is free of serum proteins is clearly important, but others are applicable to other cell types. While evidence exists in the literature for the use of proprietary media with specific cell types (see Table 9.4), commercial recipes are often a trade secret, and you can only rely on the supplier's advice or, better, screen a number of media over several subcultures with your own cells. The latter can be an extensive exercise, but is justified if you are planning long-term work with the cells.

It is important to determine the quality control performed by commercial suppliers of serum-free media. The ideal situation is for the medium to have been tested against the cells that you wish to grow (e.g., keratinocyte growth medium should have been tested on keratinocytes). Some suppliers, such as Biowhittaker (Clonetics), will supply the appropriate cells with the medium, ensuring the correct quality control, but others may have performed quality control with routine cell lines, in which case you will have to do your own quality control before purchasing the medium.

Serum Substitutes

A number of products have been developed commercially to replace all or part of the serum in conventional media. Some of these products are Ex-cyte (Bayer), Ventrex (JRH Biosciences), CPSR-1,2,3 (Sigma), Nu-serum (Becton Dickinson), and Biotain-MPS (Biowhittaker). While they may offer a degree of consistency not obtainable with regular sera, variations in batches can still occur, and the constitution of the products is not fully defined. They may be use-ful as an *ad hoc* measure or for purposes of economy, but are not a replacement for serum-free media. Nutridoma (BCL), ITS Premix, TCM, TCH (ICN), SIT (Sigma), and Excell-900 (JRH Biosciences) are defined supplements aimed at replacing serum, partially or completely [Cartwright and Shah, 1994].

The choice of serum-free media and serum replacements is now so large and diverse, that it is not possible to make individual recommendations for specific tasks. The best approach is to check the literature, contact the suppliers, obtain samples of those products that seem most relevant from previous reports, and screen the products in your own assays (e.g., for growth, survival, or special functions).

DEVELOPMENT OF SERUM-FREE MEDIUM

There are two general approaches to the development of a serum-free medium for a particular cell line or primary culture. The first is to take a known recipe for a related cell type, with or without 10–20% dialyzed serum, and alter the constituents individually or in groups, while reducing the serum, until the medium is optimized for your own particular requirement. This was the approach adopted by Ham and co-workers [Ham, 1984] and generally will provide optimal conditions. If a group of compounds is found to be effective in reducing serum supplementation, the active constituents may be identified by the systematic omission of single components and then the concentrations of the essential components optimized [Ham, 1984]. However, this is a very time-consuming and laborious process, involving growth curves and clonal growth assays at each stage, and it is not unreasonable to expect to spend at least three years developing a new medium for a new type of cell.

The time-consuming nature of the first approach has led to the second approach: supplementing existing media such as RPMI 1640 [Carney et al., 1981] or combining media such as Ham's F12 with DMEM [Barnes and Sato, 1980] and restricting the manipulation of the constituents to a shorter list of substances. Among the latter are selenium, transferrin, albumin, insulin, hydrocortisone, estrogen, triiodothyronine, ethanolamine, phosphoethanolamine, growth factors (EGF, FGF, PDGF, endothelial growth supplement, etc.), prostaglandins (PGE_1, $PGF_{2\alpha}$), and any other substances that may have special relevance. (See Table 9.3.) Selenium, transferrin, and insulin will usually be found to be essential for most cells, whereas the requirements for the other constituents will be more variable.

TABLE 9.4. Selecting a Serum-free Medium (See also Tables 9.1 and 9.2)

Cells or cell line	Serum-free medium (Table 9.1)	Refs. (Tables 9.1 and 9.2)	Commercial suppliers (of specified media or alternatives)
BHK 21			
Bronchial epithelium	LHC-9	[Pardee et al., 1984]	Biofluids; Biowhittaker
Chick embryo fibroblasts	MCDB 201, 202	(See Protocol 22.9)	Sigma
Chinese hamster ovary (CHO)	MCDB 302	[McKeehan and Ham, 1976b] [Hamilton and Ham, 1977]	Biowhittaker; Sigma; Gibco; JRH Biosciences; PromoCell
Chondrocytes		[Adolphe, 1984]	
Continuous cell lines	M199, MB752/1, CMRL 1066, MCDB media, DMEM:F12 + supplements	[Waymouth, 1984]	Biowhittaker, JRH Biosciences; ICN; Sigma
COS-1,7			Biowhittaker
Endothelium	MCDB 130, 131	[Knedler and Ham, 1987; Gupta et al., 1997; Hoheisel et al., 1998]	Biowhittaker; Sigma
Fibroblasts	MCDB 110, 202, 402	[Bettger et al., 1981; Shipley and Ham, 1983]	Biowhittaker; Sigma
Glial cells	Michler-Stucke	[Michler-Stuke and Bottenstein, 1982]	Biowhittaker; G-5; Gibco
Glioma		[Frame et al., 1980; Freshney, 1980]	Biowhittaker
HeLa cells		[Blaker et al., 1971], [Bertheussen, 1993]	Biowhittaker
Hematopoietic cells	αMEM; Iscove's	[Stanners et al., 1971; Iscove and Melchers, 1978]	Sigma; Boehringer Ingelheim
HL-60			Biowhittaker
HT-29			Biowhittaker
Human diploid fibroblasts	MCDB 110, 202	[Bettger et al., 1981; Ham, 1984]	Biowhittaker
Human leukemia	Iscove's	[Breitman et al., 1984]	
Human tumors	Brower; HITES; Masui; Bottenstein N3	Brower et al., 1986; Carney et al., 1981; Masui et al., 1986b; Bottenstein, 1984; Chopra et al., 1996]	

Cell type	Medium	Reference	Supplier
Hepatocytes, Liver epithelium	Williams E, L15	[Williams and Gunn, 1974; Mitaka et al., 1993]	Sigma
Hybridomas	Iscove's	[Iscove and Melchers, 1978; Murakami, 1984]	Sigma; Gibco; Biowhittaker; JRH Biosciences; ICN; Boehringer Mannheim; Irvine Scientific; IBF Ultroser HY: Metachem; PromoCell
Insect cells		[Peehl and Ham, 1980; Tsao et al., 1982; Boyce and Ham, 1983]	JRH Biosciences; Cell Gro
Keratinocytes	MCDB 153		Gibco; Biowhittaker
L cells (L929, LS)		[Birch and Pirt, 1970, 1971; Higuchi, 1977]	NCTC109, NCTC135, Sigma
Lymphoblastoid cell lines (human)	Iscove's	[Iscove and Melchers, 1978]	JRH Biosciences; CellGro; IBF Ultroser
Mammary epithelium	MCDB 170	[Hammond et al., 1984]	UltraMDCK, Biowhittaker
MDCK dog kidney epithelium	K-1	[Taub, 1984]	
LLC-PK$_1$ pig kidney	K-2	[Taub, 1984]	
Melanocytes	Gilchrest	(See Protocol 22.18)	Biowhittaker
Melanoma	Gilchrest	[Barnes and Sato, 1980]	Biowhittaker
Mouse embryo fibroblasts	MCDB 402	[Ham, 1984]	
Mouse leukemia		[Murakami, 1984]	
Mouse erythroleukemia	Iscove's	[Iscove and Melchers, 1978]	
Mouse myeloma		[Murakami, 1984]	Sigma, Gibco
Mouse neuroblastoma	MCDB411; DMEM:F12/N1	[Agy et al., 1981; Bottenstein, 1984]	
Neurons	DMEM:F12/N3; B27/Neurobasal	[Bottenstein, 1984; Brewer, 1995]	N2, Gibco
Skeletal muscle	F12	(See Protocol 22.12)	
3T3 cells	MCDB402	[Shipley and Ham, 1983]	
Matrix-coating products			Becton Dickinson; Gibco; Biofluids; Sigma

Further recommendations on the choice of medium can be found in McKeehan [1977], Barnes et al. [1984(a–d)], and Mather [1998].

PREPARATION OF SERUM-FREE MEDIUM

A number of recipes for serum-free media are now available—some commercially (see Trade Index)—for particular cell types [Cartwright and Shah, 1994; Mather, 1998; see also Tables 9.1 and 9.2 and Chapter 22]. The procedure for making up serum-free recipes is similar to that for preparing regular media (see Protocol 10.9; see also [Waymouth, 1984]). Ultrapure reagents and water should be used and care taken with solutions of Ca^{2+} and Fe^{2+} or Fe^{3+} to avoid precipitation. Metal salts tend to precipitate in alkaline pH in the presence of phosphate, particularly when the medium or salt solution is autoclaved, so cations in stock solutions should be kept at a low pH (below 6.5) and maintained phosphate free. They should be sterilized by autoclaving or filtration. (See Sterilization in Chapter 10.) It is often recommended that cations be added last, immediately before using the medium. Otherwise the constituents are generally made up as a series of stock solutions, minerals and vitamins 1,000×, tyrosine, tryptophan, and phenylalanine in 0.1 N HCl at 50×, essential amino acids 100× in water, salts 10× in water, and any other special cofactors, lipids, etc., 1,000× in the appropriate solvents. These are combined in the correct proportions and diluted to the final concentration, and then the pH and osmolality are checked. (See Customized Medium in Chapter 10.)

Growth factors, hormones, and cell adhesion factors are best added separately just before the medium is used, as they may need to be adjusted to suit particular experimental conditions.

CONCLUSIONS

However desirable serum-free conditions may be, there is no doubt that the relative simplicity of retaining serum, the specialized techniques required for the use of some serum-free media, the considerable investment in time, effort, and resources that go into preparing new recipes or even adapting existing ones, and the multiplicity of media required if more than one cell type is being handled all act as considerable deterrents to most laboratories to enter the serum-free arena. There is also no doubt, however, that the need for consistent and defined conditions for the investigation of regulatory processes governing growth and differentiation, the pressure from biotechnology to make the purification of products easier, and the need to eliminate all sources of potential infection will eventually force the adoption of serum-free media on a more general scale. But first, recipes must be found that are less "temperamental" than some current recipes and that can be used with equal facility and effectiveness in different laboratories.

CHAPTER 10

Preparation and Sterilization

All stocks of chemicals and glassware used in tissue culture should be reserved for that purpose alone. Traces of heavy metals or other toxic substances can be difficult to remove and are detectable only by a gradual deterioration in the culture. It follows that separate stocks imply separate washing of glassware. The requirements of tissue culture washing are higher than for general glassware, although the level of soil may be less; a special detergent may be necessary (see Selection of Detergent in this chapter), and cross-contamination from chemical glassware must be avoided. Because most laboratories propagate cells in disposable plastic flasks, the effect on growth of attached monolayers is a receding problem. Nevertheless, whenever glass is used, either for storage or for culture, the problem of chemical contaminants leaching out into media or reagents remains, and absolute cleanliness is therefore essential.

All apparatus and liquids that come in contact with cultures or other reagents must be sterilized by one of the methods described in the next section (Tables 10.1, 10.2).

APPARATUS

Glassware

Items of glassware used for dispensing and storing media and for culturing cells must be cleaned very carefully to avoid traces of toxic materials contaminating the inner surfaces and becoming incorporated into the medium. If the glass surface is to be used for cell propagation, it must not only be clean, but also carry the correct charge. Caustic alkaline detergents render the surface of the glass unsuitable for cell attachment and require subsequent neutralization with dilute HCl or H_2SO_4, but many modern detergents do not alter the glass surface and can be removed completely. The most effective washing procedure is as follows:

(1) Do not let soiled glassware dry out. A sterilizing agent, such as sodium hypochlorite, should be included in the water used to collect soiled glassware
   (a) to remove any potential biohazard, and
   (b) to prevent microbial growth in the water
(2) Select a detergent that is effective in the water of your area, that rinses off easily, and that is nontoxic. (See Selection of Detergent in this chapter.)
(3) Before drying the glassware, make sure that it has been thoroughly rinsed in tap water and deionized or distilled water.
(4) Dry glassware inverted so that it drains readily.
(5) Sterilize the glassware by dry heat to minimize the risk of depositing toxic residues from steam sterilization.

Plastic culture flasks are, on the whole, meant for single use, as detergent renders them unsuitable for cell propagation (in monolayer) and resterilization is difficult. Cells may be reseeded back into the same flask after subculturing them, but this tends to increase the risk of contamination. However, with care, cells that grow in suspension can be subcultured in the same flask for many generations.

TABLE 10.1. Sterilization of Equipment and Apparatus

Item	Sterilization
Ampules for freezer, glass	Dry heat[a]
Ampules for freezer, plastic	Autoclave[b] (usually bought sterile)
Apparatus containing glass and silicone tubing	Autoclave
Disposable tips for micropipettes	Autoclave in autoclavable trays or nylon bags
Filters, reusable	Autoclave; do not use prevacuum or postvacuum
Glassware	Dry heat
Glass bottles with screw caps	Autoclave with cap slack
Glass coverslips	Dry heat
Glass slides	Dry heat
Glass syringes	Autoclave (separate piston if PTFE)
Instruments	Dry heat
Magnetic stirrer bars	Autoclave
Pasteur pipettes, glass	Dry heat
Pipettes, glass	Dry heat
Plexiglas, Perspex, Lucite	70% EtOH (see text)
Polycarbonate	Autoclave
Repeating pipettes or syringes	Autoclave (separate PTFE pistons from glass barrels)
Screw caps	Autoclave
Silicone grease (for isolating clones)	Autoclave in glass Petri dish
Silicone tubing	Autoclave
Stoppers, rubber and silicone	Autoclave
Test tubes	Dry heat

[a]Dry heat, 160°C/l h.

[b]Autoclave, 100 kPa (1 bar/15 lb/in²), 121°C for 20 min.

Sterilization procedures are designed not just to kill replicating micro-organisms but also to eliminate the more resistant spores. Moist heat is more effective than dry heat; however, it does carry a risk of leaving a residue. Dry heat is preferable, but a minimum temperature of 160°C maintained for 1 h is required. Moist heat (for fluids and perishable items) should be maintained at 121°C for 15–20 min. (See Table 10.1, 10.2.) For moist heat to be effective, steam penetration must be assured, which means that the sterilization chamber must be evacuated prior to steam injection or the air must be completely replaced with steam by downward displacement.

Inserting Thermalog indicators (Johnsen and Jorgensen) and a temperature probe from a recording thermometer, into a sample of the load, centrally located, monitors both temperature and humidity during sterilization. The temperature of the chamber effluent is often used but this does not accurately reflect the temperature of the load. Recording thermometers have the advantage that they will create a permanent record that can be archived. Indicators (e.g., Thermalog), on the other hand, provide a visual confirmation of temperature and humidity and can be used to monitor several parts of the load simultaneously. Both are recommended.

PROTOCOL 10.1. PREPARATION AND STERILIZATION OF GLASSWARE

Materials

Disinfectant: hypochlorite, 300 ppm available chlorine, minimum, when diluted in detergent (e.g., Precept tablets, Clorox or Chloros; see Sources of Materials–Disinfectants in Trade Index)

Detergent (e.g., 7× or Decon®)

Soaking baths

Bottle brushes

Stainless-steel baskets (to collect washed and rinsed glassware for drying)

Aluminum foil

Sterility indicators (Browne's tubes for dry heat or steam sterilization; Thermalog indicators for steam sterilization only)

Sterile-indicating tape or tabs (Bennett). This is different from the sterile-indicating tape used in autoclaves, as the sterilizing temperature is higher in an oven. Most autoclave tapes tend to char and release traces of volatile material from the adhesive, which can leave a deposit on the oven or even the glassware.

Sterilizing oven, fan assisted, capable of reaching 160°C, and preferably with recording thermometer and flexible probe

TABLE 10.2. Sterilization of Liquids

Solution	Sterilization	Storage
Agar	Autoclave[a] or boil	Room temperature
Amino acids	Filter[b]	4°C
Antibiotics	Filter	−20°C
Bacto-peptone	Autoclave	Room temperature
Bovine serum albumin	Filter (use stacked filters)	4°C
Carboxylmethyl cellulose	Steam, 30 min[c]	4°C
Collagenase	Filter	−20°C
DMSO	Self-sterilizing; dispense into aliquots sterile tubes	Room temperature; keep dark, avoid contact with rubber or plastics (except polypropylene)
Drugs	Filter (check for binding; use low binding filter, e.g., Millex-GV, if necessary)	−20°C
EDTA	Autoclave	Room temperature
Glucose, 20%	Autoclave	Room temperature
Glucose, 1 2%	Filter (low concentrations; caramelizes if autoclaved)	Room temperature
Glutamine	Filter	−20°C
Glycerol	Autoclave	Room temperature
Growth factors	Filter (low protein binding)	−20°C
HEPES	Autoclave	Room temperature
HCl, 1 N	Filter	Room temperature
Lactalbumin hydrolysate	Autoclave	Room temperature
Methocel	Autoclave	4°C
NaHCO₃	Filter	Room temperature
NaOH, 1 N	Filter	Room temperature
Phenol red	Autoclave	Room temperature
Salt solutions (without glucose)	Autoclave	Room temperature
Serum	Filter; use stacked filters	−20°C
Sodium pyruvate, 100 mM	Filter	−20°C
Transferrin	Filter	−20°C
Tryptose	Autoclave	Room temperature
Trypsin	Filter	−20°C
Vitamins	Filter	−20°C
Water	Autoclave	Room temperature

[a]Autoclave, 100 kPa (15 lb/in²), 121°C for 20 min.

[b]Filter, 0.2-μm pore size.

[c]Steam, 100°C for 30 min.

Protocol

Collection and Washing of Glassware (Fig. 10.1):

1. Immediately after use, collect glassware into detergent containing disinfectant. It is important that glassware not dry before soaking, or cleaning will be much more difficult.
2. Soak overnight in detergent.
3. Rinsing:
   (a) Brush glassware by hand or with a mechanized bottle brush (Fig. 10.2) the following morning, and rinse thoroughly in four complete changes of tap water followed by three changes of deionized water. A sink spray (see inset, Fig. 3.6) is a useful accessory; otherwise bottles must be emptied and filled completely each time. Clipping bottles in a basket will help to speed up this stage.
   (b) Machine rinses should be done without detergent. If done on a spigot header (see Glassware Washing Machine in Chapter 4), this can be reduced to two rinses with tap water and one with deionized or reverse-osmosis water.
4. After rinsing thoroughly, invert bottles, etc., in stainless-steel wire baskets and dry upside down.
5. Cap bottles with aluminum foil when cool, and store.

Sterilization of Glassware:

1. Attach small square of sterile-indicating tape or other indicator label to glassware, and date.

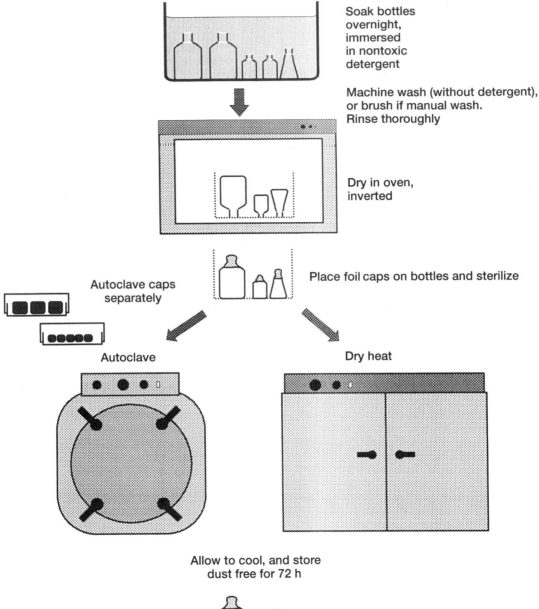

Soak bottles overnight, immersed in nontoxic detergent

Machine wash (without detergent), or brush if manual wash. Rinse thoroughly

Dry in oven, inverted

Autoclave caps separately

Place foil caps on bottles and sterilize

Autoclave

Dry heat

Allow to cool, and store dust free for 72 h

Fig. 10.1. *Washing and Sterilizing Glassware.* Procedure as in Protocol 10.1. Sterilization conditions: autoclave at 121°C for 15 min; oven, 160°C for 1 h. Caps are sterilized separately from bottles to avoid condensation forming in bottles if autoclaved with caps in place. (See Fig. 10.3.)

2. Place glassware in an oven with fan-circulated air and temperature set to 160°C.
3. To ensure that the center of the load reaches 160°C:
   (a) Place a sterility indicator in a bottle or typical item in the middle of the load.
   (b) If using a recording thermometer, place the sensor in a bottle or typical item in the middle of the load.

(c) Do not pack the load too tightly; leave room for circulation of hot air.
4. Close the oven, check that the temperature returns to 160°C, seal the oven with a strip of tape with the time recorded on it (or use automatic locking and recorder), and leave for 1 h;
5. After 1 h, switch off the oven and allow it to cool with the door closed. It is convenient to put the oven on an automatic timer so that it can be left

Fig. 10.2. Motorized Bottle Brushes. Care must be taken not to press down too hard on the brush, lest the bottle break.

to switch off on its own overnight and be accessed in the morning. This precaution allows for cooling in a sterile environment and also minimizes the heat generated during the day, when it is hardest to deal with.

6. Use glassware within 24–48 h.

Keep organic matter out of the oven. Do not use paper tape or packaging material, unless you are sure that it will not release volatile products on heating. Such products will eventually build up on the inside of the oven, making it smell when hot, and some deposition may occur inside the glassware being sterilized.

Alternatively, bottles may be loosely capped with screw caps and foil, tagged with autoclave tape, and autoclaved for 20 min at 121°C with a prevacuum and a postvacuum cycle (see Sterilizer in Chapter 4). Then the caps may be tightened when the bottles have cooled down. Caps must be very slack (loosened one complete turn) during autoclaving, so as to allow steam to enter the bottle and to prevent the liner (if one is used) from being sucked out of the cap and sealing the bottle. If a bottle becomes sealed during sterilization in an autoclave, sterilization will not be complete (Fig. 10.3). Unfortunately, misting often occurs when bottles are autoclaved, and a slight residue may be left when the mist evaporates. Also, the bottles risk becoming contaminated as they cool, by drawing in nonsterile air before they are sealed. Dry-heat sterilization is better, because it allows the bottles to cool down within the oven before they are removed.

Pipettes

Both glass and plastic pipettes are used in tissue culture. Plastic pipettes have the advantage that they are used just once and disposed of, avoiding the need for washup and sterilization. Glass pipettes are significantly cheaper, but have to be deplugged and replugged each time they are used. Glass pipettes must also be washed carefully in the pipette washer, so that they will not retain soil and become vulnerable to subsequent blockage.

△ *Safety Note.* Glass pipettes are prone to damage, and chipped ends present a severe hazard to both users and the washup staff. Discard or repair them.

PROTOCOL 10.2. PREPARATION AND STERILIZATION OF PIPETTES

Materials
Pipette cylinders (to collect used pipettes)
Disinfectant: hypochlorite, 300 ppm available chlorine, minimum, when diluted in detergent (e.g., Precept tablets, Chlorox, or Chloros)
Detergent (e.g., 7× or Decon®)
Stainless-steel baskets (to collect washed and rinsed pipettes for drying)
Sterility indicators (Browne's tubes, Thermalog)
Pipette cans (square aluminum or stainless steel with silicone cushions at either end; square cans do not roll on the bench) (Bellco, Life Sciences)
Sterile-indicating tape or tabs (Bennett). (See note in Protocol 10.1)
Sterilizing oven, fan assisted, capable of reaching 160°C, and preferably with recording thermometer and flexible probe

Protocol
Collection and Washing:
1. Place water with detergent and a disinfectant in pipette cylinder.
2. Discard pipettes, tip first, into cylinder immediately after use (Fig. 10.4).
   (a) Do not accumulate pipettes in the hood or allow pipettes to dry out.
   (b) Do not put pipettes that have been used with agar or silicones (water–repellent, antifoam, etc.) in the same cylinder as regular pipettes. Use disposable pipettes for silicones, and either rinse agar pipettes after use in hot tap water or use disposable pipettes.
3. Soak pipettes overnight or for a minimum of 2 h. If usage of pipettes is heavy, replace cylinders at intervals when full, and soak for 2 h before entering rinse cycle.
4. After soaking, remove plugs with compressed air.
5. Transfer pipettes to pipette washer (Fig. 10.5; see also Fig. 3.6), tips uppermost.

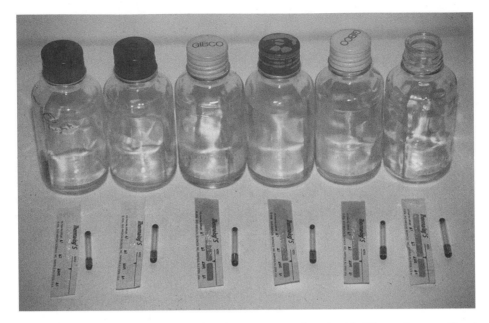

Fig. 10.3. Sterilizing Bottles. These bottles were autoclaved with Thermalog sterility indicators inside. Thermalog turns blue with high temperature and steam, and the blue area moves along the strip with time at the required sterilization conditions. The cap on the leftmost bottle was tight, and each succeeding cap was gradually slacker, until, finally, no cap was used on the bottle furthest to the right. The leftmost bottle is not sterile, because no steam entered it. The second bottle is not sterile either, because the liner drew back onto the neck and sealed it. The next three bottles are all sterile, but the brown stain on the indicator shows that there was fluid in them at the end of the cycle. Only the bottle at the far right is sterile and dry.

The glass indicators (Browne's tubes) all implied that their respective bottles were sterile.

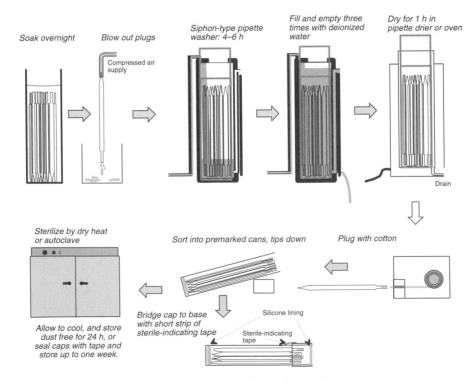

Fig. 10.4. Washing and Sterilizing Pipettes.

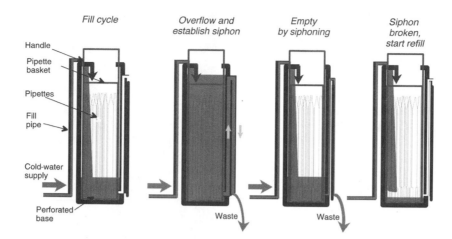

Fill cycle *Overflow and establish siphon* *Empty by siphoning* *Siphon broken, start refill*

Fig. 10.5. Siphon Pipette Washer. Connected to main water supply. A slow fill establishes a siphoning action and drains the main chamber, which then refills. This process is repeated many times over 4–6 h.

6. Rinse in tap water by siphoning action of pipette washer for a minimum of 4 h or in an automatic washing machine with a pipette adapter, but without detergent.
7. Turn valve to setting for deionized water (see Fig. 10.4; see also Fig. 3.6), or wait until last of tap water finally runs out, turn off tap water, and empty and fill washer three times with deionized water. (Use automatic deionized rinse cycle in machine.)
8. Transfer pipettes to pipette drier or drying oven, and dry with tips uppermost.
9. Plug with cotton (Fig. 10.6).
10. Sort pipettes by size and store dust free.

Sterilization:

1. Place pipettes in pipette cans. (It is useful to have both ends of the cans labeled with the size of the pipettes.)

Fig. 10.6. Semiautomatic Pipette Plugger (Bellco).

2. Fill each can with one size of pipette.
3. Fill a few cans with an assortment of 1-ml, 2-ml, 10-ml, and 25-ml pipettes (e.g., four of each size).
4. Attach sterile-indicating tape, bridging the cap to the can, and stamp date on tape.
5. Sterilize by dry heat for 1 h at 160°C. Use the smallest amount of tape possible, or replace with temperature indicator tabs (Bennett), which are small and are made of less volatile material. The temperature should be measured in the center of the load, to ensure that this, the most difficult part to reach, attains the minimum sterilizing conditions. Leave spaces between cans when loading the oven, to allow for circulation of hot air (Fig. 10.7).
6. Remove pipettes from oven, allow to cool, and transfer cans to tissue culture laboratory. If you anticipate that pipettes will lie idle for more than 48 h, seal cans around the cap with adhesive tape.

Screw Caps

There are two main types of caps that are in common use for glass bottles: (1) aluminum or phenolic plastic caps with synthetic rubber or silicone liners and (2) wadless polypropylene caps that are reusable (Duran); these caps are deeply shrouded and have ring inserts for better sealing and to improve pouring (although pouring is not recommended in sterile work). Polypropylene caps will seal only if screwed down tightly on a bottle with no chips or imperfections on the lip of the opening. Do not leave aluminum caps or any other aluminum items in alkaline detergents for more than 30 min, as they will corrode. Do not have glassware together with caps in the same detergent bath, or the aluminum may contaminate the glass. Avoid detergents that say they are made for machine washing, as they are highly caustic.

PROTOCOL 10.3. PREPARATION AND STERILIZATION OF SCREW CAPS

Materials
Disinfectant: hypochlorite, 300 ppm available chlorine (e.g., Precept tablets, Johnson & Johnson)
Detergent (e.g., 7× or Decon®)
Soaking baths
Stainless-steel baskets (to collect caps for washing and drying)
Sterility indicators (Browne's tubes, Thermalog indicators)
Glass Petri dishes (for packaging)
Autoclavable plastic film (Portex, Cedanco) or paper sterilization bags
Sterile-indicating autoclave tape
Autoclave, with recording thermometer with flexible probe that can be inserted in load

Protocol
Collection and Washing:
  Metal or phenolic caps with liners:
1. Soak 30 min (maximum) in detergent.
2. Rinse thoroughly for 2 h. (Make sure all caps are submerged.) Liners should be removed and replaced after rinsing, which may be carried out in either of two ways:
   (a) In a beaker (or pail) with running tap water led by a tube to the bottom. Stir the caps by hand every 15 min.
   (b) In a basket or, better, in a pipette-washing attachment. Rinse in an automatic washing machine, but do not use detergent in the machine.
  Polypropylene caps:
These may be washed and rinsed by hand as just described (extending the detergent soak if necessary). Because these caps may float, they must be weighted down during soaking and rinsing. For automatic washers, after soaking in detergent, use pipette-washing attachment and normal cycle without machine detergent.
  Stoppers:
Shrouded caps are preferred to stoppers, but if the latter are required, use silicone or heavy metal-free white rubber stoppers in preference to those made of natural rubber. Wash and sterilize as for caps. (There will be no problem with flotation in washing and rinsing.)

Fig. 10.7. Sterilizing Oven. Pipette cans are stacked with spaces between to allow circulation of hot air. Brown staining on front of oven shows evidence of volatile material from sterile-indicating tape.

Sterilization:
3. Place caps in a glass Petri dish with the open side down.
4. Wrap caps in cartridge paper or steam-permeable nylon film, and seal with autoclave tape (Fig. 10.8).
5. Autoclave for 20 min at 121°C and 100 kPa (1 bar, 15 lb/in²) (Fig. 10.9).

Selection of Detergent
When most culture work was done on glass, the quality (charge, chemical residue) of the glass surface was

Fig. 10.8. Packaging Screw Caps for Sterilization. The caps are enclosed in a glass Petri dish, which is then sealed in an autoclavable nylon bag (Portex).

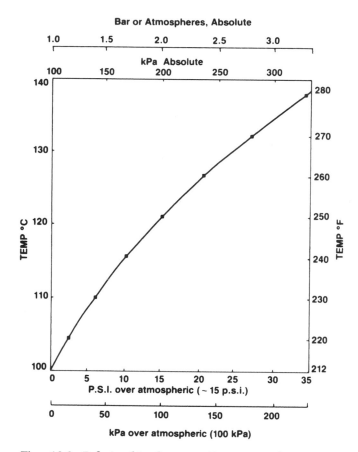

Bar or Atmospheres, Absolute

kPa Absolute

P.S.I. over atmospheric (~ 15 p.s.i.)

kPa over atmospheric (100 kPa)

Fig. 10.9. Relationship between Pressure and Temperature. 121°C and 100 kPa or 1 bar (250°C, 15 lb/in²) for 15–20 min are the conditions usually recommended.

critical. As most cell culture is now carried out on disposable plastic, the major requirements for cleaning glassware are that (a) the detergent be effective in removing residue from the glass and (b) no toxic residue be left behind to leach out into the medium or other reagents. Some tissue culture suppliers will provide a suitable detergent that has been tested with tissue culture (e.g., 7X, ICN, or Decon for manual washing), but often, machine detergents will come from a general laboratory supplier or the supplier of the machine. The washing efficiency of a detergent can be determined by washing heavily soiled glassware (e.g., a bottle of serum or a medium containing serum, that has been autoclaved). One simply uses the normal washup procedure; a visual check will then show which detergents have been effective.

The presence of a toxic residue is best determined by cloning cells (see Protocol 20.9) (e.g., on a glass Petri dish that has been washed in the detergent, rinsed as indicated in Protocol 10.1, and then sterilized by dry heat. A plastic Petri dish should be used as a control. This technique can also be employed if you are anxious about residue left on glassware after autoclaving.

Miscellaneous Equipment

Cleaning. All new apparatus and materials (silicone tubing, filter holders, instruments, etc.) should be soaked in detergent overnight, thoroughly rinsed, and dried. Anything that will corrode in the detergent —mild steel, aluminum, copper, brass, etc.—should be washed directly by hand without soaking (or with soaking for 30 min only, using detergent if necessary) and then rinsed and dried.

Used items should be rinsed in tap water and immersed in detergent immediately after use. Allow them to soak overnight, and then rinse and dry them. Again, do not expose materials that might corrode to detergent for longer than 30 min. Aluminum centrifuge buckets and rotors must never be allowed to soak in detergent.

Particular care must be taken with items treated with silicone grease or silicone fluids. These items must be treated separately and the silicone removed, if necessary, with carbon tetrachloride. Silicones are very difficult to remove if they are allowed to spread to other apparatus, particularly glassware.

Packaging. Ideally, all apparatus used for sterilization should be wrapped in a covering that will allow steam to penetrate, but that will be impermeable to dust, microorganisms, and mites. Proprietary bags, bearing sterile-indicating marks that show up after sterilization, are available from clinical sterile-supply services (e.g., Polysciences, Bio-Medical Products). Semipermeable transparent nylon film (Portex Plastics, Cedanco) is sold in rolls of flat tubes of different diameters and can be made up into bags with sterile-indicating tape. Although expensive, such film can be reused several times before becoming brittle.

Tubes and orifices should be covered with tape and paper or nylon film before packaging, and needles or other sharp points should be shrouded with a glass test tube or other appropriate guard.

Sterilization. The type of sterilization used will depend on the material. (See Table 10.1.) Metallic items are best sterilized by dry heat. Silicone rubber (which should be used in preference to natural rubber), PTFE, polycarbonate, cellulose acetate, and cellulose nitrate filters, etc., should be autoclaved for 20 min at 121°C and 100 kPa (1 bar, 15 lb/in²) with preevacuation and postevacuation steps in the cycle, except when filters are sterilized in a filter assembly. (See Protocol 10.4.) In small bench-top autoclaves and pressure cookers, make sure that the autoclave boils vigorously for 10–15 min before pressurizing to displace all the air. (Take care that enough water is put in at the start to allow for evaporation.) After sterilization, the steam is released and the items are removed to dry off in an oven or rack.

△ *Safety Note.* To avoid burns, take care releasing steam and handling hot items. Wear elbow-length insulated gloves, and keep your face well clear of escaping steam when you open doors, lids, etc. Use safety locks on autoclaves.

Reusable Sterilizing Filters

Although most laboratories now use disposable filter assemblies, some large-scale users may prefer to use stainless-steel filter housings. Filter assemblies should be made up and sterilized in accordance with Protocol 10.4.

PROTOCOL 10.4. STERILIZING FILTER ASSEMBLIES

Materials
Filter housing (Table 10.3.)
Filters to fit holder
Sterile-indicating autoclave tape

Protocol
1. After thorough washing in detergent (see Miscellaneous Equipment in this chapter), rinse assembly in water, followed by deionized water, and dry.
2. Insert support grid in filter and place filter membrane on grid. If membrane is made of polycarbonate, apply wet to counteract static electricity.
3. Place prefilters (glass fiber and others as required; see Sterile Filtration in this chapter) on top of filter.
4. Reassemble filter holder, but do not tighten up

completely. (Leave about one whole turn on bolts.)
5. Cover inlet and outlet of filter with aluminum foil.
6. Pack filter assembly in sterilizing paper or steam-permeable nylon film, and close assembly with sterile-indicating tape.
7. Autoclave at 121°C and 100 kPa (1 bar, 15 lb/in²) with no preevacuation or post-evacuation. (Use "liquids cycle" in automatic autoclaves.)
8. Remove and allow to cool.
9. Do not tighten filter holder completely until the filter is wetted at the beginning of filtration. (See Sterile Filtration in this chapter.)

Alternative Methods of Sterilization. Many plastics cannot be exposed to the temperature required for autoclaving or dry-heat sterilization. To sterilize such items, immerse them in 70% alcohol for 30 min and dry them off under UV light in a laminar-flow hood. Care must be taken with some plastics (e.g., Plexiglas, Perspex, Lucite), as they will depolymerize in alcohol or when they are exposed to UV light. Ethylene oxide may be used to sterilize plastics, but two to three weeks are required to clear it completely from the plastic surface after sterilization. The best method for sterilizing plastics is γ-irradiation, at a level of 2,000–3,000 Gy. Items should be packaged and sealed; polythene may be used and sealed by heat welding.

REAGENTS AND MEDIA

The ultimate objective in preparing reagents and media is to produce them in a pure form (1) to avoid the

TABLE 10.3. Filter Size and Fluid Volume

Filter size or designation	Disposable (D) or reusable (R)	Approximate volume that may be filtered	
		Crystalloid	Colloid
25 mm, Millex	D	1–100 ml	1–20 ml
47 mm or Sterivex cartridge	R, D	0.1–1 l	100–250 ml
90 mm	R	1–10 l	0.2–2 l
Millipak-20	D	2–10 l	200 ml–2 l
Millipak-40	D	10–20 l	2–5 l
Millipak-60	D	20–30 l	5–7 l
Millipak-100	D	30–75 l	7–10 l
Millipak-200	D	75–150 l	10–30 l
Millidisk	D	30–300 l	5–50 l
142 mm	R	10–50 l	1–5 l
293 mm	R	50–500 l	5–20 l

Examples in the table are quoted from Millipore catalogue. Similar products are available from Pall Gelman, Nalge Nunc, Sartorius, and a number of other suppliers. (See Sources of Materials in Trade Index.)

accidental inclusion of inhibitors and substances that are toxic to cell survival and growth and to the expression of specialized functions, (2) to enable the reagent to be totally defined and the functions of its constituents to be fully understood, and (3) to reduce the risk of microbial contamination.

Most reagents and media can be sterilized either by autoclaving (if they are heat stable—e.g., water, salt solutions, amino acid hydrolysates) or by membrane filtration (if they are heat labile). For autoclaving, solutions should be dispensed into borosilicate glass or polycarbonate and kept sealed to avoid evaporation and chemical pollution from the autoclave. If soda-glass bottles are used, they are better left with the caps slack to minimize breakage. The evolution of vapor will help to prevent steam entering from the autoclave, but the level of the liquid will need to be restored with sterile distilled water later. As with sterilizing apparatus, sterile indicators (e.g., Thermalog) and the probe from a recording thermometer should be placed in a mock sample in the center of the load.

Media and reagents supplied on-line to large-scale culture vessels and industrial or semi-industrial bioreactors can be sterilized on-line by short-duration ultrahigh-temperature treatment (Alfa-Laval). Adapting this process to media production might allow increased automation and ultimately reduce costs.

Water

Water used in tissue culture must be of a very high purity, particularly with serum-free media. (See also Water Purification in Chapter 4.) Because water supplies vary greatly, the degree of purification required may vary. Hard water will need a conventional, ion-exchange water softener on the supply line before entering the purification system, but this will not be necessary with soft water.

There are four main approaches to water purification: reverse osmosis, distillation, deionization, and carbon filtration. For ultrapure water (UPW), the first stage is usually reverse osmosis, but it can be replaced by distillation (Fig. 10.10; see also Fig. 4.7). Distillation has the advantage that the water is heat sterilized, but it is more expensive and the boiler needs to be cleaned out regularly. If glass distillation is used for the first stage, the still should be electric and automatically controlled, and the heating elements should be made of borosilicate glass or should be sheathed in silica. Reverse osmosis depends on the integrity of the filtration membrane; hence, the effluent must be monitored. If the costs of both power for distillation and replacement membranes are deducted directly from your budget, reverse osmosis will probably work out cheaper, but if power is supplied free or is costed independently of usage, then distillation will be

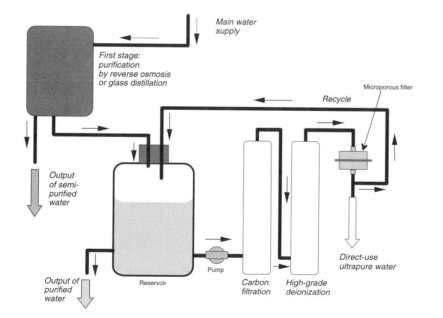

Fig. 10.10. Water Purification. Tap water is fed to a storage container via reverse osmosis or glass distillation. This semipurified water is then recycled back to the storage container via carbon filtration, deionization, and micropore filtration. Reagent-quality water is available at all times from storage; media-quality water is available from the micropore filter supply (at right of diagram). If the apparatus recycles continuously, then water of the highest purity will be collected first thing in the morning for the preparation of medium. (See also Fig. 4.7.)

cheaper. The type of reverse-osmosis cartridge used is determined by the pH of the water supply. (See manufacturer's specification for details).

The second stage is carbon filtration, which will remove both organic and inorganic colloids. The third stage is high-grade mixed-bed deionization to remove ionized inorganic material, and the final stage is micropore filtration to remove any microorganisms acquired from the system and to trap any resin that may have escaped from the deionizer. In the Millipore system, to minimize pollution during storage, the water is collected directly from the final-stage micropore filter without being stored. Water is stored only after the first stage is completed, and this stored water can be used as rinsing water. In some systems, water is recycled continuously from the micropore filter to the store, and if the supply from the first stage is turned off (e.g., overnight), then the stored water gradually "polishes" (i.e., increases in purity). In such a system, water should be used first thing in the morning for preparing media.

The quality of the deionized water should be monitored by resistivity or conductivity (the inverse of resistivity) at regular intervals and the cartridge changed when a decrease in resistance (an increase in conductivity) is observed. The ISO 3696 standard sets resistivity of type I water at a level ≥ 20 MΩcm at 25°C. The total organic carbon (TOC) should be ≤ 10 parts per billion (ppb). Conductivity meters are often supplied with water purification systems. Additionally, Millipore supply a TOC meter. Distillation can precede deionization, providing sterile water to the deionization stage. This arrangement has significant advantages, provided that care is taken to exclude the escape of resin from the deionization cartridge.

In the Millipore system, an ultrafiltration stage can be inserted between deionization and micropore filtration to produce pyrogen-free water. Millipore does not recommend storing water, so a system should be selected to give the rate of supply that you require online.

Water is sterilized by autoclaving at 121°C and 100 kPa (15 lb/in², 1 bar) for 10–15 min. Water should be dispensed in aliquots that are suitable for intended use (e.g., for media preparation from concentrates). The bottles should be sealed during sterilization, which will require borosilicate glass (Pyrex) or polycarbonate (Nalge Nunc) bottles. If bottles are unsealed (e.g., if soda glass is used, it may break if sealed), allow 10% extra volume per bottle to allow for evaporation.

Balanced Salt Solutions

The formulation of BSS has been discussed previously. (See Balanced Salt Solutions in Chapter 8.) The formula for Hanks's BSS [after Paul, 1975] contains magnesium chloride in place of some of the sulfate originally recommended; it should be autoclaved below pH 6.5 to prevent calcium and magnesium phosphates from precipitating and should be neutralized before use. Similarly, Dulbecco's PBS is made up without calcium and magnesium (PBSA), which are made up separately (PBSB) and added just before use if required. PBS is often used without the addition of the Ca^{2+} and Mg^{2+} component, and in that form it should be referred to as PBSA or PBS without Ca^{2+} and Mg^{2+}; the "PBSA" convention will be used throughout this book.

Most balanced salt solutions contain glucose, which, because it may caramelize on autoclaving, is best omitted at first and added later. If glucose is prepared as a 100× concentrate (200 g/l), caramelization during autoclaving is reduced, and it can be used at 5–25 ml/l BSS to give 1–5 g/l.

PROTOCOL 10.5. PREPARATION OF BSS

Outline
Dissolve powder with constant mixing, check pH and conductivity, dispense into aliquots, and autoclave.

Materials
Nonsterile:
BSS powder (see Sources of Materials in Trade Index)
Liquid-holding container to suit pack size of BSS:
  Clear glass or clear plastic aspirator with tap outlet at base
  or
  Erlenmeyer flask or bottle, peristaltic metering pump and tubing
Magnetic stirrer and follower
Bottles for storage, graduated and in sizes most commonly used; borosilicate glass (Pyrex, Schott)
Conductivity meter
pH meter
Autoclave tape or sterile-indicating tabs
Autoclave

Protocol
1. Add appropriate volume of ultrapure water to container
2. Place container on magnetic stirrer and set to around 200 rpm.
3. Open packet of BSS powder and add contents slowly to container while mixing.
4. Stir until powder is completely dissolved.
5. Check pH and conductivity of a sample and enter in record.

(a) pH should not vary more than 0.1 pH unit for a given BSS. (Actual pH will depend on which BSS is being prepared and how it is formulated.)

(b) Conductivity should not vary more than 5% from 150 μScm^{-1}.

(c) Discard sample; do not add back to main stock.

6. Dispense contents of container into graduated bottles.

7. Cap and seal bottles. It is useful to cover the caps with paper secured with an elastic band. The paper can be dated and labeled with the contents, using a stamp; this also avoids having to remove labels from used bottles.

8. Attach a small piece of autoclave tape or sterile-indicating tab and date.

9. Sterilize by autoclaving for 10–15 min at 121°C and 100 kPa (1 bar, 15 lb/in²), in sealed bottles. (see Fig. 10.9.)

10. Store solution at room temperature.

11. Add glucose, $NaHCO_3$, and correct pH to 7.4 as required, just before using solution.

If glucose and bicarbonate are added before sterilization, the solution should be filtered instead of autoclaved. After adding bicarbonate, and gassing with an appropriate CO_2 concentration if required (see CO_2 and Bicarbonate in Chapter 8), adjust the pH of the solution to pH 7.2.

BSS can also be made up as 10× concentrates, to save on storage space; dilute the preparation with UPW for use.

Media

During the preparation of complex solutions, care must be taken to ensure that all of the constituents dissolve and do not get filtered out during sterilization and that they remain in solution after autoclaving or storage. Concentrated media are often prepared at a low pH (between 3.5 and 5.0) to keep all the constituents in solution, but even then, some precipitation may occur. If the constituents are properly resuspended, they will usually redissolve on dilution; but if the precipitate has been formed by degradation of some of the constituents of the medium, then the quality of the medium may be reduced. If a precipitate forms, the performance of the medium should be checked by cell growth and cloning and an appropriate assay of special functions. (See Culture Testing in this chapter.)

Commercial media are supplied as (1) working-strength solutions, with or without sodium bicarbonate and glutamine; (2) 10× concentrates, usually without $NaHCO_3$ and glutamine, which are available as separate concentrates; or (3) powdered media, with or without $NaHCO_3$ and glutamine. Powdered media are the cheapest and not a great deal more expensive than making up medium from your own chemical constituents if you include time for preparation, sterilization, and quality control, the cost of raw materials of high purity, and the cost of overheads such as power and wages. Powdered media are quality controlled by the manufacturer for their growth-promoting properties, but not, of course, for sterility. They are mixed very efficiently by ball milling, so, in theory, a pack may be subdivided for use at different times. However, in practice, it is better to match the size of the pack to the volume that you intend to prepare, because once the pack is opened, the contents may deteriorate and some of the constituents may settle.

Tenfold concentrates cost about twice as much per liter of working-strength medium as powdered media, but save on sterilization costs. Buying media at working strength is the most expensive (about five times the cost of a 10× concentrate), but is the most convenient, as no further preparation is required other than the addition of serum if that is required.

PROTOCOL 10.6. PREPARATION OF MEDIUM FROM 1× STOCK

Outline

Check the formulation; if complete, it may be used directly, after adding serum if that is required. (See Selection of Medium and Serum in Chapter 8; Replacement of Medium in Chapter 12.) If the formulation is incomplete (e.g., lacking glutamine or bicarbonate), add the appropriate stock concentrate(s).

Note. A supplement (e.g., serum or antibiotics) is a constituent that is added to the medium and is not in the original formulation. It needs to be indicated in any publication. Other additions (e.g., glutamine or $NaHCO_3$) are part of the formulation and are not supplements. They need not be indicated in publications unless their concentrations are changed.

Materials

Medium stock
$NaHCO_3$, 7.5% (0.89 M)
Glutamine, 200 mM (will need to be thawed)
Serum: calf, fetal calf, etc. (will need to be thawed)
Antibiotics (not recommended for routine use):
  Penicillin, 1×10^4 U/ml
  Streptomycin, 10 mg/ml

Protocol (see also Sterile Handling in Chapter 5)

1. Check formulation of medium, and determine what additions are required.
2. Take medium to hood with any other supplement or addition that is required.
3. Unwrap bottles if polythene wrapped, and swab with 70% EtOH.
4. Uncap bottles.
5. Transfer the appropriate volume of each addition to the stock bottle to make the correct dilution; for 100 ml, use the following ingredients and amounts:

Glutamine, 200 mM	1 ml
NaHCO₃, 7.5%	2.9 ml (for 5% CO_2)
Antibiotics (if used)	0.5 ml
Serum	10 ml (for 10%)

   (a) Use a different pipette for each addition.
   (b) Move each new stock to the opposite side of the hood after it has been added, so that you will know that it has been used.
   (c) Remove all additives or supplements from the hood when the medium is complete.
6. If elevated CO_2 is used, the gas mixture may be bubbled through the medium before the serum is added. Otherwise, leave sufficient headspace to gas the air space only. Do not bubble gas through any medium containing serum, as the medium will froth and bubble out through the neck, risking contamination.
7. Recap bottles.
8. Alter labeling to record additions, date and initial.
9. Return medium to 4°C or use directly.
10. If using a new medium for the first time, pipette an aliquot into a flask or Petri dish, and incubate for at least 1 h under your standard conditions, to ensure that pH equilibrates at correct value. If it does not, readjust pH of medium and repeat, or else alter CO_2 concentration.

Note. Changing the CO_2 concentration of the incubator will affect all other culture media in the same incubator. Changing the CO_2 concentration should be regarded as a one-time adjustment and should not be used for batch-to-batch variations, which are better controlled by adding sterile acid or alkali.

PROTOCOL 10.7. PREPARATION OF MEDIUM FROM 10× CONCENTRATE

Outline

Sterilize aliquots of deionized distilled water of such a size that one aliquot, when made up to full-strength medium, will last from one to three weeks. Add concentrated medium and other constituents, adjust the pH, and use the solution or return it to the refrigerator.

Materials (for 1 litre of medium)

Premeasured aliquot(s) of sterile UPW, 885.5 ml for low NaHCO₃, 861 ml for high NaHCO₃. (Remember to allow space for all additions when choosing the container and the volume of H_2O to be sterilized.)
Medium concentrate, 10×
Glutamine, 200 mM (will need to be thawed)
NaHCO₃, 7.5% (0.89 M)
Serum: calf, fetal calf, etc. (will need to be thawed)
Antibiotics if required (not recommended for routine use):
  Penicillin, 1×10^4 U/ml
  Streptomycin, 10 mg/ml

Protocol

1. Add constituents as follows:
   (a) For sealed culture flask with gas phase of air, low HCO_3^- concentration, atmospheric CO_2 concentration, and low buffering capacity:

Constituent	Concentration of stock	Volume added	Final concentration
Medium concentrate	10×	100	1×
Glutamine	200 mM	10	2 mM
NaHCO₃	7.5%	4.5	4 mM
Water already present		885.5	
Final volume		1,000	

   (b) For sealed culture flask with gas-phase of air, atmospheric CO_2, and low HCO_3^- concentration, and with HEPES for high buffering capacity (may be vented to atmosphere by slackening cap for some cell lines at a high cell density if a lot of acid is produced):

Constituent	Concentration of stock	Volume added	Final concentration
Medium concentrate	10×	100	1×
Glutamine	200 mM	10	2 mM
NaHCO₃	7.5%	4.5	4 mM

Constituent	Concentration of stock	Volume added	Final concentration
HEPES	1.0 M	20	20 mM
Water already present		885.5	
Final volume		1,020*	

*Although this amount is higher than the correct dilution volume, it helps to compensate for the additional osmolality of the HEPES.

(c) For cultures in open vessels in a CO_2 incubator or under CO_2 in sealed flasks:

(i) with 5% CO_2 and high bicarbonate:

Constituent	Concentration of stock	Volume added	Final concentration
Medium concentrate based on Earle's salts or equivalent for high bicarbonate	10×	100	1×
Glutamine	200 mM	10	2 mM
NaHCO$_3$	7.5%	29	26 mM
Water already present		861	
Final volume		1,000	

(ii) with 2% CO_2, intermediate bicarbonate concentration, and HEPES (optional):

Constituent	Concentration of stock	Volume added	Final concentration
Medium concentrate based on Earle's salts or equivalent for high bicarbonate	10×	100	1×
Glutamine	200 mM	10	2 mM
HEPES	1.0 M	20	20 mM
NaHCO$_3$	7.5%	9	8 mM
Water already present		861	
Final volume		1,000	

2. Add 1 N NaOH to give pH 7.2 at 20°C. (See below.) When incubated, the medium will rise to pH 7.4 at 37°C, but this figure may need to be checked by a trial titration the first time the recipe is used. (see Protocol 10.6.)

3. Add any other constituents. If these are in water, remove an equivalent volume from the amount of water already present in the bottle before adding the constituents. If the constituents are in isotonic salt, they should be added to the final volume of medium. Because it is isotonic, serum should be added to the final volume, although doing so will dilute the nutrients from the medium.

Always equilibrate and check the pH at 37°C, as the solubility of CO_2 decreases with increased temperature and the pK_a of the HEPES will change.

The amount of alkali needed to neutralize 10× concentrated medium (which is made up in acid to maintain the solubility of the constituents) may vary from batch to batch and from one medium to another, and in practice, titrating the medium to pH 7.4 at 37°C can sometimes be a little difficult. When making up a new medium for the first time, add the stipulated amount of NaHCO$_3$ and allow samples with varying amounts of alkali to equilibrate overnight at 37°C in the appropriate gas phase. Check the pH the following morning, select the correct amount of alkali, and prepare the rest of the medium accordingly.

The bicarbonate concentration is important in establishing a stable equilibrium with atmospheric CO_2, but regardless of the amount of bicarbonate used, if the medium is at pH 7.4 and 37°C, the bicarbonate concentration at each concentration of CO_2 will be as in Table 8.1. (See CO_2 and Bicarbonate in Chapter 8). Some media are designed for use with a high bicarbonate concentration and elevated CO_2 in the atmosphere (e.g., Eagle's MEM with Earle's salts), while others have a low bicarbonate concentration for use with a gas phase of air (e.g., Eagle's MEM with Hanks's salts; see Tables 8.1 and 8.2). If a medium is changed and its bicarbonate concentration altered, it is important to make sure that the osmolality is still within an acceptable range. The osmolality should always be checked (see Osmolality in Chapter 8) when any significant alterations are made to a medium that are not in the original formulation.

If your consumption of medium is fairly high (>200 l/year) and you are buying the medium ready made, then it may be better to get extra constituents included in the formulation, as this practice will work out to be cheaper. HEPES in particular is very expensive to buy separately. Glutamine is often supplied separately, as it is unstable; it is best to buy it separately and store it frozen. The half-life of glutamine in medium at 4°C is about three weeks and at 37°C about 1 wk. Some dipeptides of glutamine have increased stability, while retaining the bioavailability of the glutamine. One such is Glutamax, available from Gibco.

Care should be taken with 10× concentrates to ensure that all of the constituents are in solution, or at least evenly suspended, before dilution. Some constituents (e.g., folic acid or tyrosine) can precipitate and be missed at dilution. Incubation at 37°C for several hours may overcome this problem.

Once a batch of medium has been tested for its growth-promoting and other properties, if it is found to be satisfactory, then it need not be tested each time it is made up to working strength. The sterility of a medium made by diluting a 10× concentrate should be checked, however, by incubating an aliquot of the complete medium at 37°C for 48 h before use.

Powdered Media

Instructions for the preparation of powdered media are supplied with each pack. Choose a size that you can make up all at once and use the pack within three months. Select a formulation lacking glutamine. If other unstable constituents are present, they also should be omitted and added later as a sterile concentrate just before use.

PROTOCOL 10.8. PREPARATION OF MEDIUM FROM POWDER

Outline

Dissolve the entire contents of the pack in the correct volume of high-purity water, using a magnetic stirrer and adding the powder gradually with constant mixing. When all the constituents have dissolved completely, the medium should be filtered immediately and not allowed to stand, in case any of the constituents precipitate or microbial contamination progresses. The pH is adjusted better when the final constituents (e.g., glutamine, $NaHCO_3$, or serum) have been added to the medium.

Materials

Sterile:
Graduated bottles for medium
Caps for bottles
Universal containers for contamination control
  sampling
Sterile filtration assembly (see Sterile Filtration in
  this chapter)
Nonsterile:
Powder medium. (See Sources of Materials in Trade
  Index.)

Liquid container to suit size of pack (e.g., Erlenmeyer flask or bottle)
Peristaltic metering pump
Tubing for peristaltic pump
Magnetic stirrer and follower
Conductivity meter

Protocol

1. Add appropriate volume of ultrapure water to container
2. Place container on magnetic stirrer and set to around 200 rpm.
3. Open packet of powder and add contents slowly to container while mixing.
4. Stir until powder is completely dissolved.
5. Check pH and conductivity of a sample and enter in record:
   (a) pH should be within 0.1 unit of expected level for particular medium
   (b) Conductivity should be within 2% of expected value for particular medium
   (c) Discard sample; do not add back to main stock.
6. Connect container with medium to filter via tubing and peristaltic pump. (See Sterile Filtration in this chapter and Fig. 10.11.)
7. Turn on pump and dispense medium into graduated bottles.
8. Collect samples from beginning, middle, and end of run to test for sterility. (See Sterility Testing in this chapter.)
9. Cap and seal bottles.
10. Store at 4°C.
11. Add serum, and correct pH to 7.4 as required, just before use.

For people using smaller amounts (<1.0 l/week) or several different types of medium, smaller volumes may be prepared, complete with glutamine, and filtered directly into storage bottles using a bottle-top filter sterilizer or filter flasks (Fig. 10.12c,d). With this and other negative-pressure filtration systems, some dissolved CO_2 may be lost during filtration, and the pH may rise. Provided that the correct amount of $NaHCO_3$ is in the medium to suit the gas phase (see CO_2 and Bicarbonate in Chapter 8), the medium will reequilibrate in the incubator, but this should be confirmed the first time the medium is used.

For large-scale requirements (>10 l/week), medium can be prepared in a pressure vessel, checked at intervals with a large pipette to determine whether solution is complete, and sterilized by positive pressure through an in-line disposable or reusable filter into a receiver vessel. (See Sterile Filtration in this chapter and Figs. 10.13 and 10.14.)

Fig. 10.11. Peristaltic Pump Filtration. Sterile filtration with peristaltic pump between nonsterile reservoir and sterilizing filter (Millipak, Millipore; courtesy of Millipore [U.K.], Ltd.)

Customized Medium

If one intends to explore different formulations, or if the medium is to be made up in-house from individual constituents, it is convenient to make up a number of concentrated stocks—amino acids at 50× or 100×, vitamins at 1,000×, and tyrosine and tryptophan at 50× in 0.1 N HCl, glucose at 200 g/l, and single-strength BSS. The requisite amount of each concentrate is then mixed, filtered through a sterilizing filter of 0.2 μm porosity, and diluted with the BSS. (See Sterile Filtration in this chapter, Figs. 10.11 and Fig. 10.12, and Table 8.2.)

PROTOCOL 10.9. PREPARATION OF CUSTOMIZED MEDIUM

Outline

Prepare stock concentrates (derived from Tables 8.3 and 9.1) and store frozen. Thaw and blend as required. Sterilize by filtration and store until required.

Materials

Amino acid concentrate, 100× in water, stored frozen
Tyrosine and tryptophan, 50× in 0.1 N HCl
Vitamins, 1,000× in water

Glucose, 100× (200 g/l in BSS)
Additional solutions (e.g., trace elements and nucleosides; not lipids, hormones, or growth factors, which should be added just before using the medium)
Storage bottles, selected for optimum aliquot size for subsequent dilution (see Step 5 of protocol)

Protocol

1. Thaw solutions and ensure that all solutes have redissolved.
2. Blend constituents in correct proportions:
   (a) amino acid concentrate   100 ml
   (b) tyrosine and tryptophan   200 ml
   (c) vitamins   10 ml
   (d) glucose   100 ml
3. Mix and sterilize solution by filtration. (See later.)
4. Store frozen.
5. For use, dilute 41 ml of concentrate mixture with 959 ml sterile 1× BSS.
6. Adjust to pH 7.4.
7. Store at 4°C.
8. Add serum or other supplements, such as hormones, growth factors, and lipids, just before using medium. If metals are used as trace elements, it is also better to add these just before using the

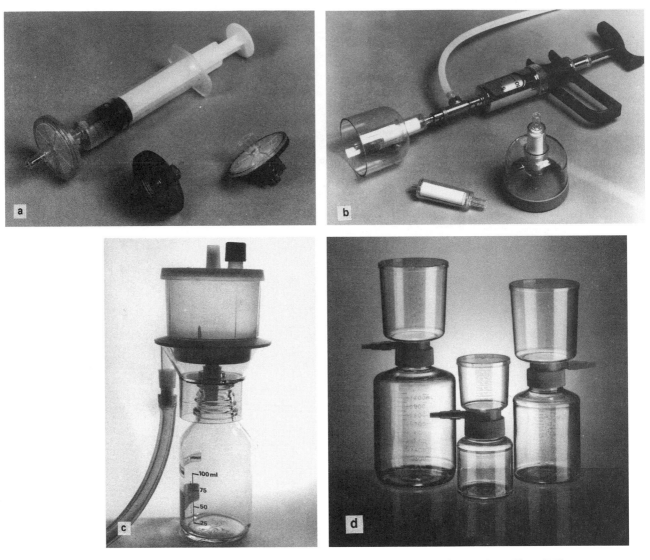

Fig. 10.12. Sterilizing Filters. Disposable filters. a. Millex 25-mm disc syringe filter (Millipore). b. Sterivex in use with repeating syringe. c. Bottle-top fitting (Becton-Dickinson). d. Filter cup and storage vessels (Stericup, Millipore). (a,b) positive pressure; (c,d) negative pressure (see also Figure 10.11). Photographs a, b, and d courtesy of Millipore (UK), Ltd.

medium, as they can precipitate in the presence of phosphate in concentrated stocks.

The advantage of this type of recipe is that it can be varied; extra nutrients (keto acids, nucleosides, minerals, etc.) can be added or the major stock solutions altered to suit requirements, but, in practice, this procedure is so laborious and time consuming that few laboratories make up their own media from basic constituents, unless they wish to alter individual constituents regularly. The reliability of commercial media depends entirely on the application of appropriate quality-control measures. Any laboratory carrying out its own preparation must make sure that appropriate quality-control measures are employed. (See Quality Control in this chapter.)

There are now several reputable suppliers of standard formulations (see Trade Index), many of whom will supply specialized, serum-free formulations and media prepared to your own formulation. It is important to ensure that the quality control these suppliers employ is relevant to the medium and the cells you wish to propagate. You might buy MCDB 153, which is tested on HeLa cell colony formation, but this test is of little relevance if you wish to grow primary keratinocytes.

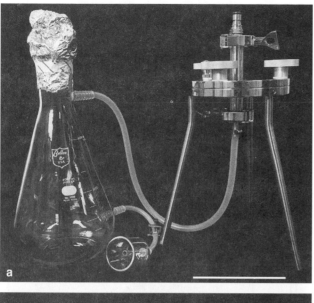

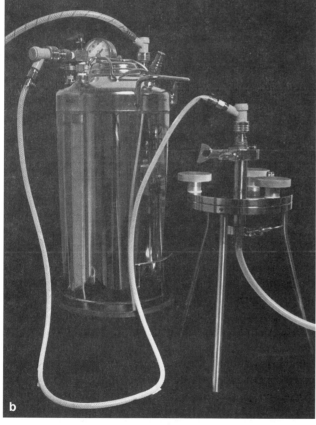

Fig. 10.13. In-Line Assembly. In-line filter assembly connected to receiver flask (a) and to pressurized reservoir (b). Only those items in (a) need be sterilized. Normally, the glass bell would be covered in protective foil; it is left off here for purposes of illustration.

Sterilization

Autoclavable Media. Some commercial suppliers offer autoclavable versions of Eagle's MEM and other media. Autoclaving is much less labor intensive, is less expensive, and has a much lower failure rate than filtration. The procedure to follow is supplied in the manufacturer's instructions and is similar to that described earlier for BSS. The medium is buffered to pH 4.25 with succinate, in order to stabilize the B vitamins during autoclaving, and is subsequently neutralized. Glutamine is replaced by glutamate or is added sterile after autoclaving.

Sterile Filtration

Filtration through 0.1–0.2-μm microporous filters is the method of choice for sterilizing heat-labile solutions (Fig. 10.15). Numerous kinds of filters are available, made from many different materials, including polyethersulphone (PES), nylon, polycarbonate, cellulose acetate, cellulose nitrate, PTFE, and ceramics, and in sizes from syringe-fitting filters (Figs. 10.12a, 10.16a) to multidisk cartridge filters (Fig. 10.16b). Low protein-binding filters are available from most suppliers (e.g., Durapore, Millipore); PES filters are generally found to be faster flowing. Polycarbonate filter membranes are absolute filters with an array of holes of a uniform porosity; the number of holes per unit area increases as the size diminishes, to maintain a uniform flow rate. Most other filters are of the mesh variety and filter by entrapment; they generally have a faster flow rate and reduced clogging, but will compress at high pressures.

Disposable filter holder designs include simple disc filters, cartridges, and hollow-fiber designs. (See Figs. 10.11 and 10.12.) Reusable filters are made up of membranous material (Fig. 10.16a) or are cartridge filters (Fig. 10.16b) housed in reusable steel holders (Figs. 10.16b and 10.13) and sterilized by autoclaving, as described in Protocol 10.4. They are usually connected to a pressure reservoir (Figs. 10.13b and 10.14) or a peristaltic pump and operate under positive pressure.

Disposable Filters. There is a wide range of disposable filters of different capacities and configurations. Syringe-tip filters are generally used for low volume filtration (for 2–20 ml) and vary in size from 13–50 mm diameter (see Fig. 10.12a). Intermediate sized filters (for 50–500 ml) can be used in-line with a peristaltic pump (see Fig. 10.11), or as bottle-top filters used with a vacuum line and a regular medium bottle (Fig. 10.12c). Intermediate sized filters can also be purchased as complete filter units for attaching to a vacuum line, with an upper chamber for the non-sterile

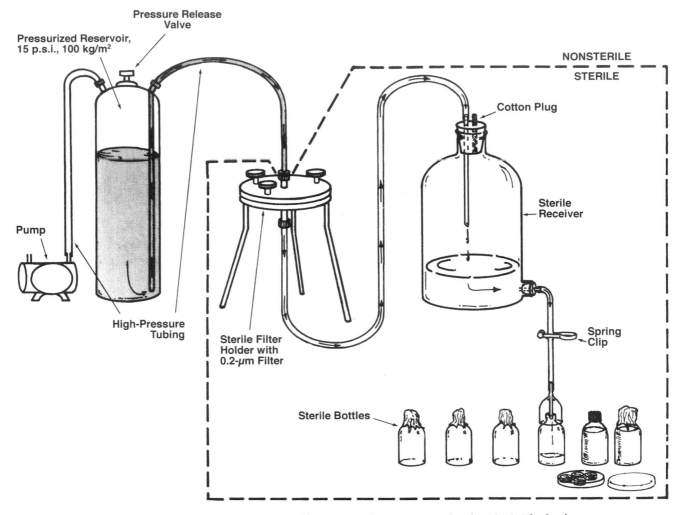

Fig. 10.14. Diagram of In-line Assembly. Setup using apparatus in Fig. 10.13. The broken line encloses those items that are sterilized. A disposable filter unit of equivalent surface area may be substituted for the metal, reusable filter holder and stand illustrated here. (See also Fig. 10.17.)

solution and a lower chamber to receive the sterile liquid and to use for storage (Fig. 10.12d). Large capacity cartridge filters (for 20–500 l) are usually operated in line under positive pressure from a reservoir (Fig. 10.17). Although disposable filters are more expensive than reusable, they are less time-consuming to use and give fewer failures.

Reusable Filters. A similar range of sizes is available as for disposable filters, but the larger sizes tend to be more commonly used, with a setup as shown in Figs. 10.13 and 10.14. When using a reusable filter holder, see Protocol 10.4 for preparation and sterilization.

The following four protocols feature disposable filters in the small sizes and a reusable filter for larger volumes, but equipment of both types is available.

PROTOCOL 10.10. STERILE FILTRATION WITH SYRINGE-TIP FILTER

Materials
Sterile:
Plastic syringe, 10–50 ml capacity
Syringe-tip filter (e.g., Pall Gelman Acrodisk or Millipore Millex)
Receiver vessel (e.g., a universal container)
Nonsterile:
Solution for sterilization (5–100 ml)

Protocol
1. Swab down hood and assemble materials.
2. Fill syringe with solution to be sterilized.
3. Uncap receiver vessel.

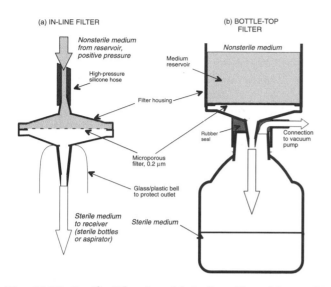

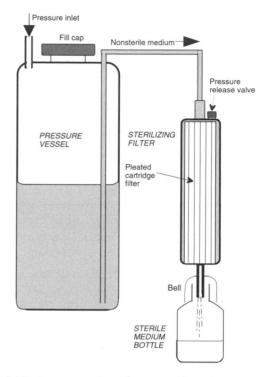

Fig. 10.15. Sterile Filtration. (a) In-line filter. Nonsterile medium from pump or pressure vessel. (b) Bottle-top filter or filter flask. (Designs are similar.) Medium added to upper chamber and collected in lower. Lower chamber can be used for storage.

Fig. 10.17. Large-capacity Filtration. Pleated cartridge filter for large-scale filtration. Several filters of decreasing porosities (e.g., 5 μm, 1 μm, 0.45 μm) can be used in series for filtering colloidal solutions such as serum.

4. Unpack filter and attach to tip of syringe, holding the sterile filter within the bottom half of the packaging while attaching to syringe.
5. Expel solution through filter into receiver vessel. Only moderate pressure is required.
6. Cap receiver vessel.
7. Discard syringe and filter.

The syringe may be refilled several times by returning the filter to the lower half of the sterile packaging, detaching it from the syringe, refilling the syringe, and reattaching the filter. If the back-pressure increases, take a new filter.

PROTOCOL 10.11. STERILE FILTRATION WITH FILTER FLASK

Materials
Sterile:
Filter flask (e.g., Corning 500 ml Vacuum Filter System or Nalgene Filter Unit)
Cap for lower chamber (if chamber is used for storage)
Sample tube or universal container for sterility test (if required)

Fig. 10.16. Reusable Filters. (a) 47-mm in-line polypropylene. (b) Millidisk range; stainless-steel housings, high-capacity cartridge-type filters. (Courtesy of Millipore, U.K., Ltd.)

Nonsterile:
Medium for sterilization (500 ml)
Vacuum pump or vacuum line

Protocol

1. Connect side arm of filter flask to vacuum pump.
2. Remove cap from bottle and lid from top chamber of filter flask.
3. Pour nonsterile medium into top chamber.
4. Switch on pump.
5. Unpack cap for lower chamber, ready for use.
6. When liquid has all been drawn into lower chamber, switch off pump, detach filter housing and top chamber, and cap lower chamber.
7. Label lower chamber, date, and initial.
8. Store at 4°C until required.

Larger and smaller filter flasks are available. If a larger volume is to be filtered and dispensed into aliquots, it may be better to use one of several bottle-top filters, which may be set up to filter material directly into standard medium bottles.

PROTOCOL 10.12. STERILE FILTRATION WITH SMALL IN-LINE FILTER

Materials

Sterile:
In-line filter with bell (see Fig. 10.11)
Graduated bottles and caps for medium
Sample tube or universal container for sterility test
Nonsterile:
Medium for sterilization (500–5,000 ml)
Peristaltic pump and tubing
Clamp stand

Protocol

1. Unpack filter and connect to outlet from peristaltic pump.
2. Clamp filter in clamp stand at a suitable height such that bottles for receiving medium can be positioned below the filter with the neck shrouded and removed easily when ready to be filled. Use one of the sterile medium bottles to set up the filter if necessary, but no not use this bottle, ultimately, for sterile collection.
3. Place inlet tubing to pump in nonsterile medium.
4. Place first medium bottle under filter bell (removing bottle used for setup).
5. Switch on pump.
6. Fill bottle to required volume.
7. Switch off pump.

8. Remove bottle, cap it, number it, and replace with fresh bottle.
9. Restart pump.
10. Repeat steps 6–10 as required, until entire batch of medium is filtered.
11. If several bottles are being prepared, remove first, middle, and last bottle for sterility testing. (See Quality Control in this chapter.) If only a few bottles are being prepared, collect small (e.g., 10 ml) intermediate samples at intervals and set aside for testing.
12. Replace the aluminum foil over the cap and neck of each bottle to keep it free from dust during storage.
13. Label, date, and initial bottles.

Store medium at 4°C until required; do not release for use until sterility test is complete. If any contamination is found in the test samples, refilter the whole batch, or discard it.

PROTOCOL 10.13. STERILE FILTRATION WITH LARGE IN-LINE FILTER

Materials (see Figs. 10.13, 10.14, and 10.17)

Sterile:
Filter (e.g., 90-mm membrane and reusable filter holder (see Table 10.3) or disposable 90-mm disk (e.g., Millipore Millex) or cartridge (e.g., Pall Gelman Capsule).
Receiver for filtrate with outlet at the base
Tubing from filter to sterile receiver
Glass bell
Bottles, foil capped (retain foil for covering cap and neck to keep bottle dust free during subsequent storage)
Caps
Nonsterile:
Pressure vessel, 5–50 l.
Pump, 100 kPa (1 bar, 15 lb/in²)
Clamps to secure filter and outlet bell
Tubing from pressure vessel to filter
Spring clip

Protocol

1. Secure the filter holder in position.
2. Connect the outlet to the receiver vessel.
3. Secure the outlet from the receiver at a suitable height such that medium bottles can be positioned below it with the neck shrouded and removed easily when the bottle filled. Use one of the sterile medium bottles to set up if necessary, but do not use this bottle, ultimately, for sterile collection.

4. Connect the pressure vessel to the filter inlet.
5. Decant the medium into the pressure vessel.
6. For reusable filters, turn on the pump just long enough to wet the filter. Stop the pump and tighten up the filter holder.
7. Switch on the pump to deliver 100 kPa (15 lb/in^2). When the receiver starts to fill, draw off aliquots into medium stock bottles of the desired volume.
8. Cap the bottles as each one is taken from the filter bell.
9. Store medium at 4°C. Do not release for general use until quality control has been performed.

Positive pressure is recommended for optimum performance of the filter and to avoid the removal of CO_2, which results from negative-pressure filtration. Positive pressure may also be applied using a peristaltic pump (Fig. 10.11) in line between the nonsterile reservoir and a disposable in-line filter, such as the Millipak (Millipore), which may be used instead of the reusable assembly. Since the disposable filter is bought sterile and no receiver is necessary, the only preparation and sterilization required is for the media bottles.

Serum

Preparing serum is one of the more difficult procedures in tissue culture, because of variations in the quality and consistency of the raw materials and because of the difficulties encountered in sterile filtration. Moreover, serum is also one of the most costly constituents of tissue culture, accounting for 20–30% of the total budget if it is bought from a commercial supplier. Buying sterile serum is certainly the best approach from the point of view of consistency and quality control, but the next protocol is suggested if the serum has to be prepared in the laboratory.

PROTOCOL 10.14. COLLECTION AND STERILIZATION OF SERUM

Outline
Collect blood, allow it to clot, and separate the serum. Filter serum, through filters of gradually reducing porosity. Bottle and freeze filtered serum.

Collection
Arrangements may be made to collect whole blood from a slaughterhouse. The blood should be collected directly from the bleeding carcass and not allowed to lie around after collection. Alternatively, blood may be withdrawn from live animals under proper veterinary supervision. The latter alternative, if performed consistently on the same group of an-imals, gives a more reproducible serum, but a lower volume for a greater expenditure of effort. If the procedure is done carefully, blood may be collected aseptically.

Clotting
Allow the blood to clot by having it stand overnight in a covered container at 4°C. This so-called natural-clot serum is superior to serum that is physically separated from the blood cells by centrifugation and defibrination, as platelets release growth factor into the serum during clotting. Separate the serum from the clot, and centrifuge the serum at 2,000 g for 1 h to remove sediment.

Sterilization
Serum is usually sterilized by filtration through a sterilizing filter of 0.1 μm porosity, but because of its viscosity and high particulate content, the serum should be passed through a graded series of fiberglass or other prefilters before passing through the final sterilizing filter. Only the last filter, a 142–350-mm in-line disc filter or equivalent disposable filter (e.g., Millipak 200), need be sterile. The prefilter assemblies may be stainless steel with replaceable cartridges, disk filter units, or a single bonded unit (Pall Gelman). The last is easiest to use, but more difficult to clean and reuse.

For Sterilizing Volumes of 5–20 l:

Materials
Sterile:
Sterilizing filter: a 200-mm, 0.1-μm-porosity filter (e.g., Millipak, Millipore). A porosity of 0.2 μm is sufficient for antibacterial and fungal sterilization, but 0.1 μm is required to remove mycoplasma.
Sterile receiving vessel with outlet at base
Sterile bottles with caps and foil
Nonsterile:
Peristaltic pump and tubing
Clamp
Nonsterile prefilters:
  Fiberglass disposable filter or 142- mm reusable filter (e.g., Pall Gelman)
  5-μm-porosity disposable or reusable filter (e.g., Pall Gelman Versapore)
  1.2-μm porosity disposable or reusable filter (e.g., Millipore Opticap)
  0.45-μm porosity disposable or reusable filter (e.g., Millipore Millipak)

Note. The preceding are examples only; contact your supplier and request a series of filters and prefilters that will suit your serum requirements. If it is a once-only activity, choose disposable filters; if collection will be repeated regularly, it will be more economical to employ reusable filter holders for the prefilter stages, while still using a disposable filter for sterilization, for added security.

Protocol

1. Insert appropriate nonsterile filters into nonsterile prefilter holders (if reusable holders are being used).
2. Connect one or more prefilters in line and upstream from a sterile disposable or reusable filter holder (Fig. 10.18) that contains a 0.1-μm-porosity filter and is connected to a sterile receiver via the peristaltic pump.
3. Place the intake of the pump into a serum container.
4. Switch on the pump, and check any reusable filter holders for leakage as they are wetted; switch off the pump and tighten filter holders as necessary.
5. Restart the pump and continue filtering, checking for leaks or blockages. Increasing the flow rate will increase the rate of filtration, but may cause the filters to become packed or clogged.
6. Collect aliquots in sterile bottles, leaving at least 20% headspace to allow for adiabatic expansion on freezing.
7. Collect samples at the beginning, middle, and end of the run to check sterility.
8. Cap and number the bottles.
9. Replace the foil over the cap and neck for storage.
10. Store serum at $-20°C$ until quality control is completed. (See Quality Control in this chapter.)

Small-Scale Serum Processing. If small amounts (<1 liter) of serum are required, then the process is similar to Protocol 10.14, but can be scaled down. After clot retraction (see Ptotocol 10.14 clotting), small volumes of serum may be centrifuged (5–10,000 g) and then filtered through a series of disposable filters (e.g., 50 mm Millipore Millex or Pall Gelman Acrodisc) and, finally, through a 50-mm, 0.1-μm-porosity sterile disposable filter.

Centrifuge very small volumes (10–20 ml) at 10,000 g, and filter the serum directly through a graded series of syringe-tip disposable 25-mm filters (e.g., Acrodisc, Pall Gelman; Millex, Millipore), finishing with a 0.1-μm sterilizing filter (e.g., Millex).

Storage. Bottle the serum in sizes that will be used up within two to three weeks after thawing. Freeze the serum as rapidly as possible, and if it is thawed, do

not refreeze it unless further prolonged storage is required.

Serum is best used within 6–12 months of preparation if it is stored at $-20°C$, but more prolonged storage may be possible at $-70°C$. Usually, however, the bulk of serum stocks makes this impractical. Polycarbonate or high-density polypropylene bottles will eliminate the risk of breakage if storage at $-70°C$ is desired. Regardless of the temperature of the freezer or the nature of the bottles, do not fill them completely; allow for the expansion of water during freezing.

Human Serum. Pooled outdated human blood or plasma from a blood bank can be used instead of or in addition to bovine or equine blood. It should be sterile and not require filtration. Titrate out the heparin or citrate anticoagulant with Ca^{2+}, allow the blood to clot overnight, and then separate and freeze the serum.

△ ***Safety Note.*** Care must be taken with human donor serum to ensure that it is screened for hepatitis, HIV, tuberculosis, and other adventitious infections.

Quality Control. Use same procedures as those for a medium. (See later.)

A major problem that is emerging with the use of serum is the possibility of viral infection. When the possibility of bacterial infection was first appreciated, it was relatively easy to devise filtration procedures to filter out anything above 1.0 μm, and a porosity of 0.45 μm became standard. Subsequently, it was learned that mycoplasma would pass through filters as low as 0.2 μm, and commercial suppliers of serum lowered the exclusion limits of their filters to 0.1 μm. This reduction in size appears to have virtually eliminated mycoplasma from serum batches used in culture, but the problem of viral contamination remains. Filtering out virus would seem to be a much more significant task, but some companies (e.g., Pall Gelman) claim that it may be possible.

Dialysis. For certian studies, the presence of constituents of low molecular weight (amino acids, glucose, nucleosides, etc.) may be undesirable. These constituents may be removed by dialysis through conventional dialysis tubing.

PROTOCOL 10.15. SERUM DIALYSIS

Materials

Sterile:
Bottles and caps
Sterilizing filter, 0.1 μm

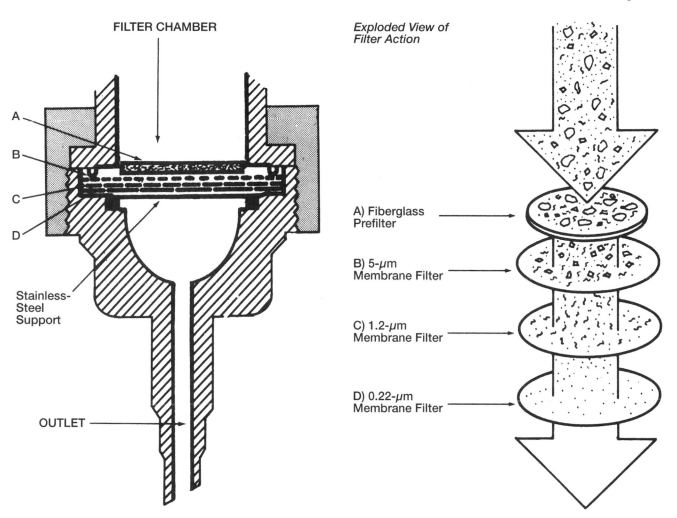

Fig. 10.18. Prefilter. Prefiltration for filtering colloidal solutions (e.g., serum) or solutions with high particulate content. Several prefilters can be connected in series, and only the final filter need be sterile. (See also Fig. 10.17.)

Prefilters: fiberglass, in sizes 5.0, 1.2, 0.45, and 0.22 μm (see Protocol 10.14)

Nonsterile:

Dialysis tubing
Beaker with ultrapure water
Bunsen burner
Tripod with wire gauze
Serum to be dialyzed
HBSS at 4°C
Measuring cylinder

Protocol

1. Boil five pieces of 30-mm × 500-mm dialysis tubing, in three changes of distilled water.
2. Transfer tubing to Hanks's balanced salt solution (HBSS), and allow to cool.
3. Tie double knots at one end of each tube.
4. Half-fill each dialysis tube with serum (20 ml).
5. Express air and knot the open end of the tube, leaving a space of about half the tube between the serum and the knot.
6. Place the tubing in 5 l of HBSS, and stir on a magnetic stirrer overnight at 4°C.
7. Change HBSS and repeat step 6 twice.
8. Collect serum into a measuring cylinder and note the volume collected. (If the volume is less than the starting volume of the serum, add HBSS to return to the starting volume. If the volume is greater than the starting volume of the serum, make due allowance when adding to the medium later.)
9. Sterilize serum through a graded series of filters. (See Protocol 10.14.)
10. Bottle and freeze the serum.

Preparation and Sterilization of Other Reagents

Individual recipes and procedures are given in the Reagent Appendix. On the whole, most reagents are sterilized by filtration if they are heat labile and by autoclaving if they are heat stable. (See Table 10.2.) Filters with low binding properties (e.g., Millex-GV) are available for sterilizing of proteins and peptides.

CONTROL, TESTING, AND STORAGE OF MEDIA

Quality Control

A medium that is prepared in the laboratory needs to be tested before use. If the medium is purchased ready made as a 1× working-strength solution, then it should be possible to rely on the quality control carried out by the supplier, other than any special requirements that you have of the medium. Likewise, if a 10× concentrate is used, the growth and sterility testing will have been done, and the only variable will be the water used for dilution. Provided that the conductivity and level of total organic carbon fall within specifications (see Water in this chapter) and no major changes have been made in the supply of water, most laboratories will accept this compliance as adequate quality control.

However, if medium is prepared from powder, it will have been sterilized in the laboratory, and quality control will be required to confirm sterility, although, given that all the constituents have dissolved, you may be prepared to accept the quality control of the supplier regarding the medium's growth-promoting activity. Media made up from basic constituents will require complete quality control, involving both sterility testing and culture testing.

Sterility Testing

Bubble Point. When positive pressure filtration is complete and all the liquid has passed through the filter, raise the pump pressure until bubbles form in the effluent from the filter. This is the *bubble point* and should occur at more than twice the pressure used for filtration. (See manufacturer's instructions.) If the filter bubbles at the sterilizing pressure (100 kPa, 15 psi) or lower, then it is perforated and should be discarded. In that case, any filtrate that has been collected should be regarded as nonsterile and refiltered. Single-use, disposable filters rarely fail the bubble point test, which is very quick and easy to perform. Reusable filters can fail, so they should be checked after every filtration run.

Incubation. Collect samples at the beginning, middle, and end of the run. If the bottle is small, this can be done by removing bottles. If the bottle is large, rather than waste medium, collect samples into smaller containers during the run. Remember, however, if you are bottling in 1,000-ml sizes, taking 1-ml samples will reduce the sensitivity by a factor of 10^3; if you are not prepared to sacrifice whole bottles at this size, then you should at least sample 100 ml. It is best not to withdraw samples from individual bottles, as this will both increase the risk of contaminating the bottles and reduce the volume in the bottle used for sampling, thereby altering the dilutions of subsequent additions to that bottle.

Incubate samples according to either of the following procedures:

(a) Incubate samples of medium at 37°C for 72 h. If any of the samples become cloudy, discard them and resterilize the batch. If there are signs of contamination in the other stored bottles, the whole batch should be discarded.
(b) For a more thorough test, and when the solution being filtered does not have its own nutrients, take samples, as discribed in this section, and dilute one-third of each into nutrient broths (e.g., L-Broth, beef-heart hydrolysate, and thioglycollate). Divide each sample in two, and incubate one at 37°C and one at 20°C for 10 days, with uninoculated controls. If there is any doubt after this incubation, mix and plate out aliquots on nutrient agar and incubate at 37°C and 20°C.

Downstream Secondary Filtration. Place a demountable 0.45-μm sterile filter in the effluent line from the main sterilizing filter. Any contamination that passes due to failure in the first filter will be trapped in the second. At the end of the run, remove the second filter and place the filter on nutrient agar. If colonies grow, discard or refilter the medium. This method has the advantage that it monitors the entire filtrate, and not just a small fraction of it, although it does not avoid risks of contamination during bottling and capping.

Autoclaved Solutions. Sterility testing of autoclaved stocks is much less essential, provided that proper monitoring (of the temperature and the time at the sterilizing temperature) of the autoclave is carried out. (See Reagents and Media in this chapter.)

Culture Testing

Media that have been produced commercially will have been tested for their capability of sustaining the growth of one or more cell lines. (If they have not, then you should change your supplier!) However, under certain circumstances, you may wish to test your own media for quality: (1) if it has been made up in the laboratory from basic constituents; (2) if any additions are made to the medium; (3) if the medium is for a special purpose that the commercial supplier is not able to test; and (4) if the medium is made up from powder and there is a risk of losing constituents during filtration.

The medium can become contaminated with toxic substances during filtration. For example, some filters are treated with traces of detergent to facilitate wetting, and the detergent may leach out into the medium as it is being filtered. Such filters should be washed by passing PBS or BSS through them before use or by discarding the first aliquot of filtrate. Polycarbonate filters (e.g., Nuclepore) are wettable without detergents and are preferred by some workers, particularly when the serum concentration in the medium is low.

There are three main types of culture test: (1) plating efficiency; (2) growth curve at regular passage densities and up to saturation density; and (3) the expression of a special function (e.g., differentiation in the presence of an inducer, viral propagation, the formation of a specific product, or the expression of a specific antigen). All of these tests should be performed on the new batch of medium with your regular medium as a control.

Plating Efficiency. The plating efficiency test (see Protocol 20.9) is the most sensitive culture test, detecting minor deficiencies and low concentrations of toxins that are not apparent at higher cell densities. Ideally, it should be performed with a limiting concentration of serum, which may otherwise mask deficiencies in the medium. To determine this concentration, do an initial plating efficiency test in different concentrations of serum, and select a concentration such that the plating efficiency is about half that of the usual concentration, but still gives countable colonies.

Growth Curve. A clonal growth assay will not always detect insufficiencies in the amount of particular constituents. For example, if the concentration of one or more amino acids is low, it may not affect clonal growth, but could influence the maximum cell concentration that is attainable.

A growth curve (see Protocols 20.7 and 20.8) gives three parameters of measurement: (1) the lag phase before cell proliferation is initiated after subculture, indicating whether the cells are having to adapt to different conditions; (2) the doubling time in the middle of the exponential growth phase, indicating the growth-promoting capacity of the medium; and (3) the terminal cell density. In cell lines whose growth is not sensitive to density (e.g., continuous cell lines; see Contact Inhibition and Density Limitation of Growth in Chapter 17), the terminal cell density indicates the total yield possible and usually reflects the total amino acid or glucose concentration. Remember that a medium that gives half the terminal cell density costs twice as much per cell produced.

Special Functions. If you are testing special functions, a standard test from the experimental system you are using (e.g., a virus titer in the medium after a set number of days) should be performed on the new medium alongside the old one.

A major implication of these tests is that they should be initiated well in advance of the exhaustion of the current stock of medium so that proper comparisons may be made and there is time to have fresh medium prepared if the medium fails any of the tests.

Storage

Opinions differ as to the shelf life of different media. As a rough guide, media made up without glutamine should last six to nine months at 4°C. Once glutamine, serum, or antibiotics are added, the storage time is reduced to two to three weeks. Hence, media that contain labile constituents should either be used within that number of weeks of preparation or be stored at -20°C.

Some forms of fluorescent lighting will cause riboflavin and tryptophan to deteriorate into toxic by-products [Wang, 1976]. Thus, incandescent lighting should be used in cold rooms where media are stored and in hot rooms where cells are cultured, and the light should be extinguished when the room is not occupied. Bottles of medium should not be exposed to fluorescent lighting for longer than a few hours; a dark freezer is recommended for long-term storage.

CHAPTER 11

Primary Culture

TYPES OF PRIMARY CELL CULTURE

A primary culture is that stage of the culture following isolation of the cells, but before the first subculture. There are three stages to consider: (1) isolation of the tissue, (2) dissection and/or disaggregation, and (3) culture following seeding into the culture vessel. Following isolation, a primary cell culture may be obtained either by allowing cells to migrate out from fragments of tissue adhering to a suitable substrate or by disaggregating the tissue mechanically or enzymatically to produce a suspension of cells, some of which will ultimately attach to the substrate. It appears to be essential for most normal untransformed cells— with the exception of hematopoietic cells—to attach to a flat surface in order to survive and proliferate with maximum efficiency. Transformed cells (see Anchorage Independence in Chapter 17), on the other hand, particularly cells from transplantable animal tumors, are often able to proliferate in suspension.

The enzymes used most frequently are crude preparations of trypsin, collagenase, elastase, hyaluronidase, DNase, pronase, dispase, alone or in various combinations. Crude preparations are often more successful than purified enzyme preparations, as the former contain other proteases as contaminants, although the latter are generally less toxic and more specific in their action. Trypsin and pronase give the most complete disaggregation, but may damage the cells. Collagenase and dispase, on the other hand, give incomplete disaggregation, but are less harmful. Hyaluronidase can be used in conjunction with collagenase to digest the intracellular matrix, and DNase is employed to disperse DNA released from lysed cells,

as it tends to impair proteolysis and promote reaggregation. (See Table 12.3.)

Although each tissue may require a different set of conditions, certain requirements are shared by most primary cultures:

(1) Fat and necrotic tissue are best removed during dissection.
(2) The tissue should be chopped finely with sharp instruments to cause minimum damage.
(3) Enzymes used for disaggregation should be removed subsequently by gentle centrifugation.
(4) The concentration of cells in the primary culture should be much higher than that normally used for subculture, since the proportion of cells from the tissue that survives in primary culture may be quite low.
(5) A rich medium, such as Ham's F12, is preferable to a simple medium, such as Eagle's MEM, and, if serum is required, fetal bovine often gives better survival than does calf or horse. Isolation of specific cell types will probably require selective media. (See Selective Media in Chapter 9 and the specific protocols in Chapter 22.)
(6) Embryonic tissue is preferable, as it disaggregates more readily, yields more viable cells, and proliferates more rapidly in primary culture than does adult tissue.

ISOLATION OF THE TISSUE

Before attempting to work with human or animal tissue, make sure that your work fits within medical eth-

149

ical rules or current legislation on experimentation with animals. For example, in the United Kingdom, the use of embryos or fetuses beyond 50% gestation or incubation is regulated under the Animal Experiments (Scientific Procedures) Act of 1986. Work with human biopsies or fetal material usually requires the consent of the local ethical committee and the patient and/or his or her relatives [Warnock, 1985; Winterton, 1989; Royal College of Physicians of London, 1990; UKCCCR, 1999; NBAC, 1999]. Taking tissue from human donors requires the consent of the donor, or of a close relative, so a suitable form (e.g., Table 11.1) should be drafted in a style readily understood by the patient, requesting his or her permission and drawing his or her attention to the use that might be made of the tissue. It is also useful to include a short summary of your project, in lay terms, explaining what you are doing, why, and what the possible outcome will be, particularly if it is seen to be of medical benefit.

△ *Safety Note.* Work with human tissue should be carried out in a Class II biological safety cabinet. (See Human Biopsy Material in Chapter 6.)

An attempt should be made to sterilize the site of the dissection with 70% alcohol if the site is likely to be contaminated (e.g., skin). Remove the tissue aseptically and transfer it to the tissue culture laboratory in BSS or medium as soon as possible. Do not dissect animals in the tissue culture laboratory, as the animals may carry microbial contamination. If a delay in transferring the tissue is unavoidable, it can be held at 4°C for up to 72 h, although a better yield will result from a quicker transfer.

Mouse Embryo Cell Culture

Mouse embryos are a convenient source of cells for undifferentiated fibroblastic cultures. They are often used as feeder layers. (See Feeder Layers in Chapter 13.)

PROTOCOL 11.1. ISOLATION OF MOUSE EMBRYOS

Outline
Remove uterus aseptically from timed pregnant mouse and dissect out embryos.

Materials
Sterile:
DBSS (see Reagent Appendix) in 50-ml sterile beaker to cool instruments after flaming
DBSS in 25–50-ml screw-capped vial or tube
Pointed forceps
Pointed scissors
Nonsterile:
Small laminar-flow hood

Timed pregnant mice (see Step 1 of this protocol)
70% alcohol in wash bottle
70% alcohol to sterilize instruments (see Fig. 6.3)
Bunsen burner

△ *Safety Note.* When sterilizing instruments by dipping them in alcohol and flaming them, take care not to return the instruments to alcohol while they are still alight!

Protocol
1. *Induction of estrus.* If males and females are housed separately, then when they are put together for mating, estrus will be induced in the female 3 d later, when the maximum number of successful matings will occur. This process enables the planned production of embryos at the appropriate time. The timing of successful matings may be determined by examining the females' vaginas each morning for a hard mucous plug.
2. *Dating the embryos.* The day of detection of a vaginal plug, or the "plug date," is noted as day zero, and the development of the embryos is timed from this date. Full term is about 19–21 d. The optimal age for preparing cultures from a whole disaggregated embryo is around 13 d, when the embryo is relatively large (Figs. 11.1, 11.2) but still contains a high proportion of undifferentiated mesenchyme, which is the main source of the culture. However, isolation and handling embryos at this stage require a license in the United Kingdom, and it may, therefore, be preferable to use embryos at 9 or 10 d, although the amount of tissue recovered from these embryos will be substantially less. Most individual organs, with the exception of the brain and the heart, begin to form at about the 9th day of gestation, but are difficult to isolate until about the 11th day. Dissection is easier at 13–14 d, and most of the organs are completely formed by the 18th day.
3. Kill the mouse by cervical dislocation (U.K. Schedule I procedure), and swab the ventral surface liberally with 70% alcohol. (Fig. 11.3a.)
4. Tear the ventral skin transversely at the median line just over the diaphragm (Fig. 11.3b), and, grasping the skin on both sides of the tear, pull in opposite directions to expose the untouched ventral surface of the abdominal wall (Fig. 11.3c).
5. Cut longitudinally along the median line of the exposed abdomen with sterile scissors, revealing the viscera. At this stage, the uteri filled with embryos are obvious in the posterior abdominal cavity.

TABLE 11.1. Donor Consent Form

CONSENT TO REMOVE TISSUE FOR DIAGNOSIS AND RESEARCH

This form requests your permission to sample one or more small pieces of tissue to be used for medical research. This tissue, or cell lines or other products derived from it, may be used by a number of different research organizations, or it may be stored for an extended period awaiting use. It is also possible that it may eventually be used by a commercial company to develop future drugs. We would like you to be aware of this and of the fact that, by signing this form, you give up any claim that you own the tissue or its components, regardless of the use that may be made of it. You should also be aware of, and agree to, the possible testing of the tissue for infectious agents, such as the AIDS virus or hepatitis.

I am willing to have tissue removed for use in medical research and development. (If the donor is too unwell to sign, a close relative should sign on his or her behalf.)

Name of donor .... *Name of relative* ....

Signature ....

Date ....

This material will be coded, and absolute confidence will be maintained. Your name will not be given to anyone other than the person taking the sample.

Do you wish to receive any information relating to your state of health?

Yes/No

Signature .... *Date* ....

Would you like any information relating to your state of health to be given to your doctor:

Yes/No

If yes, name of doctor ....

Address ....

....

....

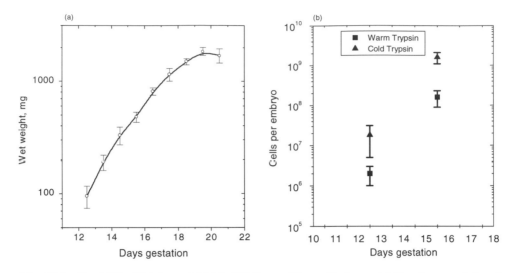

Fig. 11.1. Total Wet Weight and Yield of Cells per Mouse Embryo. (a) Total wet weight of embryo without placenta or membranes, mean ± standard deviation (squares) [From Paul et al., 1969]. (b) Cell yield per embryo after incubation in 0.25% trypsin at 37°C for 4 h (squares), or after soaking in 0.25% trypsin at 4°C for 5 h and incubation at 37°C for 30 min (triangles; see text).

6. Dissect out the uteri into a 25-ml or 50-ml screw-capped tube containing 10 or 20 ml BSS. (Fig. 11.3d–f.) Antibiotics may be added to the BSS when there is a high risk of infection. (See Reagent Appendix, Dissection BSS [DBSS].)

Note. All of the preceding steps should be done outside the tissue culture laboratory; a small laminar flow hood and rapid technique will help to maintain sterility. Do not take live animals into the tissue culture laboratory, as the animals may carry contamination. If an animal carcass must be handled in the tissue culture area, make sure that the carcass is immersed in alcohol briefly, or thoroughly swabbed, and disposed of quickly after use.

7. Take the intact uteri to the tissue culture laboratory, and transfer them to a fresh dish of sterile DBSS.
8. Dissect out the embryos (Fig. 11.3g,h):
   (a) Tear the uterus with two pairs of sterile forceps, keeping the points of the forceps close together to avoid distorting the uterus and bringing too much pressure to bear on the embryos.
   (b) Free the embryos from the membranes and placenta and place them to one side of the dish to bleed.
9. Transfer the embryos to a fresh dish. If a large number of embryos is required (i.e., more than four or five litters), it may be helpful to place the

dish on ice (for subsequent dissection and culture; see Protocols 11.4–11.8)

Chick Embryo Cell Culture

Chick embryos are easier to dissect, as they are larger than the equivalent stage of mouse embryo. Like mouse embryos, chick embryos are used to provide predominantly mesenchymal cell primary cultures for cell proliferation analysis, to provide feeder layers, and as a substrate for viral propagation. Because of their larger size, it is also easier to dissect out individual organs to generate specific cell types, such as hepatocytes, cardiac muscle, and lung epithelium. As with mouse embryos, the use of chick embryos is subject to animal legislation in the United Kingdom, and working with embryos that are more than half-term requires a license.

PROTOCOL 11.2. ISOLATION OF CHICK EMBRYOS

Outline
Remove embryo aseptically from the egg and transfer to dish.

Materials
Sterile:
Small beaker, 20–50 ml or egg cup
Forceps, straight and curved
9-cm Petri dishes
DBSS

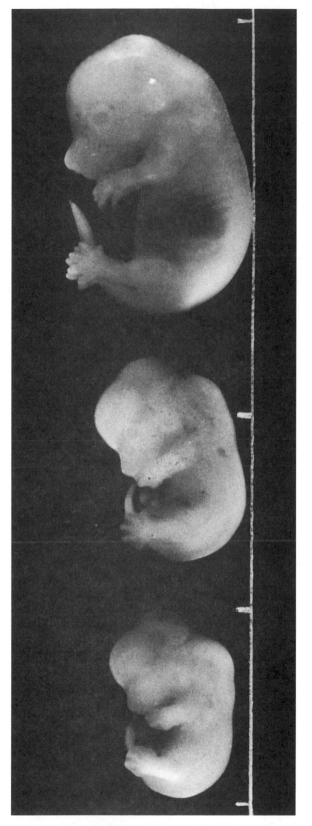

Fig. 11.2. Mouse Embryos. Embryos from the 12th, 13th, and 14th day of gestation. The 12-d embryo (bottom) came from a small litter (three) and is larger than would normally be found at this stage. Scale: 10 mm between marks.

Nonsterile:
70% alcohol
Swabs
10-d embryonated eggs
Humid incubator (no additional CO_2 above atmospheric level)

Protocol

1. Incubate the eggs at 38.5°C in a humid atmosphere, and turn the eggs through 180° daily. Although hen's eggs hatch at around 20 to 21 d, the lengths of their developmental stages are different from those of mouse embryos. For a culture of dispersed cells from the whole embryo, the egg should be taken at about 8 d, and for isolated-organ rudiments, at about 10–13 d. (10 d is the maximum in the United Kingdom without a license.)
2. Swab the egg with 70% alcohol, and place it with its blunt end facing up in a small beaker. (Fig. 11.4a.)
3. Crack the top of the shell and peel the shell off to the edge of the air sac using sterile forceps. (Fig. 11.4b.)
4. Resterilize the forceps (i.e., dip them in alcohol, burn off the alcohol, and cool the forceps in sterile BSS), and then use the forceps to peel off the white shell membrane to reveal the chorioallantoic membrane (CAM) below, with its blood vessels. (Fig. 11.4c,d.)
5. Pierce the CAM with sterile curved forceps, and lift out the embryo by grasping it gently under the head. Do not close the forceps completely, or else the neck will sever. (Fig. 11.4e–g.)
6. Transfer the embryo to a 9-cm Petri dish containing 20 ml of DBSS. (For subsequent dissection and culture, see Protocol 11.7.)

Human Biopsy Material

Handling human biopsy material presents certain problems that are not encountered with animal tissue. It usually is necessary to obtain consent (1) from the hospital ethical committee, (2) from the attending physician or surgeon, and (3) from the patient or the patient's relatives. (See Table 11.1.) Furthermore, biopsy sampling is usually performed for diagnostic purposes, and hence the needs of the pathologist must be met first. This factor is less of a problem if extensive surgical resection or nonpathological tissue (e.g., placenta or umbilical cord) is involved.

There is also the difficult problem of ownership and subsequent patent rights to deal with, and the following issues need to be addressed:

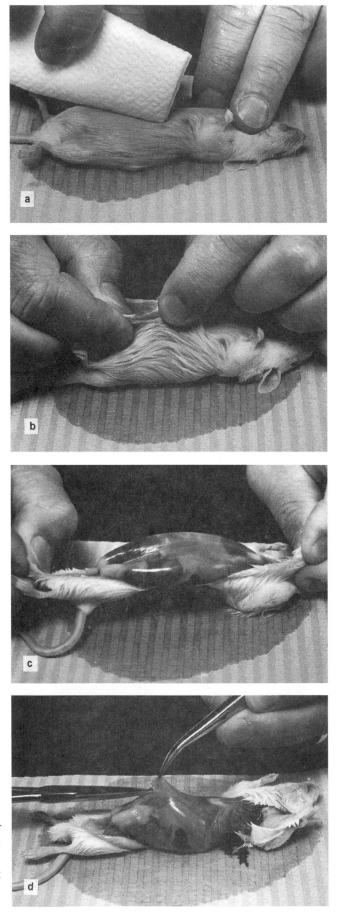

Fig. 11.3. Mouse Dissection. Stages in the aseptic removal of mouse embryos for primary culture. (See text.) (a) Swabbing the abdomen. (b), (c) Tearing the skin to expose the abdominal wall. (d) Opening the abdomen. (e) The uterus *in situ.* (f) Removing the uterus. (g), (h) Dissecting the embryos from the uterus. (i) Removing the membranes. (j) Chopping the embryos.

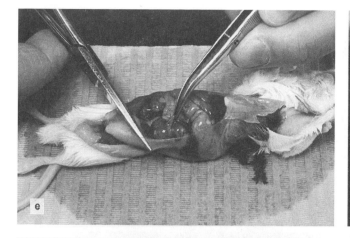

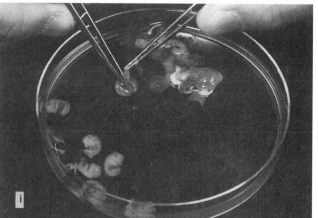

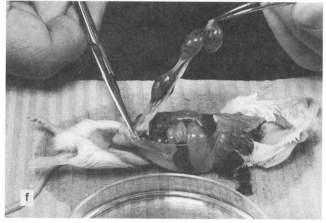

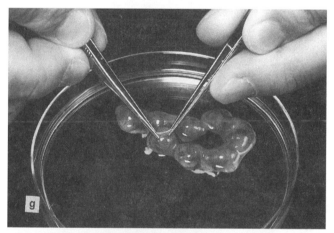

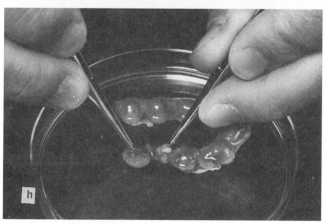

Fig. 11.3. Continued

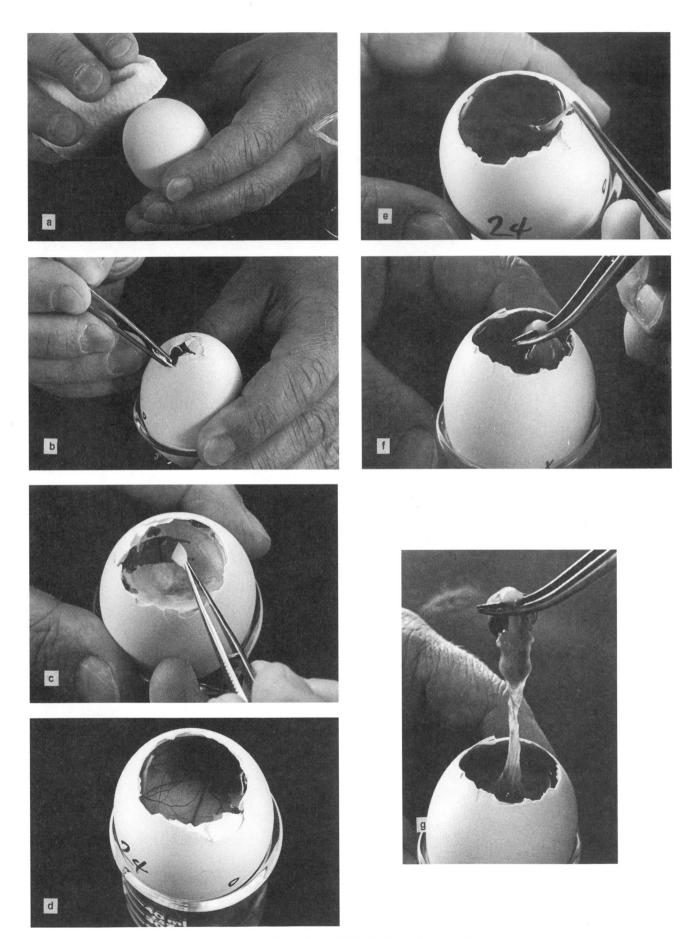

Fig. 11.4. Removing a Chick Embryo from an Egg.

(1) The patient and/or relative's informed consent to taking tissue for research purposes, over and above the clinical requirement

(2) Confidentiality of the origin of the tissue.

(3) Ownership of cell lines and their derivatives.

(4) Authority for genetic modification of the cell lines.

(5) Patent rights from any commercial collaboration.

(6) Feedback of any genetic information to the patient and/or physician.

By far, the easiest approach is to ask the donor and/or relatives to sign a disclaimer statement before the tissue is removed; otherwise, the legal aspects of ownership of the cell lines that might be derived and any future biopharmaceutical exploitation of the cell lines, their genes, and their products become exceedingly complex. Feedback of genetic information, however, is a more difficult problem; in the case of a patient in a hospital, the feedback is most likely to be directed to the doctor, but in the case of a donor that is not hospitalized, you must ask the donor if he or she wishes to know your findings and any implications that they might have.

The operation is often performed by one of the resident staff at a time that is not always convenient to the tissue culture laboratory, so some formal collection or storage system must be employed for times when you or someone on your staff cannot be there. If delivery to your lab is arranged, then there must be a system for receiving specimens, recording details of the source, tissue of origin, pathology, etc., (see Primary Records in this chapter) and alerting the person who will perform the culture that the specimens have arrived; otherwise, valuable material may be lost or spoiled.

△ *Safety Note.* Biopsy material carries a risk of infection (see Biohazards in Chapter 6), so it should be handled in a Class II biohazard cabinet, and all media and apparatus must be disinfected after use by autoclaving or immersion in a suitable disinfectant (see Disposal in Chapter 6). If possible, the tissue should be screened for infections such as hepatitis, HIV, and tuberculosis [Advisory Committee on Dangerous Pathogens, 1995b] unless the patient has already been tested for these conditions.

PROTOCOL 11.3. HUMAN BIOPSIES

Outline
Consult with hospital staff, provide labeled container(s) of medium, and arrange for collection of samples from operating room or pathologist.

Materials
Specimen tubes (15–30 ml) with leak-proof caps about one-half full with culture medium containing antibiotics (see Collection Medium in Reagent Appendix), and labeled with your name, address, and telephone number

Protocol

1. Provide containers of collection medium, clearly labeled, to the anteroom of the operating theater or to the pathology laboratory.

2. Make arrangements to be alerted when the material is ready for collection.

3. Collect the containers after surgery, or have someone send them to you immediately after collection and inform you when they have been dispatched.

4. Transfer the sample to the tissue culture laboratory. The sample should be triple wrapped (e.g., in a sealed tube within a sealed plastic bag full of absorbent tissue, in case of leakage, within a padded envelope with your name, address, and telephone number on it). Usually, if kept at 4°C, biopsy samples survive for at least 24 h and even up to 3 or 4 d, although the longer the time from surgery to culture, the more the samples will deteriorate.

5. Log receipt of sample in a hand-written record book for subsequent transfer to a computerized database.

6. *Decontamination.* A disinfectant wash is given before skin biopsy, and a parenteral antibiotic is given before gut surgery. Most surgical specimens, however, due to the needs of surgery, are sterile when removed, though problems may arise with subsequent handling. Superficial specimens (e.g., skin biopsies, melanomas, etc.) and gastrointestinal tract specimens (e.g., colon and rectal samples) are particularly prone to contamination from gut flora. It may be advantageous to consult a medical microbiologist to determine which flora to expect in a given tissue and then choose your antibiotics accordingly. If the surgical sample is large enough (i.e., 200 mg or more), then a brief dip (i.e., 30 s–1 min) in 70% alcohol will help to reduce superficial contamination without causing much harm to the center of the tissue sample.

PRIMARY CULTURE

Several techniques have been devised for primary culture of isolated tissue. These techniques can be divided into purely mechanical techniques, involving dissection with or without some form of maceration, and techniques utilizing enzymic disaggregation. (Fig. 11.5.) Briefly, dissection of primary explants is suitable when very small amounts of tissue are available, enzymic disaggregation when more tissue is available but a high recovery is required, and mechanical disaggregation when large amounts of soft tissue are available and the size of the yield is not paramount.

Primary Explant

The primary-explant technique was the original method developed by Harrison [1907], Carrel [1912], and others for initiating a tissue culture. As originally performed, a fragment of tissue was embedded in blood plasma or lymph, mixed with heterologous serum and embryo extract, and placed on a coverslip which was inverted over a concavity slide. The clotted plasma held the tissue in place, and the explant could be examined with a conventional microscope. The embryo extract and serum, together with the plasma, supplied nutrients and stimulated migration out of the explant across the solid substrate. The heterologous serum was used to promote clotting of the plasma. This technique is still used, but has been largely replaced by the simplified method described in Protocol 11.4.

PROTOCOL 11.4. PRIMARY EXPLANTS

Outline

The tissue is chopped finely and rinsed, and the pieces are seeded onto the surface of a culture flask or Petri dish in a small volume of medium with a high concentration (i.e., 40–50%) of serum, such that surface tension holds the pieces in place until they adhere spontaneously to the surface. (Fig. 11.6, see also Plate 1.) Once this is achieved, outgrowth of cells usually follows. (Fig. 11.7.)

Materials
Sterile:
Growth medium (e.g., 50:50 DMEM:F12 with 20% fetal bovine serum)
100 ml of DBSS
Petri dishes, 9 cm, non-tissue-culture grade
Forceps
Scalpels
Pipettes, 10 ml with wide tips
Centrifuge tubes, 15 or 20 ml, or universal containers
Culture flasks, 25 cm², or Petri dishes, 50 mm. The size of flasks and volume of growth medium depend on the amount of tissue—roughly five 25-cm² flasks per 100 mg of tissue, and initially 1 ml of medium per flask, building up to 5 ml per flask over the first 3–5 d.

Protocol
1. Transfer tissue to fresh, sterile BSS, and rinse.
2. Transfer the tissue to a second dish; dissect off unwanted tissue, such as fat or necrotic material; and chop finely with crossed scalpels (see Fig. 11.6) to about 1-mm cubes.
3. Transfer by pipette (10–20 ml, with wide tip) to a 15- or 50-ml sterile centrifuge tube or universal container. (Wet the inside of the pipette first with BSS, or else the pieces will stick.) Allow the pieces to settle.
4. Wash by resuspending the pieces in BSS, allowing the pieces to settle, and removing the supernatant fluid. Repeat this step two more times.
5. Transfer the pieces (remember to wet the pipette) to a culture flask, with about 20–30 pieces per 25-cm² flask.
6. Remove most of the fluid, and add about 1 ml of growth medium per 25-cm² growth surface. Tilt the flask gently to spread the pieces evenly over the growth surface.
7. Cap the flask, and place it in an incubator or hot room at 37°C for 18–24 h.
8. If the pieces have adhered, then the medium volume may be made up gradually over the next 3–5 d to 5 ml per 25 cm² and then changed weekly until a substantial outgrowth of cells is observed. (See Fig. 11.7.)
9. The explants may then be picked off from the center of the outgrowth with a scalpel and transferred by prewetted pipette to a fresh culture vessel. (Then return to step 7.)
10. Replace the medium in the first flask until the outgrowth has spread to cover at least 50% of the growth surface, at which point the cells may be subcultured. (See Protocol 12.2.)

This technique is particularly useful for small amounts of tissue, such as skin biopsies, for which there is a risk of losing cells during mechanical or enzymatic disaggregation. Its disadvantages lie in the poor adhesiveness of some tissues and the selection of cells in the outgrowth. In practice, however, most cells —fibroblasts, myoblasts, glia, and epithelia—particularly those from the embryo, migrate out successfully.

Attaching Explants. Both adherence and migration may be stimulated by placing a glass coverslip on top of the explant, with the explant near the edge of the coverslip, or the plastic dish may be scratched through the explant to attach the tissue to the flask

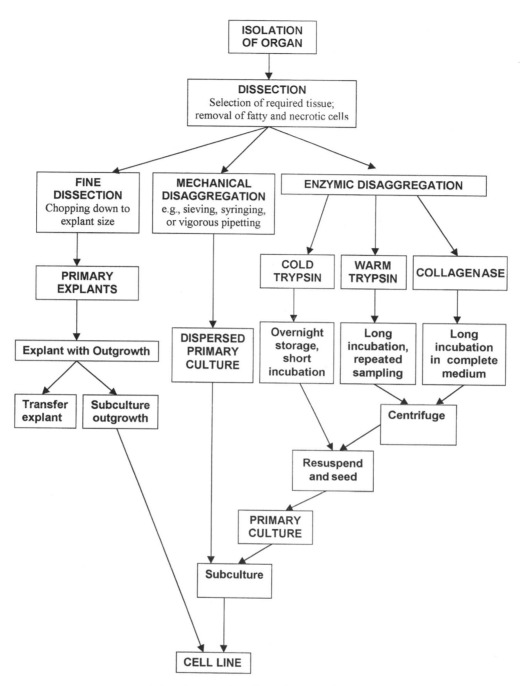

Fig. 11.5. Options for Primary Culture.

[Elliget and Lechner, 1992]. (See also Protocol 22.9.) Attachment may also be promoted by treating the plastic with polylysine or fibronectin (see Improving Clonal Growth in Chapter 13, Protocol 22.9), extracellular matrix (see Protocol 7.1), or feeder layers (see Protocol 13.3). Historically, plasma clots have been used to promote attachment. Place a drop of plasma on the plastic surface, and embed the explant in it. This should induce the plasma to clot in a few minutes, whereupon medium can be added. Alterna-

tively, purified fibrinogen and thrombin can be used [Nicosia and Ottinetti, 1990].

Enzymatic Disaggregation

Cell–cell adhesion in tissues is mediated by a variety of homotypic interacting glycopeptides (cell adhesion molecules, or CAMs) (see Cell Adhesion in Chapter 2), some of which are calcium dependent (cadherins) and hence are sensitive to chelating agents such as EDTA or EGTA. Integrins, which bind to the RGD motif in

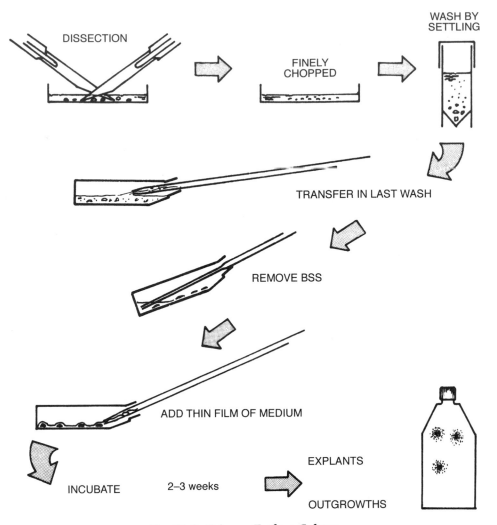

Fig. 11.6. Primary Explant Culture.

extracellular matrix, also have Ca^{2+}-binding domains and are affected by Ca^{2+} depletion. Intercellular matrix and basement membranes also contain other glycoproteins, such as fibronectin and laminin, which are protease sensitive, and proteoglycans, which are less so, and can sometimes be degraded by glycanases, such as hyaluronidase or heparinase. The easiest approach is to proceed from a simple disaggregation solution to a more complex solution (see Table 12.3) with trypsin alone or trypsin/EDTA as a starting point, adding other proteases to improve disaggregation, and deleting trypsin if necessary to increase viability. In general, increasing the purity of an enzyme will give better control and less toxicity with increased specificity, but may result in less disaggregation activity.

Mechanical and enzymatic disaggregation of the tissue avoids problems of selection by migration and yields a higher number of cells that are more representative of the whole tissue in a shorter time. However, just as the primary-explant technique selects on the basis of cell migration, dissociation techniques select cells resistant to the method of disaggregation and still capable of attachment.

Embryonic tissue disperses more readily and gives a higher yield of proliferating cells than does newborn or adult tissue. The increasing difficulty in obtaining viable proliferating cells with increasing age is due to several factors, including the onset of differentiation, an increase in fibrous connective tissue and extracellular matrix, and a reduction of the undifferentiated proliferating cell pool. When procedures of greater severity are required to disaggregate the tissue (e.g., longer trypsinization or increased agitation), the more fragile components of the tissue may be destroyed. In fibrous tumors, for example, it is very difficult to obtain complete dissociation with trypsin while still retaining viable carcinoma cells.

The choice of trypsin grade to use has always been difficult, as there are two opposing trends: (1) The purer the trypsin, the less toxic it becomes, and the

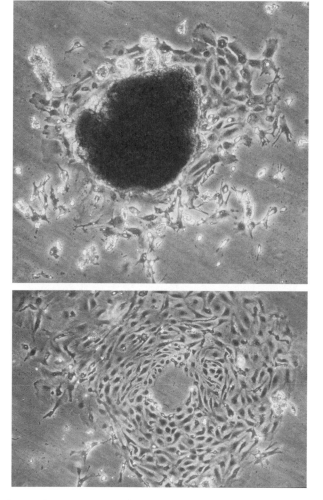

Fig. 11.7. Primary Explant. Primary explant culture from mouse squamous skin carcinoma. (a) Explant and early-stage outgrowth about 3 d after explantation. (b) Outgrowth after removal of explant, about 7 d after explantation. Olympus CK, 10× objective.

more predictable its action; (2) the cruder the trypsin, the more effective it may be, due to other proteases. In practice, a preliminary test experiment may be necessary to determine the optimum grade for viable cell yield, as the balance between sensitivity to toxic effects and disaggregation ability may be difficult to predict.

Crude trypsin is by far the most common enzyme used in tissue disaggregation [Waymouth, 1974], as it is tolerated quite well by many cells, it is effective for many tissues, and any residual activity left after washing is neutralized by the serum of the culture medium, or by a trypsin inhibitor (e.g., soya bean trypsin inhibitor, Sigma) when serum-free medium is used.

It is important to minimize the exposure of cells to active trypsin in order to preserve maximum viability. Hence, when whole tissue is being trypsinized at 37°C, dissociated cells should be collected every half hour,

and the trypsin should be removed by centrifugation and neutralized with serum in medium. Soaking the tissue for 6–18 h in trypsin at 4°C (see Protocol 11.6) allows penetration with minimal tryptic activity, and digestion may then proceed for a much shorter time (i.e., 20–30 min) at 37°C [Cole and Paul, 1966]. Although the cold-trypsin method gives a higher yield of viable cells and requires less effort, the warm-trypsin method is still used extensively and is presented here for comparison.

PROTOCOL 11.5. WARM TRYPSIN

Outline
The tissue is chopped and stirred in trypsin for a few hours. The dissociated cells are collected every half hour, centrifuged, and pooled in medium containing serum. (Fig. 11.8.)

Materials
Sterile:
Tissue, 1–5 g
DBSS, 50 ml (see Reagents & Materials)
Trypsin (crude), 2.5% in PBSA or normal saline
PBSA, 200 ml
Growth medium with serum (e.g., DMEM/F12 with 10% fetal bovine serum)
Culture flasks, 5–10/g tissue (varies depending on cellularity of tissue)
Petri dishes, 9 cm, nontissue culture grade
Two 50-ml centrifuge tubes
250-ml Erlenmeyer flask (preferably indented as in Fig. 11.9 [Bellco])
Magnetic follower, autoclaved in test tube
Curved forceps
Pipettes (Pasteur, 2 ml, 10 ml)
Nonsterile:
Magnetic stirrer
Hemocytometer or cell counter

Protocol
1. Transfer the tissue to fresh, sterile DBSS in 9-cm Petri dish, and rinse.
2. Transfer the tissue to a second dish; dissect off unwanted tissue, such as fat or necrotic material; and chop with crossed scalpels (see Fig. 11.8) to about 3-mm cubes.
3. Transfer the tissue with curved forceps to a 15- or 50-ml sterile centrifuge tube or universal container. Allow the pieces to settle.
4. Wash the tissue by resuspending the pieces in DBSS, allowing the pieces to settle, and removing the supernatant fluid. Repeat this step two more times.

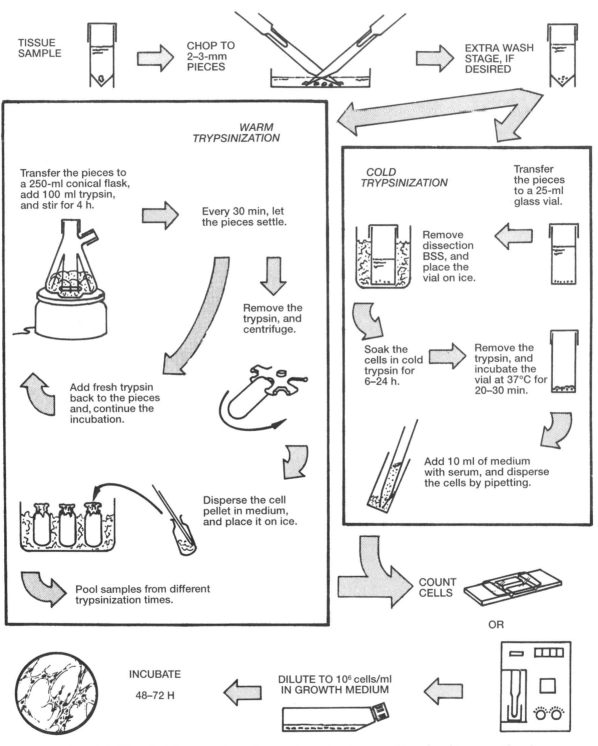

Fig. 11.8. Trypsin Disaggregation. Preparation of primary culture by disaggregation in trypsin. The warm-trypsin method is shown on the left, and the cold-trypsin method is shown on the right.

Fig. 11.9. Trypsinization Flask. The indentations in the side of the flask improve mixing, and the rim around the neck, below the side arm, allows the cell suspension to be poured off, while leaving the stirrer bar and any larger fragments behind (Bellco).

5. Transfer the pieces to the empty trypsinization flask, into which about 20 mg/ml of trypsin will be added in step 7.
6. Remove most of the residual fluid, and add 180 ml of PBSA.
7. Add 20 ml of 2.5% trypsin. (Other enzymes—e.g., collagenase, hyaluronidase, or DNase—may be added at this stage as well, if required.)
8. Add a magnetic follower to the flask.
9. Cap the flask, and place it on a magnetic stirrer in an incubator or hot room at 37°C.
10. Stir at about 100 rpm for 30 min at 37°C.
11. After 30 min, collect disaggregated cells:
    (a) Allow the pieces to settle.
    (b) Pour off the supernatant into a centrifuge tube and place it on ice.
    (c) Add fresh trypsin to the pieces remaining in the flask, and continue to stir and incubate for a further 30 min.
    (d) Centrifuge the harvested cells from step 11(b) at approximately 500 g for 5 min.
    (e) Resuspend the resulting pellet in 10 ml of medium with serum, and store the suspension on ice.
12. Repeat step 11 until complete disaggregation occurs or until no further disaggregation is apparent (usually 3–4 h).
13. Collect and pool chilled cell suspensions, count the cells by hemocytometer or electronic cell counter (see Cell Counting in Chapter 20), and check viability (see Viability in Chapter 21).

14. The cell population will be very heterogeneous; electronic cell counting will initially require confirmation with a hemocytometer, as calibration can be difficult.
15. Dilute the cell suspension to 1×10^6 per ml in growth medium, and seed as many flasks as are required, with approximately 2×10^5 cells per cm². When the survival rate is unknown or unpredictable, a cell count is of little value (e.g., in tumor biopsies, in which the proportion of necrotic cells may be quite high). In this case, set up a range of concentrations from about 5 to 25 mg of tissue per ml.
16. Change the medium at regular intervals (2–4 d as dictated by depression of pH). Check the supernatant for viable cells before discarding it, as some cells can be slow to attach or may even prefer to proliferate in suspension.

This technique is useful for the disaggregation of large amounts of tissue in a relatively short time, particularly for whole mouse embryos or chick embryos. It does not work as well with adult tissue, in which there is a lot of fibrous connective tissue, and mechanical agitation can be damaging to some of the more sensitive cell types, such as epithelium.

Trypsinization with Cold Preexposure
One of the disadvantages of using trypsin to disaggregate tissue is the damage that may result from prolonged exposure to the tissue to trypsin at 37°C—hence the need to harvest cells after 30-min incubations in the warm-trypsin method rather than have them exposed for the full time (i.e., 3 to 4 h) required to disaggregate the whole tissue. A simple method of minimizing damage to the cells during disaggregation is to soak the tissue in trypsin at 4°C to allow penetration of the enzyme with little tryptic activity. (Table 11.2.) Following this procedure, the tissue will require much shorter incubation at 37°C for disaggregation [Cole and Paul, 1966].

PROTOCOL 11.6. COLD TRYPSIN

Outline
Chop tissue and place in trypsin at 4°C for 6–18 h. Incubate after removing the trypsin, and disperse the cells in warm medium. (See Fig. 11.8.)

Materials
Sterile:
Culture medium (e.g., DMEM/F12 with 10% FBS)
DBSS

TABLE 11.2. Cell Yield by Warm and Cold Trypsinization

Duration and temperature of trypsinization		After trypsinization			After 24 h in culture	
		Cells recovered per embryo $\times 10^{-7}$	% Viability by dye exclusion (trypan blue)	Total no. of viable cells $\times 10^{-7}$	Recovered, % of total seeded	Viability, % of viable cells seeded
4°C	37°C					
0 h	4 h	1.69	86	1.45	47.2	54.9
5.5 h	0.5 h	3.32	60	1.99	74.5	124
24 h	0.5 h	3.40	75	2.55	60.3	80.2

0.25% crude trypsin in serum-free RPMI 1640 or MEM/Stirrer Salts (S-MEM)
Petri dishes, 9 cm, non-tissue culture grade
Forceps, straight and curved
Scalpels
Erlenmeyer flask(s), 25 ml, screw capped
Culture flasks, 25 cm² or 75 cm²
Pipettes (Pasteur, 2 ml, 10 ml)
Nonsterile:
Ice bath

Protocol

1. Transfer the tissue to fresh, sterile DBSS in a 9-cm Petri dish, and rinse.
2. Transfer the tissue to a second dish; dissect off unwanted tissue, such as fat or necrotic material; and chop with crossed scalpels (see Fig. 11.8) to about 3-mm cubes. Embryonic organs, if they do not exceed this size, are better left whole.
3. Transfer the tissue with curved forceps to a 15- or 50-ml sterile centrifuge tube or universal container. Allow the pieces to settle.
4. Wash the tissue by resuspending the pieces in BSS, allowing the pieces to settle, and removing the supernatant fluid. Repeat this step two more times.
5. Remove most of the residual fluid, and add 10 ml/tube/g of tissue of 0.25% trypsin in RPMI 1640 at 4°C.
6. Place the mixture at 4°C for 6–18 h.
7. Remove and discard the trypsin carefully, leaving the tissue with only the residual trypsin. (Other enzymes—e.g., collagenase, hyaluronidase, or DNase—may be added in 1–2 ml amounts at this stage, if required.)
8. Place the tube at 37°C for 20–30 min.
9. Add warm medium, approximately 1 ml for every 100 mg of original tissue, and gently pipette the mixture up and down until the tissue is completely dispersed.
10. If some tissue does not disperse, then the cell suspension may be filtered through sterile muslin or stainless steel mesh (100–200 µm), or Falcon 70-mm "Cell Strainer" (Becton Dickinson; Fig. 11.10), or the larger pieces may simply be allowed to settle. When there is a lot of tissue, increasing the volume of suspending medium to 20 ml for each gram of tissue will facilitate settling and subsequent collection of supernatant fluid. Two to three minutes should be sufficient to get rid of most of the larger pieces.
11. Determine the cell concentration in the suspension by hemocytometer or electronic cell counter (see Cell Counting in Chapter 20), and check viability (see Protocol 21.1).
12. The cell population will be very heterogeneous; electronic cell counting will initially require confirmation with a hemocytometer, as calibration can be difficult.
13. Dilute cell suspension to 1×10^6 per ml in growth medium, and seed as many flasks as are required, with approximately 2×10^5 cells per cm². When the survival rate is unknown or unpredictable, a cell count is of little value (e.g., in tumor biopsies, for which the proportion of necrotic cells may be quite high). In this case, set up a range of concentrations from about 5 to 25 mg of tissue per ml.
14. Change the medium at regular intervals (2–4 d as dictated by depression of pH). Check the supernatant for viable cells before discarding it, as some cells can be slow to attach or may even prefer to proliferate in suspension.

The cold-trypsin method usually gives a higher yield of viable cells, with improved survival after 24 h

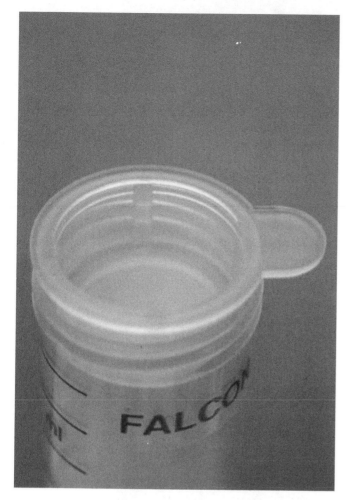

Fig. 11.10. Cell Strainer. Disposable polypropylene filter and tube for straining aggregates from primary suspensions. Can also be used for disaggregating soft tissues. (See also Fig. 11.14.)

culture (see Fig. 11.1 and Table 11.2), and preserves more different cell types than the warm method (see Plate 4). Cultures from mouse embryos contain more epithelial cells when prepared by the cold method, and erythroid cultures from 13-d fetal mouse liver respond to erythropoietin after this treatment, but not after the warm-trypsin method or mechanical disaggregation [Cole and Paul, 1966; Conkie, personal communication]. The cold-trypsin method is also convenient, as no stirring or centrifugation is required, and the incubation at 4°C may be done overnight. This method does take longer than the warm-trypsin method, however, and is not as convenient when large amounts of tissue (i.e., greater than 10 g) are being handled.

Chick Embryo Organ Rudiments

The cold-trypsin method is particularly suitable for small amounts of tissue, such as embryonic organs.

Taking a 10–13-d chick embryo as a starting point, the procedure described by Protocol 11.7 gives good reproducible cultures with evidence of several different cell types characteristic of the tissue of origin. This protocol forms a good exercise for teaching purposes.

PROTOCOL 11.7. CHICK EMBRYO ORGAN RUDIMENTS

Outline
Dissect out individual organs or tissues, and place, them, preferably whole, in cold trypsin overnight. Remove the trypsin, incubate the organs or tissue briefly, and disperse them in culture medium. Dilute and seed the cultures.

Materials
Sterile:
BSS

0.25% crude trypsin (Difco 1:250 or equivalent) in RPMI 1640 or S-MEM on ice; lower concentrations may be used with purer grades of trypsin, e.g., 0.05–0.1% Sigma crystalline or Worthington Grade IV

Culture medium (e.g., DMEM/F12 with 10% FBS), minimum of 12 ml per tissue

Petri dishes, 9 cm, non-tissue culture grade

Culture flasks, 25 cm² (two per tissue)

Scalpels (No. 11 blade for most steps)

Iridectomy knives for fine dissection

Curved and straight fine forceps

Pipettes (Pasteur, 2 ml, 10 ml)

Test tubes, preferably glass, 10–15 ml, with screw caps

Nonsterile:

Embryonated hen's eggs, 10–13 d incubation (>10 d license in the United Kingdom)

Ice bath

Protocol
1. Remove the embryo from the egg as described previously (see Protocol 11.2), and place it in sterile BSS.
2. Remove the head. (Fig. 11.11a,b.)
3. Remove an eye and open it carefully, releasing the lens and aqueous and vitreous humors. (Fig. 11.11c,d.)
4. Grasp the retina in two pairs of fine forceps and gently peel the pigmented retina off the neural retina and connective tissue. (Fig. 11.11e.) (This step requires a dissection microscope for 10-d embryos. A brief exposure to 0.25% trypsin in 1 mM EDTA will allow the two tissues to separate more easily.) Put the tissue to one side.

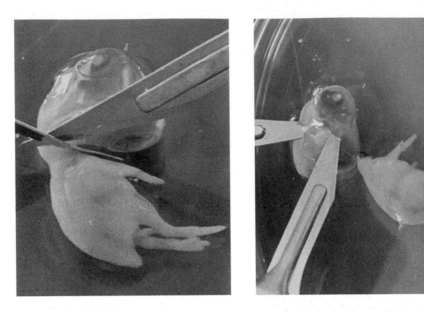

a

c

b

d

Fig. 11.11. Dissection of a Chick Embryo. (a), (b) Removing the head. (c) Removing the eye. (d) Dissecting out the lens. (e) Peeling off the retina. (f) Scooping out the brain. (g) Halving the trunk. (h) Teasing out the heart and lungs from the anterior half. (i) Teasing out the liver and gut from the posterior half. (j) Inserting the tip of the scalpel between the left kidney and the dorsal body wall. (k) Squeezing out the spinal cord. (l) Peeling the skin off the back of the trunk and hind leg. (m) Slicing muscle from the thigh. (n) Organ rudiments arranged around the periphery of the dish. From the right, clockwise, we have the following organs: brain, heart, lungs, liver, gizzard, kidneys, spinal cord, skin, and muscle.

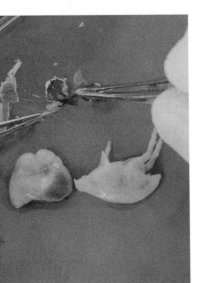

e

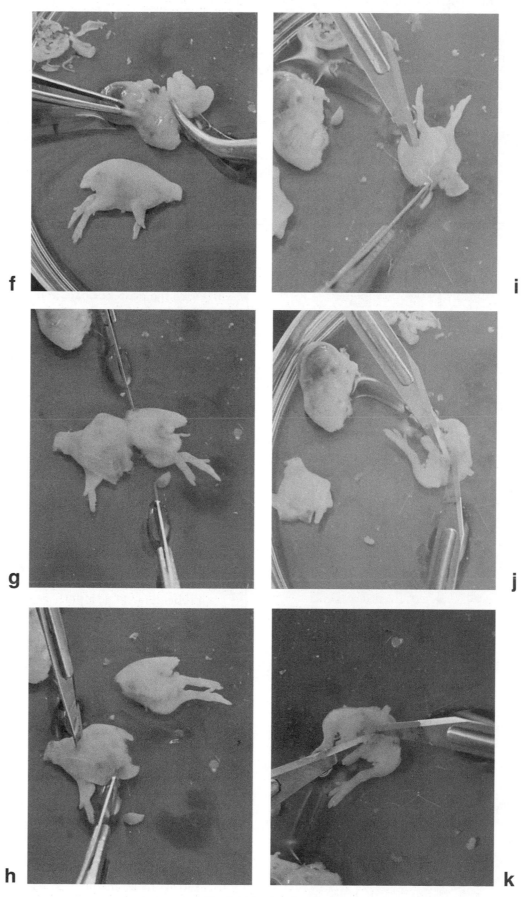

Fig. 11.11. *Continued*

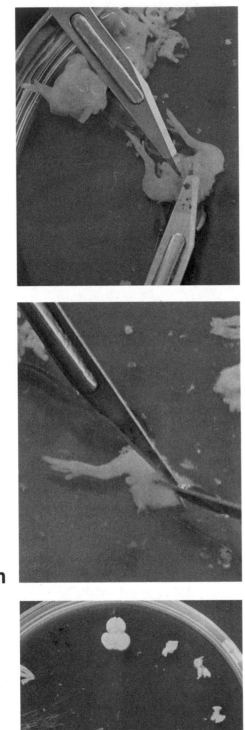

Fig. 11.11. *Continued*

5. Pierce the top of the head with curved forceps, and scoop out the brain. (Fig. 11.11f.) Place the brain with the retina at the side of the dish.

6. Halve the trunk transversely where the pink color of the liver shows through the ventral skin. (Fig. 11.11g.) If the incision is made on the line of the diaphragm, then it will pass between the heart and the liver; but sometimes the liver will go to the anterior instead of the posterior half.

7. Gently probe into the cut surface of the anterior half, and draw out the heart and lungs. (Fig. 11.11h; tease the organs out, and do not cut until you have identified them.) Separate the heart and lungs and place at the side of the dish.

8. Probe the posterior half, and draw out the liver, with the folds of the gut enclosed in between the lobes. (Fig. 11.11i.) Separate the liver from the gut and place each at the side of the dish.

9. Fold back the body wall to expose the inside of the dorsal surface of the body cavity in the posterior half. The elongated lobulated kidneys should be visible parallel to and on either side of the midline.

10. Gently slide the tip of the scalpel under each kidney and tease the kidneys away from the dorsal body wall. (Fig. 11.11j.) (This step requires a dissection microscope for 10-d embryos.) Carefully cut the kidneys free, and place them on one side.

11. Place the tips of the scalpels together on the midline at the posterior end, and, advancing the tips forward, one over the other, express the spinal cord as you would express toothpaste from a tube. (Fig. 11.11k.) (This step may be difficult with 10-d embryos.)

12. Turn the posterior trunk of the embryo over, and strip the skin off the back and upper part of the legs. (Fig. 11.11l.) Collect and place this skin on one side.

13. Dissect off muscle from each thigh, and collect this muscle together. (Fig. 11.11m.)

14. Transfer all of these tissues, and any others you may want, to separate test tubes containing 1 ml of 0.25% trypsin, and place these tubes on ice. Make sure that the tissue slides right down the tube into the trypsin.

15. Leave the test tubes for 6–18 h at 4°C.

16. Carefully remove the trypsin from the test tubes; tilting and rolling the tube slowly will help.

17. Incubate the tissue in the residual trypsin for 15–20 min at 37°C.

18. Add 4 ml of medium to each of two 25-cm² flasks for each tissue to be cultured.

19. Add 2 ml of medium to tubes containing tissues and residual trypsin, and pipette up and down gently to disperse the tissue.
20. Allow any large pieces of tissue to settle, pipette off the supernatant fluid into the first flask, mix, and transfer 1 ml of diluted suspension to the second flask. This procedure gives two flasks at different cell concentrations and avoids the need to count the cells. Experience will determine the appropriate cell concentration to use in subsequent attempts.
21. Change the medium as required (e.g., for brain, it may need to be changed after 24 h, but pigmented retina will probably last 5–7 d), and check for characteristic morphology and function.

Analysis. After 3–5 d, contracting cells may be seen in the heart cultures, colonies of pigmented cells in the pigmented retina culture, and the beginning of myotubes in skeletal muscle cultures. Culture may be fixed and stained (see Staining in Chapter 15) for future examination.

Other Enzymatic Procedures

Disaggregation in trypsin can be damaging (e.g., to some epithelial cells) or ineffective (e.g., for very fibrous tissue, such as fibrous connective tissue), so attempts have been made to utilize other enzymes. Since the extracellular matrix often contains collagen, particularly in connective tissue and muscle, collagenase has been the obvious choice [Freshney, 1972 (colon carcinoma); Lasfargues, 1973 (breast carcinoma); Chen et al., 1989 (kidney); Kralovanszky et al., 1990 (gut); Heald et al., 1991 (pancreatic islet cells)]. (See also Gastrointestinal Tract, Liver, Pancreas in Chapter 22). Other bacterial proteases, such as pronase [Schaffer et al., 1997; Glavin et al., 1996] and dispase (Boehringer-Mannheim) [Compton et al., 1998; Inamatsu et al., 1998] have also been used with varying degrees of success. The participation of carbohydrate in intracellular adhesion has led to the use of hyaluronidase [Berry and Friend, 1969] and neuraminidase in conjunction with collagenase. Other proteases continue to appear on the market (e.g., Sigma, Boehringer, Worthington). With the selection now available, screening available samples is the only option if trypsin, collagenase, dispase, pronase, hyaluronidase, and DNase, alone and in combinations, do not prove to be successful.

It is not possible to describe here all of the primary disaggregation techniques that have been used, but the method described in the next section has been found to be effective in several normal and malignant tissues.

Collagenase

This technique is very simple and effective for many tissues: embryonic, adult, normal, and malignant. It is of greatest benefit when the tissue is either too fibrous or too sensitive to allow the successful use of trypsin [Freshney, 1972]. Crude collagenase is often used and may depend, for some of its action, on contamination with other nonspecific proteases. More highly purified grades are available if nonspecific proteolytic activity is undesirable, but they may not be as effective as crude collagenase.

PROTOCOL 11.8. DISAGGREGATION IN COLLAGENASE

Outline
Place finely chopped tissue in complete medium containing collagenase and incubate. When tissue is disaggregated, remove collagenase by centrifugation, seed cells at a high concentration, and culture. (Fig. 11.12.)

Materials
Sterile:
Collagenase (2,000 units/ml), Worthington CLS or Sigma 1A
Culture medium, e.g., DMEM/F12 with 10% FBS
DBSS
Pipettes, 1 ml, 10 ml
Petri dishes, 9 cm, non-tissue culture grade
Culture flasks, 25 cm²
Centrifuge tubes, 15–50 ml, depending on the amount of tissue being processed
Scalpels
Nonsterile:
Centrifuge

Protocol
1. Transfer the tissue to fresh, sterile DBSS, and rinse.
2. Transfer the tissue to a second dish; dissect off unwanted tissue, such as fat or necrotic material; and chop finely with crossed scalpels (see Fig. 11.12) to about 1-mm cubes.
3. Transfer the tissue by pipette (10–20 ml, with wide tip) to a 15- or 50-ml sterile centrifuge tube or universal container. (Wet the inside of the pipette first with DBSS, or else the pieces will stick.) Allow the pieces to settle.
4. Wash the tissue by resuspending the pieces in DBSS, allowing the pieces to settle, and removing the supernatant fluid. Repeat this step two more times.

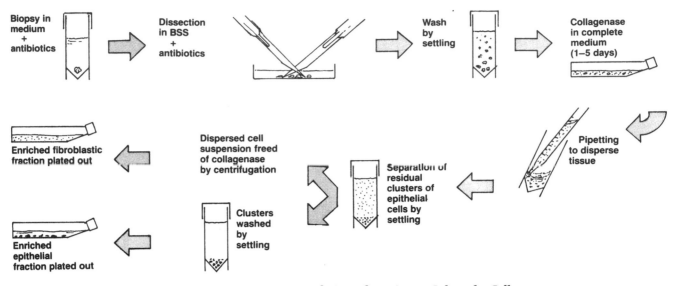

Fig. 11.12. Stages in Disaggregation of Tissue for Primary Culture by Collagenase..

5. Transfer 20–30 pieces to one 25-cm² flask and 100–200 pieces to a second flask.
6. Drain off the DBSS, and add 4.5 ml of growth medium with serum to each flask.
7. Add 0.5 ml of crude collagenase, 2,000 units/ml, to give a final concentration of 200 units/ml collagenase.
8. Incubate at 37°C for 4–48 h without agitation. Tumor tissue may be left up to 5 d or more if disaggregation is slow (e.g., in scirrhous carcinomas of the breast or the colon), although it may be necessary to centrifuge the tissue and resuspend it in fresh medium and collagenase before that amount of time has passed if an excessive drop in pH is observed (i.e., <pH 6.5).
9. Check for effective disaggregation by gently moving the flask; the pieces of tissue will "smear" on the bottom of the flask and, with moderate agitation, will break up into single cells and small clusters. (Fig. 11.13.) With some tissues (e.g., lung, kidney, and colon or breast carcinoma), small clusters of epithelial cells can be seen to resist the collagenase and may be separated from the rest by allowing them to settle for about 2 min. If these clusters are further washed with DBSS by resuspension and settling and the sediment is seeded in medium, then they will form healthy islands of epithelial cells. Epithelial cells generally survive better if they are not completely dissociated.
10. When complete disaggregation has occurred, or when the supernatant cells are collected after clusters are allowed to settle, centrifuge the mixture at 50–100 g for 3 min.
11. Discard supernatant DBSS, resuspend, combine

the pellets in 5 ml of medium, and seed in a 25-cm² flask. If the pH fell during collagenase treatment (to pH 6.5 or less by 48 h), then dilute the suspension twofold to threefold in medium after removing the collagenase.
12. Replace the medium after 48 h.

Some cells, particularly macrophages, may adhere to the first flask during the collagenase incubation. Transferring the cells to a fresh flask after collagenase treatment (and subsequent removal of the collegenase) removes many of the macrophages from the culture. The first flask may be cultured as well, if required. Light trypsinization will remove any adherent cells other than macrophages.

Disaggregation in collagenase has proved particularly suitable for the culture of human tumors, mouse kidney, human adult and fetal brain, lung, and many other tissues, particularly epithelium. The process is gentle and requires no mechanical agitation or special equipment. With more than 1 g of tissue, however, it becomes tedious at the dissection stage and can be expensive, due to the amount of collagenase required. It will also release most of the connective tissue cells, accentuating the problem of fibroblastic outgrowth, so it may need to be followed by selective culture or cell separation. (See Selective Media in Chapter 9, and Chapter 14).

The discrete clusters of epithelial cells produced by disaggregation in collagenase (see step 9 of Protocol 11.8) and by the cold-trypsin method (see Plates 2, 3) can be selected under a dissection microscope and transferred to individual wells in a microtitration plate, alone or with irradiated or mitomycin C–treated feeder cells. (See Protocols 13.3, 23.1.)

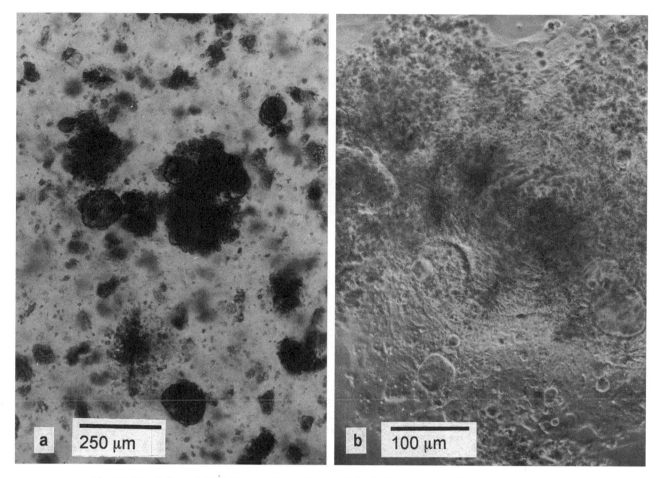

Fig. 11.13. Cells and Cell Clusters from Human Colonic Carcinoma. After 48 h dissociation in crude collagenase (Worthington CLS grade). (a) Before removal of collagenase. (b) After removal of collagenase, further disaggregation by pipetting, and culture for 48 h. The clearly defined rounded clusters in (a) form epithelial-like sheets in (b) and the more irregularly shaped clusters produce fibroblasts.

The addition of hyaluronidase aids disaggregation by attacking terminal carbohydrate residues on the surface of the cells. This combination has been found to be particularly effective for dissociating rat or rabbit liver, by perfusing the whole organ *in situ* [Berry and Friend, 1969; Seglen, 1975; Kralovansky et al., 1990] (see also Liver in Chapter 22) and completing the disaggregation by stirring the partially digested tissue in the same enzyme solution for a further 10–15 min, if necessary. This technique gives a good yield of viable hepatocytes and is a good starting point for further culture. (See Protocol 22.6.)

Collagenase may also be used in conjunction with trypsin. Cahn et al. [1967] developed a formulation that included chick serum, collagenase, and trypsin. Chick serum has a moderating effect on the activity of the trypsin, but does not inhibit it to the extent expected of other sera.

Mechanical Disaggregation

The outgrowth of cells from primary explants is a relatively slow process and can be highly selective. Enzymatic digestion, discussed earlier in this chapter, is rather more labor intensive, though, potentially, it gives a culture that is more representative of the tissue. As there is a risk of damaging cells during enzymatic digestion, many people have chosen to use the alternative of mechanical disaggregation—for example, collecting the cells that spill out when the tissue is carefully sliced [Lasfargues, 1973], pressing the dissected tissue through a series of sieves for which the mesh is gradually reduced in size (see Fig. 11.14), or, alternatively forcing the tissue fragments through a syringe and needle [Zaroff et al., 1961] or simply pipetting it repeatedly. This procedure gives a cell suspension more quickly than does enzymatic digestion, but may cause mechanical damage. The following pro-

DISAGGREGATION OF TISSUE BY SIEVING

(1) Press the tissue gently through a 100-μm sieve.

SYRINGE PISTON

SIEVE

MEDIUM OR SERUM WITH BSS, HEPES BUFFERED

PETRI DISH

(2) Lift the sieve with strong forceps or an artery clamp, and wash the dispersed cells and clumps through the sieve, leaving the debris behind.

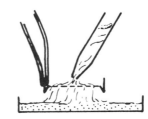

(3) Transfer the suspension of cells and clumps to a second, finer mesh sieve (≈50 μm), and repeat steps (1) and (2).

Repeat this step with a 20-μm mesh sieve, if desired. (See text.)

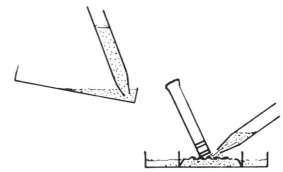

(4) Count the cells using a hemocytometer with dye exclusion viability stain.

(5) Dilute the cells in medium with serum. Seed at 10^6 viable cells per ml, and incubate.

b

tocol is one method of mechanical disaggregation that has been found to be moderately successful with soft tissues, such as brain.

PROTOCOL 11.9. MECHANICAL DISAGGREGATION BY SIEVING

Outline

The tissue in culture medium is forced through a series of sieves for which the mesh is gradually reduced in size until a reasonable suspension of single cells and small aggregates is obtained. The suspension is then diluted and cultured directly.

Materials

Sterile:
Culture medium
Forceps
Sieves (1 mm, 100 μm, 20 μm (Fig. 11.14a), or
  Falcon "Cell Strainer," (see Fig. 11.10))
9-cm Petri dishes
Scalpels
Disposable plastic syringes (2 ml or 5 ml)
Culture flasks

Protocol

1. After washing and preliminary dissection of the tissue (see steps 1 and 2 of Protocol 11.5), chop the tissue into pieces about 3–5 mm across, and place a few pieces at a time into a stainless steel or polypropylene sieve of 1-mm mesh in a 9-cm Petri dish (see Fig. 11.14b) or centrifuge tube (see Fig. 11.10).
2. Force the tissue through the mesh into medium by applying gentle pressure with the piston of a disposable plastic syringe. Pipette more medium through the sieve to wash the cells through it.
3. Pipette the partially disaggregated tissue from the Petri dish into a sieve of finer porosity, perhaps 100-μm mesh, and repeat step 2.
4. The suspension may be diluted and cultured at this stage, or it may be sieved further through 20-μm mesh if it is important to produce a single-cell suspension. In general, the more highly dispersed the cell suspension, the higher the sheer stress required and the lower the resulting viability.

5. Seed the culture flasks at 2×10^5, 1×10^6, and 2×10^6 cells/ml by diluting the cell suspension in medium.

Only soft tissues, such as spleen, embryonic liver, embryonic and adult brain, and some human and animal soft tumors, respond well to this technique. Even with brain, for which fairly complete disaggregation can be obtained easily, the viability of the resulting suspension is lower than that achieved with enzymatic digestion, although the time taken may be very much less. When the availability of tissue is not a limitation and the efficiency of the yield is unimportant, it may be possible to produce, in a shorter amount of time, as many viable cells with mechanical disaggregation as with enzymatic digestion, but at the expense of very much more tissue.

Separation of Viable and Nonviable Cells

When an adherent primary culture is prepared from dissociated cells, nonviable cells are removed at the first change of medium. With primary cultures maintained in suspension, nonviable cells are gradually diluted out when cell proliferation starts. If necessary, however, nonviable cells may be removed from the primary disaggregate by centrifuging the cells on a mixture of Ficoll and sodium metrizoate (e.g., Hypaque or Triosil) [Vries et al., 1973]. This technique is similar to the preparation of lymphocytes from peripheral blood. (See Protocol 26.1.) The dead cells will form a pellet at the bottom of the tube.

PROTOCOL 11.10 ENRICHMENT OF VIABLE CELLS

Outline

Up to 2×10^7 cells in 9 ml of medium may be layered on top of 6 ml of Ficoll-paque (Gibco, Nycomed, ICN) in a 25-ml screw-capped centrifuge bottle. The mixture is then centrifuged, and viable cells are collected from the interface.

Materials

Sterile:
Cell suspension with as few aggregates as possible
Clear centrifuge tubes or universal containers
PBSA

Fig. 11.14. Disaggregation by Sieving. (a) Stainless-steel sieves suitable for disaggregating tissue. (b) Disaggregation of tissue by sieving. (c) Falcon disposable sieve; it nests in the top of a 50-ml centrifuge tube. Can be used for mechanical disaggregation, as in (b), or for filtering aggregates out of a disaggregated suspension.

TABLE 11.3. Data Record for Primary Culture

Date............................ Time.. Operator..

Origin of tissue	Species	Record
	Race or strain	
	Age	
	Sex	
	Path. # or animal tag #	
	Tissue	
	Site	
	Stored tissue/DNA location	
Pathology		
Disaggregation agent	Trypsin, collagenase, etc.	
	Concentration	
	Duration	
	Solvent	
Cell count	Concentration after resuspension (C_I)	
	Volume (V_I)	
	Yield ($Y = C_I \times V_I$)	
	Yield per g (wet weight of tissue)	
Seeding	Number (N) and type of vessel (flask, dish, or plate wells)	
	Final concentration (C_F)	
	Volume per flask, dish, or well (V_F)	
Medium	Type	
	Batch no.	
	Serum type and concentration	
	Batch no.	
	Other additives	
	CO_2 concentration	
Matrix coating	e.g., fibronectin, Matrigel, collagen	
Other parameters		

Ficoll Hypaque or equivalent (ICN, Nygaard, Pharmacia) (Ficoll/metrizoate, adjusted to 1.077 g/cc)
Growth medium
Syringe with blunt canula or square-cut needle, Pasteur pipettes, or pipettor
Nonsterile:
Hemocytometer or cell counter
Centrifuge

Protocol

1. Allow major aggregates in the cell suspension to settle.
2. Layer 9 ml of the cell suspension onto 6 ml of the Ficoll-paque mixture. This step should be done in a wide, transparent centrifuge tube with a cap, such as the 25-ml Sterilin or Nunclon universal container, or in the clear plastic Corning 50-ml tube, using double the aforementioned volumes.
3. Centrifuge the mixture for 15 min at 400 *g* (measured at the center of the interface).
4. Carefully remove the top layer without disturbing the interface.
5. Collect the interface carefully with a syringe, Pasteur pipette, or pipettor.

△ *Safety Note.* If you are using human or other primate material, do not use a sharp needle or Pasteur pipette.

6. Dilute the mixture to 20 ml in medium (e.g., DMEM/F12/10FB).
7. Centrifuge the mixture at 70 *g* for 10 min.
8. Discard the supernatant fluid, and resuspend the pellet in 5 ml of growth medium.
9. Repeat steps 7 and 8 in order to wash cells free of density medium.
10. Count the cells using a hemocytometer on an electronic counter.
11. Seed the culture flask(s).

This procedure can be scaled up or down and works with lower ratios of density medium to a cell suspension (e.g., 5 ml of cell suspension over 1 ml of density medium).

Primary Culture in Summary

The disaggregation of tissue and preparation of the primary culture make up the first, and perhaps most vital, stage in the culture of cells with specific functions. If the required cells are lost at this stage, then the loss is irrevocable. Many different cell types may be cultured by choosing the correct techniques. (See Selective Media in Chapter 9, and Chapter 22.) In general, trypsin is more severe than collagenase, but is sometimes more effective in creating a single-cell suspension. Collagenase does not dissociate epithelial cells readily, but this characteristic can be an advantage for separating the epithelial cells from stroma cells. Mechanical disaggregation is much quicker than the procedure using collagenase, but damages more cells. The best approach is to try out the techniques described in Protocols 11.4–11.9 and select the method that works best in your system. If none of those methods is successful, try using additional enzymes, such as pronase, dispase, and DNase, and consult the literature for examples of previous work with the tissue in which you are interested.

Primary Records

Regardless of the technique used to produce a culture, it is important to keep proper records of the culture's origin and derivation, including the species, sex and tissue from which it was derived, any relevant pathology, and the procedures used for disaggregation and primary culture. (Table 11.3.) If you are working under good laboratory practice (GLP) conditions, then such records are not just desirable, but also obligatory [Food and Drug Administration, 1992; Department of Health and Social Security, 1986]. Records can be kept in a notebook or another hard-copy file of record sheets, but it is best at this stage to initiate a record in a computer database, such as Microsoft Access; this record then becomes the first step in maintaining the *provenance* of the cell line. The database record may never proceed beyond the primary culture stage, but, looking at the opposite extreme, it could be the first stage in creating an accurate record of what will become a valuable cell line. If the record becomes irrelevant, it can always be deleted, but it cannot, with any accuracy, be created later if the cell line assumes some importance.

As it may be necessary to authenticate the origin of a cell line at a later stage in its life history, particularly if cross-contamination is suspected, it is important to save a sample of tissue, blood, or DNA from the same individual at the time of isolation of tissue for the primary culture. This sample can then be used as reference material for DNA fingerprinting. (See DNA Fingerprinting in Chapter 15.)

CHAPTER 12

Cell Lines

NOMENCLATURE

The first *subculture* represents an important transition for a culture. The need to subculture implies that the primary culture has increased to occupy all of the available substrate. Hence, cell proliferation has become an important feature. While the primary culture may have a variable growth fraction (see Growth Fraction in Chapter 20), depending on the type of cells present in the culture, after the first subculture, the growth fraction is usually high (80% or more). From a very heterogeneous primary culture, containing many of the cell types present in the original tissue, a more homogeneous cell line emerges. In addition to its biological significance, this process has considerable practical importance, as the culture can now be propagated, characterized, and stored, and the potential increase in cell number and the uniformity of the cells open up a much wider range of experimental possibilities. (See Table 1.4.)

SUBCULTURE AND PROPAGATION

Once a primary culture is subcultured (or *passaged*, or *transferred*), it becomes known as a *cell line*. This term implies the presence of several cell lineages of either similar or distinct phenotypes. If one cell lineage is selected, by cloning (see Protocol 13.1), by physical cell separation (see Chapter 14), or by any other selection technique, to have certain specific properties that have been identified in the bulk of the cells in the culture, this cell line becomes known as a *cell strain* (see Glossary). Some commonly used cell lines and cell strains are listed in Table 12.1. (See also Table 2.1.) If a cell line transforms *in vitro*, it gives rise to a *continuous cell line* (see Evolution of Cell Lines in Chapter 2 and Immortalization in Chapter 17), and if selected or cloned and characterized, it is known as a *continuous cell strain*. The relative advantages and disadvantages of finite cell lines and continuous cell lines are listed in Table 12.2.

The first subculture gives rise to a *secondary* culture, the secondary to a *tertiary*, and so on, although in practice, this nomenclature is seldom used beyond the tertiary culture. Since the importance of culture lifetime was highlighted by Hayflick and others with regard to diploid fibroblasts [Hayflick and Moorhead, 1961], when each subculture divided the culture in half (the *split ratio* being 1:2), passage number has often been confused with generation number. The *passage number* is the number of times that the culture has been subcultured, while the *generation number* is the number of doublings that the cell population has undergone, given that the number of doublings in the primary culture is very approximate. When the split ratio is 1:2, then, as in Hayflick's experiments, the passage number is approximately equal to the generation number. Many laboratories subculture at split ratios greater than 1:2, and if this is the case, then it is useful to use a power of 2, e.g., 1:2, 1:4, 1:8, 1:16, to give the equivalent of one, two, three, or four doublings, respectively. None of these approximations takes account of cell loss through necrosis, apoptosis, or differentiation or premature ageing and withdrawal from cycle, which probably take place at every growth cycle between each subculture.

TABLE 12.1. Commonly Used Cell Lines

Cell line	Morphology	Origin	Species	Age	Ploidy	Characteristics	Reference
Finite, from Normal Tissue							
IMR-90	Fibroblast	Lung	Human	Embryonic	Diploid	Susceptible to human viral infection; contact inhibited	Nichols et al., 1977
MRC-5	Fibroblast	Lung	Human	Embryonic	Diploid	Susceptible to human viral infection; contact inhibited	Jacobs, 1970
MRC-9	Fibroblast	Lung	Human	Embryonic	Diploid	Susceptible to human viral infection; contact inhibited	Jacobs, 1979
WI-38	Fibroblast	Lung	Human	Embryonic	Diploid	Susceptible to human viral infection	Hayflick and Moorhead, 1961
Continuous, from Normal Tissue							
293	Epithelial	Kidney	Human	Embryonic	Aneuploid	Readily transfected.	Graham et al., 1977
3T3-A31	Fibroblast	Kidney	Mouse BALB/c	Embryonic	Aneuploid	Contact inhibited; readily transformed	Aaronson and Todaro, 1968
3T3-L1	Fibroblast		Mouse Swiss	Embryonic	Aneuploid	Adipose differentiation	Green and Kehinde, 1974
BEAS-2B	Epithelial	Lung	Human	Adult			Reddel et al., 1988
BHK21-C13	Fibroblast	Kidney	Syrian hamster	Newborn	Aneuploid	Transformable by polyoma	Macpherson and Stoker, 1962
BRL 3A	Epithelial	Liver	Rat	Newborn		Produce IGF-2	Coon, 1968
C2	Fibroblastoid	Skeletal muscle	Mouse	Embryonic		Myotubes	Morgan et al., 1992
C7	Epithelioid	Hypothalamus	Mouse			Neurophysin; vasopressin	De Vitry et al., 1974
CHO-K1	Fibroblast	Ovary	Chinese hamster	Adult	Diploid	Simple karyotype	Puck et al., 1958
COS-1, COS-7	Epithelioid	Kidney	Pig	Adult	Diploid	Good hosts for DNA transfection	Gluzman, 1981
CPAE	Endothelial	Pulmonary-artery endothelium	Cow	Adult	Diploid	Factor VIII, Angiotensin II converting enzyme	Del Vecchio and Smith, 1981
HaCaT	Epithelial	Keratinocytes	Human	Adult	Diploid	Cornification	Boukamp et al., 1988
L6	Fibroblastoid	Skeletal muscle	Rat	Embryonic	Diploid	Myotubes	Richler and Yaffe, 1970
LLC-PK1	Epithelial	Kidney	Pig	Adult	Diploid	Na(su)+-dependent glucose uptake	Hull et al., 1976; Saier, 1984
MDCK	Epithelial	Kidney	Dog	Adult	Diploid	Domes, transport	Gaush et al., 1966; Rindler et al., 1979
NRK49F	Fibroblast	Kidney	Rat	Adult	Aneuploid	Induction of suspension growth by TGf-α,β	DeLarco and Todaro, 1978
STO	Fibroblast		Mouse	Embryonic	Aneuploid	Used as feeder layer for embryonal stem cells	Bernstein, 1975
Vero	Fibroblast	Kidney	Monkey	Adult	Aneuploid	Viral substrate and assay	Hopps et al., 1963
Continuous, from Neoplastic Tissue							
A2780	Epithelial	Ovary	Human	Adult	Aneuploid	Chemosensitive with resistant variants	Tsuruo et al., 1986
A549	Epithelial	Lung	Human	Adult	Aneuploid	Synthesizes surfactant	Giard et al., 1972
A9	Fibroblast	Subcutaneous	Mouse	Adult	Aneuploid	HGPRT-ve: deriv. L929	Littlefield, 1964b
B16	Fibroblastoid	Melanoma	Mouse	Adult	Aneuploid	Melanin	Nilos and Makarski, 1978
C1300	Neuronal	Neuroblastoma	Rat	Adult	Aneuploid	Neurites	Lebermann and Sachs, 1978

Cell line	Morphology	Tissue	Species	Age	Ploidy	Characteristics	Reference
C6	Fibroblastoid	Glioma	Rat	Newborn	Aneuploid	Glial fibrillary acidic protein, GPDH	Benda et al., 1968
Caco-2	Epithelial	Colon	Human	Adult	Aneuploid	Transports ions and amino acids	Fogh, 1977
EB-3	Lymphocytic	Peripheral blood	Human	Juvenile	Diploid	EB virus +ve	Epstein and Barr, 1964
Friend	Suspension	Spleen	Mouse	Adult	Aneuploid	Hemoglobin	Scher et al., 1971
GH1, GH2, GH3	Epithelioid	Pituitary tumor	Rat	Adult	Aneuploid	Growth hormone	Buonassisi et al., 1962; Yasumura et al., 1966
H4-II-E-C3	Epithelial	Hepatoma	Rat	Adult	Aneuploid	Tyrosine aminotransferase	Pitot et al., 1964
HeLa	Epithelial	Cervix	Human	Adult	Aneuploid	G6PD Type A	Gey et al., 1952
HeLa-S3	Epithelial	Cervix	Human	Adult	Aneuploid	High plating efficiency; will grow well in suspension	Puck and Marcus, 1955
HEP-G2	Epithelioid	Hepatoma	Human	Adult	Aneuploid	Retains some microsomal metabolizing enzymes	Knowles et al., 1980
HL-60	Suspension	Myeloid leukemia	Human	Adult	Aneuploid	Phagocytosis Neotetrazolium Blue reduction	Olsson and Ologsson, 1981
HT-29	Epithelial	Colon	Human	Adult	Aneuploid	Differentiation inducible with NaBt	Fogh and Trempe, 1975
K-562	Suspension	Myeloid leukemia	Human	Adult	Aneuploid	Hemoglobin	Andersson et al., 1979a,b
L1210	Lymphocytic		Mouse	Adult	Aneuploid	Rapidly growing; suspension	Law et al., 1949
L929	Fibroblast		Mouse	Adult	Aneuploid	Clone of L-cell	Sanford et al., 1948
LS	Fibroblast		Mouse	Adult	Aneuploid	Grow in suspension; derived from L929	Paul and Struthers, personal communication
MCF-7	Epithelial	Pleural effusion from breast tumor	Human	Adult	Aneuploid	Estrogen receptor +ve, domes, α-lactalbumin	Soule et al., 1973
MCF-10	Epithelial	Fibrocystic mammary tissue	Human	Adult	Near diploid	Dome formation	Soule et al., 1990
MOG-G-CCM	Epithelioid	Glioma	Human	Adult	Aneuploid	Glutamyl synthetase	Balmforth et al., 1986
P388D1	Lymphocytic		Mouse	Adult	Aneuploid	Grow in suspension	Dawe and Potter, 1957; Koren et al., 1975
S180	Fibroblast		Mouse	Adult	Aneuploid	Cancer chemotherapy screening	Dunham and Stewart, 1953
SK/HEP-1	Endothelial	Hepatoma, endothelium	Human	Adult	Aneuploid	Factor VIII	Heffelfinger et al., 1992
WEHI-3B D+	Suspension	Marrow	Mouse	Adult	Aneuploid	IL-3 production	Nicola, 1987
ZR-75-1	Epithelial	Ascites fluid from breast tumor	Human	Adult	Aneuploid	ER-ve, EGFr+ve	Engel et al., 1978

HeLa-contaminated: cell lines shown to be predominately HeLa cells by GPDH analysis, karyotype, and DNA fingerprint

Cell line	Morphology	Tissue	Species	Age	Ploidy	Characteristics	Reference
Chang liver	Epithelial	Liver	Human	Embryonic	Aneuploid		Chang, 1954
Girardi Heart	Epithelial	Heart	Pig	Adult	Aneuploid		
Hep-2	Epithelial	Larynx	Human	Adult	Aneuploid		Moore et al., 1955
KB	Epithelial	Oral cavity	Human	Adult	Aneuploid		Eagle, 1955
WISH	Epithelial	Amnion	Human	Newborn	Aneuploid		Hayflick, 1961

TABLE 12.2. Properties of Finite and Continuous Cell Lines

Properties	Finite	Continuous (transformed)
Ploidy	Euploid, eiploid	Aneuploid, heteroploid
Transformation	Normal	Immortal, growth-control altered, and tumorigenic
Anchorage dependence	Yes	No
Contact inhibition	Yes	No
Density limitation of cell proliferation	Yes	Reduced or lost
Mode of growth	Monolayer	Monolayer or suspension
Maintenance	Cyclic	Steady state possible
Serum requirement	High	Low
Cloning efficiency	Low	High
Markers	Tissue specific	Chromosomal, enzymic, antigenic
Special functions (e.g., virus susceptibility, differentiation)	May be retained	Often lost
Growth rate	Slow (T_D of 24–96 h)	Rapid (T_D of 12–24 h)
Yield	Low	High
Control parameters	Generation no.; Tissue-specific markers	Stain characteristics

IMMORTALIZATION OF CELL LINES

Cell lines with limited culture lifespans are known as *finite* cell lines and behave in a fairly reproducible fashion. (See Senescence in Chapter 2.) They grow through a limited number of cell generations, usually between 20 and 80 population doublings, before extinction. The actual number of doublings depends on species and cell lineage differences, clonal variation, and culture conditions, but is consistent for one cell line grown under the same conditions. It is, therefore, important that reference to a cell line should express the approximate generation number or number of doublings since explanation; I say "approximate" because the number of generations that have elapsed in the primary culture is difficult to assess.

CELL LINE DESIGNATIONS

Cell lines should also be given a code or designation (e.g., normal human brain (NHB)); a cell-strain or cell-line number (if several cell lines were derived from the same source; e.g., NHB1, NHB2, etc.); and, if cloned, a clone number (e.g., NHB2-1, NHB2-2, etc.). It is useful to keep a log book or computer database file where the receipt of biopsies or specimens is recorded prior to initiation of a culture. In such a case, the accession number in the log book or database file, perhaps linked to an identifier letter code, can be used to establish the cell-line designation; for example, LT156 would be lung tumor biopsy number 156. This method is less likely to generate ambiguities, such as the same letter code being used for two different cell lines, and gives automatic reference to the record of accession of the line.

For finite cell lines, the number of population doublings should be estimated and indicated after a slash, e.g., NHB2/2, and increases by one for a split ratio of 1:2 (e.g., NHB2/2, NHB2/3, etc.), by two for a split ratio of 1:4 (e.g., NHB2/2, NHB2/4, etc.), and so on. This procedure is more difficult when dealing with a continuous cell line, for which the generation number is very high and often unknown. If the line is immortal, then the generation number is less informative, and the number after the slash may be used to indicate the number of generations or, more simply, the number of passages since the last thaw from the freezer. (See Serial Replacement in Chapter 19).

When referenced in publications or reports, each cell line should be prefixed with a code designating the laboratory in which it was derived (e.g., WI for Wistar Institute, NCI for National Cancer Institute, SK for Sloan Kettering) [Federoff, 1975]. In publications or reports, the cell line should be given its full designation the first time it is mentioned, with the abbreviated or truncated version following in parentheses, and in the Materials and Methods section of the paper and thereafter, its abbreviation may be used.

It is important that, whatever nomenclature is adopted, the cell line designation is unique, or else confusion will arise in subsequent reports in the literature. Cell banks deal with this problem by giving each cell line an accession number; when reporting on cell lines acquired from a cell bank, you should give this accession number in the Materials and Methods. section. Punctuation can also give rise to problems when one is searching for a cell line in a database, so always adhere to a standard syntax, and do not use apostrophes or spaces.

SELECTION OF CELL LINE

Apart from specific functional requirements, there are a number of general parameters to consider in selecting a cell line:

(1) **Finite vs. continuous.** Is there a continuous cell line that expresses the right functions? A continuous cell line generally is easier to maintain, grows faster, clones more easily, and produces a higher cell yield per flask. (See Table 12.2.)

(2) **Normal or Transformed.** Is it important whether the line is malignantly transformed or not? If it is, then it might be possible to obtain an immortalized line that is not tumorigenic.

(3) **Species.** Is species important? Nonhuman cell lines have fewer biohazard restrictions and have the advantage that the original tissue may be more accessible.

(4) **Growth characteristics.** What do you require in terms of growth rate, yield, plating efficiency, and ease of harvesting? You will need to consider the following parameters:
   ◆ Population-doubling time
   ◆ Saturation density (yield per flask)
   ◆ Cloning efficiency
   ◆ Growth fraction
   ◆ Ability to grow in suspension

(5) **Availability.** If you have to use a finite cell line, are there sufficient stocks available, or will you have to generate your own line(s)? If you choose a continuous cell line, are authenticated stocks available?

(6) **Validation.** How well characterized is the line, if it exists already, or, if not, can you do the necessary characterization? (See Authentication, Ch. 15)

(7) **Phenotypic expression.** Can the line be made to express the right characteristics?

(8) **Control cell line.** If you are using a mutant, transfected, transformed, or abnormal cell line, is there a normal equivalent available, should it be required?

(9) **Stability.** How stable is the cell line? Has it been cloned? If not, can you clone it, and how long would this cloning process take to generate sufficient frozen and usable stocks?

ROUTINE MAINTENANCE

Once a culture is initiated, whether it is a primary culture or a subculture of a cell line, it will need a periodic medium change, or "feeding," followed eventually by subculture if the cells are proliferating. In nonproliferating cultures, the medium will still need to be changed periodically, as the cells will still metabolize and some constituents of the medium will become exhausted or will degrade spontaneously. Intervals between medium changes and between subcultures vary from one cell line to another, depending on the rate of growth and metabolism; rapidly growing transformed cell lines, such as HeLa, are usually subcultured once per week, and the medium should be changed four days later. More slowly growing, particularly nontransformed, cell lines may need to be subcultured only every two, three, or even four weeks, and the medium should be changed weekly between subcultures. (For a more detailed discussion of the growth cycle, see Split Ratios and Growth Cycle in this chapter and Growth Cycle in Chapter 20.)

Cell Morphology

Whatever procedure is undertaken, it is vital that the culture be examined regularly to confirm the absence of contamination (see Visible Microbial Contamination in Chapter 18; see Fig. 18.1), the healthy status of the cells, and the lack of any signs of deterioration, such as granularity around the nucleus, cytoplasmic vacuolation, and rounding up of the cells with detachment from the substrate. (Fig. 12.1.) Such signs may imply that the culture requires a medium change, or may indicate a more serious problem, e.g., inadequate or toxic medium or serum, microbial contamination, or senescence of the cell line. During routine maintenance, the medium change or subculture frequency should prevent such deterioration, as it is often difficult to reverse.

Replacement of Medium

Four factors indicate the need for the replacement of culture medium:

(1) **A drop in pH.** The rate of fall and absolute level should be considered. Most cells stop growing as the pH falls from pH 7.0 to pH 6.5 and start to lose viability between pH 6.5 and pH 6.0, so if the medium goes from red through orange to yellow, the medium should be changed. Try to estimate the rate of fall; a culture at pH 7.0 that falls 0.1 pH units in one day will not come to harm if left a day or two longer before feeding, but a culture that falls 0.4 pH units in one day will need to be

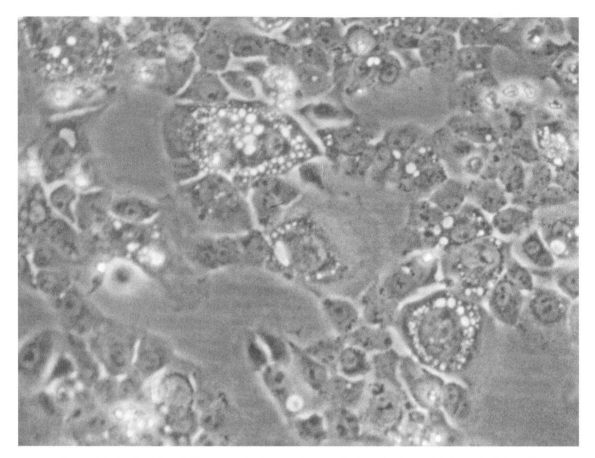

Fig. 12.1. Unhealthy Cells. Vacuolation and granulation in bronchial epithelial cells (BEAS-2B). The cytoplasm of the cells becomes granular, particularly around the nucleus, and vacuolation occurs. The cells may become more refractile at the edge if cell spreading is impaired.

fed within 24–48 hr and cannot be left over a weekend without feeding.

(2) **Cell concentration.** Cultures at a high cell concentration exhaust the medium faster than those at a low concentration. This factor is usually evident in the rate of change of pH, but not always.

(3) **Cell type.** Normal cells (e.g., diploid fibroblasts) usually stop dividing at a high cell density (see Contact Inhibition and Density Limitation of Growth in Chapter 17), due to cell crowding, growth factor depletion, and other reasons. The cells block in the G1 phase of the cell cycle and deteriorate very little, even if left for two to three weeks or longer. Transformed cells, continuous cell lines, and some embryonic cells, however, deteriorate rapidly at high cell densities unless the medium is changed daily or they are subcultured.

(4) **Morphological deterioration.** (See Cell Morphology in this chapter.) This factor must be anticipated by regular examination and familiarity with the cell line. If deterioration is allowed to progress too far, it will be irreversible.

Volume, Depth, and Surface Area

The usual ratio of medium volume to surface area is $0.2–0.5$ ml/cm². The upper limit is set by gaseous diffusion through the liquid layer, and the optimum ratio depends on the oxygen requirement of the cells. Cells with a high O_2 requirement do better in shallow medium (e.g., 2 mm), and those with a low requirement may do better in deep medium (e.g., 5 mm). If the depth of the medium is greater than 5 mm, then gaseous diffusion may become limiting. With monolayer cultures, this problem can be overcome by rolling the bottle or perfusing the culture with medium and arranging for gas exchange in an intermediate reservoir. (See Hollow Fibers in Chapter 24 and Perfusion Systems in Chapter 25.)

PROTOCOL 12.1. CHANGING THE MEDIUM OR FEEDING

Outline

Examine the culture by eye and on an inverted microscope. If indicated, e.g. by a fall in pH, remove the old medium and add fresh medium. Return the culture to the incubator.

Materials

Sterile:
For a list of standard materials, see Protocol 5.1
Pipettes
Medium

Protocol

1. Prepare the hood, bring the reagents and materials necessary for procedure, and place those required immediately in the hood, after swabbing it and them with 70% alcohol. (See Protocol 5.1.)
2. Examine the culture carefully for signs of contamination or deterioration. (See Figs. 12.1 and 18.1.)
3. Check the previously described criteria—pH and cell density or concentration—and, based on your knowledge of the behavior of the culture, decide whether or not to replace the medium. If feeding is required, proceed as follows.
4. Take the culture to the sterile work area.
5. Remove and discard the medium.
6. Add the same volume of fresh medium, prewarmed to 37°C if it is important that there be no check in cell growth.
7. Return the culture to the incubator.

Note. When a culture is at a low density and growing slowly, it may be preferable to half-feed it—i.e., to remove only half of the medium at step 3 and replace it in step 4 with the same volume as was removed.

Holding Medium

A *holding medium* may be used when stimulation of mitosis, which usually accompanies a medium change, even at high cell densities, is undesirable. Holding media are usually regular media with the serum concentration reduced to 0.5 or 2% or eliminated completely. For serum-free media, growth factors and other mitogens are omitted. This omission inhibits mitosis in most untransformed cells. Transformed cell lines are unsuitable for this procedure, as either they may continue to divide successfully or the culture may deteriorate, because transformed cells do not block in a regulated fashion in G_1 of the cell cycle. (See Cell Cycle in Chapter 2).

Holding media are used to maintain cell lines with a finite lifespan without using up the limited number of cell generations available to them. (See Senescence in Chapter 2.) Reduction of serum and cessation of cell proliferation also promote expression of the differentiated phenotype in some cells [Maltese and Volpe, 1979; Schousboe et al., 1979]. Media used for the collection of biopsy samples can also be referred to as holding media.

Subculture

The growth of cells in culture usually follows a standard pattern. (Fig. 12.2.) A lag following seeding is followed by a period of exponential growth, called the log phase. When the *cell density* (cells/cm² substrate) reaches a level such that all of the available substrate is occupied, or when the *cell concentration* (cells/ml medium) exceeds the capacity of the medium, growth ceases or is greatly reduced. (See Plates 5–8.) Then either the medium must be changed more frequently or the culture must be divided. For an adherent cell line, dividing a culture, or *subculture* as it is called, usually involves removal of the medium and dissociation of the cells in the monolayer with trypsin, although some loosely adherent cells (e.g., HeLa-S₃) may be subcultured by shaking the bottle, collecting the cells in the medium, and diluting as appropriate in fresh medium in new bottles. Exceptionally, some cell monolayers cannot be dissociated in trypsin and require the action of alternative proteases, such as pronase, dispase, and collagenase. (Table 12.3.) Of these proteases, pronase is the most effective, but can be toxic to some cells. Dispase and collagenase are generally less toxic than trypsin, but may not give complete dissociation of epithelial cells. The severity of the treatment depends on the cell type, and a protocol should be selected with the least severity that is compatible with the generation of a single-cell suspension of high viability.

The attachment of cells to each other and to the culture substrate is mediated by cell surface glycoproteins, and Ca^{2+}. (See Cell Adhesion in Chapter 2.) Other proteins, and proteoglycans, derived from the cells and from the serum, become associated with the cell surface and the surface of the substrate and facilitate cell adhesion. Subculture usually requires chelation of Ca^{2+} and degradation of extracellular matrix and, potentially, cell adhesion molecules.

Criteria for Subculture

The need to subculture a monolayer is determined by the following criteria:

(1) **Density of culture.** Normal cells should be subcultured as soon as they reach confluence. If left more than 24 h, they will withdraw from the cycle and take longer to recover when reseeded. Transformed cells will continue to proliferate beyond confluence, but will start to deteriorate after about two doublings, and reseeding efficiency will de-

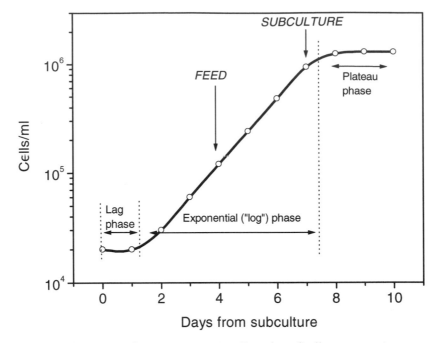

Fig. 12.2. *Growth Curve and Maintenance.* Semilog plot of cell concentration versus time from subculture, showing the lag phase, the exponential phase, and a plateau, and indicating times at which subculture and feeding should be performed.

cline. These cells also should be subcultured on reaching confluence.

(2) **Exhaustion of medium.** Exhaustion of the medium (see Replacement of Medium in this chapter) usually indicates that the medium requires replacement, but if a fall in pH occurs so rapidly that the medium must be changed more frequently, then subculture may be required. Usually, a drop in pH is accompanied by an increase in cell density, which is the prime indicator of the need to subculture.

(3) **Time since last subculture.** Routine subculture is best performed according to a strict schedule, so that reproducible behavior is achieved and monitored. If cells have not reached a high-enough density (i.e., they are not confluent) by the appropriate time, then increase the seeding density, or if they reach confluence too soon, then reduce the seeding density. Once this routine is established, the recurrent growth should be consistent in duration and cell yield from a given seeding density. Deviations from this pattern then signify a departure from normal conditions or indicate deterioration of the cells. Ideally, a cell concentration should be found that allows for the cells to be subcultured after 7 d, with the medium being changed after 3–4 d.

(4) **Requirements for Other Procedures.** When cells are required for purposes other than routine propagation, they also have to be subcultured, in order to increase the stock or to change the type of culture vessel or medium. Ideally, this procedure should be done at the regular subculture time, when it will be known that the culture is performing routinely, what the reseeding conditions should be, and what outcome can be expected. However, demands for cells do not always fit the established routine for maintenance, and compromises have to be made, but (a) cells should not be subcultured while still within the lag period, and (b) cells should always be taken between the middle of the log phase and the time before which they have entered the plateau phase of a previous subculture (unless there is a specific requirement for plateau phase cells, in which case they will need frequent feeding or continuous perfusion).

PROTOCOL 12.2. SUBCULTURE

Outline
Remove the medium. Expose the cells briefly to trypsin. Incubate the cells. Disperse the cells in medium. Count the cells. Dilute and reseed the subculture.

Materials
For a list of standard materials, see Protocol 5.1.
Sterile:
Complete growth medium

Trypsin, 0.25% in PBSA, or 1 mM EDTA (see Table 12.3)
PBSA
Culture flasks
Nonsterile:
Hemocytometer or electronic cell counter

Protocol

1. Prepare the hood, and bring the reagents and materials to the hood to begin the procedure. (See Standard Procedure in Chapter 5.)
2. Examine the culture carefully for signs of contamination or deterioration. (See Figs. 12.1 and 18.1.)

3. Check the criteria, (see Criteria for Subculture in this chapter) and, based on your knowledge of the behavior of the culture, decide whether or not to subculture. If subculture is required, proceed as follows.
4. Take the culture to a sterile work area, and remove and discard the medium.
5. Add PBSA prewash (0.2 ml/cm²) to the side of the flask opposite the cells so as to avoid dislodging cells, rinse the cells, and discard the rinse. This step is designed to remove traces of serum that would inhibit the action of the trypsin.
6. Add trypsin (0.1 ml/cm²) to the side of the flask

TABLE 12.3. Cell Dissociation Procedures

Procedure	Pretreatment	Dissociation agent	Medium	Applicable to
Shake-off	None	Gentle mechanical shaking, rocking, or vigorous pipetting	Culture medium	Mitotic or other loosely adherent cells
Scraping	None	Cell scraper	Culture medium	Cell lines for which proteases are to be avoided (e.g., receptor or cell-surface protein analysis); can damage some cells and rarely gives a single-cell suspension
Trypsin[a] alone	Remove medium completely	0.01–0.5% crude trypsin; usually 0.25%	PBSA, CMF, or saline citrate	Most continuous cell lines
Prewash + trypsin	PBSA	0.25% crude trypsin	PBSA	Some strongly adherent continuous cell lines and many early-passage cells
Prewash + trypsin	1 mM EDTA in PBSA	0.25% crude trypsin	PBSA	Strongly adherent early-passage cell lines
Prewash + trypsin	1 mM EDTA in PBSA	0.25% crude trypsin	PBSA + 1 mM EDTA	Many epithelial cells, but some can be sensitive to EDTA; EGTA can be used
Trypsin + collagenase	1 mM EDTA in PBSA	0.25% crude trypsin; 200 U/ml crude collagenase	PBSA + 1 mM EDTA	Dense cultures and multilayers, particularly with fibroblasts
Dispase	None	0.1–1.0 mg/ml Dispase	Culture medium	Removal of epithelium in sheets (does not dissociate epithelium)
Pronase	None	0.1–1.0 mg/ml Pronase	Culture medium	Provision of good single-cell suspensions, but may be harmful to some cells
DNase	PBSA or 1 mM EDTA in PBSA	2–10 µg/ml crystalline DNase	Culture medium	Use of other dissociation agents which damage cells and release DNA

[a]Digestive enzymes are available (Difco, Worthington, Boehringer Mannheim, Sigma) in varying degrees of purity. Crude preparations—e.g., Difco trypsin, 1:250, or Worthington CLS–grade collagenase—contain other proteases that may be helpful in dissociating some cells, but may be toxic to other cells. Start with a crude preparation, and progress to purer grades if necessary. Purer grades are often used at a lower concentration (mg/ml), as their specific activities (enzyme units/g) are higher. Purified trypsin at 4°C has been recommended for cells grown in low-serum concentrations or in the absence of serum [McKeehan, 1977] and is generally found to be more consistent. Batch testing and reservation, as for serum, may be necessary for some applications.

opposite the cells. Turn the flask over and lay it down. Ensure that the monolayer is completely covered. Leave the flask stationary for 15–30 s, and then withdraw all but a few drops of the trypsin, making sure that the monolayer has not detached. Using trypsin at 4°C helps to prevent premature detachment.

7. Incubate, with the flask lying flat, until the cells round up (Figs. 12.3 and 12.4 and see Plates 8–11); when the bottle is tilted, the monolayer should slide down the surface. (This usually occurs after 5–15 min.) Do not leave the flask longer than necessary, but on the other hand, do not force the cells to detach before they are ready to do so, or else clumping may result.

Note. In each case, the main dissociating agent, be it trypsin or EDTA, is present only briefly, and the incubation is performed in the residue after most of the dissociating agent has been removed. If you encounter difficulty in getting cells to detach and, subsequently, in preparing a single-cell suspension, you may employ alternative procedures, such as those described in Table 12.3.

8. Add medium (0.1–0.2 ml/cm²), and disperse the cells by repeated pipetting over the surface bearing the monolayer. Finally, pipette the cell suspension up and down a few times, with the tip of the pipette resting on the bottom corner of bottle, taking care not to create a foam. The degree of pipetting required will vary from one cell line to another; some cell lines disperse easily, while others require vigorous pipetting in order to disperse them. Almost all cells incur mechanical damage from shearing forces if pipetted too vigorously; primary suspensions and early-passage cultures are particularly prone to damage, due partly to their greater fragility and partly to their larger size, but continuous cell lines are usually more resilient and require vigorous pipetting for complete disaggregation. Pipette the suspension up and down sufficiently to disperse the cells into a single-cell suspension. If this step is difficult, apply a more aggressive dissociating agent. (See Table 12.3.)

9. A single-cell suspension is desirable at subculture to ensure an accurate cell count and uniform growth on reseeding. It is essential if quantitative estimates of cell proliferation or of plating efficiency are being made and if cells are to be isolated as clones.

10. Count the cells using a hemocytometer or an electronic particle counter. (See Cell Counting in Chapter 20.)

11. Dilute the suspension to the appropriate seeding concentration

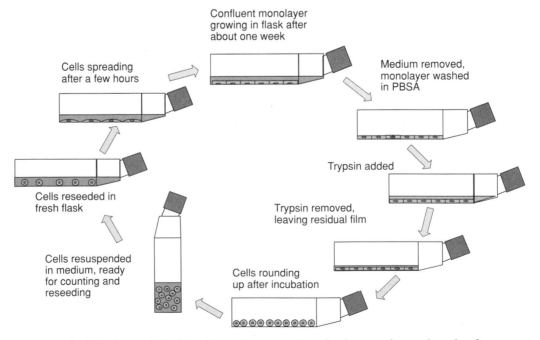

Fig. 12.3. Subculture of the Monolayer. Stages in the subculture and growth cycle of monolayer cells following trypsinization. (See also Plates 5–8.)

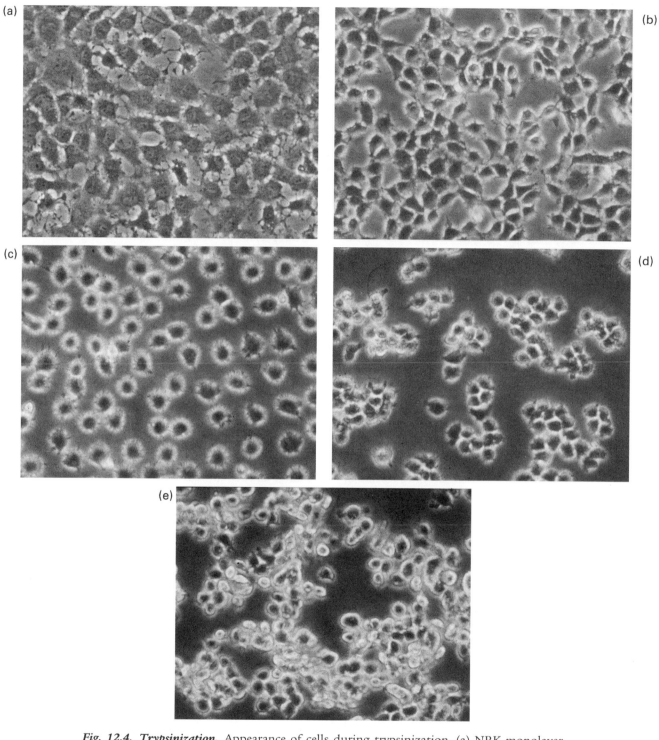

Fig. 12.4. Trypsinization. Appearance of cells during trypsinization. (a) NRK monolayer after PBSA/EDTA prewash and before trypsinization. Some retraction of the cells from the effect of the EDTA is already visible. (b) NRK monolayer immediately after the removal of trypsin. (c) NRK monolayer 1 min after the removal of trypsin. The cells are almost completely rounded up. (d) NRK monolayer 5 min after the removal of trypsin. The cells are rounding up and detaching, as evidenced by their tendency to form clumps. (e) NRK monolayer 15 min after the removal of trypsin. The cells are now brightly refractile, indicating that they are fully rounded up and detached. (See also Plates 8–11.)

(a) by adding the appropriate volume of cells to a premeasured volume of medium in a culture flask or

(b) by diluting the cells to the total volume required and distributing that volume among several flasks.

Procedure (a) is useful for routine subculture when only a few flasks are used and precise cell counts and reproducibility are not critical, but procedure (b) is preferable when setting up several replicates, because the total number of manipulations is reduced and the concentrations of cells in each flask will be identical.

12. If the cells are grown in elevated CO_2, gas the flask by blowing the correct gas mixture from a premixed cylinder, or a gas blender, through a filtered line into the flask above the medium. Do not bubble gas through the medium, as doing so will generate bubbles, which can denature some constituents of the medium and increase the risk of contamination. If the normal gas phase is air, as with Eagle's medium with Hanks' salts, this step may be omitted.

13. Cap the flask(s), and return them to the incubator. Check after about 1 h for a change in the pH. If the pH rises in a medium with a gas phase of air, then return the flask(s) to the aseptic area and gas the culture briefly (1–2 s) with 5% CO_2. Since each culture will behave predictably in the same medium, you eventually will know which cells to gas when they are reseeded, without having to incubate them first. If the pH rises in medium that already has a 5%-CO_2 gas phase, either increase the CO_2 to 7% or 10% or add sterile 0.1 N of HCl.

Note. The procedure in step 13 should not become a long-term solution to the problem of high pH after subculture. If the problem persists, then reduce the pH of the medium at the time it is made up, and check the pH of the medium in the incubator or in a gassed flask.

The expansion of air inside plastic flasks causes larger flasks to swell and prevents them from lying flat. Release the pressure 30 min after placing the flask in the incubator by briefly slackening the cap, or, alternatively, this problem may be prevented by compressing the top and bottom of large flasks before sealing them. Incubation then restores the correct shape. Care must be taken not to crack the flasks.

Split Ratios and Growth Cycle

When handling a cell line for the first time, or when using an early-passage culture with which you have little experience, it is good practice to subculture the cell line to a split ratio of 2 or 4 at the first attempt. As you gain experience and the cell line seems established in the laboratory, it may be possible to increase the split ratio—i.e., to reduce the cell concentration after subculture—but always keep a flask at a low split ratio when attempting to increase its split ratio.

For finite cell lines, it is convenient to reduce the cell concentration at subculture by two-, four-, eight-, or sixteenfold (i.e., a split ratio of 2, 4, 8, and 16, respectively), making the calculation of the number of population doublings easier (i.e., respectively, a split ratio of 2 corresponds to 1 population doubling, 4 to 2, 8 to 3, and 16 to 4); for example, a culture divided eightfold requires three doublings to return to the same cell density. A fragile or slowly growing line should also be split 1:2, while a vigorous, rapidly growing line should be split 1:8 or 1:16. Once a cell line becomes continuous (usually taken as beyond 150 or 200 generations), the generation number is disregarded, and the culture should simply be cut back to between 10^4 and 10^5 cells/ml. The split ratio, or dilution, is also chosen to establish a convenient subculture interval, perhaps 1 or 2 weeks, and to ensure that the cells (1) are not diluted below the concentration that permits them to reenter the growth cycle with a lag period of 24 h or less and (2) do not enter a plateau before the next subculture.

Even when a split ratio is employed, the number of cells per flask should be recorded after trypsinization and at reseeding, so that the growth rate can be estimated at each subculture and the consistency can be monitored. (See Protocols 20.7 and 20.8.) Otherwise, minor alterations will not be detected for several passages.

Routine passage leads to the repetition of a standard growth cycle. (See Fig. 12.5.) It is essential to become familiar with this cycle for each cell line that is handled, as this cycle controls the seeding concentration, the duration of growth before subculture, the duration of experiments, and the appropriate times for sampling to give greatest consistency. Cells at different phases of the growth cycle behave differently with respect to proliferation, enzyme activity, glycolysis and respiration, synthesis of specialized products, and many other properties.

Cell Concentration at Subculture

The ideal method of determining the correct seeding density is to perform a growth curve at different seeding concentrations (see Protocols 20.7 and 20.8) and thereby determine the minimum concentration that will give a short lag period and early entry in rapid logarithmic growth (i.e., a short population-doubling time), but will reach the top of the exponential phase at a time that is convenient for the next subculture.

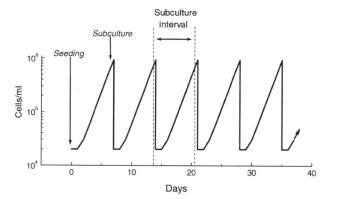

Fig. 12.5. Serial Subculture. Repeated growth cycle over several weeks. If the cells are growing correctly, then they should reach the same concentration after the same time in each cycle, given that the seeding concentration and subculture interval remain constant.

TABLE 12.4. Monolayer vs. Suspension Culture

Monolayer	Suspension
Culture requirements	
Cyclic maintenance	Steady state
Trypsin passage	Dilution
Limited by surface area	Volume (gas exchange)
Growth properties	
Contact inhibition	Homogeneous suspension
Cell interaction	
Diffusion boundary	
Useful for	
Cytology	Bulk production
Mitotic shake-off	Batch harvesting
In situ extractions	
Continuous product harvesting	
Applicable to	
Most cell types, including primaries	Only transformed cells

As a general rule, most continuous cell lines subculture satisfactorily at a seeding concentration of between 1×10^4 and 5×10^4 cells/ml, finite fibroblast cell lines subculture at about the same concentration, and more fragile cultures, such as endothelium and some early-passage epithelia, subculture at around 1×10^5/ml. For a new culture, start at a high seeding concentration and gradually reduce until a convenient growth cycle is achieved without any deterioration in the culture.

Propagation in Suspension

Protocol 12.2 refers to the subculture of monolayers, since most primary cultures and continuous lines grow in this way. Cells that grow continuously in suspension, either because they are nonadhesive (e.g., many leukemias and murine ascites tumors) or because they have been kept in suspension mechanically, or selected, may be subcultured like bacteria or yeast. These cells have a number of advantages (Table 12.4); for example, trypsin treatment is not required, the whole process is quicker and less traumatic for the cells, and scale-up is easier. (See also Suspension in Chapter 25). Replacement of the medium is not usually carried out with suspension cultures, and instead, the culture is either diluted and expanded, or some cell suspension is withdrawn and the residue is diluted back to an appropriate seeding concentration. In either case, a growth cycle will result, similar to that for monolayer cells, but usually with a shorter lag period.

Cells that grow spontaneously in suspension can be maintained in regular culture flasks, which need not be tissue-culture treated (although they must be sterile, of course). The rules regarding the depth of medium in static cultures are as for monolayers—i.e., 2–5 mm to allow for gas exchange. When the depth of

a suspension culture is increased—e.g., if it is expanded—the medium requires agitation, which is best achieved with a suspended rotating magnetic pendulum (Techne), with the culture flask placed on a magnetic stirrer. (Fig. 12.6; see also Suspension in Chapter 25.) Roller bottles rotating on a rack can also be used to agitate suspension cultures. (See Table 7.1 and Fig. 25.11.)

Subculture of Suspension Culture

Protocol 12.3 describes routine subculture of a suspension culture into a fresh vessel. A continuous culture can be maintained in the same vessel, but the probability of contamination gradually increases with any buildup of minor spillage on the neck of the flask during dilution. The criteria for subculture are similar to those for monolayers:

(1) **Cell concentration**, which should not exceed 1×10^6/ml for most suspension-growing cells.
(2) **pH**, which is linked to cell concentration.
(3) **Time since last subculture**, which, as for monolayers, should fit a regular schedule.
(4) **Cell production requirements** for experimental or production purposes.

PROTOCOL 12.3. SUBCULTURE IN SUSPENSION

Outline

Withdraw a sample of the cell suspension, count the cells, and seed an appropriate volume of the cell

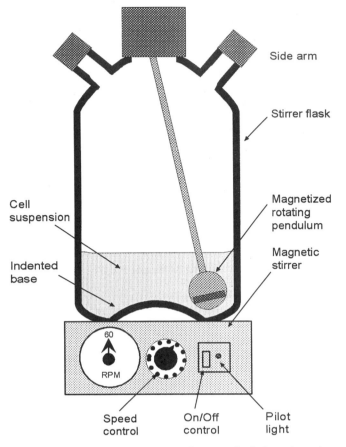

Fig. 12.6. Stirrer Culture. A small stirrer flask, based on the Techne design, with a capacity of 250–1000 ml. The cell suspension is stirred by a rotating pendulum, which rotates in an annular depression in the base of the flask.

suspension into fresh medium in a new flask, restoring the cell concentration to the starting level.

Materials
For a list of standard materials, see Protocol 5.1.
Sterile:
Starter culture
Growth medium (e.g., Eagle's MEM with Spinner Salts (S-MEM) or RPMI 1640 plus 5% calf serum)
Stirrer flask with magnetic pendulum stirrer (Techne, Bellco)
Nonsterile:
Magnetic stirrer
Hemocytometer or electronic cell counter

Protocol
1. Prepare the hood, and bring the reagents and materials to the hood to begin the procedure. (See Standard Procedure in Chapter 5).
2. Examine the culture carefully for signs of contamination or deterioration. This step is more difficult with suspension cultures than with monolayer cells, but cells that are in poor condition are indicated by shrinkage and/or granularity. Healthy cells should look clear and hyaline, with the nucleus visible on phase contrast, and are often found in small clumps in a static culture.
3. Take the culture to the sterile work area, remove a sample from it, and count the cells in the sample.
4. Based on the previously described criteria under Subculture of Suspension Culture and on your knowledge of the behavior of the culture, decide whether or not to subculture. If subculture is required, proceed as follows.
5. Mix the cell suspension, and disperse any clumps by pipetting the cell suspension up and down.
6. Add the appropriate volume of medium to a fresh flask.
7. Add a sufficient number of cells to give a final concentration of 1×10^5 cells/ml for slow-growing cells (36–48 h doubling time) or 2×10^4/ml for rapidly growing cells (12–24 h doubling time).
8. Gas the culture if it is CO_2-buffered.
9. Cap the flask, and take it to incubator.
10. Regular culture flasks should be laid flat, as for monolayer cultures. Stirrer bottles should be placed on a magnetic stirrer and stirred at 60–100 rpm. Take care that the stirrer motor does not overheat the culture. Insert a polystyrene foam mat under the bottle if necessary. Induction-drive stirrers generate less heat and have no moving parts.

Suspension cultures have a number of advantages. (See Table 12.4.) First, the production and harvesting

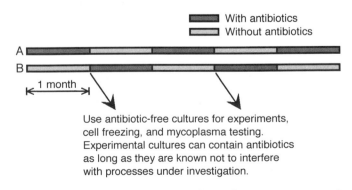

Fig. 12.7. Parallel Cultures and Antibiotics. A suggested scheme for maintaining parallel cell cultures with and without antibiotics, such that each culture always spends part of the time out of antibiotics.

of large quantities of cells may be achieved without increasing the surface area of the substrate. (See Suspension in Chapter 25.) Furthermore, if dilution of the culture is continuous and the cell concentration is kept constant, then a steady state can be achieved; this steady state is not readily achieved in monolayer cultures. Maintenance of monolayer cultures is essentially cyclic, with the result that growth rate and metabolism vary, depending on the phase of the growth cycle.

Use of Antibiotics

The continuous use of antibiotics encourages cryptic contaminations, particularly mycoplasma, and the development of antibiotic-resistant organisms. (See Antibiotics in Chapter 8.) It may also interfere with cellular processes under investigation. However, there may be circumstances for which contamination is particularly prevalent or a particularly valuable cell line is being carried, and in these cases, antibiotics may be used. If they are used, then it is important to maintain

TABLE 12.5. Data Record, Feeding

Date .... Time .... Operator ....

	Date:				
Cell line	Designation				
	Generation or pass no.				
Status	Phase of growth cycle				
	Appearance of cells				
	Density of cells				
	pH of medium (approx.)				
	Clarity of medium				
Medium	Type				
	Batch no.				
	Serum type and concentration				
	Batch no.				
	Other additives				
	CO_2 concentration				
Other parameters					

TABLE 12.6. Data Record, Subculture

Date .... Time .... Operator....................

	Date:				
Cell line	Designation				
	Generation or pass no.				
Status before subculture	Phase of growth cycle				
	Appearance of cells				
	Density of cells				
	pH of medium (approx.)				
	Clarity of medium				
Dissociation agent	Prewash				
	Trypsin				
	EDTA				
	Other				
	Mechanical				
Cell count	Concentration after resuspension (C_I)				
	Volume (V_I)				
	Yield ($Y = C_I \times V_I$)				
	Yield per flask				
Seeding	Number (N) & type of vessel (flask, dish, or plate wells)				
	Final concentration (C_F)				
	Volume per flask, dish, or well (V_F)				
	Split ratio ($Y/C_F \times V_F \times N$), or Number of flasks seeded ÷ Number of flasks trypsinized, where the flasks are of same size				
Medium/serum	Type				
	Batch no.				
	Serum type and concentration				
	Batch no.				
	Other additives				
	CO_2 concentration				
Matrix coating	e.g., fibronectin, Matrigel, collagen				
Other parameters					

some antibiotic-free stocks in order to reveal any cryptic contaminations; these stocks can be maintained in parallel, and stock may be alternated in and out of antibiotics (see Fig. 12.7) until antibiotic-free culture is possible. It is not advisable to adopt this procedure as a permanent regime, and, if a chronic contamination is suspected, the cells should be discarded or the contamination eradicated (see Persistent Contamination in Chapter 18 and Contamination in Chapter 28), and then you may revert to antibiotic-free maintenance.

Maintenance Records

Details of routine maintenance, including feeding and subculture, should be kept (Tables 12.5 and 12.6), and deviations or changes should be added to the database record for that cell line. Such records are required for GLP [Food and Drug Administration, 1992; Department of Health and Social Security, 1986], as for primary-culture records, but are also good practice in any laboratory. If standard procedures are defined, then entries need say only "Fed" or "Subcultured," with the date, a note of the cell numbers, and any comments on visual assessment, and any deviation from the standard procedure should be recorded as well.

This set of records forms part of the continuing provenance of the cell line, and all data, or at least any major event—e.g., if the line is cloned or changed to serum-free medium—should be entered in the database.

CHAPTER 13

Cloning and Selection

It can be seen from the preceding two chapters that a major recurrent problem in tissue culture is the preservation of a specific cell type and its specialized properties. While environmental conditions undoubtedly play a significant role in maintaining the differentiated properties of specialized cells in a culture (see Induction of Differentiation in Chapter 16), the selective overgrowth of unspecialized cells and cells of the wrong lineage is still a major problem.

CLONING

The traditional microbiological approach to the problem of culture heterogeneity is to isolate pure cell strains by cloning, but, while this technique is relatively easy for continuous cell lines, its success in most primary cultures is limited by poor cloning efficiencies. However, the cloning of primary cultures can be successful. Clark and Pateman [1978] isolated a Kupffer cell line from Chinese hamster liver by cloning the primary culture; Sertoli cells have been cloned [Zwain et al., 1991]; and juxtaglomerular [Muirhead et al., 1990] and glomerular cells [Troyer and Kreisberg, 1990] from kidney have also been isolated by cloning.

A further problem of cultures derived from normal tissue is that they may survive only for a limited number of generations (see Senescence in Chapter 2), and by the time that a clone has produced a usable number of cells, it may already be near to senescence. (Fig. 13.1.) Although cloning of continuous cell lines is more successful than cloning finite cell lines considerable heterogeneity may still arise within the clone as

it is grown up for use. (See Genetic Instability in Chapter 17.) Nevertheless, cloning may help to reduce the heterogeneity of a culture.

Cloning is also used as a survival assay (see Plating Efficiency in Chapter 20) for optimizing growth conditions (selection of medium and serum) and for determining chemosensitivity and radiosensitivity.

Cloning may be carried out in monolayer culture, using Petri dishes, multiwell plates, or flasks. Since the cells remain attached, it is relatively easy to discern individual colonies. Micromanipulation is the only conclusive method for determining genuine clonality of a colony (i.e., that the colony was derived from one cell), but when symmetrical colonies are derived from a single-cell suspension, particularly if colony formation is monitored at the early stages, then it is possible to say that the probability of the colonies being clones is high.

Cloning can also be carried out in suspension by seeding cells into a gel, such as agar, or a viscous solution, such as Methocel. The stability of the gel, or viscosity of the Methocel ensures that daughter cells do not break away from the colony as it forms. Hematopoietic cells are usually cloned in suspension; depending on the cells and growth factors used, the colony generates undifferentiated cells with high repopulation efficiency, *in vivo* or *in vitro*, or may mature into colonies of differentiated hematopoietic cells with very little repopulation efficiency. Cloning then becomes an assay for reproductive potential and stem cell identity.

Continuous cell lines generally have a high plating efficiency in monolayer and in suspension due to their

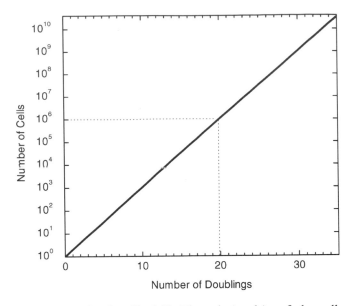

Fig. 13.1. Clonal Cell Yield. The relationship of the cell yield in a clone to the number of population doublings; e.g., 20 doublings are required to produce 10^6 cells.

transformed status, while normal cells, which may have a moderately high cloning efficiency in monolayer, have a very low cloning efficiency in suspension. This distinction has allowed suspension cloning to be used as an assay for transformation. (See Anchorage Independence in Chapter 17.)

Dilution cloning [Puck and Marcus, 1955], is the technique that is used most widely, based on the observation that cells diluted below a certain density form discrete colonies.

PROTOCOL 13.1. DILUTION CLONING

Outline
Seed the cells at low density. Incubate them until colonies form. Stain the cells (if they used as a survival assay) or isolate them, and propagate the cells into a cell strain if they are being used for selection. (Fig. 13.2)

Materials
Sterile:
Medium, 5% CO_2 equilibrated (as normally used for the cell line being used; the concentration of serum may need to be increased if the cells are serum dependent, and fetal bovine is usually better than calf)
Trypsin, 0.25%, 1:250, or equivalent
Pipettes, 1, 5, 10, and 25 ml
Petri dishes, 6 or 9 cm

Tubes, or universal containers, for dilution
Nonsterile:
Hemocytometer or electronic cell counter

Protocol
1. Trypsinize the cells (see Protocol 12.2) to produce a single-cell suspension. Under-trypsinizing will produce clumps and overtrypsinizing will reduce the viability of the cells, but it is fundamental to the concept of cloning that the cells be singly suspended. It may be necessary when cloning a new cell line for the first time to try different lengths of trypsinization and different recipes (see Table 12.3), to give the optimum plating efficiency from a good single-cell suspension.
2. While the cells are trypsinizing, number the dishes (on the side of the base), and measure out medium for the dilution steps. Up to four dilution steps may be necessary to reduce a regular monolayer to a concentration suitable for cloning.
3. When the cells round up and start to detach, disperse the monolayer in medium containing serum or trypsin inhibitor
4. Count the cells, and dilute the cell suspension to the desired seeding concentration. If cloning the cells for the first time, choose a range of 10, 50, 100, and 200 cells/ml (Table 13.1).
5. Seed the Petri dishes with the requisite amount of medium containing cells (see Volume, Depth and Surface Area in Chapter 12).
6. Place the dishes in a transparent plastic sandwich or cake box.
7. Put the box in a humid CO_2 incubator or gassed sealed container (2–10% CO_2, see CO_2 and Bicarbonate in Chapter 8).
8. Leave the cultures untouched for 1 week. If colonies have formed:
   (a) For plating efficiency assay, stain and count the colonies (see Protocol 15.3; see also Plates 12, 13).
   (b) For clonal selection, isolate individual colonies (see Protocol 13.6). If no colonies are visible, replace medium and continue to culture for another week. Feed the dishes again and culture them for a third week if necessary. If no colonies appear by 3 weeks, then it is unlikely that they will appear at all.

Feeding. As the density of cells during cloning is very low, the need to feed the dishes after one week is debatable. Feeding mainly counteracts the loss of nutrients (such as glutamine), which are unstable, and replaces growth factors that have degraded. However, it also increases the risk of contamination, so it is reasonable to leave dishes for two weeks without feed-

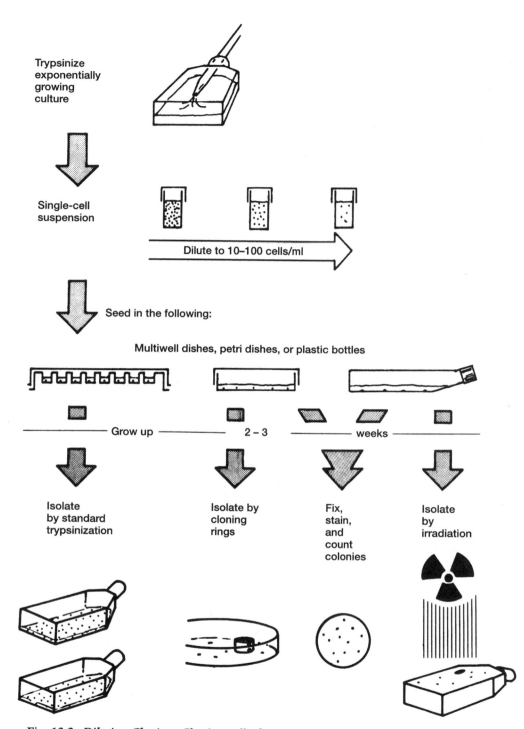

Trypsinize
exponentially
growing
culture

Single-cell
suspension

Dilute to 10–100 cells/ml

Seed in the following:

Multiwell dishes, petri dishes, or plastic bottles

——— Grow up ——— 2 – 3 ——— weeks ———

Isolate
by standard
trypsinization

Isolate by
cloning
rings

Fix,
stain,
and
count
colonies

Isolate
by
irradiation

Fig. 13.2. Dilution Cloning. Cloning cells from a monolayer culture. When clones form,
they may be isolated directly from multiwell dishes (center left and lower left of the figure),
by the cloning ring technique (center and lower center), or by irradiating the flask while
shielding one colony (center right and lower right). If isolation is not required and the
cloning is being performed for quantitative assay (see Protocols 20.9 and 21.3), then the
colonies are fixed, stained, and counted.

TABLE 13.1. Relationship of Seeding Density to Plating Efficiency

Expected plating efficiency (%)	Optimal cell number to be seeded			
	Per ml	Per cm²	Per dish, 6 cm	Per dish, 9 cm
0.1	1×10^4	2×10^3	40,000	100,000
1.0	1×10^3	200	4,000	10,000
10	100	20	400	1,000
50	20	4	80	200
100	10	2	40	100

ing. If it is necessary to leave the dishes for a third week, then the medium should be replaced, or at least half of it.

The preferential formation of colonies at the center of the plate can be due to incorrect seeding, either from seeding the cells into the center of a plate that already contains medium or from swirling the plate such that the cells tend to focus in the center, but it can also be due to resonance in the incubator. (See Incubator in Chapter 4.)

Microtitration Plates. If the prime purpose of cloning is to isolate colonies, then seeding into microtitration plates can be an advantage. When the clones grow up, isolation is easy, but the plates have to be monitored at the early stages in order to mark which wells genuinely have single clones. The statistical probability of a well having a single clone can be increased by reducing the seeding density to a level such that only 1 in 5 or 10 wells would be expected to have a colony; e.g., from Table 13.1, 100 cells/ml at 10% plating efficiency would give 10 colonies/ml, or 1 colony/0.1 ml, as added to a microtitration plate—i.e., 1 colony/well. If the seeding concentration is reduced to 10 cells/ml, then, theoretically, only 1 in 10 wells will contain a colony, and the probability of wells containing more than one colony is very low.

STIMULATION OF PLATING EFFICIENCY

When cells are plated at low densities, the rate of survival falls in all but a few cell lines. This does not usually present a severe problem with continuous cell lines, for which the plating efficiency seldom drops below 10%, but with primary cultures and finite cell lines, the plating efficiency may be quite low—0.5–5%, or even zero. Numerous attempts have been made to improve plating efficiencies, based on the assumption either that cells require a greater range of nutrients at low densities, because of loss by leakage, or that cell-derived diffusible signals or conditioning factors are

present in high-density cultures and are absent or too dilute at low densities. The intracellular metabolic pool of a leaky cell in a dense population will soon reach equilibrium with the surrounding medium, while that of an isolated cell never will. This principle was the basis of the capillary technique of Sanford et al. [1948], by which the L929 clone of L-cells was first produced. The confines of the capillary tube allowed the cell to create a locally enriched environment that mimicked the higher cell-density state. In microdrop techniques developed later, the cells were seeded as a microdrop under liquid paraffin. By keeping one colony separate from another, as in the capillary techniques, colonies could be isolated subsequently. As media improved, however, plating efficiencies increased, and Puck and Marcus [1955] were able to show that cloning cells by simple dilution (as described in Protocol 13.1) in association with a feeder layer of irradiated mouse embryo fibroblasts (see Protocol 13.3) gave acceptable cloning efficiencies, although subsequent isolation required trypsinization from within a collar placed over each colony.

Improving Clonal Growth

(1) **Medium.** Choose a rich medium, such as Ham's F12, or a medium that has been optimized for the cell type in use (e.g., MCDB 110 [Ham, 1984] for human fibroblasts, Ham's F12 or MCDB 302 for CHO [Ham, 1963; Hamilton and Ham, 1977]). (See Tables 9.1 and 9.2 and Chapter 22.)

(2) **Serum.** When serum is required, fetal bovine is generally better than calf or horse. Select a batch for cloning experiments that gives a high plating efficiency during tests.

(3) **Hormones.** Insulin, 1×10^{-10} IU/ml, has been found to increase the plating efficiency of several cell types [Hamilton and Ham, 1977]. Dexamethasone, 2.5×10^{-5} M, 10 µg/ml, a soluble synthetic hydrocortisone analogue, improves the plating efficiency of chick myoblasts and human normal glia, glioma, fibroblasts, and melanoma and

gives increased clonal growth (colony size) if removed five days after plating [Freshney et al., 1980a,b]. Lower concentrations (e.g., 10^{-7} M) have been found to be preferable for epithelial cells. (See Epidermis, Breast, Cervix in Chapter 22).

(4) **Intermediary metabolites.** Keto acids—e.g., pyruvate or α-ketoglutarate [Griffiths and Pirt, 1967; McKeehan and McKeehan, 1979] and nucleosides [α-MEM; Stanners et al., 1971]—have been used to supplement media and are already included in the formulation of a rich medium, such as Ham's F12. Pyruvate is also added to Dulbecco's modification of Eagle's MEM [Dulbecco and Freeman, 1959; Morton, 1970].

(5) **Carbon dioxide.** CO_2 is essential for obtaining maximum cloning efficiency for most cells. While 5% CO_2 is usually used, 2% is sufficient for many cells and may even be slightly better for human glia and fibroblasts. HEPES (20 mM) may be used with 2% CO_2, protecting the cells against pH fluctuations during feeding and in the event of failure of the CO_2 supply. (Using 2% CO_2 also cuts down on the consumption of CO_2.) At the other extreme, Dulbecco's modification of Eagle's MEM is normally equilibrated with 10% CO_2 and is frequently used for cloning myeloma hybrids for monoclonal antibody production. The concentration of bicarbonate must be adjusted if the CO_2 tension is altered, so that equilibrium is reached at pH 7.4. (See Table 8.1.)

(6) **Treatment of substrate.** Polylysine improves the plating efficiency of human fibroblasts in low serum concentrations [McKeehan and Ham, 1976a] (see Matrix Coating in Chapter 7):

(a) Add 1 mg/ml of poly-D-lysine in UPW to the plates (~5 ml/25 cm²).

(b) Remove and wash the plates with 5 ml of PBSA per 25 cm². The plates may be used immediately or stored for several weeks before use.

Fibronectin also improves the plating of many cells [Barnes and Sato, 1980]. The plates may be pretreated with 5 μg/ml of fibronectin incorporated in the medium.

(7) **Trypsin.** Purified (twice recrystallized) trypsin used at 0.05 μg/ml may be preferable to crude trypsin, but opinions vary. McKeehan [1977] noted a marked improvement in plating efficiency when trypsinization (using pure trypsin) was carried out at 4°C.

Conditioned Medium

Medium that has been used for the growth of other cells acquires metabolites, growth factors, and matrix products from these cells. This conditioned medium can improve the plating efficiency of some cells if it is diluted into the regular growth medium.

PROTOCOL 13.2. PREPARATION OF CONDITIONED MEDIUM

Outline

Harvest medium from homologous cells, or a different cell line, from the late log phase. Filter the medium, and dilute it with fresh medium as required.

Materials

Cells for conditioning: same cells, another cell line (e.g., 3T3 cells), or mouse embryo fibroblasts
Cloning medium: Ham's F12, with 10% FBS, or as appropriate for the cells to be cloned.
Sterilizing filter: 0.45 μm or 0.22 μm, filter flask

Protocol

1. Grow conditioning cells to 50% of confluence.
2. Change the medium, and incubate the cells for a further 48 h.
3. Collect the medium.
4. Centrifuge the medium at 1,000 g for 10 min.
5. Filter the medium through a 0.45-μm sterilizing filter. (The medium may need to be clarified first by prefiltration through 5-μm and 1.2-μm filters; see Protocol 10.14.)
6. Store the medium frozen at −20°C.
7. Thaw the medium before use, and add it to cloning medium in the following proportion: 1 part conditioned medium to 2 parts cloning medium.

Note. Centrifugation, freezing and thawing, and filtration steps all help to avoid the risk of carrying any cells over from the conditioning cells. If the same cells are used for conditioning as for cloning, then this problem is less important, but better cloning may be obtained by using a different cell line or primary mouse fibroblasts. If a cell strain is derived by this method, then its identity must be confirmed (e.g., see Protocols 15.10 and 15.11) to preclude cross-contamination from the conditioned medium.

Feeder Layers

The reason that some cells do not clone well is related to their inability to survive at low densities. One way to maintain cells at clonogenic densities, but, at the same time, to mimic high cell densities, is to clone the cells onto a growth-arrested feeder layer. (Fig. 13.3.) The feeder cells may provide nutrients, growth factors, and matrix constituents that enable the cloned cells to survive more readily.

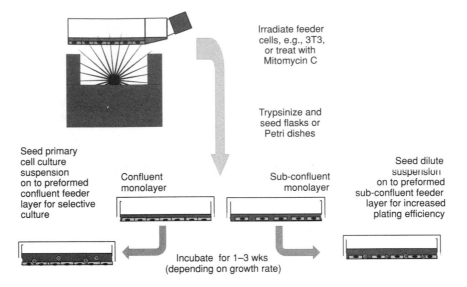

Fig. 13.3. Feeder Layers. Cells are irradiated and trypsinized (or may be trypsinized first and then irradiated in suspension) and plated at a low density to enhance cloning efficiency or at high density to provide a confluent monolayer for selective growth.

PROTOCOL 13.3. PREPARATION OF FEEDER LAYERS

Outline
Plant homologous or heterologous cells—e.g., from mouse embryo—rendered nonproliferative by irradiation or drug treatment at medium density before the cloning of test cells.

Materials
Sterile or aseptically prepared:
Secondary culture of 13-d mouse embryo fibroblasts (see Protocols 11.6 and 12.2)
Culture medium for cells to be cloned
Mitomycin C, 5 μg/ml, in BSS or serum-free medium
Nonsterile:
X-ray or ^{60}Co source capable of delivering 30 Gy in 30 min or less (instead of mitomycin C)

Protocol
1. Trypsinize embryo fibroblasts, from the primary culture (see Protocols 11.6 and 12.2), and reseed the cells at 10^5 cells/ml.
2. At 50% confluence, add mitomycin C, 2 μg/10^6 cells, 0.25 μg/ml, overnight [Macpherson and Bryden, 1971], or irradiate the culture with 30 Gy.
3. Change the medium after treatment, and after a further 24 h, trypsinize the cells and reseed them in fresh medium at 5×10^4 cells/ml (10^4 cells/cm^2).
4. Incubate the culture for a further 24–48 h, and then seed the cells for cloning.

The feeder cells will remain viable for up to 3 weeks, but will eventually die out and are not carried over if the colonies are isolated. Other cell lines or homologous cells may be used to improve the plating efficiency, but heterologous cells have the advantage that if clones are to be isolated later, chromosome analysis will rule out accidental contamination from the feeder layer. Other cell lines that have been used for feeder cells include 3T3, MRC-5, and STO cells. Early-passage mouse embryo cells probably produce more matrix components than do established cell lines, but screening different cells is the only way to be sure which type of cell is best.

SUSPENSION CLONING

Some cells, particularly hematopoietic stem cells and virally transformed fibroblasts, clone readily in suspension. To hold the colony together and prevent mixing, the cells are suspended in agar or Methocel and plated out over an agar underlay or into dishes that need not be treated for tissue culture.

PROTOCOL 13.4. CLONING IN AGAR
This protocol has been submitted by Mary Freshney, Beatson Institute for Cancer Research, Garscube Estate, Switchback Road, Bearsden, Glasgow, G61

1BD, United Kingdom. (See Fig. 13.4 and Protocol 22.20.)

Outline

Agar is a liquid at high temperatures, but is a gel at 37°C. Cells are suspended in warm agar medium and, when incubated after the agar gels, form discrete colonies that may be easily isolated.

Materials

Sterile:

Noble agar, Difco

Medium at double strength (i.e., Ham's F12, RPMI 1640, Dulbecco's MEM, or CMRL 1066). Prepare the medium from 10× concentrate to half the recommended final volume, and add twice the normal concentration of serum.

Fetal bovine serum

Growth medium, 1×, for cell dilutions

Sterile ultrapure water (UPW)

Sterile conical flask

Pipettes, including sterile plastic disposable pipettes for agar solutions

Universal containers, bijoux bottles, or tubes

35-mm Petri dishes, non-tissue-culture grade

Nonsterile:

Bunsen burner and tripod

Water bath at 55°C

Water bath at 37°C

Electronic cell counter or hemocytometer

Tray

Note. Before preparing the medium and the cells, work out the cell dilutions and label the Petri dishes. For an assay to measure the cloning efficiency of a cell line, prepare to set up three dishes for each cell dilution. Convenient cell numbers per 35-mm dish are 1,000, 333, 111, and 37—i.e., one-third dilution of the cell suspension. If any growth factors, hormones, or other supplements are to be added to the dishes, they should be added to the 0.6% agar underlay.

Protocol

1. Number or label the Petri dishes. It is convenient to place them on a tray.
2. Prepare 2× medium containing 40% FBS, and keep it at 37°C.
3. Weigh out 1.2 g of agar.
4. Measure 100 ml of sterile UPW into a sterile conical flask and another 100 ml into a sterile bottle. Add 1.2 g of agar to the flask. Cover the flask, and boil the solution for 2 min. Alternatively, the agar may be sterilized in the autoclave in advance, but, if subsequently stored, it will

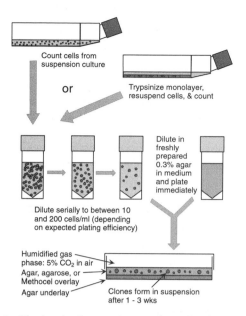

Fig. 13.4. Cloning in Suspension. Cultured cells or primary suspensions from bone marrow or tumors, suspended in agar or low-melting-temperature agarose, which is then allowed to gel, form colonies in suspension. Use of an underlay prevents attachment to the base of the dish.

still need to be boiled, in order to melt it for use.

5. Transfer the boiled agar and the bottle of sterile UPW to a water bath at 55°C.
6. Prepare a 0.6% agar underlay by combining an equal volume of 2× medium and 1.2% agar. Keep the underlay at 37°C. (Fig. 13.5.)

If any growth factors, hormones or other supplements are being used, they should be added to the underlay medium at this point.

Note. If a titration of growth factors is being carried out or a selection of different factors is being

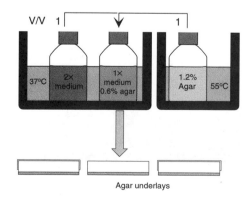

Fig. 13.5. Agar Underlay. Agar, 1.2%, at 55°C is mixed with 2× medium at 37°C and dispensed immediately into dishes, where it is allowed to gel at room temperature or 4°C.

used, then add the required amount to the Petri dishes before the underlay is added.

7. Add 1 ml of 0.6% agar medium to the dishes, mix, and ensure that the medium covers the base of the dish. Leave the dishes at room temperature to set.
8. Prepare the cell suspension, and count the cells.
9. Prepare 0.3% agar medium, and keep it at 37°C. This medium may be prepared by diluting 2× medium at 37°C with 1.2% agar at 55°C and UPW at 55°C in the respective proportions of 2:1:1. (See Fig. 13.6.)
10. Prepare the following cell dilutions, making the top concentration of cells 1×10^5/ml:
    (a) 1×10^5/ml.
    (b) Dilute 1×10^5/ml by 1/3 to give 3.3×10^4/ml.
    (c) Dilute 3.3×10^4/ml by 1/3 to give 1.1×10^4/ml.
    (d) Dilute 1.1×10^4/ml by 1/3 to give 3.7×10^3/ml.
11. Label four bijoux bottles or tubes one for each dilution and pipette 40 μl of each cell dilution, including the 1×10^5/ml concentration, into the respective container. Add 4 ml of 0.3% agar medium at 37°C to each container, mix, and pipette 1 ml from each container onto each of three Petri dishes. (Fig. 13.7.) This will give final concentrations as follows:
    (a) 1×10^3/ml/dish
    (b) 330/ml/dish
    (c) 110/ml/dish
    (d) 37/ml/dish.

Note. Always be sure that the agar medium for the top layer has had adequate time to cool to 37°C before adding the cells to it.

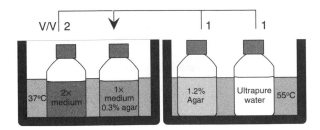

Fig. 13.6. Agar Dilution. Agar, 1.2%, and UPW are maintained at 55°C and mixed with 2× medium to give 0.3% agar for cloning. The use of low-melting-point agarose allows all solutions to be maintained at 37°C, but this agarose can be more difficult to gel.

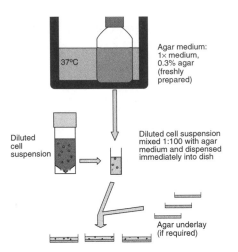

Fig. 13.7. Cloning in Agar. Determination of plating efficiency in suspension. Freshly prepared agar medium is mixed with a series of cell dilutions, prepared as for dilution cloning, diluted at a proportion of 1:100 in agar medium, and plated immediately into dishes, with or without an underlay. Non-tissue-culture-grade plastic should be used if there is no underlay.

12. Allow the solution in the Petri dishes to gel at room temperature.
13. Put the Petri dishes into a clean plastic box with a lid, and incubate them at 37°C in a humid incubator for 10 d.

Agarose, which has reduced sulphated polysaccharides, can be substituted for agar. Some types of agarose have a lower gelling temperature and can be manipulated more easily at 37°C. They are gelled at 4°C and then are returned to 37°C.

Because of the complexity of handling melted agar with cells, and due to the impurities that may be present in agar, some laboratories prefer to use Methocel, which is a viscous solution and not a gel [Buick et al., 1979]. It has a higher viscosity when warm. Because it is a sol and not a gel, cells will sediment through it slowly. It is, therefore, essential to use an underlay with Methocel. Colonies form at the interface between the Methocel and the agar (or agarose) underlay, placing themselves in the same focal plane and making analysis and photography easier.

PROTOCOL 13.5. CLONING IN METHOCEL

Outline

Suspend the cells in medium containing Methocel, and seed the cells into dishes containing gelled agar medium. (Fig. 13.8.)

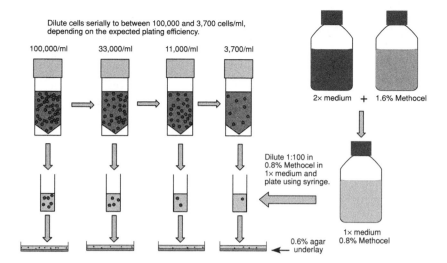

Fig. 13.8. Cloning in Methocel. Determination of plating efficiency in suspension. A series of cell dilutions is prepared as for agar cloning, diluted 1:100 in Methocel medium, and plated into non-tissue-culture-grade dishes or dishes with an agar underlay.

Materials

Sterile:

Noble agar, Difco

Medium at double strength (i.e., Ham's F12, RPMI 1640, Dulbecco's MEM, or CMRL 1066). Prepare the medium from 10× concentrate to half the recommended final volume, and add twice the normal concentration of serum.

Fetal bovine serum

Growth medium, 1×, for cell dilutions

1.6% Methocel, 4 Ns m^{-2} (4,000 cps), in UPW; place on ice (see Reagent Appendix)

Sterile ultrapure water (UPW)

Sterile conical flask

Pipettes, including sterile plastic disposable pipettes for agar solutions

Universal containers, bijoux bottles, or tubes

35-mm Petri dishes, non-tissue-culture grade

Syringes to dispense Methocel (Methocel tends to cling to the inside of pipettes, making dispensing inaccurate)

Nonsterile:

Bunsen burner and tripod

Water bath at 55°C

Water bath at 37°C

Electronic cell counter or hemocytometer

Tray

Protocol

1. Prepare agar underlays as in Protocol 13.4, steps 1–7.
2. Dilute the Methocel to 0.8% with an equal volume of 2× medium. Mix it well, and keep it on ice.
3. Trypsinize monolayer cells, or collect cells from suspension culture and count them.
4. Prepare the following cell dilutions, making the top concentration of cells 1×10^5/ml:
   (a) 1×10^5/ml.
   (b) Dilute 1×10^5/ml by 1/3 to give 3.3×10^4/ml.
   (c) Dilute 3.3×10^4/ml by 1/3 to give 1.1×10^4/ml.
   (d) Dilute 1.1×10^4/ml by 1/3 to give 3.7×10^3/ml.
   Methocel is viscous, so manipulations are easier to perform using a syringe.
5. Label four bijoux bottles or tubes one for each dilution, and pipette 40 μl of each cell dilution, including the 1×10^5/ml concentration, into the respective container. Add 4 ml of 0.8% Methocel medium to each container, and mix well with a vortex or, if the cells are known to be particularly fragile, by sucking the solution gently up and down with a syringe several times. Then use a syringe to add 1 ml from each container to each of three Petri dishes. (See Fig. 13.8.) This will give final concentrations as follows:
   (a) 1×10^3/ml/dish
   (b) 330/ml/dish
   (c) 110/ml/dish
   (d) 37/ml/dish.
6. Incubate the dishes in a humid incubator until colonies form. Since the colonies form at the interface between the agar and the Methocel, fresh medium may be added, 1 ml per dish or well, after 1 week and then removed and replaced with

more fresh medium after 2 weeks without disturbing the colonies.

Many of the recommendations that apply to medium supplementation for monolayer cloning also apply to suspension cloning. In addition, sulfhydryl compounds, such as mercaptoethanol (5×10^{-5} M), glutathione (1 mM), and (α-thioglycerol (7.5×10^{-5} M) [Iscove et al., 1980], are sometimes used. Macpherson [1973] found that the inclusion of DEAE dextran was beneficial for cloning, a finding that was later confirmed by Hamburger et al. [1978], who also found that macrophages enhanced the cloning of tumor cells, although others have found them to be detrimental. Courtenay et al. [1978] incorporated rat red blood cells into the medium and demonstrated that a low oxygen tension enhanced cloning.

Most cell types clone in suspension with a lower efficiency than in monolayer, some cells by two or three orders of magnitude. The isolation of colonies is, however, much easier.

ISOLATION OF CLONES

When cloning is used for the selection of specific cell strains, the colonies that form (see Plate 13) need to be isolated for further propagation. If monolayer cells

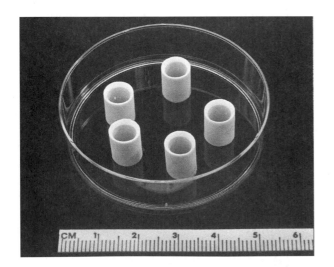

Fig. 13.9. Cloning Rings. Porcelain rings (Fisher) are illustrated, but thick-walled glass (Scientific Laboratory Supplies), stainless steel rings (e.g., roller bearings), or plastic (e.g., cut from nylon, silicone, or Teflon thick-walled tubing) can be used. Whatever the material, the base must be smooth in order to seal with silicone grease onto the base of the Petri dish, and the internal diameter must be just wide enough to enclose one whole clone without the external diameter overlapping adjacent clones.

are cloned directly into multiwell plates (see Protocol 13.1—Microtitration Plates), then colonies may be isolated by trypsinizing individual wells. It is necessary to confirm the clonal origin of the colony during its formation by regular microscopic observation. If, however, cloning is performed in Petri dishes, then there is no physical separation between colonies. This separation must be created by removing the medium and placing a stainless steel or ceramic ring around the colony to be isolated. (Fig. 13.9.)

PROTOCOL 13.6. ISOLATION OF CLONES WITH CLONING RINGS

Outline
The colony is trypsinized from within a porcelain, glass, PTFE, or stainless steel ring and transferred to one of the wells of a 24- or 12-well plate, or directly to a 25-cm² flask. (Fig. 13.10.)

Materials
Sterile:
Cloning rings (Bellco, Scientific Laboratory Supplies, Fisher); sterilize in a glass Petri dish by dry heat or autoclave
Silicone grease; sterilize in a glass Petri dish by dry heat, 160°C for 1 h
Yellow tips for pipettor, or Pasteur pipettes with a bent end (Bellco #1273)
0.25% trypsin in PBSA
Growth medium
24-well plate and/or 25-cm² flasks
Sterile forceps
Nonsterile:
Pipettor, 50–100 μl
Felt-tip pen or, preferably, a Nikon ring marker that fits into the objective nosepiece of a microscope in place of one of the objectives

Protocol
1. Examine the clones, and mark those that you wish to isolate with a felt-tip marker on the underside of the dish, or use a ring marker (Nikon).
2. Remove the medium from the dish, and rinse the clones gently with PBSA.
3. Using sterile forceps, take one cloning ring, dip it in silicone grease, and press it down on the dish alongside the silicone grease, to spread the grease around the base of the ring.
4. Place the ring around the desired colony.
5. Repeat steps 4 and 5 for two or three other colonies in the same dish.

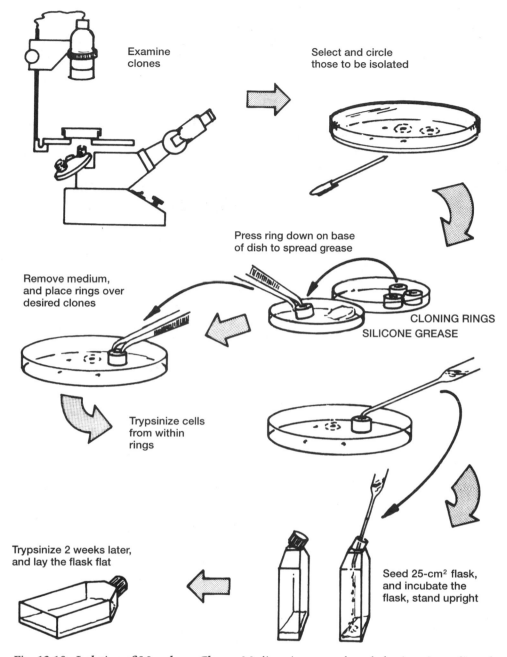

Fig. 13.10. Isolation of Monolayer Clones. Medium is removed, and cloning rings, dipped in silicone grease, are placed around each colony, which is then trypsinized from within the ring.

6. Add a sufficient amount of 0.25% trypsin to fill the hole in the ring (0.1–0.4 ml, depending on the internal diameter of the ring). Leave the trypsin for 20 s, and then remove it.
7. Close the dish, and incubate it for 15 min at 37°C.
8. Add 0.1–0.4 ml of medium to each ring.
9. Taking each clone in turn, pipette the medium up and down to disperse the cells, and transfer the medium to a well of a 24-well plate or to a 25-cm² flask standing on end. Use a separate pipette, or a separate pipettor tip, for each clone.
10. Wash out the ring with another 0.1–0.4 ml of medium, and transfer the medium to the same well or flask.

Note. The dish will dry out if left open for too long. Either limit the number of clones isolated or recover the dish between manipulations.

11. Make up the medium in the wells to 1.0 ml, close the plate, and incubate it. If you are using flasks, then add 1 ml of medium to each flask, stand each flask on end, and incubate the flasks.
12. When the clone grows to fill the well, transfer up to a 25-cm² flask, incubated conventionally with 5 ml of medium. If you are using the up-ended flask technique, then remove the medium when the end of a flask is confluent, trypsinize the cells, resuspend them in 5 ml of medium, and lay the flask down flat. Continue the incubation.

Flasks are available with a removable top film (Nalge Nunc) that may be peeled off to allow harvesting of clones. Alternatively, where an irradiation source is available, clones may be isolated by shielding one and irradiating the rest of the monolayer (30 Gy).

PROTOCOL 13.7. ISOLATING CELL COLONIES BY IRRADIATION

Outline
Invert the flask under an X-ray machine or ⁶⁰Co source, screening the desired colony with lead.

△ *Safety Note.* X-ray machines and ⁶⁰Co sources must be used under strict supervision and with appropriate monitoring to safeguard your own exposure and that of others (see High Energy Sources in Chapter 6). Contact your local radioprotection officer before setting up this type of experiment.

Materials
Sterile:
Growth medium
0.25% trypsin
PBSA
Nonsterile:
X-ray or cobalt source
Pieces of lead cut from a 2-mm-thick sheet and of a size from about 2–5 mm in diameter

Protocol
1. Select the desired colony, and mark it with a felt-tip pen or a Nikon ring marker.
2. Select a piece of lead of appropriate size.
3. Take the flask to the radiation source.
4. Invert the flask under the source.
5. Cover the colony with a 2-mm-thick piece of lead.
6. Irradiate the flask with 30 Gy.

7. Return the flask to a sterile area. Remove the medium, trypsinize the cells, and allow the cells to reestablish in the same bottle, using the irradiated cells as a feeder layer.

If irradiation and trypsinization are carried out when the colony is about 100 cells in size, then the trypsinized cells will reclone. Three serial clonings may be performed within six weeks by this method.

Other Isolation Techniques for Monolayer Clones

(1) Distribute small coverslips or broken fragments of coverslips on the bottom of a Petri dish. When plated out at the correct density, some colonies are found to be singly distributed on a piece of glass and may be transferred to a fresh dish or multiwell plate.
(2) Use the capillary technique of Sanford et al. [1948]. A dilute cell suspension is drawn into a sterile glass capillary tube (e.g., a 50-μl Drummond Microcap), allowing colonies to form inside the tube. The tube is then carefully broken on either side of a colony and transferred to a fresh plate. This technique was exploited by Echarti and Maurer [1989, 1991] for clonogenic assay of hematopoietic cells and tumor cells, for which the colony-forming efficiency can be quantified by scanning the capillary in a densitometer.

Suspension Clones
The isolation of colonies growing in suspension is simple, but requires a dissection microscope.

PROTOCOL 13.8 ISOLATION OF SUSPENSION CLONES

Outline
Draw the colony into a pipettor or Pasteur pipette, and transfer the colony to a flask or the well of a multiwell plate (Fig. 13.11).

Materials
Sterile:
Growth medium in universal container
Multiwell plates, 24 well
Culture flask, 25 cm²
Yellow tips for the pipettor
Nonsterile:
Dissecting microscope, 20–50× magnification
Pipettor, 100 μl
Felt-tip pen or Nikon ring marker

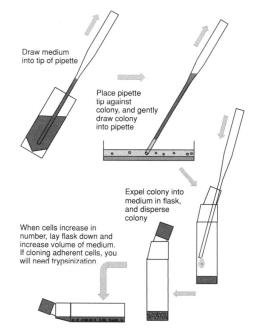

Fig. 13.11. Isolation of Suspension Clones. Mark the colony as for monolayer, then draw the colony into a Pasteur pipette. Transfer the colony to a culture flask, disperse it in medium, and incubate it. Make up medium when cells start to grow.

Protocol

1. Examine the dishes on an inverted microscope, and mark the clones with a felt-tip pen or a Nikon ring marker.
2. Pipette 1 ml of medium into each well of a 24-well plate.
3. Pick the colonies using a dissecting microscope:
   (a) Use a separate pipettor tip for each clone.
   (b) Set the pipettor to 100 μl.
   (c) Draw approximately 50 μl of medium into the pipettor tip, place the tip of the pipettor against the colony to be isolated, and gently draw in the colony in the remaining 50 μl of the stroke of the pipette.
4. Transfer the contents of the pipettor tip to a 24-well dish, and flush out the colony with medium. If the medium is made with Methocel, then the colony will settle, adhere, and grow out. If the medium is made from agar, then you may need to pipette the colony up and down a few times in the well to disperse the agar. Clones may also be seeded directly into a 25-cm² plastic flask that is standing on end. (See Protocol 13.6.)

REPLICA PLATING

Bacterial colonies can be replated by pressing a moist pad gently down onto the colonies growing on a nu-trient agar plate and transferring the pad to a second, fresh agar plate. Various attempts have been made to adapt this technique to cell culture, usually by placing a mesh screen or filter over monolayer clones and transferring it to a fresh dish after a few days [Hornsby et al., 1992]. For clones that have been developed in microtitration plates, there are a number of transfer devices available—e.g., the Corning Costar Transtar (see Figure 4.19), which can be used with suspension cultures directly after agitating the culture or with monolayer cultures following trypsinization and resuspension.

SELECTIVE INHIBITORS

Manipulating the conditions of a culture by using a selective medium is a standard method for selecting microorganisms. Its application to animal cells in culture is limited, however, by the basic metabolic similarities of most cells isolated from one animal, in terms of their nutritional requirements. The problem is accentuated by the effect of serum, which tends to mask the selective properties of different media. Most selective media that have been shown to be generally successful have been serum-free formulations. (See Selective Media in Chapter 9.) A number of metabolic inhibitors, however, have had recurrent success. Gilbert and Migeon [1975, 1977] replaced the L-valine in the culture medium with D-valine and demonstrated that cells possessing D-amino acid oxidase would grow preferentially. Kidney tubular epithelia [Gross et al., 1992], bovine mammary epithelia [Sordillo et al., 1988], endothelial cells from rat brain [Abbott et al., 1992], and Schwann cell cultures [Armati and Bonner, 1990] have been selected in this way. However this technique appears not to be effective against human fibroblasts [Masson et al., 1993].

Much of the effort in developing selective conditions has been aimed at suppressing fibroblastic overgrowth. Kao and Prockop [1977] used cis-OH-proline, although it can prove toxic to other cells. Fry and Bridges [1979] found that phenobarbitone inhibited fibroblastic overgrowth in cultures of hepatocytes, and Braaten et al. [1974] were able to reduce the fibroblastic contamination of neonatal pancreas by treating the culture with sodium ethylmercurithiosalicylate. Fibroblasts also tend to be more sensitive to geneticin (G418) at 100 μg/ml [Levin et al., 1995].

One of the more successful approaches was the development of a monoclonal antibody to the stromal cells of a human breast carcinoma [Edwards et al., 1980]. Used with complement, this antibody proved to be cytotoxic to fibroblasts from several tumors and helped to purify a number of malignant cell lines. Attempts have also been made to kill cells selectively with drug- or toxin-conjugated antibodies, but this approach is complicated by the need to get a stably bound drug or toxin that will become dissociated at the target cell. This has been difficult to achieve. Selective antibodies are now more successfully used in "panning" or magnetizable bead separation techniques. (See Magnetic Sorting in Chapter 14.)

Selective media are also commonly used to isolate hybrid clones from somatic hybridization experiments. HAT medium, a combination of hypoxanthine, aminopterin, and thymidine (see HAT Medium in reagent appendix), selects hybrids with both hypoxanthine guanine phosphoribosyltransferase and thymidine kinase from parental cells deficient in one or the other enzyme (see Selection of Hybrid Clones in Chapter 27) [Littlefield, 1964a].

Transfected cells are also selected by resistance to a number of drugs, such as neomycin, its analogue Geneticin (G418), hygromycin, and methotrexate, by including a resistance-conferring gene in the construct used for transfection (e.g., *neo* (aminoglycoside phosphotransferase), *hph* (hygromycin B phosphotransferase), or *dhfr* (dihydrofolate reductase; see DNA Transfer in Chapter 27). Culture in the correct concentration of the selective marker, determined by titration against the transfected and nontransfected controls, selects for stable transfectants. Selection with methotrexate has the additional advantage that increasing the methotrexate concentration leads to amplification of the *dhfr* gene and can coamplify other genes in the construct.

Negative selection is also possible by using the Herpes simplex virus (HSV) *TK* gene, which activates Ganciclovir (Syntex) into a cytotoxic product. Transfected cells will be sensitized to the drug.

When a mixture of cells shows different responses to growth factors, it is possible to stimulate one cell type with the appropriate growth factor and then, taking advantage of the increased sensitivity of the more rapidly growing cells, kill the cells selectively with irradiation or cytosine arabinoside (ara-c). (See Neurons in Chapter 22). Alternatively, if an inhibitor is known or a growth factor is removed, which will take one population out of the cycle, then remaining cycling cells can be killed with ara-c or irradiation.

ISOLATION OF GENETIC VARIANTS

The following protocol for the development of mutant cell lines that amplify the dihydrofolate reductase (DHFR) gene was contributed by June Biedler, Memorial Sloan-Kettering Cancer Center, New York, New York.

PROTOCOL 13.9. METHOTREXATE RESISTANCE AND DHFR AMPLIFICATION

Principle

Cells exposed to gradually increasing concentrations of folic acid antagonists, such as methotrexate (MTX), over a prolonged period of time will develop resistance to the toxic effects of the drug [Biedler et al., 1972]. Resistance resulting from amplification of the DHFR gene generally develops the most rapidly, although other mechanisms—e.g., alteration in antifolate transport and/or mutations affecting enzyme structure or affinity—may confer part or all of the resistant phenotype.

Outline

Expose the cells to a graded series of concentrations of MTX for weeks, periodically replacing the medium with fresh medium containing the same drug concentration. Select for subculturing those flasks in which a small percentage of cells survive and form colonies. Repeatedly subculture such cells in the same and in two- to tenfold higher MTX concentrations until the cells acquire the desired degree of resistance.

Materials

Sterile:
Chinese hamster cells or rapidly growing human or mouse cell lines
Methotrexate Sodium Parenteral (Lederle Laboratories)
0.15 M of NaCl
Tissue-culture flasks
Pipettes
Culture medium that does not contain thymidine and hypoxanthine (e.g., Eagle's MEM with 10% fetal bovine serum)

Nonsterile:
Inverted microscope
Ultralow-temperature cabinet or liquid-nitrogen freezer

Protocol

1. Clone the parental cell line to obtain a rapidly growing, genotypically uniform population to be used for selection.
2. Dilute the MTX with sterile 0.15 M (0.85%) NaCl. The drug packaged for use in the clinic is in solution at 2.5 mg/ml.
3. Inoculate 2.5×10^5 cells into replicate 25-cm^2 flasks containing no drug or 0.01, 0.02, 0.05, and 0.1 μg/ml of MTX in complete tissue-culture medium. Adjust the pH of each solution to pH 7.4, and incubate the flasks at 37°C for 5–7 d.

4. Observe the cultures with an inverted microscope. Replace the medium with fresh medium containing the same amount of MTX in cultures showing clonal growth of a small proportion of cells amid a background of enlarged, substrate-adherent, and probably dying cells, and reincubate those cultures.

5. Allow the cells to grow for another 5–7 d changing the growth medium as necessary, but continuously exposing the cells to MTX. When the cell density has reached $2–10 \times 10^6$ cells/flask, subculture the cells at 2.5×10^5 cells/flask into new flasks containing the same and two- to tenfold higher drug concentrations.

6. After another 5–7 d, observe the new passage flasks as well as the cultures from the previous passage that had been exposed to higher drug concentrations; change the medium and select for viable cells as before.

7. Continue the selection with progressively higher drug concentrations at each subculture step until the desired level of resistance is obtained: 2–3 months for Chinese hamster cells with low to moderate levels of resistance, increase in DHFR

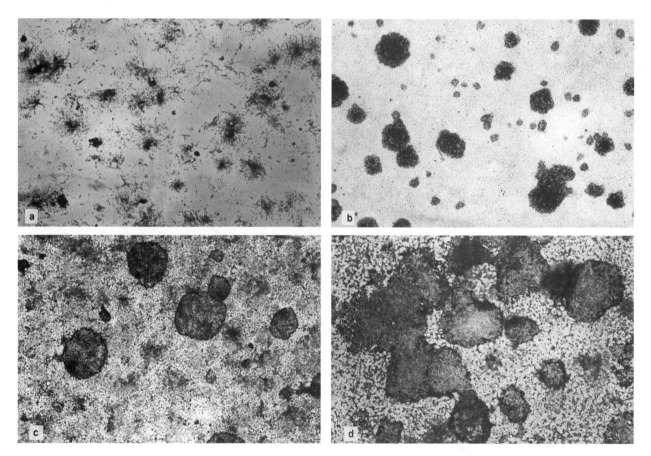

Fig. 13.12. Selective Feeder Layers. Selective cloning of breast epithelium on a confluent feeder layer. (a) Colonies forming on plastic alone after seeding 4,000 cells/cm² (2×10^4 cells/ml) from a breast carcinoma culture. Small, dense colonies are epithelial cells, and larger, stellate colonies are fibroblasts. (b) Colonies of cells from the same culture, seeded at 400 cells/cm² (2,000 cells/ml) on a confluent feeder layer of FHS74Int cells [Owens et al., 1974]. The epithelial colonies are much larger than those in (a), the plating efficiency is higher, and there are no fibroblastic colonies. (c) Colonies from a different breast carcinoma culture plated onto the same feeder layer. Note the different colony morphology with a lighter stained center and ring at the point of interaction with the feeder layer. (d) Colonies from normal breast culture seeded onto FHI cells (fetal human intestine; similar to FHS74Int). There are a few small, fibroblastic colonies present in (c) and (d). (From a technique described by Dr. A. J. Hackett, personal communication.)

activity, and/or transport alteration; 4–6 months or more for high levels of resistance and enzyme overproduction, for Chinese hamster, mouse, or fast-growing human cells.

8. Periodically freeze samples of the developing lines at −70°C or in liquid nitrogen. (See Protocol 19.1.)

Analysis. Characterize resistant cells for levels of resistance to the drug in a clonal growth assay (see Protocol 21.3), for increase in activity or amount of DHFR by biochemical or gel electrophoresis techniques [Albrecht et al., 1972; Melera et al., 1980], and/or for increase in mRNA and copy number of the reductase gene by Northern, Southern, or dot blots [Scotto et al., 1986] with DHFR-specific probes to determine the mechanism(s) of resistance.

Variations. Cell culture media other than Eagle's MEM can be used; the composition of the medium (e.g., folic acid content) can be expected to influence the rate and type of MTX resistance development. Media containing thymidine, hypoxanthine, and glycine (see Tables 8.3 and 9.1) prevent the develop-

ment of antifolate resistance and should be avoided. Cells can be treated with chemical mutagens prior to selection [Thompson and Baker, 1973]; this treatment may also alter the rate and type of mutant selection.

Selection can also be done using cells plated in the drug at low density in 100-mm tissue-culture dishes (with the isolation of individual colonies using cloning rings; see Protocol 13.6), using single cells in 96-well cluster dishes, or in soft agar, to enable the isolation of one or multiple clonal populations at each or any step during resistance development.

Cells can be made to be resistant to a number of other agents, such as antibiotics, other antimetabolites, toxic metals, and so on, by similar techniques; differences in the mechanism of action or degree of toxicity of the agents, however, may require that treatment with the agent be intermittent rather than continuous and may increase the time necessary for selection.

Cell lines of different species or with slower growth rates, such as some human tumor cell lines, may require different (usually lower) initial drug concentrations, longer exposure times at each concentration,

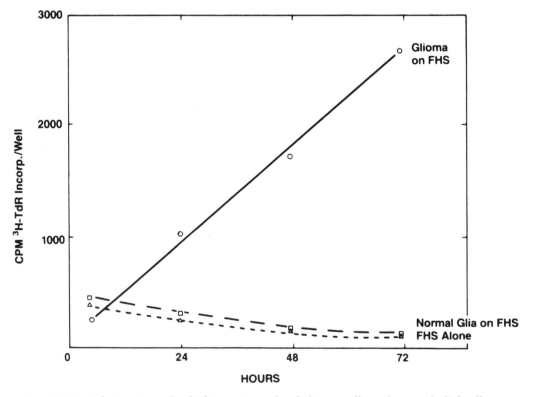

Fig. 13.13. Selective Growth of Glioma. Growth of glioma cells and normal glial cells on a confluent feeder layer. The cells were seeded at 2×10^4/ml (4×10^3/cm²) onto confluent, mitomycin C–treated feeder layers (see text) of FHS74Int cells [Owens et al., 1974] and labeled at intervals thereafter with [³H]-thymidine (see text). The cells were then extracted and counted.

and smaller increases in the concentration between selection steps.

Solubilization of MTX other than the Lederle product will require the addition of equimolar amounts of NaOH and sterilization through a 0.2-μm filter.

INTERACTION WITH SUBSTRATE

Selective Adhesion

Different cell types have different affinities for the culture substrate and attach at different rates. If a primary cell suspension is seeded into one flask and transferred to a second flask after 30 min, a third flask after 1 h, and so on for up to 24 h, then the most adhesive cells will be found in the first flask and the least adhesive in the last. Macrophages will tend to remain in the first flask, fibroblasts in the next few flasks, epithelial cells in the next few flasks, and, finally, hematopoietic cells in the last flask.

If collagenase in complete medium is used for primary disaggregation of the tissue (see Protocol 11.8), then most of the cells that are released will not attach within 48 h unless the collagenase is removed. However, macrophages migrate out of the fragments of tissue and attach during this period and can be removed from other cells by transferring the disaggregate to a fresh flask after 48–72 h of treatment with collagenase. This technique works well during disaggregation of biopsy specimens from human tumors.

Selective Detachment

Treatment of a heterogeneous monolayer with trypsin or collagenase will remove some cells more rapidly than others. Periodic brief exposure to trypsin removed fibroblasts from cultures of fetal human intestine [Owens et al., 1974] and skin [Milo et al., 1980]. Lasfargues [1973] found that the exposure of cultures of breast tissue to collagenase for a few days at a time removed fibroblasts and left the epithelial cells. EDTA, on the other hand, may release epithelial cells more readily than it will release fibroblasts [Paul, 1975].

Dispase II (Boehringer Mannheim) selectively dislodges sheets of epithelium from human cervical cultures grown on feeder layers of 3T3 cells without dislodging the 3T3 cells (see Protocol 22.5). This technique may be effective in subculturing epithelial cells from other sources, excluding stromal fibroblasts.

Nature of Substrate

The hydrophilic nature of most culture substrates (see Substrate Materials in Chapter 7) appears to be necessary for cell attachment, but little is known about variations in charge distribution on the cell surface

Fig. 13.14. Growth of Melanoma, Fibroblasts, and Glia in Suspension. Cells were plated out at 5×10^5 per 35-mm dish (2.5×10^5 cells/ml) in 1.5% Methocel over a 1.25% agar underlay. Colonies were photographed after 3 wk. (a) Melanoma. (b) Human normal embryonic skin fibroblasts. (c) Human normal adult glia. (d) The colony-forming efficiency of normal and malignant glial cells in suspension. Unshaded bars, colonies of >8 cells; cross-hatched bars, colonies of >16 cells; stippled bars, colonies of >32 cells (approximate). Colony counts were done on an Artek Colony Counter (see Fig. 18.6c) at different threshold settings. *Illustration continued on following page*

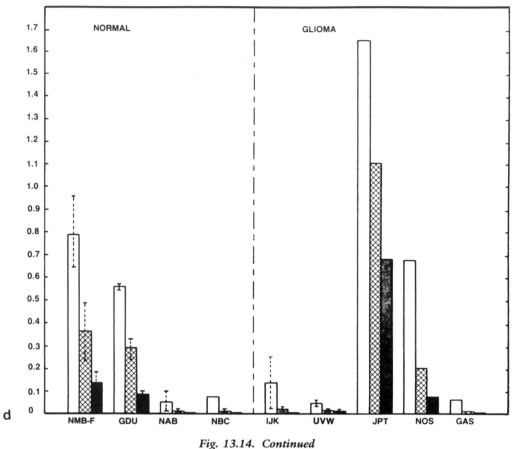

Fig. 13.14. Continued

and how different receptor arrays might interact with a complex charge array on the substrate. Since cell sorting in the embryo is a highly selective process and relates to differences in specific receptor sites on the cell surface and the distribution of ligands on adjacent cells and/or the extracellular matrix, qualitative and quantitative variations in substrate affinity should be anticipated in cultured cells. (See also Cell Adhesion in Chapter 2.)

While several sources of ECM are now available (e.g., Matrigel, Collagen [Becton Dickinson]), the emphasis so far has been on promoting cell survival or differentiation, and little has been made of the potential for selectivity in "designer" matrices, although collagen has been reported to favor epithelial proliferation [Kibbey et al., 1992; Kinsella et al., 1992] and Matrigel also favors epithelial survival and differentiation [Bissell et al., 1987; Ghosh et al., 1991; Kibbey et al., 1992]. Since the constituents are now better understood, mixing various collagens with proteoglycans, laminin, and/or fibronectin could be used to create more selective substrates.

The selective effect of substrates on growth may depend on both differential rates of attachment and growth, although in practice the two are indistin-

guishable. Macrophages also attach to Teflon, but do not proliferate. Collagen has been used in gel form to favor epithelial cell growth [Lillie et al., 1980] and in its denatured form to support endothelial outgrowth from aorta into a fibrin clot [Nicosia and Leighton, 1981].

Primaria plastics (Becton Dickinson) have a different charge on the plastic surface from conventional tissue-culture plastics designed to enhance epithelial growth relative to fibroblasts. Becton Dickinson and others also supply (see Sources of Materials in section titled Matrix) plastics coated in natural or synthetic matrices that may facilitate growth of more fastidious cell types, but are probably not selective.

Selective Feeder Layers

As well as conditioning the substrate (see Feeder Layers in this chapter), feeder layers can also be used for the selective growth of epidermal cells [Rheinwald and Green, 1975] and for repressing stromal overgrowth in cultures of breast (Fig. 13.12; see also Plates 14, 15) and colon carcinoma (see Protocol 22.5) [Freshney et al., 1982b]. The role of the feeder layer is probably quite complex; it provides not only extracellular matrix for adhesion of the epithelium, but also positively

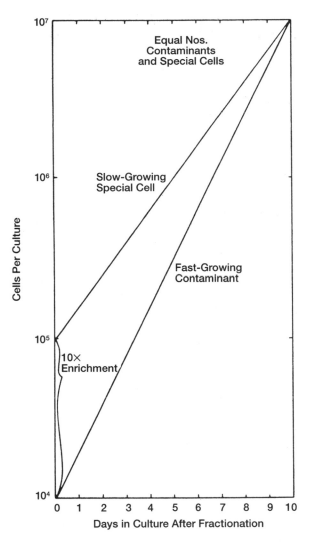

Fig. 13.15. Mixed Culture Overgrowth. Overgrowth of a slow-growing cell line by a rapidly growing contaminant. This figure portrays a hypothetical example, but it demonstrates that a 10% contamination with a cell population that doubles every 24 h will reach equal proportions with a cell population that doubles every 36 h after only 10 d of growth.

acting growth factors and negative regulators that inactivate TGF-β [Maas-Szabowski and Fusenig, 1996]. Human glioma will grow on confluent feeder layers of normal glia, while cells derived from normal brain will not [MacDonald et al., 1985]. (Fig. 13.13 and Protocol 23.1.)

Semisolid Media

The transformation of many fibroblast cultures reduces the anchorage dependence of cell proliferation (see Anchorage Independence in Chapter 17) [Macpherson and Montagnier, 1964]. By culturing the cells in agar (see Protocol 13.4) after viral transformation, it is possible to isolate colonies of transformed cells and exclude most of the normal cells. Normal cells will not form colonies in suspension with the high efficiency of virally transformed cells, although they will often do so with low plating efficiencies. The difference between virally transformed fibroblasts and untransformed cells is not seen as clearly in attempts

at selective culture of spontaneously arising tumors: Normal glia and fetal skin fibroblasts will form colonies in suspension just as readily as do glioma and melanoma. (Fig. 13.14.)

Cell cloning and the use of selective conditions have a significant advantage over physical cell separation techniques (see Chapter 14), in that contaminating cells are either eliminated entirely by clonal selection or repressed by constant or repeated application of selective conditions. Even the best physical cell separation techniques still allow some overlap between cell populations, such that overgrowth recurs. A steady state cannot be achieved, and the constitution of the culture changes continuously. From Fig. 13.15, it can be seen that a 90%-pure culture of cell line A will be 50% overgrown by a 10% contamination with cell line B in 10 days, given that B grows 50% faster than A. For continued culture, therefore, selective conditions are required in addition to, or in place of, physical separation techniques.

CHAPTER 14

Cell Separation

While cloning and employing selective culture conditions are the preferred methods for purifying a culture (see Isolation of Clones in Chapter 13), there are occasions when cells do not grow with a high enough plating efficiency to make cloning possible or when appropriate selection conditions are not available. It may then be necessary to resort to a separation technique, such as density sedimentation or flow cytometry. Separation techniques have the advantage that they give a high yield more quickly than cloning does, although not with the same purity.

The more successful separation techniques depend on differences in (1) cell density (specific gravity), (2) affinity of antibodies to cell surface epitopes, (3) cell size, and (4) light scatter, or the fluorescent emission of labeled cells sorted electronically by flow cytometry. The first two techniques involve a relatively low level of technology and are inexpensive, while the second two call for high technology with a significant outlay of capital.

CELL DENSITY AND
ISOPYKNIC SEDIMENTATION

Separation of cells by density can be performed by centrifugation at low g using conventional equipment [Pretlow and Pretlow, 1989; Sharpe, 1988]. The cells are deposited in a density gradient to an equilibrium position equivalent to their own density (isopyknic sedimentation). The density medium should be nontoxic and nonviscous at high densities (1.10 g/ml),

and should exert little osmotic pressure in solution. Serum albumin [Turner et al., 1967], dextran [Schulman, 1968], Ficoll (Pharmacia) [Sykes et al., 1970], metrizamide (Nygaard) [Munthe-Kaas and Seglen, 1974], and Percoll (Amersham Pharmacia) [Pertoft & Laurent, 1982] have all been used successfully. (See also Protocol 26.1.) Percoll (colloidal silica) and the radiopaque iodinated compounds metrizamide and metrizoate are among the more effective media currently used.

PROTOCOL 14.1. CELL SEPARATION BY DENSITY GRADIENT

Outline
Form a gradient (1) by layering different densities of Percoll, (2) by high-speed spin, or (3) with a special gradient former. Centrifuge cells through the gradient, collect fractions, dilute the medium, and culture the cells (Fig. 14.1).

Materials
Sterile:
Growth medium
Growth medium + 20% Percoll
Centrifuge tubes, 25 ml
PBSA
Trypsin, 0.25%
Syringe or gradient harvester
Plates, 24 well or microtitration
Nonsterile:
Refractometer or density meter

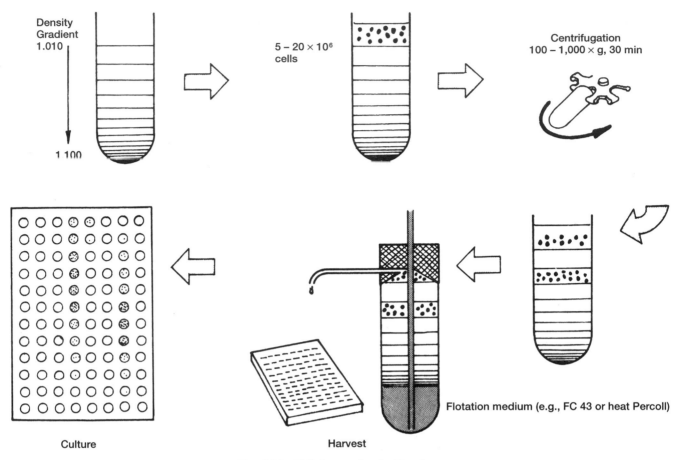

Fig. 14.1. Cell Separation by Density.

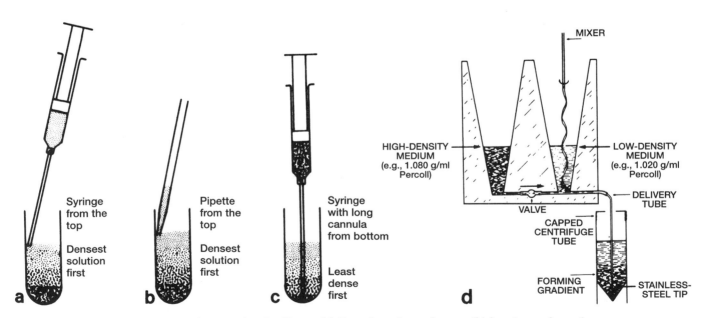

Fig. 14.2. Layering Density Gradients. (a) By syringe from the top, (b) by pipette from the top, (c) by syringe from the bottom, and (d) by a gradient mixing device (Buchler).

Hemocytometer or cell counter
Low-speed centrifuge

Protocol

1. Prepare the gradient by one of the following techniques:
   (a) Layering:
   (i) Adjust the density of the Percoll medium to 1.10 g/ml and its osmotic strength to 290 mOsm/kg.
   (ii) Mix the Percoll and regular media in varying proportions to give the desired density range (e.g., 1.020–1.100 g/ml) in 10 or 20 steps.
   (iii) Layer one step over another, building up a stepwise density gradient in a 25-ml centrifuge tube. (See Fig. 14.2.)
   Gradients may be used immediately or left overnight.
   (b) Centrifugation:
   (i) Place the medium containing Percoll at density 1.085 g/ml in a tube.
   (ii) Centrifuge at 20,000 g for 1 h.
   (iii) Centrifugation generates a sigmoid gradient (Fig. 14.3), the shape of which is determined by the starting concentration of Percoll, the duration and centrifugal force of the centrifugation, the shape of the tube, and the type of rotor.
   (c) With a gradient former:
   A continuous linear gradient may be produced by mixing, for example, 1.020 g/ml with 1.08 g/ml Percoll in a gradient former (Fisons, Pharmacia, Buchler; cl12.5. 14.2d).

2. Trypsinize cells and resuspend them in the medium plus serum or a trypsin inhibitor. Check to make sure that the cells are singly suspended.

3. Using a syringe, pipettor, or fine-tipped pipette, layer up to 2×10^7 cells in 2 ml of medium on top of the gradient.

4. The tube may be allowed to stand on the bench for 4 h and will sediment under 1 g; or it may be centrifuged for 20 min between 100 and 1,000 g. If the latter procedure is used, increase centrifuge speed gradually at start of run and do not apply brake at end of run.

5. Collect fractions using a syringe or a gradient harvester (Fisons; Fig. 14.4). Fractions of 1 ml may be collected into a 24-well plate or 0.1 ml into microtitration plates. Samples should be taken at intervals for cell counting and for determining the density (ρ) of the gradient medium. Density may be measured on a refractometer (Hilger) or density meter (Paar).

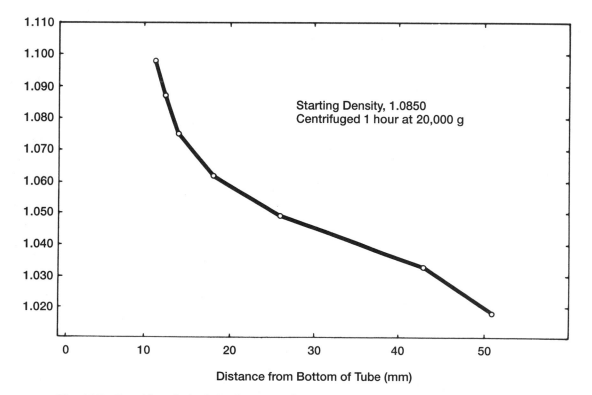

Fig. 14.3. Centrifuge-derived Gradient. Gradient generated by spinning Percoll at 20,000 g for 1 h.

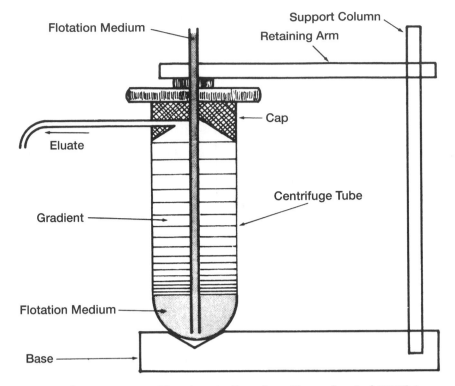

Fig. 14.4. Gradient Harvester. Flotation medium (e.g., Fluorochemical FC43) is pumped down the inlet tube to the bottom of the gradient, displacing the gradient and cells upward and out through the delivery tube. (After an original design by Dr. G. D. Birnie.)

6. Add an equal volume of medium to each well, and mix to ensure that the cells settle to the bottom of the well. Change the medium to remove the Percoll after 24–48 h incubation.

Variations

Position of Cells. Cells may be incorporated into the gradient during its formation by centrifugation. Only one spin is required, although spinning the cells at such a high g force may damage them. There is evidence that Percoll may be incorporated into cells [Pertoft and Laurent, 1982], so for this and other media (see below), it is preferable to layer cells on top of the gradient.

Other Media. Ficoll is one of the most popular media, since, like Percoll, it can be autoclaved. Ficoll is a little more viscous than Percoll at high densities and may cause some cells to agglutinate. Metrizamide (Nygaard), a nonionic derivative of metrizoate, which is a radio-opaque iodinated substance used in radiography (Isopaque, Hypaque, Renografin) and in lymphocyte purification (see Protocol 26.1), is less viscous than Ficoll at high densities [Rickwood and Birnie, 1975], but may be incorporated into some cells (Fig. 14.5), as may Isopaque [Splinter et al., 1978]. When

such media are used, cells should always be layered on top of the gradient and not mixed in during its formation.

Marker Beads. Amersham Pharmacia manufactures colored marker beads of standard densities that may be used to determine the density of regions of the gradient.

Isopyknic sedimentation is quicker than velocity sedimentation at unit gravity (see Cell Size and Sedimentation Velocity, this chapter) and gives a higher yield of cells for a given volume of gradient. It is ideal when clear differences in density ($\geq$0.02 g/cc) exist between cells. Cell density may be affected by the medium used to attain the gradient (e.g., metrizamide; see Fig. 14.5), by the position of the cells in the growth cycle, and by serum (Fig. 14.6). Because high g forces are not required, isopyknic sedimentation can be done on any centrifuge and can even be performed at 1 g.

ANTIBODY-BASED TECHNIQUES

Immune Panning

The attachment of cells to dishes coated with antibodies, a process called *immune panning*, has been used successfully with a number of different cell types [Sil-

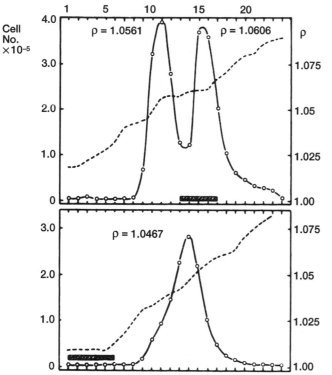

Fig. 14.5. Metrizamide Gradients. Incorporation of metrizamide (Nygaard) into cells during isopyknic centrifugation. MRC-5 cells (human diploid embryonic lung fibroblasts) were layered in a metrizamide-containing medium of the appropriate density in the center of the gradient or in the medium alone at the top of the gradient. After centrifugation, the cells were eluted and counted, and samples were taken from each fraction to determine the density. [Reproduced from Freshney, 1976, with permission of the publisher.] Slashed bars: position of cells at start.

vestri et al., 1991; Murphy et al., 1992]. The vast majority of panning methods are derived from the work of Wysocki and Sata [1978], who developed the technique for separating lymphocyte populations. A cell-type-specific antibody raised against a cell surface epitope is conjugated to the bottom of a Petri dish, and when the mixed cell population is added to the dish, the cells to which the antibody is directed attach rapidly to the bottom of the dish. The remainder can then be removed. Immune panning can be used positively, to select a specific subset of cells that can be released subsequently by mechanical detachment or light trypsinization, or negatively, to remove unwanted cells.

Magnetic Sorting

Magnetic sorting can be achieved by conjugating a specific antibody, raised against a cell surface epitope, to ferritin beads (Dynabeads, from Dynal), mixing the cell suspension with the beads, and then running the suspension past a magnet which draws the cells that have attached to the beads onto the side of the separating chamber. The cells and beads are released when the current is switched off, and the cells may be separated from the beads by trypsinization or vigorous pipetting. Several cell types have been separated by this method, including stem cell purification by negative sorting [Bertoncello et al., 1991] (Fig. 14.7), purging bone marrow of leukemic cells [Trickett et al., 1990], and kidney tubular epithelium [Pizzonia et al.,

1991]. Protocols for the use of Dynabeads are available on the Dynal Web site. (See Suppliers and Resources in Trade Index.)

The use of immunomagnetic microbeads [Gaudernack et al., 1986] (Fig. 14.8) allows the cells to be cultured or processed through further sorting procedures, without the need to remove the beads (Miltenyi). The method is, therefore, particularly useful for a positive sort. The following protocol has been abstracted from the Myltenyi instruction sheet.

PROTOCOL 14.2. MAGNETIC-ACTIVATED CELL SORTING (MACS)

Outline

Buffy coat or another mixed cell suspension is mixed with antibody-conjugated microbeads, diluted, and placed in a magnetic separation column. Cells bound to microbeads migrate to the sides of the column, while unbound cells flow through. Bound cells are released from the column when it is removed from the separator magnet and are purged from the column with a syringe piston.

Materials

Sterile:

Buffer: PBSA, pH 7.2, with 0.5% BSA and 2 mM EDTA

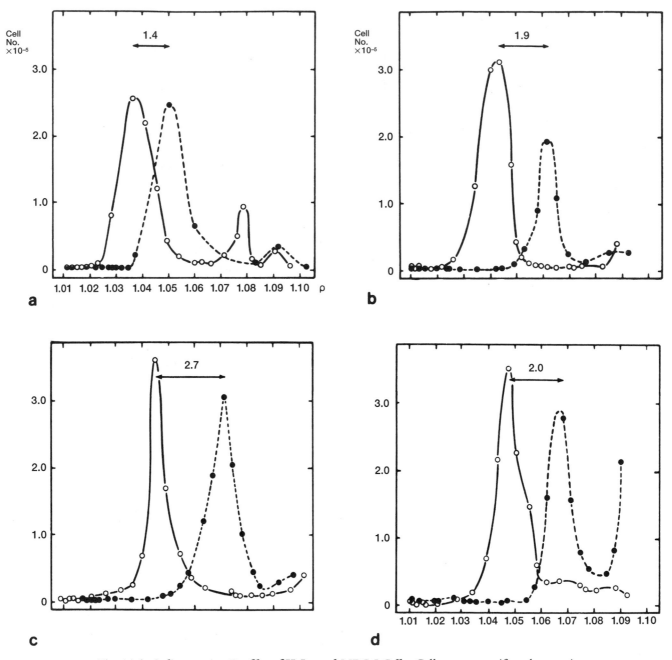

Fig. 14.6. Sedimentation Profiles of HeLa and MRC-5 Cells. Cells were centrifuged to equilibrium in gradients of metrizamide in a culture medium. a,b. Cells taken from log phase of growth. c,d. Cells taken from plateau phase. Gradients in (a) and (c) contained no serum; those in (b) and (d) contained 10% fetal bovine serum. Numbers over the arrows are the differences in density between the peaks, multiplied by 100. Solid circles, HeLa; open circles, MRC-5. [From Freshney, 1976, with permission of the publisher.]

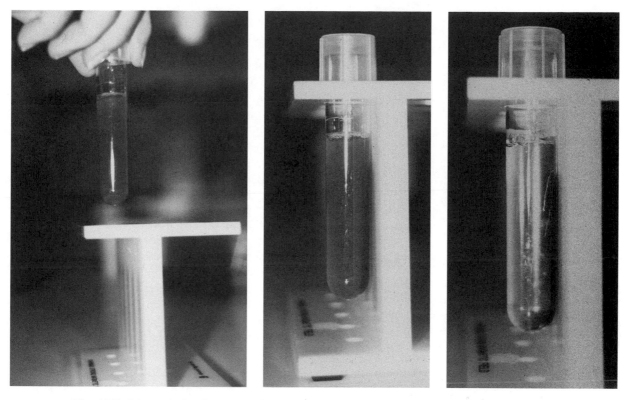

Fig. 14.7. Magnetic Sorting. Negative sort. Committed progenitor cells from bone marrow suspension bound to Dynal paramagnetic beads with antibodies to lineage markers. Lineage negative (stem) cells are not bound and remain in the suspension ready for sorting by flow cytometry. (a) Inserting the tube into the magnetic holder; (b) tube immediately after being placed in magnetic holder; (c) tube 30 seconds after placement in magnetic holder.

Magnetic cell separator: MiniMACS, MidiMACS, VarioMACS, or SuperMACS (Mylteni)

RS^+ or VS^+ column adaptors

Positive selection column: MS^+/RS^+ for up to 1×10^7 cells; LS^+/VS^+ for up to 1×10^8 cells

Magnetizable microbeads conjugated to antibody raised against cell surface antigen of cells to be collected.

Collection tube to match volume being collected: MS^+/RS^+ 1 ml; LS^+/VS^+ 5 ml

Protocol

Labeling:

1. Isolate peripheral blood mononuclear cells by the standard method (see Protocol 26.1), or prepare a cell suspension by trypsinization or an alternative procedure (see Protocols 12.2, 11.5, and 11.8).

2. Remove dead cells with Ficoll-paque. (See Protocol 11.10.)

3. Remove clumps by passing cells through 30 μm of nylon mesh or filter (Mylteni, #414:07). (Wet the filter with a buffer before use.)

4. Wash the cells in the buffer by centrifugation.

5. Resuspend pellet from centrifugation in 80 μl of buffer per 10^7 total cells (80 μl minimum volume, even for $<1 \times 10^7$ cells).

6. Add 20 μl of MACS microbeads per 1×10^7 cells.

7. Mix and incubate suspension for 15 min at 6–12°C. (For fewer cells, use the same volume.)

8. Dilute suspension by adding 10–20× the labeling volume of buffer.

9. Centrifuge 300 g for 10 min.

10. Remove the supernatant and resuspend cells plus microbeads in 500 μl buffer per 1×10^8 total cells.

Positive magnetic separation:

11. Place the column in the magnetic field of the MACS separator.

12. Wash the column: MS^+/RS^+ 500 μl; LS^+/VS^+ 3 ml.

13. Apply the cell suspension to the column: MS^+/RS^+ 500–1000 μl; LS^+/VS^+ 1–10 ml.

14. Allow negative cells to pass through the column.

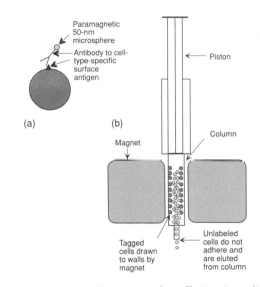

Fig. 14.8. Magnetically Activated Cell Sorting (MACS). Positive sort. (a) Cells are preincubated with antibodies, raised against a cell-type-specific surface antigen and conjugated to paramagnetic microspheres. (b) When the cells are introduced to the column, cells bound to microspheres migrate to the outer wall, and unlabeled cells go straight through. The magnetically bound cells, released when the column is removed from the magnet, are flushed out with the piston. Diagram based on Mylteni MACS sorting.

15. Rinse the column with buffer: MS$^+$/RS$^+$ 3 × 500 μl; LS$^+$/VS$^+$ 3 × 3 ml.
16. Remove the column from the separator and place the column on the collection tube.
17. Pipette buffer into the column (MS$^+$/RS$^+$ 1 ml, LS$^+$/VS$^+$ 5 ml), and flush out positive cells using the plunger supplied with the column.
18. Count the cells, adjusting the concentration in the growth medium:
    1 × 10^5–1 × 10^6 cells/ml and seed culture flasks for primary culture.
    10–1000 cells per ml and seed Petri dishes for cloning.

CELL SIZE AND SEDIMENTATION VELOCITY

The relationship between the particle size and sedimentation rate at 1 g, although complex for submicron-sized particles, is fairly simple for cells and can be expressed approximately as

$$v \approx \frac{r^2}{4} \qquad (1)$$

[Miller and Phillips, 1969], where v is the sedimentation rate in mm/h and r is the radius of the cell in μm. (See Table 20.1.)

Layering cells over a serum gradient in medium, will allow the cells to settle through the medium according to equation (1). However, unit gravity sedimentation is unable to handle large numbers of cells (~1 × 10^6 cells/cm^2 of surface area at the top of the gradient) and does not give particularly good separations unless the mean cell sizes are very different and the cell populations are homogeneous in size. Most cell separations based on cell size use either centrifugal elutriation (giving moderate resolution, but a high yield) or a cell sorter (high resolution with a low yield).

Centrifugal Elutriation

The centrifugal elutriator (Beckman) is a device for increasing the sedimentation rate and improving the yield and resolution of cell separation by performing the separation in a specially designed centrifuge and rotor (Fig. 14.9) [Lutz et al., 1992]. Cells in the suspending medium are pumped into the separation chamber in the rotor while it is turning. While the cells are in the chamber, centrifugal force tends to push the cells to the outer edge of the rotor (Fig. 14.10). Meanwhile, the medium is pumped through the chamber such that the centripetal flow rate approximates the sedimentation rate of the cells. If the cells were uniform, they would remain stationary, but since they vary in size, density, and cell surface configuration, they tend to sediment at different rates. Because the sedimentation chamber is tapered, the flow rate increases toward the edge of the rotor, and a continuous range of flow rates is generated. Cells of differing sedimentation rates will, therefore, reach equilibrium at different positions in the chamber. The sedimentation chamber is illuminated by a stroboscopic light and can be observed through a viewing port. When the cells are seen to reach equilibrium, the flow rate is increased and the cells are pumped out into receiving vessels. The separation can be performed in a complete medium and the cells cultured directly afterward.

The procedure comprises four phases: (1) setting up and sterilizing the apparatus, (2) calibration, (3) loading the sample and establishing the equilibrium conditions, and (4) harvesting fractions. The details of the protocol are provided in the operating manual for the elutriator (Beckman–Coulter). Equilibrium is reached in a few minutes, and the whole run may take 30 min. On each run, 1 × 10^8 cells may be separated, and the run may be repeated as often as necessary. The apparatus is, however, fairly expensive, and a considerable amount of experience is required before effective separations may be made. A number of cell types have been separated by this method [Teofili et al., 1996; Lag et al., 1996; Yoshioka et al., 1997], as have cells of

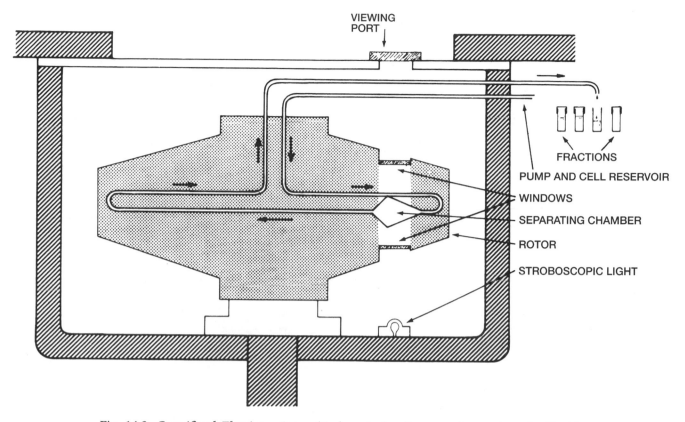

Fig. 14.9. Centrifugal Elutriator Rotor (Beckman). A cell suspension and carrier liquid enter at the center of the rotor and are pumped to the periphery and then into the outer end of the separating chamber. The return loop is via the opposite side of the rotor, to maintain balance.

different phases of the cell cycle [Breder et al., 1996; Mikulits et al., 1997].

FLUORESCENCE-ACTIVATED CELL SORTING

Fluorescence-activated cell sorting [Vaughan and Milner, 1989] operates by projecting a single stream of cells through a laser beam in such a way that the light scattered from the cells is detected by one or more photomultipliers and recorded (Figs. 14.11, 14.12). If the cells are pretreated with a fluorescent stain (e.g., propidium iodide or chromomycin A_3 for DNA) or a fluorescent antibody, the fluorescence emission excited by the laser is detected by a second photomultiplier tube. The information obtained is then processed and displayed as a two- (Fig. 14.13) or three-dimensional graph on the monitor.

A *flow cytometer* is an analytical instrument that processes the output of the photomultipliers to analyze the constitution of a cell population (e.g., to determine the proportion of cells in different phases of

the cell cycle, measured by a combination of DNA fluorescence and cell size measurements).

A *fluorescence-activated cell sorter (FACS)* is an instrument that uses the emission signals from each cell to sort the cell into one of two sample collection tubes and a waste reservoir. If specific coordinates are set to delineate sections of the display, the cell sorter will

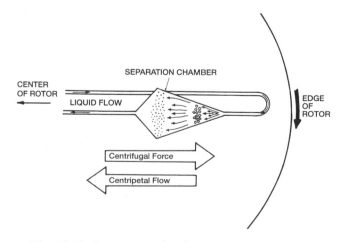

Fig. 14.10. Separation Chamber of Elutriator Rotor.

Fig. 14.11. Fluorescence-Activated Cell Sorter (FACS). (a) FACStar analytical flow cytometer. (b) Close-up of flow chamber compartment of cell sorter version of the earlier
FACS IV (Becton Dickinson; see also Fig. 14.12).

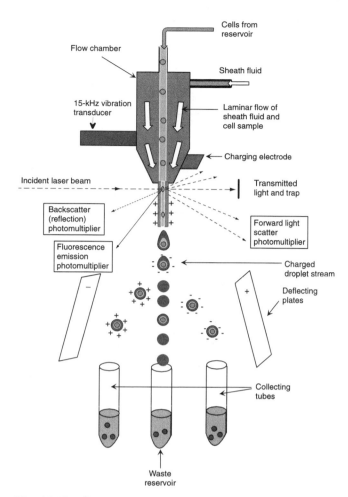

Fig. 14.12. Flow Cytometry. Principle of operation of flow cytophotometer. Cell stream in PBSA enters at the top, and sheath liquid, also PBSA, is injected around the cell stream to generate a laminar flow within the flow chamber. When the cell stream exits the chamber, it cuts a laser beam, and the signal generated triggers the charging electrode, thereby charging the cell stream. The cell stream then breaks up into droplets, induced by the 15-kHz vibration transducer attached to the flow chamber. The droplets carry the charge briefly applied to the exiting cell stream and are deflected by the electrode plates below the flow chamber. The charge is applied to the plates with a sufficient delay to allow for the transit time from cutting the beam to entering the space between the plates. (See also Fig. 14.11b.)

divert those cells with properties that would place them within these coordinates (e.g., high or low light scatter, high or low fluorescence) into the appropriate receiver tube, placed below the cell stream. The stream itself is broken up into droplets by a high-frequency vibration applied to the flow chamber, and the droplets containing single cells with specific attributes are charged as they leave the chamber. These droplets are deflected, left or right according to the charge applied, as they pass between two oppositely charged plates.

The charge is applied briefly and at a set time after the cell has cut the laser beam such that the droplet containing one specifically marked cell is deflected into the receiver. The concentration in the cell stream must be low enough that the gap between cells is sufficient to prevent two cells from inhabiting one droplet [Vaughan and Milner, 1989].

All cells having similar properties are collected into the same tube. A second set of coordinates may be established and a second group of cells collected simultaneously into another tube by changing the polarity of the cell stream and deflecting the cells in the opposite direction. All remaining cells are collected in a central waste reservoir.

This method may be used to separate cells according to any differences that may be detected by light scatter (e.g., cell size) or fluorescence (e.g., DNA, RNA, or protein content; enzyme activity; specific antigens) and has been applied to a wide range of cell types. It has probably been utilized most extensively for hematopoietic cells [Battye and Shortman, 1991; Pipia and Long, 1997], for which disaggregation into the obligatory single-cell suspension is relatively simple, but has also been used for solid tissues (e.g., lung [Aitken et al., 1991], skin [Swope et al., 1997], and gut [Boxberger et al., 1997; see also Protocol 22.17]. It is an extremely powerful tool, but is limited by the cell yield (about 10^7 cells is a reasonable maximum number of cells that can be processed at one time) and the very high cost of the instrument (approximately $200,000). It also requires a full-time skilled operator.

OTHER TECHNIQUES

The many other techniques that have been used successfully to separate cells are too numerous to describe in detail. They are summarized below and listed in Table 14.1.

Electrophoresis is performed either in a Ficoll gradient [Platsoucas et al., 1979] or by curtain electrophoresis; the second technique is probably more effective and has been used to separate kidney tubular epithelium [Kreisberg et al., 1977].

Affinity chromatography is used on antibodies [Varon and Manthorpe, 1980; Au and Varon, 1979] or plant lectins [Pereira and Kabat, 1979] that are bound to nylon fiber [Edelman, 1973] or Sephadex (Amersham Pharmacia). This technique appears to be useful for fresh blood cells, but less so for cultured cells.

Countercurrent distribution [Walter, 1975, 1977; Sharpe, 1988] has been utilized to purify murine ascites tumor cells with reasonable viability.

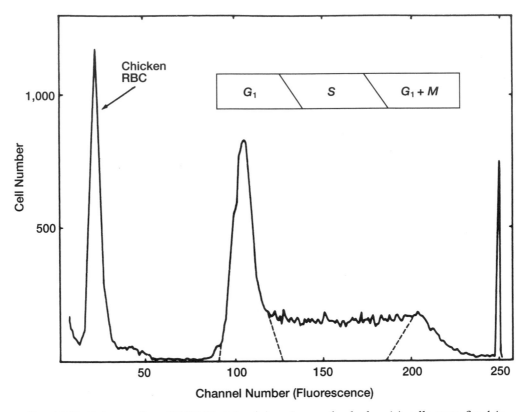

Fig. 14.13. Printout from FACS II. Friend (murine erythroleukemia) cells were fixed in methanol as a single-cell suspension and stained with chromomycin A_3. The suspension was then mixed with chicken erythrocytes, fixed, stained in the same way, and run through a FACS II. The printout plots cell number on the vertical axis and channel number (fluorescence) on the horizontal axis. Since fluorescence is directly proportional to the amount of DNA per cell, the trace gives a distribution analysis of the cell population by DNA content. The lowest DNA content is found in the chicken erythrocytes, included as a standard. The major peak around channel 100 represents those cells as the G_1 phase of the cell cycle. Cells around channel 200, therefore, represent G_2 and metaphase cells (with double the amount of DNA per cell), and cells in intermediate channels are in the S (DNA synthetic) phase. Cells accumulated in channel 250 are those for which the DNA value is off the scale or those that have formed aggregates. (Courtesy of Prof. B. D. Young.)

BEGINNER'S APPROACH TO CELL SEPARATION

It is best to start with a simple technique such as density gradient centrifugation or, if a specific cell surface phenotype can be predicted, panning on coated dishes or MACS. If the resolution or yield is insufficient, then it may be necessary to employ FACS or centrifugal elutriation. Centrifugal elutriation is useful for the rapid sorting of large numbers of cells but FACS will probably give the purest cell population, based on the combined application of two or more stringent criteria.

When a high-purity cell suspension is required and selective culture is not an option, it will probably be necessary to employ at least a two-step fractionation, in a manner analogous to the purification of proteins. In many such procedures, density gradient separation is used as a first step, with panning or MACS as a second, and the final purification is performed by FACS as has been used in the isolation of hematopoietic stem cells [Cooper and Broxmeyer, 1994].

TABLE 14.1. Cell Separation Methods

Method	Basis for separation	Equipment	Comments	Reference
Sedimentation velocity at 1 g	Cell size	Custom-made separating funnel and baffles	Simple technique, but not very high resolution or yield	[Miller and Phillips, 1969]
Isopyknic sedimentation	Cell density	Centrifuge	Simple and rapid	[Pertoft and Laurent, 1982]; see Protocol 14.1
MACS	Surface antibodies	Simple magnet and flow chamber	Specific, given a highly specific surface antibody	[Gaudernack et al., 1986]; see Protocol 14.2
Immune panning	Surface antibodies	Antibody-coated dishes	Simple, low technology with precoated plates available, but also depends on specific surface antibody	[Wysocki and Sata, 1978]
Centrifugal elutriation	Cell size, density, and surface configuration	Special centrifuge and elutriator rotor	Rapid, high cell yield, but quite complex process	[Lutz et al., 1992]
FACS	Cell surface area, fluorescent markers, fluorogenic enzyme substrates, multiparameter	Flow cytometer	Complex technology and expensive; very effective; high resolution, but low yield	[Vaughan and Milner, 1989]
Affinity chromatography	Cell surface antigens, cell surface carbohydrate	Sterilizable chromatography column	Elutriation of cells from columns is difficult, better in free suspension	[Edelman, 1973]
Countercurrent distribution	Affinity of cell surface constituents for solvent phase	Shaker	Some cells may suffer loss of viability, but method is quite successful for others	[Walter, 1977]
Electrophoresis in gradient or curtain	Surface charge	Curtain electrophoresis apparatus		[Kreisberg et al., 1977; Platsoucas et al., 1979]

CHAPTER 15

Characterization

THE NEED FOR CHARACTERIZATION

When a new cell line is derived, either from a primary culture or from an existing cell line, it is difficult to assess its future value. Often, it is only after a period of use and dissemination that the true importance of the cell line becomes apparent, and, at that point, details of its origin are required. However, by that time, it is too late to collect information retrospectively. It is, therefore, vital that adequate records are kept, from the time of isolation of the tissue, or of the receipt of a new cell line, detailing the origin and handling of the cell line. These records form the *provenance* of the cell line, and, as with fine art objects and antiques, where the term originated, the more detailed the provenance, the more valuable the cell line. (See Primary Records in Chapter 11; Maintenance Records in Chapter 12).

This aspect of cell culture has become particularly important with the widespread dissemination of cell lines through cell banks and personal contacts to research laboratories and commercial companies far removed from their origin. In particular, if a cell line becomes incorporated into a procedure that requires validation of its components, then the authentication of the cell line becomes crucial. Authentication requires that the cell line be characterized on receipt, and periodically during use, and that these data are compatible with, and added to, the existing provenance.

Special attention must be paid to the possibility that the cell line has become cross-contaminated. The demonstration that the majority of continuous cell lines in use in the United States in the late 1960s had become cross-contaminated with HeLa cells [Gartler, 1967; Nelson-Rees and Flandermeyer, 1977; Lavappa, 1978] first brought this serious problem to light, but the continued use of the lines 30 years later indicates that many people are still unaware, or are unwilling to accept, that many lines in common use (e.g., Hep-2, KB, Girardi Heart, WISH, Chang Liver, etc.) are not authentic. (See Fig. 15.12 and Table 12.1.). The use of HeLa-contaminated lines, without proper acknowledgement of the contamination, is still a major problem [Hay, 1991; UKCCCR, 1999; Stacey et al., 1999; MacLeod et al., 1999] and emphasizes that any vigorously growing cell line can overgrow another, more slowly growing line [Masters et al., 1988; van Helden et al., 1988; Christensen et al., 1993; Gignac et al., 1993].

Some of the methods in general use for cell line characterization are listed in Table 15.1. (See also Hay et al. [2000].)

There are six main requirements for cell line characterization:

(1) Confirmation of the species of origin;
(2) Correlation with the tissue of origin, which comprises the following charcteristics:
   (a) identification of the lineage to which the cell belongs and
   (b) position of the cells within that lineage (i.e., the stem, precursor, or differentiated status;
(3) Determination of whether the cell line is transformed or not;
   (a) is the cell line finite or continuous?

TABLE 15.1. Characterization of Cell Lines and Cell Strains

Criterion	Method	Reference	Protocol No.
Karyotype	Chromosome spread with banding	Rothfels and Siminovitch [1958], Rooney and Czepulkowski [1986]	15.8
Isoenzyme analysis	Agar gel electrophoresis	Hay [1992]	15.11
Cell surface antigens	Immunohistochemistry	Hay [1992], Burchell et al. [1983, 1987]	15.12
Cytoskeleton	Immunocytochemistry with antibodies to specific cytokeratins	Lane [1982], Moll et al. [1982]	15.12
DNA fingerprint	Restriction enzyme digest; PAGE; satellite DNA probes	Hay [1992], Jeffreys et al. [1985]	15.10

(b) does it express properties associated with malignancy? (See Chapter 16)

(4) Confirmation of the absence of cross-contamination (see Cross-contamination in Chapter 18);

(5) Indication of whether the cell line is prone to genetic instability, transformation, and phenotypic variation (see Dedifferentiation in Chapter 16; Genetic Instability in Chapter 17);

(6) Identification of specific cell lines within a group from the same origin, selected cell strains, or hybrid cell lines, all of which require demonstration of features unique to that cell line or cell strain.

Species Identification

Chromosomal analysis (see Protocol 15.9) is one of the best methods for distinguishing between species. Isoenzyme electrophoresis (see Protocol 15.11) is also a good diagnostic test and is quicker than chromosomal analysis, but requires the appropriate apparatus and reagents. In practice, a combination of the two methods is often used and gives unambiguous results [Hay, 2000]. Recently, techniques have been introduced for "chromosome painting"—i.e., using combinations of specific molecular probes to hybridize to individual chromosomes. (See Chromosome Painting in this chapter.) These probes identify individual chromosome pairs and are species specific. The availability of probes is limited to a few species at present, and most are either mouse or human, but chromosome painting is a good method for distinguishing between human and mouse chromosomes in potential cross-contaminations and interspecific hybrids.

Certain chromosome banding patterns can also be used to distinguish human and mouse chromosomes. (See Fig. 15.9.)

Lineage or Tissue Markers

Cell Surface Antigens. These markers are particularly useful in sorting hematopoietic cells [Visser and De Vries, 1990] and have also been effective in discriminating epithelium from stroma with antibodies such as anti-EMA [Heyderman et al., 1979] and anti-HMFG 1 and 2 [Burchell and Taylor-Papadimitriou, 1989], and neuroectodermally derived cells (e.g., anti-A2B5) [Dickson et al., 1983] from cells derived from other germ layers.

Intermediate Filament Proteins. These are among the most widely used lineage or tissue markers [Lane, 1982; Ramaekers et al., 1982]. Glial fibrillary acidic protein (GFAP) for astrocytes [Bignami et al., 1980] (see Plate 25) and desmin [Bochaton-Piallat et al., 1992; Brouty-Boyé et al., 1992] for muscle are the most specific, while cytokeratin marks epithelial cells and mesothelium [Lane et al., 1982; Moll et al., 1982] (see Plate 24). Vimentin, though usually restricted to mesodermally derived cells *in vivo*, can appear in other cell types *in vitro*.

Differentiated Products and Functions. Hemoglobin for erythroid cells, myosin or tropomyosin for muscle, melanin for melanocytes, and serum albumin for hepatocytes are among the best examples of specific cell type markers, but, like all differentiation markers, they depend on the complete expression of the differentiated phenotype.

Transport of inorganic ions, and the resultant transfer of water, is characteristic of some epithelia; grown as monolayers, they will produce *domes*, which are hemicysts in the monolayer caused by accumulation of water on the underside of the monolayer [Rabito et al., 1980]. (Fig. 15.1 and see Plate 23.) These domes are found in many types of absorptive and secretory epithelia [Abaza et al., 1974; Lever, 1986]. Other specific functions that can be expressed *in vitro* include muscle contraction and depolarization of nerve cell membrane.

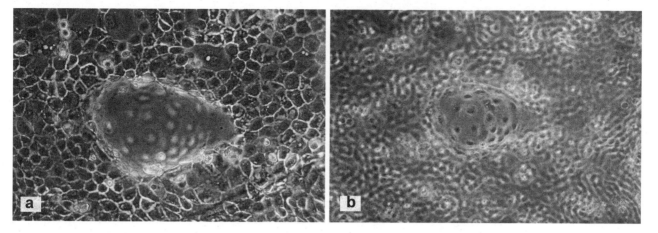

Fig. 15.1. Domes. (a) Dome, or hemicyst, formed in an epithelial monolayer by downward transport of ions and water, lower focus (on monolayer). (b) Upper focus (top of dome). (See also Plate 23.)

Enzymes. Three parameters are available in enzymic characterization: (1) the constitutive level (i.e., in the absence of inducers or repressors); (2) the response to inducers and repressors; and (3) isoenzyme polymorphisms (see Isoenzymes in this chapter). Creatine kinase BB isoenzyme is characteristic of neuronal and neuroendocrine cells, as is neuron-specific enolase; lactic dehydrogenase is present in most tissues, but as different isoenzymes, and a high level of tyrosine aminotransferase, inducible by dexamethasone, is generally regarded as specific to hepatocytes (Table 15.2) [Granner et al., 1968].

Regulation. Although differentiation is usually regarded as an irreversible process, the level of expression of many differentiated products is under the regulatory control of environmental influences, such as hormones, the matrix, and adjacent cells. (See Induction of Differentiation in Chapter 16). Hence, the measurement of specific lineage markers may require preincubation of the cells in, for example, a hormone such as hydrocortisone, specific growth factors, or growth of the cells on extracellular matrix of the correct type. Maximum expression of both tyrosine aminotransferase in liver cells and glutamine synthetase in glia requires prior induction with dexamethasone. Glutamine synthetase is also repressed by glutamine, so glutamate should be substituted in the medium 48 h before assay [DeMars, 1957].

Lineage Fidelity. Although many of the markers described above have been claimed as lineage markers, they are more properly regarded as tissue or cell type markers, as they are often more characteristic of the function of the cell than its embryologic origin. Cytokeratins occur in mesothelium and kidney epithe-

lium, although both of these tissues derive from the mesoderm. Neuron-specific enolase and creatine kinase BB are expressed in neuroendocrine cells of the lung, although these cells are now recognized to derive from the endoderm and not from neuroectoderm, as one might expect of neuroendocrine-type cells.

Unique Markers

Unique markers include specific chromosomal aberrations (e.g., deletions, translocations, polysomy); major histocompatibility (MHC) group antigens (e.g., HLA in human), which are highly polymorphic; and DNA fingerprinting (see Protocol 15.10). Enzymic deficiencies (e.g., thymidine kinase deficiency (TK^-)) and drug resistance (e.g., vinblastine resistance, usually coupled to the expression of the P-glycoprotein by one of the *mdr* genes, which code for the efflux protein) are not truly unique, but may be used to distinguish among cell lines from the same tissues but different donors.

Transformation

Transformation forms a major element in cell line characterization and is dealt with separately. (See Chapter 17.)

MORPHOLOGY

Observation of morphology is the simplest and most direct technique used to identify cells. It has, however, certain shortcomings that should be recognized. Most of these are related to the plasticity of cellular morphology in response to different culture conditions; for example, epithelial cells growing in the center of a confluent sheet are usually regular, polygonal, and

TABLE 15.2. Enzymic Markers

Enzyme	Cell Type	Inducer	Repressor	Reference
Glutamyl synthetase	Astroglia (brain)	Hydrocortisone	Glutamine	Hallermeyer and Hamprecht, [1984]
Tyrosine aminotransferase	Hepatocytes	Hydrocortisone		Granner et al., [1968]
Sucrase	Enterocytes	NaBt		Pignata et al., [1994]; Vachon et al., [1996]
Alkaline phosphatase	Type II pneumocyte (in lung alveolus)	Dexamethasone, oncostatin, IL-6	TGF-β	Edelson et al., [1988]; McCormick & Freshney [2000]
Alkaline phosphatase	Enterocytes	Dexamethasone, NaBt		Vachon et al., [1996]
Nonspecific esterase	Macrophages	PMA, Vitamin D$_3$		Murao et al., [1983]
Angiotensin-converting enzyme	Endothelium	Collagen, Matrigel		Del Vecchio and Smith, [1981]
Neurone-specific enolase	Neurons, neuroendocrine cells			Hansson et al., [1984]
Tyrosinase	Melanocytes	cAMP		Park et al., [1993, 1999]
DOPA-decarboxylase	Neurons, SCLC			Dow et al., [1995]; Gazdar et al., [1980]
Creatine kinase MM	Muscle cells	IGF-II	FGF-1,2,7	Stewart et al., [1996]
Creatine kinase BB	Neurons, neuroendocrine cells, SCLC			Gazdar et al., [1981]

with a clearly defined edge, while the same cells growing at the edge of a patch may be more irregular and distended and, if transformed, may break away from the patch and become fibroblast-like in shape.

Subconfluent fibroblasts from hamster kidney or human lung or skin assume multipolar or bipolar shapes (see Plate 16) and are well spread on the culture surface, but at confluence they are bipolar and less well spread (see Plate 17). They also form characteristic parallel arrays and whorls that are visible to the naked eye. Mouse 3T3 cells and human glial cells grow like multipolar fibroblasts at low cell density, but become epithelial-like at confluence. (Fig. 15.2 and see Plates 16–22.) Alterations in the substrate [Gospoda-rowicz et al., 1978b; Freshney, 1980], and the constitution of the medium (see Soluble Inducers in Chapter 16), can also affect cellular morphology. Hence, comparative observations of cells should always be made at the same stage of growth and cell density in the same medium, and growing on the same substrate.

The terms "fibroblastic" and "epithelial" are used rather loosely in tissue culture and often describe the appearance rather than the origin of the cells. Thus, a bipolar or multipolar migratory cell, the length of which is usually more than twice its width, would be called "fibroblastic," while a monolayer cell that is polygonal, with more regular dimensions, and that grows in a discrete patch along with other cells is usu-

Fig. 15.2. Cell Morphology. (a) BHK-21 (baby hamster kidney fibroblasts), clone 13, in log growth. The culture is not confluent, and the cells are well spread and randomly oriented (although some orientation is beginning to appear). (b) Cells of an epithelial-like morphology from fetal human intestine (FHI). (c) Astrocytes from human astrocytoma. This pattern is quite characteristic, but is lost as the cells are passaged, and a morphology not unlike that shown in (b), (e), or (f) develops. (d) Plateau-phase BHK-21 C13 cells. The cells are smaller, more highly condensed, and have assumed a parallel orientation with each other. (e) Bovine aortic endothelium. This morphology has a regular appearance similar to that shown in (b), though here the cells are more closely packed. (f) Again, a

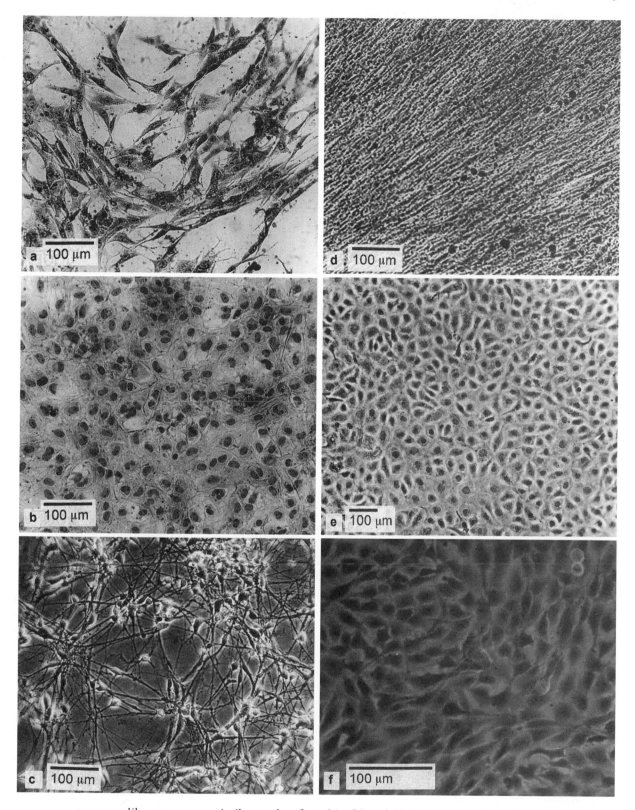

pavementlike appearance similar to that found in (b) and (c), but now produced by 3T3 cells, mouse fibroblasts. With experience, these cell types may be distinguished, but their similarity underlines the need for criteria for identification other than morphology. (a,b,d) Giemsa stained. (c,e–h) Phase contrast. (e) By courtesy of Dr. P. Del Vecchio. (See also Plates 16–22).

ally regarded as "epithelial." However, when the identity of the cells has not been confirmed, the terms "fibroblast-like" (or "fibroblastoid") and "epithelial-like" (or "epithelioid") should be used.

Frequent brief observations of living cultures, preferably with phase-contrast optics, are more valuable than infrequent stained preparations that are studied at length. The former give a more general impression of the cell's morphology and its plasticity and also reveal differences in granularity and vacuolation that bear on the health of the culture. Unhealthy cells often become granular and then display vacuolation around the nucleus. (See Fig. 12.1.)

Microscopy

The inverted microscope is one of the most important tools in the tissue culture laboratory, but is often used incorrectly. The following protocol provides brief general instructions for the use of an inverted microscope.

PROTOCOL 15.1. USING AN INVERTED MICROSCOPE

Outline
Place the culture on the microscope stage, switch on the power, and select the correct optics. Focus on the specimen and center the condenser and phase ring, if necessary.

Materials
Nonsterile:
Inverted microscope (see Inverted Microscope in Chapter 4) with phase contrast on 10× and 20× or 40× objectives and condenser with appropriate phase rings
Lens tissue

Protocol
1. Make sure that the microscope is clean. (Wipe the stage with 70% alcohol, and clean the objectives with lens tissue, if necessary.)
2. Switch on the power, bringing the lamp intensity up from its lowest to the correct intensity, instead of switching the lamp straight to bright illumination.
3. Check the alignment of the condenser and the light source:
   (a) Place a stained slide, flask, or dish on the stage.
   (b) Close down the field aperture.
   (c) Focus on the image of the iris diaphragm.
   (d) Center the image of the iris in the field of view.

4. Center the phase ring:
   (a) If a phase telescope is provided, insert it in place of the standard eyepiece. Then focus on the phase ring, and check that the condenser ring (white on black) is concentric with the objective ring (black on white).
   (b) If no phase telescope provided, then replace the stained specimen with an unstained culture (living or fixed), refocus, and move the phase ring adjustment until optimum contrast is obtained.
5. Examine the cells. A 10× phase contrast objective will give sufficient detail for routine examination. A 4× phase contrast objective will not give sufficient detail, and higher magnifications than 10× restrict the area scanned.
   (a) Note the growth phase (e.g., sparse, subconfluent, confluent, dense).
   (b) Note the state of the cells (e.g., clear and hyaline or granular, vacuolated, etc.).
6. Check the clarity of the medium (e.g., absence of debris, floating granules, signs of contamination) (see Visible Microbial Contamination in Chapter 18), selecting a higher magnification objective as required.
7. Record your observations.
8. Turn the rheostat down and switch off the power to the microscope.
9. Return the culture to the incubator, or take it to a hood.

Cultures taken straight from the incubator will develop condensation on the upper surface of the flask or dish. This condensation can be cleared in a flask by inverting the flask and running medium over the inside of the top of the flask (without allowing it to run into the neck) and standing the flask vertically for 10–20 sec to allow the liquid to drain down. Clearing the condensation is more difficult for Petri dishes and may require replacement of the lid. If the dish is correctly labeled, i.e., on one side of the base, this should not prejudice identification of the culture.

It is useful to keep a set of photographs for each cell line (see Protocols 15.6 and 15.7), prepared shortly after acquisition and at intervals thereafter, as a record in case a morphological change is subsequently suspected. This record can be supplemented with photographs of stained preparations and a scanned autoradiograph from a DNA fingerprint.

Staining

A polychromatic blood stain, such as Giemsa, provides a convenient method of preparing a stained culture. The recommended procedure is as follows.

PROTOCOL 15.2. STAINING WITH GIEMSA

Principle

Giemsa is a polychromatic blood stain. It is usually combined with May–Grünwald stain when staining blood, but not when staining cells. It stains the nucleus pink or magenta, the nucleoli dark blue, and the cytoplasm pale gray-blue. It stains cells fixed in alcohol or formaldehyde, and will not work correctly unless the preparation is completely anhydrous.

Outline

Fix the culture in methanol, and stain it directly with Giemsa. Wash the culture, and examine it wet.

Materials

Nonsterile:
PBSA
PBSA:methanol, 1:1
Undiluted Giemsa stain
Methanol
Deionized water

Protocol

This protocol assumes a cell monolayer is being used, but fixed cell suspensions (see Preparation of Suspension Cultures for Cytology in this chapter) can also be used, starting at Step 6.

1. Remove and discard the medium.
2. Rinse the monolayer with PBSA, and discard the rinse.
3. Add 5 ml of PBSA/methanol per 25 cm². Leave it for 2 min and then discard the PBSA/methanol.
4. Add 5 ml of fresh methanol, and leave it for 10 min.
5. Discard the methanol, and replace it with fresh anhydrous methanol. Rinse the monolayer, and then discard the methanol.
6. At this point, the flask may be dried and stored or stained directly. It is important that staining be done directly from fresh anhydrous methanol, even with a dry flask. If the methanol is poured off and the flask is left for some time, water will be absorbed by the residual methanol and will inhibit subsequent staining. Even "dry" monolayers can absorb moisture from the air.
7. Add neat Giemsa stain, 2 ml per 25 cm², making sure that the entire monolayer is covered and remains covered.
8. After 2 min, dilute the stain with 8 ml of water, and agitate it gently for a further 2 min.
9. Displace the stain with water so that the scum that forms is floated off and not left behind to coat the cells. Wash the cells gently in running tap water until any pink cloudy background stain (precipitate) is removed, but stain is not leached out of cells.
10. Pour off the water, rinse the monolayer in deionized water, and examine the cells on the microscope while the monolayer is still wet. Store the cells dry, and rewet them to re-examine.

Note. Giemsa staining is a simple procedure that gives a good high-contrast polychromatic stain, but precipitated stain may give a spotted appearance to the cells. This precipitate forms as a scum at the surface of the staining solution, due to oxidation, and throughout the solution, particularly on the surface of the slide, when water is added. Washing off the stain by upward displacement, rather than pouring it off or removing the slides, is designed to prevent the cells from coming in contact with the scum. Extensive washing at the end of the procedure is designed to remove any precipitate left on the preparation.

Fixed cell preparations can also be stained with crystal violet. Crystal violet is a monochromatic stain and stains the nucleus dark blue and the cytoplasm light blue. It is not as good as Giemsa for morphological observations, but it is easier to use and can be reused. It is a convenient stain to use for staining clones for counting, as it does not have the precipitation problems of Giemsa and is easier to wash off, giving a clearer background.

PROTOCOL 15.3. STAINING WITH CRYSTAL VIOLET

Materials

Nonsterile:
PBSA
PBSA/MeOH: 1:1 PBSA:methanol
Methanol
Crystal violet
Filter funnel and filter paper of a size appropriate to take 5 ml of stain from each dish
Bottle for recycled stain

Protocol

1. Remove and discard the medium.
2. Rinse the monolayer with PBSA, and discard the rinse.
3. Add 5 ml of PBSA/MeOH per 25 cm². Leave it for 2 min and then discard the PBSA/MeOH.
4. Add 5 ml of fresh methanol, and leave it for 10 min.
5. Discard the methanol. Drain the dishes, and allow them to dry.

6. Add 5 ml of crystal violet per 25 cm², and leave for 10 min.
7. Place the filter funnel in the bottle.
8. Discard the stain into the filter funnel.
9. Rinse the dish in tap water and then in deionized water, and allow the dish to dry.

Culture Vessels for Cytology: Monolayer Cultures

The following is a list of culture vessels which have been found suitable for cytology:

(1) Regular 25-cm² flasks or 50-mm Petri dishes;
(2) Coverslips (glass or Thermanox, Lux, Bayer, ICN) in multiwell dishes, Petri dishes, or Leighton tubes (Fig. 15.3a; see also Figs. 7.2 and 7.3);
(3) Microscope slides in 90-mm Petri dishes or with attached multiwell chambers (see Sources of Materials–Chamber Slides in Trade Index; Fig. 15.3b);
(4) The Gabridge chamber/dish (Bionique), which can be used with a variety of different plastic membranes or standard glass or plastic coverslips, allowing the culture to be placed directly on a thin, optically clear surface for high-resolution microscopy (Fig. 15.3c);
(5) Petriperm dishes (Heraeus), cellulose acetate or polycarbonate filters, Thermanox, and PTFE-coated coverslips, which have all been used for EM cytology studies (some pretreatment of filters or PTFE may be required—e.g., gelatin, collagen, fibronectin, Matrigel, or serum coating—see Improving Clonal Growth in Chapter 13).

Preparation of Suspension Culture for Cytology

In this section, four ways of preparing cytological specimens from suspension cultures are described.

Smear Preparation. This is the type of preparation commonly used to prepare a blood smear for cytological observation.

PROTOCOL 15.4. SMEAR PREPARATION

Materials
Sterile:
Concentrated cell suspension in 50–100% serum
Serum
Nonsterile:
Microscope slides
Methanol
Fan

Protocol
1. Place a drop of serum on one end of a slide. Dip the end of a second slide into the drop and move the second slide up the first slide, distributing a thin film of serum on the slide. (Fig. 15.4.)
2. Dry off the serum with a fan.
3. Place a drop of cell suspension 10⁶ cells/ml or more, in 50–100% serum, on one end of the slide. Dip the end of the second slide into the drop and move the second slide up the first slide (as for the serum), distributing a thin film of cells on the slide. (See Fig. 15.4.)
4. Dry off the slide quickly, and fix the smear in methanol.

△ ***Safety Note.*** Gloves should be worn if human or other primate cells are being handled.

Centrifugation. Centrifuges (usually called *cytocentrifuges*) are available with special slide carriers for spinning cells onto a slide (e.g., the Cytospin (Shandon; or the Cytotek (Miles Scientific; Figs. 15.5, 15.6)). The carriers have sample compartments leading to an orifice that is placed against a microscope slide and are located within a specially designed, enclosed rotor. Coating the slides with serum helps to prevent the cells from bursting when they hit the slide.

PROTOCOL 15.5 CYTOCENTRIFUGE

Outline
Dispense a small volume of concentrated cells into the sample compartment in situ and placed against a slide. Centrifuge the cells onto the slide.

Materials
Nonsterile:
Cell suspension, 5 × 10⁵ cells/ml, in 50–100% serum
Serum
Methanol
Cytocentrifuge (Shandon, Miles)
Microscope slides
Slide carriers for centrifuge
Fan

Protocol
1. Coat the slides in serum by dipping or smearing as in Protocol 15.4.
2. Dry the slides with a fan.
3. Label the slides.
4. Place the slides on slide carriers, and insert the carriers in the rotor.

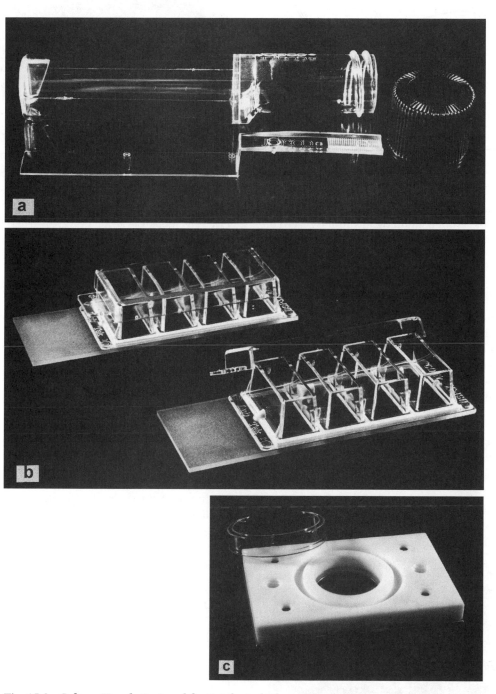

Fig. 15.3. Culture Vessels Designed for Cytological Observation. (a) Costar disposable Leighton tube with coverslip with handle, for easy retrieval. (b) Lab-Tek plastic chambers on a regular microscope slide. One, 2, 4, 8, and 16 chambers per slide are available (Nalge Nunc). (c) Reusable chamber to take a regular glass coverslip (Bionique). (Courtesy of Dr. M. Gabridge.)

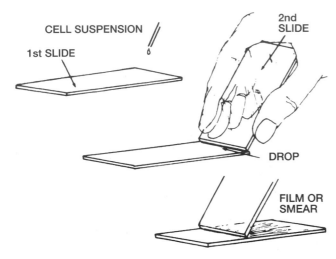

CELL SUSPENSION
1st SLIDE
2nd SLIDE
DROP
FILM OR SMEAR

Fig. 15.4. Preparing a Cell Smear. Top: A drop of cell suspension in serum or serum-containing medium is placed on a slide. Center and bottom: A second slide is used to spread the drop.

5. Place approximately $1-2 \times 10^5$ cells in 200–400 μl of medium in at least two sample blocks.
6. Switch on the centrifuge, and spin the cells down onto the slides at 100 g for 5 min.
7. Dry off the slides quickly, and fix them in MeOH for 10 min.

Note. Some cytocentrifuges require that fixation is performed *in situ*.

Fig. 15.5. Cytotek Centrifuge.

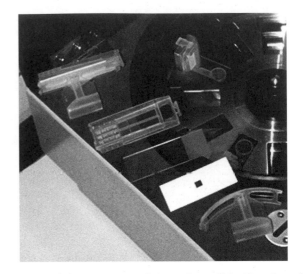

Fig. 15.6. Slide Carriers. Polypropylene slide chambers for Cytotek; they prevent contact between the slide, wick, and cell suspension until centrifugation commences (Miles Scientific).

△ **Safety Note.** The rotor must be sealed during spinning if human or other primate cells are being used.

Drop Technique. This procedure is the same as used for chromosome preparation (see Chromosome Content in this chapter), but without the colcemid and hypotonic treatments.

Filtration. This technique is sometimes used in exfoliative cytology. (See the instructions of the following manufacturers for further details: Pall Gelman, Millipore, Sartorius.)

PROTOCOL 15.6. FILTRATION CYTOLOGY

Outline
Gently draw a low-concentration cell suspension through a filter. Wash the cells and fix it *in situ*.

Materials
Nonsterile:
PBSA, 20 ml
Methanol, 50 ml
Giemsa stain
Mountant (DPX or Permount)
Transparent filters (e.g., 25-mm TPX or polycarbonate, 0.5-μm porosity, Millipore, Pall Gelman, Sartorius)
Filter holder (e.g., Millipore, Pall Gelman, Sartorius)
Vacuum flask (Millipore)

8. Stain the filter in Giemsa, and let it dry.
9. Mount the filter on a slide in DPX or Permount by pressing the coverslip down with a heavy weight to flatten the filter.

Photography

When using conventional film, there are two major frame sizes you may wish to consider: 35 mm, which is best for routine color transparencies and high-volume black-and-white photographs, and $3\frac{1}{2} \times 4\frac{1}{2}$ in (9×12 cm) Polaroid with positive/negative film (type 665), which is good for low-volume black-and-white photographs. Instant film, made by Fuji, may also be used. With instant film, you obtain an immediate result and know that you have the record without having to develop and print a film. The unit cost is high, but there is a considerable saving in time. However, some instant films have relatively low contrast, and their color saturation is not always as good as that of conventional film. Instant films also tend to have a shorter shelf life.

Specific instructions are supplied with microscope cameras, but a few general guidelines may be useful for photographing living cultures.

PROTOCOL 15.7. MICROSCOPE PHOTOGRAPHY ON FILM

Outline
Load the camera with film. Set up the microscope, check the alignment of the optics, and focus on the relevant area of the specimen. Divert the light beam to the camera, focus, and take the photograph.

Materials
Sterile or Aseptically Prepared:
Culture for examination, which should be taken to a hood (make sure that the culture is free of debris —e.g., change the medium on a primary culture before photography)
For flask cultures:
    Loosen the cap on the flask and allow the culture to cool to room temperature;
    Tighten the cap on the flask when the contents of the flask are at room temperature, in order to avoid any distortion of the flask as it cools;
    Rinse the medium over the inside of the top of the flask in order to remove any condensation that may have formed;
    Allow the culture to drain by standing it vertically.
For dishes and plates:
    Allow the culture to cool to room temperature;

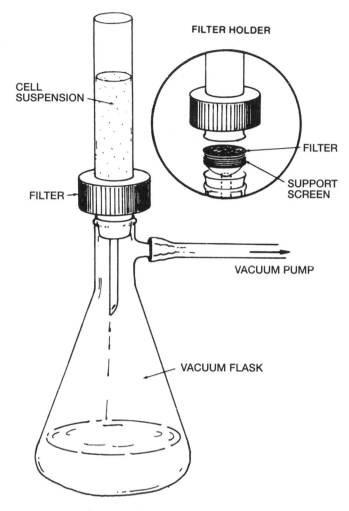

Fig. 15.7. Filter Cytology. Filter assembly for cytological preparation. (See text.)

Cell suspension (~10^6 cells in 5–10 ml medium with 20% serum)
Vacuum pump

Protocol
1. Set up the filter assembly (Fig. 15.7) with a 25-mm-diameter, 0.5-μm-porosity transparent filter.
2. Draw the cell suspension onto the filter using a vacuum pump. Do not let all of the medium run through.
3. Gently add 10 ml of PBSA when the cell suspension is down to 2 ml.
4. Repeat step 3 when the PBSA is down to 2 ml.
5. Add 10 ml of methanol to the PBSA, and keep adding methanol until pure methanol is being drawn through the filter.
6. Switch off the vacuum before all of the methanol runs through.
7. Lift out the filter, and air dry it.

Replace the lid of the dish or plate (make sure that the sample is labeled on the base of the dish or plate).

Nonsterile:

Inverted microscope (see Inverted Microscope in Chapter 4) with the following attributes:

Phase contrast on 10× and 20× or 40× objectives

Phase condenser with phase rings for 10× and 20× or 40× objectives

Trinocular head with camera eyepiece for 35-mm or Polaroid film

Neutral density filters, 2× and 4×

Lens tissue

Record pad for camera

Micrometer slide

Protocol

1. Choose and load the film, and set the film speed on the exposure meter before bringing out the culture:
   (a) For color, use a high-resolution print film (e.g., Fuji Superia or Kodak Royal Gold 100 ISO), or tungsten film for transparencies (e.g., Kodak Ektachrome 160 ISO, Tungsten).
   (b) For black-and-white film, choose high-contrast, fine-grain film (e.g., Kodak Micro-neg Pan).
2. Set the camera or exposure meter to the film speed. (The camera back will do this automatically if it and the film are DX coded.)
3. Set the camera on automatic exposure control with center-spot metering.
4. Select the field and magnification (10× is usually best for a representative shot of a cell monolayer, 20× for cellular detail, and 4× for clones), making sure that the center spot is over an area of typical brightness, and avoiding imperfections or marks on the flask. (Always label the side of the flask or dish, and not the top or bottom.)
5. Check the focus:
   (a) Focus the binocular eyepieces to your own eyes, using the frame or graticule in the eyepiece.
   (b) Focus the microscope.
   (c) Switch the beam-splitter prism to the camera.
   (d) Focus the camera eyepiece to your eye, if a separate eyepiece is provided.
   (e) Finally, critically refocus the microscope, using the camera eyepiece.
6. Turn the light to maximum brightness (necessary for color film to achieve the correct color temperature; less important for black-and-white film); check the focus and exposure; take a photograph; and wind on (usually automatic). When using color film, if the image is too bright for the exposure to be controlled, then reduce the light intensity with neutral gray filters (not by turning down the rheostat, which changes the color temperature).
7. Repeat step 6 on half and double exposure if you are uncertain of the correct exposure.
8. Turn down the light immediately in order to avoid overheating the culture.

Note. An infrared filter may be incorporated to minimize overheating.

9. Repeat step 6 with micrometer slide to give the scale of magnification. If this photograph is processed in the same manner as the other shots, it can be used to generate a scale bar.
10. Return the culture to the incubator.
11. Label Polaroid prints immediately. If you are using 35-mm film, then keep a record against the frame number; otherwise, the prints or slides will be difficult to identify.
12. File the photographs in a readily accessible way (e.g., albums or filing cabinet sheets with transparent pockets).

Reproduction of photographs for publication is best done on high-contrast, glossy, double-weight paper. Reproduction for theses and reports, for which several copies are required, can be achieved quite satisfactorily by laser photocopying. If two to four prints are blocked, the unit cost compares very favorably with that for photographic printing, the time taken is much less, the quality is good, and binding can be achieved without the problems of mounted photographs.

Electronic Image Recording

Although the initial capital outlay is higher for electronic image recording than for a conventional camera, there are significant advantages in making records by CCD camera. Its resolution, although still not quite as high as that for film, is more than adequate for publication, and its images may be stored electronically, with the additional facility for searching, electronic transmission to other sites, formatting, and editing that electronic storage provides. The image is viewed on-screen at the time it is taken, confirming that it is satisfactory, and can be printed instantly if required, in black and white or in color. In addition to the cost of the equipment, the only disadvantages at present are the inability to produce good-quality enlargements and the relatively large footprint required for the equipment in a location where space may already be at a premium. Also, as each image may take up around 2 Mb, electronic storage can become a problem.

PROTOCOL 15.8. ELECTRONIC IMAGES FROM MICROSCOPE

Outline
Set up the microscope, checking the alignment of the optics, and focusing on the relevant area of the specimen. Check the connections of the camera to the computer, divert the light beam to the camera, focus, and save the image.

Materials
Sterile or Aseptically Prepared:
Culture for examination, which should be taken to a hood (make sure that the culture is free of debris—e.g., change the medium on a primary culture before photography)
For flask cultures: Loosen the cap on the flask and, allow the culture to cool to room temperature;
Tighten the cap on the flask only when the contents of the flask are at room temperature, in order to avoid any distortion of the flask as it cools;
Rinse the medium over the inside of the top of the flask in order to remove any condensation that may have formed;
Allow the culture to drain by standing vertically.
For dishes and plates:
Allow the culture to cool to room temperature;
Replace the lid of the dish or plate (make sure that the sample is labeled on the base of the dish or plate)
Nonsterile:
Inverted microscope (see Inverted Microscope in Chapter 4) with the following attributes:
Phase contrast on 10× and 20× or 40× objectives
Phase condenser with phase rings for 10× and 20× or 40× objectives
Trinocular head or other port with adapter with CCD camera
Neutral density filters, 2× and 4×
Lens tissue
Record pad for images
Micrometer slide

Protocol
1. Select the field and magnification (10× is usually best for a representative shot of a cell monolayer, 20× for cellular detail, and 4× for clones), avoiding imperfections or marks on the flask. (Always label the side of the flask or dish, and not the top or bottom.)
2. Check the focus:
   (a) Focus the binocular eyepieces to your own eyes, using the frame or graticule in the eyepiece.
   (b) Focus the microscope.
3. Switch to the monitor, and check the density and contrast of the image on screen, adjusting the light intensity with the rheostat for a black-and-white camera and with neutral density filters for a color camera (although the image can later be reprocessed electronically if the color temperature is incorrect).
4. Refocus the microscope if necessary.
5. Save the image at a resolution appropriate to the image's ultimate use and the electronic storage available (e.g., 940 × 740 will require 2 Mb storage space and provide reasonable quality if reproduced at 5 × 7 in, 12 × 18 cm.)
6. Turn down the light in order to avoid overheating the culture.

Note. An infrared filter may be incorporated to minimize overheating.

7. Repeat steps 4 and 5 with a micrometer slide at the same magnification to give the scale of magnification. If this image is processed in the same manner as the other shots, it can be used to generate a scale bar.
8. Return the culture to the incubator.
9. Complete a record against the filename; otherwise, the images will be difficult to identify.
10. Saved images can be viewed, edited, and formatted using programs such as Adobe Photo-Shop and printed by color laser printer or Kodak ColorEase printer.

CHROMOSOME ANALYSIS

Chromosome content is one of the most characteristic and well-defined criteria for identifying cell lines and relating them to the species and sex from which they were derived. (See Committee on Standardized Genetic Nomenclature for Mice [1972] for information on mouse karyotype, Committee for a Standardized Karyotype of *Rattus norvegicus* [1973] and Shepel et al. [1994] for rat, and Paris Conference [1971 (1975)] or An International System for Human Cytogenetic Nomenclature [Mitelman, 1995] for human. There is also the *Atlas of Mammalian Chromosomes* [Hsu and Benirschke,

1967], although it predates chromosome banding; see Chromosome Banding in this chapter. Chromosome analysis can also distinguish between normal and malignant cells, since the chromosome number is more stable in normal cells (except in mice, where the chromosome complement of normal cells can change quite rapidly after explanation into culture).

PROTOCOL 15.9. CHROMOSOME PREPARATIONS

Outline
Fix cells arrested in metaphase and swollen in hypotonic medium. Drop the cells on a slide, stain and examined them (Fig. 15.8) [Rothfels and Siminovitch, 1958; Rooney and Czepulkowski, 1986].

Materials
Sterile or Aseptically Prepared:
Culture of cells in log phase
Colcemid, 10^{-5} M in PBSA
PBSA
Trypsin, 0.25% crude
Nonsterile:
Hypotonic solution: 0.04 M KCl, 0.025 M sodium citrate
Acetic methanol fixative: 1 part glacial acetic acid plus 3 parts anhydrous methanol or ethanol, made up fresh and kept on ice
Giemsa stain
DPX or Permount mountant
Ice
Centrifuge tubes
Pasteur pipettes
Slides
Coverslips, #00
Slide dishes
Low-speed centrifuge
Vortex mixer

Protocol
1. Set up a 75-cm² flask culture at between 2 × 10⁴ and 5 × 10⁴ cells/ml (4 × 10³ and 1 × 10⁴ cells/cm²) in 20 ml.
2. Approximately 3–5 d later, when the cells are in the log phase of growth, add 1:100 V:V, 1 × 10⁻⁵ M colcemid (1 × 10⁻⁷ M, final) to the medium already in the flask.
3. After 4–6 h, remove the medium gently, add 5 ml of 0.25% trypsin, and incubate the culture for 10 min.
4. Centrifuge the cells in trypsin, and discard the supernatant trypsin.

5. Resuspend the cells in 5 ml of hypotonic solution, and leave them for 20 min at 37°C.
6. Add an equal volume of freshly prepared, ice-cold acetic methanol, mixing constantly, and then centrifuge the cells at 100 g for 2 min.
7. Discard the supernatant mixture, "buzz" the pellet on a vortex mixer, and slowly add fresh acetic methanol with constant mixing.
8. Leave the cells for 10 min on ice.
9. Centrifuge the cells for 2 min at 100 g.
10. Discard the supernatant acetic methanol, and resuspend the pellet by "buzzing" (e.g., holding the bottom of the tube against a vortex mixer) in 0.2 ml of acetic methanol, to give a finely dispersed cell suspension.
11. Draw one drop of the suspension into the tip of a Pasteur pipette, and drop from around 12 in (30 cm) onto a cold slide. Tilt the slide and let the drop run down the slide as it spreads.
12. Dry off the slide rapidly over a beaker of boiling water, and examine it on the microscope with phase contrast. If the cells are evenly spread and not touching, then prepare more slides at the same cell concentration. If the cells are piled up and overlapping, then dilute the suspension two- to fourfold and make a further drop preparation. If the cells from the diluted suspension are satisfactory, then prepare more slides. If not, then dilute the suspension further and repeat this step.
13. Stain the cells with Giemsa: (a) Immerse the slides in neat stain for 2 min; (b) Place the dish in the sink, and add approximately 10 V of water, allowing surplus stain to overflow from the top of the slide dish. (c) Leave the slides for a further 2 min; (d) Displace the remaining stain with running water, and finish by running the slides individually under tap water to remove precipitated stain (which gives a pink, cloudly appearance on slides). (e) Check the staining under the microscope. If it is satisfactory, then dry the slide thoroughly and mount the #00 coverslip in DPX or Permount.

Chromosome Banding

This group of techniques [see Rooney and Czepulkowski, 1986] was devised to enable individual chromosome pairs to be identified when there is little morphological difference between them [Wang and Fedoroff, 1972, 1973]. For Giemsa banding, the chromosomal proteins are partially digested by crude trypsin, producing a banded appearance on subsequent staining. Trypsinization is not required for quinacrine banding. The banding pattern is characteristic for each chromosome pair. (Fig. 15.9.)

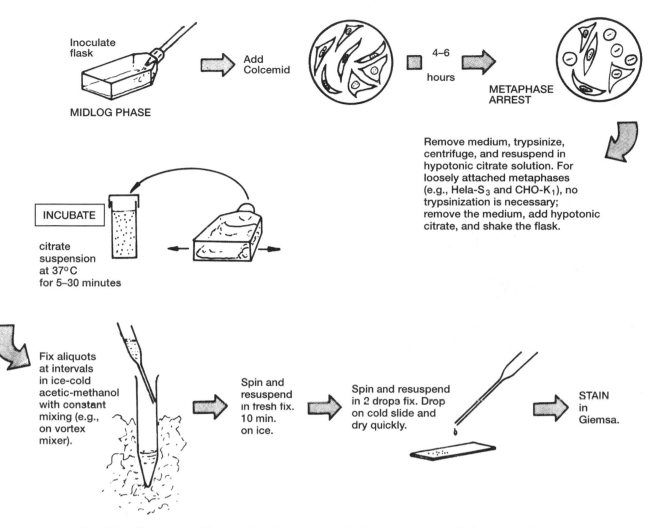

Fig. 15.8. Chromosome Preparation. Preparation of chromosome spreads from monolayer cultures by the drop technique.

A protocol for chromosome banding has been contributed by Mike Griffiths. (See Protocol 26.5.)

Chromosome Analysis
The following are methods by which the chromosome complement may be analyzed:

(1) **Chromosome Count.** Count the chromosome number per spread for between 50 and 100 spreads. (The chromosomes need not be banded.) Closed-circuit television or a camera lucida attachment may help. You should attempt to count all of the mitoses that you see and classify them (a) by chromosome number or (b), if counting is impossible, as "near diploid uncountable" or "polyploid uncountable." Plot the results as a histogram. (See Fig. 2.6.)
(2) **Karyotype.** Photograph or save about 10 or 20 good spreads of banded chromosomes, and print

on 20 × 25-cm high-contrast paper. Cut out the chromosomes, sort them into sequence, and stick them down on paper (Figs. 15.10 and 15.11.) If the image is recorded with a CCD camera, or scanned from a slide or print, then chromosome sorting may be done on an image editor, such as Adobe PhotoShop, or Chantal (Leica).

Variations

Metaphase Block. (1) Vinblastine, 10^{-6} M, may be used instead of colcemid. (2) Duration of the metaphase block may be increased to give more metaphases for chromosome counting, but chromosome condensation will increase, making banding very difficult.

Collection of Mitosis. Some cells—e.g., CHO and HeLa—detach readily when in metaphase if the flask is shaken ("shake-off" technique), eliminating trypsin-

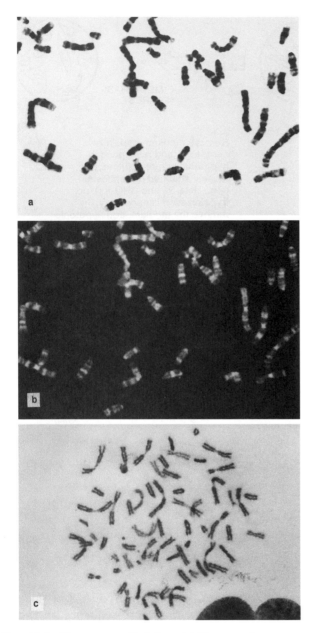

Fig. 15.9. Chromosome Staining. (a) Human chromosomes banded by the standard trypsin–Giemsa technique. (b) The same preparation as in (a), but stained with Hoechst 33258. (c) Human/mouse hybrid stained with Giemsa at pH 11. Human chromosomes are less intensely stained than mouse chromosomes. Several human/mouse chromosomal translocations can be seen. (Courtesy of Dr. R.L. Church.)

Hypotonic Treatment. Substitute 0.075 M KCl alone or HBSS diluted to 50% with distilled water for the hypotonic citrate used in Protocol 15.9. The duration of the hypotonic treatment may be varied from 5 min to 30 min to reduce lysis or increase spreading.

Spreading. There are perhaps more variations at this stage than any other, all designed to improve the degree and flatness of the spread. They include (1) dropping cells onto a slide from a greater height (clamp the pipette, and mark the position of the slide using a trial run with fixative alone); (2) flame drying (dry the slide after dropping the cells, by heating the slide over a flame, or actually burn off the fixative by igniting the drop on the slide as it spreads; however, the latter may make subsequent banding more difficult); (3) making the slide ultracold (chill the slide on solid CO_2 before dropping the cells on to it); (4) refrigerating the fixed cell suspension overnight before dropping the cells onto a slide); (5) dropping cells on a chilled slide (e.g., steep the slide in cold alcohol and dry it off) and then placing the slide over a beaker of boiling water; and (6) tilting the slide or blowing the drop across the slide as the drop spreads.

Banding. The methods for banding include the following: (1) Giemsa-banding (use trypsin and EDTA rather than trypsin alone); (2) Q-banding [Caspersson et al., 1968] (stain the cells in 5% (w/v) quinacrine dihydrochloride in 45% acetic acid, rinse the slide, and mount it in deionized water at pH 4.5 [Lin and Uchida, 1973; Uchida and Lin, 1974]); (3) C-banding (this technique emphasizes the centromeric regions; the fixed preparations are pretreated for 15 min with 0.2 M of HCl and 2 min with 0.07 M NaOH and then are treated overnight with SSC (either 0.03 M sodium citrate, 0.3 M NaCl or 0.09 M sodium citrate, 0.9 M NaCl) before staining with Giemsa stain [Arrighi and Hsu, 1974] (see Protocol 26.5).

Techniques have been developed for discriminating between human and mouse chromosomes, principally to aid the karyotypic analysis of human and mouse hybrids. These methods include fluorescent staining with Hoechst 33258, which causes mouse centromeres to fluoresce more brightly than human centromeres, [Hilwig and Gropp, 1972; Lin et al., 1974], alkaline staining with Giemsa ("Giemsa-11") [Bobrow et al., 1972; Friend et al., 1976] (see Fig. 15.9c), and *in situ* hybridization with *chromosome paints* (see later and Protocol 27.2).

Chromosome counting and karyotyping allow species identification of the cells and, when banding is used, distinguish cell line variation and marker chromosomes. However, banding and karyotyping is time-consuming, and chromosome counting with a quick

ization. The procedure is as follows: (1) add colcemid, (2) remove the colcemid carefully and replace it with hypotonic citrate/KCl, (3) shake the flask to dislodge cells in metaphase either before or after incubation in hypotonic medium, and (4) fix the cells as previously described.

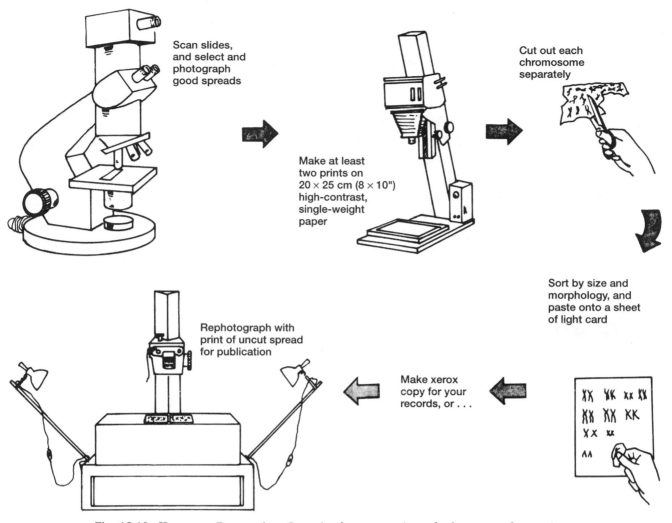

Fig. 15.10. Karyotype Preparation. Steps in the preparation of a karyotype from microphotographs of metaphase spreads.

check on gross chromosome morphology may be sufficient to confirm or exclude a suspected cross-contamination.

Chromosome Painting

With the advent of fluorescently labeled probes that bind to specific regions, and even specific genes, on chromosomes, it has become possible to locate genes, identify translocations, and determine the species of origin of chromosomes. This technique employs *in situ* hybridization technology, which can also be used for extrachromosomal and cytoplasmic localization of specific nucleic acid sequences, such as might be used to localize specific messenger RNA species. (See Applied Imaging; Cambio; Oncer; Protocols 27.1 and 27.2.)

DNA CONTENT

The amount of DNA per cell is relatively stable and is characteristic to the species in normal cell lines, such as human, chick, and hamster fibroblasts, but varies in cell lines from the mouse and from many neoplasms. DNA can be measured by propidium iodide fluorescence with a CCD camera or flow cytometry [Vaughan and Milner, 1989] (see Fluorescence-Activated Cell Sorting in Chapter 14; Flow Cytometry in Chapter 20), although the generation of the necessary single-cell suspension will, of course, destroy the topography of the specimen. Analysis of DNA content is particularly useful in the characterization of transformed cells that are often aneuploid and heteroploid. (See Genetic Instability in Chapter 17.)

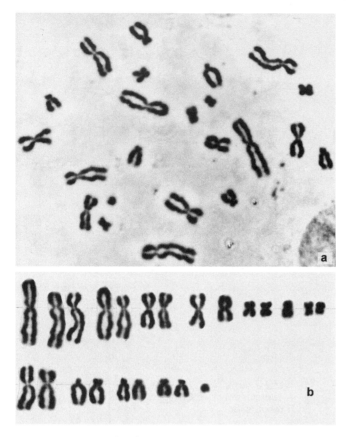

Fig. 15.11. *Example of Karyotype.* Chinese hamster cells recloned from the Y-5 strain of Yerganian and Leonard [1961]. (Chromosomes stained by acetic-orcein.)

DNA Hybridization

It is possible to hybridize specific molecular probes to unique DNA sequences (Southern blotting) and detect the hybrids by radioisotopic, fluorescent, or luminescent labels [Sambrook et al., 1989; Ausubel et al., 1996]. Extracted DNA, whole or cut with restriction endonucleases, electrophoresed, blotted onto nitrocellulose, and hybridized to a particular probe or set of probes, provides information about species-specific regions, amplified regions of the DNA, or altered base sequences that are characteristic to that cell line. Thus, strain-specific gene amplifications, such as amplification of the dihydrofolate reductase (DHFR) gene, may be detected in cell lines selected for resistance to methotrexate [Biedler et al., 1972]; amplification of the MDR gene in vinblastine-resistant cells [Schoenlein et al., 1992]; overexpression of a specific oncogene, or oncogenes in transformed cell lines [Bishop, 1991; Hames and Glover, 1991; Weinberg, 1989]; or deletion, or loss of heterozygosity in suppressor genes [Witkowski, 1990; Marshall, 1991].

It is also possible to label a cell strain for future identification by transfecting in a reporter gene, such as β-galactosidase, and then to detect it by Southern blotting or by chromogenic assay for the gene product.

DNA Fingerprinting

DNA contains regions known as satellite DNA that are apparently not transcribed. The functions of these regions are not understood, and they may be purely structural or may provide a reservoir of potentially coded regions for further genetic evolution. Regardless of their function, however, these regions are not highly conserved, since they are not transcribed, and they give rise to regions of hypervariability. When the DNA is cut with specific endonucleases, it may be probed with cDNAs that hybridize to these hypervariable regions. These probes were originated by Jeffreys et al. [1985], but more are now becoming available. (See Protocol 15.10 and Stacey et al. [1992].) Electrophoresis reveals variations in fragment length (restriction fragment length polymorphism, RFLP) that are specific to the individual from which the DNA was derived. When analyzed by polyacrylamide electrophoresis, each individual's DNA gives a specific hybridization pattern as revealed by autoradiography with radioactive probes. These patterns have come to be known as DNA fingerprints and are cell line specific, except if more than one cell line has been derived from one individual or if highly inbred donor animals have been used.

DNA fingerprints appear to be quite stable in culture, and cell lines from the same origin, but maintained separately in different laboratories for many years, still retain the same or very similar DNA fingerprints. DNA fingerprinting has become a very powerful tool to determine the origin of a cell line, if the original cell line or DNA from it or from the individual has been retained, and emphasizes the need to retain a tissue, blood, or DNA sample when tissue is isolated for primary culture. (See Primary Records in Chapter 11.) Furthermore, if a cross-contamination is suspected, this can be confirmed or denied by fingerprinting the cells and all potential contaminants. DNA fingerprinting has confirmed earlier isoenzyme and karyotypic [Gartler, 1967; Nelson-Rees and Flandermeyer, 1977; Lavappa, 1978; Nelson-Rees et al., 1981] data indicating that many commonly used cell lines are cross-contaminated with HeLa. (See Fig. 15.12 and Table 12.1.)

The following protocol for DNA fingerprinting has been provided by Glyn Stacey, NIBSC, South Mimms, Herts EN6 3QG, England, United Kingdom.

Principle

Since the first report of multilocus DNA fingerprinting by Jeffreys et al. [1985], many probes have been developed for analysis of polymorphic loci called variable number tandem repeats, or VNTRs. In multilocus DNA fingerprinting, the probe is used under conditions that allow cross-hybridization with different repetitive DNA families. Thus, the final result represents polymorphic information from many parts of the genome and is therefore potentially sensitive to change in a cell line's genomic DNA complement. The use of repetitive sequences in the M13 phage genome as a probe for identity testing has been demonstrated in animals, plants, and bacteria and therefore represents a very wide-ranging authentication technique that has the added advantage that pure probe DNA is available commercially at low cost.

PROTOCOL 15.10. MULTILOCUS DNA FINGERPRINTING OF CELL LINES

(a) Preparation of Southern Blots of Cell Line DNA

Outline
Extract DNA from cells digested with a restriction enzyme, immobilize the digest on a nylon membrane by the Southern blot technique [Ausubel et al., 1996], and hybridize the DNA with labeled DNA from the M13 phage.

Materials
Sterile:
PBSA: phosphate-buffered saline, without Ca^{2+} and Mg^{2+} (pH 7.2)
Hinfl enzyme and buffer (e.g., Life Technologies)
Nonsterile:
Agarose gels: 0.7% agarose (Type 1A, Sigma) in 1× TBE, 1 mg/l ethidium bromide

△ *Safety Note.* Ethidium bromide is a potential carcinogen.

TBE, 5× (stock solution, pH 8.2): 54 g of Trisbase (Sigma), 27.5 g of boric acid (Sigma), and 20 ml of 0.5 M EDTA, disodium salt (Sigma)
Loading buffer, 6×: 0.25% (w/v) bromophenol blue, 0.25% (w/v) xylene cyanol, and 40% (w/v) sucrose in distilled water
Acid wash: 500 ml of 0.1 M HCl
Alkaline wash: 500 ml of 0.2 M sodium hydroxide, 0.6 M NaCl
Neutralizing wash: 500 ml of 0.5 M Tris, pH 7.6, 1.5 M NaCl
SSC, 2×: 1 in 10 dilution of stock 20× SSC
SSC, 20× (stock solution): 175.3 g of NaCl (Analar, Merck) and 88.2 g of sodium citrate (Analar,

Merck) dissolved in distilled water to a final volume of 1.0 l at final pH 7.2.

Protocol
1. Wash approximately 10^7 cells (harvested by scraping or trypsinization) twice in PBSA, and recover the cells as a pellet.
2. Extract the high-molecular-weight genomic DNA, and check its integrity by minigel electrophoresis of a small aliquot, as described in step 6 of this Protocol. A method for DNA extraction that has been used routinely for cell lines and that avoids the use of phenol is described in Stacey et al. [1992].
3. Quantify the DNA by UV spectroscopy using a 1-in-50 dilution of DNA [Sambrook et al., 1989].
4. Mix 5 μg of DNA with 40 U of Hinfl enzyme, 3 μl of 10× enzyme buffer, and sterile distilled water to a total volume of 30 μl.
5. Incubate the mixture at 38°C overnight. Remix, and microfuge the digest after 1 h.
6. Test that the digestion is complete by agarose minigel electrophoresis of 1 μl of digest in 9 μl of electrophoresis buffer (1× TBE) and 2 μl of 6× loading buffer. Hinfl digests should show a low-molecular-weight smear of DNA fragments (<5 kb) with no residual genomic DNA band (20–30 kb).
7. To the remainder of the digest, add 6 μl of 6× loading mix and load it into an analytical agarose gel (0.7% in 1× TBE, 1 mg/l ethidium bromide at least 20 cm long) in parallel with a HindIII digest marker and a digest of 5 μg of DNA from a standard cell line (e.g., HeLa). Electrophorese the mixture at approximately 3 V cm⁻¹ until the 2.3-kb HindIII fragment has run the length of the gel (i.e., after 17–20 h). Photograph the gel against a white ruler over a UV transilluminator as a record of migration distance.
8. Treat the separated DNA fragments in the gel with acid wash (15 min), alkaline wash (30 min), and finally a neutralizing wash (30 min). The agarose gel can then be blotted onto a nylon membrane (Hybond N, Amersham) most simply by capillary action using 20× SSC transfer buffer, as described by Sambrook et al. [1989]. Then rinse the membrane in 2× SSC, dry it, and wrap it in Saran Wrap (Dow) or equivalent.
9. Fix DNA fragments to the membrane by exposing the DNA to ultraviolet light (302 nm) for 5 min using a UV-transilluminator (e.g., TM20, UVP Inc).

10. Fixed membranes should be stored in the dark in dry polythene bags prior to hybridization.

(b) Preparation of Labeled M13 Phage DNA

Outline

Label M13mp9 DNA by random primer extension [Feinberg and Vogelstein, 1983], and purify the DNA on a Sephadex column.

Materials

Sterile:

Template: M13mp9 ssDNA (0.25 ng/μl Boehringer Mannheim) diluted 1:4 in distilled water

HEPES, 1 M, pH 6.6

DTM reagent: dATP, dCTP, and dTTP, at 100 μM each in 250 mM of Tris-HCl, 25 mM of MgCl$_2$ and 50 mM of 2-mercaptoethanol (pH 8.0)

△ *Safety Note.* 2-mercaptoethanol is volatile and toxic.

OL reagent: 90 O.D. units of oligodeoxyribonucleotides in 1 mM of Tris and 1 mM of EDTA (pH 7.5)

Labeling buffer: LS buffer (100:100:28 of 1 M HEPES:DTM:OL)

Bovine serum albumin (20 mg/ml of molecular biology grade, Life Technologies)

α-[^{32}P]-dGTP (3 Ci/mMol Amersham international plc)

△ *Safety Note.* ^{32}P is a source of high-energy β particles, to which exposure should be minimized. Always use gloves and 1-cm Perspex screens.

Klenow enzyme (1 μg/μl, sequencing grade, Life Technologies)

Sephadex column (e.g., NICK column, Amersham Pharmacia)

Column buffer: 10 mM Tris-HCl, pH 7.5, 1 mM EDTA

Protocol

1. Mix 2 μl of diluted M13 DNA with 11 μl of distilled water in a sterile microtube.
2. Place the tube in a boiling water bath for 2 min.
3. Transfer the tube directly to ice for 5 min.
4. Add 11.4 μl of LS buffer and 0.5 μl of BSA to the tube.
5. Behind a Perspex screen, add 10 μl of [^{32}P]-dGTP and 2 μl of Klenow enzyme to the tube.
6. Mix the contents of the tube, and briefly centrifuge the tube.
7. Incubate the tube at room temperature overnight behind a Perspex screen.

8. Carefully pipette the contents of the tube onto a Sephadex column previously equilibrated with 10 ml of column buffer.
9. Apply 2× 400 μl of column buffer, collecting the eluate from the second 400 μl in a microtube as the purified probe.

△ *Safety Note.* Radioactivity in the column will remain high, due to retention of unincorporated nucleotide.

10. Boil the purified probe for 2 min and chill it on ice before mixing it with hybridization solution.

(c) Hybridization

Outline

Hybridize labeled M13mp9 DNA to Southern-blotted cell DNA, and, after stringency washing, use autoradiography to visualize the pattern of fragments binding the M13 sequences [Westneat et al., 1988].

Materials

Nonsterile:

Prehybridization/hybridization solutions: 0.263 M disodium hydrogen phosphate, 7% (w/v) sodium dodecyl sulfate, 1 mM EDTA, 1% (w/v) BSA fraction V (Boehringer Mannheim GmbH)

Stringency wash: SSC, 2×

SSC, 20× (stock solution): 175.3 g of sodium chloride (Analar Merck) and 88.2 g of sodium citrate (Analar Merck) dissolved in distilled water to a final colume of 1 l at final pH 7.2.

Protocol

1. Add membranes to 200 ml (for up to three membranes) of prewarmed prehybridization solution at 55°C in a polythene sandwich box.
2. Shake the box at 55°C for 4 h.
3. Transfer the membranes to 150 ml of prewarmed hybridization solution in a second sandwich box at 55°C, with the boiled and chilled probe added.
4. Shake the box at 55°C overnight, for a minimum of 16 h.
5. Transfer up to two membranes to 1 l of stringency wash, and shake the mixture at room temperature for 15 min.
6. Repeat step 5
7. Replace the wash with 1 l of prewarmed (55°C) stringency wash plus 0.1% SDS, and shake the box at 55°C for 15 min.

8. Rinse the membranes in 1× SSC at room temperature.

9. Allow the membranes to air dry on filter paper until just moist; then wrap them in Saran Wrap and measure their surface counts using a Geiger-Muller monitor. Apply autoradiography at −80°C to permit an estimation of the time for X-ray film exposure. Two films (e.g., Fuji RX) should be applied to the hybridized Southern membrane in an autoradiography cassette.

10. Develop the first film after overnight exposure in order to enable more accurate esitmation of the final exposure time required. Use standard X-ray-developing chemicals (e.g., 1:5 v/v of CD15 developer for 5 min, and CD40 fixer for 5 min).

Figure 15.12 shows typical DNA fingerprints produced by this method.

△ *Safety Note.* Procedures involving radioactive materials should be performed according to national and local regulations. Always consult your local biological safety officer before beginning such work.

Discussion. This simple method of fingerprinting using M13 phage DNA (see Fig. 15.12) provides useful fingerprint patterns for cell lines from a wide range of animals, including insects. Note that some cultures may show fingerprints with high-molecular-weight bands that are poorly resolved, and it may be helpful in such cases to interpret only those bands in a molecular-weight range of about 3–15 kb. The method may be refined by using the 0.78-kb Cla1–Bsm1 M13 DNA fragment, which contains the most useful probe sequences. Nonradioactive probe-labeling methods can also be applied [Moreno, 1990].

Other useful multilocus fingerprint probes include the probes 33.15 and 33.6 [Jeffreys et al., 1985] and oligonucleotide probes, such as $(GTG)_5$ [Ali et al., 1986], both of which are available commercially from Zeneca-Cellmark Diagnostics (Abingdon, UK) and Fresnius (Germany), respectively. Nevertheless, M13 phage DNA is the most readily available and the cheapest multilocus fingerprinting probe that can be usefully applied to a wide range of species.

A variety of single-locus probes and polymerase chain reaction (PCR) methods for VNTR analysis are also proving very useful for the identification of cell cultures. For human cell lines, multiplex PCR kits readily give profiles that are specific to the individual of origin (e.g., Promega). However, experience with similar systems indicates that STR analysis may not be able to differentiate between clones of common origin (Y. Reid, personal communication, American Type Culture Collection), although this can be achieved using multilocus DNA fingerprinting [Stacey et al., 1993]. Techniques based on random amplified polymorphic DNA analysis using single, short, random sequence primers may prove problematic in terms of reproducibility. However, single-primer methods based on mini- or microsatellite sequences appear to be more promising and can have a very wide range of application [Meyer et al., 1997; Stacey, unpublished data]. PCR methods are also available that can determine the species of origin of a cell line [e.g., Stacey et al., 1997]. Also in Stacey et al., the authors validated a simple method of species identification that does not require the expense and time required for PCR and isoenzyme analysis. In this method, agarose electrophoresis of restriction digests is used to directly visualize species-specific DNA fragments that enable identification of most species encountered in cell culture.

Multilocus methods have the potential to detect changes in genetic structure at many loci throughout the genome [Thacker et al., 1988] and are generally more versatile for the analysis of multiple species. The fingerprinting method described in Protocol 15.10 provides a cheap and straightforward route to useful quality control for cell lines from a wide range of species. Such testing enables simultaneous confirmation of identity and exclusion of cross-contamination. These are essential criteria to ensure consistent experimental data and to avoid wasted time and money due to mislabeled or cross-contaminated cultures.

RNA AND PROTEIN

Cells of a particular characteristic phenotype can be recognized by analysis of gene expression by Northern blotting [Sambrook et al., 1989; Ausubel et al., 1996] using radioactive, fluorescent, or luminescent probes. This procedure can be carried out at the cellular level, as in FISH (see Protocol 27.2), or by *in situ* hybridization with radioactive probes (see Protocol 27.1).

Qualitative analysis of total cell protein reveals differences between cells when whole cells, or cell membrane extracts, are run on two-dimensional gels [O'Farrell, 1975]. This technique produces a characteristic "fingerprint" similar to polypeptide maps of protein hydrolyzates, but contains so much information that interpretation can be difficult. It is possible to scan these gels by computerized densitometry (Molecular Dynamics), which attributes quantitative values to each spot and can normalize each fingerprint to set standards and interpret positional differences in spots. Labeling the cells with [32P], [35S]-methionine, or a combination of [14C]-labeled amino acids, followed by autoradiography, may make analy-

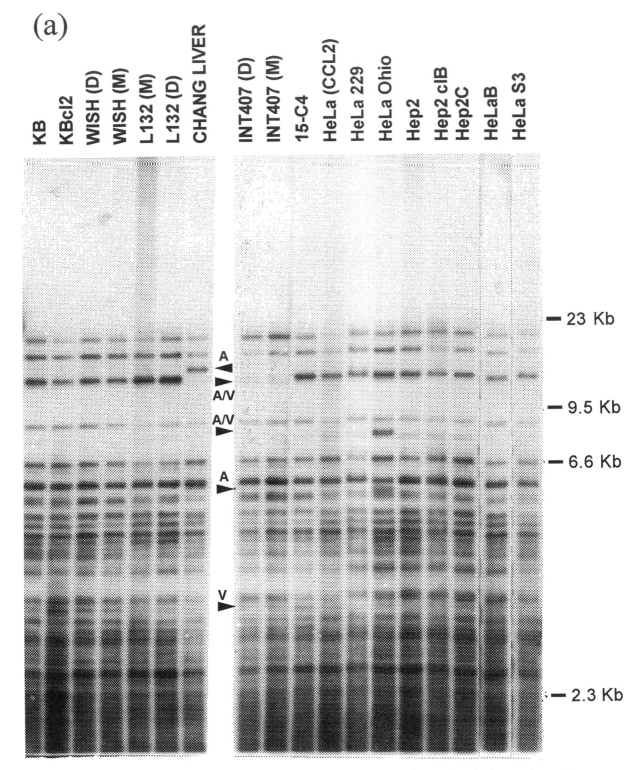

Fig. 15.12. DNA Fingerprints. Southern blots of cell line DNA digested with the Hinf I restriction enzyme and hybridized with the minisatellite probe 33.15 [Jeffreys et al., 1985]. (a) DNA fingerprints of HeLa-cell-contaminated cell lines and subclones of the HeLa cell line. Banding patterns are identical in cases when master banks (M) and their derivative working or distribution cell banks (D) were analyzed. Fingerprint patterns were generally consistent between the cell lines, but some cases showed additional bands (A) (e.g., HeLa Ohio, INT 407, and Chang Liver). (Photograph courtesy of G. Stacey, NIBSC, UK.) (b) Four human glioma cell lines—MOG-G-CCM, U-251MG, SB-18, and MOG-G-UVW (three separate freezings)—and duplicate lanes of BHK-21 controls. (Fingerprints courtesy of G. Stacey, NIBSC, UK.)

(b)

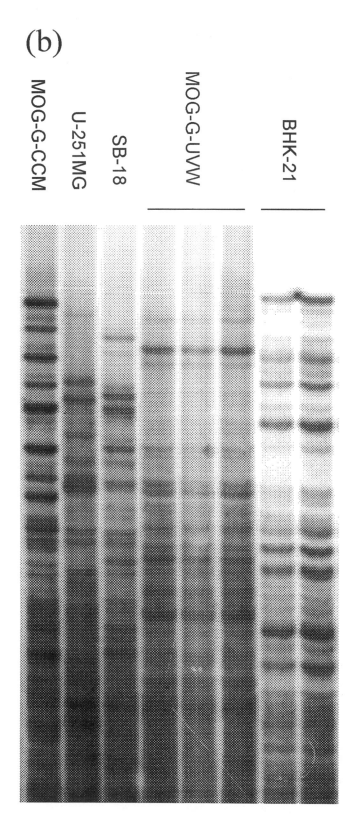

MOG-G-CCM
U-251MG
SB-18
MOG-G-UVW
BHK-21

Fig. 15.12. Continued

sis easier, but it is not a technique suitable for routine use unless the technology for preparing two-dimensional gels is currently in use in the laboratory.

ENZYME ACTIVITY

Specialized functions *in vivo* are often expressed in the activity of specific enzymes—e.g., urea cycle enzymes in liver, angiotensin-converting enzyme in endothelium. Unfortunately, many enzyme activities are lost *in vitro* (see Dedifferentiation in Chapter 2) and are no longer available as markers of tissue specificity—e.g., liver parenchyma loses arginase activity within a few days of culture. However, some cell lines do express specific enzymes, such as tyrosine aminotransferase in the rat hepatoma HTC cell lines [Granner et al., 1968]. When looking for specific marker enzymes, the constitutive (uninduced) level and the induced level should be measured and compared with a number of control cell lines. Glutamyl synthetase activity, for instance, characteristic of astroglia in brain, is increased severalfold when the cells are cultured in the presence of glutamate instead of glutamine [DeMars, 1957]. Induction of enzyme activity requires specialized conditions for each enzyme, and information on these conditions may be obtained from the literature. Common inducers are glucocorticoid hormones, such as dexamethasone; polypeptide hormones, such as insulin and glucagon; and alteration in substrate or product concentrations in the medium, as in the aforementioned example with glutamyl synthetase.

Isoenzymes

Enzyme activities can also be compared qualitatively between cell strains, due to enzyme protein polymorphisms among species and, sometimes, among races, individuals, and tissues within a species. These so-called *isoenzymes*, or *isozymes*, may be separated chromatographically or electrophoretically, and the distribution patterns (zymograms) may be found to be characteristic of species or tissue. (Fig. 15.13.) Nims et al. [1998] determined that interspecies cell line cross-contamination can be detected using isoenzyme analysis if the contaminating cells represent at least 10% of the total cell population.

Electrophoresis media include agarose, cellulose acetate, starch, and polyacrylamide. In each case, a crude enzyme extract is applied to one point in the gel, and a potential difference is applied across the gel. The different isoenzymes migrate at different rates and can be detected later by staining with chromogenic substrates. Stained gels can be read directly by eye and photographed, or scanned with a densitometer.

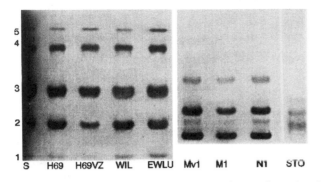

Fig. 15.13. Isoenzyme Patterns. Agarose electrophoresis of lactate dehydrogenase isoenzymes in four human cell lines (NCI-H69; H69VZ, an adherent derivative of NCI-H69; WIL, a human lung adenocarcinoma; and EWLU, human fetal lung fibroblasts), two from mink lung—MV1Lu (MV1) and M1, a *myc* transfected derivative of MV1Lu—and STO mouse fibroblasts.

The following protocol for isoenzyme analysis using an agarose gel system has been contributed by Jane L. Caputo, American Type Culture Collection, 10801 University Blvd., Manassas, Virginia 20110.

Principle

The species of origin of a cell line can be determined with the Authentikit gel electrophoresis system, which can be used to determine the mobility of seven isoenzymes: nucleoside phosphorylase (NP; E.C. 2.4.2.1); glucose-6-phosphate dehydrogenase (G6PD; E.C. 1.1.1.49); malate dehydrogenase (MD, E.C. 1.1.1.37); lactate dehydrogenase (LD; E.C. 1.1.1.27); aspartate aminotransferase (AST, E.C. 2.6.1.1); mannose-6-phosphate isomerase (MPI, E.C. 5.3.1.8); and peptidase B (Pep B, E.C. 3.4.11.4). This system allows easy screening of up to six cells lines per gel for seven genetic markers in less than three hours [Hay, 2000]. In most cases, the species of origin can be determined by using only four of the seven isoenzymes listed: nucleoside phosphorylase, glucose-6-phosphate dehydrogenase, malate dehydrogenase, and lactate dehydrogenase. Similarly, interspecies cell line cross-contamination can be detected in most cases using these four isoenzymes [Nims et al., [1998], although peptidase B was also needed to analyze a contamination of a mouse cell line with a hamster cell line.

PROTOCOL 15.11. ISOENZYME ANALYSIS

Outline

Harvest the cells, wash them in PBSA, resuspend them at 5×10^7 cells/ml, and prepare the cell extract. Store the extract at −70°C. Prepare the gel apparatus. Apply 1–2 μl of the cell extract, stan-

dard, and control to the agarose gels. Fill the chambers with water, place the agarose gels in the chambers, and run electrophoresis for 25 min. Apply enzyme reagents, incubate the gels for 5–20 min, rinse the gels, dry them, and examine the finished gels showing enzyme zones.

Materials

Sterile or Aseptically Prepared:
Cells
Extracts (Innovative Chemistry, Inc.):
  Standard, L929 extract
  Control, HeLa S3 extract
Cell Extraction Buffer (Innovative Chemistry, Inc.) or Triton X-100 Extract Solution: 1:15, V/V, Triton X-100 in 0.9% NaCl, pH 7.1, containing 6.6 × 10^{-4}M EDTA (store at 4°C)
Nonsterile:
Authentikit apparatus and reagents (all from Innovative Chemistry, Inc.):
  Agarose gel films, SAB 8.6, one for each enzyme tested
  Incubation tray liners, one for each enzyme tested
  Incubation trays
  Stain and wash trays
  Temperature-controlled chamber cover and base
  Power supply ((160 v DC) with timer
  Safety interconnector
  Incubation chamber/dryer
  Magnetic stirrer with 1-inch-long stir bar
  Buffer, SAB 8.6
  Enzymatic substrates, 1 vial of each:
    Nucleoside phosphorylase (NP)
    Glucose-6-phosphate dehydrogenase (G6PD)
    Malate dehydrogenase (MD)
    Lactate dehydrogenase (LD)
    Aspartate aminotransferase (AST)
    Mannose-6-phosphate isomerase (MPI)
    Peptidase B (Pep B)
  Sample applicator Teflon tips®
  Gel documentation form
  Enzyme migration data form
  Cell I.D. final analysis form
Microliter syringe
Microcentrifuge, Eppendorf (Brinkmann)
Eppendorf tubes, 1.5 ml
Pipettor, 1.0 ml
Pipettor tips
Pipettes, 5.0 ml
Graduated cylinder, 100 ml
Marking pen
Deionized water
Distilled water
PBSA

Protective gloves
Container for disposal of pipettor tips

Protocol

Preparation of extract:

1. Grow cells in the conventional manner to 2×10^7 viable cells.
2. Harvest cells according to the procedure recommended for the particular cell line.
3. Resuspend the cell pellet in PBSA, and count the number of viable cells.
4. Centrifuge the suspension at 300 g for 5–10 min to pellet the cells and decant the supernatant.
5. Repeat steps 3 and 4 for a total of three washes.
6. Add 100 μl of extract solution to the cell pellet.
7. Gently draw the cell suspension into a small-pore pipette, and pipette the suspension up and down until all cells are lysed.
8. Transfer the cell lysate to Eppendorf tubes (1.5 ml), and pipette the lysate up and down for several minutes until the cell membranes become cloudy and clump together. Spin the lysate at top speed ($\sim$9,000 g) for 2 min in a microfuge. Recover the supernatant, divide it into aliquots in the desired volume, and store it at $-70°C$.

Setup of electrophoresis apparatus:

1. Turn on the incubation chamber.
2. Place one tray liner in each incubator tray.
3. Add 6.0 ml of deionized water to each incubator tray, and place the trays at 37°C for 20 min.
4. Place 95 ml of SAB 8.6 buffer in each chamber of the electrophoresis cell base (a total of 190 ml of buffer per base). This buffer should not be reused, since the pH changes significantly during the electrophoretic run.
5. Connect the filled cell base to the safety interconnector which is plugged into a voltage-regulated power supply. The power supply should be plugged into a grounded electrical outlet.
6. Fill each temperature-controlled electrophoresis cell cover with 500 ml of cold water (4–10°C). The cell cover must be filled when it is in a vertical position, or else the water will run out. Cool the water with ice, or store a sufficient supply of cold water in a standard refrigerator. Replace this water before each run.
7. Place 500 ml of deionized or distilled water in each stain, and wash the tray with it. This water must not be reused.

Electrophoresis procedure:

1. Label each agarose gel to be used. This can be done with labels available from Innovative Chemistry, Inc., or by writing on the back of the gel with a permanent marking pen.
2. Place the gel on the counter, with the plastic nipples down. Orient the gel so that the sample application wells are closest to you.
3. Label each well with the identity of the sample to be tested (e.g., standard, control, unknown 1, unknown 2, etc.). Six unknowns can be tested on each gel.
4. Gently peel the agarose gel from its hard plastic cover. Be careful to handle the gel only by its edges. Discard the hard plastic cover.
5. Add the cell extracts to the sample wells. Use the dispenser with a Teflon tip attached to dispense exactly 1 μl of cell extract into each well. Use a fresh tip for each sample. Dispense the standard extract into track 1, the control into track 2, and the unknowns in tracks 3 to 8. Only the drop of cell extract should touch the well, not the tip, in order to avoid damaging the agarose. If 2 μl of cell extract are required, allow the first 1 μl to diffuse into the agarose before applying the second.
6. Insert each loaded agarose film into a cell cover. The agarose side must face outward. Match the anode ($+$) side of the agarose film with the anode ($+$) side of the cell cover. It may be necessary to bend the agarose gel film slightly to insert it into the cell cover.
7. Place each cell cover on a cell base. The black end with the magnet inside must be nearest to the power supply. Turn on the power supply (160 v), and set the timer for 25 min.
8. Remove each reagent substrate vial from the refrigerator about 5 min before the end of the run, and allow it to warm to room temperature.
9. Add 0.5–1 ml of deionized water to each reagent vial immediately before use. Swirl the vial gently to dissolve the reagent.
10. When the electrophoresis is finished, remove the cell cover from the cell base and place it on absorbent paper.

Staining procedure:

1. To remove the agarose gel film from the cell cover, grasp the film by its edges, squeeze inward, and remove it from the cover.
2. Place the agarose gel film, gel side up, on absorbent paper on a flat surface, with the wells placed toward you.
3. Carefully blot the residual buffer from both ends of the agarose film with a lint-free tissue.
4. Place a 5-ml pipette along the lower edge of the agarose gel film.
5. Pour the reconstituted substrate evenly onto the

agarose film along the leading edge of the pipette.

6. With a single, smooth motion, push the pipette across the surface of the agarose. Drag the pipette back toward you, and push it across the surface one more time. Roll it off the end of the agarose, removing the excess substrate in the process. Be very careful not to damage the agarose. No pressure is necessary to perform this step.

7. Place the agarose gel film, agarose side up, into a prewarmed incubator tray, and place the tray in a 37°C incubator for approximately 5 to 20 min.

8. Following incubation, wash the agarose gel film two times in 500 ml of double-distilled or deionized water for 15 min each time, with agitation provided by a magnetic stirring bar. After the first 15 min, remove each gel from the water, discard the water, add 500 ml of fresh water to the dish, and immerse the gel in the water. Cover the gel to protect it from light. Be sure that the gel is immersed and not floating on top of the wash water.

9. Remove the gel from the water, and place it on a drying rack in the drying chamber of the incubator or oven. Dry it for 30 min or until the agarose is dry. Alternatively, agarose gel films will dry at room temperature overnight.

10. To clean up, pour out the buffer from the cell base, and rinse the cell base with distilled water. Pour the water out of the cell cover, rinse the inside of the cell cover, and allow it to dry.

11. Evaluate you results. The bands are permanent, and the films can be kept for future reference. If a background staining develops over time, the gels were not sufficiently washed.

Analysis

Attach the dried gel to the gel documentation form; line up the sample wells at the origin, and measure them. The bright yellow background of the forms are millimeter lined and give maximum contrast and enhance the purple bands on the dry gel. Measure the enzyme migration distance by measuring from the middle of the application zone to the middle of the enzyme zone. Record measured distances for the standard, the control, and the unknowns on the enzyme migration data form. Transfer the enzyme migration data to the cell I.D. final analysis form to confirm the species identification. Detailed instructions for identifying unknowns are included with the forms. Keep these forms in your notebook as a permanent record of your results. A detailed discussion of each enzyme can be found in the *Handbook for Cell Authentication*

and Identification, available from Innovative Chemistry, Inc.

Variations

Cell extracts can be prepared by ultracentrifugation, freezing and thawing rapidly three times, or treatment with octyl alcohol [Macy, 1978]. A cell extraction buffer is also available from Innovative Chemistry, Inc.

The control and standard may be prepared using Protocol 15.11 for preparation of extract. The standard is prepared from the mouse L929 cell line (ATCC CCL-1), and the control is prepared from the human HeLa cell line (ATCC CCL 2). O'Brien et al. [1980] reported that samples could be stored at −70°C for up to a year without substantial loss of enzyme activity.

ANTIGENIC MARKERS

As a result of the abundance of antibodies and kits from commercial suppliers (see Sources of Materials, Antibodies in the Trade Index), immunostaining and Elisa assays are among the most useful techniques available for cell line characterization. (Table 15.3.) Regardless of the source of the antibody, however, it is essential to be certain of its specificity by using appropriate control material. This is true for monoclonal antibodies and polyclonal antisera alike; a monoclonal antibody is highly specific for a particular epitope, but the generality or specificity of the expression of the epitope must still be demonstrated.

Antibody localization is determined by fluorescence, wherein the antibody is conjugated to a fluorochrome, such as fluorescein or rhodamine, or by the deposition of a precipitated product from the activity of horseradish peroxidase or alkaline phosphatase conjugated to the antibody. Immunological staining may be direct—i.e., the specific antibody is itself conjugated to the fluorochrome or enzyme and used to stain the specimen directly. Usually, however, an indirect method is used wherein the primary (specific) antibody is used in its native form to bind to the antigen in the specimen, and this procedure is followed by treatment with a second antibody, raised against the immunoglobulin of the first antibody. The second antibody may be conjugated to a fluorochrome [Johnson, 1989]. (Fig. 15.14.)

Various methods have been employed to enhance the sensitivity of detection of these methods, particularly the peroxidase-linked methods. In the peroxidase-antiperoxidase (PAP) technique, a further amplifying tier is added by reaction with peroxidase conjugated to anti-peroxidase antibody from the same species as the primary antibody [Jones and Gregory,

TABLE 15.3. Antibodies Used in Cell Line Recognition

Cell type	Antibody to	Localization	Specificity	Comments
All cells	Actin	Microfilament cytoskeleton	Low	Specific antibodies available for myocytes
All cells	Integrins	Cell surface	Low	Specific subtypes on some cells
Anterior pituitary	hGH	Golgi and secreted	High	
Astroglia	GFAP	Intermediate-filament cytoskeleton	High	See also Table 22.2
B-cells	CD22	Cell surface		
Breast epithelium	α-lactalbumin	Golgi and secreted	High	Needs differentiation induction
Colorectal and lung adenocarcinoma	CEA	Cell surface and Golgi	Intermediate	Cell adhesion molecule
Endothelium	Factor VIII	Weibl–Palade bodies	High	
Epithelium	L-CAM	Cell surface	High	
Epithelium	Cytokeratin	Intermediate-filament cytoskeleton	High	Some antibodies also stain mesothelium; specific antibodies for different epithelia
Epithelium	EMA	Cell surface	High	
Epithelium	HMFG I & II	Cell surface	High	Same antigen as EMA
Fetal hepatocytes	AFP	Golgi and secreted	High	Also expressed by hepatomas
Hematopoietic stem cells	CD34	Cell surface	High	Also in precursors but not mature cells
Hepatocytes	Albumin	Golgi and secreted	High	
Immature B-cells	CD20	Cell surface	High	Not in plasma cells
Mesodermal cells	Vimentin	Intermediate-filament cytoskeleton	Low	Also expressed in epithelial and glial cells in culture
Mesothelium	Mesothelial cell antigen	Cell surface	High	
Monocytes and macrophages	CD14	Cell surface	High	
Myeloid cells	CD13	Cell surface	High	
Myocytes	Desmin	Intermediate-filament cytoskeleton	High	
Neural and neuroendocrine cells	NSE	Cytoplasmic	Intermediate	Can be expressed in SCLC
Neural cells	N-CAM	Cell surface	Intermediate	Also expressed in SCLC
Neuroendocrine lung and stomach	GRP	Golgi and secreted	Intermediate	Also expressed in SCLC
Neuronal cells	Neorofilament protein	Intermediate-filament cytoskeleton	High	Also stains SCLC
Oligodendrocytes	Myelin basic protein	Cell surface	High	Also stains peripheral neurones
Oligodendrocytes	Gal-C	Cell surface	High	See also Table 22.2
Placental epithelium	hCG	Golgi and secreted	Low	Also expressed in some lung tumors
Prostatic epithelium	PSA	Golgi and secreted	High	
T-cells	CD3	Cell surface	High	
T-cells and endothelium	I-CAM	Cell surface	Intermediate	

Abbreviations: GFAP, glial fibrillary acidic protein; EMA, epithelial membrane antigen; HMFG, human milk fat globule protein; AFP, α-fetoprotein; CEA, carcinoembryonic antigen; SCLC, small cell lung cancer; hCG, human chorionic gonadotropin; I-CAM, intercellular cell adhesion molecule; NSE, neurone specific enolase; PSA, prostate-specific antigen; Gal-C, galactocerebroside; GRP, gastrin-releasing peptide.

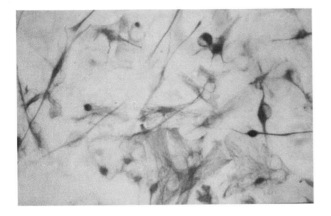

Fig. 15.14. GFAP. Indirect immunoperoxidase staining of glial fibrillary acidic protein (GFAP) in a human anaplastic astrocytoma, MOG-G-CCM. Reichert Polyvar, 10× objective. (See also Plate 25.)

1989]. Even greater sensitivity has been obtained by using a biotin-conjugated second antibody with streptavidin conjugated to peroxidase or alkaline phosphatase (Amersham) or gold-conjugated second antibody with subsequent silver intensification (Janssen).

PROTOCOL 15.12. INDIRECT IMMUNOFLUORESCENCE

Outline
Fix the cells, and treat them sequentially with first and second antibodies. Examine the cells by UV light.

Materials
Sterile or Aseptically Prepared:
Culture grown on glass coverslip, chambered slide (GIBCO-Life Techologies), or polystyrene Petri dish
Nonsterile:
Freshly prepared fixative: 5% acetic acid in ethanol (place at −20°C)
Primary antibody diluted 1:100–1:1,000 in culture medium with 10% FBS
Second antibody raised against the species of the first (e.g., if the first antibody was raised in rabbit, then the second should be from a different species, such as, goat anti-rabbit immunoglobulin); the second antibody should be conjugated to fluorescein or rhodamine
Swine serum or another blocking agent
PBSA
Mountant: 50% glycerol in PBSA and containing a fluorescence-quenching inhibitor (Vecta)

Protocol
1. Wash the coverslip with cells in PBSA, and place it in a suitable dish—e.g., a 24-well plate for a 13-mm coverslip.

2. Place the dish at −20°C for 10 min, add cold fixative to it, and leave it for 20 min.
3. Remove the fixative, wash the coverslip in PBSA, add 1 ml of normal swine serum, and leave the dish at room temperature for 20 min.
4. Rinse the coverslip in PBSA, drain it on paper tissue, and place the coverslip, inverted, on a 50-μl drop of diluted primary antibody.
5. Place the coverslip at 37°C for 30 min at room temperature for 1–3 h or overnight at 4°C. For the 4°C incubation the antibody may be diluted 1:1,000.
6. Rinse the coverslip in PBSA, and transfer it to the second antibody, diluted 1:20, for 20 min at 37°C.
7. Rinse the coverslip in PBSA, and mount it on a slide in 50% glycerol in PBSA with fluorescence bleaching retardant (Vecta).
8. Examine the slide on a fluorescence microscope.

Variations

Fixation. For the cell surface or, particularly, for fixation-sensitive antigens, treat the cells with antibodies first, and then postfix as in step 2. When a glass substrate is used, cold acetone may be substituted for acid ethanol.

Indirect Peroxidase. Substitute peroxidase-conjugated antibody for the fluorescent antibody at stage 6, and then transfer it to peroxidase substrate. (Diaminobenzidine-stained preparations can be dehydrated and mounted in DPX, but ethyl carbazole must be mounted in glycerol as in step 7.

PAP. Use peroxidase conjugated second antibody at stage 6, and then transfer the coverslip to diluted PAP complex (1:100) (most immunobiological suppliers—e.g., Dako, Vecta) in PBSA for 20 min. Rinse it, and add peroxidase substrate as for indirect peroxidase. Incubate, wash, and mount the coverslip.

Cell Surface Markers. Specific cell surface antigens are usually stained in living cells (at 4°C in the presence of sodium azide to inhibit pinocytosis), while intracellular antigens are stained in fixed cells, sometimes requiring light trypsinization to permit access of the antibody to the antigen.

HLA and blood-group antigens can be demonstrated on many human cell lines and serve as useful characterization tools. HLA polymorphisms are particularly valuable, especially when the donor patient profile is known [Espmark and Ahlqvist-Roth, 1978; Stoner et al., 1981; Pollack et al., 1981].

DIFFERENTIATION

Many of the characteristics described under antigenic markers or enzyme activities may also be regarded as markers of differentiation, and as such, they can help to correlate cell lines with their tissue of origin. Other examples of differentiation, and thus specific markers of cell line identity, are given in Table 2.1, and the appropriate assays for these properties may be derived from the references cited in the table. (See also Markers of Differentiation in Chapter 16; Chapter 22.)

AUTHENTICATION

Characterization of a cell line is vital, not only in determining its functionality, but also in proving its authenticity. Cell culture has been plagued since the 1960s with repeated examples of cross contamination, mostly, but by no means invariably, with HeLa cells. (See DNA Fingerprinting in this chapter, Fig. 15.12 and Table 12.1.) While some of the work with cell lines may remain perfectly valid—e.g., if it is a molecular process of interest regardless of the origin of the cells

—much of the work, which attempts to correlate cell behavior with the tissue or tumor of origin, is totally invalidated by cross-contamination. Characterization studies, particularly with continuous cell lines, have also become a process of authentication that is vital to the validation of the data derived from these cells.

The nature of the technique used for characterization depends on the type of work being carried out; for example, if molecular technology is readily available, then expression analysis by Northern blots coupled with DNA fingerprinting is likely to be of most use, while a cytology laboratory may prefer to use chromosome analysis coupled with FISH and chromosome painting, and a laboratory with immunological expertise may prefer to use MHC analysis (e.g., HLA typing) coupled with lineage-specific markers.

Regardless of the intrinsic ability of the laboratory, certain techniques deserve general acclaim, either because of their specificity or their simplicity. Such techniques are DNA fingerprinting and multiple isoenzyme analysis by agarose gels. (See Protocols 15.10 and 15.11.) Combined with a functional assay related to your own interests, these procedures should provide sufficient data to authenticate a cell line.

CHAPTER 16

Differentiation

EXPRESSION OF *IN VIVO* PHENOTYPE

The phenotype of cells cultured and propagated as a cell line is often different from the characteristics that predominate in the tissue from which it was derived. (See Dedifferentiation in Chapter 2.) This is due to several factors that regulate the geometry, growth, and function *in vivo*, but that are absent from the *in vitro* microenvironment. *Differentiation* is the process leading to the expression of phenotypic properties characteristic of the functionally mature cell *in vivo*. This is not to say that differentiation is either complete or irreversible. There are processes that are not normally reversible, such as the cessation of DNA synthesis in the erythroblast nucleus, but other processes are reversible, such as the induction of albumin synthesis in differentiated hepatocytes, which is often lost in culture, but can be reinduced.

Differentiation as used in this text describes the combination of *constitutive* (stably expressed without induction) and *adaptive* (subject to positive and negative regulation of expression) properties found in the mature cell. *Commitment*, on the other hand, implies an irreversible transition from a stem cell to a particular defined lineage endowing the cell with the potential to express a limited repertoire of properties, either constitutively or when induced to do so.

Terminal differentiation implies that a cell has progressed down a particular lineage to a point at which the mature phenotype is fully expressed and beyond which the cell cannot progress. In principle, this definition need not exclude cells, such as fibrocytes, that can revert to a less differentiated phenotype and resume proliferation but in practice, the term tends to

be reserved for cells like neurons, skeletal muscle, or keratinized squames, for which differentiation is irreversible.

Dedifferentiation

Dedifferentiation has been used to describe the loss of the differentiated properties of a tissue when it becomes malignant or when it is grown in culture. As dedifferentiation comprises complex processes with several contributory factors, including cell death, selective overgrowth, and adaptive responses, the term should be used with caution. When used correctly, dedifferentiation means the loss by a cell of the specific phenotypic properties associated with the mature cell. When dedifferentiation occurs, it is, most probably, either an adaptive process, implying that the differentiated phenotype may be regained given the right inducers (see also Differentiation in Chapter 2), or a selective process, implying that a progenitor cell has been selected because of its greater proliferative potential. The latter does not preclude the possibility that the progenitor cell may be induced to mature to the fully differentiated cell, given the correct environmental conditions. If the wrong lineage has been selected (e.g., stromal fibroblasts from liver instead of hepatocytes), no amount of induction can bring back the required phenotype; in the past, this deficit has often been erroneously attributed to dedifferentiation.

STAGES OF COMMITMENT AND DIFFERENTIATION

There are two main pathways to differentiation in the adult organism. In constant renewal tissues, like the epidermis, intestinal mucosa, and blood, a small population of totipotent or pluripotent undifferentiated stem cells, capable of self-renewal, gives rise, on demand, to committed progenitor cells that will proliferate and progress toward terminal differentiation, losing their capacity to divide as they reach the terminal stages. (See Fig. 2.4.) This process gives rise to mature, differentiated cells that normally will not divide. Proliferation in the progenitor compartment is regulated by feedback to generate the correct size of the differentiated cell pool.

In tissues that do not turn over rapidly, but replenish themselves in response to trauma, the resting tissue shows little proliferation; however, the mature cells may reenter division. In connective tissue, for example, cells such as fibrocytes may respond to a local reduction in cell density and/or the presence of one or more growth factors by losing their differentiated properties (e.g., collagen synthesis) and reentering the cell cycle. When the tissue has regained the appropriate cell density by division, cell proliferation stops and differentiation is reinduced. This type of renewal is rapid, because a relatively large population of cells is recruited.

It is not clear whether the cells that reenter the cell cycle to regenerate the tissue are phenotypically identical to the bulk of the differentiated cell population, or whether they represent a subset of reversibly differentiated cells. In liver, which responds to damage by regeneration, it appears that the mature hepatocyte can reenter the cell cycle, while in skeletal muscle, where terminally differentiated cells cannot reenter the cycle, regenerating cells are derived from the satellite cells, which appear to form a quiescent stem cell population. Mature fibrocytes, blood vessel endothelial cells, and glial cells appear to be able to reenter the cell cycle to regenerate, but the possibility of a regenerative subset within the total population cannot be ruled out.

PROLIFERATION AND DIFFERENTIATION

As differentiation progresses, cell division is reduced and eventually lost. In most cell systems, cell proliferation is incompatible with the expression of differentiated properties. Tumor cells can sometimes break this restriction, and in melanoma, for example, melanin continues to be synthesized while the cells are proliferating. Even in these cases, however, synthesis of the differentiated product increases when division stops.

There are severe implications for this relationship in culture, where expansion and propagation are often the main requirements, and it is, therefore, not surprising to find that the majority of cell lines do not express fully differentiated properties. This fact was noted many years ago by the exponents of organ culture (see Organ Culture in Chapter 24), who set out to retain three-dimensional, high-cell-density tissue architecture and to prevent dissociation and selective overgrowth of undifferentiated cells. However, although of considerable value in elucidating cellular interactions regulating differentiation, organ culture has always suffered from the inability to propagate large numbers of identical cultures, particularly if large numbers of cells are required, and the heterogeneity of the sample, assumed to be essential for the maintenance of the tissue phenotype, has in itself made the ultimate biochemical analysis of pure cell populations and of their responses extremely difficult.

Hence, in recent years, there have been many attempts to reinduce the differentiated phenotype in pure populations of cells by re-creating the correct environment and, by doing so, to define individual influences exerted on the induction and maintenance of differentiation. This process usually implies the cessation of cell division and the creation of an interactive, high-density cell population, as in histotypic or organotypic culture. This subject is discussed in greater detail in Chapter 24.

COMMITMENT AND LINEAGE

Progression from a stem cell to a particular pathway of differentiation usually implies a rapid increase in commitment, with advancing stages of progression. (See Fig. 2.4.) A hematopoietic stem cell, after commitment to lymphocytic differentiation, will not change lineage at a later stage and adopt myeloid or erythrocytic characteristics. Similarly, a primitive neuroectodermal stem cell, once committed to become a neuron, will not change to a glial cell. This is not to say that, if the inductive environment is altered early enough (i.e., before commitment), a cell can alter its destiny, or even adopt a mixed phenotype under artifactual or pathological circumstances. Commitment may therefore be regarded as the point between the stem cell and a particular progenitor stage, where a cell or its progeny can no longer transfer to a separate lineage.

Many claims have been made in the past regarding cells transferring from one lineage to another. Perhaps the most substantiated of these claims is that of the regeneration of the amphibian lens by recruitment of cells from the iris [Clayton et al., 1980; Cioni et al., 1986]. Since the iris can be fully differentiated and still regenerate lens, this claim has been proposed as *transdifferentiation*. It is, however, one of the few examples, and most other claims have been from tumor cell systems for which the origin of the tumor population may not be clear. For example, small-cell carcinoma of the lung has been found to change to squamous or large-cell carcinoma following relapse after the completion of chemotherapy. Whether this implies that one cell type, the Kulchitsky cell [de Leij et al., 1985], presumed to give rise to small-cell lung carcinoma, changed its commitment, or whether it implies that the tumor originally derived from a multipotent stem cell and on recurrence progressed down a different route, is still not clear [Gazdar et al., 1983; Goodwin et al., 1983; Terasaki et al., 1984].

Similarly, the K562 cell line was isolated from a myeloid leukemia, but subsequently was shown to be capable of erythroid differentiation [Andersson et al., 1979b]. Rather than being a committed myeloid progenitor converting to erythroid, the tumor probably arose in the common stem cell known to give rise to both erythroid and myeloid lineages. For some reason, as yet unknown, continued culture favored erythroid differentiation rather than the myeloid features seen in the original tumor and early culture. In some cases, again in cultures derived from tumors, a mixed phenotype may be generated. For instance, the C_6 glioma of rat expresses both astrocytic and oligodendrocytic features, and these features may be demonstrated simultaneously in the same cells.

In general, however, these cases are unusual and are restricted to tumor cultures. Most cultures from normal tissues, although they may differentiate in different directions, will not alter to a different lineage once they are committed. This raises the question of the actual status of cell lines derived from normal tissues. Most cultures are derived from (1) stem cells, or progenitor cells, which may differentiate in one or more different directions (e.g., lung mucosa, which can become squamous or mucin secreting, depending on the stimuli); (2) committed progenitor cells, which stay true to the lineage; or (3) differentiated cells, such as fibrocytes, which may dedifferentiate and proliferate, but still retain lineage fidelity (see Origin of Cultured Cells in Chapter 2). Some mouse embryo cultures, loosely called fibroblasts (e.g., the various cell lines designated 3T3), probably more correctly belong to category (1), as they can be induced to become adipocytes, muscle cells, and endothelium, as well as fibrocytes.

Many cell lines may have different degrees of commitment, depending on the "stemness" or progenitor status of the cells from which they were derived, but unless the correct environmental conditions are reestablished and proliferation is discouraged, they will remain at the same position in the lineage. Conversely, some cell populations—e.g., bronchial epithelium—will spontaneously mature in regular serum-containing media and require defined serum-free conditions to remain proliferative.

There are now some well-described examples in which progenitor cells (e.g., the O2A common progenitor of the oligodendrocyte and type 2 astrocyte in the brain, which remains a proliferating progenitor cell in a mixture of PDGF and bFGF), will differentiate into an oligodendrocyte in the absence of growth factors or serum, or into a type 2 astrocyte in fetal bovine serum or a combination of ciliary neurotropic factor (CNTF) and bFGF [Raff et al., 1988; Raff, 1990]. (See Table 22.2.) Similarly, cardiac muscle cells remain undifferentiated and proliferative in serum and bFGF, but differentiate in the absence of serum [Goldman and Wurzel, 1992], and primitive embryonal stem cell cultures differentiate spontaneously unless kept in the undifferentiated proliferative phase by bFGF, SCF (stem cell factor, Steel factor, kit ligand), and LIF (lymphocyte inhibitory factor) [Matsui et al., 1992].

Hence, with the advent of more defined media, it is gradually becoming possible to define the correct inducer environment that will maintain cells in a stem-like, or progenitor, status or that will induce the cells to differentiate.

MARKERS OF DIFFERENTIATION

Markers expressed early and retained throughout subsequent maturation stages are generally regarded as lineage markers—e.g., intermediate filament proteins, such as the cytokeratins (epithelium) [Moll et al., 1982], or glial fibrillary acidic protein (astrocytes) [Eng and Bigbee, 1979; Bignami et al., 1980]. Markers of the mature phenotype representing terminal differentiation are more usually specific cell products or enzymes involved in the synthesis of these products— e.g., hemoglobin in an erythrocyte, serum albumin in a hepatocyte, transglutaminase [Schmidt et al., 1985] or involucrin [Parkinson and Yeudall, 1992] in a differentiating keratinocyte [Schmidt et al., 1985], and glycerol phosphate dehydrogenase in an oligodendrocyte [Breen and De Vellis, 1974]. (See Table 2.1.) These properties are often expressed well after commitment and are more likely to be reversible and under adaptive control by hormones, nutrients, matrix constituents and cell-cell interaction. (See Fig. 16.1.)

As the genes coding for a large number of differentiated products have now been identified and sequenced, it is possible to look for the expression of differentiation marker proteins by RT-PCR [Ausubel et al., 1996], which identifies the expression of specific mRNAs. This method is not particularly good as a quantitative estimation and will not necessarily con-

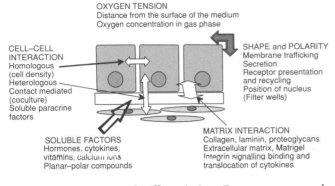

Fig. 16.1. Regulation of Differentiation. Parameters controlling the expression of differentiation *in vitro*.

firm synthesis of the final product, but, with densitometry of the spots on macroautoradiographs, along with appropriate loading controls, it is possible to distinguish between very low levels (or no expression) and high levels of expression of specific gene transcripts. Given the availability of the correct primers and cDNAs, expression of several genes can be screened simultaneously, which is a considerable advantage over other methods of determining product expression.

Differentiation should be regarded as the expression of one, or preferably more than one, marker associated with terminal differentiation. While lineage markers are helpful in confirming cell identity, the expression of the functional properties of the mature cells is the best criterion for terminal differentiation.

INDUCTION OF DIFFERENTIATION

There are four main parameters that govern the control of differentiation. (Fig. 16.1.)

Soluble Inducers

Physiological Inducers. (Table 16.1.) This category includes hormones such as hydrocortisone, glucagon, and thyroxin (or triiodothyronine); paracrine factors that are released by one cell and influence adjacent cells [e.g., KGF in prostate; Planz et al., 1998]; alveolar maturation factor [Post et al., 1984] and other potential candidates, such as IL-6 and oncostatin M in lung maturation [McCormick et al., 1996, McCormick and Freshney, 2000], glia maturation factor in brain [Keles et al., 1992] (see Plates 27 and 28; Paracrine Growth Factors in this chapter), and interferons [e.g., Pfeffer and Eisenkraft, 1991]; vitamins such as vitamin D_3 [Jeng et al., 1994; Rattner et al., 1997] and retinoic acid [Saunders et al., 1993; Hafny et al., 1996; Ghigo et al., 1998]; and inorganic ions, particularly Ca^{2+},

since high Ca^{2+} promotes keratinocyte differentiation [Cho and Bikle, 1997], for example (see Epidermis in Chapter 22).

Nonphysiological Inducers. Rossi and Friend [1967] observed that mouse erythroleukemia cells treated with dimethyl sulfoxide (DMSO), to induce the production of Friend leukemia virus, turned red due to the production of hemoglobin. (See Plates 26 and 37.) Subsequently, it was demonstrated that many other cells—e.g., neuroblastoma, myeloma, and mammary carcinoma—also responded to DMSO by differentiating. Many other compounds have now been added to this list of nonphysiological inducers: hexamethylene bisacetamide (HMBA); N-methyl acetamide; sodium butyrate; benzodiazepines, whose action may be related to that of DMSO; and a range of cytotoxic drugs, such as methotrexate, cytosine arabinoside, and mitomycin C. (Table 16.2.)

The action of these compounds is unclear, but may be mediated by changes in membrane fluidity (particularly for polar solvents, like DMSO, and anesthetics and tranquillizers); by their influence as lipid intercalators on enzymes of signal transduction, such as protein kinase C (PKC) and phospholipase D (PLD), which tend to relocate from the soluble cytoplasm to the endoplasmic reticulum when activated; or by alterations in DNA methylation or histone acetylation. The induction of differentiation by polar solvents, such as DMSO, may be phenotypically normal, but the induction by cytotoxic drugs may also induce gene expression unrelated to differentiation [McLean et al., 1986].

Tumor promoters, such as phorbol meristate acetate (PMA), have been shown to induce squamous differentiation in bronchial mucosa, although not in bronchial carcinoma [Willey et al., 1984; Mascui et al., 1986; Saunders et al., 1993]. Although these tumor promoters are not normal regulators *in vivo*, they bind to specific receptors and activate signal transduction, such as by the activation of PKC [Dotto et al., 1985].

Cell Interaction

Homologous. Homologous cell interaction occurs at high cell density. It may involve gap junctional communication [Finbow and Pitts, 1981], where metabolites, second messengers, such as cyclic AMP, diacylglycerol (DAG), Ca^{2+}, or electrical charge may be communicated between cells. This interaction probably harmonizes the expression of differentiation within a population of similar cells, rather than initiating its expression.

The presence of homotypic cell–cell adhesion molecules, such as the CAMs or cadherins, which are cal-

Plate 1. Human Lung. *Outgrowth from primary explant from squamous cell carcinoma of human lung, MOG-DAN. Giemsa stained. Low magnification.*

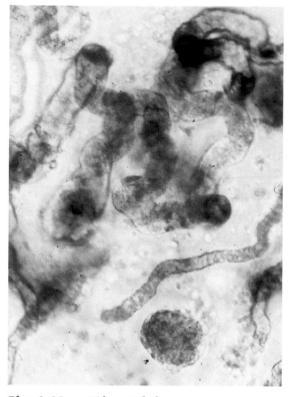

Plate 2. Mouse Kidney Tubules. *Disaggregation of newborn-mouse kidney by cold trypsin method. Connective tissue has dissociated, but fragments of tubules and glomeruli remain intact. Also seen with collagenase digestion. Normal bright field illumination.*

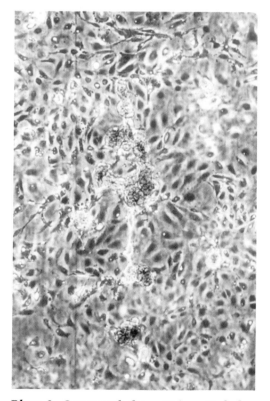

Plate 3. Outgrowth from Kidney Tubules. *Outgrowth of cells from attached fragments following dissagregation by cold trypsin method. Phase contrast.*

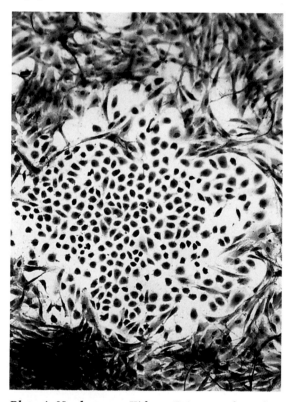

Plate 4. Newborn-rat Kidney. *Primary culture from trypsinized (cold method) newborn-rat kidney. Giemsa stained. (Courtesy of M.G. Freshney.)*

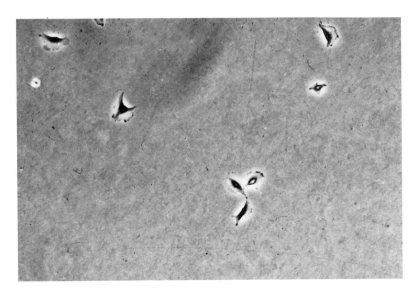

Plate 5. Newly Subcultured Monolayer. *NRK cells 24 h after subculture.*

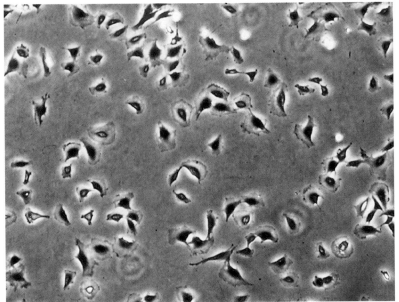

Plate 6. Midlog Phase Cells. *NRK cells 3d after subculture; ready for medium change.*

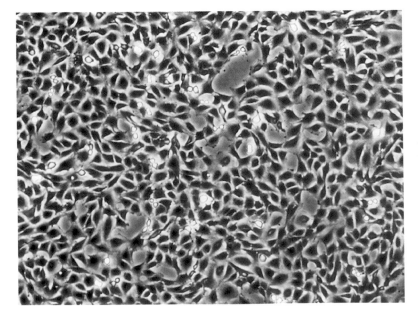

Plate 7. Early-Plateau-phase Cells. *NRK cells 7d after subculture and ready for subculture again.*

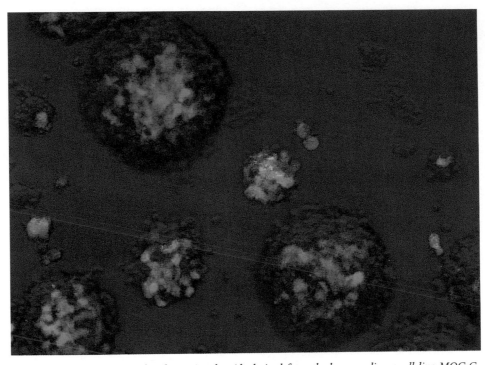

Plate 33. Spheroids. *Transfected mosaic spheroids derived from the human glioma cell line MOG-G-UVW. The spheroids, ranging in size from 100 to 500 µm in diameter, are composed of mixtures of cells transfected with the GFP gene (green). The relative proportions of the cell lines present in the spheroids reflect the proportions of the cells added to the stirrer flask. (Courtesy of Tom Wheldon and Marie Boyd)*

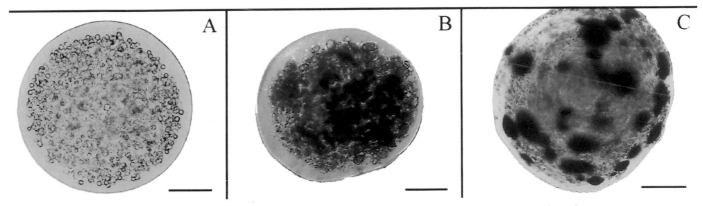

Plate 34. Alginate Encapsulation. *Light-microscopic images of cells encapsulated in alginate after 2 h (A), 3 weeks (B), and 4 months (C) in vitro. Within the alginate beads both cell death and cell proliferation occur, and for many cell lines, multicellular spheroids will form inside the beads (scale bar = 70µm). (Courtesy of Tracy-Ann Read and Rolf Bjerkuig.)*

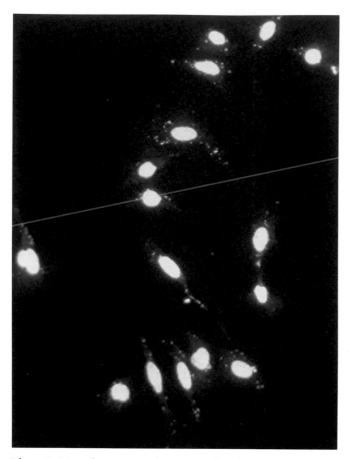

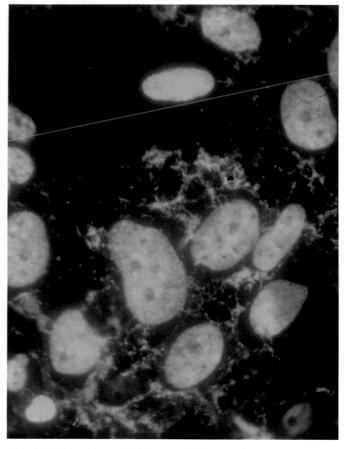

Plate 30. Mycoplasma. *Mycoplasma stained with hoechst 33258. 50 x objective.*

Plate 31. Mycoplasma, High Power. *High-power fluorescence photomicrograph (100 x objective) of Hoechst 33258-stained cells with mycoplasma infection.*

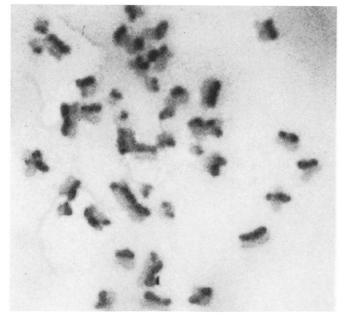

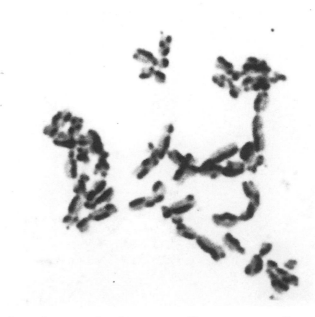

Plate 32a. Control SCE in A2780 cells. *Untreated cells: A2780/Cp70 with an additional human chromosome 2 transferred (A2780/cp70+chr2). (Courtesy of Robert Brown and Maureen Illand).*

Plate 32b. SCE induced in A2780 cells. *A2780/Cp70 cells treated with 10μM Cisplatin for 1h, showing extensive SCE: dark Giemsa staining alternating between strands within individual chromosomes. (Courtesy of Robert Brown and Maureen Illand).*

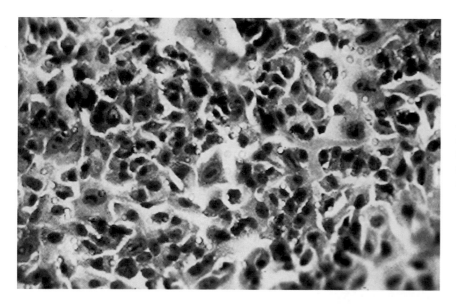

Plate 35a. *Culture on Filter Well.* *A549 human lung adenocarcinoma grown on polycarbonate filter. Holes in the filter (15 μm) are visible, particularly in the top right-hand corner space. Giemsa stained.*

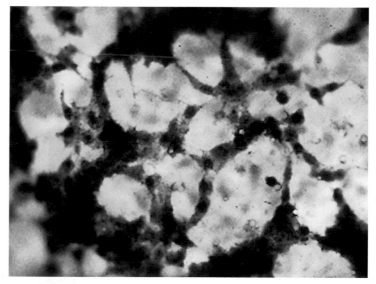

Plate 35b. *A549 cells growing on filter with human fetal lung fibroblasts on other side of filter.*

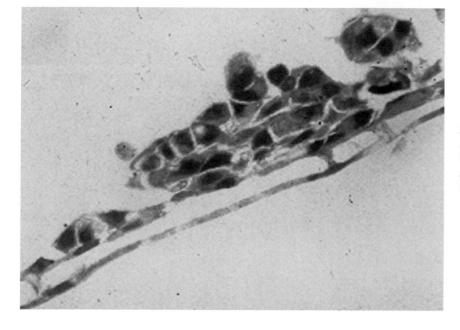

Plate 35c. *Section through filter with A549 cells and fibroblasts. Both cell types were seeded on top, but fibroblasts migrated through first, although some remain on the top. Three cells are seen traveling through the pores. H xE stain; x 40 objective.*

Plate 36. Microcarriers. *Vero cells growing on microcarriers. Courtesy of ICN Biochemicals.)*

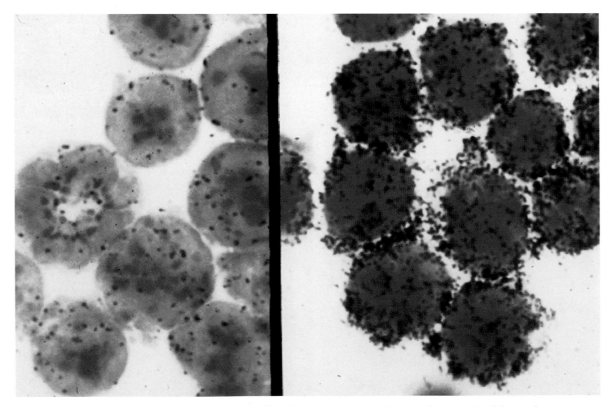

Plate 37. In Situ **Labeling of Friend Cells.** *Effect of DMSO on induction of mRNA for globin. Left-hand figure, control; right hand, induced with 2% DMSO. (Courtesy of David Conkie.)*

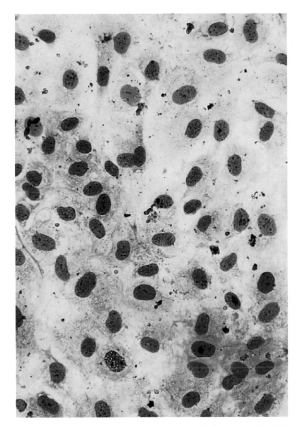

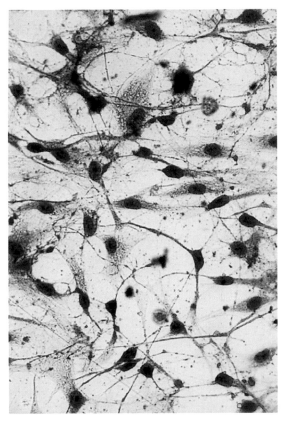

Plate 27. Normal Human Glial. *Undifferentiated glial cells from normal human brain. Giemsa stained.*

Plate 28. Differentiated Human Glial Cells. *Morphological differentiation induced in glial cells from normal human brain by glia maturation factor. Giemsa stained.*

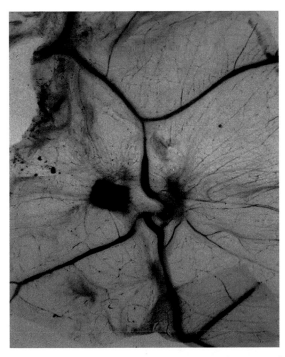

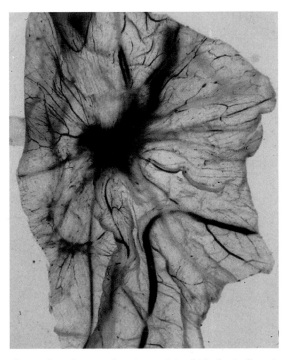

Plate 29a. Induction of Angiogenesis. *A crude extract of normal glial cells was placed on the chorioallantoic membrane at 10 d incubation, and the membrane was removed 2 weeks later. (Courtesy of Margaret Frame).*

Plate 29b. *Induction of angiogenesis in chick chorioallantoic membrane by crude extract of Walker 256 carcinoma cells placed on the chorioallantoic membrane at 10 d incubation, and the membrane was removed 2 weeks later. (Courtesy of Margaret Frame.)*

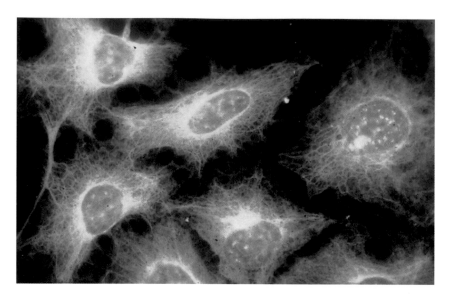

Plate 24. Cytokeratin. *Immunofluorescent staining of cytokeratin in A549 cells. 100 x objective Pan cytokeratin antibody (Courtesy of N. E. Fusenig.)*

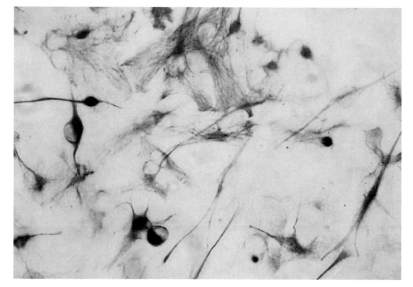

Plate 25. GFAP in Glial Cells. *Immunoperoxidase staining of glial fibrillary acidic protein in MOG-G-CCM, an anaplastic astrocytoma.*

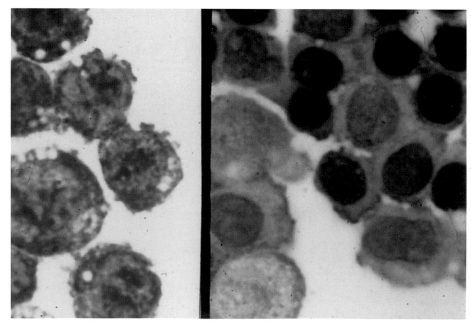

Plate 26. Hemoglobin in Friend Cells. *Benzidine staining of hemoglobin induced in Friend erythroleukemia cells by DMSO. (Photo courtesy of David Conkie.)*

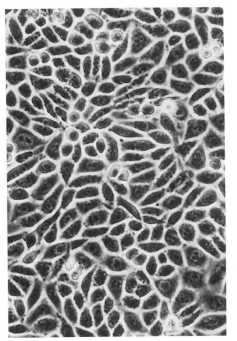

Plate 20. CHO-K1 Cells. *Continuous cell line from hamster ovary. Phase contrast.*

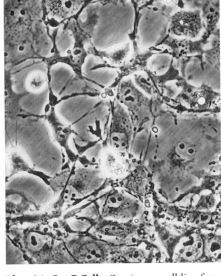

Plate 21. Cos-7 Cells. *Continuous cell line from monkey kidney. Phase contrast.*

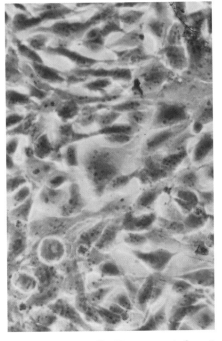

Plate 22. HeLa Cells. *Human cervical carcinoma. Phase contrast.*

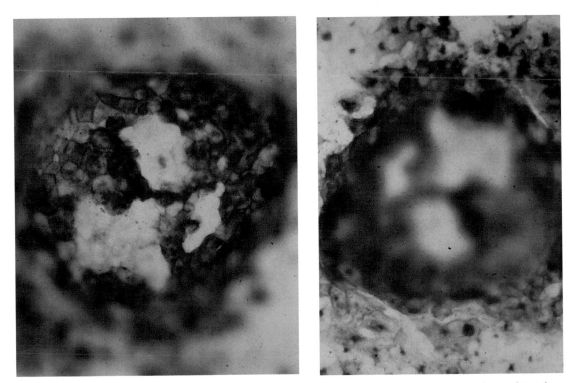

Plate 23a. Domes. *Dome forming in monolayer of WIL lung adenocarcinoma. Mosaic of CEA-positive and CEA-negative cells. Focused on top of dome.*

Plate 23b. *Dome forming in monolayer of WIL lung adenocarcinoma. Mosaic of CEA-positive and CEA-negative cells. Focused on monolayer.*

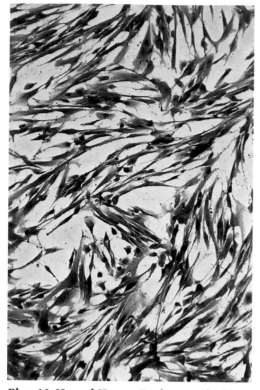

Plate 16. Normal Human Fetal Lung Fibroblasts. *Subconfluent normal fetal human lung fibroblasts. Giemsa stained.*

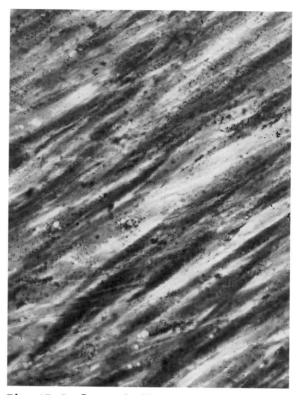

Plate 17. Confluent Fibroblasts. *Confluent normal fetal human lung fibroblasts. Phase contrast.*

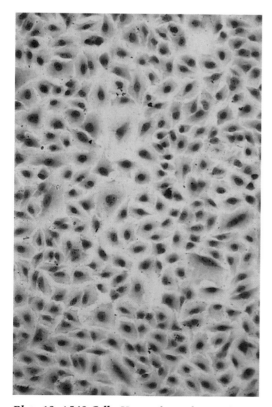

Plate 18. A549 Cells. *Human lung adenocarcinoma. Giemsa stained.*

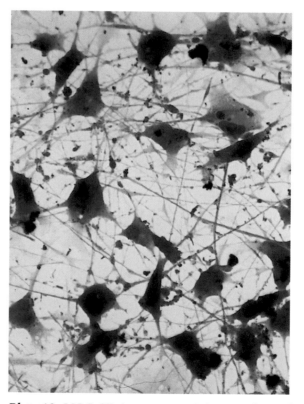

Plate 19. MOG-GP Astrocytoma. *Primary culture of astroglial cells from an astrocytoma. Giemsa.*

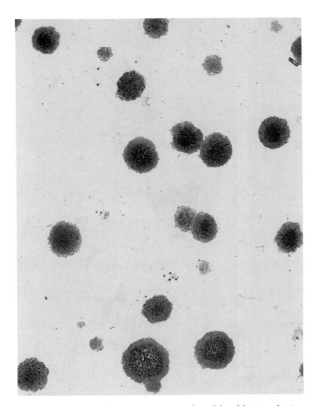

Plate 12. HeLa Clones. *HeLa-S3 cloned by dilution cloning and stained with Giemsa after 3 weeks.*

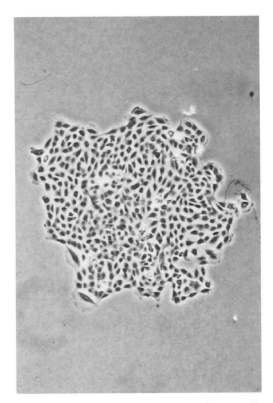

Plate 13. NRK Clone. *Small clone of NRK cells, cloned by dilution cloning. Phase contrast, 4 x objective.*

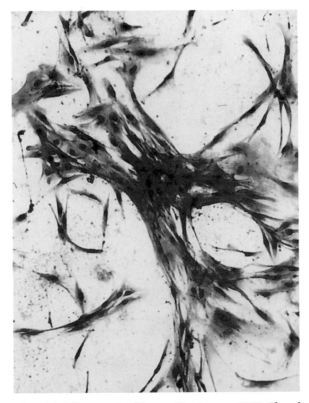

Plate 14. Micrograph of Breast Carcinoma, JUW, Cloned on plastic. *Micrograph of secondary culture from human breast carcinoma, 4000 cells/cm2, growing on plastic. Mainly fibroblasts. Giemsa stain.*

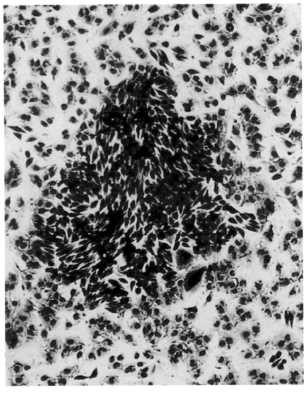

Plate 15. Microphoto of Breast Carcinoma on Feeder Layer. *Microphotograph of epithelial colony from human breast carcinoma cells, 400 cells/cm2, growing on confluent feeder layer of FHS74Int human fetal intestinal cells. Giemsa stain.*

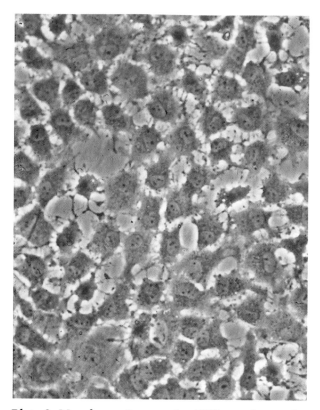

Plate 8. Monolayer, Pre-trypsin. *NRK monolayer, after PBSA/EDTA prewash and before trypsinization.*

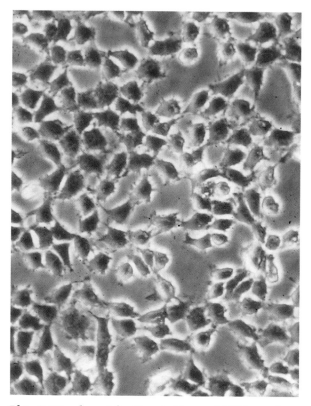

Plate 9. Monolayer After Trypsin Removal. *NRK monolayer immediately after removal of trypsin.*

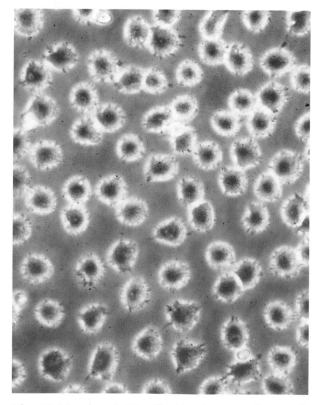

Plate 10. Monolayer 1 Min After Trypsin Removal. *NRK cells rounding up after 1 min incubation.*

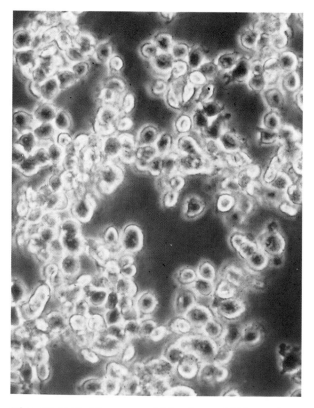

Plate 11. Fully Disaggregated Monolayer. *NRK monolayer 15 min after removal of trypsin.*

TABLE 16.1. Soluble Inducers of Differentiation: Physiological

Inducer	Cell type	Reference
Steroid and related:		
Hydrocortisone	Glia, glioma	[McLean et al., 1986]
	Lung alveolar type II cells	[Rooney et al., 1995; McCormick et al., 1996]
	Hepatocytes	[Granner et al., 1968]
	Mammary epithelium	[Marte et al., 1994]
	Myeloid leukemia	[Sachs, 1978]
Retinoids	Tracheobronchial epithelium	[Kaartinen et al., 1993]
	Endothelium	[Lechardeur et al., 1995; Hafny et al., 1996]
	Enterocytes (Caco-2)	[McCormack et al., 1996]
	Embryonal carcinoma	[Mills et al., 1996]
	Melanoma	[Lotan and Lotan, 1980; Meyskens and Fuller, 1980]
	Myeloid leukemia	[Degos, 1997]
	Neuroblastoma	[Ghigo et al., 1998]
Peptide hormones:		
Melanotropin	Melanocytes	[Goding and Fisher, 1997]
Thyrotropin	Thyroid	[Chambard et al., 1983]
Erythropoietin	Erythroblasts	[Goldwasser, 1975]
Prolactin	Mammary epithelium	[Takahashi et al., 1991; Rudland, 1992; Marte et al., 1994]
Insulin	Mammary epithelium	[Marte et al., 1994; Rudland, 1992]
Cytokines:		
Nerve growth factor	Neurons	[Levi-Montalcini, 1979]
Glia maturation factor	Glial cells	[Keles et al., 1992]
Epimorphin	Kidney epithelium	[Hirai et al., 1992]
Fibrocyte-pneumocyte factor	Type II pneumocytes	[Post et al., 1984]
Interferon-α, β	A549 cells	[McCormick et al., 1996]
	HL60, myeloid leukemia	[Kohlhepp et al., 1987]
Interferon-γ	Neuroblastoma	[Wuarin et al., 1991]
CNTF	Type 2 astrocytes	[Raff, 1990]
IL-6	A549 cells	[McCormick et al., 1996]
Oncostatin M	A549 cells	[McCormick et al., 1999]
KGF	Keratinocytes	[Aaronson et al., 1991]
	Prostatic epithelium	[Thomson et al., 1997; Yan et al., 1992]
HGF	Kidney (MDCK)	[Bhargava et al., 1992; Li et al., 1992]
	Hepatocytes	[Montesano et al., 1991]
TGFβ	Bronchial epithelium	[Masui et al., 1986a]
	Melanocytes	[Fuller and Meyskens, 1981]
Vitamins:		
Vitamin E	Neuroblastoma	[Prasad et al., 1980]
Vitamin D_3	Monocytes (U937)	[Yen et al., 1993]
	Myeloma	[Murao et al., 1983]
	Osteoblasts	[Vilamitjana-Amedee et al., 1993]
	Enterocytes (IEC-6)	[Jeng et al., 1994]
Vitamin K	Hepatoma	[Bouzahzah et al., 1995]
	Kidney epithelium	[Cancela et al., 1997]
Retinoids (see earlier in this table)		
Minerals:		
Ca^{2+}	Keratinocytes	[Boyce and Ham, 1983]

TABLE 16.2. Soluble Inducers of Differentiation: Nonphysiological

Inducer	Cell type	Reference
Planar–polar compounds:		
DMSO	Murine erythroleukemia	[Rossi and Friend, 1967; Dinnen and Ebisuzaki, 1990]
	Myeloma	[Tarella et al., 1982]
	Neuroblastoma	[Kimhi et al., 1976]
	Mammary epithelium	[Rudland et al., 1992]
	Hepatocytes	[Mitaka et al., 1993; Hino et al., 1999]
Sodium butyrate	Erythroleukemia	[Andersson et al., 1979b]
	Colon cancer	[Velcich et al., 1995]
N-methyl acetamide	Glioma	[McLean et al., 1986]
N-methyl formamide, dimethyl formamide	Colon cancer	[Dexter et al., 1979]
HMBA	Erythroleukemia	[Osborne et al., 1982; Marks et al., 1994]
Benzodiazepines	Erythroleukemia	[Clarke and Ryan, 1980]
Cytotoxic drugs:		
Genistein	Erythroleukemia	[Watanabe et al., 1991]
Cytosine arabinoside	Myeloid leukemia	[Takeda et al., 1982]
Mitomycin C; anthracyclines	Melanoma	[Raz, 1982]
Methotrexate	Colorectal carcinoma	[Lesuffleur et al., 1990]
Signal transduction modifiers:		
Glioma	Oat cell cancer	[Frame et al., 1984]
PMA	Bronchial epithelium	[Willey et al., 1984; Masui et al., 1986a]
	Mammary epithelium	[Wada et al., 1994]
	Colon (HT29, Caco-2)	[Velcich et al., 1995; Pignata et al., 1994]
	Monocyte (U937)	[Hass et al., 1993]
	Erythroleukemia (K562)	[Kujoth and Fahl, 1997]
	Neuroblastoma	[Spinelli et al., 1982]

Abbreviations: HMBA, Hexamethylene bisacetamide; PMA (TPA), phorbol meristate acetate.

cium-dependent, provides another mechanism by which contacting cells may interact. These adhesion molecules promote interaction primarily between like cells via identical, reciprocally acting, extracellular domains, and they appear to have signal transduction potential via phosphorylation of the intracellular domains [Doherty et al., 1991; Gumbiner 1995].

Heterologous. Heterologous cell interaction—e.g., between mesodermally and endodermally or ectodermally derived cells—is responsible for initiating and promoting differentiation. During and immediately following gastrulation in the embryo, and later during organogenesis, mutual interaction between cells originating in different germ layers promotes differentiation [Yamada et al., 1991; Hirai et al., 1992; Hemmati-Brivaniou et al., 1994; Muller et al., 1997]. For example, when endodermal cells form a diverticulum from the gut and proliferate within adjacent mesoderm, the mesoderm induces the formation of alveoli and bronchiolar ducts and is itself induced to become

elastic tissue [Hardman et al., 1990; Caniggia et al., 1991].

The extent to which this process is continued in the adult is not clear, but evidence from epidermal maturation suggests that the underlying dermis is required for the formation of keratinized squames with fully cross-linked keratin [Fusenig, 1994a; Limat et al., 1995]. Conditioned medium from lung fibroblasts induces differentiation in A549 and NCI-H441 cells derived from lung tumors of the Clara cell or Type II cell lineage [McCormick and Freshney 2000; McCormick et al., 1996], implying transmission of paracrine factors.

Paracrine Growth Factors

Some growth factors act as morphogens [Gumbiner, 1992] (e.g., epimorphin [Hirai et al., 1992] and hepatocyte growth factor (HGF), one of the family of heparin-binding growth factors (HBGF) (see Table 9.3), shown to be released from fibroblasts, such as the MRC-5 [Kenworthy et al., 1992]). These factors induce tubule formation in the MDCK continuous cell line from dog kidney [Orellana et al., 1996; Montesano et al., 1997], salivary gland epithelium [Furue and Saito, 1997], and mammary epithelium [Soriano et al., 1995]. KGF, also a paracrine factor, is produced by dermal fibroblasts. It influences epidermal differentiation [Aaronson et al., 1991; Gumbiner, 1992], and regulates differentiation in prostatic epithelium [Planz et al., 1998; Thomson et al., 1997]. Growth factors such as FGF-1,2,3, KGF (FGF-7), TGF-β, and activin are also active in embryonic induction [Jessell and Melton, 1992].

HGF and KGF are both produced only by fibroblasts, bind to receptors found only on other cells (principally epithelium), and are classic examples of paracrine growth factors. The release of paracrine factors may be under the control of systemic hormones in some cases. It has been demonstrated that type II alveolar cells in the lung produce surfactant in response to dexamethasone *in vivo*. *In vitro* experiments have shown that this induction of surfactant synthesis is dependent on the steroid binding to receptors in the stroma, which then releases a peptide to activate the alveolar cells [Post et al., 1984]. Similarly, the response of epithelial cells in the mouse prostate to androgens is mediated by the stroma [Thomson et al., 1997]. KGF has been shown to be at least one component of the interaction in the prostate [Yan et al., 1992].

The differentiation of the intestinal enterocyte, which is stimulated by hydrocortisone, also requires underlying stromal fibroblasts [Kédinger et al., 1987], and, in this case, modification of the extracellular matrix between the two cell types is implicated [Simon-Assman et al., 1986]. Matrix modification is also implicated in the response of alveolar Type II cells to paracrine factors (see later) [Yevdokimova and Freshney, 1997]. Differentiation in the hematopoietic system is under control of several positively acting lineage-specific growth factors, such as IL-1, IL-6, G-CSF, and GM-CSF (see Tables 9.3 and 22.3), the last of which is also dependent on the matrix for activation [de Wynter et al., 1993]. Matrix heparan sulphates are also implicated in FGF activation [Klagsbrun and Baird, 1991; Fernig and Gallagher, 1994].

Negatively Acting Paracrine Factors

Negatively acting factors, such as MIP-1α, maintain the stem cell phenotype [Graham et al., 1992], and, similarly, PDGF and FGF-2 promote the growth and self-renewal of the O-2A progenitor cell, but inhibit its differentiation [Bögler et al., 1990]. Other negative regulators of differentiation include LIF in ES cell differentiation [Smith et al., 1988] and TGF-β in alveolar type II cell differentiation [Torday and Kourembanas,

1990; McCormick and Freshney, 2000; McCormick et al., 1996].

Cell–Matrix Interactions

Surrounding the surface of most cells is a complex mixture of glycoproteins and proteoglycans that is almost certainly highly specific for each tissue, and even for parts of a tissue. Reid [1990] showed that the construction of artificial matrices from different constituents can regulate gene expression. For example, addition of liver-derived matrix material induced expression of the albumin gene in hepatocytes. Furthermore, collagen has been found to be essential for the functional expression of many epithelial cells [Berdichevsky et al., 1992; Flynn et al., 1982; Burwen and Pitelka, 1980] and for endothelium to mature into capillaries [Folkman and Haudenschild, 1980]. The RGD motif (arginine-glycine-aspartic acid) in matrix molecules appears to be the receptor interactive moiety in many cases [Yamada and Geiger, 1997]. Small polypeptides containing this sequence effectively block matrix-induced differentiation, implying that the intact matrix molecule is required [Pignatelli and Bodmer, 1988].

Attempts to mimic matrix effects by use of synthetic macromolecules have been partially successful using poly-D-lysine to promote neurite extension in neuronal cultures (see Neurons in Chapter 22), but it seems that there is still a great deal to learn about the specificity of matrix interactions. It is unlikely that charge alone is sufficient to mimic the more complex signals demonstrated in many different types of matrix interaction, but charge alterations probably allow cell attachment and spreading, and under these conditions, the cells may be capable of producing their own matrix.

It has been shown that endothelial cells [Kinsella et al., 1992; Garrido et al., 1995] and many epithelial cells differentiate more effectively on Matrigel [Kibbey et al., 1992; Darcy et al., 1995], a matrix material produced by the Engelberth Holm Swarm (EHS) sarcoma and made up predominantly of laminin, but also of collagen and proteoglycans. This technique is useful, but has the problem of introducing another biological variable to the system. Defined matrices are required; however, as yet, they must be made in the laboratory and are not commercially available. While fibronectin, laminin, collagen, and a number of other matrix constituents are available commercially, the specificity probably lies largely in the proteoglycan moiety, within which there is the potential for wide variability, particularly in the number, type, and distribution of the sulfated glycosaminoglycans, such as heparan sulfate [Fernig and Gallagher 1986].

The extracellular matrix may also play a role in the modulation of growth factor activity. It has been suggested that the matrix proteoglycans, particularly heparan sulfate proteoglycans (HSPGs), may bind certain growth factors, such as GM-CSF [Damon et al., 1989; Luikart et al., 1990], and make them more available to adjacent cells. Transmembrane HSPGs may also act as low-affinity receptors for growth factors and transport these growth factors to the high-affinity receptors [Klagsbrun and Baird, 1991; Fernig and Gallagher, 1994]. A549 cells require glucocorticoid to respond to differentiation inducers such as oncostatin M, IL-6 and lung fibroblast-conditioned medium. The glucocorticoid, in this case dexamethasone (DX), induces the A549 cells to produce a low-charge density fraction of heparan sulfate, which, when partially purified, was shown to substitute for the DX and activate OSM, IL-6, and fibroblast-conditioned medium [Yevdokimova and Freshney, 1997].

Polarity and Cell Shape

Studies with hepatocytes [Sattler et al., 1978] showed that full maturation required the growth of the cells on collagen gel and the subsequent release of the gel from the bottom of the dish by using a spatula or bent Pasteur pipette. This process allowed shrinkage of the gel and an alteration in the shape of the cell from flattened to cuboidal, or even columnar. Accompanying or following the shape change, and also possibly due to access to medium through the gel, the cells developed polarity, visible by electron microscopy; when the nucleus became asymmetrically distributed, nearer to the bottom of the cell, an active Golgi complex formed and secretion toward the apical surface was observed.

A similar establishment of polarity has been demonstrated in thyroid epithelium [Chambard et al., 1983], using a filter well assembly. In this case, the lower (basal) surface generated receptors for thyroid-stimulating hormone (TSH) and secreted triiodothyronine, and the upper (apical) surface released thyroglobulin. Studies with hepatocytes [Guguen-Guillouzo and Guillouzo, 1986] (see Liver in Chapter 22) and bronchial epithelium [Sanders et al., 1993] suggest that floating collagen may not be essential, but the success of filter well culture (see Filter-Well Inserts in Chapter 24) confirms that access to medium from below helps to establish polarity.

DIFFERENTIATION AND MALIGNANCY

It is frequently observed that, with increasing progression of cancer, histology of a tumor indicates poorer differentiation, and from a prognostic standpoint, patients with poorly differentiated tumors generally have a lower survival rate than patients with differentiated tumors. It has also been stated that cancer is principally a failure of cells to differentiate normally. It is therefore surprising to find that many tumors grown in tissue culture can be induced to differentiate. (See Table 16.2.) Indeed, much of the fundamental data on cellular differentiation has been derived from the Friend murine leukemia, mouse and human myeloma, hepatoma, and neuroblastoma. Nevertheless, there appears to be an inverse relationship between the expression of differentiated properties and the expression of malignancy-associated properties, even to the extent that the induction of differentiation has often been proposed as a mode of therapy [Spremulli and Dexter, 1984; Freshney, 1985].

PRACTICAL ASPECTS

It is clear that, given the correct environmental conditions, and assuming that the appropriate cells are present, partial—or even complete—differentiation is achievable in cell culture. As a general approach to promoting differentiation, as opposed to cell proliferation and propagation, the following may be suggested:

(1) Select the correct cell type by use of appropriate isolation conditions and a selective medium. (See Selective Media in Chapter 9 and Chapter 22).

(2) Grow the cell to a high cell density ($>10^5$ cells/cm^2) on the appropriate matrix. The matrix may be collagen of a type that is appropriate to the site of origin of the cells, with or without fibronectin or laminin, or it may be more complex, tissue-derived or cell-derived (see Protocol 7.1), e.g. Matrigel (see Cell–Matrix Interactions in this chapter) or a synthetic matrix (e.g., poly-D-lysine for neurons).

(3) Change the cells to a differentiating medium rather than a propagation medium—e.g., for epidermis, increase Ca^{2+} to around 3 mM, and for bronchial mucosa, increase the serum concentration (see Protocol 22.9). For other cell types, this step may require defining of the growth factors appropriate to maintaining cell proliferation and those responsible for inducing differentiation.

(4) Add differentiation-inducing agents, such as glucocorticoids; retinoids; vitamin D_3; DMSO; HMBA; prostaglandins; and cytokines, such as bFGF, EGF, KGF, HGF, IL-6, OSM, TGFβ, interferons, NGF, and melanocyte-stimulating hormone (MSH), as appropriate for the type of cell. (See Tables 16.1 and 16.2.)

(5) Add the interacting cell type during the growth phase (step (2) in this procedure), the induction phase (steps (3) and (4) in this procedure), or both phases. Selection of the correct cell type is not always obvious, but lung fibroblasts for lung epithelial maturation [Post et al., 1984; Speirs et al., 1991], glial cells for neuronal maturation [Seifert and Müller, 1984], and bone marrow adipocytes for hematopoietic cells (see Protocol 22.19) are some of the better characterized examples.

(6) Elevating the culture in a filter well [Chambard et al., 1983] may be advantageous, particularly for certain epithelia, as it provides access for the basal surface to nutrients and ligands, the opportunity to establish polarity, and regulation of the nutrient and oxygen concentration at the apical surface, by adjusting the composition and depth of overlying medium.

Not all of these factors may be required, and the sequence that they are presented in is meant to imply some degree of priority. Scheduling may also be important; for example, the matrix generally turns over slowly, so prolonged exposure to matrix inducing conditions may be important, while some hormones may be effective in relatively short exposures. Furthermore, the response to hormones may depend on the presence of the appropriate extracellular matrix, cell density, or heterologous cell interaction.

CHAPTER 17

Transformation

ROLE IN CELL LINE CHARACTERIZATION

Transformation is seen as a particular event or series of events that depends on and promotes genetic instability. It alters many of the cell line's properties, including growth rate, mode of growth, specialized product formation, longevity, and tumorigenicity (see Table 17.1). It is therefore important that these characteristics are included when a cell line is validated to determine whether the cell line originates from neoplastic cells or has undergone transformation in culture. The transformation status is a vital characteristic that is required when culturing cells from tumors, in order to confirm that the cells are derived from the neoplastic component of the tumor, rather than from normal infiltrating fibroblasts, blood vessel cells, and inflammatory cells.

More than one criterion is necessary to confirm neoplastic status, as most of the aforementioned characteristics are expressed in normal cells at particular stages of development. The exceptions are gross aneuploidy, heteroploidy, and tumorigenicity, which are regarded as conclusive positive indicators of malignant transformation. However, some tumor cells can be near euploid and nontumorigenic, and other criteria are therefore required.

These general principles of transformation are discussed throughout this chapter, with protocols included as appropriate. The criteria that determine whether cells are transformed are listed in Table 17.1.

WHAT IS TRANSFORMATION?

In microbiology, where the term was first employed in this context, *transformation* implies a change in phenotype that is dependent on the uptake of new genetic material. Although this process is now possible in mammalian cells (see DNA Transfer in Chapter 27), it is called *transfection* or *DNA transfer* in this case to distinguish it from transformation. Transformation of cultured cells implies a spontaneous or induced permanent phenotypic change resulting from a heritable change in DNA and gene expression. Although transformation can arise from infection with a transforming virus, such as polyoma, or from gene transfection, it can also arise spontaneously or following exposure to ionizing radiation or chemical carcinogens.

Transformation is associated with *genetic instability* and three major classes of phenotypic change, one or all of which may be expressed in one cell strain: (1) *immortalization*, the acquisition of an infinite life span, (2) *aberrant growth control*, the loss of contact inhibition and anchorage dependence, and (3) *malignancy*, as evidenced by the tumorigenic potential of the cells. The term *transformation* is used here to imply all three of these processes. The acquisition of an infinite life span alone is referred to as *immortalization*, since it can be achieved without grossly aberrant growth control and malignancy, which are usually correlated.

GENETIC INSTABILITY

The characteristics of a cell line do not always remain stable. In addition to the selective and adaptive processes already described (see Dedifferentiation in Chapter 2; Dedifferentiation in Chapter 16) cell lines are also prone to genetic instability. Normal, human finite cell lines are usually genetically stable, but cell lines from other species, particularly the mouse, are

TABLE 17.1. Properties of Transformed Cells

Property	Assay	Protocol, Figure, or Reference
Growth		
Immortal	Grow beyond 100 pd	[Hayflick and Moorhead, 1961]
Anchorage independent	Clone in agar; may grow in stirred suspension	Protocols 13.4, 13.5, 12.3
Loss of contact inhibition	Microscopic observation; time lapse	Figs. 17.2, 17.3
Growth on confluent monolayers of homologous cells	Focus formation	Fig. 17.2
Reduced density limitation of growth	High saturation density; high growth fraction at saturation density	Protocols 17.2, 20.10, 20.11
Low serum requirement	Clone in limiting serum	Protocols 20.9, 21.3
Growth factor independent	Clone in limiting serum	Protocols 20.9, 21.3
Production of autocrine growth factors	Immunostaining; clone in limiting serum with conditioned medium; receptor-blocking antibody or peptide inhibitor	Protocols 15.12, 20.9, 21.3
Transforming growth factor production	Suspension cloning of NRK	Protocols 13.4, 13.5
High plating efficiency	Clone in limiting serum	Protocols 20.9, 21.3
Shorter population-doubling time	Growth curve	Protocols 20.7, 20.8
Genetic		
High spontaneous mutation rate	Sister chromatid exchange	Protocol 21.5
Aneuploid	Chromosome content	Protocol 15.9
Heteroploid	Chromosome content	Protocol 15.9
Overexpressed or mutated oncogenes	Southern blot; FISH, immunostaining	[Ausubel et al., 1996]; Protocols 27.2, 15.12
Deleted or mutated suppressor genes	Southern blot; FISH, immunostaining	[Ausubel et al. 1996]; Protocols 27.2, 15.12
Gene and chromosomal translocations	FISH, chromosome paints	Protocol 27.2
Structural		
Modified actin cytoskeleton	Immunostaining	Protocol 15.12
Loss of cell-surface-associated fibronectin	Immunostaining	Protocol 15.12
Modified extracellular matrix	Immunostaining; DEAE chromatography	Protocol 15.12; [Yevdokimova and Freshney, 1997]
Altered expression of cell adhesion molecules (CAMs cadherins, integrins)	Immunostaining	Protocol 15.12
Disruption in cell polarity	Immunostaining; polarized transport in filter wells	Protocol 15.12; [Halleux and Schneider, 1994]
Neoplastic		
Tumorigenic	Xenograft in nude or scid mice	[Giovanella et al., 1974]
Angiogenic	CAM assay; filter wells	Fig. 17.7; [Ment et al., 1997]
Enhanced protease secretion (e.g., plasminogen activator)	Plasminogen activator assay	Fig. 17.8; [Whur et al., 1980]; [Boxman et al., 1995]
Invasive	Organoid confrontation; filter-well invasion assay	Figs. 17.5, 17.6; [Mareel et al., 1979]; [Brunton et al., 1997]

genetically unstable and transform quite readily. Continuous cell lines, particularly from tumors of all species, are very unstable, not surprisingly, as this instability was a major reason for their undergoing the necessary mutations to become continuous.

Evidence of genetic rearrangement can be seen in chromosome counts (see Fig. 2.6) and karyotype analysis. While the mouse karyotype is made up exclusively of small telocentric chromosomes, several metacentrics are apparent in many continuous murine cell lines (e.g., Robertsonian fusion). Furthermore, while virtually every cell in the animal has the normal diploid set, this condition is more variable in culture. In extreme cases—e.g., continuous cell strains, such as HeLa-S3—less than half of the cells will have exactly the same karyotype; i.e., they are *heteroploid*.

Most continuous cell strains, even after cloning, contain a range of genotypes that are constantly changing. Because transformation often involves chromosomal rearrangement, it is probable that it can occur only in cells with the capacity for chromosomal alterations. Additionally, transformation may cause genetic instability to arise in a previously stable genotype. Hence, transformed continuous lines retain a capacity for genetic variation that is not apparent *in vivo* or in many finite cell lines.

There are two main causes of genetic variation: (1) The spontaneous mutation rate appears to be higher *in vitro*, associated, perhaps, with the high rate of cell proliferation, and (2) mutant cells are not eliminated unless their growth capacity is impaired. It is not surprising that phenotypic variation will arise as a result of this genetic variation. Minimal-deviation rat hepatoma cells, grown in culture, express tyrosine aminotransferase activity constitutively and may be induced further by dexamethasone [Granner et al., 1968], but subclones of a cloned strain of H4-II-E-C3 [Pitot et al., 1964] differed both in the constitutive level of the enzyme and in its capacity to be induced by dexamethasone. (Fig. 17.1.)

Chromosomal Aberrations

Both variations in ploidy and increases in the frequency of individual chromosomal aberrations can be found [Biedler, 1976; Croce, 1991], and variations in chromosome number are found in most tumor cultures [see Fig. 2.6; Protocol 15.9; Sandberg, 1982]. The incidence of genetic instability and frequency of chromosomal rearrangement can be determined by the sister chromatid exchange assay [Venitt, 1984]. (See Protocol 21.5 and Fig. 21.9.)

Some specific aberrations are associated with particular types of malignancy [Croce, 1991]. The first of these aberrations to be documented was the Philadelphia chromosome in chronic myeloid leukemia (trisomy 13). Subsequently, translocations of the long arms of chromosomes 8 and 14 were found in Burkitt's lymphoma [Lebeau and Rowley, 1984]. Several other leukemias also express other translocations [Mark, 1971]. Meningiomas often have consistent aberrations, and small-cell lung cancer frequently has a 3p2 deletion [Wurster-Hill et al., 1984]. These aberrations constitute tumor-specific markers that can be extremely valuable in cell line characterization and confirmation of neoplasia.

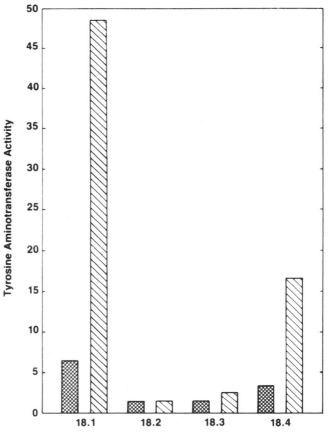

Fig. 17.1. Clonal Variation. Variation in tyrosine aminotransferase activity among four subclones of clone 18 of a rat minimal-deviation hepatoma cell strain, H-4-II-E-C3. Cells were cloned; clone 18 was isolated, grown up, and recloned, and the second-generation clones were assayed for tyrosine aminotransferase activity, with and without pretreatment of the culture with dexamethasone. Crosshatched bars, basal level; hatched bars, induced level. Data provided by J. Somerville.

DNA Content

Flow cytometry [Traganos et al., 1977] shows that the DNA content of tumor cells mimics chromosomal aberrations—i.e., it may vary from the normal somatic cell DNA content and show marked heterogeneity within a population. DNA analysis does not substitute for chromosome analysis, however, as cells with an apparently normal DNA content can still have an aneuploid karyotype. Deletions and polysomy may cancel out, or translocations may occur without net loss of DNA.

IMMORTALIZATION

Most normal cells have a finite life span of 20–100 generations (see Senescence in Chapter 2), but some cells, notably those from rodents and from most tumors, can produce continuous cell lines with an infinite life span. The rodent cells are karyotypically normal at isolation and appear to go through a crisis after about 12 generations; most of the cells die out in this crisis, but a few survive with an enhanced growth rate and give rise to a continuous cell line.

If continuous cell lines from mouse embryos (e.g., the various 3T3 cell lines) are maintained at a low cell density and are not allowed to remain at confluence for any length of time, they remain sensitive to contact inhibition and density limitation of growth [Todaro and Green, 1963]. If, however, they are allowed to remain at confluence for extended periods, foci of cells appear with reduced contact inhibition, begin to pile up, and will ultimately overgrow. (Fig. 17.2b.)

The fact that these cells are not apparent at low densities or when confluence is first reached suggests that they arise *de novo*, by a further transformation event. They appear to have a growth advantage, and subsequent subcultures will rapidly be overgrown by the randomly growing cell. This cell type is often found to be *tumorigenic*.

Control of Senescence

The finite life span of cells in culture is regulated by a group of 10 or more dominantly acting senescence genes, the products of which negatively regulate cell cycle progression [Pereira-Smith and Smith, 1988; Goldstein et al., 1989; Sasaki et al., 1996]. Somatic hybridization experiments between finite and immortal cell lines usually generate hybrids with a finite life span, suggesting that the senescence genes are dominant. It is likely that one or more of these genes negatively regulate the expression of telomerase [Holt et al., 1996; Greider and Blackburn, 1996; Smith and de Lange, 1997; Bryan and Reddel, 1997], required for the

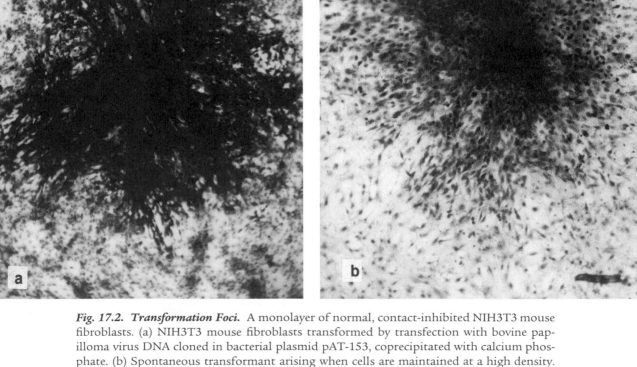

Fig. 17.2. Transformation Foci. A monolayer of normal, contact-inhibited NIH3T3 mouse fibroblasts. (a) NIH3T3 mouse fibroblasts transformed by transfection with bovine papilloma virus DNA cloned in bacterial plasmid pAT-153, coprecipitated with calcium phosphate. (b) Spontaneous transformant arising when cells are maintained at a high density. (Courtesy of D. Spandidos; photographs by M. Freshney.)

terminal synthesis of telomeric DNA, which otherwise becomes progressively shorter during a finite life span, until the chromosomal DNA can no longer replicate. Telomerase is expressed in germ cells and has moderate activity in stem cells, but is absent from somatic cells. Deletions and/or mutations within senescence genes, or overexpression or mutation of one or more oncogenes that override the action of the senescence genes, can allow cells to escape from the negative control of the cell cycle and reexpress telomerase.

It has been assumed that immortalization is a multistep process involving the inactivation of a number of cell cycle regulatory genes, such as Rb and p53. The SV40 LT gene is often used to induce immortalization. The product of this gene, T antigen, is known to bind Rb and p53. By doing so, it not only allows an extended proliferative life span, but also restricts the DNA surveillance activity of genes like p53, thereby allowing an increase in genomic instability and an increased chance of generating further mutations favorable to immortalization (e.g., the up regulation of telomerase or the down regulation of one of the telomerase inhibitors). Recent studies have shown that transfection of the telomerase gene with a regulatable promoter is sufficient to immortalize cells [Bodnar et al., 1998; Vaziri and Benchimol, 1998].

Immortalization per se does not imply the development of aberrant growth control and malignancy, since a number of immortal cell lines, such as 3T3 cells and BHK21-C13, retain contact inhibition of cell motility, density limitation of cell proliferation, and anchorage dependence, and are not tumorigenic. It must be assumed, however, that some aspects of growth control are abnormal and that there is a likely increase in genomic instability. Furthermore, immortalized cell lines often lose the ability to differentiate.

Immortalization with Viral Genes

A number of viral genes have been used to immortalize cells. (Table 17.2.) It has been recognized for some time that SV40 can be used to immortalize cells, and the gene responsible for this appears to be the large T (LT) gene [Mayne et al., 1996]. Other viral genes that have been used to immortalize cells are adenovirus E1a [Seigel, 1996], human papilloma virus (HPV) E6 and E7 [Peters et al., 1996; Le Poole et al., 1997], and Epstein Barr virus (EBV; usually the whole virus is used) [Bolton and Spurr, 1996]. Most of these genes probably act by blocking the inhibition of cell cycle progression by inhibiting the activity of genes such as CIP-1/WAF-1/p21, Rb, p53, and p16, thus giving an increased life span and enhanced opportunity for further mutations. Those genes that have been used most extensively are EBV for lymphoblastoid cells [Bolton and Spurr, 1996] and SV40LT for adherent cells such as fibroblasts [Mayne et al., 1996], keratinocytes [Steinberg, 1996], and endothelial cells [Punchard et al., 1996]. Endothelial cells have also been immortalized by fusion with A549 cells [Suggs et al., 1986] and by irradiation [Punchard et al., 1996].

Typically, cells are transfected or retrovirally infected with the immortalizing gene before they enter senescence. This extends their proliferative life span for another 20–30 population doublings, whereupon the cells cease proliferation and enter *crisis*. After a variable period in crisis (up to several months), a

TABLE 17.2. Genes Used in Immortalization

Gene	Insertion	Cell type	Reference
SV40LT	Lipofection	Keratinocytes	[Steinberg, 1996]
	Calcium phosphate transfection	Fibroblasts	[Mayne et al., 1996]
	Calcium phosphate transfection	Astroglial cells	[Burke et al., 1996]
	Adenovirus infection	Oesophageal epithelium	[Inokuchi et al., 1995]
	Microinjection	Rat brain endothelium	[Lechardeur et al., 1995]
	Transfection	Prostate epithelium	[Rundlett et al., 1992]
	Transfection	Mammary epithelium	[Shay et al., 1993]
	Strontium phosphate transfection	Bronchial epithelium	[De Silva et al., 1996]
	Strontium phosphate transfection	Mesothelial cells	[Duncan et al., 1996]
HPV16 E6/E7		Cervical epithelium	[Demers et al., 1994]
	Transfection	Keratinocytes	[Bryan et al., 1995]
	Strontium phosphate transfection	Mesothelial cells	[De Silva et al., 1994]
	Strontium phosphate transfection	Bronchial epithelium	[De Silva et al., 1994]
	Retroviral infection	Ovarian surface epithelium	[Tsao et al., 1995]
Ad5 E1a		Epithelial cells	[Douglas and Quinlan, 1994]
hTRT	Transfection	Pigmented retinal epithelium	[Bodnar et al., 1998]
hTRT	Transfection	Foreskin fibroblasts	[Bodnar et al., 1998]

subset of immortal cells overgrows. The proportion of cells that eventually immortalize can be 1×10^{-5} to 1×10^{-9}.

Immortalization of Human Fibroblasts

By far, the most successful and most frequently used method for deriving immortal human fibroblasts is through the expression of SV40 T antigen, which does not lead directly to immortalization, but initiates a chain of events that results in an immortalized derivative appearing with a low probability, estimated at about 1 in 10^7 [Shay and Wright, 1989; Huschtcha and Holliday, 1983]. SV40-transfected cells are selected directly under appropriate culture conditions, and the surviving cells are subcultured to give rise to a precrisis SV40-transformed cell population. These cells are cultivated continuously until they reach the end of their proliferative life span, when they inevitably enter crisis. They must then be nurtured with care, and sufficient cells must be cultured, to give a reasonable chance for an immortalized derivative to appear.

The choice of T antigen expression vectors depends on the choice of the dominant selectable marker gene, which is the source of the promoter that drives T antigen expression and alternative forms of the T antigen itself. While, in my experience, selection for *gpt* is effective in human fibroblasts, G418 (*neo*) and hygromycin (*hygB*) are much more effective and easier to use. The majority of human SV40-immortalized fibroblast cell lines have been established with either SV40 virus or constructs, such as pSV3*neo*, that express T antigen from the endogenous promoter. We recommend the use of pSV3*neo* [Southern and Berg, 1982; Mayne et al., 1986] for the constitutive expression of T antigen.

Cells should be used between passages 7 and 15. Trypsinize the cells 24–48 h before transfection, and seed ≤ 2–2.5×10^5 cells per 9-cm dish or ≤ 5.5–6.8×10^5 cells per 175-cm² flask. Cells should be 70–80% confluent when transfected, in a final volume of medium of 10 ml/9-cm dish or 30 ml/175-cm² flask.

The calcium phosphate precipitation method relies on the formation of a DNA precipitate in the presence of calcium and phosphate ions. The DNA is first sterilized by precipitation in ethanol and resuspension in sterile buffer. It is then mixed carefully with calcium, and the resulting solution is added very slowly, with mixing, to a phosphate solution. When making the precipitate, it is important to note that optimal gene transfer occurs when the final concentration of DNA in the precipitate is 20 μg/ml. The volume of DNA precipitate applied to the cultures should never exceed one-tenth of the total volume. It is necessary to leave

the mixture to develop for 30 min before adding it to the cell cultures.

The following protocol and preceding introduction for the immortalization of fibroblasts has been condensed from Mayne et al. [1996].

PROTOCOL 17.1. FIBROBLAST IMMORTALIZATION

Materials
Sterile:
HEPES buffer: HEPES, 12.5 mM; pH 7.12
10× CaHEPES: CaCl$_2$, 1.25 M; HEPES, 125 mM; pH 7.12
2× HEPES-buffered phosphate (2× HBP): Na$_2$HPO$_4$, 1.5 mM; NaCl, 280 mM; HEPES, 25 mM; pH 7.12
NaOAc: NaOAc, 3 M; pH 5.5
Tris-buffered EDTA (TBE): Tris HCl, 2 mM; EDTA, 0.1 mM; pH 7.12
Absolute ethanol
Eagle's MEM/15% FCS
G418 (Gibco): 20 mg/ml in HEPES buffer, pH 7.5, sterilized by filtration through a 0.2-μm membrane and stored in small aliquots at $-20°$C
Hygromycin (Boehringer Mannheim): 2 mg/ml in UPW, filter sterilized with a 0.2-μm membrane and stored in small aliquots at $-20°$C.

Protocol
1. Estimate the amount of DNA required for the transfection:
   (a) Use a maximum of 20 μg of your T antigen vector (without carrier DNA) for each plate and 60 μg of the vector DNA per flask.
   (b) Include an additional 20 μg of DNA, as it is not always possible to recover the full expected amount after preparation of the precipitate.
2. Prepare a sterile solution of the vector DNA in a microcentrifuge tube:
   (a) Precipitate the DNA with one-tenth of a volume of 3.0 M NaOAc, pH 5.5, and 2.5 volumes of ethanol.
   (b) Mix well, ensuring that the entire inside of the tube has come into contact with the ethanol solution,
   (c) Leave the tube on ice briefly (about five minutes).
   (d) Centrifuge the tube for 15 min at 15,000 rpm in a microcentrifuge to collect the precipitate.
   (e) Gently remove the tube from the centrifuge, and open it in a sterile cell culture cabinet.

(f) Remove the supernatant by aspiration, taking care not to disturb the pellet. Ensure that the ethanol is well drained.

(g) Allow the pellet to air dry in the cell culture hood until all traces of ethanol have evaporated.

(h) Resuspend the DNA pellet in TBE to give a final concentration of 0.5 mg/ml. It may be necessary to vortex the tube in order to release the pellet from the side of the tube.

(i) Incubate the tube at 37°C for 5–10 min, with occasional vortexing to ensure that the pellet is well resuspended.

Note. Do not use higher TBE concentrations for resuspending your DNA, as doing so can interfere with the formation of the DNA precipitate.

3. Prepare the DNA calcium phosphate precipitate:

(a) Calculate the total volume of precipitate required. The final concentration of the DNA in the precipitate should be 20 μg/ml, and you will need 1 ml for each 9-cm plate and 3 ml for each 175-cm² flask. Remember to make an extra 1 ml of precipitate in order to ensure recovery of sufficient volume for all of your cultures, as some loss of volume will occur during preparation of the precipitate.

(b) Dilute the DNA/TBE mix in 12.5 mM HEPES to give 20 μg/ml in the final mix.

(c) Add 10× CaCl$_2$, one-tenth of the volume of the final mix.

(d) Add the solution dropwise, while mixing, to an equal volume of 2× HBP. For example, for nine 9-cm dishes at 1 ml/dish, plus 1 ml to spare (i.e., 10 ml of mix), we have the following:

(i) DNA/TBE	0.5 ml;
(ii) 12.5 mM HEPES	3.5 ml;
(iii) 10× CaCl$_2$	1.0 ml;
(iv) Add dropwise to 2× HBP	5.0 ml.

The final concentrations in the mix are as follows:

DNA	20 μg/ml;
CaCl$_2$	0.125 M;
Na$_2$PO$_4$	0.75 mM;
NaCl	140 mM;
HEPES	12.5 mM.

The pH is 7.12

Note. Use plastic pipettes and tubes when preparing DNA calcium phosphate precipitates, as the precipitates stick very firmly to glass. For mixing, we rec-

ommend the use of two pipettes in two handheld pipette aids, one for blowing bubbles of sterile air into the mixture, and the other for carefully adding the DNA/calcium mix in a dropwise fashion to the phosphate solution; good mixing results in an even precipitate. Use 1–5-ml pipettes, depending on the volume of precipitate being made. As the DNA/calcium solution is added to the phosphate solution, a light, even precipitate will begin to form. This precipitate is quite obvious and gives a milky appearance when complete.

(e) After all of the DNA/calcium solution has been added to the phosphate, replace the lid on the tube, invert the tube gently once or twice, and leave the tube to stand at room temperature for 30 min.

(f) It is important to make a mock precipitate without DNA. This allows you to assess the effectiveness of your selection conditions. A mock precipitate can be made exactly as described for the regular precipitate, but the DNA for the mock precipitate is replaced with additional Hepes buffer.

4. Add 1 ml of the DNA or mock precipitate to each plate or 3 ml to each flask. Make sure that the volume of precipitate is no more than one-tenth of the total volume of the culture medium already on the cells.

5. Leave the precipitate on the cells for a minimum of 6 h, but not more than overnight ($\approx$16 h). For fibroblasts from some individuals, exposure to calcium and phosphate for more than 6 h may be toxic.

6. Remove the calcium phosphate precipitate by aspiration. There is no need to wash the cells further, or, in our experience, to further treat the cells with either DMSO or glycerol.

7. Add medium to the cultures, and incubate them until 48 h from the start of the experiment; then add selective agents to the cultures. The agent that you add will depend on the vector used for transfection. Vectors carrying the *neo* gene confer resistance to G418 (Gibco, Geneticin), and vectors carrying the *hyg b* gene confer resistance to hygromycin B (Boehringer-Mannheim). All fibroblasts, in our experience, require 100–200 μg/ml of G418 or 10–20 μg/ml of hygromycin B to kill the cells gradually over a period of a week.

8. Change the medium on the transfected plates:

(a) Dispense the total volume of medium required into a suitably sized sterile bottle, and add G418 or hygromycin B from the concentrated stocks to give the correct final concentration.

(b) Gently swirl the solution to mix it.

(c) Aspirate the medium from the plates or flasks, and replace it with the selective medium.

(d) Return the plates or flasks to the incubator.

9. Monitor the effects of the selective medium on a daily basis by examining the culture under the microscope. When a significant number of cells have lifted and died, replace the medium with fresh selective medium. Selection should be maintained at all times.

10. Continue to replace the medium until the background of cells has lifted and died. The mock transfected plates that have not been transfected with DNA should have no viable cells remaining after 7–10 d. If there are cells remaining, then the selection has not worked adequately, and it may be necessary to raise the concentration of the selective agent.

11. Once the background of cells has died, it is no longer necessary to routinely change the medium on the cells. The cells should then be left undisturbed in the incubator for four to six weeks to allow the transfected cells to grow and form colonies.

12. Once colonies arise, pick out individual colonies by using cloning rings (see Protocol 13.6) or bulk the colonies together by trypsinizing the whole dish or flask.

13. Freeze aliquots of cells at the earliest opportunity and regularly thereafter. Once transfectants have been expanded into cultures and ample stocks frozen in liquid nitrogen, it is necessary to keep the culture going for an extended period of time until it reaches crisis. To minimize the risk from fungal contaminants, add Amphotericin B (Fungizone, Gibco) at 2.5 μg/ml.

14. Subculture the cultures routinely until they reach the end of their *in vitro* life span. As the cells approach crisis, the growth rate often slows. As the cultures begin to degenerate and cell division ceases, it is no longer necessary to subculture the cells. However, if heavy cell debris begins to cling to the remaining viable cells, it may be advisable to trypsinize the cells to remove the debris. Either return all of the cells to the same vessel, or use a smaller vessel to compensate for the cell death that is occurring. In general, the cells grow and survive better if they are not too sparse. With patience, care, and the culture of sufficient cells from your freezer stocks, you should, in most cases, obtain a postcrisis line.

15. When healthy cells begin to emerge, allow the colonies to grow to a reasonable size before subculturing, and then begin to subculture the colonies again. Do not be tempted to put too few cells into a large flask.

16. Freeze an ampoule of cells at the earliest opportunity, and continue to build up a freezer stock before using the culture.

17. To check that your postcrisis line is truly immortal, we recommend selecting and expanding individual clones from the culture. This procedure has the additional benefit of providing a homogeneous culture derived from a single cell.

Posttransfection Care of Cultures

(1) The level of selective agent is chosen to produce a gentle kill over a period of about a week. The majority of cells on the DNA-treated plates should die within seven days, and those cells that remain should be the successful transfectants.

(2) It is advisable to freeze ampoules of cells routinely both to build up a stock of transfected cells and to save time if cultures are lost due to contamination.

(3) In most cases, crisis is a marked event, with the majority of cells showing signs of deterioration and a net loss of viable cells. On average, crisis lasts from 3–6 months, and the culture may deteriorate to the point where very few, if any, obviously healthy cells are present.

(4) Once a culture has entered crisis, you can then reliably predict the timing of crisis for parallel cultures stored in liquid nitrogen. This prediction allows one to retrieve ampoules from parallel cultures from the freezer and to build up a number of flasks sitting at the threshold of crisis. As these parallel cultures enter crisis and begin to lose viability, these flasks can be pooled. In some cases, there will be an adequate cover of cells in the flask, but high levels of cellular debris. In these cases, replate the cells. The cells should be trypsinized and centrifuged, and the pellet should be returned to the original flask. The flask may be rinsed several times with trypsin to remove any adhering cell debris before returning the cells to it.

(5) The first sign of a culture emerging from crisis is usually the appearance of one or more foci of apparently healthy, robust cells with the typical appearance of SV40-transformed cells. On subculturing, these foci expand to give a healthy, regenerating culture. In many cases, the early-emerging postcrisis cells grow poorly, but the growth properties improve with further subculturing, and subcloning helps to select individual clones with better growth properties. It is important to freeze an ampoule of your new cell line as

early as possible. Our working definition for an immortal line is that the culture has undergone a minimum of 100 population doublings posttransfection and has survived subsequent subcloning.

(6) As the appearance of an immortal derivative within these cultures is a relatively rare event, it is essential that any postcrisis cell lines which emerge are checked to ensure that they were derived from the original starting material and are not the result of cross-contamination from other immortal cell lines in the laboratory. (See Authentication in Chapter 15; Cross-contamination in Chapter 18; Protocol 15.10.)

Telomerase-Induced Immortalization

The primary cause of senescence appears to be telomeric shortening, followed by telomeric fusion and the formation of dicentric chromosomes and subsequent apoptosis. Transfecting cells with the telomerase gene *htrt* extends the life span of the cell line, and a proportion of these cells become immortal, but not malignantly transformed [Bodnar et al., 1998]. As a high proportion of the *htrt*$^+$ clones become immortal, this appears to be a promising technique for immortalization, although the functionality of these lines has yet to be demonstrated.

Transgenic Mouse

The transgenic mouse Immortomouse (*H-2K^b-tsA58* SV40 large T) carries the temperature-sensitive SV40LT gene. A number of tissues from this mouse give rise to immortal cell lines, including colonic epithelium [Whitehead and Joseph, 1994] and brain astroglia [Noble and Barnett, 1996].

ABERRANT GROWTH CONTROL

Cells cultured from tumors, as well as cultures that have transformed *in vitro*, show aberrations in growth control, such as growth to higher saturation densities, clonogenicity in agar, and growth on confluent monolayers of homologous cells [Aaronson et al., 1970]. These cell lines exhibit lower serum or growth factor dependence, usually form clones with a higher efficiency, and are assumed to have acquired some degree of autonomous growth control by overexpression of oncogenes or by deletion of suppressor genes. Growth control is often autocrine—i.e., the cells secrete mitogens for which they possess receptors, or the cells express receptors or stages in signal transduction that are permanently active and unregulated. Although immortalization does not necessarily imply a loss of growth control, many cells progress readily from immortalization to aberrant growth, perhaps due to ge-

netic instability that is intrinsic to the immortalized genotype.

Anchorage Independence

Many of the properties associated with neoplastic transformation *in vitro* are the result of cell surface modifications [Hynes, 1974; Nicolson, 1976; Bruynell et al., 1990]—e.g., changes in the binding of plant lectins [Laferte and Loh, 1992] and in cell surface glycoproteins [Bruynell et al., 1990; Carraway et al., 1992] —which may be correlated with the development of invasion and metastasis *in vivo*. Fibronectin (large extracellular transformation-sensitive [LETS] protein) is lost from the surface of transformed fibroblasts [Hynes, 1973; Vaheri et al., 1976] due to alterations in integrins. This loss may contribute to a decrease in cell–cell and cell–substrate adhesion [Yamada, 1991; Reeves, 1992] and to a decreased requirement for attachment and spreading for the cells to proliferate.

Transformed cells may lack specific CAMs (e.g., L-CAM), which, when transfected back into the cell, regenerate the normal, noninvasive phenotype [Mege et al., 1989], and, as such, they may be recognized as tumor suppressor genes. Other CAMs may be overexpressed, such as N-CAM in small-cell lung cancer [Patel et al., 1989], wherein the extracellular domain is subject to alternative splicing [Rygaard et al., 1992]. The expression of and degree of phosphorylation of integrins may also change [Watt, 1991], potentially altering cytoskeletal interactions, the regulation of gene transcription, the substrate adhesion of the cells, and the relationship between cell spreading and cell proliferation.

In addition, the loss of cell–cell recognition, a product of reduced adhesion, leads to a disorganized growth pattern and the loss of contact inhibition of cell motility and density limitation of cell proliferation. (See Contact Inhibition in this chapter.) Cells can grow detached from the substrate, either in stirred suspension culture or suspended in semisolid media, such as agar or Methocel. There is an obvious analogy between altered cell adhesion in culture and detachment from the tissue in which a tumor arises and the subsequent formation of metastases in foreign sites, but the validity of this analogy not clear.

Suspension Cloning. Macpherson and Montagnier [1964] were able to demonstrate that polyoma-transformed BHK21 cells could be grown preferentially in soft agar, while untransformed cells cloned very poorly. Subsequently, it has been shown that colony formation in suspension is frequently enhanced following viral transformation. The situation regarding spontaneous tumors is less clear, however, in spite of the fact that Freedman and Shin [1974; Kahn and

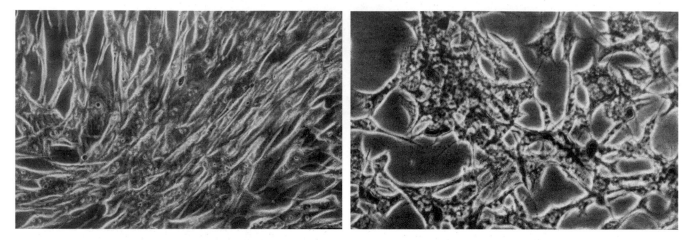

Fig. 17.3. Contact Inhibition. Late log-phase cultures of (a) BHK21-C13 and (b) the polyoma-transformed clone BHK21-PyY. C13 cells tend to assume a parallel orientation and will not overgrow each other, while PyY cells show no recognition of each other and grow randomly over each other or the substrate. Olympus CK2 microscope 10× objective.

Shin, 1979] demonstrated a close correlation between tumorigenicity and suspension cloning in Methocel. Although Hamburger and Salmon [1977] showed that many human tumors contain a small percentage of cells (< 1.0%) that are clonogenic in agar, a number of normal cells will also clone in suspension [Laug et al., 1980; Freshney and Hart, 1982] (see Fig. 13.14) with equivalent efficiency. Since normal fibroblasts are among these cells, the value of this technique for assaying for the presence of tumor cells in short-term cultures from human tumors is in some doubt. However, it remains a valuable technique for assaying neoplastic transformation *in vitro* by tumor viruses and was used extensively by Styles [1977] to assay for carcinogenesis.

Techniques for cloning in suspension are described in Chapter 13. (See Protocols 13.4 and 13.5.) Variations with particular relevance to the assay of neoplastic cells lie in the choice of the suspending medium. It has been suggested [Neugut and Weinstein, 1979] that agar may allow only the most highly transformed cells to clone, while agarose (which lacks sulfated polysaccharides) is less selective. Montagnier [1968] was able to show that untransformed BHK21 cells, which would grow in agarose but not in agar, could be prevented from growing in agarose by the addition of dextran sulfate.

Contact Inhibition

The loss of contact inhibition may be detected morphologically by the formation of a disoriented monolayer of cells (Fig. 17.3) or rounded cells in foci within the regular pattern of normal surrounding cells (see Fig. 17.2). Cultures of human glioma show a disorganized growth pattern and exhibit reduced density

limitation of growth by growing to a higher saturation density than that of normal glial cell lines (Fig. 17.4) [Freshney et al., 1980a,b]. Since variations in cell size influence the saturation density, the increase in the labeling index with [3H]-thymidine at saturation density (see Protocol 20.10) is a better measure of the reduced density limitation of growth. Human glioma, labeled for 24 h at saturation density with [3H]-thymidine, gave a labeling index of 8%, while normal glial cells gave 2% [Guner et al., 1977].

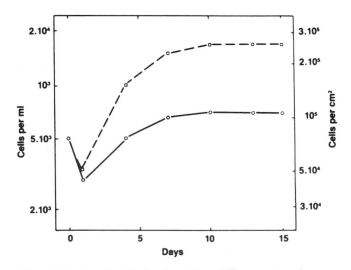

Fig. 17.4. Density Limitation. The difference in plateaux (saturation densities) attained by cultures from normal brain (circles, solid line) and a glioma (squares, broken line). Cells were seeded onto 13-mm coverslips, and 48 h later, the coverslips were transferred to 9-cm Petri dishes containing 20 ml of growth medium, to minimize exhaustion of the medium.

PROTOCOL 17.2. DENSITY LIMITATION OF CELL PROLIFERATION

Outline
Grow the culture to saturation density in nonlimiting medium conditions, and determine, autographically, the percentage of cells labeling with [³H]-thymidine.

Materials
Sterile or Aseptically Prepared:
Culture of cells ready for subculture
Growth medium
Maintenance medium (no serum or growth factors) containing 37 KBq/ml (1.0 μCi/ml) of [³H]-thymidine, 74 GBq/mmol (2Ci/mmol)
PBSA
0.25% trypsin
24-well plates containing coverslips 13 mm in diameter
9-cm Petri dishes (bacteriological grade, one per coverslip)

Protocol
1. Trypsinize the cells, and seed 10^5 cells/ml into a 24-well plate, 1 ml/well; each well should contain a 13-mm-diameter coverslip.
2. Incubate the cells in a humidified CO_2 incubator for 1–3 d.
3. Transfer the coverslips to 9-cm bacteriological grade Petri dishes, each containing 20 ml of medium, and place the dishes in the CO_2 incubator.
4. Continue culturing, changing the medium every 2 d once the cells become confluent on the coverslips. Trypsinize and count the cells from two coverslips every 3–4 d. As the cells become denser on the coverslip, it may be necessary to add 200–500 units/ml of crude collagenase to the trypsin in order to achieve complete dissociation of the cells for counting.
5. When cell growth ceases—i.e., two sequential counts show no significant increase—add 2.0 ml 37 KBq/ml (1.0 μCi/ml) of [³H]-thymidine, 74 GBq/mmol (2 Ci/mmol), and incubate the cells for a further 24 h.

△ *Safety Note.* Handle [³H]-thymidine with care. Although it is a low-energy β emitter, it localizes to DNA and can induce radiolytic damage. Wear gloves, do not handle [³H]-thymidine in a horizontal laminar flow hood, but instead in a biohazard or cytotoxic drug handling hood (see Radiation in Chapter 6), and discard waste liquids and solids by the appropriate route specified in the local rules governing the handling of radioisotopes.

6. Transfer the coverslips back to a 24-well plate, and trypsinize the cells for autoradiography. (See Autoradiography in Chapter 26.) The cells may be fixed in suspension and dropped on a slide as for chromosome preparations (without the hypotonic treatment), centrifuged onto a slide using a cytocentrifuge (see Protocol 15.5), or trapped on filters by vacuum filtration (see Protocol 15.6).

Note. It is necessary to trypsinize high-density cultures for autoradiography, because of their thickness and the weak penetration of β emission from ³H. (The mean path length of β particles in water is approximately 1 μm.) Labeled cells in the underlying layers will not be detected by the radiosensitive emulsion, due to the absorption of the β particles by the overlying cells. If the cells remain as a monolayer at saturation density, this step may be omitted, and autoradiographs may be prepared by mounting the coverslips, cells uppermost, on a microscope slide.

Analysis. Count the number of labeled cells as a percentage of the total number of cells. Scan the autoradiographs under the microscope, and count the total number of cells and the proportion of cells labeled in representative parts of the slide. (See Fig. 20.7.)

Variations. Cells in DNA synthesis may also be labeled with bromodeoxyuridine (BUdR) and subsequently detected by antibody to BUdR-labeled DNA (Dako). Human cycling cells can also be labeled with the Ki67 monoclonal antibody (Dako) against DNA polymerase, or with anti-PCNA against proliferating cell nuclear antigen (PCNA). While there is generally good agreement between [³H]-thymidine and BUdR labeling, Ki67 and PCNA will label more cells, as the antigen is present throughout the cycle and is not restricted to the S-phase.

Growth of cells at high density in nonlimiting medium can also be achieved by growing the cells in a filter well (Falcon, Costar, Millipore), choosing a filter diameter substantially below that of the dish (e.g., the Corning Costar 8-mm filter in a 24-well plate), and counting the number of cells in plateau, performing an autoradiograph or immunostaining with Ki67 or anti-PCNA.

Serum Dependence
Transformed cells have a lower serum dependence than that of their normal counterparts [Temin, 1966, Eagle et al., 1970], due, in part, to the secretion of growth factors by tumors [Todaro and DeLarco, 1978]. These factors have been collectively described

as autocrine growth factors. Implicit in this definition is that (1) the cell produces the factor; (2) the cell has receptors for the factor; and (3) the cell responds to the factor by entering mitosis. Some of these factors may have an apparent transforming activity on normal cells (e.g., TGFα) binding to the EGF receptor and inducing mitosis [Richmond et al., 1985], although, unlike true transformation (see What is Transformation in this chapter), this type is probably reversible. These factors also cause nontransformed cells to adopt a transformed phenotype and grow in suspension [Todaro and DeLarco, 1978]. This effect can be assayed by treating NRK cells with conditioned medium from the test cell and cloning them in suspension. (See Protocols 13.2, 13.4, and 13.5.)

Tumor cells can also produce many hemopoietic growth factors, such as interleukins 1, 2, and 3, along with colony-stimulating factor (CSF) [Fontana et al., 1984; Metcalf, 1990]. It has been proposed [Cuttitta et al., 1985] that some factors, such as gastrin-releasing peptide and vasoactive intestinal peptide (VIP), hitherto believed to be ectopic hormones produced by lung carcinomas, may in fact be autocrine growth factors. Autocrine growth factors can be detected by immunostaining (see Protocol 15.12), but their value as transformation markers is limited, because many normal cells—e.g., glia, fibroblasts, and endothelial cells—produce autocrine factors when proliferating.

Autonomous growth control is also achieved in transformed cells by the expression of modified receptors, such as the *erb*-B2 oncogene product, and the modified G protein, such as mutant *ras*, or by the overexpression of genes regulating stages in signal transduction (e.g., *src* kinase) or transcriptional control (e.g., *myc*, *fos*, and *jun*) [Bishop, 1991]. In many cases, the gene product is permanently active and is unable to be regulated. Overexpression of the genes can be detected by immunostaining (see Protocol 15.12) or immunoblotting for the protein product or Northern blotting for mRNA. In some cases, the oncogene product (e.g., *erb*-B2, activated Ha-*ras*) can be distinguished from the normal product (e.g., EGF receptor, normal *ras*, respectively) qualitatively as well as quantitatively, by specific antibodies.

TUMORIGENICITY

Transformation is a multistep process that often culminates in the production of neoplastic cells [Quintanilla et al., 1986]. However, cell lines derived from malignant tumors, presumably already transformed, can undergo further transformation with an increased growth rate, a reduced anchorage dependence, more pronounced aneuploidy, and immortalization. This suggests that a series of steps, not necessarily coordinated or interdependent and not necessarily individually tumorigenic, is required for malignant transformation. Furthermore, all cell lineages present within a tumor need not have the same transformed properties, and the same set of properties need not be expressed in every tumor. Progression may imply the expression of new properties or the deletion of old ones that may induce metastasis or even spontaneous remission.

There are, therefore, several steps in transformation, the sequence of which may be determined by environmental selective pressure. *In vitro*, where little restriction on growth is imposed, the events need not necessarily follow in the same sequence as *in vivo*.

Malignancy

Malignancy implies that the cells have developed the capacity to generate invasive tumors if implanted *in vivo* into an isologous host or if transplanted as a xenograft into an immune-deprived animal. While the development of malignancy can be recognized as a discrete phenotypic event, it often accompanies the development of aberrant growth control, suggesting that some of the lesions responsible for aberrant growth control also cause malignancy. An obvious candidate for such lesions is a deficit in cell–cell interaction that deprives the cell of control of proliferation (density limitation of cell proliferation) and of motility control (contact inhibition).

Two approaches have been used to explore malignancy-associated properties: (1) Cells have been cultured from malignant tumors and characterized; (2) transformation *in vitro* with a virus or a chemical carcinogen, or transfection with oncogenes, has produced cells that were tumorigenic and that could be compared with the untransformed cells. The second approach provides transformed clones of the same lineage, which can be shown to be malignant, and these clones can be compared with untransformed clones, which are not malignant. Unfortunately, many of the characteristics of cells transformed *in vitro* have not been found in cells derived from spontaneous tumors. Ideally, tumor cells and equivalent normal cells should be isolated and characterized. Unfortunately, there have been relatively few instances for which this arrangement has been possible, and even then, although the cells may belong to the same lineage, their position in that lineage is not always clear, and thus comparison is not strictly justified.

There are a number of properties that can be recognized in cells transformed *in vitro*, or cells cultured from *in vivo* tumors, and associated with malignant invasion. The first, and most important, property is tumorigenesis, which is universally accepted as a reli-

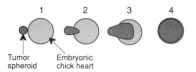

Fig. 17.5. Chick Heart Assay. Tumor spheroids (see Protocol 24.2) are cocultured with healed fragments of 8-d embryonic chick heart. (1) The spheroid adheres to the heart after a few hours; (2) after 24–48 h, it starts to penetrate the chick heart. (3) It spreads within the heart fragment, (4) by 8–10 d it has completely replaced the heart tissue.

able criterion for malignant transformation, but is subject to a significant number of false negatives, due to problems with the implantation technique and rejection by the immune system, even in immunocompromised animals.

Tumorigenesis

The only generally accepted sign of malignancy is the demonstration of the formation of invasive or metastasizing tumors *in vivo.* Transplantable tumor cells (~1×10^6) injected into isogeneic hosts will produce invasive tumors in a high proportion of cases, while 10^6 normal cells of similar origin will not. Models have been developed, using immune-suppressed or immune-deficient host animals, to study the tumorigenicity of human tumors. The genetically athymic "nude" mouse [Giovanella et al., 1974] and thymectomized irradiated mice [Bradley et al., 1978; Selby et al., 1980] have both been used extensively as hosts for xenografts. The take rate of the grafts varies, however, and many clearly defined tumor cell lines and tumor biopsies fail to produce tumors as xenografts; those that do take frequently fail to metastasize, although they may be invasive locally. Take rates can be im-

proved by sublethal irradiation of the host nude mouse (30–60 Gy), by using asplenic athymic (*scid*) mice, or by implanting the cells in Matrigel [Pretlow et al., 1991]. In spite of the frequency of false negatives, tumorigenesis remains a good indicator of malignancy.

Invasiveness

Tumorigenesis assays should always be accompanied by histology of the tumor to confirm its histpathological similarity to the original tumor and to demonstrate that it is invasive. However, if the cells are not tumorigenic, or if transplantation facilities are not available or not considered desirable, then it is possible to utilize a number of *in vitro* assays. Some of these assays also provide models that are more readily quantified as experimental systems.

Chick Chorioallantoic Membrane. The chorioallantoic membrane (CAM) assay can be performed on chick embryos *in ovo* or on explanted CAM *in vitro.* Easty and Easty [1974] showed that invasion of the CAM could be demonstrated in organ culture, and others [Hart and Fidler, 1978] attempted, with some limited success, to construct a chamber capable of quantifying the penetration of tumor cells across the CAM. An advantage of the CAM assay *in ovo* is that it may also show angiogenesis (see Angiogensis in this chapter), and the subsequent histology may reveal whether the tumor cells have penetrated the underlying basement membrane.

Organoid Confrontation. Mareel et al. [1979] developed an *in vitro* model for invasion, using chick embryo heart fragments cocultured with reaggregated clusters of tumor cells. (Fig. 17.5.) Invasion appears to

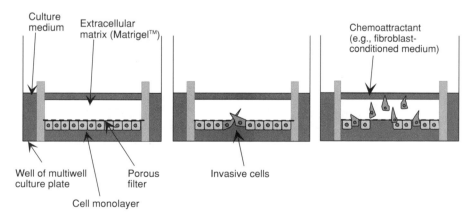

Fig. 17.6. Filter-Well Invasion. Cells plated on the underside of the filter migrate through the filter into growth-factor-depleted Matrigel in the well of the filter insert, encouraged by the addition of a chemoattractant, such as fibroblast-conditioned medium, to the upper side of the Matrigel. (After Brunton et al., [1997].)

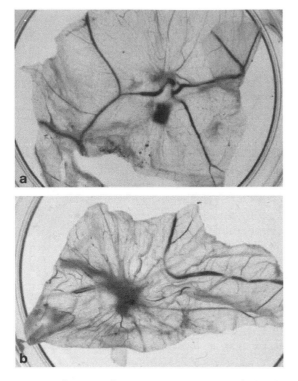

Fig. 17.7. Induction of Angiogenesis. Neovascularization induced in chick chorioallantoic membrane by tumor cell extract. A crude extract of Walker 256 carcinoma cells absorbed into sterile filter paper was placed on the chorioallantoic membrane of 9 d chick embryos. The membrane was removed 6 d later. (a) Control. (b) Walker 256 extract. (Courtesy of Margaret Frame.) (See also Plate 29.)

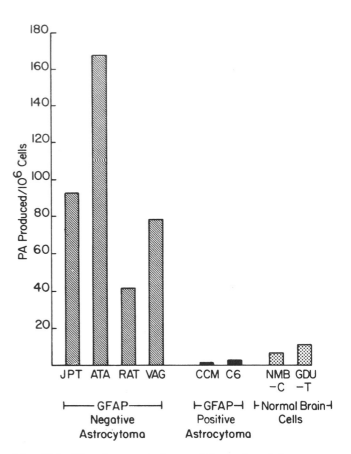

Fig. 17.8. Plasminogen Activator. PA produced by tumor cells *in vitro*. (The units are arbitrary.) The PA activities of the four gliomas—JPT, ATA, RAT, and VAG—were all higher than cells cultured from normal brain (NMB-C, GDU-T). It was also found that the only cells to produce the differentiated glial marker glial fibrillary acidic protein—CCM and C6—had the lowest PA of all.

be correlated with the malignant origin of the cells, is progressive, and causes destruction of the host tissue. The application of this technique to human tumor cells shows a good correlation between malignancy and invasiveness in the assay [de Ridder and Calliauw, 1990]. This technique has been used extensively, but is difficult to quantify and requires skilled histological interpretation.

Filter Wells. A number of filter-well techniques have been developed, based on the penetration of filters coated with Matrigel or some other extracellular matrix constituent (Fig. 17.6) [Repesh, 1989; Schlechte et al., 1990; Brunton et al., 1997; Lamb et al., 1997]. The degree of penetration into the gel, or through to the distal side of the filter, is rated as invasiveness and is determined histologically by the number of cells and the distance moved, or by prelabeling the cells with ^{125}I and counting the radioactivity on the distal side of the filter or bottom of the dish. It is more readily quantified, but lacks the presence of normal host cells in the barrier, normally associated with invasion *in vivo*. The penetration into Matrigel is likely to be a measure of the matrix degradation and reflects

the production of proteases or glycosidases by the cells. It is not clear how cells that do not make their own degradative enzymes, but instead rely on the production of proteases induced in the stroma, will perform in these assays.

Angiogenesis

Tumor cells release factors, including VEGF [Joukov et al., 1997], FGF-2 [Thomas et al., 1997], and angiogenin [Hu et al., 1997], that are capable of inducing neovascularization [Folkman, 1992; Skobe and Fusenig, 1998]. Fragments of tumor, pellets of cultured cells, or cell extracts, implanted on the surface of the CAM of a hen's egg, promote an increase in vascularization that is apparent to the naked eye 6–8 d later. (Fig. 17.7 and Plate 29.) Since this assay is not readily quantified, the stimulation of cell migration [Gullino, 1985], cell proliferation [Freshney, 1985], or morphogenesis [Chen et al., 1997; Ment et al., 1997; Jain et al., 1997] in filter-well cultures of vascular endothelium may provide the basis for more quantitative assays.

Plasminogen Activator

Other products that tend to be increased in transformed cells are proteolytic enzymes [Mahdavi and Hynes, 1979], long since associated with theories of invasive growth [Liotta, 1987]. Since proteolytic activity may be associated with the cell surface of many normal cells and is absent on some tumor cells, an equivalent normal cell must be used as a control when using this criterion. Plasminogen activator (PA) is higher in some cultures from human glioma than in cultures from normal brain [Hince and Roscoe, 1980]

(Fig. 17.8), and other cultures have previously shown that PA is associated with many different tumors [de Vries et al., 1995; del Vecchio et al., 1993; Schwartz et al., 1991]. PA may be measured by clarification of a fibrin clot or by release of free soluble ^{125}I from [^{125}I] fibrin [Unkless et al., 1974; Strickland and Beers, 1976]. In addition, a simple chromogenic assay has been developed by Whur et al. [1980].

It has been proposed that, for some carcinomas, soluble urokinase-like PA (uPA) is elevated more than tissue-type PA (tPA) [Markus et al., 1980; Duffy et al., 1990], so it is informative to couple the chromogenic assay with zymogram analysis [Davies et al., 1993; Boxman et al., 1995] and then immunoblot to determine the proportion of each type of PA.

CHAPTER 18

Contamination

SOURCES OF CONTAMINATION

Maintaining asepsis is still one of the most difficult challenges to the newcomer to tissue culture. Part of the difficulty, of course, is due to awkwardness during early training, and experience will eventually cure the problem. However, in certain situations, even the most experienced worker will suffer from contamination. There are several potential routes to contamination (Table 18.1) including failure in the sterilization procedures for glassware and pipettes, turbulence and particulates (dust and spores) in the air in the room, poorly maintained incubators and refrigerators, faulty laminar-flow hoods, the importation of contaminated cell lines or biopsies, and poor technique. The last of these is probably the most significant.

Operator Technique

If reagents are sterile and equipment is in proper working order, contamination depends on the interaction of the operator's technique with environmental conditions. If the skill and level of care of the operator is high and the atmosphere is clean, free of dust, and still, contamination as a result of manipulation will be rare. If the environment deteriorates (e.g., as a result of construction work or a seasonal increase in humidity), or if the operator's technique declines (through the omission of one or more apparently unnecessary precautions), the probability of infection increases. If both happen simultaneously or sequentially, then the results can be catastrophic.

Let us consider this conjunction of events in graphic form. (See Fig. 5.1.) The maintenance of good technique may be represented by the top graph, with occasional lapses shown as a downward peak, and the quality of the environment may be depicted by the bottom graph, with occasional sporadic increases in risk, such as a contaminated Petri dish opened accidentally in the area or dust generated from equipment maintenance. Provided that the two curves (of good technique and a high-quality environment) are kept well apart, the coincidence of a lapse in technique and an environmental breakdown will be rare. If, however, there is a progressive decline in technique or in the environment, the frequency of contamination will increase, and if both conditions deteriorate, contamination will be regular and widespread.

Environment

It is fairly obvious that the environment in which tissue culture is carried out must be as clean as possible and free from disturbance and through traffic. (See Sterile Handling Area in Chapter 3.) Equipment brought in from storage and air currents from doors, refrigerators, centrifuges, and the movement of staff increase the risk of contamination. Accordingly, one should keep the area clean and free of through traffic and wipe down anything that is brought in.

Use and Maintenance of Laminar-flow Hood

The most common form of poor technique is improper use of the laminar-flow hood. If it becomes overcrowded with bottles and equipment (Figs. 5.2, 5.3), the laminar airflow is disrupted, and the protective boundary layer between operator and room is lost.

TABLE 18.1. Routes to Contamination

Route or cause	Prevention
Technique	
Manipulations, pipetting, dispensing, etc.	
Nonsterile surfaces and equipment.	Clear work area of items not in immediate use.
Spillage on necks and outside of bottles and on work surface.	Swab regularly with 70% alcohol. Do not pour liquids. Dispense or transfer by pipette, autodispenser, or transfer device. If pouring is unavoidable: (1) do so in one smooth movement, (2) discard the bottle that you pour from, and (3) wipe up any spillage.
Touching or holding pipettes too low down, touching necks of bottles, inside screw caps.	Hold pipettes above graduations.
	Do not work over open vessels.
Splash-back from waste beaker.	Discard waste into a beaker with a funnel or, preferably, by drawing off the waste into a reservoir by means of a vacuum pump.
Sedimentary dust or particles of skin settling on the culture or bottle. Hands or apparatus held over an open dish or bottle.	Do not work over (vertical laminar flow and open bench) or behind and over (horizontal laminar flow) an open bottle or dish.
Work surface	
Dust and spillage.	Swab the surface with 70% alcohol before, after, and during work.
	Mop up spillage immediately.
Operator hair, hands, breath, clothing	
Dust from skin, hair, or clothing dropped or blown into the culture.	Wash hands thoroughly or wear gloves. Wear a lint-free lab coat with tight cuffs and gloves overlapping them.
Aerosols from talking, coughing, sneezing, etc.	Keep talking to a minimum and face away from work when you talk.
	Avoid working with a cold or throat infection, or wear a mask.
	Tie back long hair or wear a cap.
	Wear a lab coat different from the one you wear in the general lab area or animal house.
Materials and reagents	
Solutions	
Nonsterile reagents and media.	Filter or autoclave solutions before using them.
Dirty storage conditions.	Clean up storage areas and disinfect regularly.
Inadequate sterilization procedures.	Monitor the performance of the autoclave with a recording thermometer or sterility indicator. (See trade index and Chapter 10.)
	Check the integrity of filters with a bubble-point or microbial assay after using them.
	Test all solutions after sterilization.
Poor commercial supplier.	Test solutions; change suppliers.
Glassware and screw caps	
Dust and spores from storage.	Shroud caps with foil. Wipe bottles with EtOH before taking them into the hood.
	Replace stocks from the back of the shelf. Do not store anything unsealed for more than 24 h.
Ineffective sterilization (e.g., an overfilled oven or sealed bottles, preventing the ingress of steam).	Check the temperature of the load throughout the cycle. In the autoclave, keep caps slack on empty bottles. Stack oven and autoclave correctly. (See Chapter 10.)
Instruments, pipettes	
Ineffective sterilization.	Sterilize items by dry heat before using them.
	Monitor the performance of the oven.
Contact with a nonsterile surface or some other material.	Resterilize instruments. (Use 70% alcohol; burn and cool off the instruments.)
	Do not grasp any part of an instrument or pipette that will pass into a culture vessel.
Invasion by insects, mites, or dust.	Do not store instruments for more than 24 h, unless sealed with tape.
Culture flasks and media bottles in use	
Dust and spores from incubator or refrigerator.	Use screw caps instead of stoppers. Swab bottles before placing in hood. Box plates and dishes.

TABLE 18.1. (*Continued*)

Route or cause	Prevention
Dirty storage or incubation conditions.	Cover caps and necks of bottles with aluminum foil during storage or incubation.
	Wipe flasks and bottles with 70% alcohol before using them.
	Clean out stores and incubators regularly.
Media under the cap and spreading to the outside of the bottle.	Discard all bottles that show spillage on the outside of the neck. Do not pour.
Equipment and Facilities	
Room air	
Drafts, eddies, turbulence, dust, aerosols.	Clean filtered air.
	Reduce traffic and extraneous activity.
	Wipe the floor and work surfaces regularly.
Laminar-Flow Hoods	
Perforated filter.	Check filters regularly for holes and leaks.
Change of filter needed.	Check the pressure drop across the filter.
Spillages, particularly in crevices or below a work surface.	Clear around and below the work surface regularly. Let alcohol run into crevices.
Dry incubators	
Growth of molds and bacteria on spillages.	Wipe up any spillage with 70% alcohol on a swab.
	Clean out incubators regularly.
CO_2, humidified incubators	
Growth of molds and bacteria on walls and shelves in a humid atmosphere.	Clean out with detergent followed by 70% alcohol. (See Protocol 18.1.)
Spores, etc., carried on forced-air circulation.	Enclose open dishes in plastic boxes with close-fitting lids. (But do not seal the lids.)
	Swab boxes with 70% alcohol before opening them.
	Put a fungicide or bacteriocide in humidifying water. (But check first for toxicity.)
Mites, insects, and other infestations in wooden furniture, or benches, in incubators, and on mice, etc., taken from the animal house.	
Entry of mites, etc., into sterile packages.	Seal all sterile packs.
	Avoid wooden furniture if possible; use plastic laminate, one-piece, or stainless-steel bench tops.
	If wooden furniture is used, seal it with polyurethane varnish or wax polish, and wash it regularly with disinfectant.
	Keep animals out of the tissue culture lab.
Importation of Biological Materials	
Tissue samples	
Infected at source or duing dissection.	Do not bring animals into the tissue culture lab.
	Incorporate antibiotics into the dissection fluid. (See Isolation of the Tissue in Chapter 11.)
	Dip all potentially infected large-tissue samples in 70% alcohol for 30 s.
Incoming cell lines	
Contaminated at the source or during transit.	Handle these cell lines alone, preferably in quarantine, after all other sterile work is finished. Swab down the bench or hood after use with 2% phenolic disinfectant in 70% alcohol, and do not use it until the next morning.
	Check for contamination by growing a culture for two weeks without antibiotics. (Keep a duplicate culture in antibiotics at the first subculture.)
	Check for contamination visually, by phase-contrast microscopy and Hoechst stain for mycoplasma. Using indicator cells allows screening before first subculture.

Note: No one-to-one relationship between prevention and cause is intended throughout this table; preventative measures are interactive and may relate to more than one cause.

This in turn leads to the entry of nonsterile air into the hood and the release of potentially biohazardous materials into the room. In addition, the risk of collision between sterile pipettes and nonsterile surfaces of bottles, etc., increases. One should bring into the hood only those items that are directly involved in the current operation.

Laminar-flow hoods also must be maintained regularly, and the integrity of their filters and the security of their cabinets, which are dependent on both the internal air velocity and outside turbulence, should be checked at least twice a year by a competent engineer.

Humid Incubators

A major source of contamination stems from the use of humid incubators. (See Incubators in Chapter 5.) High humidity is not required, unless open vessels are being used; sealed flasks are better kept in a dry incubator or the hot room. If there is a need to gas flasks with CO_2, this is better done from a cylinder and the flasks sealed and placed in a normal incubator. Using permeable caps (see Venting in Chapter 7) minimizes the risk of contamination, but increases the unit cost and still exposes the flask to a higher risk atmosphere than in a dry incubator. Permeable caps may be sealed with a secondary rubber cap for transfer to a nongassed incubator. (Becton Dickinson.) If flasks are maintained in a CO_2 incubator with slack or permeable caps, it is possible to keep the incubator dry and use a different incubator for open plates. The CO_2-monitoring system will need to be recalibrated if the incubator is used dry, and the flasks will need to be checked for evaporation.

There are, however, many situations in which a humid incubator must be used. To reduce the risk of contamination, an incubator should be selected with an interior, all of which is readily accessible and can be cleaned easily. Cultures placed in the incubator should be enclosed in a plastic box. (See Boxed Cultures in Chapter 5.)

Fungicides. Copper-lined incubators have reduced fungal growth, but are usually about 20–30% more expensive than conventional ones. Placing copper foil in the humidifier tray also inhibits the growth of fungus, but only in the tray, and will not protect the walls of the incubator. A number of fungal retardants are in common use, including copper sulfate, riboflavin, sodium dodecyl sulfate (SDS), and Roccall, a proprietary fungicidal cleaner used in a 2% solution. A comparison of colony formation in incubators with and without Roccall shows no toxic effect. Many of these retardants are detergents, so it is important not to have a CO_2 or an air line bubbling through liquid containing them, or the liquid will foam. Remember,

a fungicide will only protect the tray; there is no substitute for regular cleaning!

Cleaning Incubators. Cleaning should be carried out regularly using 10% Roccall or an equivalent nontoxic antifungal cleaner. The frequency will depend on where the incubator is located; monthly may suffice for a clean area with filtered room air, but a shorter interval will be required for a rural site, where the spore count is higher, or during construction work or renovation. The frequency of access will also influence the buildup of fungal contamination. When the incubator is in use, any spillage must be mopped up immediately and contaminated cultures removed as soon as they are detected.

PROTOCOL 18.1. CLEANING INCUBATORS

Outline
Remove cultures to an alternative incubator, switch off the empty incubator, wash it out with detergent and alcohol, switch on the heat, and allow the incubator to dry. Replenish the water in the tray and restore the CO_2.

Materials
Nonsterile:
Water containing 2% Roccall or an equivalent fungal inhibitor, to refill the water tray
Alternative CO_2 incubator or plastic box and sealing tape
Detergent: 10% Roccall, or another fungicidal detergent
Ethanol, 70%

Protocol
1. Remove all cultures to another CO_2 incubator, or box the cultures up, gas them with CO_2, and place them in a regular incubator or hot room.
2. Switch off the incubator that is to be cleaned.
3. Remove all the shelves, the water tray, and any demountable panels from the incubator.
4. Wash the inside of the incubator with detergent solution; try to reach all corners and crevices.
5. Rinse with water.
6. Wipe the interior of the incubator with 70% ethanol.
7. Restore heat (not CO_2), and leave the door open until the chamber is dry.
8. Wash the shelves and panels in detergent, rinse them in water, and wipe them with 70% ethanol.
9. Return the panels and shelves to the incubator.

10. Replace the water tray and fill it with water containing 2% Roccall.
11. Close the door and restore CO_2.
12. When the temperature and CO_2 have stabilized, return the cultures to the incubator.

Cold Stores

Refrigerators and cold rooms also tend to build up fungal contamination on the walls in a humid climate, due to condensation that forms every time the door is opened, admitting moist air. The moist air increases the risk of deposition of spores on stored bottles; hence, they should be swabbed with alcohol before being placed in the hood. (See Standard Procedure in Chapter 5.) The cold store should be cleared, and the walls and shelving should be washed down with disinfectant every few months.

Sterile Materials

There should be no risk of contamination from sterile equipment and reagents if the appropriate quality control is carried out, either in-house or by the supplier. However, occasionally items can slip through if, for example, the load is too tightly packed, there is a cold spot in the sterilizing oven or autoclave, packaging is punctured, or a low-titer contamination has escaped detection despite quality control. It is difficult to guard against this eventuality, other than by following the correct procedures (see Quality Control in Chapter 10) and checking to make sure that all packing is correctly sealed.

Imported Cell Lines and Biopsies

Any biological material brought into the laboratory runs the risk of being contaminated. All such material should be maintained in quarantine (see next section) until it is shown to be clear of contamination, at which point it can join other stocks in general use. (See Cultures in Chapter 5.) Whenever possible, all cell lines should be acquired via a reputable cell bank, which will have screened for contamination. Cell lines from any other source, as well as biopsies from all animal and human donors, should be regarded as contaminated until shown to be otherwise.

Quarantine

Any culture that is suspected of being contaminated and any imported material that is awaiting testing should be kept in quarantine. Preferably, quarantine should take place in a separate room with its own hood and incubator, but if this is not feasible, one of the hoods that are in general use may be employed. The hood should be used last thing in the day and should be sprayed and swabbed down with 70% alcohol containing 2% phenolic disinfectant after use.

Then it should be withdrawn from service until the following day.

TYPES OF MICROBIAL CONTAMINATION

Bacteria, yeasts, fungi, molds, and mycoplasmas all appear as contaminants in tissue culture, and if protozoology is carried on in the same laboratory, some protozoa can infect cell lines. Usually, the species or type of infection is not important, unless it becomes a frequent occurrence. It is only necessary to note the general kind of contaminant (e.g., bacterial rods or cocci, yeast, etc.), how it was detected, the location where the culture was last handled, and the operator's name. If a particular type of infection recurs frequently, it may be beneficial to identify it in order to find its origin. [For more detailed screening procedures for microbial contamination, see European Pharmacopoeia, 1980; United States Pharmacopeia, 1985; Doyle et al., 1990; Doyle and Bolton, 1994; and Hay and Cour, 1997.]

MONITORING FOR CONTAMINATION

Potential sources of contamination are listed in Table 18.1, along with the precautions that should be taken to avoid them. Even in the best laboratories, however, contaminations do arise, so the following procedure is recommended:

(1) Check for contamination by eye and with a microscope at each handling of a culture. Check for mycoplasma every month.
(2) If it is suspected, but not obvious, that a culture is contaminated, but the fact cannot be confirmed *in situ*, clear the hood or bench of everything except your suspected culture and one can of Pasteur pipettes. Because of the potential risk to other cultures, this is best done after all your other culture work is finished. Remove a sample from the culture and place it on a microscope slide. (Kovaslides are convenient for this, as they do not require a coverslip.) Check the culture with a microscope, preferably by phase contrast. If it is confirmed that the culture is contaminated, discard the pipettes, swab the hood or bench with 70% alcohol containing a phenolic disinfectant, and do not use the hood or bench until the next day.
(3) Record the nature of the contamination.
(4) If the contamination is new and is not widespread, discard the culture, the bottle with the medium used to feed it, and any other bottle (e.g., trypsin) that has been used in conjunction with the cul-

ture. Discard all these into disinfectant, preferably in a fume hood and outside the tissue culture area.

(5) If the contamination is new and widespread (i.e., in at least two different cultures), discard all media, stock solutions, trypsin, etc.

(6) If the same kind of contamination has occurred before check stock solutions for contamination (a) by incubation alone or in nutrient broth (see Sterility Testing in Chapter 10) or (b) by plating out the solution on nutrient agar (Oxoid, Difco). If (a) and (b) prove negative, but contamination is still suspected, incubate 100 ml of solution, filter it through a 0.2-μm filter, and plate out filter on nutrient agar with an uninoculated control.

(7) If the contamination is widespread, multispecific, and repeated, check the laboratory's sterilization procedures (e.g., the temperatures of ovens and autoclaves, particularly in the center of the load, the durations of the sterilization cycle, the packaging, the storage practices, (e.g., unsealed glassware should be resterilized every 24 hr), and the integrity of the aseptic room and the filters on the laminar-flow hood filters).

(8) Do not attempt to decontaminate cultures unless they are irreplaceable.

Visible Microbial Contamination

Characteristic features of microbial contamination are as follows:

(1) A sudden change in pH, usually a decrease with most bacterial infections, very little change with yeast until the contamination is heavy, and sometimes an increase in pH with fungal contamination.

(2) Cloudiness in the medium, sometimes with a slight film or scum on the surface or spots on the growth surface that dissipate when the flask is moved.

(3) Under a low-power microscope ($\sim \times 100$), spaces between cells will appear granular and may shimmer with bacterial contamination (Fig. 18.1a). Yeasts appear as separate round or ovoid particles that may bud off smaller particles (Fig. 18.1b). Fungi produce thin filamentous mycelia (Fig. 18.1c) and, sometimes, denser clumps of spores. With toxic infection, some deterioration of the cells will be apparent.

(4) Under high-power microscopy ($\sim \times 400$), it may be possible to resolve individual bacteria and distinguish between rods and cocci. At this magnification, the shimmering that is visible in some infections will be seen to be due to mobility of the bacteria. Some bacteria form clumps or associate with the cultured cells.

(5) With a slide preparation, the morphology of the bacteria can be resolved at $\times 1,000$, but this is not usually necessary. Microbial infection may be confused with precipitates of media constituents (particularly protein) or with cell debris, but can be distinguished by their regular morphology. Precipitates may be crystalline or globular and irregular and are not usually as uniform in size. If you are in doubt, plate out a sample of medium on nutrient agar. (See Sterility Testing in Chapter 10).

Mycoplasma

Mycoplasmal infections (Fig. 18.1d–f) cannot be detected by the naked eye other than through signs of deterioration in the culture. The culture must be tested specially by fluorescent staining, PCR, Elisa assay, immunostaining, autoradiography, or microbiological assay. (See Microbiological Culture later in this chapter.) Fluorescent staining of DNA by Hoechst 33258 [Chen, 1977] is the easiest and most reliable method (see Protocol 18.2) and reveals mycoplasmal infections as a fine particulate or filamentous staining over the cytoplasm at $\times 500$ magnification (Fig. 18.2; Plates 30, 31). The nuclei of the cultured cells are also brightly stained by this method and thereby act as a positive control for the staining procedure. Most other microbial contaminations will also show up with fluorescence staining, so low levels of contamination or particularly small organisms such as micrococci can also be detected.

It is important to appreciate the fact that mycoplasmas do not always reveal their presence by means of macroscopic alterations of the cells or media. Many mycoplasma contaminants, particularly in continuous cell lines, grow slowly and do not destroy host cells. However, they can alter the metabolism of the culture in many different ways. Because mycoplasmas take up thymidine from the medium, infected cultures show abnormal labeling with [3H]-thymidine. Immunological studies can also be totally frustrated by mycoplasmal contamination, as attempts to produce antibodies against the cell surface may raise antimycoplasma antibodies. Mycoplasmas can alter cell behavior and metabolism in many other ways [McGarrity, 1982; Doyle et al., 1990; Izutsu et al., 1996; Dorazio et al., 1996; Giron et al., 1996; Paddenberg, 1996], so there is an absolute requirement for routine, periodic assays to detect possible covert contamination of all cell cultures, particularly continuous cell lines.

Monitoring Cultures for Mycoplasmas. Superficial signs of chronic mycoplasmal infection include a diminished rate of cell proliferation, reduced satu-

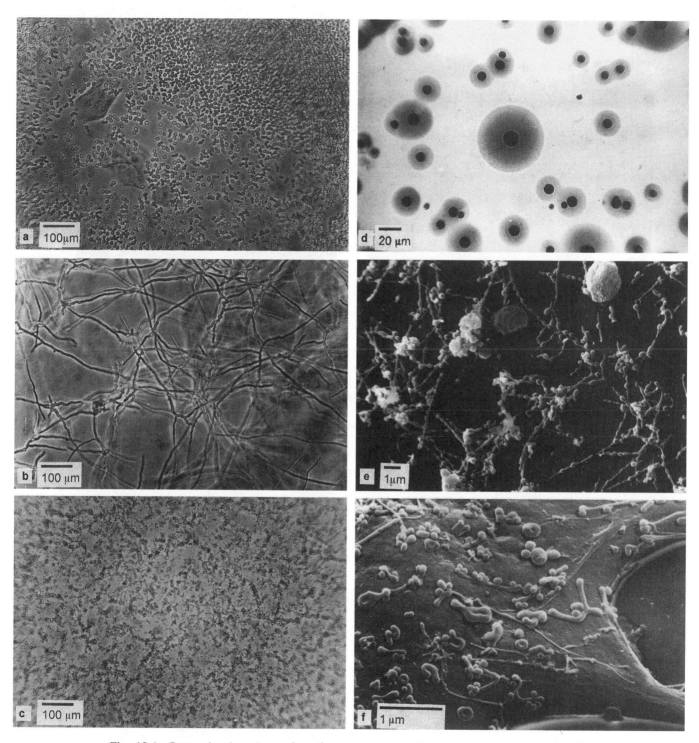

Fig. 18.1. Contamination. Examples of microorganisms found to contaminate cell cultures. (a) Yeast. (b) Mold. (c) Bacteria. (d) Mycoplasma colonies growing on special nutrient agar (not as seen in cell culture [see Fig. 18.2]). (e,f) Scanning electron micrograph of mycoplasma growing on the surface of cultured cells (d–f, courtesy of Dr. M. Gabridge; a–c, 10× objective).

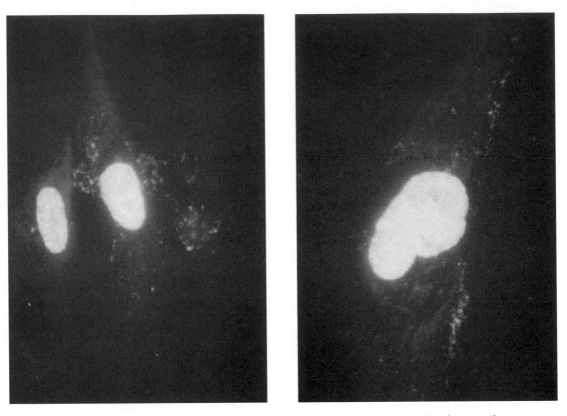

Fig. 18.2. Mycoplasma. Human normal diploid lung fibroblasts infected with mycoplasma and stained with Hoechst 33258. The nuclei fluoresce brightly from cellular DNA, and extranuclear fluorescence due to mycoplasma is also apparent. Light cytoplasmic staining is present in these preparations, but its diffuse nature makes it easily distinguishable from the bright particulate or filamentous staining of the mycoplasma. Usually, there is no cytoplasmic background. The lack of nuclear detail is due to the overexposure of the photographs that is necessary to reveal the mycoplasma. 50× water immersion objective (Leica). (See also Plates 30, 31.)

ration density [Stanbridge and Doersen, 1978], and agglutination during growth in suspension [Giron et al., 1996]. Acute infection causes total deterioration, with perhaps a few resistant colonies, although these and any resulting cell lines are not necessarily free of contamination and may carry a chronic infection.

Principle

The cultures are stained with Hoechst 33258, a fluorescent dye that binds specifically to DNA [Chen, 1977]. Since mycoplasmas contain DNA, they can be detected readily by their characteristic particulate or filamentous pattern of fluorescence on the cell surface and, if the contamination is heavy, in surrounding areas. Monolayer cell cultures can be fixed and stained directly, but following centrifugation, the medium from cells growing in suspension will need to be added to an indicator cell (i.e., another monolayer) known to be free of mycoplasma, but also known to be a good host for mycoplasma and to spread well in culture, with adequate cytoplasm to reveal any adher-

ent mycoplasma. Vero cells, 3T6, NRK, and A549 are all suitable.

The use of an indicator cell is recommended in the following protocol as it also helps to avoid problems with false positives arising from debris when cells are assayed just after thawing or from primary cultures, which you might not want to sacrifice anyway. If there is a lot of debris, the medium can be filtered through a sterile 5-μm filter or centrifuged at 100 *g*.

PROTOCOL 18.2. FLUORESCENCE DETECTION OF MYCOPLASMA

Outline

Culture cells in the absence of antibiotics for at least one week, transfer the supernatant medium to an early log phase-indicator culture, incubate 3–5 d, fix and stain the cells, and look for fluorescence other than in the nucleus.

Materials

Sterile or Aseptically Prepared:

Supernatant medium, free of antibiotics, from 7 d-monolayer or centrifuged suspension cell culture

Indicator cells (e.g., 3T6, NRK, Vero, A549)

Petri dishes, 60 mm

Nonsterile:

Hoechst 33258 stain, 50 ng/ml in BSS without phenol red (BSS-PR) or PBSA

PBSA

Deionized water

Fixative: freshly prepared acetic methanol (1:3, cold)

Mountant: see Reagent Appendix

Protocol

1. Seed indicator cells into Petri dishes without using antibiotics; seed enough to give 50–60% confluence in 4–5 d (e.g., 2×10^4 NRK or 1×10^5 A549, in 5 ml of medium).
2. Add 1.5 ml of medium from the test culture.
3. Incubate the solution until the cells reach 50–60% confluence.

Note. If cultures reach confluence by the end of the assay, staining will be inhibited, and the subsequent visualization of mycoplasma will be impaired.

4. Remove the medium and discard it.
5. Rinse the monolayer with BSS-PR or PBSA, and discard the rinse.
6. Add fresh BSS-PR or PBSA diluted 50:50 with fixative, rinse the monolayer, and discard the rinse.
7. Add pure fixative, rinse, and discard the rinse.
8. Add more fixative ($\sim$0.5 ml/cm²), and fix for 10 min.
9. Remove and discard the fixative.
10. Dry the monolayer completely if it is to be stored. (Samples may be accumulated at this stage and stained later.)
11. If you are proceeding directly with staining, wash off the fixative with deionized water and discard the wash.
12. Add Hoechst 33258 in BSS-PR or PBSA, and stain 10 min at room temperature.
13. Remove and discard the stain.
14. Rinse the monolayer with water and discard the rinse.
15. Mount a coverslip in a drop of mountant, and blot off any surplus from the edges of the coverslip.
16. Examine the monolayer by epifluorescence with a 330/380-nm excitation filter and an LP 440-nm barrier filter.

Analysis. Check for extranuclear fluorescence. Mycoplasmas give pinpoints or filaments of fluorescence over the cytoplasm and, sometimes, in intercellular spaces. The pinpoints are close to the limits of resolution with a 50× objective (0.1–1.0 μm) and are usually regular in size and shape. Not all of the cells will necessarily be infected, so as much as possible of the preparation should be scanned before you declare the culture uninfected.

Sometimes a light, uniform staining of the cytoplasm is observed, probably due to RNA. This fluorescence tends to fade on storage of the preparation, and examination the next day (after storing dry and in the dark) usually gives clearer results. This artefact never has the sharp punctate or filamentous appearance of mycoplasma and can be distinguished fairly readily with further experience in observation.

If there is any doubt regarding the interpretation of the fluorescence test, it should be repeated after generating a further subculture of the test cells in the absence of antibiotics. If results are still equivocal, adopt another assay, such as PCR or Elisa.

Alternative Methods. Among other methods that have been reported for the detection of mycoplasmal infections are methods that detect mycoplasma-specific enzymes such as arginine deiminase or nucleoside phosphorylase [see Schneider and Stanbridge, 1975; Levine and Becker, 1977] and those that detect toxicity with 6-methylpurine deoxyriboside (Mycotect Gibco). PCR or microbiological culture are also widely used.

Microbiological Culture. This is a very sensitive method, widely employed in quality control and validation procedures. However, it is best not to use it, unless facilities and training for the culture of mycoplasma is available, as these microorganisms are quite fastidious. The cultured cells are seeded into mycoplasma broth [Doyle et al., 1990], grown for 6 d, and plated out onto special nutrient agar [Hay, 2000]. Colonies form in about 8 d and can be recognized by their size ($\sim$200 μm diameter) and their characteristic "fried-egg" morphology-dense center with a lighter periphery (Fig. 18.1d). Commercial kits for microbiological detection (Mycotrim) are available from NEN.

If the microbiological culture method is used to detect mycoplasma, it is necessary to grow a known strain of mycoplasma at the same time as a positive control. This requirement discourages most laboratories, which do not wish to introduce live mycoplasma deliberately, unless that can be done in a separate laboratory from the rest of tissue culture work.

While using selective culture conditions and examining the morphology of a colony enables the species of mycoplasma to be identified, the microbiological culture method is much slower and more difficult to perform than the fluorescence technique. Commercial screening for mycoplasma by microbiological culture is available (e.g., from ICN, MA Bioservices, and Microbiological Associates). Specific monoclonal antibodies now allow the characterization of mycoplasma contaminations.

Molecular Hybridization. Molecular probes specific to mycoplasmal DNA can be used in Southern blot analysis to detect infections by conventional molecular hybridization techniques. A kit is available that employs similar probes and a simple single-step separation of hybrid DNA for subsequent detection by scintillation counting (GenProbe, Lab Impex Ltd.).

PCR. Detection of mycoplasma by PCR [Hopert et al., 1993; Pruckler and Ades, 1995; Otto et al., 1996; Toji et al., 1998] offers a high-sensitivity method with few false positives. PCR is probably the preferred method for a molecular biology laboratory, especially if no fluorescence microscope is available.

[³H]-Thymidine Incorporation. One other method that has been used quite successfully is autoradiography with [³H]-thymidine [Nardone et al., 1965]. The culture is incubated overnight with 4 KBq/ml (~0.1 μCi/ml) of [³H]-thymidine with high specific activity, and an autoradiograph is prepared. (See Protocol 26.3.) Grains over the cytoplasm are indicative of contamination (see Fig. 26.2), which can be accompanied by a lack of nuclear labeling due to trapping of the thymidine at the cell surface by the mycoplasma.

Viral Contamination

Incoming cell lines, natural products, such as serum, in media and enzymes such as trypsin, used for subculture, are all potential sources of viral contamination. A number of reagents are screened by manufacturers against a limited range of viruses, and claims have been made that the larger viruses can be filtered out during processing, but there is no certain way at present to avoid viral contamination.

Detection of Viral Contamination. Screening with a panel of antibodies by immunostaining (see Protocol 15.12) or Elisa assays is probably the best way of detecting viral infection. Alternatively, one may use PCR with the appropriate viral primers. Some com-

mercial companies (Microbiological Associates, MA Bioservices) offer viral screening.

ERADICATION OF CONTAMINATION

Bacteria, Fungi, and Yeasts

Decontamination should be attempted only in extreme situations, under quarantine, and with expert supervision.

PROTOCOL 18.3. ERADICATION OF MICROBIAL CONTAMINATION

Outline
Wash the culture several times in a high concentration of antibiotics by rinsing the monolayer or by centrifugation of cells in suspension. Then grow the culture for three subcultures with, and three without, antibiotics. Test for contamination after each subculture.

Materials
Sterile:
DBSS (see Dissection BSS in Reagents and Materials)
High-antibiotic medium (see Collection Medium in Reagents and Materials and Table 8.4; see next section for mycoplasma)
Materials for subculture (see Protocol 12.2)
Nonsterile:
Phase-contrast microscope, preferably with 40× and 100× objectives
Materials for staining mycoplasma (see Protocol 18.2)

Protocol
1. Collect the contaminated medium carefully. If possible, the organism should be tested for sensitivity to a range of individual antibiotics. If not, autoclave the medium or add hypochlorite. (See Disposal in Chapter 6.)
2. Wash the cells in DBSS:
   (a) For monolayers, rinse the culture three times, trypsinize, and wash the cells twice more by centrifugation.
   (b) For suspension cultures, wash the culture five times by centrifugation
3. Reseed a fresh flask.
4. Add high-antibiotic medium and change the culture every 2 d.
5. Subculture in a high-antibiotic medium.
6. Repeat steps 1–4 for three subcultures.

7. Remove the antibiotics, and culture the cells without them for a further three subcultures.
8. Check the cultures by phase-contrast microscopy and Hoechst staining. (See Protocol 18.2.)
9. Culture the medium for a further two months without antibiotics, and check to make sure that all contamination has been eliminated. (See Sterility Testing in Chapter 10.)

The general rule should be that contaminated cultures are discarded and that decontamination is not attempted unless it is absolutely vital to retain the cell strain. In any event, complete decontamination is difficult to achieve, particularly with yeast, and attempts to do so may produce hardier, antibiotic-resistant strains.

Eradication of Mycoplasma

If mycoplasma is detected in a culture, the first and overriding rule, as with other forms of contamination, is that the culture should be discarded for autoclaving or incineration. In exceptional cases (e.g., if the contaminated line is irreplaceable), one may attempt to decontaminate the culture. Decontamination should be done, however, only by an experienced operator, and the work must be carried out under conditions of quarantine.

Several agents are active against mycoplasma, including kanamycin, gentamycin, tylosin [Friend et al., 1966], polyanethol sulfonate [Mardh, 1975], and 5-bromouracil in combination with Hoechst 33258 and UV light [Marcus et al., 1980]. Coculturing with macrophages [Schimmelpfeng et al., 1968], animal passage [Van Diggelen et al., 1977], and cytotoxic antibodies [Pollock and Kenney, 1963] can also be effective in some cases. However, the most successful agents have been tylosin (ICN), Mycoplasma Removal Agent [MRA, also from ICN; Drexler et al., 1994], ciprofloxacin [Mowles, 1988; Hlubinova et al., 1994], and BM-Cycline (Boehringer-Mannheim).

Suspected cultures should be treated as in Protocol 18.3, using MRA, BM-Cycline, or tylosin at the manufacturer's recommended concentration, in place of the usual antibiotics in DBSS and the collection medium. However, this operation should not be undertaken unless it is absolutely essential, and even, then it must be preformed in experienced hands and in isolation. It is far safer to discard infected cultures.

Eradication of Viral Contamination

There are no reliable methods for eliminating viruses from a culture at present; disposal or tolerance are the only options.

Persistent Contamination

Many laboratories have suffered from periods of contamination which seems to be refractory to all the remedies suggested in Table 18.1. There is no easy resolution to this problem, other than to follow the previous recommendations in a logical and analytical fashion, paying particular attention to changes in technique, new staff, new suppliers, new equipment, and insufficient maintenance of laminar-flow hoods or other equipment. Typically, an increase in the contamination rate stems from deterioration in aseptic technique, an increased spore count in the atmosphere, poorly maintained incubators, a contaminated cold room or refrigerator, or a minor, intermittent fault in a sterilizing oven or autoclave.

The constant use of antibiotics also favors the development of chronic contamination. Many organisms are inhibited, but not killed, by antibiotics. They will, therefore, persist in the culture, undetected for most of the time, but periodically surfacing when conditions change or when there are intrinsic host-parasite-type population fluctuations. It is essential that your cultures be maintained in antibiotic-free conditions for at least part of the time, and preferably all the time; otherwise cryptic contaminations will persist, their origins will be difficult to determine, and eliminating them will be impossible.

A slight change in practices, the introduction of new staff, or an increase in activity as more people use a facility can all contribute to an increase in the rate of contamination. Procedures must remain stringent, even if the reason is not always obvious to the operator, and alterations in routine should not be made casually. If strict practices are maintained, contamination may not be eliminated entirely, but it will be detected early.

CROSS-CONTAMINATION

During the history of tissue culture, a number of cell strains have evolved with very short doubling times and high plating efficiencies. Although these properties make such cell lines valuable experimental material, they also make them potentially hazardous for cross-infecting other cell lines [Nelson-Rees and Flandermeyer, 1977; Lavappa, 1978; Nelson-Rees et al., 1981; Kneuchel and Masters, 1999]. The extensive cross-contamination of many cell lines with HeLa and other rapidly growing cell lines is now clearly established [Stacey et al., 1999; Macleod et al., 1999], but many operators are still unaware of the seriousness of the risk. (See also The Need For Characterization, DNA Fingerprinting, and Authentication in Chapter 15.)

The following practices help avoid cross-contamination:

(1) Obtain cell lines from a reputable cell bank that has performed appropriate characterization of the cells (see Cell Banks in Chapter 19; Sources of Materials—Cell Banks in Trade Index), or perform the necessary characterization yourself as soon as possible (see Chapter 15).

(2) Do not have media bottles or culture flasks with more than one cell line open simultaneously.

(3) Handle rapidly growing lines, such as HeLa, on their own and after other cultures.

(4) Never use the same pipette for different cell lines.

(5) Never use the same bottle of medium, trypsin, etc., for different cell lines.

(6) Do not put a pipette back into a bottle of medium, trypsin, etc., after it has been in a culture flask containing cells.

(7) Add medium and any other reagents to the flask first, and then add the cells last.

(8) Do not use unplugged pipettes, or pipettors without plugged tips, for routine maintenance.

(9) Check the characteristics of the culture regularly, and suspect any sudden change in morphology, growth rate, etc. Cross-contamination or its absence may be confirmed by DNA fingerprinting (see Protocol 15.10) [Stacey et al., 1992], karyotype [Nelson-Rees and Flandermeyer, 1977] (see Protocols 15.9 and 27.2), or isoenzyme analysis [O'Brien et al., 1980] (see Protocol 15.11).

CONCLUSIONS

• Check living cultures regularly for contamination by using normal and phase-contrast microscopy and for mycoplasmas by employing fluorescent staining of fixed preparations.

• Do not maintain all cultures routinely in antibiotics: Grow at least one set of cultures of each cell line without antibiotics for a minimum of two weeks at a time, and preferably continuously, in order to allow cryptic contaminations to become overt.

• Do not attempt to decontaminate a culture unless it is irreplaceable, and then do so only under strict quarantine.

• Quarantine all new lines that come into your laboratory until you are sure that they are uncontaminated.

• Do not share media or other solutions among cell lines or among operators, and check cell line characteristics (see Chapter 15) periodically to guard against cross-contamination.

• New cell lines should be characterized, preferably by DNA fingerprinting, as soon after isolation as possible.

It cannot be overemphasized that cross-contaminations can and do occur. It is essential that the preceding precautions be taken and that cell strain characteristics be checked regularly.

CHAPTER 19

Cryopreservation

NEED FOR CRYOPRESERVATION

The adoption of newly developed cell lines or imported cells lines into regular use implies an investment in time and resources that increases, often exponentially, with continued use. The use of a cell line adds to its provenance, and the work of the laboratory becomes increasingly dependent on it. The cell line becomes a valuable resource replacement of which would be expensive and time-consuming. If unique, it might be impossible to replace. It is, therefore, essential to protect this considerable investment by preserving the cell line.

Rationale for Freezing

Cell lines in continuous culture are prone to variation, due to selection in early-passage culture (see Evolution of Cell Lines in Chapter 2), senescence in finite cell lines (see Control of Senescence in Chapter 17), and genetic instability (see Genetic Instability in Chapter 17) in continuous cell lines. In addition, even the best-run laboratory is prone to equipment failure and contamination. Cross-contamination also continues to occur at an alarming frequency. (See Cross-Contamination in Chapter 18.) There are many reasons, therefore, for freezing down a stock of cells; these reasons can be summarized as follows:

(1) Genotypic drift due to genetic instability
(2) Senescence
(3) Transformation
(4) Phenotypic instability due to selection and dedifferentiation
(5) Contamination by microorganisms

(6) Cross-contamination by other cell lines
(7) Incubator failure
(8) Saving time and materials maintaining lines not in immediate use
(9) Need for distribution to other users.

PRESERVATION

Selection of Cell Line

A cell line is selected with the required properties. If it is a finite cell line, it is grown to around the fifth population doubling in order to create a sufficient bulk of cells for freezing. Continuous cell lines should be cloned (see Cloning in Chapter 13) and an appropriate clone selected and grown up to sufficient bulk to freeze. Prior to freezing, the cells should be maintained under standardized culture conditions, characterized (see Chapter 13), and checked for contamination (see Monitoring for Contamination in Chapter 18), particularly cross-contamination. (Table 19.1.)

Continuous cell lines have the advantages that they survive indefinitely, grow more rapidly, and can be cloned more easily, but they may be less stable genetically. Finite cell lines are usually diploid or close to it and are stable between certain passage levels, but they are harder to clone, grow more slowly, and eventually die out or transform. (See Table 12.2.)

Standardization of Culture Conditions

Medium. The type of medium used will influence the selection of different cell types and regulate their phenotypic expression. (See Chapters 8 and 9.) Con-

TABLE 19.1. Requirements before Freezing

Acquisition	Finite cell line	Freeze at early passage (<5 subcultures)
	Continuous cell line	Clone, select, and characterize; amplify
Standardization	Medium	Select optimal medium, and adhere to this medium
	Serum (if used)	Select a batch for use at all stages (see Batch Reservation in Chapter 8)
	Substrate	Standardize on one type and supplier, though not necessarily on one size or configuration
Validation	Provenance	Record details of life history and properties
	Authentication	Check cell line characteristics against provenance (see Chapters 15 and 16)
	Transformation	Determine transformed status (see Table 17.1)
	Contamination	Microbial (see Visible Microbial Contamination in Chapter 18 and Sterility Testing in Chapter 10)
		Mycoplasma (see Protocol 18.2)
	Cross-contamination	Criteria to confirm identity (e.g., DNA fingerprint; see Protocol 15.10)

sequently, once a medium has been selected, standardize on that medium, and, preferably, on one supplier, if the medium is being purchased ready made.

Serum. The best method of eliminating serum variation is to convert to a serum-free medium (see Chapter 9), although, unfortunately, serum-free formulations are not yet available for all cell types, and the conversion may be costly and time consuming. Serum substitutes (see Serum Substitutes in Chapter 9) may offer greater consistency and are generally cheaper than serum or growth factor supplementation, but do not offer the control over the physiological environment afforded by serum-free medium. If serum is required, select a batch (see Batch Reservation in Chapter 8), and use that batch throughout each stage of cryopreservation.

Stages of Cryopreservation

As soon as a small surplus of cells becomes available, from subculturing a primary culture or newly acquired cell line, a few ampules should be frozen as what is called a *token freeze*. When a propagated cell line has been produced or a cloned cell strain selected, with the desired characteristics and shown to be free of contamination, then a *seed stock* should be stored frozen (Table 19.2).

For most cell lines, particularly those in constant and general use, the seed stock should be protected and not be made available for general issue. When an ampule is thawed to check the viability of the seed stock freezing, if usage of the cell line is anticipated in the near future, a *distribution stock* should be frozen and ampules from this stock issued to individuals as required. Individuals requiring stocks over a prolonged period should then freeze down their own *user stocks*, which should be discarded when the work is

finished. When the distribution stock becomes depleted, it may be replenished from the seed stock. When the seed stock falls below five ampules, it should be replenished before any other ampules are issued, and with the minimum increase in generation number from the first freezing.

Storage in liquid nitrogen (Figs. 19.1 and Fig. 19.2; see also Cryostorage Container in Chapter 4) is currently the most satisfactory method of preserving cultured cells [Farrant, 1980]. The cell suspension, preferably at a high concentration, should be frozen slowly, at ~1°C per min [Leibo and Mazur, 1971; Harris and Griffiths, 1977], in the presence of a preservative such as glycerol or dimethyl sulfoxide [Lovelock and Bishop, 1959]. The frozen cells are transferred rapidly to liquid nitrogen when they are at or below −70°C; it is critical that they do not warm up above −50°C, when they will start to deteriorate. The ampules are then stored immersed in liquid nitrogen or in the gas phase above the liquid.

It must be realized that the cooling rate is proportional to the difference in temperature between the ampules and the ambient air. If the ampules are placed in a freezer at −70°C, they will cool rapidly to around −50°C, but the cooling rate falls off significantly after that. (Fig. 19.3.) Hence, the time that the ampules spend in the −70°C freezer needs to be longer than the amount of time projected by a 1°C cooling rate, as the bottom of the curve is asymptotic. It is safer to leave the ampules at −70°C overnight before transferring them to liquid nitrogen. Furthermore, when removed from the insulated tube or container, they will heat up at a rate of ~10°C/min, so the transfer to liquid nitrogen must take significantly less than two minutes.

When required, the cells are thawed rapidly and reseeded at a relatively high concentration to optimize

TABLE 19.2. Stages in Cell Line Preservation

Stage	Source	No. of ampules	Distribution	Validation
Token Freeze	Originator	1–3	None	Provenance only
Seed Stock	Original stock or token freeze	12	None (replenishment of distribution stock only)	Viability Authentication Transformation Contamination Cross-contamination
Distribution Stock	Test thaw from seed stock	50–100 (or more as required)	Users, including other laboratories	Viability Contamination Cross-contamination
User Stock	Distribution stock		None	Viability Contamination Cross-contamination

recovery. If liquid nitrogen storage is not available, the cells may be stored in a conventional freezer. The temperature in this freezer should be as low as possible; little deterioration has been found at $-196°C$ [Green et al., 1967], but significant deterioration (5–10% per annum) may occur at $-70°C$.

PROTOCOL 19.1. FREEZING CELLS

Outline
Grow the culture to late log phase, prepare a high-cell-density suspension, and freeze it slowly with a preservative. (See Fig. 19.2.)

Materials
Sterile or Aseptically Prepared:
Cultures to be frozen
If monolayer: PBSA and 0.25% crude trypsin
Growth medium (Serum improves survival of the cells after freezing; up to 50%, or even pure, serum has been used. If serum is being used with serum-free cultures, it should be washed off after thawing.)
Preservative, free of impurities: DMSO or glycerol (DMSO should be colorless, and it needs to be stored in glass or polypropylene, as it is a powerful solvent and will leach impurities out of rubber and some plastics. Glycerol should be not more than one year old, as it may become toxic after prolonged storage, and should be dispensed into a universal container.)
Syringe, 1–5 ml, for dispensing glycerol
Plastic ampules, 1.2 ml
Nonsterile:
Hemocytometer or electronic cell counter

Canes or racks for storage
Insulated container for freezing: polystyrene box lined with cotton wool, or plastic foam insulation tube (or controlled rate freezer if available)
Curved forceps
Protective gloves, nitrile

Protocol
1. Check the culture for the following (see Table 19.1):
   (a) healthy growth
   (b) freedom from contamination
   (c) specific characteristics.
2. Grow the culture up to the late log phase and, if you are using a monolayer, trysinize and count the cells. (See Protocol 12.2.) If you are using a suspension, count and centrifuge the cells. (See Protocol 12.3.)
3. Dilute one of the preservatives in growth medium to make freezing medium:
   (a) Add dimethyl sulfoxide (DMSO) to 5–10% or
   (b) Add glycerol to 10–15%.
4. Resuspend the cells in freezing medium at approximately $1 \times 10^6 - 1 \times 10^7$ cells/ml. It is not advisable to place ampules on ice in an attempt to minimize deterioration of the cells. A delay of up to 30 min at room temperature is not harmful when using DMSO and is beneficial when using glycerol.

△ *Safety Note.* DMSO can penetrate many synthetic and natural membranes, including *skin* and rubber gloves [Horita and Weber, 1964]. Consequently, any potentially harmful substances in regular use (e.g., carcinogens) may well be carried into the circulation through the skin and even through

Fig. 19.1. Liquid-Nitrogen Freezers. (a) Narrow-necked freezer with storage on canes in canisters. (b) Interior of the narrow-necked freezer, looking down on canes in canisters. (c) Wide-necked freezer with storage in triangular drawers. (d) Narrow-necked freezer with storage in square drawers. (See also Fig. 19.8.)

Fig. 19.4. Ampules on Canes. Plastic ampules clipped onto aluminum canes, enclosed in a cardboard tube, and placed inside an insulating foam tube.

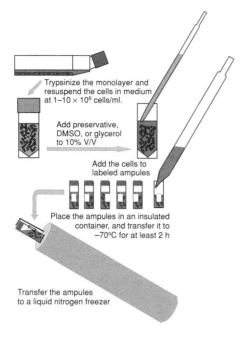

Fig. 19.2. Freezing Cells.

Trypsinize the monolayer and resuspend the cells in medium at $1-10 \times 10^6$ cells/ml.

Add preservative, DMSO, or glycerol to 10% V/V

Add the cells to labeled ampules

Place the ampules in an insulated container, and transfer it to $-70°C$ for at least 2 h

Transfer the ampules to a liquid nitrogen freezer

rubber gloves. DMSO should always be handled with caution, particularly in the presence of any toxic substances.

5. Dispense the cell suspensions into prelabeled ampules, and seal the ampules.

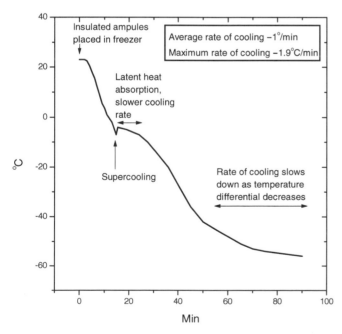

Fig. 19.3. Freezing Curve. Record of the fall in temperature in an ampule containing medium, clipped with five other ampules on an aluminum cane, enclosed in a cardboard tube, placed within a polyurea-foam tube (see Fig. 19.4), and placed in a freezer at $-70°C$.

Insulated ampules placed in freezer

Average rate of cooling $-1°/min$
Maximum rate of cooling $-1.9°C/min$

Latent heat absorption, slower cooling rate

Supercooling

Rate of cooling slows down as temperature differential decreases

6. Place the ampules on canes for canister storage (Fig. 19.4), or leave them loose for drawer storage (see Fig. 19.1c,d).

7. Freeze the ampules by one of the following methods:

(a) Lay the ampules on cotton wool in a polystyrene foam box with a wall thickness of ~15 mm. This box, plus the cotton wool, should provide sufficient insulation such that the ampules will cool at 1°C/min when the box is placed at $-70°C$ or $-90°C$ in a regular deep freeze or insulated container with solid CO_2.

(b) Insert the canes in tubular foam pipe insulation, with a wall thickness of ~15 mm (Fig. 19.4), and place the insulation at $-70°C$ or $-90°C$ in a regular deep freeze or insulated container with solid CO_2.

(c) Place the cells in a Taylor Wharton freezer neck plug (Fig. 19.5) or a Nalge Nunc freezing container (Fig. 19.6).

(d) Use a controlled-rate freezer programmed to freeze at 1°C/min, with accelerated freezing through the eutectic point. (See Fig. 19.3.)

8. When the ampules have reached $-70°C$ (a minimum of 4–6 h after placing them at $-70°C$ if starting from 20°C ambient (see Fig. 19.3 and above), but preferably overnight), transfer them to a liquid N_2 freezer. This transfer must be done quickly, as the ampules will reheat at ~10°C/min, and the cells will deteriorate rapidly if the temperature rises above $-50°C$.

△ **Safety Note.** Protective gloves and a face mask should be used when handling liquid nitrogen.

9. When the ampules are safely located in the freezer, make sure that the appropriate entries are made in the freezer index. (See Tables 19.3 and 19.4.)

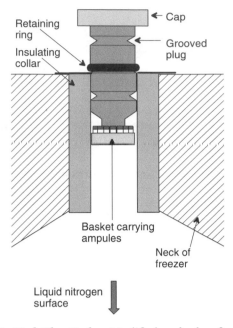

Fig. 19.5. Neck-Plug Cooler. Modified neck plug for narrow-necked freezers, allowing controlled cooling at different rates (Taylor Wharton). Shown is the section of the freezer neck with the modified neck plug in place. The "O" ring is used to set the height of the ampules within the neck of the freezer. The lower the height, the faster the cooling.

△ *Safety Note.* Inexperienced users should avoid using glass ampules as they have a serious risk of explosion when thawed. If glass ampules are used, they must be perfectly and quickly sealed in a gas-oxygen flame. If sealing takes too long, the cells will heat up and die, and the air in the ampule will expand and blow a hole in the top of the ampule. If the ampule is not perfectly sealed, it may inspire liquid nitrogen during storage in the liquid phase of the nitrogen

Fig. 19.6. Nalge Nunc Cooler. Plastic holder with fluid-filled base. The specific heat of the coolant in the base insulates the container and gives a cooling rate of ~1°C/min in the ampules.

freezer and will subsequently explode violently on thawing.

It is possible to check for leakage by placing glass ampules in a dish of 1% methylene blue in 70% alcohol at 4°C for 10 min before freezing. If the ampules are not properly sealed, the methylene blue will be drawn into the ampule, and the ampule should be discarded. The ampules may need to be relabeled after this procedure.

Ampules. Plastic ampules are unbreakable, but must be sealed with the correct torsion on the screw cap, or else they may leak. They are of a larger diameter and taller than equivalent glass ampules. Due to this size difference, check that they will fit in the canes or racks used for storage. (Special canes for plastic ampules are available.)

Plastic ampules are preferred for the average experimental and teaching laboratory, as they are safer and more convenient, but repositories and cell banks generally prefer glass ampules, since the long-term storage properties of glass are well characterized and, when correctly performed, sealing is absolute.

The ampules should also carry a label with the cell strain designation and, preferably, the date and user's initials, although the latter is not always feasible in the available space. Remember, cell cultures stored in liquid nitrogen may well outlive you! They can easily outlive your stay in a particular laboratory. The record therefore should be readily interpreted by others and sufficiently comprehensive so that the cells may be of use to others.

Cooling Rate

Most cultured cells survive best if they are cooled at 1°C/min, but if recovery is low, try changing the freezing rate (i.e., use more or less insulation) [Leibo and Mazur, 1971], or use a programmable freezer (Fig. 19.7) with a probe, which senses the temperature of the ampule and adds liquid nitrogen to the freezing chamber at the correct rate to achieve a preprogrammed cooling rate. These freezers are, however, relatively expensive, compared to the simple devices described above, and have few advantages unless you wish to vary the cooling rate [e.g., Foreman and Pegg, 1979] or alter the shape of the cooling curve.

Cryofreezers

These are four main types of liquid-nitrogen storage systems (Fig. 19.8), based on whether the storage vessel is wide-necked or narrow-necked, and whether storage is in the vapor or liquid phase. Wide-necked freezers are chosen for ease of access and maximum capacity, and narrow-necked freezers for economy (since they have a slow evaporation rate). Storage in

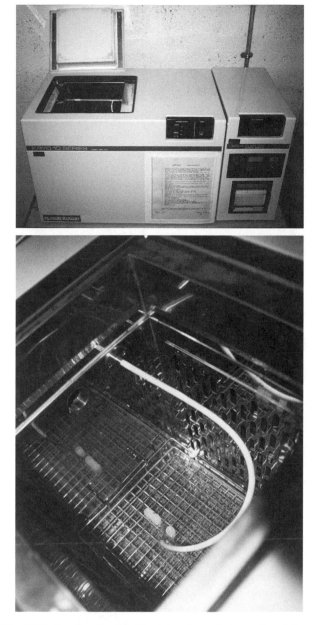

Fig. 19.7. Programmable Freezer. Ampules are placed in an insulated chamber, and the cooling rate is regulated by injecting liquid nitrogen into the chamber at a rate determined by a sensor on the rack with the ampules and a preset program in the console unit (Planer Biomed). (a) Control unit and freezing chamber (lid open). (b) Close-up of a freezing chamber with four ampules, one with a probe in it.

the vapor phase eliminates the risk of explosion with sealed ampules, while storage in the liquid phase means that the container can be filled and the liquid nitrogen will therefore last longer. Storage capacities and static holding times (i.e., the time to boil dry without being opened) are given in the manufacturer's literature for each freezer.

It is possible to effect a compromise by selecting a relatively narrow-necked freezer while still using a tray system for storage. Freezers are available with inventory control based on square-array storage trays; these trays are mounted on racks that are accessed by the same system as the cane and canister of conventional narrow-necked freezers. These freezers do not have the storage capacity of the cane and canister, but have equivalent holding times and a honeycomb storage array that many people prefer.

△ *Safety Note.* Biohazardous material *must* be stored in the gas phase, and teaching and demonstrating are best done with gas-phase storage. Above all, if liquid-phase storage is used, the user must be made aware of the explosion hazard of both glass and plastic and must wear a face shield or goggles.

Freezer Records

Records should provide (a) an inventory showing what is in each part of the freezer and (b) a cell strain index, describing the cell line, its designation, what its special characteristics are, and where it is located. This record may be kept on a conventional card index, but a computerized database will give superior data storage and retrieval. This type of data can be provided by separate tables within the same database used for the provenance of the cell line. (Tables 19.3 and 19.4; Primary Records in Chapter 11; Maintenance Records in Chapter 12). Material stored on disks or tape must have back-up copies on disk or tape or must have a hard-copy printout.

Using a computerized database requires that the curator of the freezers manages this database. If entries are to be made by users, user stocks can have both read and write access, while access to seed stock and distribution stock should be read only. Alternatively, the whole file can be read only and updated by the curator from paper entries on cards or a log book. (See Tables 19.3 and 19.4.)

PROTOCOL 19.2. THAWING FROZEN CELLS

Outline
Thaw the cells rapidly, dilute them slowly, and reseed them at a high cell density. (Fig. 19.9.)
Sterile:
Culture flask
Centrifuge tube (if centrifugation is required)
Growth medium
Pipettes, 1 ml, 10 ml
Syringe and 19-g needle (if you are using glass ampules)

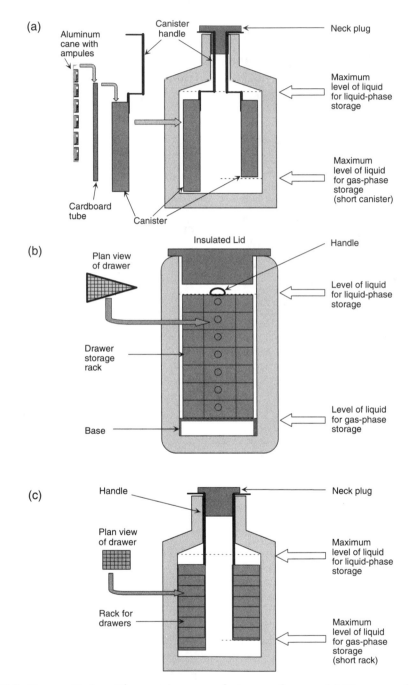

Fig. 19.8. Freezer Design. Three main types of nitrogen freezers: (a) Narrow-necked with ampules on canes in canisters (high capacity, low boil-off rate). (b) Wide-necked with ampules in triangular racks (high capacity, high boil-off rate). (c) Narrow-necked with ampules in square racks (moderate capacity, low boil-off rate).

TABLE 19.3. Cell Line Record

Cell line	Freeze date:						
	Location:						

Species Normal/ Adult/ neoplastic Fetal/NB	**Mycoplasma:** Method Date of test Result	**Freeze instructions:** Rate Preservative, %
Tissue Site Pass./ Gen. No	**Authentication:** Date of test Method	**Thaw instructions:** Thaw rapidly to 37°C Dilute to: 5 ml 10 ml 20 ml 50 ml
Author Ref Mode of growth	**Normal Maintenance:** Subculture Seeding Agent frequency conc.	Centrifuge to remove Yes/No preservative?
Special characteristics: 	Medium change Type Serum, frequency etc. .... % Gas phase Buffer pH	Special requirements **Biohazard precautions:**
Person completing card Date	Any other special conditions	

Nonsterile:
Protective gloves and face mask
Bucket of water at 37°C with lid
Forceps
70% alcohol
Swab
Naphthalene black (amido black), 1%, or trypan
  blue, 0.4%.

Protocol
1. Check the index for the location of the ampule to be thawed.
2. Collect all materials, prepare the medium, and label the culture flask.
3. Retrieve the ampule from the freezer, check that it is the correct one, and place it in 10 cm of water at 37°C in a 5-l bucket with a lid.

TABLE 19.4. Freezer Record

Position: Freezer no. ....   Canister/Section no. ....   Tube/drawer no. ....

Cell strain/line: ....   Freeze date ....   Frozen by ....

No. of ampules frozen ....   No. of cells/ampoule ....   in .... ml

Growth medium ....   Serum .... Conc. ....   Freeze medium ....

Method of cooling ....   Cooling rate ....

Thawing record:

Thaw date	No. of amp.	No. left	Seeding			Viability*			Notes
			Conc.	Vol.	Medium	Dye exclusion	~% attached by 24 h	Cloning efficiency	

*Only one parameter need be used here.

△ *Safety Note.* A face shield, or protective goggles, and gloves must be worn. If the ampule has been stored in the gas phase, a lid for the bucket is not necessary. However, ampules, even plastic, stored under liquid nitrogen may inspire the liquid and, on thawing, will explode violently. A plastic bucket with a lid is therefore essential to contain any explosion.

4. When the ampule is thawed, check the label to confirm the identity of the cells; then swab the ampule thoroughly with 70% alcohol, and open it.
5. Transfer the contents of the ampule to a culture flask.
6. Add medium slowly to the cell suspension: 10 ml over about 2 min added dropwise at the start, and then a little faster, gradually diluting the cells and preservative. This gradual process is partic-

ularly important with DMSO, with which sudden dilution can cause severe osmotic damage and reduce cell survival by half.
7. The dregs in the ampule may be stained with naphthalene black or trypan blue to determine their viability. (See Protocol 21.1.)

The number of cells frozen should be sufficient to allow for 1:10 or 1:20 dilution on thawing to dilute out the preservative but still keep the cell concentration higher than at normal passage; for example, for cells subcultured normally at 1×10^5/ml, 1×10^7 should be frozen in 1 ml of medium, and, after thawing the cells, the whole 1 ml should be diluted to 20 ml of medium, giving 5×10^5 cells/ml (five times the normal seeding concentration). This dilutes the preservative from 10% to 0.5%, at which concentration it is less likely to be toxic. Residual preservative may be diluted out as soon as the cells start to grow (for sus-

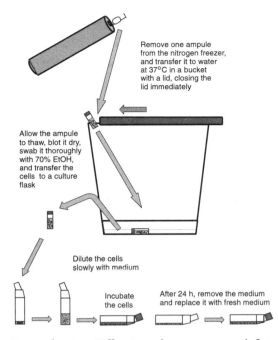

Fig. 19.9. Thawing Cells. Ampules are removed from the freezer and thawed rapidly in warm water, under cover, to avoid the risk of explosion if the ampules have been stored in the liquid phase.

pension cultures) or the medium changed as soon as the cells have attached (for monolayers).

The dye exclusion viability and the approximate take (e.g., the proportion of cells attached versus those still floating after 24 h) should be recorded on the appropriate record card or file to assist in future thawing. One ampule should be thawed from each batch as it is frozen, to check that the operation was successful.

Variations. In some cases—e.g., L5178Y mouse lymphoma with DMSO—the preservative may be too toxic for the cells to survive on thawing (M. Freshney, personal communication). In these cases, dilute the cells slowly, as in Step 6, Protocol 19.2, centrifuge them for 2 min at 100 *g*, discard the supernatant medium with the preservative, and resuspend the cells in fresh growth medium for culture.

Whole flasks may be frozen by growing the cells to late log phase, adding 5–10% DMSO to the smallest volume of medium which will effectively cover the monolayer, and placing the flask in an expanded polystyrene container of 15-mm wall thickness [Ohno et al., 1991]. The insulated container is placed in a −70°C to −90°C freezer and will freeze at approximately 1°C/min. Survival is good for several months, as long as the flask in its container is not removed from the freezer. Twenty-four-well plates may also be

frozen in the same manner [Ure et al., 1992] with about 150 μl freezing medium per well and can be used to store large numbers of clones during evaluation procedures.

Serial Replacement

Stock cultures should be replaced from the freezer at regular intervals to minimize the effects of genetic drift and phenotypic variation. After a cell line has been in culture for 2½ months, thaw out another vial, check its characteristics, make sure that is is free from contamination (see Monitoring for Contamination in Chapter 18), and grow it up to replace existing stocks. Discard the existing stocks when they have been out of the freezer for 3 months, and move on to the new stock. Repeat this process every 3 months with cells that have a population-doubling time (PDT) of approximately 24 h; cell lines with shorter or longer PDTs may need shorter or longer replacement intervals, respectively.

CELL BANKS

Several cell banks exist (see Sources of Materials—Cell Banks in the Trade Index) for the secure storage and distribution of validated cell lines. Since many cell lines may come under patent restrictions, particularly hybridomas and other genitically modified cell lines, it has also been necessary to provide patent repositories with limited access.

As a general rule, it is preferable to obtain your initial seed stock from a reputable cell bank, where the necessary characterization and quality control will have been done. Furthermore, it is highly recommended that you submit valuable cultures to a cell bank in addition to maintaining your own frozen stock, as the former will protect you against loss of your own lines and allow their distribution to others. If you feel that your cells should not be distributed, then they can be banked with that restriction placed on them.

In addition to the provision of frozen cell stocks, some cell banks also make their catalogue information available on line or on disk. There are also data banks that can be accessed on-line and that maintain information on cell lines held by subscribers in their own laboratories. (See Reagents and Materials—Data Banks in the Trade Index.) This type of data bank provides a vast increase in the amount of material that is potentially available, but it must be remembered that cell lines obtained from other laboratories will vary significantly in the amount of characterization that they have had and in the quality of their maintenance.

TRANSPORTING CELLS

Cultures may be transferred from one laboratory to another as frozen ampules or as living cultures. In either case,

(1) advise the recipient as to when the cells are to be shipped;
(2) fax or e-mail instructions on the following:
   (a) what to do on receipt,
   (b) medium or serum required,
   (c) any special supplements, and
   (d) subculture regimen;
(3) tape the data sheet for the cells and a copy of the instructions to the outside of the box.

Frozen Ampules

Ship frozen ampules in solid carbon dioxide, in a thick-walled polystyrene foam container. The carrier must be informed when cells are shipped in solid CO_2. Usually, cells will remain frozen for up to 3 days if properly packed, but if they thaw slowly, their viability will decline rapidly.

Wrap the ampule in absorbent paper tissue (in case of leakage), and place it in a polypropylene tube with a tight cap. Place the tube on solid CO_2 in an insulated box. The box should be about $300 \times 300 \times 300$ mm, with a wall thickness of 50 mm and a central space of about $200 \times 200 \times 200$ mm. You will need about 5 kg of solid CO_2 to fill the central space.

Transfer the ampule from the liquid nitrogen freezer to the solid CO_2 as quickly as possible, as it will warm up at $\sim 10-20°C/min$ and must not rise above $-50°C$.

On arrival, the ampule should be thawed and seeded as normal. (See Protocol 19.2.)

Living Cultures

Alternatively, cells may be shipped as a growing culture. The cells should be at the mid to late log phase; confluent or postconfluent cultures will exhaust the medium more rapidly and may tend to detach in transit. The flask should be filled to the top with medium, taped securely around the neck with a stretch-type waterproof adhesive tape, and sealed in a small polythene bag. The bagged flask is then packed within a rigid container filled with absorbent packing. This container is then sealed in a polythene bag and then in plastic foam or bubble wrap. Place on the box a label that says "fragile" and, in large letters, the following instructions: DO NOT FREEZE!

On receipt, most of the medium is removed, leaving only the normal amount for culture—e.g., 5 ml for a 25-cm^2 flask. The culture can be weaned onto new medium when it is ready for the first feed, but keep the original shipping medium in case there are any problems of adaptation to the new stock.

It is usually better to ship cells via a courier, who should be briefed as to the contents of the package and the urgency of delivery. International mailings will require negotiation with customs controls, and a competent nominated agent who is familiar with this type of importation can help to move the package through customs. Cells shipped to the United States are required to be quarantined and tested by the Bureau of Agriculture, so it is better to arrange this shipment through one of the cell banks, ECACC or ATCC.

Warn the recipient before shipping, enclose details of the cell line and requirements for its maintenance, and fax these details to the recipient before the cells are shipped. Enclose copies of these details, including instructions on what to do on arrival with the cells in the package.

CHAPTER 20

Quantitation

Quantitation covers the determination of cell number, cellular constituents, cell activities (e.g., gene expression, enzyme activity, motility, etc.), and cell proliferation kinetics. Proper quantitation is required for consistency of maintenance (see Subculture in Chapter 12) and for the analysis of the response of cells to environmental variations or experimental conditions.

CELL COUNTING

While estimates can be made of the stage of growth of a culture from its appearance on the microscope, the standardization of culture conditions and proper quantitative experiments are difficult unless the cells are counted before and after, and preferably during, each experiment.

Hemocytometer

The concentration of a cell suspension may be determined by placing the cells in an optically flat chamber under a microscope. (Fig. 20.1.) The cell number within a defined area of known depth is counted and the cell concentration derived from the count.

PROTOCOL 20.1. CELL COUNTING BY HEMOCYTOMETER

Outline
Trypsinize a monolayer culture, or take a sample from suspension culture, prepare a slide, and insert the cells into a hemocytometer chamber. Count the cells on the microscope and calculate cell concentration.

Materials
Sterile:
PBSA
0.25% crude trypsin
Growth medium
Pasteur pipettes or pipettor yellow tips
Nonsterile:
Pipettor, 20 μl or adjustable
Hemocytometer (Improved Neubauer)
Tally counter
Microscope

Protocol
1. Sample the cells:
   (a) For a monolayer,
      (i) Trypsinize the monolayer (see Protocol 12.2) and resuspend in medium to give an estimated 1×10^6/ml.
      (ii) Mix the suspension thoroughly to disperse the cells, and transfer a small sample (~1 ml) to a vial or universal container.
   (b) For a suspension culture,
      (i) Mix the suspension thoroughly to disperse any clumps;
      (ii) Transfer 1 ml of the suspension to a vial or universal container.
      A minimum of approximately 1×10^6 cells/ml is required for this method, so the suspension may need to be concen-

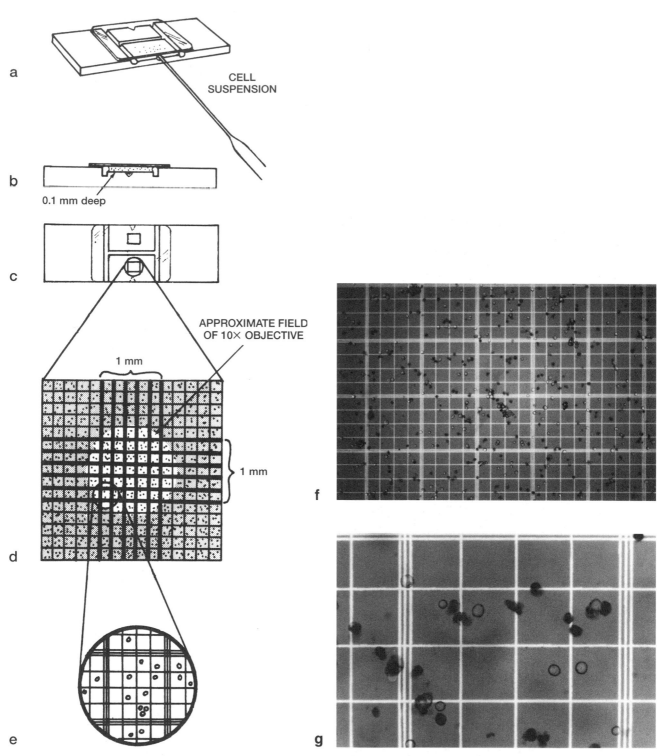

Fig. 20.1. Hemocytometer slide. Improved Neubauer pattern. (a) Adding a cell suspension to an assembled slide. (b) Longitudinal section of the slide, showing the position of the cell sample in a 0.1-mm-deep chamber. (c) Top view of the slide. (d) Magnified view of the total area of the grid. The light central area is that area which would be covered by the average 10× objective (depending on the field of view of the eyepiece). This area covers approximately the central 1 mm² of the grid. (e) Magnified view of one of the 25 smaller squares, bounded by the triple parallel lines that make up the 1-mm² central area. This view is subdivided by single grid lines into the 16 smallest squares to aid counting. (f) Low-power (10× objective) microphotograph showing 20 of the 25 smaller squares of a slide loaded with cells pretreated with naphthalene black (amido black). (See the text.) Viable cells are unstained and clear, with a refractile ring around them; nonviable cells are dark and have no refractile ring. (g) High-power (40× objective) microphotograph of one of the smaller squares, bounded by three parallel lines and containing 16 of the smallest squares. The distance between each set of triple lines is 200 μm.

trated by centrifuging it (at 100 g for 2 min) and resuspending it in a measured smaller volume.

2. Prepare the slide:
   (a) Clean the surface of the slide with 70% alcohol, taking care not to scratch the semi-silvered surface.
   (b) Clean the coverslip, and, wetting the edges very slightly, press it down over the grooves and semisilvered counting area. (See Fig. 20.1.) The appearance of interference patterns ("Newton's rings," or rainbow colors between the coverslip and the slide, like the rings formed by oil on water) indicates that the coverslip is properly attached, thereby determining the depth of the counting chamber.

3. Mix the cell sample thoroughly, pipetting vigorously to disperse any clumps, and collect about 20 μl into the tip of a Pasteur pipette or pipettor. Do not let the fluid rise in a Pasteur pipette, or else cells will be lost in the upper part of the stem.

4. Transfer the cell suspension immediately to the edge of the hemocytometer chamber, and let the suspension run out of the pipette and be drawn under the coverslip by capillarity. Do not overfill or underfill the chamber, or else its dimensions may change, due to alterations in the surface tension; the fluid should run only to the edges of the grooves.

5. Reload the pipette, and fill the second chamber if there is one.

6. Blot off any surplus fluid (without drawing from under the coverslip), and transfer the slide to the microscope stage.

7. Select a 10× objective, and focus on the grid lines in the chamber. (See Fig. 20.1.) If focusing is difficult because of poor contrast, close down the field iris, or make the lighting slightly oblique by offsetting the condenser.

8. Move the slide so that the field you see is the central area of the grid and is the largest area that you can see bounded by three parallel lines. This area is 1 mm². With a standard 10× objective, this area will almost fill the field, or the corners will be slightly outside the field, depending on the field of view. (See Fig. 20.1.)

9. Count the cells lying within this 1-mm² area, using the subdivisions (also bounded by three parallel lines) and single grid lines as an aid for counting. Count cells that lie on the top and left-hand lines of each square, but not those on the bottom or right-hand lines, in order to avoid counting the same cell twice. For routine sub-

culture, attempt to count between 100 and 300 cells per mm²; the more cells that are counted, the more accurate the count becomes. For more precise quantitative experiments, 500–1,000 cells should be counted.
   (a) If there are very few cells (<100/mm²), count one or more additional squares (each 1 mm²) surrounding the central square.
   (b) If there are too many cells (>1,000/mm²), count only five small squares (each bounded by three parallel lines) across the diagonal of the larger (1-mm²) square.

10. If the slide has two chambers, move to the second chamber and do a second count. If not, rinse the slide and repeat the count with a fresh sample.

Analysis. Calculate the average of the two counts, and derive the concentration of your sample using the formula

$$c = n/v$$

where c is the cell concentration (cells/ml), n is the number of cells counted, and v is the volume counted (ml). For the Improved Neubauer slide, the depth of the chamber is 0.1 mm, and, assuming that only the central 1 mm² is used, v is 0.1 mm³, or 1×10^{-4} ml. The formula then becomes

$$c = n/10^{-4}, \text{ or } c = n \times 10^4.$$

If the cell concentration is high and only the 5 diagonal squares within the central 1 mm² were counted (i.e., 1/5 of the total), this equation becomes
$$c = n \times 5 \times 10^4$$

If the cell concentration is low and the other eight 1 mm² squares were counted in addition to the central 1 mm² (i.e., making nine squares total), this expression becomes

$$c = \frac{n \times 10^4}{9}.$$

Hemocytometer counting is cheap and gives you the opportunity to see what you are counting. If the cells were previously mixed with an equal volume of a viability stain (see Viability in Chapter 21; Fig. 20.1f,g), a viability determination may be performed at the same time. (Remember to compensate for the additional dilution to obtain an absolute count.) However, the procedure is rather slow and prone to error both in the method of sampling and the total number of cells counted; it also requires a minimum of 10⁶ cells/ml.

Most of the errors in this procedure occur by incorrect sampling and transfer of cells to the chamber. Make sure that the cell suspension is properly mixed before you take a sample, and do not allow the cells time to settle or adhere in the tip of the pipette before transferring them to the chamber. Ensure also that you have a single-cell suspension, as aggregates make counting inaccurate. Larger aggregates may enter the chamber more slowly or not at all. If aggregation cannot be eliminated during preparation of the cell suspension (see Table 12.3), lyse the cells in 0.1 M citric acid containing 0.1% crystal violet at 37°C for 1 hr and then count the nuclei [Sanford et al., 1951].

Electronic Counting

Although a number of different automatic methods have been developed for counting cells in suspension, the system devised originally by Coulter Electronics is the one most widely used. Briefly, in this method, cells drawn through a fine orifice increase the electrical resistance to the current flowing through the orifice, producing a series of pulses that are sorted and counted.

Electronic Cell Counter. (See Fig. 20.2.) There are three main components of the electronic cell counter (see Fig. 20.3): (1) an orifice tube connected to a metering pump (Fig. 20.3a), with a 150-μm orifice drilled through an industrial ruby (Fig. 20.3b), (2) an amplifier, pulse-height analyzer, and scaler connected to two electrodes, one in the orifice tube and one in the sample beaker, and (3) an analogue and a digital readout showing the cell count and a number of other parameters, such as cell volume and size distribution, depending on the model purchased (Schärfe, Beckman Coulter).

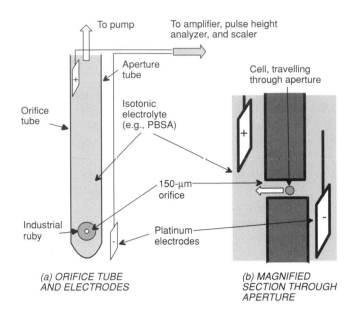

(a) ORIFICE TUBE AND ELECTRODES

(b) MAGNIFIED SECTION THROUGH APERTURE

Fig. 20.3. Cell Counter Orifice Tube. Principle of operation of an electronic particle counter. The electrolyte (PBSA) in an orifice tube is connected to a pump and draws a measured volume of the cell sample from a beaker. A cell passing through the orifice alters the flow of current and generates a signal, the amplitude of which is proportional to the volume of the cell.

When the count is initiated, a measured volume of cell suspension is drawn through the orifice. (See Fig. 20.3b.) As each cell passes through the orifice, it changes the resistance to the current flowing through the orifice by an amount proportional to the volume of the cell. This change in resistance generates a pulse (amps^{-1}) that is amplified and counted. Since the size of the pulse is proportional to the volume of the cell (or any other particle) passing through the orifice, a series of signals of varying pulse height are generated. The lower threshold (in cell diameter) is set to eliminate electronic noise and fine particulate debris, but to retain pulses derived from cells. (See Calibration, below.) The upper threshold is either set to infinity or is adjustable, depending on the model. These thresholds determine which range of particles sizes is counted. The display on the counter shows a histogram depicting the distribution of cell size.

PROTOCOL 20.2. ELECTRONIC CELL COUNTING

Outline

Dilute a sample of cells in electrolyte (physiological saline or PBSA), place the diluted sample under the

Fig. 20.2. Electronic Cell Counter. CASY 1 (Schärfe Systems) cell counter, also suitable for cell sizing.

orifice tube, and count the cells by drawing 0.5 ml of the diluted sample through the counter.

Materials
Sterile or Aseptically Prepared:
Cell culture
PBSA
0.25% crude trypsin
Growth medium
Nonsterile:
Counting cups

Protocol
1. Trypsinize the cell monolayer, or collect a sample from the suspension culture. The cells must be well mixed and singly suspended.
2. Dilute the sample of cell suspension to 1:50 in 20 ml of counting fluid in a 25-ml beaker or disposable sample cup. An automatic dispenser will speed up this dilution and improve reproducibility.

Note. Dispensing counting fluid rapidly can generate air bubbles that will be counted as they pass through the orifice. Consequently, the counting fluid should stand for a few moments before counting. If the fluid is dispensed first and the cells added second, this problem is minimized.

3. Mix the suspension well, and place it under the tip of the orifice tube, ensuring that the orifice is covered and that the external electrode lies submerged in the counting fluid in the sample beaker.
4. Check the program settings:
   (a) Threshold setting(s) (minimum cell size, usually 7.0 μm)
   (b) Volume to be counted (usually 0.5 ml)
   (c) Background subtraction (if used)
   (d) Dilution settings (e.g., 50 if 0.4 ml is counted in 20 ml of PBSA).
5. Check the visual analogue display
   (a) to ensure that all cells fall within the threshold setting(s);
   (b) to check for viability or cell debris (indicated by a shoulder on the curve or histogram falling below the normal lower threshold setting);
   (c) to check for aggregation (indicated by particles appearing above the normal size range).
6. Initiate the count sequence.
7. When the count cycle is complete, the size distribution will appear on the analogue screen.

Switching to the digital screen will give the cell count per ml.

Analysis. If the background and dilution are set correctly, the cell concentration will be that of the starting suspension before dilution in counting fluid. If not, then, if the counter is set to take 0.5 ml of a 1:50 dilution, the final count on the readout should be multiplied by 100 to give the concentration per ml of the original cell suspension. However, some counters automatically compensate for the volume sampled, so check the instruction manual.

Calibration. Older counters required calibration by counting cells at increasing increments of the lower threshold (upper threshold set to infinity). The plot of these counts generated a plateau, the center of which gave the correct threshold setting. In modern counters, with an analogue display, the lower threshold may be set manually on the read-out, usually at 7.0 μm, while the upper is set to 30 μm or infinity.

Cell Sizing
Most counters, by virtue of their design, will do a cell size analysis, and differences are found mainly in the display and downloading capabilities. A graphical display, either LCD or VDU, is a significant advantage to routine counting and is available on most machines.

Electronic cell counting is rapid and has a low inherent error, due to the high number of cells counted. It is prone to misinterpretation, however, because cell aggregates, dead cells, and particles of debris of the correct size will all be counted indiscriminately. The cell suspension should be examined carefully before dilution and counting.

Electronic particle counters are expensive, but if used correctly, are very convenient and give greater speed and accuracy to cell counting.

Stained Monolayers
There are occasions when cells cannot be harvested for counting or are too few to count in suspension (e.g., at low cell concentrations in microtitration plates). In these cases, the cells may be fixed and stained *in situ* and counted by eye on a microscope. Since this procedure is tedious and subject to high operator error, isotopic labeling or the estimation of the total amount of DNA (see Protocol 20.3) or protein (see Protocol 20.4) is preferable, though these measurements may not correlate directly with the cell number—e.g., if the ploidy of the cell varies. A rough estimate of the cell number per well can also be obtained by staining the cells with crystal violet and measuring the absorption on a densitometer. This method has also been used to calculate the number of cells per colony in clonal

TABLE 20.1. Relationship Between Cell Size, Volume, and Mass

Cell type	Diameter (μm)	Volume (μm³)	Cells/g $\times$ 10^{-6}	
			Calculated	Measured
Murine leukemia (e.g., L5178Y or Friend)	11–12	800	1,250	1,000
HeLa	14–16	1,200	800	250
Human diploid fibroblasts	16–18	2,500	400	180

growth assays [McKeehan et al., 1977]. Staining cells with Coomassie blue, sulforhodamine B [Boyd, 1989; Skehan et al., 1989], or MTT [Plumb et al., 1989] also gives an estimation of the cell number, given that linearity has been demonstrated previously in a standard plot of absorption against cell number. (See Protocol 21.4.) MTT staining has the advantage that it stains only viable cells.

CELL WEIGHT

Wet weight is seldom used unless very large cell numbers are involved, as the amount of adherent extracellular liquid gives a high error. As a rough guide, however, there are about 2.5×10^8 HeLa cells (14–16 μm in diameter) per gram wet weight, about $8–10 \times 10^8$ cells/g for murine leukemias (e.g., L5178Y murine lymphoma, Friend murine erythroleukemia, myelomas and hybridomas; 11–12 μm in diameter), and about 1.8×10^8 cells/g for human diploid fibroblasts (16–18 μm in diameter). (See Table 20.1.) Similarly, dry weight is seldom used, since salt derived from the medium contributes to the weight of unfixed cells, and fixed cells lose some of their low-molecular-weight intracellular constituents and lipids. However, an estimate of dry weight can be derived by interferometry [Brown and Dunn, 1989].

DNA CONTENT

In practice, besides the cell number, DNA and protein are the two most useful measurements for quantifying the amount of cellular material. DNA may be assayed by several fluorescence methods, including reaction with DAPI [Brunk et al., 1979] or Hoechst 33258 [Labarca and Paigen, 1980]. DNA can also be measured by its absorbance at 260 nm, where 50 μg/ml has an optical density (O.D.) of 1.0. Because of interference from other cellular constituents, the direct absorbance method is useful only for purified DNA. The following protocol is a relatively simple and straightforward assay for DNA.

PROTOCOL 20.3. DNA ESTIMATION BY HOECHST 33258

Principle
The fluorescence emission of Hoechst 33258 at 458 nm is increased by interaction of the dye with DNA at pH 7.4 and in high salt to dissociate the chromatin protein. This method gives a sensitivity of 10 ng/ml, but requires intact double-stranded DNA.

Outline
Homogenize cells or tissue in buffer, and then sonicate the homogenate. Mix aliquots of the culture with H33258, and measure the fluorescence.

Materials
Nonsterile:
Buffer: 0.05 M NaPO$_4$; 2.0 M NaCl, pH 7.4, containing 2×10^{-3} M EDTA
Hoechst 33258: in buffer, 1 μg/ml for DNA above 100 ng/ml and 0.1 μg/ml for 10–100 ng/ml

Protocol
1. Homogenize the cells in buffer, 1×10^5 cells/ml for 1 min, using a Potter homogenizer.
2. Sonicate the cells for 30 sec
3. Dilute the cells 1:10 in Hoechst 33258 and buffer.
4. Read fluorescence emission at 492 nm with excitation at 356 nm, using calf thymus DNA as a standard.

PROTEIN

The protein content of cells is widely used for estimating total cellular material and can be used in growth experiments or as a denominator in expressions of the specific activity of enzymes, the receptor content, or intracellular metabolite concentrations. The amount of proteins in solubilized cells can be estimated directly by measuring the absorbance at 280 nm, with minimal interference from nucleic acids and other constituents. The absorbance at 280 nm can detect down to 100 μg of protein, or about 2×10^5 cells.

Colorimetric assays are more sensitive than measurements of absorption, and among these assays, the Bradford reaction with Coomassie blue [Bradford, 1976] is the most widely used.

Solubilization of Sample

Since most assays rely on a final colorimetric step, they must be carried out on clear solutions. Cell monolayers and cell pellets may be dissolved in 0.5–1.0 M NaOH by heating them to 100°C for 30 min or leaving them overnight at room temperature. Alternatively, using 0.3 M NaOH and 1% sodium lauryl sulfate, the solution is complete after 30 min at room temperature.

The protein content can be estimated colorimetrically by the Bradford method, which is not dependent on specific amino acids and is quite sensitive, requiring 50–100,000 cells. Color is generated in one step after a short incubation and read within 30 min.

PROTOCOL 20.4. PROTEIN ESTIMATION BY BRADFORD METHOD

Principle
Coomassie blue undergoes a spectral change on binding to protein in acidic solution.

Outline
Dissolve protein, mix it with a color reagent, and read the O.D. after 10 min.

Materials
Sodium lauryl sulfate (SLS, SDS), 3.5 mM in water or 0.3 M NaOH

Coomassie Brilliant Blue G-250, 0.12 mM (0.01%), in 4.7% EtOH and 85% (w/v) phosphoric acid: Dissolve 100 mg of Coomassie blue in 50 ml of 95% EtOH; add 100 ml of 85% phosphoric acid; and dilute the solution to 1 l with water

Protein standard solution (e.g., BSA, 10 μg/ml); on first setting up the assay and at intervals of 1–2 months, perform a standard curve with BSA or a similar standard protein (1–50 μg/ml)

Protocol
1. Solubilize the protein (1–20 μg) or cells (around 1×10^6) in 100 μl of 3.5 mM sodium dodecyl sulfate in water or 0.3 M NaOH.
2. Add 100 μl of reagent blank (SLS), 100 μl of test protein solution, and 100 μl of BSA standard to separate, triplicate tubes.

3. Add 1.0 ml of Coomassie blue. Mix the solution, and let it stand for 10 min.
4. Read the tests and the BSA standard on a spectrophotometer at 595 nm against the reagent blank.

Variations. Reagents for the Bradford assay are available in kits from BioRad. Sulforhodamine B [Skehan et al., 1990] and the reagent in the kit available from Pierce (see Suppliers List) are very sensitive and are suitable for small numbers of cells ($\sim 1 \times 10^3$).

RATES OF SYNTHESIS

DNA Synthesis
Measurements of DNA synthesis are often taken to be representative of the amount of cell proliferation. (See also Metabolic Assays in Chapter 21.) [³H]-thymidine ([³H]-TdR) or [³H]-deoxycytidine is the usual precursor that is employed. Exposure to one of these precursors may be for short periods (0.5–1 h) for rate estimations or for longer periods (24 h or more) to measure accumulated DNA synthesis when the basal rate is low (e.g., in high-density cultures). [³H]-TdR should not be used for incubations longer than 24 h or at high specific activities, as radiolysis of DNA will occur, due to the short path length of β-emission (~ 1 μm) from decaying tritium; the β-emission releases energy within the nucleus and causes DNA strand breaks. If prolonged incubations or high specific activities are required, [¹⁴C]-TdR or ³²P should be used.

PROTOCOL 20.5. DNA SYNTHESIS

Materials
Sterile or Aseptically Prepared:
Cells at a suitable stage (grown in glass vials or test tubes if the hot 2 M PCA method is being used; otherwise, grown in conventional plastic)

[³H]-TdR, 0.4 MBq/ml ($\sim$10 μCi/ml), 100 μl for each ml of culture medium

Note. Some media—e.g., Ham's F10 and F12—contain thymidine, which will ultimately determine the specific activity of added [³H]-TdR. Allowance will have to be made for this factor when judging the amount of isotope to add. While 40 KBq/ml ($\sim$1.0 Ci/ml) of isotope may be sufficient for most media, 0.2 MBq/ml ($\sim$5 μCi/ml) should be used with F10 or F12.

Nonsterile:
Trichloracetic acid (TCA), 0.6 M (on ice), 6 ml for each 1 ml of culture
HBSS or PBSA, ice cold, 2 ml per ml of culture
Perchloric acid, 2 M, or 0.3 M NaOH with 35 mM sodium lauryl sulphate (SLS, SDS), 0.5 ml per 1 ml of culture
MeOH, 1 ml per 1 ml of culture
Scintillation vials
Scintillant (10× volume of perchloric acid or NaOH/SDS)

Outline
Label the cells with [³H]-TdR, extract the DNA, and determine the level of radioactivity by means of a scintillation counter.

Protocol
1. Grow the culture to the desired density (usually the midlog phase for maximum DNA synthesis or the plateau for density-limited DNA synthesis; see Contact Inhibition in Chapter 17).
2. Add [³H]-TdR, 40 KBq/ml (~1.0 μCi/ml), or 2 MBq/mmol (~50 Ci/mol) in HBSS.
3. Incubate the cells for 1–24 h as required.
4. Remove the radioactive medium carefully, and discard it into the proper container for liquid radioactive waste.
5. Wash the cells carefully with 2 ml of HBSS or PBSA, and add 2 ml of ice-cold 0.6 M TCA for 10 min. Fix the cells in MeOH first if they are loosely adherent. (See Protein Synthesis in this chapter.)
6. Repeat the trichloracetic acid wash twice, for 5 min each time.
7. Add 0.5 ml of 2 M perchloric acid, place the solution on a hot plate at 60°C for 30 min, and allow the solution to cool. Alternatively, add 0.5 ml of SLS in NaOH, and incubate the solution for 30 min at 37°C, or overnight at room temperature.
8. Collect the solubilized pellet, transfer it to the scintillant, and determine the radioactivity on a scintillation counter.

Note. If you are using the perchloric acid method, the residue may be dissolved in alkali for protein determination (see Protocol 20.4). Replicate cultures should be set up to provide cell counts to allow calculation of the DNA synthesis by cell number.

For suspension cultures, spin the cells at 100 g for 10 min in steps 4, 5, and 6. Mix the cells on a vortex mixer to disperse the pellet before each wash (in 0.6 M trichloracetic acid and 2 M perchloric acid) in step 7. Spin the cells after step 7 at 1,000 g for 10 min

to separate the precipitate (for protein estimation, if required) and the supernatant (for scintillation counting).

△ *Safety Note.* [³H]-thymidine represents a particular hazard, because it induces radiolytic damage in DNA, as mentioned previously. Take care to avoid accidental ingestion, injection, or inhalation of aerosols. When using [³H]-thymidine, work in a biohazard cabinet or on an open bench, but do not use horizontal laminar flow, or else aerosols will be blown directly at you.

Incubation with an isotopic precursor can provide several different types of data, depending on the incubation conditions and subsequent processing. Incubation followed by a short wash in ice-cold BSS and extraction into 0.6 M TCA will give a measure of the uptake and, if carried out over a few minutes' duration, will give a fair measure of the unidirectional flux. In experiments measuring uptake, the incorporation of precursors into acid-insoluble molecules, such as protein and DNA, is assumed to be minimal, due to the short incubation time, and only the acid-soluble pools are counted by extraction into cold 0.6 M TCA. In longer incubations (i.e., 2–24 h), it is assumed that the precursor pools become saturated. Equilibrium levels may be measured by cold trichloracetic acid extraction, and the incorporation into polymers may be measured by extraction with hot 2 M perchloric acid (DNA), cold dilute alkali (RNA), or hot 1.0 M NaOH (protein).

Protein Synthesis
Colorimetric assays measure the total amount of protein present at any one time. Sequential observations over a period of time may be used to measure the net protein accumulation or loss (i.e., protein synthesized − protein degraded), while the rate of protein synthesis may be determined by incubating cells with a radioisotopically labeled amino acid, such as [³H]-leucine or [³⁵S]-methionine, and measuring (e.g., by scintillation counting) the amount of radioactivity incorporated per 10⁶ cells or per milligram of protein over a set period of time.

△ *Safety Note.* Radioisotopes must be handled with care and according to local regulations governing permitted amounts, authorized work areas and disposal, and so forth.

PROTOCOL 20.6. PROTEIN SYNTHESIS

Materials
Sterile:
Cell culture, 1 × 10⁴–1 × 10⁶ cells, in, for example, a 24-well plate

[³H]-leucine, 2 MBq/ml (~50 μCi/ml) in a culture medium without serum (the specific activity is unimportant, as it will be determined by the leucine concentration in the medium)

Nonsterile:

Sodium lauryl sulfate, 35 mM in 0.3 N NaOH

Trichloracetic acid (TCA)

Scintillation vials

Eppendorf tubes

Scintillation fluid with a minimum of 10% water tolerance

Protocol

1. Incubate the culture to the required cell density.
2. Remove the culture from the incubator, and add a prewarmed solution of radioisotope in medium or BSS diluting 1:10 (e.g., 100 μl per 1 ml/well)
3. Return the culture to the incubator as rapidly as possible.
4. Incubate the culture for 4–24 h.

Note. Different proteins turn over at different rates. This protocol is not aimed at any specific subset of proteins, but at the total protein in rapidly proliferating cells. When assaying protein synthesis in a cell line for the first time, check that the rate of synthesis is linear over the chosen incubation time. A lag may be encountered if the amino acid pool is slow to saturate.

5. Remove the culture from the incubator, and withdraw the medium carefully from the wells into the proper container for radioactive liquid waste.
6. Wash the cells gently with cold HBSS or PBSA.

Note. Some monolayer cultures—particularly some loosely adherent continuous cell lines, such as HeLa-S₃—may detach during washing. In this case, remove the isotope and add methanol to fix monolayer. Leave the culture for 10 min, carefully remove the methanol, and dry the monolayer.

7. Replace the plates on ice. Add 0.6 M TCA, at 4°C, for 10 min, to remove any unincorporated precursor.
8. Repeat step 7 twice, but for only five minutes each time.
9. Wash the culture with MeOH, and then dry the plates.
10. Add 0.5 ml of 0.3 M NaOH, containing 35 mM SLS, and leave for 30 min at room temperature.
11. Mix the contents of each well, and transfer them to separate scintillation vials.

12. Add 5 ml of scintillant, and count on a scintillation counter.

Note. Biodegradable scintillants—e.g., Ecoscint—are preferred over toluene- or xylene-based scintillants, as the former are less toxic to handle and can be poured down the sink with excess water provided that the levels of radioactivity fall within the legal limits.

For suspension cultures, spin the cells at 1,000 *g* for 10 min in step 5 to remove the medium, and at steps 6, 7 and 8. Also, omit step 9.

Plates from step 9 can also be quantified directly on a phosphorimager.

PREPARATION OF SAMPLES FOR ENZYME ASSAY AND IMMUNOASSAY

As the amount of cellular material available from cultures is often too small for efficient homogenization, other methods of lysis are required to release soluble products and enzymes for assay. It is convenient either to set up cultures of the necessary cell number in sample tubes or multiwell plates (see Microtitration Assays in Chapter 21) or to trypsinize a bulk culture and place aliquots of cells into assay tubes. In either case, the cells should be washed in HBSS or PBSA to remove the serum, and lysis buffer should be added to the cells. The lysis buffer should be chosen to suit the assay, but, if the particular lysis buffer is unimportant, 0.15 M NaCl or PBSA may be used. If the product to be measured is membrane bound, add 1% detergent (Na deoxycholate, Nonidet P40) to the lysis buffer. If the cells are pelleted, resuspend them in the buffer by vortex mixing. Freeze and thaw the preparation three times by placing it in EtOH containing solid CO_2 (~−90°C) for 1 min and then in 37°C water for 2 min (longer for samples greater than 1 ml). Finally, spin the preparation at 10,000 *g* for 1 min (e.g., in an Eppendorf centrifuge), and collect the supernatant for assay.

Alternatively, the whole extract may be assayed for enzyme activity, and the insoluble material may be removed by centrifugation later if necessary.

CYTOMETRY

In Situ Labeling

Fluorescence labeling, either directly with a fluorescent dye (e.g., Hoechst 33258 for DNA) or with a conjugated antibody for detection of an antigen or molecular probe (see Protocols 27.2 and 27.1), can measure the amounts of enzyme, DNA, RNA, protein, or other cellular constituents *in situ* using a CCD camera. This process allows qualitative as well as quantitative analyses to be made, but is slow if large numbers of cells are to be scanned.

Flow Cytometry

Flow cytometry of a cell suspension [Vaughan and Miller, 1989], (see also Fluorescence Activated Cell Sorting in Chapter 14) while losing the relationship between cytochemistry and morphology, samples up to 1×10^7 cells, can measure multiple cellular constituents and activities [Kurtz and Wells, 1979; Klingel et al., 1994] and enables correlation of these measurements with other cellular parameters, such as cell size, lineage, DNA content or viability [Al-Rubeai, 1997].

REPLICATE SAMPLING

Since, in most cases, cultured cells can be prepared in a uniform suspension, the provision of large numbers of replicates for statistical analysis is often unnecessary. Usually, three replicates are sufficient, and for many simple observations (e.g., cell counts), duplicates may be sufficient.

Many types of culture vessel are available for replicate monolayer cultures (see Sampling and Analysis in Chapter 7), and the choice of which vessel to use is determined (1) by the number of cells required in each sample and (2) by the frequency or type of sampling. For example, if the incubation time is not a variable, replicate sampling is most readily performed in multiwell plates, such as microtitration plates and 24-well plates. If, however, samples are collected over a period of time (e.g., daily for 5 days), then the constant removal of a plate for daily processing may impair growth in the rest of the wells. In this case, the replicates are best prepared in individual tubes or 4-well plates. Plain glass or tissue-culture-treated plastic test tubes may be used, though Leighton tubes are superior, as they provide a flat growth surface. Alternatively, if the optical quality of the tubes is not critical, flat-bottomed glass specimen tubes and even glass scintillation vials may be good containers. If glass vials or tubes are used, they must be washed as tissue culture glassware (see Glassware in Chapter 10); they cannot be used for tissue culture after use with scintillant.

Sealing large numbers of vials or tubes can become tedious, so many people seal tubes with vinyl tape rather than screw caps. Such tape can also be color coded to identify different treatments.

Handling suspension cultures is generally easier than dealing with monolayer cultures, because the shape of the container and its surface charge are less important. Multiple sampling can also be performed on one culture when using suspension cultures. This sampling is done conveniently by sealing the bottle containing the culture with a silicone rubber membrane closure (Pierce) and then sampling using a syringe and needle. (Remember to replace the volume of culture removed with an equal volume of air.)

Data Acquisition

Analysis of data from cultured cells is not necessarily different from the way that data from any other system are handled. However the production of large amounts of data in cell culture experiments is relatively easy, particularly when dealing with microtitration plates, with which several hundred data points can be generated without a great deal of effort, or even several thousand by using robotics. Handling tissue culture-derived data will depend on how they are generated, on the scale of the experiment, and the number of parameters. While cell counting is the accepted method for generating data to construct a growth curve with one cell line under two or three sets of conditions, this does not lend itself to expansion to, say, determining multiple growth curves to measure the response of several cell lines to combinations of growth factors. In this example, if a colorimetric endpoint is chosen, then absorbance (e.g., MTT assay; see Protocol 21.4) or fluorescence emission (e.g., sulforhodamine [Boyd, 1989]) can be used, the plates analyzed on an enzyme-linked immunosorbent assay (ELISA) plate reader, and data reduction achieved by using the appropriate software.

A radiometric end point (e.g., [³H]-thymidine) used in conjunction with microtitration plates can be determined by simultaneous measurement of the whole plate in a microtitration plate scintillation counter (Packard, Pharmacia Wallac). Similarly, large numbers of sample tubes can be read automatically by γ or β counting, using robotic systems (Beckman).

Data Analysis

The ability to generate large amounts of data has been made possible by the creation of multiple replicate analysis systems, such as microtitration. As with ELISA analysis, the rate-limiting step is no longer the generation of the data, but is instead its analysis. It is important, therefore, when choosing a parameter of measurement to suit the culture system, that some thought be given to the amount of data to be generated and how that data will be handled. The easiest approach is to direct the data into a computer, either via a network or to a PC dedicated to that project. A number of companies now market computer programs that display and analyze data from microtitration plate assays (not just ELISA). These programs include titration curves, enzyme kinetics, and binding assays. With the necessary skills, and using a spreadsheet for importation of the data, you may be able to set up this type of program for yourself; alternatively, consultant advice is often available from the suppliers of plate readers (Molecular Devices, Alpha Laboratories, Lab Systems, Biorad).

CELL PROLIFERATION

Measurements of cell proliferation rates are often used to determine the response of cells to a particular stimulus or toxin. (See Protocols 20.7 and 20.8.) Quantitation of culture growth is also important in routine maintenance, as it is a crucial element for monitoring the consistency of the culture and knowing the best time to subculture (see Subculture in Chapter 12), the optimum dilution, and the estimated plating efficiency at different cell densities. Testing medium, serum, new culture vessels or substrates, and so forth all require quantitative assessment.

Experimental Design

Knowledge of the growth state of a culture, and its kinetic parameters, is critical in the design of cell culture experiments. Cultures vary significantly in many of their properties between the lag phase, the period of exponential growth (log phase), and the stationary phase (plateau). It is, therefore, important to take account of the status of the culture both at the initiation of an experiment and at the time of sampling, in order to determine whether is it proliferating or not and, if it is, the duration of the population doubling time (PDT) and the cell cycle time. Cells that have entered the plateau phase have a greatly reduced growth fraction, a different morphology, may be more differentiated, and may become polarized. They generally tend to secrete more extracellular matrix and may be more difficult to disaggregate. Generally, cell cultures are most consistent and uniform in the log phase and sampling at the end of the log phase gives the highest yield and greatest reproducibility.

It is also important to consider the effects of the duration of an experiment on the transition from one state to another. Adding a drug in the middle of the exponential phase and assaying later may give different results, depending on whether the culture is still in exponential growth when it is harvested or whether it has entered the plateau phase. Scheduling treatment and sampling requires a detailed knowledge of the parameters of the growth cycle.

Growth Cycle

As described previously (see Subculture in Chapter 12), following subculture, cells progress through a characteristic growth pattern of the lag phase; the exponential, or log phase; and the stationary, or plateau phase. (Fig. 20.4.) The log and plateau phases give vital information about the cell line, the PDT during exponential growth, and the maximum cell density achieved in the plateau phase (i.e., the saturation density). The measurement of the PDT is used to quantify the response of the cells to different inhibitory or stimulatory culture conditions, such as variations in nutrient concentration, hormonal effects, or toxic drugs. It is also a good monitor of the culture during serial passage and enables the calculation of cell yields and the dilution factor required at subculture.

Single time points are unsatisfactory for monitoring growth if the shape of the growth curve is not known. A reduced cell count after, say, 5 days could be caused by a reduced growth rate of some or all of the cells; a longer lag period, implying adaptation or cell loss (it is difficult to distinguish between the two); or a reduction in saturation density. This is not to say that growth curves are of no value. They can be useful for testing media, sera, growth factors and some drugs, and once the response being monitored is fully characterized and the type of response is predictable (e.g., a change in the PDT), then single time-point observations may be sufficient. Growth curves are particularly useful for the determination of the saturation density, although the amount of growth at saturation density should be assessed by the labeling index with [^{3}H]-thymidine. (See Protocol 20.11 and Contact Inhibition in Chapter 17.)

The PDT derived from a growth curve should not be confused with the cell cycle or generation time. The PDT is an average figure that applies to the whole population and it describes the net result of a wide range of division rates, including a rate of zero, within the culture. The *cell cycle time* or *generation time* is measured from one point in the cell cycle until the same point is reached again (see Cell Cycle Time in this chapter) and refers only to the dividing cells in the population, while the PDT is influenced by nongrowing and dying cells as well. PDTs vary from 12–15 h in rapidly growing mouse leukemias, like the L1210, to 24–36 h in many adherent continuous cell lines, and up to 60 or 72 h in slow-growing finite cell lines.

A new growth cycle begins each time the culture is subcultured and can be analyzed in more detail, as described in the following protocol.

PROTOCOL 20.7. GROWTH CURVE, MONOLAYER

Outline

Set up a series of cultures at three different cell concentrations, and count the cells at daily intervals until they reach the plateau phase.

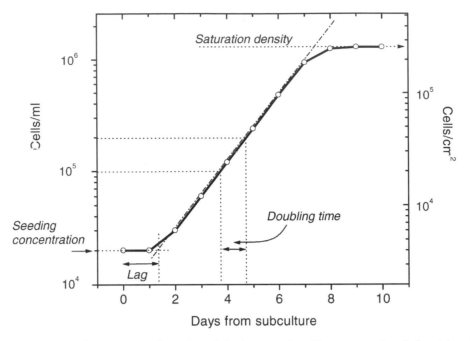

Fig. 20.4. Growth Curve. Semilog plot of the increase in cell concentration (left axis) and cell density (right axis) following subculture, from cultures sampled daily and counted. It should be possible to draw a straight line through the part of the plot that represents the exponential phase and derive the population-doubling time (PDT) from the middle region of this best-fit line. The time at the intercept with the seeding concentration is the lag time, and the saturation density is found at the plateau (at least three linear points without an increase in cell concentration) at the top end of the curve.

Materials

Sterile or Aseptically Prepared:
Monolayer cell culture
Trypsin, 0.25%, crude, with 10 mM EDTA
Growth medium, 100 ml, with 2 mM $NaHCO_3$
PBSA (prewash and for cell counting)
Plates, 24 well
Nonsterile:
Plastic box to hold the plates
CO_2 incubator or CO_2 supply to purge the box with 5% CO_2

Protocol

1. Trypsinize the cells as for a regular subculture. (See Protocol 12.2.)
2. Dilute the cell suspension to 1×10^5 cells/ml, 3×10^4 cells/ml, and 1×10^4 cells/ml, in 25 ml of medium for each concentration.
3. Seed three 24-well plates, one at each cell concentration, with 1 ml of the appropriate cell suspension per well. Add the cell suspension slowly from the center of the well, so that it does not swirl around the well. Similarly, do not shake the plate to mix the cells, as the circular movement of the medium will concentrate the cells in the middle of the well.

4. Place the plates in a humid CO_2 incubator or a sealed box gassed with 5% CO_2.
5. After 24 h, remove the plates from the incubator, and count the cells in three wells of each plate:
   (a) Remove the medium completely from the three wells containing cells to be counted
   (b) Add 1 ml of trypsin/EDTA to each of the three wells.
   (c) Incubate the plate for 15 min.
   (d) Disperse the cells in the trypsin/EDTA and transfer 0.4 ml of the suspension to 19.6 ml of PBSA for counting. Count the cells on an electronic cell counter.

Note. A hemocytometer may be used to count the cells, but may be difficult to use for lower cell concentrations. If you use a hemocytometer, reduce the volume of trypsin to 0.1 ml, and disperse the cells carefully without frothing the trypsin using a pipettor. Transfer the cells to the hemocytometer.

6. Return the plate to the incubator as soon as the cell samples in trypsin are removed. The plate must be out of the incubator for the minimum

length of time, in order to avoid disrupting normal cell growth.

7. Repeat sampling at 48 and 72 h, as in steps 5 and 6.
8. Change the medium at 72 h, or sooner, if indicated by a drop in the pH. (See Protocol 12.1.)
9. Continue sampling daily for rapidly growing cells (i.e., cells with a PDT of 12–14 h), but reduce the frequency of sampling to every 2 d for slowly growing cells (i.e., cells with a PDT >24 h) until the plateau phase is reached.
10. Keep changing the medium every 1, 2, or 3 d, indicated by the pH.

Analysis. (1) Calculate the cell number per well, per ml of culture medium (see Fig. 20.4), and per cm² of available growth surface in the well. (Stain one or two wells (see Staining in Chapter 15), at each density in order to determine whether the distribution of cells in the wells is uniform and whether the cells are growing up the sides of the well.) (2) Plot the cell density (cells per cm²) and the cell concentration (cells per ml), both on a log scale, against time on a linear scale. (See Fig. 20.4.) (3) Determine the lag time, PDT, and plateau density. (See Analysis of the Growth Cycle in this chapter and Fig. 20.4.) (4) Establish the appropriate starting density for routine passage. Repeat the growth curve at intermediate cell concentrations if necessary. (5) Compare growth curves under different conditions (Fig. 20.5), and try to interpret the data (see the legend for Fig. 20.5.)

Variations. Use different culture vessels—e.g., 25-cm² flasks, although more cells and medium will be required—or flat-bottomed glass sample tubes. Individual tubes have the advantage that the rest of the tubes are not disturbed when samples are removed for counting. The frequency of changing the medium may also be altered. This type of assay is useful for comparing different media or supplements.

Suspension Cultures

A growth curve can also be generated from cells growing in suspension, although the medium is not replaced in such cases.

PROTOCOL 20.8. GROWTH CURVE, SUSPENSION

Outline

Set up a series of cultures at three different cell concentrations, and count the cells daily until they reach the plateau phase.

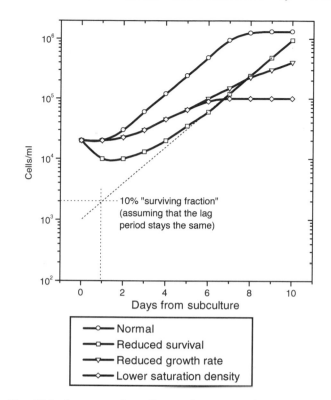

Fig. 20.5. Interpretation of Growth Curves. Changes in the shape of a growth curve can be interpreted in a number of different ways, but the labels in the key of this plot indicate what would normally be deduced from these curves.

Materials

Sterile or Aseptically Prepared:
Suspension cell culture
Growth medium, 100 ml
PBSA (for cell counting)
Plates, 24-well
Nonsterile:
Plastic box to hold the plates
CO_2 incubator or 5% CO_2 supply to purge the box with 5% CO_2.

Protocol

1. Add the cell suspension in growth medium to wells at a range of concentrations as for monolayer cultures. (See Protocol 20.7.)
2. Sample 0.4 ml of the culture from triplicate wells at intervals, ensuring that the cells are well mixed and completely disaggregated

Analysis. Calculate the cell concentration per sample, and plot it on a log scale against time on a linear scale as for monolayer growth. (See the analysis for Protocol 20.7.)

Variations. Seed two 75-cm² flasks with 20 ml of cell suspension in growth medium for each cell concentration, and sample 0.4 ml of culture from each flask daily or as required. Mix the culture well before sampling it, and keep the flasks out of the incubator for the minimum length of time. Do not feed the cultures during the growth curve.

Analysis of the Growth Cycle

The growth cycle (see Fig. 20.4) is conventionally divided into three phases:

The Lag Phase. This phase is the time following subculture and reseeding during which there is little evidence of an increase in the cell number. It is a period of adaptation during which the cell replaces elements of the cell surface and extracellular matrix lost during trypsinization, attaches to the substrate, and spreads out. During spreading, the cytoskeleton reappears, an integral part of the spreading process. The amount of enzymes, such as DNA polymerase, increases, followed by the synthesis of new DNA and structural proteins. Some specialized cell products may disappear and not reappear until the cessation of cell proliferation at a high cell density.

The Log Phase. This phase is the period of exponential increase in the cell number following the lag period and terminating in one or two population doublings after confluence is reached. The length of the log phase depends on the seeding density, the growth rate of the cells, and the density that inhibits cell proliferation. In the log phase, the growth fraction is high (usually 90–100%), and the culture is in its most reproducible form. It is the optimal time for sampling, since the population is at its most uniform and the viability is high. However, the cells are randomly distributed in the cell cycle and, for some purposes, may need to be synchronized. (See Cell Synchrony in Chapter 26.)

The Plateau Phase. Toward the end of the log phase, the culture becomes confluent—i.e., all of the available growth surface is occupied and all of the cells are in contact with surrounding cells. Following confluence, the growth rate of the culture is reduced, and in some cases, cell proliferation ceases almost completely after one or two further population doublings. At this stage, the culture enters the plateau, or stationary, phase, and the growth fraction falls to between 0 and 10%. The cells may become less motile; some fibroblasts become oriented with respect to one another, forming a typical parallel array of cells. "Ruffling" of the plasma membrane is reduced, and the cell both occupies less surface area of substrate and presents less of its own surface to the medium.

There may be a relative increase in the synthesis of specialized versus structural proteins, and the constitution and charge of the cell surface may be changed.

The cessation of motility, membrane ruffling, and growth following contact of cells at confluence was originally described by Abercrombie and Heaysman [1954] and was designated *contact inhibition*. It has since been realized that the reduction in the growth of normal cells after confluence is reached is not due solely to contact, but may also involve reduced cell spreading [Stoker et al., 1968; Folkman and Moscona, 1978]; depletion of nutrients; and, particularly, growth factors [Dulbecco and Elkington, 1973; Stoker, 1973; Westermark and Wasteson, 1975] in the medium [Holley et al., 1978]. The term *density limitation* (of cell proliferation) has therefore been used instead, to remove the implication that cell–cell contact is the major limiting factor [Stoker and Rubin, 1967], and the term *contact inhibition* is best reserved for those events resulting directly from cell contact (i.e., reduced cell motility and membrane ruffling, resulting in the formation of a strict monolayer and orientation of the cells with respect to each other).

Cultures of normal simple epithelial and endothelial cells stop growing after reaching confluence and remain as a monolayer. Most cultures, however, with regular replenishment of medium, will continue to proliferate (although at a reduced rate) well beyond confluence, resulting in multilayers of cells. Human embryonic lung and adult skin fibroblasts, which express contact inhibition of movement, will continue to proliferate, laying down layers of collagen between the cell layers until multilayers of six or more cells can be reached under optimal conditions [Kruse et al., 1970]. These fibroblasts still retain an ordered parallel array, however. Therefore, the terms "plateau" and "stationary" are not strictly accurate and should be used with caution.

Cultures that have transformed spontaneously or have been transformed by virus or chemical carcinogens will usually reach a higher cell density in the plateau phase than their normal counterparts [Westermark, 1974]. (Fig. 20.6.) This higher cell density is accompanied by a higher growth fraction and the loss of density limitation of cell proliferation. The plateau phase for these cultures is an equilibrium between cell proliferation and cell loss. These cultures are often *anchorage independent* for growth—i.e., they can easily be made to grow in suspension. (See Contact Inhibition in Chapter 17).

Derivatives from the Growth Cycle

The construction of a growth curve from cell counts performed at intervals after subculture enables the measurement of a number of parameters that should

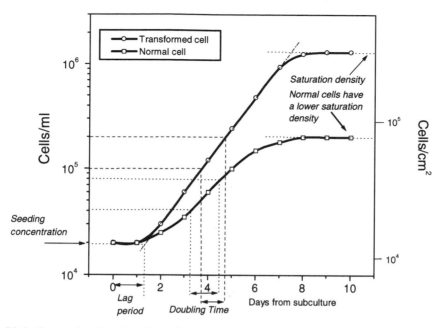

Fig. 20.6. Saturation Density. Transformation produces an increase in the saturation density of transformed cells, relative to that found in the equivalent normal cells. This increase is often accompanied by a shorter PDT.

be found to be characteristic of the cell line under a given set of culture conditions. The first of these parameters is the duration of the lag period, or *lag time*, obtained by extrapolating a line drawn through the points for the exponential phase until it intersects the seeding concentration (see Fig. 20.4) and then reading off the elapsed time since seeding equivalent to that intercept. The second parameter is the *population doubling time* (PDT)—i.e., the time taken for the culture to increase twofold in the middle of the exponential, or log, phase of growth. This parameter should not be confused with the generation time or cell cycle time (see Cell Cycle Time in this chapter), which are determined by measuring the transit of a population of cells from one point in the cell cycle until they return to the same point.

The last of the commonly derived measurements from the growth cycle are the *plateau level* and *saturation density*. The plateau level is the cell concentration (cells/ml of medium) in the plateau phase and is dependent on the cell type and the frequency with which the medium is replenished. The saturation density is the density of the cells (cells/cm² of growth surface) in the plateau phase. Saturation density and plateau level are difficult to measure accurately, as a steady state is not easily achieved in the plateau phase. Ideally, the culture should be perfused, to avoid nutrient limitation or growth factor depletion, but a reasonable compromise is to grow the cells on a restricted area, say a 15-mm diameter coverslip or filter well, in a 90-mm diameter Petri dish with 20 ml of medium that

is replaced daily (see Protocol 17.2). Under these conditions, the limitation of growth by the medium is minimal, and the cell density exerts the major effect. A count of the cells under these conditions is a more accurate and reproducible measurement than a cell count in plateau under conventional culture conditions. Note that the term "plateau" does not imply the complete cessation of cell proliferation, but instead represents a steady state in which cell division is balanced by cell loss.

With normal cells, a steady state may be achievable by not replenishing the growth factors in the medium. In this case, the cells are seeded and grown and the plateau reached without changing the medium. Clearly, the conditions used to attain the plateau phase must be carefully defined.

PLATING EFFICIENCY

Colony formation at low cell density, or *plating efficiency*, is the preferred method for analyzing cell proliferation and survival. (See Protocol 21.3.) This technique reveals differences in the growth rate within a population and distinguishes between alterations in the growth rate (colony size) and cell survival (colony number). It should be remembered, however, that cells may grow differently as isolated colonies at low cell densities. In this situation fewer cells will survive, even under ideal conditions, and all cell interaction is lost until the colony starts to form. Heterogeneity in clonal growth rates reflects differences in the capacity for cell proliferation between lineages within a population, but these differences are not necessarily expressed in an interacting monolayer at higher densities, when cell communication is possible.

When cells are plated out as a single-cell suspension at low cell densities (2–50 cells/cm²), they grow as discrete colonies. (See Dilution Cloning in Chapter 13 and Plate 12.) The number of these colonies can be used to express the plating efficiency:

$$\frac{\text{No. of colonies formed}}{\text{No. of cells seeded}} \times 100 = \text{Plating efficiency.}$$

If it can be confirmed that each colony grew from a single cell, then this term becomes the *cloning efficiency*. Measurements of the plating efficiency are derived by counting the number of colonies over a certain size (usually around 50 cells) growing from a low inoculum of cells, and this term should not be used for the recovery of adherent cells after seeding at higher cell densities (e.g., 2×10^4 cells/cm²). Survival at higher densities is more properly referred to as the *seeding efficiency*:

$$\frac{\text{No. of cells recovered}}{\text{No. of cells seeded}} \times 100 = \text{Seeding efficiency.}$$

It should be measured at a time when the maximum number of cells has attached, but before mitosis starts. This time is a difficult point to define, as the window between maximum cell attachment and the initiation of mitosis may be quite narrow, and the events may even overlap; however, it still provides a crude measurement of recovery in, for example, routine cell freezing or primary culture.

PROTOCOL 20.9. PLATING EFFICIENCY

Outline
Seed the cells at a low density and incubate the cultures until colonies form; stain and count the colonies. (See Protocol 13.1.)

Materials
Sterile:
Growth medium
Trypsin, 0.25%, crude
Petri dishes, 6 or 9 cm
Tubes, or universal containers, for dilution
Nonsterile:
Hemocytometer or electronic cell counter
Fixative: anhydrous methanol
PBSA
Stain: crystal violet

Filter funnel and filter paper (to recycle the stain)

Protocol
1. Trypsinize the cells (see Protocol 12.2) to produce a single-cell suspension.
2. While the cells are trypsinizing,
   (a) Number the dishes on the side of the base;
   (b) Measure out medium for the dilution steps. The final step should be performed at the cloning concentration (20–1,000 cells/ml, depending on the expected plating efficiency). There should be more than enough medium for three replicates in the final dilution.
3. When the cells round up and start to detach,
   (a) Disperse the monolayer in medium containing serum or a trypsin inhibitor;
   (b) Count the cells;
   (c) Dilute the cells to the desired seeding concentration.
4. Seed the Petri dishes with the requisite amount of medium containing cells.
5. Place the dishes in a transparent plastic box in a humid CO_2 incubator.
6. Incubate the dishes for 1–3 weeks. (If no colonies have appeared by the end of the third week, it is unlikely that they will appear at all.)
7. Stain the cells with crystal violet:
   (a) Remove the medium from the cells;
   (b) Rinse the cells with PBSA, and discard the rinse;
   (c) Fix the cells in methanol for 10 min;
   (d) Discard the methanol, and add crystal violet, neat, 2–3 ml per 5-cm dish, or 10 ml per 9-cm dish, making sure that the whole of the growth surface is covered;
   (e) Stain for 10 min;
   (f) Remove the stain, and return it to the stock bottle via a filter.

Analysis. Count the colonies and the calculate plating efficiency. (See calculation in introduction to this section.) Magnifying viewers make counting the colonies easier. It will be necessary to define a threshold, above which colonies will be counted. If the majority of the colonies are between a hundred and a few thousands, then set the threshold at 50 cells per colony. However, if the colonies are very small (<100 cells), then set the threshold at 16 cells per colony. Below 16 cells, equivalent to 4 cell consecutive divisions, it would be hard to presume continued cell proliferation. The size distribution of the colonies may also be determined (e.g., to assay the growth-promoting ability of a test medium or serum; see Culture Testing in Chapter 10) by counting the number of cells per colony by eye or estimating it by densitometry. To do so, after fixing and staining the colonies with crystal violet, measure absorption on a densitometer [McKeehan et al., 1977], gel scanner or automatic colony counter (see next section).

Automatic Colony Counting

If the colonies are uniform in shape and quite discrete, they may be counted on an *automatic colony counter* (see Sources of Materials—Colony Counters in Trade Index), which scans the plate with a CCD camera and analyzes the image to give an instantaneous readout of the number of colonies. A size discriminator gives an analysis of the size of the colonies, based on the average colony diameter but this is not always proportional to the cell number, as cells may pile up in the center of a colony.

Though expensive, these instruments can save a great deal of time and make colony counting more objective. However, they do not work well with colonies that overlap more than ~20% or that have irregular outlines.

LABELING INDEX

If a culture is labeled with [³H]-TdR, then cells that are synthesizing DNA will incorporate the isotope. (See DNA Synthesis in this chapter.) The percentage of labeled cells, determined by autoradiography [e.g., Westermark, 1974; Macieira-Coelho, 1973; see Protocol 26.3 and Fig. 26.2a, is known as the labeling index (LI). Measurement of the LI after a 30-min labeling period with [³H]-TdR shows a large difference between exponentially growing cells (LI = 10–20%) and cells at the plateau phase (LI ~ 1%). Since the LI is very low in the plateau phase, the duration of the [³H]-TdR labeling period may have to be increased to 24 h. With this length of labeling period, normal cells can be shown to have a lower LI with [³H]-TdR than that of neoplastic cells. (See Protocol 17.2.)

PROTOCOL 20.10. LABELING INDEX WITH [³H]-THYMIDINE

Outline
Grow cells to an appropriate density. Label the cells with [³H]-thymidine for 30 min. Wash and fix the cells. Remove any unincorporated precursor from the cells and prepare autoradiographs.

Materials
Sterile or Aseptically Prepared:
Culture of cells
Growth medium
Trypsin, 0.25%, crude
PBSA
[³H]-thymidine, 2.0 MBq/ml (~50 μCi/ml), 75 GBq/mmol (~2 Ci/mmol)

△ *Safety Note.* Handle [³H]-thymidine with care, as it is radioactive! Follow local guidelines for its use. (See Radiation in Chapter 6.)

Multiwell plate(s) containing 13-mm Thermanox coverslips (Lux)
Nonsterile:
Hemocytometer or electronic cell counter.
PBSA
Acetic methanol (1 part glacial acetic acid to 3 parts methanol), ice cold, freshly prepared
Microscope slides
Mountant (e.g., DPX or Permount)
Trichloracetic acid (TCA), 0.6 M, ice cold
Deionized water
Methanol

Protocol
1. Set up the cultures at 2 × 10⁴ cells/ml–5 × 10⁴ cells/ml in 24-well plates containing coverslips.
2. Allow the cells to attach, start to proliferate (48–72 h), and grow to the desired cell density.
3. Add [³H]-thymidine to the medium, 100 KBq/ml (~5 μCi/ml), and incubate for the cultures 30 min.

Note. Some media—e.g., Ham's F10 and F12—contain thymidine. (See Tables 8.3, 9.1, and 9.2.) In these cases, the concentration of radioactive thymidine must be increased to give the same specific activity in the medium. In prolonged exposure to high specific activity, [³H]-thymidine causes radiolysis of the DNA. This effect can be reduced by using [¹⁴C]-thymidine, or [³H]-thymidine of a lower specific activity.

4. Remove the labeled medium, and discard it into a designated container for radioactive waste.
5. Wash the coverslips three times with PBSA. Lift the coverslips off the bottom of the wells (but not right out of them) at each wash in order to allow removal of the isotope from underneath.
6. Add 1:1 PBSA: acetic methanol, 1 ml per well, and then remove it immediately.
7. Add 1 ml of acetic methanol at 4°C to each well, and leave the cultures for 10 min.
8. Remove the coverslips, and dry them with a fan.
9. Mount the coverslip on a microscope slide with the cells uppermost.
10. Leave the mountant to dry overnight.

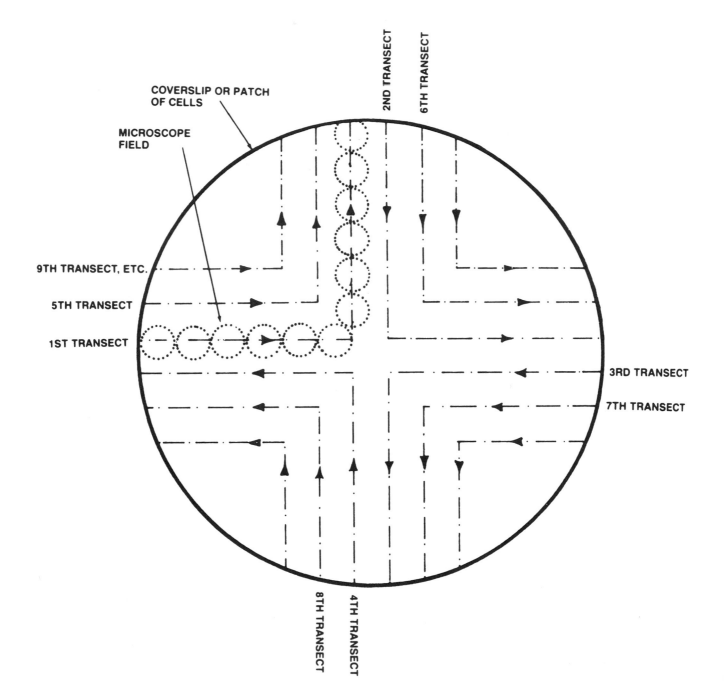

Fig. 20.7. Scanning Slides or Dishes. Scanning pattern for the analysis of cytological preparations on slides or dishes. Each dotted circle represents one microscope field, and the large circle represents the extent of the specimen (e.g., a coverslip, culture dish, or well, or a spot of cells on a slide). Guide lines can be drawn with a nylon-tipped pen.

11. Place the slides in 0.6 M TCA at 4°C in a staining dish, and leave them for 10 min. Replace the TCA twice during this extraction, thereby removing unincorporated precursors.
12. Rinse the slides in deionized water, then in methanol, and dry the slides.
13. Prepare an autoradiograph. (See Protocol 26.3.)

Analysis. Count the percentage of labeled cells. To cover a representative area, follow the scanning pattern illustrated in Fig. 20.7.

Growth Fraction

If cells are labeled with [³H]-thymidine for varying lengths of time up to 48 h, then the plot of the LI against time increases rapidly over the first few hours and then flattens out to a very low gradient, almost a plateau. (Fig. 20.8.) The level of this plateau, read against the vertical axis, is the growth fraction of the culture—i.e., the proportion of the cells in cycle at the time of labeling.

PROTOCOL 20.11. DETERMINATION OF GROWTH FRACTION

Outline
Label the culture continuously for 48 h, sampling at intervals for autoradiography.

Protocol
Follow Protocol 20.10, except that at step 3, incubation should be carried out for 15 min, 30 min, and 1, 2, 4, 8, 24, and 48 h.

Analysis. Count the number of labeled cells as a percentage of the total number of cells, using the scanning pattern from Protocol. 20.10. Plot the LI against time. (See Fig. 20.8.)

Note. Autoradiographs with ³H can be prepared only when the cells remain as a monolayer. If they form a multilayer, then they must be trypsinized after labeling, and slides must be prepared by the drop technique (see Protocol 15.9, without the hypotonic step) or by cytocentrifugation (see Protocol 15.5), as the energy of β-emission from ³H is too low to penetrate an overlying layer of cells.

The LI can also be determined by labeling cells with BUdR, which becomes incorporated into DNA. This effect can be detected subsequently by immunostaining with an anti-BUdR antibody (Dako). Results from this method are generally in agreement with the results of using [³H]-thymidine [Khan et al., 1991].

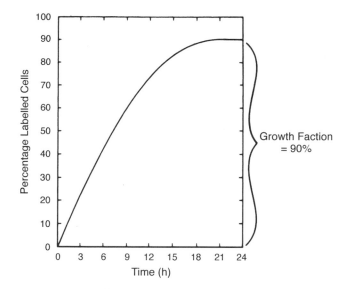

Fig. 20.8. Growth Fraction. To determine the growth fraction, cells are labeled continuously with [³H]-thymidine, and the percentage of labeled cells is determined at intervals by autoradiography. (See text.)

Mitotic Index

The mitotic index is the fraction or percentage of cells in mitosis and is determined by counting mitoses in stained cultures as a proportion of the whole population of cells. The scanning pattern should be as in Fig. 20.7.

Division Index

A number of antibodies, such as Ki67 or anti-PCNA, are able to stain cells in the division cycle. These antibodies are raised against proteins expressed during the cell cycle, but that are not expressed in resting cells. Some of the antibodies are directed against DNA polymerase, but the epitopes for others are not known.

The cells are stained by immunofluorescence or immunoperoxidase (see Protocol 15.12), and the proportion of stained cells is determined. This figure gives a higher index than that of either mitotic counting or DNA labeling, as the whole cycle is responsive to the stain. It gives a particularly useful indication of the growth fraction as well.

CELL CYCLE TIME

To determine the length of the cell cycle (generation time) and its constituent phases, cells are labeled continuously with BUdR, and the incorporation of the label is detected at intervals after labeling by immunofluorescence microscopy or flow cytometry. For flow cytometry, the cells are also stained with propidium iodide and analyzed for BUdR incorporation versus DNA content, to follow the progression of cells around the cycle [Dolbeare and Selden, 1994; Poot et al., 1994].

CELL MIGRATION

Cells in culture, particularly fibroblasts, are motile and can migrate significant distances across the substrate, dependent on cell density and the presence of stimu-lants such as growth factors. Motility is evidenced by ruffling of the cell membrane as visualized by time-lapse video. The amount of motility is difficult to quantify, but the quantification of cell migration can be achieved by detailed analysis of time-lapse video sequences. (See Protocol 26.5.), or by image analysis of tracks made by the cell's phagocytosis in dishes coated with colloidal gold [Kawa et al., 1997].

Migration can also be assayed by the movement of cells through a porous membrane, as in chemotaxis assays in a Boyden chamber [Schor, 1994] or invasion assays in a filter well (see Invasion in Chapter 17).

CHAPTER 21

Cytotoxicity

INTRODUCTION

Once a cell is explanted from its normal *in vivo* environment, the question of viability, particularly in the course of experimental manipulations, becomes fundamental. Previous chapters have dealt with the status of the cultured material relative to the tissue of origin and how to quantify changes in growth and phenotypic expression. However, none of these data are acceptable unless the great majority of the cells are shown to be viable. Furthermore, many experiments carried out *in vitro* are for the sole purpose of determining the potential cytotoxicity of the compounds being studied, either because the compounds are being used as pharmaceuticals or cosmetics and must be shown to be nontoxic, or because they are designed as anticancer agents and cytotoxicity may be crucial to their action.

New drugs, cosmetics, food additives, and so on go through extensive cytotoxicity testing before they are released for use by the public. This testing usually involves a large number of animal experiments, although, in Europe, these experiments will be subject to new legislation, due to be introduced in 1999, but deferred. There is much pressure, both humane and economic, to perform at least part of cytotoxicity testing *in vitro*. The introduction of specialized cell lines and interactive organotypic cultures, and the continued use of long-established cultures, may make this a reasonable proposition.

Toxicity is a complex event *in vivo*, where there may be direct cellular damage, as with a cytotoxic anticancer drug, physiological effects, such as membrane transport in the kidney or neurotoxicity in the brain, inflammatory effects, both at the site of application and at other sites, and other systemic effects. Currently, it is difficult to monitor systemic and physiological effects *in vitro*, so most assays determine effects at the cellular level, or *cytotoxicity*. Definitions of cytotoxicity vary, depending on the nature of the study and whether cells are killed or simply have their metabolism altered. While an anticancer agent may be required to kill cells, demonstrating the lack of toxicity of other pharmaceuticals may require a more subtle analysis of minor metabolic changes or an alteration in cell–cell signaling, such as might give rise to an inflammatory or allergic response.

All of these assays oversimplify the events that they measure and are employed because they are cheap, easily quantified, and reproducible. However, it has become increasingly apparent that they are inadequate for modern drug development, which requires greater emphasis on molecular target specificity and precise metabolic regulation. Gross tests of cytotoxicity are still required, but there is a growing need to supplement them with more subtle tests of metabolic perturbation. Perhaps the most obvious of these tests is the induction of an inflammatory or allergic response, which need not imply cytotoxicity of the allergen or inflammatory agent, but is still one of the hardest results to demonstrate *in vitro*.

It is not within the scope of this text to define all of the requirements of a cytotoxicity assay, many of which may be quite specialized, so instead I will concentrate on those aspects that influence cell growth or survival. Cell growth is generally taken to be the regenerative potential of cells, as measured by clonal growth (e.g., in a plating efficiency assay; see Protocol 20.9), net change in population size (e.g., in a growth curve; see Protocol 20.7 and 20.8), or a change in cell mass (total protein or DNA) or metabolic activity (e.g., DNA, RNA, or protein synthesis; MTT reduction).

IN VITRO LIMITATIONS

It is important that any *in vitro* measurement can be interpreted in terms of the *in vivo* response of the cells, or at least that the differences which exist between *in vitro* and *in vivo* measurements are clearly understood.

Pharmacokinetics. The measurement of toxicity *in vitro* is generally a cellular event. For example, it would be very difficult to re-create the complex pharmacokinetics of drug exposure *in vitro*, and between *in vitro* and *in vivo* experiments, there usually are significant differences in exposure time to and concentration of the drug, rate of change of the concentration, metabolism, tissue penetration, clearance, and excretion. Although it may be possible to simulate these parameters—e.g., using multicellular tumor spheroids for drug penetration—most studies concentrate on a direct cellular response, thereby gaining simplicity and reproducibility.

Metabolism. Many nontoxic substances become toxic after being metabolized by the liver; in addition, many substances that are toxic *in vitro* may be detoxified by liver enzymes. For testing *in vitro* to be accepted as an alternative to animal testing, it must be demonstrated that potential toxins reach the cells *in vitro* in the same form as they would *in vivo*. This proof may require additional processing by purified liver microsomal enzyme preparations [McGregor et al., 1988], coculture with activated hepatocytes [Guillouzo, 1989; Frazier, 1992], or genetic modification of the target cells with the introduction of genes for metabolizing enzymes under the control of a regulatable promoter [Macé et al., 1994].

Tissue and Systemic Responses. The nature of the response must also be considered carefully. A toxic response *in vitro* may be measured by changes in cell survival (see Protocol 21.3) or metabolism (see Metabolic Assays in this chapter), while the major problem *in vivo* may be a tissue response (e.g., an inflammatory reaction, fibrosis, kidney transport) or a systemic response (e.g., pyrexia, vascular dilatation). For *in vitro* testing to be more effective, models of these responses must be constructed, perhaps utilizing organotypic

cultures reassembled from several different cell types and maintained in the appropriate hormonal milieu.

It should not be assumed that complex tissue and even systemic reactions cannot be simulated *in vitro*. Assays for inflammatory responses, teratogenic disorders, and neurological dysfunctions may be feasible *in vitro*, given a proper understanding of cell–cell interaction and the interplay of endocrine hormones with local paracrine and autocrine factors.

NATURE OF THE ASSAY

The choice of assay will depend on the agent under study, the nature of the response, and the particular target cell. Assays can be divided into five major classes:

(1) *Viability:* An immediate or short-term response, such as an alteration in membrane permeability or a perturbation of a particular metabolic pathway.
(2) *Survival:* The long-term retention of self-renewal capacity (5–10 generations or more).
(3) *Metabolic:* Assays, usually microtitration based, of intermediate duration that can either measure a metabolic response (e.g., dehydrogenase activity; DNA, RNA, or protein synthesis) at the time of, or shortly after, exposure, or measure the same parameter two or three population doublings after exposure, when it is more likely to reflect cell growth potential and/or survival.
(4) *Transformation:* Survival in an altered state (e.g., a state expressing genetic mutation(s) or malignant transformation).
(5) *Irritancy:* A response analogous to inflammation, allergy, or irritation *in vivo*; as yet difficult to model *in vitro*, but may be possible to assay by monitoring cytokine release in organotypic cultures.

Viability

Viability assays are used to measure the proportion of viable cells following a potentially traumatic procedure, such as primary disaggregation, cell separation, or freezing and thawing.

Most viability tests rely on a breakdown in membrane integrity that is determined by the uptake of a dye to which the cell is normally impermeable (e.g., trypan blue, erythrosin, or naphthalene black) or the release of a dye normally taken up and retained by viable cells (e.g., diacetyl fluorescein or neutral red). However, this effect is immediate and does not always predict ultimate survival. Furthermore, dye exclusion tends to overestimate viability—e.g., 90% of cells thawed from liquid nitrogen may exclude trypan blue, but only 60% prove to be capable of attachment 24 h later.

PROTOCOL 21.1. ESTIMATION OF VIABILITY BY DYE EXCLUSION

Principle
Viable cells are impermeable to naphthalene black, trypan blue, and a number of other dyes [Kaltenbach et al., 1958].

Outline
Mix a cell suspension with stain, and examine it by low-power microscopy.

Materials
Sterile or Aseptically Prepared:
Cells
Growth medium
0.25% trypsin
PBSA
Nonsterile:
Hemocytometer
Viability stain (e.g., 0.4% trypan blue or 1% naphthalene black in PBSA or HBSS)
Pasteur pipettes
Microscope
Tally counter

Protocol
1. Prepare a cell suspension at a high concentration ($\sim 10^6$ cells/ml) by trypsinization or by centrifugation and resuspension.
2. Take a clean hemocytometer slide and fix the coverslip in place. (See Protocol 20.1.)
3. Mix one drop of cell suspension with one drop (trypan blue) or four drops (naphthalene black) of stain.
4. Load the counting chamber of the hemocytometer. (See Protocol 20.1.)
5. Leave the slide for 1–2 min (Do not leave them for a longer period of time, or else viable cells will deteriorate and take up the stain.)
6. Place the slide on the microscope, and use a 10× objective to look at the counting grid (see Fig. 20.1).
7. Count the total number of cells and the number of stained cells.
8. Wash the hemocytometer, and return it to its box.

Analysis
Calculate the percentage of unstained cells. This figure is the percentage viability in this method. If the respective volumes of cell suspension and stain are measured accurately at step 3, then this method of viability determination can be incorporated into Protocol 20.1.

PROTOCOL 21.2. ESTIMATION OF VIABILITY BY DYE UPTAKE

Principle
Viable cells take up diacetyl fluorescein and hydrolyze it to fluorescein, to which the cell membrane of live cells is impermeable [Rotman and Papermaster, 1966]. Live cells fluoresce green; dead cells do not. Nonviable cells may be stained with propidium iodide and subsequently fluoresce red. Viability is expressed as the percentage of cells fluorescing green. This method may be applied to CCD analysis or flow cytometry. (See Cytometry in Chapter 20.)

Outline
Stain a cell suspension in a mixture of propidium iodide and diacetyl fluorescein, and examine the cells by fluorescence microscopy or flow cytometry.

Materials
Sterile or Aseptically Prepared:
Single-cell suspension
Fluorescein diacetate, 10 μg/ml, in HBSS
Propidium iodide, 500 μg/ml
Nonsterile:
Fluorescence microscope
Filters:
    fluorescein: excitation 450/590 nm, emission LP 515 nm
    propidium iodide: excitation 488 nm, emission 615 nm

Protocol
1. Prepare the cell suspension as for dye exclusion (see Protocol 21.1), but in medium without phenol red.
2. Add the fluorescent dye mixture at a proportion of 1:10 to give a final concentration of 1 μg/ml of diacetyl fluorescein and 50 μg/ml of propidium iodide.
3. Incubate the cells at 37°C for 10 min.
4. Place a drop of the cells on a microscope slide, add a coverslip, and examine the cells by fluorescence microscopy.

Analysis
Cells that fluoresce green are viable, while those that fluoresce red are nonviable. Viability may be expressed as the percentage of cells that fluoresce green as a proportion of the total number of cells.

The stained cell suspension can also be analyzed by flow cytometry. (See Flow Cytometry in Chapter 20.)

Neutral-Red Uptake. Living cells take up neutral red, 40 μg/ml in culture medium, and sequester it in the lysosomes. However, neutral red is not retained by nonviable cells. Uptake of neutral red is quantified by fixing the cells in formaldehyde and solubilizing the stain in acetic ethanol, allowing the plate to be read on an Elisa plate reader at 570 nm [Borenfreund et al., 1990; Babich and Borenfreund, 1990]. Neutral red tends to precipitate, so the medium with stain is usually incubated overnight and centrifuged before use. This assay does not measure the total number of cells, but it does show a reduction in the absorbance related to loss of viable cells and is readily automated.

Survival

While short-term tests are convenient and usually are quick and easy to perform, they reveal only cells that are dead (i.e., permeable) at the time of the assay. Frequently, however, cells that have been subjected to toxic influences (e.g., irradiation, antineoplastic drugs) show an effect several hours, or even days, later. The nature of the tests required to measure viability in these cases is necessarily different, since by the time the measurement is made, the dead cells may have disappeared. Therefore, long-term tests are used to demonstrate survival rather than short-term toxicity, which may be reversible. Survival implies the retention of regenerative capacity and is usually measured by plating efficiency.

PROTOCOL 21.3. CLONOGENIC ASSAY

Principle
Plating efficiency (see Protocol 20.9) measures survival by demonstrating proliferative capacity for several cell generations, provided that the cells plate with a high-enough efficiency that the colonies can be considered representative of the entire cell population. Though not ideal, an efficiency over 10% is usually acceptable.

Outline
Treat the cells with experimental agent at a range of concentrations for 24 h. Trypsinize the cells, seed them at a low cell density, and incubate them for 1–3 weeks. Stain the cells, and count the number of colonies. (See Fig. 21.1.)

Materials
Sterile:
Growth medium

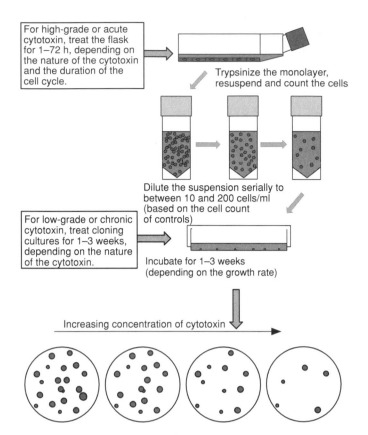

For high-grade or acute cytotoxin, treat the flask for 1–72 h, depending on the nature of the cytotoxin and the duration of the cell cycle.

Trypsinize the monolayer, resuspend and count the cells

Dilute the suspension serially to between 10 and 200 cells/ml (based on the cell count of controls)

For low-grade or chronic cytotoxin, treat cloning cultures for 1–3 weeks, depending on the nature of the cytotoxin.

Incubate for 1–3 weeks (depending on the growth rate)

Increasing concentration of cytotoxin

Fig. 21.1. Clonogenic Assay.

PBSA
Trypsin, 0.25%, crude
Compound to be tested at 10✕ the maximum concentration to be used, dissolved in serum-free medium; check the pH and the osmolality of the test solution, and adjust if necessary
Flasks, 25 cm²
Petri dishes, 6 or 9 cm, labeled on the side of the base
Nonsterile:
PBSA
Methanol
Crystal violet, 1%
Hemocytometer or electronic cell counter

Protocol
1. Prepare a series of cultures in 25-cm² flasks, three for each of six agent concentrations, and three controls. Seed the cells at 5 ✕ 10⁴ cells/ml in 4.5 ml of growth medium, and incubate them for 48 h, by which time the cultures will have progressed into the log phase. (See Growth Cycle in Chapter 20.)
2. Prepare a serial dilution of the compound to give 2 ml at 10✕ of each of the final concentrations required:

(a) If you are testing a compound for the first time, use 5-fold dilutions over a range of 3–5 logs.

(b) If you can predict the approximate toxic concentration then select a narrower, arithmetic range over one or two decades.

3. Dilute each concentration 1:10 by adding 0.5 ml of 10✕ concentrate to each of three flasks for each concentration, starting with control medium (no compound added) and progressing from lowest to highest concentration.

4. Return the flasks to the incubator.

5. If the compound is slow acting or partially reversible, repeat step 3 twice; that is, expose the cultures to the agent for 3 d, replacing the medium and compound daily by changing the medium. With fast-acting compounds, 1 h exposure will be sufficient.

6. Remove the medium from each group of three flasks in turn (working from lowest concentration of compound to highest), and trypsinize the cells.

7. Dilute and seed the cells into Petri dishes at the required density for clonal growth (see Protocol 13.1), diluting all of the cultures by the same amount as the control. Work from the flasks exposed to the highest concentration down to the lowest and then the control.

8. Incubate the cultures until colonies form

9. Fix the cultures in absolute methanol, and stain them for 10 min in 1% crystal violet. (See Protocol 15.3.)

10. Wash the dishes in tap water, drain them, stand them in an inverted position to dry.

11. Count the colonies with >50 cells (>5 generations)

Analysis. Plot the survival curve, which is a graph of the relative plating efficiency (the plating efficiency as a fraction of the control) against the log or linear drug concentration. (See Fig. 21.2.) Determine the IC_{50} or IC_{90}, which is the concentration of compound promoting 50% or 90% inhibition of colony formation, respectively. As this is a semilog plot, the IC_{90} is more appropriate, as it is more likely to fall on the linear part of the curve, while the IC_{50} tends to fall on the knee of the curve, giving a less stable value. Differences in sensitivity may be determined by the slope of the curve and the length of the knee, which influences the IC_{50} and the IC_{90}, although a more significant difference can be observed in the IC_{90}. (See Fig. 21.3.) A fraction of resistant cells is indicated by a flattening of the lower end of the curve. Total resistance is indicated by the lack of a gradient on the curve. Complex survival curves may be compared by calculating the area under the

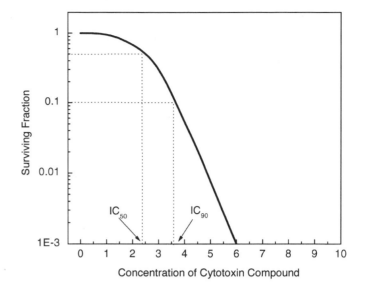

Fig. 21.2. Survival Curve. Semilog plot of the surviving fraction of cells (colonies forming from test cells/colonies forming from control cells) against the concentration of cytotoxin. Typically, the curve has a "knee," and the IC_{90} lies in the linear range of the curve. The IC_{50}, falling on the knee, is a less stable value.

curve, but this is done for expediency and is not mathematically valid.

Variations

Concentration of Agent. A wide range of concentrations in log increments (e.g., 1×10^{-6} M, 1×10^{-5} M, 1×10^{-4} M, 1×10^{-3} M, 0) should be used for the first attempt and a narrower range (log or linear), based on the results from the first range, for subsequent attempts.

Invariate Agent Concentrations. Some conditions that are tested cannot easily be varied—e.g., the quality of medium, water, or an insoluble plastic. In these cases, the serum concentrations can be varied, since serum may have a masking effect on minor toxic effects.

Duration of Exposure to Agent. Some agents act rapidly, while others act more slowly. Exposure to ionizing radiation, for example, need last only a matter of minutes to achieve the required dose, while testing some cycle-dependent antimetabolic drugs may take several days to achieve a measurable effect. Duration of exposure (T) and drug concentration (C) are related, although $C \times T$ is not always a constant. Prolonging exposure can increase sensitivity beyond that predicted by $C \times T$, due to cell cycle effects and cumulative damage.

Time of Exposure to Agent. When the agent is soluble and expected to be toxic, the procedure in Pro-

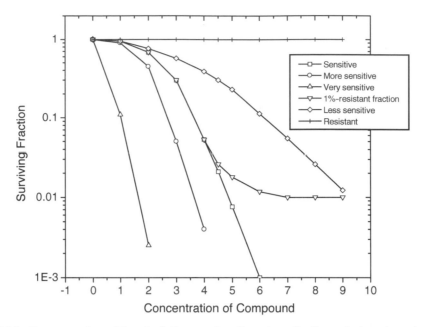

Fig. 21.3. Interpretation of Survival Curves. Semilog plot of cell survival against the concentration of cytotoxin. The slope increases with increasing sensitivity and decreases with reduced sensitivity until it becomes totally flat for complete resistance. Partial resistance can be expressed as the resistant fraction shown by the curve flattening out at the lower end.

tocol 21.3 should be followed, but when the quality of the agent is unknown, stimulation is expected, or only a minor effect is expected (e.g., 20% inhibition rather than severalfold), the agent may be incorporated during clonal growth rather than at preincubation. Confirmation of anticipated toxicity—e.g., for a cytotoxic drug—requires a conservative assay, with minimal drug exposure as compared to that *in vivo*, to be applied before cloning. Confirmation of the lack of toxicity—e.g., for tap water or a nontoxic pharmaceutical—requires a more stringent assay, with prolonged exposure during cloning.

Cell Density during Exposure. The density of the cells during exposure to an agent can alter the response of the cells and the agent; for example, HeLa cells are less sensitive to the alkylating agent mustine at high cell densities [Freshney et al., 1975]. To determine the effects of cell density it should be varied in the preincubation phase, during exposure to the drug.

Cell Density during Cloning. The number of colonies may fall at high concentrations of a toxic agent, but it is possible to compensate for this effect by seeding more cells so that approximately the same number of colonies form at each concentration. This procedure removes the risk of a low clonal density influencing survival and improves statistical reliability, but is prone to the error that cells from higher drug concentrations are plated at a higher cell con-

centration, a factor that may also influence survival. It is preferable to plate cells on a preformed feeder layer, the density of which (5×10^3 cells/cm^2) greatly exceeds that of the cloning cells. This step ensures that the cell density is uniform regardless of clonal survival, which contributes little to the total cell density. Note that cloning on a feeder layer can sometimes reveal a resistant fraction of cells that is not apparent without the feeder layer. (Fig. 21.4.).

Colony Size. Some agents are cytostatic (i.e., they inhibit cell proliferation), but not cytotoxic, and during continuous exposure, they may reduce the size of colonies without reducing the number of colonies. In this case, the size of the colonies should be determined by densitometry [McKeehan et al., 1977], automatic colony counting, or by visually counting the number of cells per colony. (See also Protocol 20.9.)

For *colony counting*, the threshold number of cells per colony (e.g., 50 as in Analysis of Protocol 20.9) is purely arbitrary, and it is assumed that most of the colonies are greatly in excess of this number. Colonies should be grown until they are quite large ($>10^3$ cells), when the growth of larger colonies tends to slow down and smaller, but still viable, colonies tend to catch up with these larger colonies. For *colony sizing*, stain the cultures earlier, before the growth rate of larger colonies has slowed down, and score all of the colonies.

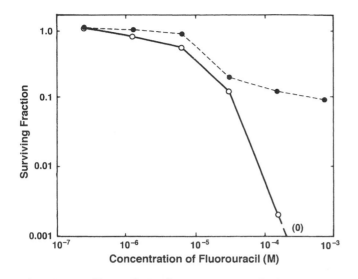

Fig. 21.4. Effect of Feeder Layer on Resistant Fraction. Human glioma cells were plated out in the presence (dotted line) and absence (solid line) of a feeder layer after treatment with various concentrations of 5-fluorouracil. A 10%-resistant fraction is apparent at 10^{-4} M drug only in the presence of a feeder layer. In the absence of the feeder layer, the small number of colonies making up the resistant fraction were unable to survive alone.

Solvents. Some agents to be tested have low solubilities in aqueous media, and it may be necessary to use an organic solvent to dissolve them. Ethanol, propylene glycol, and dimethyl sulfoxide have been used for this purpose, but may themselves be toxic to cells. Therefore, you should use the minimum concentration of solvent to obtain a solution. The agent may be made up at a high concentration in, for example, 100% ethanol; then it should be diluted gradually with BSS and finally diluted into medium. The final concentration of solvent should be <0.5%, and a *solvent control* must be included (i.e., a control with the same final concentration of solvent but without the agent being tested).

Take care when using organic solvents with plastics or rubber. It is better to use glass with undiluted solvents and to use plastic only when the solvent concentration is <10%.

While calculating the plating efficiency is one of the best methods for testing cell survival rates, it should be used only when the cloning efficiency is high enough for colonies that form to be representative of the whole cell population. Ideally, this means that controls should plate at 100% efficiency. In practice, however, this is seldom possible, and control plating efficiencies of 20% or less are often accepted.

Cell Proliferation Assays

Cell counts after a few days in culture can also be used to determine the effect of various compounds on cell proliferation, but, at least in the early stages of testing, a complete growth curve is required (see Protocols 20.7 and 20.8), because cell counts at a single point in time can be ambiguous. (See Fig. 20.5, d 7.) Growth curve analyses, using cell counting, are feasible only with relatively small numbers of samples, as they become cumbersome in a large screen. In cases for which there are many samples, a single point in time—e.g., the number of cells three days after exposure—can be used, but any significant effect should be backed up with a complete growth curve over the whole growth cycle or by an alternative assay, such as a survival curve by clonogenic assay (see Protocol 21.3) or MTT assay (see Protocol 21.4).

Metabolic Assays

Plating efficiency tests are labor intensive and time consuming to set up and analyze, particularly when a large number of samples is involved, and the duration of each experiment may be anywhere from 2 to 4 weeks. Furthermore, some cell lines have poor plating efficiencies, particularly freshly isolated normal cells, so a number of alternatives have been devised for assaying cells at higher densities (e.g., in microtitration plates). None of these tests measures survival directly, however. Instead, the net increase in the number of cells (i.e., the growth yield), the increase in the total amount of protein or DNA, or the residual ability to synthesize protein or DNA is determined. Survival in these cases is defined as the retention of metabolic or proliferative ability by the cell population as a whole some time after removal of the toxic influence. However, such assays cannot discriminate between a reduction in metabolic or proliferative activity per cell and a reduced number of cells, and therefore any novel or exceptional observation should be confirmed by clonogenic survival assay.

Microtitration Assays

The introduction of multiwell plates revolutionized the approach to replicate sampling in tissue culture. These plates are economical to use, lend themselves to automated handling, and can be of good optical quality. The most popular is the 96-well microtitration plate, each well having 28–32 mm² of growth area and the capacity for 0.1 or 0.2 ml medium and up to 1×10^5 cells. Microtitration offers a method whereby large numbers of samples may be handled simultaneously, but with relatively few cells per sample. With this method, the whole population is exposed to the agent, and viability is determined subsequently.

The end point of a microtitration assay is usually an estimate of the number of cells. While this result can be achieved directly by cell counts or by indirect methods, such as isotope incorporation, cell viability as measured by MTT reduction [Mosmann, 1983] is now widely chosen as the optimal end point [Cole, 1986; Alley et al., 1988]. MTT is a yellow water-soluble tetrazolium dye that is reduced by live, but not dead, cells to a purple formazan product that is insoluble in aqueous solutions. However, a number of factors can influence the reduction of MTT [Vistica et al., 1991]. The assay described in Protocol 21.4, provided by Jane Plumb of the CRC Department of Medical Oncology, University of Glasgow, Scotland, United Kingdom, has been shown to give the same results as a standard clonogenic assay [Plumb et al., 1989]. It illustrates the use of microtitration in the assay of anticancer drugs, but would be applicable, with minor modifications, to any cytotoxicity assay.

PROTOCOL 21.4. MTT-BASED CYTOTOXICITY ASSAY

Principle
Cells in the exponential phase of growth are exposed to a cytotoxic drug. The duration of exposure is usually determined as the time required for maximal damage to occur, but is also influenced by the stability of the drug. After removal of the drug, the cells are allowed to proliferate for two to three population-doubling times (PDTs) in order to distinguish between cells that remain viable and are capable of proliferation and those that remain viable but cannot proliferate. The number of surviving cells is then determined indirectly by MTT dye reduction. The amount of MTT–formazan produced can be determined spectrophotometrically once the MTT–formazan has been dissolved in a suitable solvent.

Outline
Incubate monolayer cultures in microtitration plates in a range of drug concentrations. (Fig. 21.5.) Remove the drug, and feed the plates daily for two to three PDTs; then feed the plates again, and add MTT to each well. Incubate the plates in the dark for 4 h, and then remove the medium and MTT. Dissolve the water-insoluble MTT–formazan crystals in DMSO, add a buffer to adjust the final pH, and record the absorbance in an ELISA plate reader.

Materials
Sterile:
Growth medium
Trypsin (0.25% + EDTA, 1 mM, in PBSA)

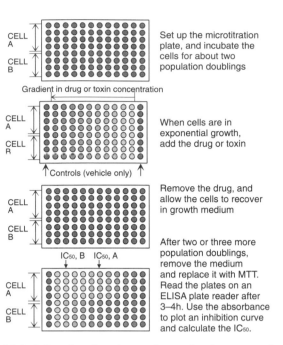

Set up the microtitration plate, and incubate the cells for about two population doublings

Gradient in drug or toxin concentration

When cells are in exponential growth, add the drug or toxin

Controls (vehicle only)

Remove the drug, and allow the cells to recover in growth medium

IC_{50}, B IC_{50}, A

After two or three more population doublings, remove the medium and replace it with MTT. Read the plates on an ELISA plate reader after 3–4h. Use the absorbance to plot an inhibition curve and calculate the IC_{50}.

Fig. 21.5. Microtitration Assay. Stages in the assay of two different cell lines exposed to a range of concentrations of the same drug and then allowed to recover before the estimation of survival by the MTT reaction. (See Protocol 21.4.)

MTT: 3-(4,5-dimethylthiazol-2-yl)-2,5-diphenyltetrazolium bromide (Sigma), 50 mg/ml, filter sterilized
Sorensen's glycine buffer (0.1 M glycine, 0.1 M NaCl adjusted to pH 10.5 with 1 M NaOH)
Microtitration plates (Linbro, ICN)
Pipettor tips, preferably in an autoclavable tip box
Petri dishes (non-TC-treated), 5 cm and 9 cm or reservoir (Corning Costar)
Universal containers, 30 ml and 100 ml
Nonsterile:
Plastic box (clear polystyrene, to hold plates)
Multichannel pipettor
Dimethyl sulfoxide (DMSO)
DMSO dispenser (optional; Well-fill, Life Sciences)
ELISA plate reader (see Sources of Materials—Plate Readers in Trade Index)
Plate carrier for centrifuge (for cells growing in suspension)

Protocol
Plating out cells:
1. Trypsinize a subconfluent monolayer culture, and collect the cells in growth medium containing serum.
2. Centrifuge the suspension (5 min at 200 g) to pellet the cells. Resuspend the cells in growth medium, and count them.
3. Dilute the cells to $2.5–50 \times 10^3$ cells/ml, depending on the growth rate of the cell line and

allowing 20 ml of cell suspension per microtitration plate.

4. Transfer the cell suspension to a 9-cm Petri dish, and, with a multichannel pipette, add 200 μl of the suspension into each well of the central 10 columns of a flat-bottomed 96-well plate (80 wells per plate), starting with column 2 and ending with column 11 and placing 0.5–10 $\times$ 10^3 cells into each well.

5. Add 200 μl of growth medium to the eight wells in columns 1 and 12. Column 1 will be used to blank the plate reader; column 12 helps to maintain the humidity for column 11 and minimizes the "edge effect."

6. Put the plates in a plastic lunch box, and incubate in a humidified atmosphere at 37°C for 1–3 d, such that the cells are in the exponential phase of growth at the time that drug is added.

7. For nonadherent cells, prepare a suspension in fresh growth medium. Dilute the cells to 5–100 $\times$ 10^3 cells/ml, and plate out only 100 μl of the suspension into round-bottomed 96-well plates. Add drug immediately to these plates.

Drug addition:

8. Prepare a serial fivefold dilution of the cytotoxic drug in growth medium to give eight concentrations. This set of concentrations should be chosen such that the highest concentration kills most of the cells and the lowest kills none of the cells. Once the toxicity of a drug is known, a smaller range of concentrations can be used. Normally, three plates are used for each drug to give triplicate determinations within one experiment.

9. For adherent cells, remove the medium from the wells in columns 2 to 11. This can be achieved with a hypodermic needle attached to a suction line.

10. Feed the cells in the eight wells in columns 2 and 11 with 200 μl of fresh growth medium; these cells are the controls.

11. Add the cytotoxic drug to the cells in columns 3 to 10. Only four wells are needed for each drug concentration, such that rows A–D can be used for one drug and rows E–H for a second drug.

12. Transfer the drug solutions to 5-cm Petri dishes, and add 200 μl to each group of four wells with a four-tip pipettor.

13. Return the plates to the plastic box, and incubate them for a defined exposure period. For nonadherent cells, prepare the drug dilution at twice the desired final concentration, and add 100 μl to the 100 μl of cells already in the wells.

Growth period:

14. At the end of the drug exposure period, remove the medium from all of the wells containing cells, and feed the cells with 200 μl of fresh medium. Centrifuge plates containing nonadherent cells (5 min at 200 *g*) to pellet the cells. Then remove the medium, using a fine-gauge needle to prevent disturbance of the cell pellet.

15. Feed the plates daily for 2–3 PDTs.

Estimation of surviving cell numbers:

16. Feed the plate with 200 μl of fresh medium at the end of the growth period, and add 50 μl of MTT to all of the wells in columns 1 to 11.

17. Wrap the plates in aluminum foil, and incubate them for 4 h in a humidified atmosphere at 37°C. Note that 4 h is a minimum incubation time, and plates can be left for up to 8 h.

18. Remove the medium and MTT from the wells (centrifuge for nonadherent cells), and dissolve the remaining MTT–formazan crystals by adding 200 μl of DMSO to all of the wells in columns 1 to 11.

19. Add glycine buffer (25 μl per well) to all of the wells containing DMSO.

20. Record absorbance at 570 nm immediately, since the product is unstable. Use the wells in column 1, which contain medium and MTT but no cells, to blank the plate reader.

Analysis

Plot a graph of the absorbance (*y*-axis) against the concentration of drug (*x*-axis). The mean absorbance reading from the wells in columns 2 and 11 is used as the control absorbance, and the IC$_{50}$ concentration is determined as the drug concentration that is required to reduce the absorbance to half that of the control. The absolute value of the absorbance should be plotted so that control values may be compared, but the data can then be converted to a percentage-inhibition curve (Fig. 21.6), to normalize a series of curves. The absorbance values in columns 2 and 11 should be the same. Occasionally, they are not, however, and this is taken to indicate uneven plating of cells across the plate.

Variations

Other Applications. A similar assay has also been used to determine cellular radiosensitivity [Carmi-

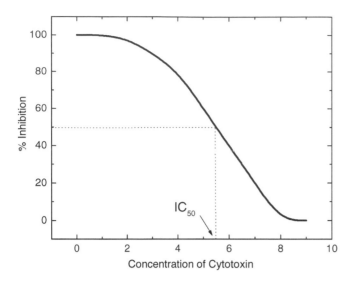

Fig. 21.6. Percentage-Inhibition Curve. Percentage inhibition ([absorbance of test wells/absorbance of control wells] × 100) plotted against the concentration of cytotoxin. Typically, a sigmoid curve is obtained, and, ideally, the IC_{50} will lie in the center of the inflexion of the curve (although is often not the case).

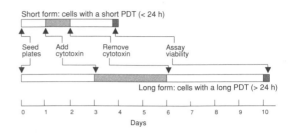

Fig. 21.7. Assay Duration. Pattern for short-form and long-form assays. The upper diagram represents an assay that is suitable for cell with a PDT < 24 h, and the bottom diagram represents an assay that is suitable for cells with a PDT >24 h, although intermediate time scales are also possible.

chael et al., 1987b]. MTT can be used to determine the number of cells after a variety of treatments other than cytotoxic drug exposure, such as growth factor stimulation. However, in each case, it is essential to ensure that the treatment itself does not affect the ability of the cell to reduce the dye.

Duration of Exposure. As with clonogenic assays (see Protocol 21.3), some agents may act more quickly, and the exposure period and recovery may be shortened. The cells must remain in exponential growth throughout (see Subculture in Chapter 12; and Growth Cycle in Chapter 20), and the cell concentration at the end should still be within the linear range of the MTT densitometric assay. When using a cell line for the first time, parallel plates should be set up for cell counts to generate a growth curve (see Protocols 20.7 and 20.8) and for MTT–formazan absorbance to ensure that absorbance is still proportional to the number of cells. If the growth curve shows that the cells are moving into the stationary phase or the absorbance is nonlinear when plotted against cell concentration, shorten the assay and proceed directly to step 16 of Protocol 21.4.

Duration of exposure is related to the number of cell cycles that the cells have gone through during exposure and recovery. Not only will the cell density increase more rapidly during exposure, but in addition the response to cycle-dependent drugs will be quicker. Cell cycle time will influence the choice between a short-form and long-form assay. (Figs. 21.7

and 21.8.) When first trying an assay, it may be desirable to sample on each day of drug exposure and recovery. If a stable IC_{50} is reached earlier, then the assay may be shortened.

End Point. Sulforhodamine is a fluorescent dye that stains protein and can also be used to estimate the amount of protein (i.e., cells) per well on a plate reader with fluorescence detection [Boyd, 1989]. It stains all cells and does not discriminate between live and dead cells.

Labeling with [3H]-thymidine (DNA synthesis), [3H]-uridine (RNA synthesis), or other isotopes can be substituted for MTT reduction. Quantitation is achieved by microtitration plate scintillation counting. Two types of scintillation counting are available: (1) The cellular contents may be aspirated onto filters by trypsinization and onto glass fiber filters by suction transfer, and the filters may be dried and counted in scintillant, or (2) the whole plate may be counted on a specially adapted counter (Pharmacia-Wallac, Packard) or phosphoimager (BioRad).

In practice, it may not matter which criterion is used for determining viability or survival at the end of an assay; it is, rather, the design of the assay (e.g., drug exposure, recovery, cell density, growth rate, etc.) that is most important.

Handling. A variety of automated handling techniques are available [i.e., autodispensers, diluters, cell harvesters, and densitometers; see Figs. 4.16–4.18] to reduce the time required per sample.

Comparison of Microtitration with Cloning

The volume of medium required per sample for microtitration is less than one-fiftieth of that required for cloning, though the number of cells is approximately the same for both techniques. Microtitration, however, is unable to distinguish between differential responses between cells within a population and the

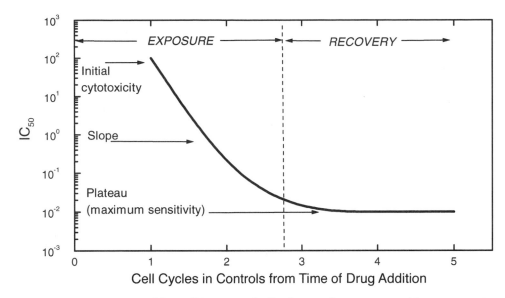

Fig. 21.8. Time Course of the Fall in ID$_{50}$. Idealized curve for an agent with a progressive increase in cytotoxicity with time, but eventually reaching a maximum effect after three cell cycles. Not all cytotoxic drugs will conform to this pattern.

degree of response in each cell—e.g., a 50% inhibition could mean that 50% of the cells respond or that each cell is inhibited by 50%.

A comparison of IC$_{50}$'s derived by microtitration and plating efficiency assays showed a strong corre-

lation between the two methods (Fig. 21.9) for the assay of antineoplastic drugs.

A significant feature of microtitration assays, particularly with a colorimetric or radiometric end point, is the generation of large amounts of data, often in a

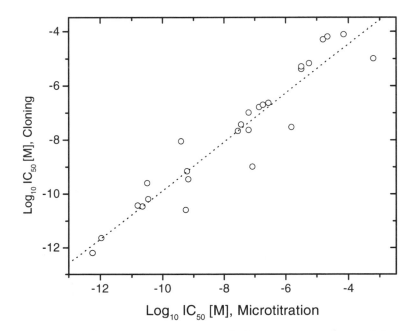

Fig. 21.9. Correlation between Microtitration and Cloning. Measurement of the IC$_{50}$'s of a group of five cell lines from human glioma and six drugs (vincristine, bleomycin, VM-26 epidophyllotoxin, 5-fluorouracil, methyl CCNU, mithramycin). Most of the outlying points were derived from one cell line that later proved to be a mixture of cell types. The broken line is the regression, with the data points from the heterogeneous cell line omitted. [From Freshney et al., 1982a.]

format that is readily analyzed by computer. As discussed in Data Analysis in Chapter 20, there are now a number of different software programs (Packard, Wallac, Molecular Devices, Biorad, Lab Systems) that can be used to generate tabular and graphical output from your data. It is important, however, that a system be used with which the operator can scan the raw data as well as the data-reduced end point. Computer analysis saves a great deal of time, but it makes different assumptions or corrections to deal with aberrant data points, which are not apparent unless the raw data is available for scrutiny.

Drug Interaction

The investigation of cytotoxicity often involves the study of the interaction of different drugs; drug interaction is readily determined by microtitration systems, in which several different ratios of interacting drugs can be examined simultaneously. Analysis of drug interaction can be performed using an isobologram to interpret the data [Steel, 1979; Berenbaum, 1985]. A rectilinear plot implies an additive response, while a curvilinear plot implies synergy if the curve dips below the predicted line and antagonism if it goes above.

ANTICANCER DRUG SCREENING

Drug screening for the identification of new anticancer drugs can be a tedious and often inefficient method of discovering new active compounds. The trend is now more toward monitoring effects on specific molecular targets. However, there have been attempts to improve screening by adopting rapid, easily automated assays, like those based on the determination of the number of viable cells by staining the cells with MTT [Mosmann, 1983; Carmichael et al., 1987a; Plumb et al., 1989]. To further cut down on manipulations, the MTT incubation step is omitted, and the end point is determined by measuring the amount of total protein using sulforhodamine B [Boyd, 1989]. While this method is quicker and easier than the MTT assay, it should be remembered that nonviable, and certainly nonreplicating, cells will still stain, so the assay should be confirmed when activity is detected, using a more reliable indicator, such as clonogenicity or MTT reduction.

Predictive Testing

The possibility has often been considered that measurement of the chemosensitivity of cells derived from a patient's tumor might be used in designing a chemotherapeutic regime for the patient [Freshney, 1978]. This technique has never been exhaustively tested, although the results of small-scale trials have been encouraging [Hamburger and Salmon, 1977; Bateman et al., 1979; Hill, 1983; Thomas et al., 1985; Von Hoff et al., 1986]. What is required now is the development of reliable and reproducible culture techniques for the most common tumors (e.g., breast, lung, colon): such that cultures of pure tumor cells capable of cell proliferation over several cell cycles may be prepared routinely. Assays might then be performed in a high proportion of cases, within 2 weeks of receipt of the biopsy.

The major problem, however, is one of logistics. The number of patients with tumors that will grow *in vitro* sufficiently to be tested, that can be expected to respond, and that will produce a response which can be followed up, is extremely small. Hence, it has proved difficult to use any *in vitro* test as a predictor of response or even to verify the reliability of the assay. The correlation of insensitivity *in vitro* with nonresponders is high, but few clinicians would withhold chemotherapy because of an *in vitro* test, particularly when the agent in question would probably not be used alone.

TRANSFORMATION

Commonly used *in vitro* assays for transformation include anchorage independence (see Protocol 13.4), reduced density limitation of cell proliferation (see Protocol 17.2), and evidence of mutagenesis. Mutagenesis can be assayed by sister chromatid exchange; this procedure is described in the following protocol, which was contributed by Maureen Illand and Robert Brown of the CRC Department of Medical Oncology, University of Glasgow, CRC Beatson Laboratories, Garscube Estate, Bearsden, Glasgow G61 1BD, Scotland, United Kingdom.

Background

Sister chromatid exchanges (SCEs) are reciprocal exchanges of DNA segments between sister chromatids at identical loci during the S-phase of the cell cycle. As SCEs are more sensitive indicators of mutagenic activity than chromosome breaks, they have become a major tool in mutagenesis research [Latt, 1981].

With the development of the thymidine analogue bromodeoxyuridine (BUdR) and its subsequent use in DNA labeling experiments, the resolution of SCEs greatly improved in comparison to previous methods, which involved the incorporation of radioactive nucleotides into replicating DNA [Taylor, 1958]. Later, the fluorescence plus Giemsa (FPG) technique of Perry and Wolf [1974] for the scoring of SCEs was enhanced, and for the first time, permanent staining of

SCEs was demonstrated. Previously, during the scoring process, rapid bleaching of the fluorescent stain occurred [Latt, 1981].

PROTOCOL 21.5. SISTER CHROMATID EXCHANGE

Principle

The FPG method involves two distinct steps: (1) Cells are labeled with BUdR for two complete cycles and then treated with colcemid to block the cells in metaphase. Following BUdR exposure, the DNA of one chromatid of each chromosome contains bromouracil in one strand, while the DNA of its sister chromatid contains bromouracil in both strands. (2) Chromosomes are then prepared from these cells, stained with the fluorescent dye Hoechst 33258, and then the BUdR is photodegraded using ultraviolet light, followed by Giemsa staining. These final steps highlight the differential incorporation of bromouracil into the sister chromatids. DNA that contains bromouracil quenches the fluorescence of Hoechst–DNA complexes. Therefore, the chromatid containing bromouracil substituted in both strands fluoresces weakly and stains weakly with Giemsa, while the chromatid containing bromouracil in only one strand fluoresces more intensely, degrades the

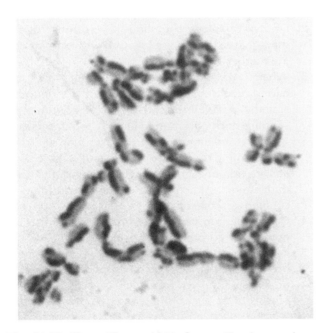

Fig. 21.10. Sister Chromatid Exchange. Ovarian carcinoma cells A2780/Cp70's with an additional human chromosome 2 transferred (A2780/cp70 +chr2), treated with 10 μM Cisplatin for one hour. (See also Plate 32.)

BUdR, and subsequently stains darkly with Giemsa. (Fig. 21.10 and see Plate 32.) If any sister chromatid exchanges occur, this staining pattern produces "harlequin chromosomes."

Outline

Trypsinize metaphase-arrested cells that are BUdR labeled for two cell cycles, incubate the cells in hypotonic buffer, and then fix the cells. Prepare slides of the cells, after treating the cells with Hoechst 33258, and photodegrade chromosome spreads. Stain the chromosomes with Giemsa, and visualize on a light microscope under oil immersion.

Materials

Sterile:
PBSA
PE: 10 mM EDTA in PBS
Trypsin: 0.12% in PE
BUdR (Sigma): Prepare 1 mM of stock in sterile ultrapure water
Karyomax: Colcemid, 10 μg/ml (Gibco)
Growth medium
Nonsterile:
SSC, 2×: 1:10 dilution of 20× SSC (see Reagents Appendix)
Hypotonic buffer: 0.075 M KCl
Sorensen's buffer: phosphate buffer, 0.066 M, pH 6.8 (tablets from Merck)
Methanol: acetic acid, 3:1, ice cold and freshly prepared
Giemsa solution, 0.76%: place 1 g of Giemsa powder (Merck) in 66 ml of glycerol and heat in a waterbath at 56–60°C for $1\frac{1}{2}$–2 h. Cool the solution, and add 66 ml of absolute alcohol
Giemsa, diluted to 3.5% in Sorensens buffer, pH 6.8
Hoechst 33258 (Sigma), 20 μg/ml, in UPW.

△ **Safety Note.** Hoechst 33258 is carcinogenic; weigh it out and dissolve it in a fume hood.

Protocol

Pretreatment:
1. Seed the cells at the appropriate density (e.g., 1 × 10⁶ cells per 75-cm² flask), and incubate for 2 d at 37°C.
2. Add BUdR to the growth medium at a final concentration of 10 μM.
3. Incubate the cells in the dark at 37°C for a further 48 h (~2 cell cycles).
4. Add colcemid to the cells 1–6 h before harvesting, depending on the cycling time of the cells. For human cell lines, the final concentration of colcemid should be 0.01 μg/ml.

Harvesting cells:

5. Wash the cells with PBS, and trypsinize them using 1 ml of 0.12% trypsin in PE.

6. Resuspend the cells in 10 ml of growth medium. Transfer the cell suspension to 50-ml Falcon tubes.

7. Centrifuge the suspension at 1500 rpm for 5 min.

8. Remove the supernatant, leaving approximately 1.2 ml above the pellet. Flick the side of the tube to resuspend the pellet.

9. Slowly add 10 ml of hypotonic buffer (pre-warmed to 37°C), and incubate the cell suspension for 10–15 min at room temperature.

10. Spin the cells in a benchtop centrifuge at 1500 rpm for 5 min.

11. Remove the supernatant, leaving 1.2 ml above the pellet. Flick the side of the tube to resuspend the pellet.

12. Add 10 ml of ice-cold fixative, initially drop by drop, mixing well after each addition. Leave the tube on ice for 10 min.

13. Repeat steps 9–11 once more, letting the cells remain in fixative overnight at 4°C, to improve slide preparations.

14. Spin the fixed cells at 1500 rpm for 5 min, and resuspend the cells in 3–5 ml of methanol/acetic acid. Store the cells at −20°C.

Slide preparation:

15. Slides should be clean and grease free before use, so wipe them with absolute alcohol.

16. Using a short glass Pasteur pipette, take up approximately 500 μl of the fixed cells.

17. Hold the slide at a downward angle, and, holding the Pasteur pipette at least 15 cm (6 in) above the slide, drop 3 drops of the cell suspension onto the slide.

18. Air dry the slide in the dark.

19. Check the slide under phase contrast to ensure that there are an even number of metaphase spreads across the slide and that the chromosomes are well separated.

Harlequin staining:

20. Immerse the slides in a Coplin jar of Hoechst 33258 at a concentration of 20 μg/ml for 10 min. (Wear gloves, as Hoechst is a COSHH3 chemical.)

21. Transfer the slides to a slide rack (Life Sciences), and drop 500 μl of 2× SSC onto each slide.

22. Cover the slides with a 22-mm × 50-mm coverslip, and seal the edges with a tem-porary seal, such as Cowgum, to prevent evaporation.

23. Place the slides in the slide rack, coverslips facing downwards, and place the slide rack on a shortwave UV box. Maintain a distance of approximately 4 cm between the slides and the UV source. The longer the slides are exposed to UV, the paler the pale chromatid will become; expose the slides for about 25–60 min.

24. Remove the coverslips from the slides, and wash the slides three times in UPW, 5 min per wash. Cover the slide holder with aluminium foil.

25. Air dry the slides in the dark.

26. Stain the slides in a Coplin jar containing 3.5% Giemsa solution in Sorensen's buffer, pH 6.8, for 3–5 mins.

27. Carefully rinse the slides in tap water, and drain them using a paper tissue.

28. Air dry the slides on the bench for 1 h. Dip each slide into xylene, drop 4 drops of DPX mountant (BDH) onto the slide, and lower a 22-mm × 50-mm coverslip, expressing any air bubbles with tissue. (Carry out this final step in a fume hood, as xylene fumes are toxic. Also, wear gloves.)

29. Air dry the slides in a fume hood overnight.

Analysis

1. Under the 40× objective of a light microscope, scan the slides for metaphase spreads.

2. Find an area on the slides where most of the metaphase spreads are located, and examine this area under oil immersion.

3. When no sister chromatid exchanges have occurred, each chromosome has one continuously staining pale chromatid and one continuously staining dark chromatid. One SCE has occurred when there is one area of dark staining and then light staining on one chromatid and, on the sister chromatid, one area of light staining and then dark staining. Each point of the discontinuity in staining is scored as one SCE.

4. Count the number of SCEs per cell and also the number of chromosomes per cell.

5. Larger chromosomes usually have a greater number of SCEs than smaller ones, and the incidence of SCEs may vary from cell to cell. Therefore, scoring SCEs per chromosome is a more accurate measure of SCE rate. SCE score is calculated by the following formula:

$$\frac{\text{mean no. of SCEs cell}}{\text{mean no. of chromosomes/cell}}.$$

Aim to score approximately fifty spreads per cell line being studied.

Variations

Pulse labeling of cells with BUdR and subsequent staining, as described previously, can detect differences in early and late replicating regions of chromosomes during the cell cycle. When cells are labeled with BUdR at the latter part of the cell cycle, DNA that replicates early will stain darkly with Giemsa, due to very little BUdR incorporation, and for regions of the chromosome that are pulsed at the earlier stages of the cell cycle, only those regions that replicate their DNA early will stain faintly with Giemsa [Latt, 1973].

Additionally, cells that have undergone only one cell cycle of continuous BUdR labeling show differential staining of chromatids only at certain bands (lateral asymmetry), due to the differences in thymine content of the DNA [Brito Babapulle, 1981]. Following photodegradation of BUdR, the fluorescent dye acridine orange can also be used to stain SCEs. With this dye, green fluorescence is observed at regions that have double-stranded DNA and thus will have little BUdR incorporation, and red fluorescence is observed at regions that have single-stranded DNA, which will have incorporated the BUdR. Consequently, the red fluorescence is equivalent to lighter staining with Giemsa, and the green fluorescence is equivalent to the darker staining with Giemsa [Karenberg and Freelander, 1974].

Carcinogenicity

The potential for *in vitro* testing for carcinogenesis is considerable [Berky and Sherrod, 1977; Grafstrom, 1990a; Grafstrom, 1990b; Zhu et al., 1991], but this is one area in which *in vivo* testing is far from adequate; the models are poor, and the tests often take weeks, or even months, to perform. The development of a satisfactory *in vitro* test is hampered (1) by the lack of a universally acceptable criterion for malignant transformation *in vitro* and (2) by the inherent stability of human cells used as targets.

The most generally accepted tests so far assume that carcinogenesis, in most cases, is related to mutagenesis. (See Genetic Instability in Chapter 17.) This assumption is the basis of the Ames tests [Ames, 1980], wherein bacteria are used as targets and activation can be carried out using liver microsomal enzyme preparations. This test has a high predictive value, but nevertheless, dissimilarities in uptake, susceptibility, and the type of cellular response have led to the introduction of alternative tests using mammalian and human cells as targets.

Some of these tests are also mutagenesis assays, using suspensions of L5178Y lymphoma cells as targets [Cole et al., 1990] and the induction of mutations or reversion, or cytological evidence of sister chromatid exchange (see Protocol 21.5), as evidence of mutagenesis. Others [Styles, 1977] have used transformation as an end point, assaying clonogenicity in suspension (see Anchorage Independence in Chapter 17) as a criterion for transformation. Critics of these systems say that both use cells which are already partially transformed as targets; even the BHK21-C13 cell used by some workers is a continuous cell line and may not be regarded as completely normal. Furthermore, the bulk of the common cancers arise in epithelial tissues, which have so far been difficult to grow routinely, and not in connective tissue cells.

The demonstration of increased oncogene expression or amplification, or the presence of increased or altered oncogene products, may provide reliable criteria, in some cases functionally related to the carcinogen. However, unless a total genomic screen is employed, this method means predicting which oncogenes will be altered. Likewise, the deletion or mutation of suppressor genes is open to molecular analysis, where deletions or mutations in the p53, Rb, p16, and L-CAM (E-cadherin) genes would cover a high proportion of malignant transformation events. As the proposals gain in complexity, the resistance to relinquishing the Ames test increases, and *in vitro* carcinogenesis assays remain the experimental tool of those studying the mechanism of carcinogenesis.

INFLAMMATION

There is an increasing need for tissue culture testing to reveal the inflammatory responses that are likely to be induced by pharmaceuticals and cosmetics with topical application or by xenobiotics that may be inhaled or ingested and may be responsible for many forms of allergy. This is an area that is only at the early stages of development, but that bears great promise for the future. It is a sensitive topic in more ways than one. Animal-rights groups are naturally incensed at the needless use of large numbers of animals to test new cosmetics that have little benefit except commercial advantage to the manufacturer, particularly when the testing of substances (such as shampoos) involves the Draze test, in which the compound is added to a rabbit's eye. More important, clinically, is the apparent increase in allergenic responses produced by pharmaceuticals and xenobiotics. These responses are little understood and poorly controlled, largely due to the absence of a simple reproducible *in vitro* test.

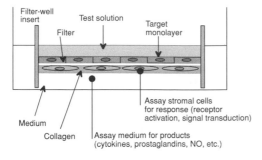

Fig. 21.11. Organotypic Assay. A hypothetical assay system for exposing one cell layer (e.g., epidermal keratinocytes) co-cultured with another associated cell type (e.g., skin fibroblasts in collagen gel) to an irritant and measuring the response by cytokine release (as proposed by Organogenesis, Inc., and others).

Since the advent of filter-well technology, several models for skin and cornea have appeared [Braa and Triglia, 1991; Triglia et al., 1991; Fusenig, 1994b; Roguet et al., 1994; Kondo et al., 1997], utilizing the facility for coculture of different cell types that the filter-well system provides. In these systems, the interaction of an allergen or irritant with a primary target (e.g., epidermis) is presumed to initiate a paracrine response, which triggers the release of a cytokine from a second, stromal component (e.g., dermis). (Fig. 21.11.) This cytokine can then be measured by ELISA technology to monitor the degree of the response. Although still in the early stages of development, kits for the measurement of irritant responses are available (Epiderm (MatTek) [Koschier at al., 1997]; Episkin (Saduc) [Cohen et al., 1997]). A protocol for corneal culture suitable for modeling irritant responses in the eye was provided by Carolyn Cahn. (See Protocol 22.2.)

It would seem that this type of system may be a major area of development, with the real prospect that allergen screening from patients' own skin may become possible and that, ultimately, analysis of the GI tract will reveal responsible allergens for irritable bowel syndrome. In each case, and in many others, there is the possibility of specific mechanistic studies into the processes of abnormal cell interaction that typify many allergic and degenerative diseases.

CHAPTER 22

Specialized Cells

The expression of specialized functions in a culture is controlled by the nutritional constitution of the medium, the presence of hormones and other inducer or repressor substances, and the interaction of the cells with the substrate and other cells. (See Chapter 16.) There are many reviews of culture techniques for specific cell types [Doyle et al., 1990–1999; Pollard and Walker, 1990; Freshney, 1992; Freshney et al., 1994; Leigh and Watt, 1994; Freshney and Freshney, 1996; Ravid and Freshney, 1998; Haynes, 1999; see also Table 2.1].

The development of techniques for cell line immortalization [Freshney and Freshney, 1996; see also Immortalization in Chapter 17] has meant that it has been possible to generate continuous cell lines from a number of finite lines from untransformed tissue [Chang et al., 1982; Freshney et al., 1994; Klein et al., 1990; Steele et al., 1992; Wyllie et al., 1992]. In many cases the differentiated properties are lost, but by using a switchable promoter (e.g., temperature sensitivity), it may prove possible to recover the differentiated phenotype. The development of the transgenic mouse carrying the large T gene of SV40 has opened up a wide range of possibilities [Yanai et al., 1991], because cells cultured from these animals are already immortalized, but still retain some differentiated functions.

Many specialized cells (e.g., epidermal keratinocytes, melanocytes, endothelial cells, smooth muscle cells, dermal fibroblasts, melanocytes, and mammary epithelium) are commercially available. (See Sources of Materials—Specialized Cells in Trade Index.) The cost is naturally very high, but as demand increases, the cost may fall. Skin cultures are also available for cy-

totoxicity and inflammation research, such as Episkin (Saduc) and Epiderm (MatTek) and from Advanced Tissue Sciences. These products are prepared in filter wells by combining keratinocytes with dermal fibroblasts and collagen supported by a nylon net in a so-called "skin equivalent." Other tissues—in particular, the cornea—have also been prepared in a similar way for toxicity studies. (See Protocol 22.2.)

A number of specialized procedures have now been devised, and some representative examples have been contributed by experts in various areas. The protocols in this chapter assume that the basic prerequisites of the cell biology laboratory, as specified in Chapter 4, will be available. Consequently, items such as inverted microscopes, bench centrifuges, and water baths will not be among the materials listed.

EPITHELIAL CELLS

Epithelial cells are responsible for the recognized functions of many organs (e.g., controlled absorption in the kidney and gut, secretion in the liver and pancreas, gas exchange in the lung, and barrier protection in skin). Epithelial cells are also of interest as models of differentiation and stem cell kinetics (e.g., epidermal keratinocytes) and are among the principal tissues in which common cancers arise. Consequently, the culture of various epithelial cells has been a focus of attention for many years. The major problem in the culture of pure epithelium has been the overgrowth of the culture by stromal cells, such as connective tissue fibroblasts and vascular endothelium. Most of the variations in technique are aimed at preventing such overgrowth by nutritional manipulation of the medium or alterations in the culture substrate (Table 22.1) that promote the growth of the undifferentiated epidermis and, preferably, the stem cells. Subsequent modifications may then be employed to enhance epithelial differentiation, although perhaps at the expense of proliferation.

Factors contained in serum—many of them derived from platelets—have a strong mitogenic effect on fibroblasts and tend to inhibit epithelial proliferation by inducing terminal differentiation. Consequently, one of the most significant events in the isolation and propagation of specialized cell cultures has been the development of selective, serum-free media, supplemented with specific growth factors as appropriate.

The isolation of epithelial cells from donor tissue is best performed with collagenase (see Collagenase in Chapter 11), which disperses the stroma, but leaves the epithelial cells in small clusters, favoring their subsequent survival.

Epidermis

This protocol for the culture of epidermal keratinocytes was contributed by Norbert E. Fusenig, Division of Carcinogenesis and Differentiation, German Cancer Research Center, Im Neuenheimer Feld 280, 6900 Heidelberg, Germany.

Because of the progress in basic cell culture technology and in our understanding of the culture requirements of the various epithelial tissues, keratinocytes of most stratified epithelia can now be grown and studied in cell culture. Mostly, the squamous epithelia of the skin and their isolated epithelial cells, the keratinocytes, have been used to study their physiology and pathology *in vitro*. In addition, cells of the oral mucosa [Tomakidi et al., 1997], as well as of skin appendages such as the hair follicle [for a review see Fusenig et al., 1994], have been isolated and cultured under various conditions, and reconstructed tissues have been formed in culture as well as in transplants [Limat et al., 1995].

In order to avoid overgrowth by mesenchymal cells, the epithelial compartment has to be separated from the connective tissue and dispersed into single cells, which are then cultured on different substrata by using different media formulations. The most commonly used and reliable method is the feeder layer technique first established by Rheinwald and Green [1975], who employed postmitotic (irradiated or mitomycin-C-treated) 3T3 cells as mesenchymal feeders. Primary fibroblasts derived from dermis or submucosa and rendered postmitotic can be used instead [Limat et al., 1989; Tomakidi et al., 1997]. The feeder cells, formerly called "lethally irradiated," recently were demonstrated to maintain a broad spectrum of physiological functions [Maas-Szabowski and Fusenig, 1996].

Alternatively, keratinocytes can be grown without feeder cells in serum-free media supplemented with pituitary extract or a variety of growth factors and other supplements, either at low (0.03–0.06 mM) Ca^{2+} concentrations to inhibit stratification or at normal (1.2–1.4 mM) Ca^{2+} concentrations to allow stratifica-

TABLE 22.1. Inhibition of Fibroblastic Overgrowth

Method	Agent	Tissue	Reference
Selective detachment	Trypsin	Fetal intestine, cardiac muscle, epidermis	Owens et al. [1974], Milo et al. [1980]
	Collagenase	Breast carcinoma	Freshney [1972]; Lasfargues [1973]
Confluent feeder layers	Mouse 3T3 cells	Epidermis	Rheinwald and Green [1975]
	Fetal human intestine	Normal and malignant breast epithelium	Stampfer et al. [1980]
		Colon carcinoma	Freshney et al. [1982b]
Selective inhibitors	D-valine	Kidney epithelium	Gilbert and Migeon [1975, 1977]
	Cis-OH-proline	Cell lines	Kao and Prockop [1977]
	Ethylmercurithiosalicylate	Neonatal pancreas	Braaten et al. [1974]
	Phenobarbitone	Liver	Fry and Bridges [1979]
	Antimesodermal antibody	Squamous carcinomas	Edwards et al. [1980]
		Colonic adenoma	Paraskeva et al. [1985]
	Geneticin	Melanocytes, melanoma	Levin et al. [1995]
Selective media	MCDB 153	Epidermis	Boyce and Ham [1983]
	MCDB 170	Breast	Hammond et al. [1984]
	Low Ca^{2+}	Epidermal melanocytes	Naeyaert et al. [1991]

tion and keratinization. [For a review, see Holbrock and Hennings, 1983.] Although maintenance at high cell density in subcultures, in the presence of feeder cells (e.g., irradiated 3T3 cells), or culture in a low-calcium, serum-free medium helps to reduce fibroblast contamination, the mesenchymal cells are not completely eliminated by these procedures.

PROTOCOL 22.1 EPIDERMAL KERATINOCYTES

Principle

Separation of the epithelial compartment from the underlying connective tissue is usually done by enzymatic digestion using trypsin [e.g., see Smola et al., 1993], dispase type II (2.4 U/ml Boehringer Mannheim, [Tomakidi et al., 1997]), or thermolysin [Germain et al., 1993]. The isolated epithelium is further dispersed by additional incubation in trypsin or, mechanically, by pipetting, after which it is filtered through nylon gauze and propagated in a serum-free, low-calcium medium or on growth-arrested feeder cells by using different media formulations. Subpopulations of keratinocytes with stem cell characteristics can be isolated due to their selective attachment to basement membrane constituents. [See Bickenbach and Chism, 1998.]

Outline

Separate the epidermis from the dermis enzymatically, disaggregate the keratinocytes, and seed them in a serum-free medium or on a growth-arrested feeder layer.

Materials

Sterile:

Prepared feeder layers, 3-d-old. (See Protocol 22.4.) Irradiate 3T3 at 30 Gy and human fibroblasts at 70 Gy

FAD: high-calcium medium for feeder layer cultures; Ham's F12 and DMEM with additives, as formulated by Wu et al. [1982]

Ham's F12:DMEM, 1:3, supplemented with adenine (1.8×10^{-4} M) cholera toxin (10^{-10} M) EGF (1 to 10 ng/l), hydrocortisone (0.4 μg/ml), and 5–10% FCS, as well as antibiotics (penicillin, 100 U/ml), and streptomycin (50 μg/ml)

Keratinocyte growth medium (KGM): serum-free medium (Clonetics, Promocell, Gibco) based on the original formulation of the medium MCDB 153 by Boyce and Ham [1983], supplemented with bovine pituitary extract with low Ca^{2+} (0.03–0.5 mM), which can be increased by adding $CaCl_2$

Keratinocyte-defined medium (KDM): serum-free medium (Clonetics, Promocell, Gibco) based on the original formulation of the medium MCDB 153 by Boyce and Ham [1983]; fully defined medium, tested as a fully competent medium in organotypic cultures [Stark et al., 1999]

Supplemented KDM (SKDM):

(a) Keratinocyte-defined medium (KDM) without pituitary extract (Promocell, Clonetics)

(b) Supplements: insulin (5 μg/ml), hydrocortisone (0.5 μg/ml), rh-EGF (0.1 ng/ml), transferrin (20 μg/ml), 0.1% highly purified bovine serum albumin (free of endotoxins and fatty acids), and L-ascorbic acid (50 μg/ml)

(c) Adjust Ca^{2+} to 1.3 mM.

(d) Primary keratinocyte cultures can be maintained on uncoated plastic dishes in SKDM, although with low proliferative activity. Growth is improved when keratinocytes are plated on collagen (type I)-coated dishes.

In organotypic cultures on lifted collagen gels populated with fibroblasts, this defined medium provided epidermal growth and differentiation indistinguishable from that obtained in FAD [Stark et al., 1999].

PBSA

EDTA (0.5%)

Glycerol (analytical grade)

FAD medium with 20% FCS and 10% glycerol

Enzymes:

(a) Thermolysin (T-7902, Sigma)

(b) Trypsin (1:250; 0.2 and 0.6%)

(c) Dispase (grade II, 2.4 U/ml, Boehringer Mannheim)

Cryopreservation ampules

Centrifuge tube, 50 ml

Nylon gauze

"Cell Strainer" (Becton Dickinson)

Petri dishes, bacteriological grade, 100 mm

Scalpels, curved forceps

Protocol

Specimens:

1. Obtain foreskin (neonatal as well as juvenile), the most frequent laboratory source for human skin, or trunk skin obtained from surgery or post mortem (up to 48 h). Keratinocytes derived from foreskin seem to attach and proliferate better than cells obtained from adult skin.

2. Incubate skin biopsies for up to 30 min in betadine solution (10% iodine), to prevent infection. (This does not visibly decrease keratinocyte cell viability.)

3. Rinse twice in PBSA for 10 min each.

Epidermal Separation:
Split-thickness skin is optimal.
4. Slice full-thickness skin with a Castroviejo dermatome (Storz Instrument) set to 0.1–0.2 mm. Alternatively, subcutaneous tissue and part of the dermis can be eliminated with curved scissors.
5. Dissect skin into 1 × 2-cm pieces with scalpels.
6. Rinse the tissue two to five times in PBSA.
7. Incubate tissue with protease by one of the following steps:
   (a) Trypsin: Float skin samples on 0.6% trypsin in PBSA (pH 7.4) for 20–30 min at 37°C. Alternatively (and this works particularly well also with full-thickness skin):
   (b) Cold trypsin: Float the samples on ice-cold 0.2% trypsin at 4°C for 15–24 h. The pH of the trypsin solution has to be monitored with phenol red to avoid a shift leading to altered enzyme activity and cell viability.
   (c) Dispase (grade II): Incubate with 2.4 U/ml for 30 min at 37°C.
   (d) Thermolysin (0.5 mg/ml) [Germain et al., 1993]: Incubate for 12–16 h at 4°C.
8. Monitor the separation of the epidermis carefully. When the first detachment of the epidermis is visible at the cut edges of skin samples, place the pieces (dermis side down) in 100-mm plastic Petri dishes and irrigate with 5 ml complete culture medium including serum.
9. Peel off the epidermis with two fine curved forceps, and collect it in a 50-ml centrifuge tube containing 20 ml of complete culture medium.
   (a) When separation has been performed with trypsin, viable keratinocytes are detached from the epidermal parts by gently pipetting and sieving through nylon gauze of 100-μm mesh ("Cell Strainer," Becton Dickinson).
   (b) To obtain single cells from epidermis separated by dispase or thermolysin, additional incubation with trypsin and EDTA (0.2% and 0.05%, respectively) for 10 min at 37°C is required.
10. After splitting the skin in trypsin, gently scrape off the loosely attached epidermal cells on the remaining dermal part, and add the cells to the epidermal suspension. This can lead to a higher cell yield if the purity of the keratinocyte population is not a major issue, but the amount of mesenchymal cell contamination in the resulting keratinocyte cultures is significantly higher.
11. Wash the isolated epidermal cells twice in culture medium by centrifugation at 100 g for 10 min, and count the total number of cells and the viable cells (those that do not absorb trypan blue; see Protocol 21.1).

Primary Culture:
12. Seed cells at 37°C in FAD or KGM medium at desired densities:
   (a) In FAD with feeder cells (dishes seeded 3 d previously with postmitotic 3T3 cells [Rheinwald and Green, 1975] or skin fibroblasts [Limat et al., 1989]), seed keratinocytes at 2–5 × 10⁴ cells/cm².
   (b) For culture free of fibroblasts, seed primary keratinocytes at high density (1–5 × 10⁵ cells/cm²) in FAD, and maintain them at high density in subcultures.
   (c) Alternatively, seed cells at low (10³ cells/cm²) and high (5 × 10⁴ cells/cm²) density in KGM in low Ca²⁺ (0.03–0.1 mM), and subculture in the same medium.
   (d) Cells will attach and grow, but at a lower rate, in SKDM, so this medium is advised for experimental conditions when low proliferative activities are required.
13. When cells have attached (after 1–3 d), rinse cultures extensively with medium to eliminate nonattached dead and differentiated cells, and continue cultivation in either FAD or KGM. Stratification and slowing down of growth can be achieved by shifting the Ca²⁺ concentration in KGM.

Subculture:
14. Subculture as follows:
   (a) Cultures in FAD:
      (i) Incubate in 0.05–0.1% EDTA for 5–15 min to initiate cell detachment, which is visible by the enlargement of intercellular spaces.
      (ii) Incubate in 0.1% trypsin and 1.3 mM (0.05%) EDTA at 37°C for 5–10 min, followed by gentle pipetting, to completely detach the cells.
   (b) Cultures in KGM:
      (i) EDTA pretreatment is not required, due to the low Ca²⁺ concentration.
      (ii) Incubate in 0.1% trypsin with 1.3 mM EDTA, as with FAD cultures.

Cryopreservation:
15. Cryopreservation is often necessary to maintain large quantities of cells derived from the same tissue sample; the best results are reported when cells from preconfluent primary cultures are used.
   (a) Trypsinize cells as before, and centrifuge at 100 g for 10 min.
   (b) Resuspend cells in complete culture medium with serum, and count.
   (c) Dispense aliquots of 2 × 10⁶ cells/ml in FAD medium with 20% FCS and 10% glycerol into cryopreservation tubes.

(d) Equilibrate at 4°C for 1–2 h.

(e) Freeze cells with a programmed freezing apparatus (Planer) at a cooling rate of 1°C per min.

(f) To recover cells:

(i) Thaw cryotubes quickly in a 37°C water bath.

(ii) Dilute cells tenfold with medium.

(iii) Centrifuge cells and resuspend them at an appropriate concentration in the desired culture medium, and seed culture vessel.

Human and mouse cells can be grown in all three media for several months. While mouse cells can be subcultured only once or twice, human cells can be passaged four and seven times in FAD and KGM, respectively.

Characterization

Cultured cells have to be characterized for their epidermal (epithelial) phenotype to exclude contamination by mesenchymal cells. This is best achieved using cytokeratin-specific antibodies for the epithelial cells. Contaminating endothelial cells can be identified by antibodies against CD31 or factor VIII-related antigen. Identifying fibroblasts unequivocally is difficult, because the use of antibodies against vimentin (the mesenchymal cytoskeletal element) is not specific; keratinocytes *in vitro* may initiate vimentin synthesis at frequencies that depend on culture conditions [Smola et al., in preparation]. As a practical assessment for mesenchymal cell contamination, cells should be plated at clonal densities (1–5 $\times$ 10^2 cells/cm^2) on feeder cells, and clone morphology should be identified at low magnification following fixation and hemalium and eosin (H&E) staining of 10- to 14-d cultures. A more specific and highly sensitive method to identify contaminating fibroblasts is the analysis of expression of keratinocyte growth factor (KGF) by RT-PCR. Since this factor is produced in fibroblasts and not in keratinocytes, it represents a selective marker. Moreover, KGF expression is enhanced by cocultured keratinocytes so that a minority of contaminating fibroblasts will be detected by this assay [Maas-Szabowski et al. 1999].

Variations

Keratinocytes can be obtained from the outer root sheaths (ORSs) of plucked scalp hair follicles by dissociating cells from the dissected follicle [Limat et al., 1989]. Up to 5 $\times$ 10^3 cells can be obtained from three hair follicles and plated on fibroblast feeder cells, resulting in about 1 $\times$ 10^6 cells within 15 d. Like interfollicular keratinocytes, ORS cells can be subcultured on feeder layer dishes and cryopreserved. These ORS-derived keratinocytes are similar in culture

to interfollicular cells, form a regular neoepidermis when transplanted onto nude mice, and are used clinically to cover chronic wounds [Limat et al., 1996].

Keratinocytes can also be grown at clonal density to study clonal cell populations and their different proliferation potentials. For this purpose, keratinocytes are cocultured with X-irradiated 3T3 cells [Rheinwald and Green, 1975], at reduced Ca^{2+} concentrations with fibroblast-conditioned medium [Yuspa et al., 1981] or in defined serum-free medium [Boyce and Ham, 1983].

In order to provide more *in vivo*-like conditions and to study the regulation of skin physiology by epithelial–mesenchymal interactions, organotypic coculture systems have been developed by seeding the cells on collagen gels (populated with fibroblasts) or pieces of dermis and lifting the supports to the air–medium interphase. [For a review of this procedure, see Fusenig, 1994a.] Under these improved growth conditions, keratinocytes express many aspects of growth and differentiation of the epidermis *in vivo*, including ultrastructural features and a complete basement membrane, features that are absent or less pronounced in submerged cultures on plastic [Smola et al., 1998]. Such organotypic cultures can now also be established and maintained for three weeks in defined SKDM medium, allowing the molecular interaction between epithelial and mesenchymal cells to be analyzed without disturbing influences from serum or other undefined tissue extracts [Stark et al., in press].

Such organotypic keratinocyte cultures are also available commercially, either as cocultures with fibroblasts or monocultures of keratinocytes growing on polycarbonate filter inserts (Falcon #3501), and are increasingly used for *in vitro* toxicity and biocompatibility testing of skin-related products [Fusenig, 1994b].

Optimal growth and differentiation of isolated keratinocytes are obtained under *in vivo* conditions when the cells are transplanted as intact cultures or in suspension onto nude mice. [For review, see Fusenig 1994a.] This leads to an almost complete expression of cells, differentiation characteristics of normal epidermis, and all biochemical and ultrastructural features of a completely keratinized epithelium, including the formation of a basement membrane [Breitkreutz et al., 1997]. Combined cultures of epidermis and stroma, mounted on mesh filters, are now commercially available (Episkin, Epiderm) and have been suggested as models for irritation and inflammation research.

Cornea

There have been a number of attempts to replace the rabbit's eye (Draze) test by using cultured corneal epithelium, so there has been considerable interest in culturing cornea. The following protocol for the culture of normal human corneal cells in a serum-free medium was provided by Carolyn Cahn, Gillette Medical Evaluation Laboratories, 401 Professional Drive, Gaithersburg, MD 20879.

PROTOCOL 22.2. CORNEAL EPITHELIAL CELLS

Principle

The corneal epithelium contacts the external environment directly and is the first tissue compromised in ocular injury. Animal models are most frequently used to model the human ocular surface. The system to be described is being developed for *in vitro* toxicological investigations, which can complement *in vivo* studies.

The corneal epithelium *in vivo* is oxygenated by diffusion from the tear film; in addition, it undergoes frequent mitosis. These two characteristics, coupled with the availability of donor corneal tissue, combine to make the corneal epithelium a good starting material for the generation of primary cultures.

Among the sources for the acquisition of corneal tissue is a sophisticated eye-banking system that provide donor corneas for transplantation. Eye-bank tissue is downgraded and made available to researchers after three days of storage, due to the limited *in vitro* viability of the endothelial cell layer. While the endothelium is labile, the epithelium retains its generative capacity for extended periods of time in storage at 4°C, making expired tissue suitable as a source of viable epithelium.

Outline

Corneal tissue is placed on collagen and allowed to adhere, and the outgrowth is expanded and propagated on fibronectin-collagen-coated surfaces. Early-passage cultures are transfected with SV40 early-region genes or infected with intact Ad12–SV40 hybrid virus. Large T-antigens apparently confer an extended life span. (See Protocol 17.1.)

Materials

Sterile or Aseptically Prepared:
Keratinocyte serum-free medium (KGM, Clonetics) containing 0.15 mM of calcium, human epidermal growth factor (0.1 ng/ml), insulin (5 μg/ml), hydrocortisone (0.5 μg/ml), and bovine pituitary extract (30 μg/ml)
Eagle's Minimal Essential Medium
PBSA
Fetal bovine serum
Trypsin-EDTA: Trypsin, 0.05%, EDTA, 0.5 mM

Fibronectin/Collagen (FNC) (Bethesda Research Faculty and Facility, Ijamsville, MD); this solution consists of fibronectin, 10 μg/ml, collagen, 35 μg/ml, with bovine serum albumin (BSA), 100 μg/ml, added as a stabilizer
Biocoat six-well plate precoated with rat-tail collagen, type I (Collaborative Research)

Protocol

Primary Cultures:

1. Place donor corneas epithelial side up on a sterile surface, and cut them into 12 triangular-shaped wedges, using a single cut of the scalpel and avoiding any sawing motion. Careful handling of the cornea in this manner decreases damage to the collagen matrix of the stroma and minimizes the liberation of fibroblasts.

2. Turn each corneal segment epithelial side down, and place four segments in each well of a six-well tray (precoated with rat-tail collagen, type I, Biocoat, Becton Dickinson).

3. Press each segment down gently with forceps to ensure good contact between the tissue and the tissue culture surface. Allow the tissue to dry for 20 min.

4. Place one drop of keratinocyte serum-free medium carefully upon each segment, and incubate the culture overnight at 37°C in 5% CO_2. Although the donor corneas received from the eye bank are stored in antibiotic-containing medium (either McCarey–Kauffman or Dexsol), all manipulations are performed under antibiotic-free conditions.

5. The next day, add 1 ml of medium to each well. During the initial culture period, cells are observed to emigrate only from the limbal region of the cornea. No cells are observed to migrate away from the central cornea or the sclera. Fibroblast outgrowth is minimized by utilizing a serum-free medium that is low in calcium (0.15 mM) and that minimizes disruption of the collagen matrix.

6. Remove the tissue segment with forceps 5 d after the explantation of the donor cornea slice, and add 3 ml of medium. After removal of the donor tissue, adherent cells continue to proliferate, and within 2 weeks from the time of establishment of the culture, confluent monolayers form, displaying the typical cobblestone morphology associated with epithelia. The yield is approximately 6 × 10⁶ cells/cornea.

Propagation:

7. Following the initial outgrowth period, feed the cultures twice per week.

8. At 70–80% confluence, rinse the cells in Dulbecco's phosphate-buffered saline (PBSA), and

release with trypsin/EDTA (0.05% trypsin, 0.53 mM EDTA) for 4 min at 37°C.

9. Stop the reaction with 10% FBS in PBSA.

10. Wash the cells (centrifugation followed by resuspension in KGM), count them, and plate at 1×10^4 cells/cm² onto tissue culture surfaces coated with FNC.

11. Incubate the culture at 37°C in 95% air and 5% CO_2.

12. Exchange the culture medium with fresh medium 1 d after trypsinization and reseeding.

Immediately after passage, cells appear more spindle shaped, are refractile, and are highly migratory. Within 7 d, control cultures become 70–80% confluent, continue to display a cobblestone morphology, and, if allowed to become postconfluent, retain the ability to stratify in discrete areas.

Although corneal epithelial cultures can be subcultured up to five times (approximately 9–10 population doublings), most of the proliferation occurs between the first and third passages. Approximate yields are $1-2 \times 10^6$ cells/cornea. Senescence always ensues by P5.

Development of Continuous Cell Lines

Lines of human corneal epithelium (HCE) with an extended life span have been developed (see Protocol 17.1), to provide a larger supply of cells for experimental purposes.

Long-Term Storage of Cells

Cells can be stored frozen in liquid nitrogen. (See Protocol 19.1.)

Phenotypic Development in Vitro. Both primary cultures and HCE lines retain phenotypic characteristics of corneal epithelium *in situ*. They continue to synthesize collagenase and express EGF receptors and corneal-specific cytokeratins, although the level of expression under current culture conditions is less than that observed *in situ*. When corneal epithelium is cultured upon collagen membranes at air–liquid interfaces, its morphology and barrier function are fairly well preserved. Stratified membranes develop that are able to inhibit the diffusion of Na-fluorescein. Air–liquid interface cultures survive for 2 weeks *in vitro* and have been used to investigate injury and repair mechanisms. Studies are underway to reconstitute cells into more complete three-dimensional tissue models by supplying corneal fibroblasts in a collagen gel.

Human corneal epithelial cells propagated *in vitro* may provide a suitable model for exploring basic cell biological mechanisms as well as toxicological phenomena. Although the primary cultures are adequate for *in vitro* studies, HCE lines with an extended life span provide a reliable source of material that can be shared among laboratories.

Breast

Milk [Buehring, 1972; Ceriani et al., 1979] and reduction mammoplasty are suitable sources of normal ductal epithelium from the breast; the former gives purer cultures of epithelial cells. Disaggregation in collagenase [Speirs et al., 1996] is preferred for primary disaggregation, growth on confluent feeder layers of fetal human intestine [Stampfer et al., 1980; Freshney et al., 1982b] represses stromal contamination of both normal and malignant tissue (see Fig. 13.12), and optimization of the medium [Stampfer et al., 1980; Smith et al., 1981; Hammond et al., 1984] enables serial passage and cloning of the epithelial cells. Cultivation in collagen gel allows three-dimensional structures to form that correlate well with the histology of the original donor tissue [Berdichevsky et al., 1992; Gomm et al., 1997].

As with epidermis, cholera toxin [Taylor-Papadimitriou et al., 1980] and EGF [Osborne et al., 1980] stimulate the growth of epithelioid cells from normal breast tissue *in vitro*.

The hormonal picture is more complex. Many epithelial cells survive better with insulin, (1–10 IU/ml) and hydrocortisone ($\sim 1 \times 10^{-8}$ M), added to the culture. The differentiation of acinar breast epithelium in organ culture requires hydrocortisone, insulin, and prolactin [Darcy, et al., 1995], and estrogen, progesterone, and growth hormone have also been shown to be required in cell culture [Klevjer-Anderson and Buehring, 1980].

The following protocol for the culture of cells from human milk was contributed by Joyce Taylor-Papadimitriou, ICRF Breast Cancer Biology Group, Guy's Hospital, Thomas Guy House, London SE1 9RT, UK.

PROTOCOL 22.3. MAMMARY EPITHELIUM

Principle

Milk from early lactation or after weaning gives the highest cell yield and contains clumps of epithelium that can proliferate in culture [Buehring, 1972; Taylor-Papadimitriou et al., 1980]. Primary cultures grown in a hormone-supplemented human-serum-containing medium give cell lines of limited life span, but that are clonogenic [Stoker et al., 1982]. These lines are eventually overtaken by nonepithelial "late milk" cells [McKay and Taylor-Papadimitriou, 1981].

Outline

Cells centrifuged from early-lactation milk are grown in the presence of endogenous macrophages in an enriched medium and may be subcultured with a mixed protease chelating solution.

Materials

Sterile:

Growth medium RPMI 1640

Fetal calf serum (FCS, ICN)

Human serum (HuS; outdated pooled serum from blood banks; Australia antigen negative)

Nunc plastic dishes, 5 cm

Universal containers or 20–50-ml centrifuge tubes

Stock solutions:

  (a) Insulin (Sigma), 1 mg/ml in 6 mM HCl

  (b) Hydrocortisone, 0.5 mg/ml in physiological saline

  (c) Cholera toxin (Schwartz-Mann), 50 μg/ml in physiological saline

Note. Serum and stock solutions of insulin, hydrocortisone, cholera toxin, pancreatin, and trypsin should be kept at $-20°C$

Trypsinization solution (TEGPED):

  (d) EGTA (Sigma, ethylene glycol-bis-(β-aminoethylether) N'N' tetraacetic acid), 13 mM in PBSA ....10 ml

  (e) EDTA (Sigma, diaminoethane tetraacetic acid), 7 mM in PBSA ....4 ml

  (f) Trypsin (Difco), 0.2% in HBSS ....4 ml

  (g) Pancreatin (Difco), 1.0% in HBSS...........2 ml

Growth medium: RPMI 1640 containing 15% FCS; 10% HuS; cholera toxin, 50 ng/ml; hydrocortisone, 0.5 μg/ml; insulin, 1 μg/ml

Protocol

Milk (2–7 d postpartum) can best be collected on hospital wards. The breast is swabbed with sterile H_2O and the milk manually expressed into a sterile container. Five to 20 ml are usually obtained per patient. The milks are pooled and diluted 1:1 with RPMI 1640 medium to facilitate centrifugation.

Primary Cultures:

1. Spin diluted milk at 600–1,000 g for 20 min. Carefully remove the supernatant, leaving some liquid so as not to disturb the pellet.
2. Wash the pelleted cells two to four times with RPMI containing 5% FCS until the supernatant is not turbid.
3. Resuspend the packed cell volume in growth medium, and plate 50 μl of packed cells in 5-cm dishes (Nunc) in a 6-ml growth medium. Incubate the cultures at 37°C in 5% CO_2.
4. Change the medium after 3–5 d and thereafter twice weekly. Colonies appear around 6–8 d and expand to push off the milk macrophages, which initially act as feeders.

Subculture of Milk Cells:

5. Incubate the cells in TEGPED (1.5 ml per 5-cm plate) at 37°C for 5–15 min, depending on the age of the culture, to produce a single-cell suspension.
6. Centrifuge cells at 100 g for 5 min and resuspend in 6-ml of medium.
7. Divide the suspension, 2 ml into each of three fresh 5-cm dishes or 25-cm² flasks.
8. Dilute threefold by adding 4 ml of fresh medium.

Variations

It is convenient to use the macrophages that are already present in the milk as feeders, but they are gradually lost as the epithelial colonies expand. However, macrophages can be removed by absorption to glass, and in that case other feeders must be added. Irradiated or mitomycin-treated 3T6 cells (see Protocols 13.3, 22.4) show the best growth-promoting activity [Taylor-Papadimitriou et al., 1977]. Analogues of cyclic AMP can be used to replace the cholera toxin [Taylor-Papadimitriou et al., 1980], although this is not possible with macrophage feeders, which are killed by the analogues.

Uses and Application of Milk Epithelial Cell Culture. Milk cultures provide cells from the fully functioning gland and allow the definition of phenotypes by immunological markers [Chang and Taylor-Papadimitriou, 1983]. Milk cells have been successfully transformed by SV40 virus [Chang et al., 1982] and provide an important source of normal cells for comparison with breast cancer cell lines and for transfection with oncogenes.

Cervix

This protocol for the culture of epithelial cells from cervical biopsy samples was contributed by Margaret Stanley, Department of Pathology, University of Cambridge, UK.

PROTOCOL 22.4. CERVICAL EPITHELIUM

(a) With 3T3 Feeder Layers

Principle

Cervical keratinocytes can be grown in serial culture at clonal density by using a modification [Stanley and Parkinson, 1979] of the method described for epidermal keratinocytes [Rheinwald and Green, 1975].

Outline

Single-cell suspensions from enzymatically disaggregated epithelium from punch or wedge cervical biopsies are inoculated into flasks or plates together with lethally inactivated Swiss 3T3 cells and are grown in a medium supplemented with serum and growth factor. When the keratinocyte colonies that arise contain 1,000 cells or more, the cultures can be trypsinized and passaged again with the use of fibroblast feeder support.

Materials

Sterile:

Transport medium: Dulbecco's modification of Eagle's medium (DMEM), supplemented with 10% fetal calf serum, 100 μg/ml of gentamicin sulfate, and 10 μg/ml of amphotericin

Trypsin EDTA solution for disaggregation of epithelial cells: 0.25% trypsin (v/v), 0.25 mM (0.01%) EDTA as the disodium salt with pH 7.4 in PBSA

Keratinocyte culture medium (KCM): DMEM supplemented with 10% fetal bovine serum, 0.5 μg/ml of hydrocortisone, and 1 $\times$ 10^{-10} M cholera toxin

Fetal bovine serum: not all batches support growth adequately. Serum samples should be tested and a large batch of suitable quality bought.

Epidermal growth factor (EGF): Sigma 100 μg dissolved in 1 ml sterile UPW. Aliquot in 100 μl aliquots and store at $-20°$C. Prepare working stocks by diluting 100 μl of EGF at 100 μg/ml in 10 ml medium with serum. Store at 4°C. Dilute 1:100 in medium with serum for use at a final concentration of 10 ng/ml.

Cholera toxin (CT): Sigma. Add 1.18 ml of sterile UPW to 1 mg of CT to give a 10^{-5}-M solution. Store at +4°C. Working stocks should be 100 μl of 10^{-5}-M CT in 10 ml of medium with serum. Filter sterilize the culture and keep it at +4°C. Use at a final concentration of 10^{-10} M.

Hydrocortisone: Sigma. Dissolve 1 mg of hydrocortisone in 1 ml of 50% ethanol/water (v/v). Dispense into 100-μl aliquots and keep at $-20°$C. Use at a final concentration of 0.5 μg/ml.

Petri dishes for tissue culture, 60 mm and 90 mm diameter

Forceps, rat toothed

Curved iris scissors

Disposable scalpels, No. 22 blade

Pipettes

Centrifuge tubes

Nonsterile:

Hemocytometer

Radiation source (e.g., X rays or ^{60}Co)

Swiss 3T3 Fibroblasts:

(a) A large master stock of cells should be prepared and frozen in individual ampules of 1 $\times$ 10^6 cells. Cells should not be used for more than 20 passages.

(i) Grow 3T3s in DMEM/10% calf serum in 175-cm^2 tissue culture flasks. Inoculate cells at 1.5 $\times$ 10^4 cells/cm^2. Change the medium after 2 d. Subculture every 4–5 d.

(ii) To avoid low-level contamination, maintain one master flask of cells on antibiotic-free medium; these cells are then used at each passage to inoculate the flasks required for that week's feeder cells.

(b) Feeder layers are inactivated by irradiation with 60 Gy (6,000 rad), either from an X-ray or ^{60}Co source. Irradiated cells (XR-3T3) may be kept at 4°C for 3–4 d.

(c) In the absence of a source of irradiation, inactivate feeder cells with mitomycin C.

(i) Expose 3T3 cells growing in monolayer to 400 μg/ml of mitomycin C for 1 h at 37°C.

(ii) Trypsinize the treated cells, resuspend and wash the cell pellet twice with fresh medium with serum, resuspend the cells at a suitable concentration in complete medium, and use.

Protocol

Primary Culture:

1. Remove the cervical biopsy from the transport medium, and wash the cells two to three times with 5 ml of sterile PBSA containing gentamicin sulfate, 50 μg/ml, and amphotericin, 5 μg/ml.

2. Place the biopsy, epithelial surface down, on a sterile culture dish.

3. Using a disposable scalpel fitted with a No. 22 blade, cut and scrape away as much of the muscle and stroma as possible, leaving a thin, opaque epithelial strip.

4. Mince the epithelial strip finely with curved iris scissors.

5. Add 10 ml of trypsin/EDTA (prewarmed to 37°C) to the epithelial mince, and transfer the tissue to a sterile glass universal containing a small plastic-coated magnetic stirrer bar.

6. Add a further 5–10 ml of trypsin/EDTA.

7. Place the universal on a magnetic stirrer in an incubator or a hot room at 37°C, and stir slowly for 30–40 min.

8. Allow the suspension to stand at room temperature for 2–3 min.

9. Remove the supernatant containing single cells, and filter it through a stainless steel or plastic mesh into a 50-ml centrifuge tube. (See Fig. 11.10.) Add 10 ml of complete medium to this suspension.

10. Add a further 15 ml of warm trypsin/EDTA to the fragments in the universal, and repeat the steps 5–9 twice. Combine the trypsin superna-

tants and spin the suspension in a bench centrifuge at 1,000 rpm (80 g) for 5 min.

11. Remove the supernatant; add 10 ml of complete medium to the pellet, resuspend the cells vigorously to give a single-cell suspension, and count the cells with a hemocytometer. Assess cell viability with trypan blue exclusion.

12. Dilute the cervical cell suspension with KCM, and plate cells out at 2×10^4 cells/cm^2 together with 1×10^5 cells/cm^2 of lethally inactivated 3T3 cells (i.e., for 1×10^5 cervical cells, 5×10^5 XR-3T3/60-mm dish).

13. Incubate the cultures at 37°C in 5% CO$_2$.

14. Seventy-two hours after the initial plating, replace the medium with a complete medium supplemented with EGF at 10 ng/ml. Check the cultures microscopically to ensure that the feeder layer is adequate. Add further feeder cells if necessary.

15. Change the medium twice weekly; EGF should be present in the medium, except when the cells are initially plated. Keratinocyte colonies become visible on the microscope by days 8–12 and should be visible to the naked eye by days 14–16. Cultures should be passaged at this time.

Subculture:

1. Remove the medium from the cell layer, and remove the feeders by rinsing rapidly with 0.01% EDTA. Wash twice with PBSA.

2. To each culture dish, add enough prewarmed trypsin/EDTA to cover the cell sheet. Leave the cultures at 37°C until the keratinocytes have detached; check for detachment with a microscope. Do not leave the cells in trypsin for more than 20 min.

3. Remove the cell suspension from the plate and transfer it to a sterile centrifuge tube.

4. Rinse the growth surface with complete medium and add to the suspension. Mix and dispense the suspension with a 10-ml pipette.

5. Spin the cells at 1,000 rpm for 5 min.

6. Remove the supernatant, add 10 ml of complete medium, and resuspend the cells vigorously with a 10-ml pipette to achieve a single-cell suspension.

7. Count the cells with a hemocytometer.

8. Cells may be replated on inactivated 3T3 cells and grown as just described or frozen for later recovery.

(b) Serum-free Medium

Cervical keratinocytes can also be grown in a serum-free keratinocyte growth medium, without feeder support, by using modifications of the methods originally described for epidermal keratinocytes (Boyce and Ham, 1983).

Outline

Single-cell suspensions from enzymatically disaggregated epithelium from punch or wedge cervical biopsies are inoculated into flasks or plates and grown in a keratinocyte growth medium. Confluent sheets of keratinocytes can be trypsinized and passaged again.

Materials

Sterile:

Keratinocyte growth medium: KGM (Bio-Whittaker), Keratinocyte-SFM (Gibco)

Tissue culture dishes, 60 mm and 90 mm in diameter

Forceps, rat toothed

Curved iris scissors

Disposable scalpels with no. 22 blades

Pipettes

Centrifuge tubes

Nonsterile:

Hemocytometer

Protocol

Primary Culture:

1. Follow steps 1–11 of the above protocol with 3T3 feeder layers to obtain a single-cell suspension in medium supplemented with serum.

2. Spin the suspension at 1,000 rpm (80 g) for 5 min. Remove the supernatant and resuspend the cells in 10 ml of PBSA.

3. Spin the suspension again and wash the cells once more with PBSA. Resuspend the cells in 10 ml of keratinocyte growth medium (prepared as per the manufacturer's instructions), and seed into culture dishes or flasks at a density of 2×10^5 cells/cm^2 (10^6 cells/50-mm Petri dish, 4×10^6/90-mm Petri dish).

4. Incubate the cultures at 37°C in 5% CO$_2$.

5. Change the medium after 72 h and twice weekly thereafter. Cultures become confluent within 14–20 d and should be passaged at that time.

Subculture:

1. Follow steps 1–7 of the above subculture protocol with 3T3 feeder layers.

2. Spin the suspension at 1,000 rpm (80 g) for 5 min. Remove the supernatant and resuspend the cells in 10 ml of PBSA.

3. Spin the suspension again and wash the cells once more with PBSA. Resuspend the cells in KGM and plate them onto culture dishes at 10^5 cells/cm^2 (5×10^5 cells/50-mm Petri dish, 2×10^6 cells/90-mm Petri dish).

4. Cells may also be frozen at this stage for recovery at a later date.

Cervical keratinocytes grown as just described can be used for a range of purposes, including investigations of papillomavirus carcinogenesis, differentiation studies, and examination of the response of these cells to mutagens and carcinogens.

Gastrointestinal Tract

The culture of normal epithelium from the gut lining has not been extensively reported, although there are numerous reports in the literature of continuous lines from human colon carcinoma. (See Colon in Chapter 23). Owens et al. [1974] were able to culture cells from fetal human intestine as a finite cell line (FHS 74 Int), and extensive use has been made of the rat intestinal cell line IEC-6 [Rak et al., 1995; Bedrin et al., 1997]. This protocol has been condensed from Paraskeva and Williams [1992], which should be consulted for greater detail and alternative methodology and applications.

PROTOCOL 22.5. COLORECTAL EPITHELIUM

The preparation of tissue specimens for cell culture is usually started within 1–2 h of removal from the patient. If this is impossible, fine cutting of the tissue into small pieces (1–2 mm) with scalpels and storage overnight at 4°C in washing medium can also prove successful.

Outline

In general, adenomas are usually digested enzymatically, and carcinomas can be dealt with simply by cutting with surgical blades. If a well-differentiated colorectal cancer does not readily release tumor cells when the specimen is cut, it can be digested enzymatically.

Materials

Sterile:

Growth medium: DMEM containing 2 mM glutamine and supplemented with 20% fetal bovine serum (batch selection is essential), hydrocortisone sodium succinate (1 μg/ml), insulin (0.2 u/ml), 2 mM glutamine, penicillin (100 u/ml), and streptomycin (100 μg/ml).

Washing medium: growth medium with 5% FBS, 200 u/ml of penicillin, 200 μg/ml of streptomycin, and 50 μg/ml of gentamycin. Gentamycin is kept in the primary culture for at least the first week and then is removed.

Digestion solution: DMEM and antibiotics, as described for the washing solution, with collagenase (1.5 mg/ml, Worthington type IV) hyaluronidase (0.25 mg/ml, Sigma type I) and 2.5–5% FBS. Although we use Worthington collagenase, Sigma culture grades can also be tried.

Dispase for subculture: Prepare Dispase (a neutral protease, Boehringer, grade 1) at 2 U/ml in DMEM containing 10% FBS, glutamine, penicillin, and streptomycin. Sterile filter the solution and store at −20°C. If a precipitate forms on thawing, the solution should be centrifuged and the active supernatant should be removed and used.

Protocol

(a) Colorectal tumors
Enzyme Digestion:

1. Wash tumor specimens four times in washing medium, and mince in a small volume of the same medium, just enough to cover the tissue. (Do not allow the tissue to dry out!) Mince the tissues with crossed surgical blades or sharp scissors to fragments of approximately 1 mm³.

2. After cutting, wash the tissue again four times (the number of washings can be varied with experience, depending on whether contamination is a factor) by bench centrifugation (300 g for 3 min) and resuspension.

3. After washing the tumor fragments, put them into the digestion solution, and rotate at 37°C, usually overnight (approximately 12–16 h). The time the specimens are left in the solution is not critical, because digestion is a mild process, but it is important not to let the digestion medium become acid during the procedure. A low pH indicates that the specimens were left in the medium for too long or too much tissue was put in the volume of the digestion mixture. Approximately 1 cm³ of tumor tissue is put into 20–40 ml of digestion solution.

Nonenzymic Tissue Preparation:

Adenomas almost invariably need digestion with enzymes. However, with carcinomas, it is often found that during the cutting of the tumor with blades into 1-mm³ pieces, small clumps of tumor cells are released from the tumor tissue into the washing medium. In this case, the following procedure can be carried out.

1. Remove the washing medium containing released clumps of cells, and separate the cells into large and small clumps by allowing them to settle by gravity for a few minutes in a centrifuge tube. Remove the supernatant phase.

2. Either put the remaining tissue pieces directly into the culture, or rotate the tissue gently for 30–60 min in washing medium to release more small clumps, which can then be removed, plated, and put into culture separately from the remaining larger pieces of tissue.

3. Wash all samples three times before putting them into culture.

Standard primary culture conditions. Growth medium in 25-cm² flasks coated with collagen type IV in the presence of Swiss 3T3 feeder cells (approximately 1×10^4 cells/cm²; see Protocol 22.4) at 37°C in a 5%-CO_2 incubator. Change the culture medium twice weekly.

4. Inoculate epithelial tubules and clumps of cells derived from tissue specimens into flasks in 4 ml of medium per 25 cm². The tubules and cells start to attach to the substratum, and epithelial cells migrate out within 1–2 d. Most of the tubules and small clumps of epithelium attach within 7 d, but the larger organoids can take up to 6 weeks to attach, although they will remain viable all that time.

The attachment of epithelium during primary culture and subculture is more reproducible and efficient when cells are inoculated onto collagen-coated flasks (Biocoat, Becton Dickinson; see also Protocols 22.2 and 22.9), and significantly better growth is obtained with 3T3 feeders than without.

When the epithelial colonies expand to several hundred cells per colony, they become less dependent on 3T3 feeders, and no further addition of feeders is necessary.

Subculture and Propagation

Most colorectal adenoma primary cultures and adenoma-derived cell lines cannot at present be passaged by routine trypsin/EDTA procedures [Paraskeva et al., 1984, 1985]. Disaggregation to single cells of the cultured adenoma cells with 0.1% trypsin in 0.25 mM (0.1%) EDTA will result in extremely poor or even zero growth, so Dispase is used instead.

1. Add Dispase to the cell monolayer, just enough to cover the cells (~2.5 ml/25-cm² flask), and leave the solution to stand for 40–60 min for primary cultures and 20–40 min for cell lines.

2. Once the epithelial layers begin to detach (they do so as sheets rather than single cells), pipette to help detachment and disaggregation into smaller clumps.

3. Wash and replate the cells under standard culture conditions. It may take several days for

clumps to attach, so replace the medium carefully when feeding.

(b) Normal Colon

One of the most reproducible and successful methods for the short-term primary culture of normal adult colonic epithelium is the following, described by Buset et al. [1987]:

1. Biopsies of about 1–3 mm³ are taken with biopsy forceps to sample only the mucosal layer and not the muscle layer.

2. After fine mincing with two scalpels, digest the tissue in a serum-free medium containing bovine serum albumin, neuraminidase, hyaluronidase, and collagenase for 10 min at 37°C and a further 1 h at room temperature.

3. Wash the digested tissue and seed cells in a serum-free NCTC 168 medium (JRH Biologicals) supplemented with ethanolamine, phospho-ethanolamine, hydrocortisone, ascorbic acid, transferrin, insulin, epidermal growth factor, pentagastrin, and deoxycholic acid.

The best substrate for cell attachment is a mixture of ungelled collagen I and bovine serum albumin. However, even under these conditions, normal adult colonic epithelial cells still have a very short life of no more than 4 d.

(c) Fibroblast Contamination

One or a combination of the following techniques can be employed to deal with fibroblast contamination:

(1) Physically remove a well-isolated fibroblast colony by scraping it with a sterile blunt instrument (e.g., a cell scraper). Care has to be taken to wash the culture up to six times to remove any fibroblasts that have detached in order to prevent them from reseeding and reattaching to the flask.

(2) Differential trypsinization can be attempted with the carcinomas [Kirkland and Bailey, 1986].

(3) Dispase preferentially (but not exclusively) removes the epithelium during passaging and leaves behind most of the fibroblastic cells attached to the culture vessel [Paraskeva et al., 1984]. During subculture, cells that have been removed with dispase can be preincubated in plastic Petri dishes for 2–6 h to allow the preferential attachment of any fibroblasts that may have been removed together with the epithelium. Clumps of epithelial cells still floating can be transferred to new flasks under standard culture conditions. This technique takes advantage of the fact that fibroblasts in general attach much more quickly to plastic than do clumps of

epithelial cells, so that a partial purification step is possible.

(4) Use a conjugate between anti-Thy-1 monoclonal antibody and the toxin ricin [Paraskeva et al., 1985]. Thy-1 antigen is present on colorectal fibroblasts, but not colorectal epithelial cells; therefore, the conjugate kills contaminating fibroblasts, but shows no signs of toxicity toward the epithelium, whether derived from an adenoma or a carcinoma.

(5) Reduce the concentration of serum to about 2.5–5% if there are heavy concentrations of fibroblastic cells. It is worth remembering that normal fibroblasts have a finite growth span *in vitro* and that using any or all of the preceding techniques will eventually push the cells through so many divisions that any fibroblasts will senesce.

Liver

Although cultures from adult liver do not express all the properties of liver parenchyma, there is little doubt that the correct lineage of cells may be cultured. So far, attempts at generating proliferating cell lines have not been particularly successful, but functional hepatocytes can be cultured under the correct conditions [Guguen-Guillouzo, 1992].

Some of the most useful continuous liver cell lines were derived from Reuber H35 [Pitot et al., 1964] and Morris [Granner et al., 1968] minimal-deviation hepatomas of the rat. Inducing tyrosine aminotransferase in these cell lines with dexamethasone proved to be a valuable model for studying the regulation of enzyme adaptation in mammalian cells [Granner et al., 1968; Reel and Kenney, 1968]. Cell lines such as Hep-G2 have also been generated from human hepatoma and retain some of the metabolizing properties of normal liver [Knowles et al., 1980].

The following protocol for the culture of isolated adult hepatocytes was contributed by Christiane Guguen-Guillouzo, Hôpital de Pontchaillou, INSERM U49, Rue Henri le Guilloux, 35033 Rennes, France.

PROTOCOL 22.6. ISOLATION OF RAT HEPATOCYTES

Principle

When perfused into the liver through the vessels and capillaries at an adequate flow rate, proteolytic enzymes such as collagenase, which are relatively noncytotoxic, will disrupt intercellular junctions and will digest the connective framework within 15 min if the liver is previously cleared of blood and depleted of Ca^{2+} by washing with calcium-free buffer [Berry and Friend, 1969]. Hepatocytes are selected from the cell suspension by two or three differential centrifugations.

Outline

Introduce a cannula in the portal vein or a portal branch, wash the liver with a calcium-free buffer (15 min), perfuse the liver with the enzymatic solution (15 min), collect and wash the cells, and count the viable hepatocytes [Guguen-Guillouzo and Guillouzo, 1986].

Materials

Sterile:
L-15 Leibovitz medium
Calcium-free HEPES buffer pH 7.65: 160.8 mM NaCl; 3.15 mM KCl: 0.7 mM $Na_2HPO_4 \cdot 12H_2O$, 33 mM HEPES, sterilization by 0.22-μm Millipore filter, and storage at 4°C (2 months)
Collagenase (Sigma grade I; Boehringer 103578)
Collagenase solution: 0.025% collagenase; 0.075% $CaCl_2 \cdot 2H_2O$ in calcium-free HEPES buffer pH 7.65; preparation and sterilization by filtration just before use
Nembutal (Abbot, 5%)
Heparin (Roche)
Tygon tube (ID, 3.0 mm; OD, 5.0 mm)
Disposable scalp vein infusion needles, 20G (Dubernard Hospital Laboratory, Bordeaux, France)
Sewing thread for cannulation
Graduated bottles and Petri dishes
Surgical instruments (sharp, straight, and curved scissors and clips)
2 × 1-ml disposable syringes
Nonsterile:
Chronometer
Peristaltic pump (10 to 200 rpm)
Water bath

Protocol

1. Warm the washing HEPES buffer and collagenase solution in a water bath (usually approximately 38–39°C to achieve 37°C in the liver). Oxygenation is not necessary.
2. Set the pump flow rate at 30 ml/min.
3. Anesthetize the rat (180–200 g) by intraperitoneal injection of nembutal (100 μl/100 g), and inject heparin into the femoral vein (1,000 IU).
4. Open the abdomen, place a loosely tied ligature around the portal vein approximately 5 mm from the liver, insert the cannula up to the liver, and ligate.
5. Rapidly incise the subhepatic vessels to avoid excess pressure, and start the perfusion with 500 ml of calcium-free HEPES buffer at a flow

rate of 30 ml/min; verify that the liver whitens within a few seconds.

6. Perfuse 300 ml of the collagenase solution at a flow rate of 15 ml/min for 20 min. The liver becomes swollen.

7. Remove the liver and wash it with HEPES buffer; after disrupting the Glisson capsule, disperse the cells in 100 ml of L-15 Leibovitz medium.

8. Filter the suspension through two-layer gauze or 60–80-μm nylon mesh, allow the viable cells to sediment for 20 min (usually at room temperature), and discard the supernatant (60 ml) containing debris and dead cells.

9. Wash the cells three times by slow centrifugations (50 g for 40 s) to remove collagenase, damaged cells, and nonparenchymal cells.

10. Collect the hepatocytes in Ham's F12 or Williams' E medium enriched with 0.2% bovine albumin (grade V, Sigma) and 10 μg/ml of bovine insulin (80–100 ml).

Analysis

Determine the cell yield and viability by the well-preserved refringent shape or the trypan blue exclusion test (0.2% w/v; usually 4–6 × 10^8 cells with a viability of more than 95%).

Variations

Isolation of Hepatocytes from Other Species. The basic two-step perfusion procedure [Seglen, 1975] can be used for obtaining hepatocytes from various rodents, including mouse, rabbit, guinea pig, or woodchuck, by adapting the volume and the flow rate of the perfused solutions to the size of the liver. The technique has been adapted for the human liver [Guguen-Guillouzo et al., 1982] by perfusing a portion of the whole liver (usually with 1.5 l HEPES buffer and 1 l collagenase solution at 70 and 30 ml/min, respectively) or by taking biopsies (15–30 ml/min, depending on the size of the tissue sample). A complete isolation into a single-cell suspension can be obtained by an additional collagenase incubation at 37°C under gentle stirring for 10 to 20 min (especially for human liver) [Guguen-Guillouzo and Guillouzo, 1986]. Fish hepatocytes can be obtained by cannulating the intestinal vein and incising the heart to avoid excess pressure. Perfusion is performed at room temperature at a flow rate of 12 ml/min.

Maintenance and Differentiation. Isolated parenchymal cells can be maintained in suspension for 4–6 h and used for short-term experiments. When seeded in nutrient medium supplemented with 1 × 10^{-6} M dexamethasone on plastic culture dishes (7 × 10^5 viable cells/ml) the cells survive for a few days.

Survival for several weeks is obtained by seeding the cells onto a biomatrix [Rojkind et al., 1980]. However, the cells rapidly lose their specific differentiated functions. Higher stability (2 months) can be obtained by coculturing hepatocytes with rat liver epithelial cells presumed to derive from primitive biliary cells [Guguen-Guillouzo et al., 1983]. Hepatocytes are also more stable when seeded on Matrigel [Bissell et al., 1987] and are capable of undergoing from one to three rounds of cell division when they are cultured in a medium supplemented with EGF and pyruvate [McGowan, 1986] or with nicotinamide and EGF [Mitaka et al., 1991]. A combination of EGF and DMSO promotes differentiation of hepatocytes [Mitaka et al., 1993; Hino et al., 1999].

Pancreas

Both acinar [Bosco et al., 1994] and islet cells [Vonen et al., 1992; Cao et al., 1997] have been grown from pancreatic tissue, and lines have been established from the "immortomouse" [Blouin et al., 1997]. The conversion of adenoma to carcinoma has generated particular interest [Perl et al., 1998].

The following protocol for the culture of pancreatic acinar cells was contributed by Robert J. Hay and Maria das Gracas Miranda, American Type Culture Collection, 10801 University Boulevard, Manassas, VA 20110-2209.

PROTOCOL 22.7. PANCREATIC EPITHELIUM

Principle

Fractionated populations of pancreatic acinar epithelia are inoculated onto collagen-coated culture dishes, and two-dimensional aggregate colonies are allowed to develop for subsequent study [Hay, 1979; Ruoff and Hay, 1979].

Outline

Dissociate guinea pig pancreatic tissue, filter the mixed-cell suspension through cheesecloth or nylon sieves, and layer the suspension over a BSA solution. After three sequential centrifugation and fractionation steps, dispense the cell pellet, and seed the cells collagen-coated culture vessels.

Materials

Sterile:
Conical flask, 25 ml (Bellco), siliconized
Pipettes, 5 or 10 ml, wide bore (Bellco)
Büchner funnel
Dialysis tubing (Spectro-Por)
Sidearm flask, 250 ml
Polypropylene centrifuge tubes, 50 ml

Cheesecloth

F12K tissue culture medium with 20% bovine calf serum (F12K-CS20)

Collagenase (Gibco)

Trypsin 1:250 (Difco)

Dextrose (Difco, No. 0155-174)

HBSS

HBSS without Ca^{2+} and Mg^{2+} (HBSS-DVC)

Bovine serum albumin, fraction V (Sigma)

Collagen-coated culture dishes (Biocoat, Beckton Dickinson)

Collagenase solution: Dissolve collagenase in 1× HBSS-DVC (pH 7.2) to give 1,800 U/ml. Adjust the solution to pH 7.2, and dialyze using a 12-KDa exclusion membrane for 4 h at 4°C, against 1× HBSS-DVC containing 0.2% glucose. The HBSS is discarded, and this step is repeated for 16 to 18 h with fresh HBSS-DVC. Filter, sterilize, and store the solution in 10–20-ml aliquots at −70°C or below.

Trypsin solution: Trypsin 1:250, 0.25% in citrate buffer; 3 g/l of trisodium citrate; 6 g/l of NaCl, 5 g/l dextrose; 0.02 g/l of phenol red; pH 7.6.

Nonsterile:

Agitating water bath (Backer model M5B#11-22-1)

Centrifuge

Protocol

1. Make up the dissociation fluid by mixing 1 part of collagenase with 2 parts of trypsin solution.

2. Aseptically remove the entire pancreas (0.5–1.0 g), and place the organ in F12K-CS20. Trim away mesenteric membranes and other extraneous matter, and mince the pancreas into 1–3-mm fragments.

3. Transfer the fragments to a siliconized, 25-ml conical flask in 5 ml of prewarmed dissociation fluid. Agitate the solution at about 120 rpm for 15 min at 37°C in a shaker bath. Repeat this dissociation step two or three times with fresh fluid, until most of the tissue has been dispersed.

4. After each dissociation, allow large fragments to settle, and transfer the supernatant to a 50-ml polypropylene centrifuge tube with approximately 12 ml of cold F12K-CS20 to neutralize the dissociation fluid. Spin the suspension at 600 rpm for 5 min, and resuspend the pellet in 5 to 10 ml of cold F12K-CS20. Keep the cell suspension in ice.

5. Pool the cell suspensions and pass through several layers of sterile cheesecloth in a Büchner funnel. Apply light suction by inserting the funnel stem into a vacuum flask during filtration. Take an aliquot for cell quantitation.

6. Generally, 1 to 2 × 10^5 cells/mg of tissue with 90% to 95% viability are obtained at this step.

7. Layer 5 × 10^7 to 5 × 10^8 cells from the resulting fluid onto the surface of 2–4 columns consisting of 35 ml of 4% cold BSA in HBSS-DVC (pH 7.2) in polypropylene 50-ml centrifuge tubes. Centrifuge the tubes at 600 rpm for 5 min. This step is critical to achieving a good separation of acinar cells from islet, ductal, and stromal cells of the pancreas. Discard the supernatant, and repeat this fractionation step twice more, pooling and resuspending the pellets in cold F12K-CS20 after each step.

8. Collect the cell pellets in 5 to 10 ml of cold F12K-CS20. Take an aliquot for counting. Yields of 2–5 × 10^4 cells/mg of tissue are obtained, with 80–95% of the total being acinar cells. Inoculation densities can be 3 × 10^5/cm². Cells adhere as aggregate colonies within 72 h.

Variations

After cells have adhered, F12K-C20 can be replaced by serum-free MEM with insulin (10 ng/ml), transferrin (5 ng/ml), selenium (9 ng/ml), and EGF (21 ng/ml). This medium supports pancreatic cell growth for at least 15 days without altering the cell morphology.

The preceding method has been applied to studies with guinea pig and human (transplant donor) tissues. The addition of irradiated human lung fibroblasts as feeder layer produces a marked stimulation (up to 500%) in the incorporation of [³H]-thymidine and prolongs survival by at least a factor of two.

Kidney

The kidney is a structurally complex organ in which the system of nephrons and collecting ducts is made up of numerous functionally and phenotypically distinct segments. This segmental heterogeneity is compounded by a cellular diversity that has yet to be fully characterized. Some tubular segments possess several morphologically distinct cell types. In addition, evidence points to rapid adaptive changes in cell ultrastructure that may correlate with changes in cell function [Stanton et al., 1981]. The structural and cellular heterogeneity presents a challenge to the cell culturist who is interested in isolating pure or highly enriched cell populations. The difficulty of the problem is further compounded in studies of the human kidney, with which form and access to the specimen may make some manipulations, such as vascular perfusion, difficult or impossible.

Several approaches have been used successfully to culture the cells of specific tubular segments. Density gradient methods are now commonly employed to isolate enriched populations of enzyme-digested tubule segments and are particularly effective in establishing proximal tubule cell cultures from experimental animals [Taub et al., 1989]. Specific nephron or collecting duct segments can also be isolated by microdissection and then explanted to the culture substrate. This method, developed with the use of experimental animals [Horster, 1979], has been applied to the culture of human kidneys [Wilson et al., 1985] and the cyst wall epithelium of polycystic kidneys [Wilson et al., 1986]. Immunodissection [Smith and Garcia-Perez, 1985] and immunomagnetic separation [Pizzonia et al., 1991] methods have also been developed to isolate specific nephron cell types on the basis of their expression of cell-type-specific ectoantigens. These elegant methods hold considerable promise for the study of specific kidney cell types in health and disease, but as yet have been applied almost exclusively to studies on experimental animals. The limited (and unscheduled) availability and inconsistent form (e.g., excised pieces, damaged vasculature, lengthy postnephrectomy period) of human donor kidneys make progressive enzymatic dissociation a more practical means for isolating human kidney cells for culture.

A number of methods for the primary culture of human kidney tissue have been reported [Detrisac et al., 1984; States et al., 1986; McAteer et al., 1991]. The following protocol and preceding introduction was contributed by James McAteer, Stephen Kempson, and Andrew Evan, Indiana University School of Medicine, Indianapolis, IN 46202-5120.

PROTOCOL 22.8. KIDNEY EPITHELIUM

Principle

The primary culture of tissue fragments excised from the outer cortex of the human kidney provides a means of isolating cells that express many of the functional characteristics of the proximal tubule [Kempson et al., 1989]. Progressive enzymatic dissociation and crude filtration yields single cells and small aggregates of cells that, when seeded at high density, give rise to a heterogeneous epithelial-enriched population. Large numbers of cells can be harvested, making it practical to establish multiple replicate primary cultures or to propagate cells for frozen storage. Experience with the method shows that the functional characteristics of such primary and subcultured cells are reproducible for kidneys from different donors and that, with proper handling of specimens, good cultures can be derived from kidneys following even a lengthy postnephrectomy period.

Outline

Tissue fragments are excised from the outer cortex, minced, washed, and incubated (with agitation) in a collagenase–trypsin solution. The tissue is periodically pipette triturated, and the dissociated cells are collected and pooled. The cell suspension is filtered through a size-limiting screen to remove undigested fragments. The cells are then washed free of enzyme, suspended in medium supplemented with serum, and seeded on culture-grade plastic.

Materials

Sterile:
Basal medium DMEM/F12: 50:50 mixture of DMEM and Ham's F12
Fetal bovine serum (Sterile Systems)
Complete culture medium: DMEM/F12 plus 10% fetal bovine serum
BSS
Collagenase (Worthington type IV), 0.1%, trypsin (1:250, Sigma), 0.1% solution in 0.15 M of NaCl
DNase, 0.5 mg/ml in saline
Trypsin–EDTA solution: trypsin (1:250), 0.1%, EDTA (culture grade, Sigma) 1 mM
Scalpels, tissue forceps
Nitex screen (160 μm) (Tetko)
Culture dishes, flasks
Tubes, sterile, 50 ml
Nonsterile:
Orbital shaker (Bellco)
Centrifuge, Centra 4-B (IEC)

Protocol

Tissue Dissociation:
1. Cut 5–10-mm-thick coronal slices of kidney, and wash the fragments in chilled basal medium.
2. Excise fragments from the outer cortex, and use crossed blades to mince the tissue into 1–2-mm pieces.
3. Transfer approximately 5 ml of tissue fragments to a 50-ml tube containing 20 ml of the warm collagenase–trypsin solution. Secure the tube to the platform of an orbital shaker within an incubator at 37°C.
4. Incubate the tissue with gentle agitation for 1 h.
5. Discard and replenish the enzyme solution.
6. Subsequently, collect the supernatant at 20-min intervals following the addition of 20 ml of basal medium containing 0.05 mg/ml DNase and gentle trituration through a 10-ml pipette (10 cycles).
7. Dilute the supernatant that has been collected with an equal volume of complete culture medium, and hold the resulting suspension on ice.

8. Repeat this cell collection procedure five or more times, until the fragments are fully dispersed.

9. Pool the harvested supernatants and dispense them into aliquots in 50-ml tubes.

10. Sediment the culture by centrifugation (1,200 rpm (100 g), 15 min), and resuspend each pellet with 45 ml of complete medium containing DNAase.

11. Filter the suspension through Nitex cloth (160 μm).

Cell Plating, Subculture, and Cryopreservation

The isolation protocol yields cell aggregates as well as single cells. Hence, counting cells is unreliable. To propagate cells for subculture or cryopreservation, use 15 ml to seed one 75-cm² flask. Subculture NHK-C cells by routine methods. Dissociation with trypsin-EDTA produces a monodisperse suspension that is suitable for counting with a hemocytometer. Seed culture-grade dishes at approximately 1×10^5 cells per cm². Cryopreservation is performed by routine methods using complete medium containing 10% dimethyl sulfoxide. (See Protocol 19.1.)

Comments and Safety Precautions.

A prominent variable in this method is the quality of the kidney specimen itself. Procedures for procuring tissue are impossible to standardize. Donor tissue commonly includes segments of kidney collected at surgical nephrectomy (e.g., for renal cell carcinoma) and intact kidney tissue that is judged unsuitable for transplantation (e.g., because it has an anomalous vasculature). These specimens are collected under different conditions (e.g., with or without perfusion, with varying periods of warm ischemia, and for different durations of the postnephrectomy period). In addition, differences in the age and health of the donor ensure that no two specimens are equivalent. Although not all kidneys yield satisfactory cultures, the foregoing protocol has been effective for a wide variety of specimens, including fresh surgical specimens and intact, perfused kidneys held on wet ice for nearly 100 h.

△ *Safety Note.* Work involving any tissue or fluid specimen of human origin requires precautions to avoid unprotected contact with bloodborne pathogens. (See Biohazards in Chapter 6.)

Applications. NHK-C cells exhibit predominantly the functional characteristics of the renal proximal tubule, including a parathyroid hormone (PTH)-inhibitable Na⁺-dependent inorganic phosphate transport system [Kempson et al., 1989]. They also show phlor-

izin-sensitive Na⁺-dependent hexose transport, exhibit a proximal tubule-like pattern of hormonal stimulation of cAMP (PTH responsive, vasopressin insensitive), and express several proximal tubule brush border enzymes (maltase, leucine aminopeptidase, gamma-glutamyl transpetidase). In addition, NHK-C cells have been used to demonstrate the specificity of phosphonoformic acid, and inhibitor of Na⁺/Pi cotransport [Yusufi et al., 1986], and they serve as a model of oxidant injury in the renal tubule [Andreoli and McAteer, 1990].

Bronchial and Tracheal Epithelium

The following protocol for the isolation and culture of normal human bronchial epithelial cells from autopsy tissue was contributed by Moira A. LaVeck and John F. Lechner, National Institutes of Health, Bethesda, MD. John Lechner's current address is Inhalation Toxicology Research Institute, P.O. Box 5890, Albuquerque, NM 87185, USA.

PROTOCOL 22.9. BRONCHIAL AND TRACHEAL EPITHELIUM

Principle

In the presence of serum, normal human bronchial epithelial (NHBE) cells cease to divide and, furthermore, terminally differentiate. The serum-free medium that has been optimized for NHBE cell growth does not support lung fibroblast cell replication, thus permitting the establishment of pure NHBE cell cultures [Lechner and LaVeck, 1985].

Outline

This procedure involves explanting fragments of large airway tissue in a serum-free medium (LHC-9) in order to initiate and subsequently propagate fibroblast-free outgrowths of NHBE cells; four subculturings and 30 population doublings are routine.

Materials
Sterile:
Culture medium:
  (a) L-15 (Gibco)
  (b) LHC-9 [Lechner and LaVeck, 1985] (see Table 9.1)
  (c) HB [Lechner and LaVeck, 1985] (see Reagent Appendix)
Bronchial tissue from autopsy of noncancerous donors
Plastic tissue culture dishes (60 and 100 mm)
FN/V/BSA: Mixture of human fibronectin, 10 μg/ml (Collaborative Research), collagen (Vitrogen 100, 30 μg/ml; Collagen Corp.), and crystallized bovine

serum albumin (BSA), 10 μg/ml (Miles Biochemical) in LHC basal medium (Biofluids, Inc.) [Lechner and LaVeck, 1985]

Scalpels No. 1621 (Becton Dickinson)

Surgical scissors

Half-curved microdissecting forceps

Pipettes (10 and 25 ml)

Trypsin (Cooper Biomedical), 0.02%, EGTA (Sigma), 0.5 mM, and polyvinylpyrrolidine (USB), 1% solution [Lechner and LaVeck, 1985]

High-O_2 gas mixture (50% O_2, 45% N_2, 5% CO_2)

Nonsterile:

Gloves (human tissue can be contaminated with biologically hazardous agents)

Controlled atmosphere chamber (Bellco No. 7741)

Rocker platform (Bellco No. 7740)

Phase-contrast inverted microscope

Protocol

1. Coat a culture dish with 1 ml of the FN/V/BSA mixture per 60-mm dish, and incubate the dish in a humidified CO_2 incubator at 37°C for at least 2 h (not to exceed 48 h). Vacuum aspirate the mixture and fill the dish with 5 ml of culture medium.

2. Aseptically dissected lung tissue from noncancerous donors autopsied within the previous 12 h is placed into ice-cold L-15 medium for transport to the laboratory, where the bronchus is further dissected from the peripheral lung tissues.

3. Before culturing, scratch an area of one square centimeter at one edge of the surface of the 60-mm culture dishes with a scalpel blade.

4. Open the airways (submerged in the L-15 medium) with surgical scissors, and cut (slice, do not saw) the tissue with a scalpel into two pieces, 20 × 30 mm.

5. Using a scooping motion to prevent damage to the epithelium, pick up the moist fragments and place them epithelium side up onto the scratched area of the 60-mm dish. Remove the medium, and incubate the fragments at room temperature for 3 to 5 min to allow time for them to adhere to the scratched areas of the dishes.

6. Add 3 ml of HB medium to each dish and place them in a controlled-atmosphere chamber. Flush the chamber with a high-O_2 gas mixture and place it on a rocker platform. Rock the chamber at 10 cycles per minute, causing the medium to flow intermittently over the epithelial surface. Incubate rocking tissue fragments at 37°C, changing the medium and atmosphere after 1 d and at 2 d intervals for 6–8 d. This step

improves subsequent explant cultures by reversing any ischemic damage to the epithelium that occurred from time of death of the donor until the tissue was placed in the ice-cold L-15 medium.

7. Before explanting, scratch seven areas of the surface of each 100-mm culture dish with a scalpel. Coat the surfaces of the scratched culture dishes with the FN/V/BSA mixture solution, and aspirate the surplus solution as before.

8. Cut the moist ischemia-reversed fragments into 7 × 7-mm pieces, and explant the pieces epithelium side up on the scratched areas. Incubate the pieces at room temperature without medium for 3–5 min, as before.

9. Add 10 ml of LHC-9 medium to each dish, and incubate explants at 37°C in a humidified 5% air/CO_2 incubator. Replace spent medium with fresh medium every 3 to 4 d.

10. After 8 to 11 d of incubation, when epithelial cell outgrowths radiate from the tissue explants more than 0.5 cm, transfer the explants to new culture dishes scratched and coated with FN/V/BSA to produce new outgrowths of epithelial cells. This step can be repeated up to seven times with high yields of NHBE cells.

11. Incubate the postexplant outgrowth cultures in LHC-9 medium for an additional 2 to 4 d before trypsinizing (with the trypsin/EGTA/PVP solution) for subculture or for experimental use.

Prostate

A number of reports have used cells cultured from prostate tissue [Cronauer et al., 1997] with a particular interest in regulating cell proliferation and differentiation by paracrine growth factors [Yan et al., 1992; Chopra et al., 1996; Thomson et al., 1997]. The following protocol for the primary culture of rat prostate epithelial cells was contributed by W. L. McKeehan, W. Alton Jones Cell Science Center, Lake Placid, NY.

PROTOCOL 22.10. PROSTATIC EPITHELIUM

Principle

Isolated normal rat prostate epithelial cells are a valuable model for studying the cell biology of maintenance, growth, and functioning of normal prostate under investigator-controlled and -defined conditions [McKeehan et al., 1982, 1984; Chaproniere and McKeehan, 1986]. Normal cells serve as the control cell type out of which prostate adenocarcinoma cells arise. Conditions similar to those described here have also been useful for the primary culture of epithelial cells for transplantable rat tumors. Key features of this method are the specific support of epithelial cells by an improved nutrient medium and hormonelike growth factors [Yan et al., 1992], while concurrently inhibiting fibroblast outgrowth due to deficient or inhibitory properties of the medium.

Outline

Remove and prepare prostates for cell culture (30 min). Incubate prostatic tissue with collagenase (1 h), and collect single cells and small aggregates of cells (1 h). Inoculate and culture cells to confluent monolayer (7 d).

Materials

Sterile:

Growth medium WAJC404 [McKeehan et al., 1984; Chaproniere and McKeehan, 1986] (see Table 9.1)

Media salt solution (MSS) [McKeehan et al., 1984]:

    (a) NaCl ....12 mM
    (b) KCl ....2 mM
    (c) KH_2PO_4 ....1 mM
    (d) HEPES, pH 7.6 ....30 mM
    (e) Glucose...4 mM
    (f) Phenol red.......................................3.3 μM

Fetal calf or horse serum

Penicillin

Kanamycin

Cholera toxin

Dexamethasone

Epidermal growth factor (EGF)

Ovine or rat prolactin

Insulin

Partially purified [McKeehan et al., 1984; Chaproniere and MeKeehan, 1986] or purified [Crabb et al., 1986] prostatropin (prostate epithelial cell growth factor)

Collagenase, type I (Sigma)

Petri dishes (glass or plastic), 60 mm

Scissors

Syringe and 14-G cannula

Erlenmeyer flasks, 25 ml

Wire screen, 1-mm mesh, made to fit a 50-ml conical centrifuge tube

Conical centrifuge tubes, plastic, 50 ml

Nylon screen filters of 250, 150, 100, and 40 μm to fit 50-ml conical tubes

25-well plastic tissue culture dishes

Nonsterile:

Shaking water bath at 37°C

Centrifuge

Hemocytometer or Coulter counter

Rats, 10–12 weeks old, male

Protocol

1. Aseptically remove desired lobes of the prostate and place them in a sterile 60-mm Petri dish.
2. Trim fat from the lobes and weigh them.
3. Add 2 ml of collagenase at 675 U/ml in MSS containing 100 U/ml of penicillin and 100 μg/ml of kanamycin.
4. Mince the tissue with scissors to approximately 1-mm pieces, small enough to fit through a 14-G cannula.
5. Using a syringe and 14-G cannula, transfer the minced tissue fragments to a sterile 25-ml Erlenmeyer flask. Add collagenase to 1 ml per 0.1 g of original wet tissue weight.
6. Incubate for 1 h at 37°C on a shaking water bath.
7. Aspirate the digested suspension three times through a 14-G cannula, and then pass the suspension through a coarse (1-mm) wire mesh screen fitted to 50-ml plastic conical tubes to remove debris and undigested material. Rinse the suspension with an equal volume of MSS containing 5% whole fetal calf serum or horse serum.
8. Collect cells by centrifugation at 100 g for 5 min at 4°C.
9. Resuspend the cell pellet in 5 ml of MSS plus 5% serum, and pass the suspension successively through nylon screen filters of mesh sizes 250, 150, 100, and 40 μm. Wash each screen with 5 ml of MSS plus 5% serum.
10. Collect cells by centrifugation, and resuspend them in 5 ml of nutrient medium WAJC404.
11. Count the cells and adjust the concentration to 4 × 10⁶ cells/ml.
12. Inoculate 50 μl containing 2 × 10⁵ cells into each well of a 24-well plate (area of well ∼2 cm²) containing 1 ml of medium WAJC404, 10 ng/ml of cholera toxin, 1 μM dexamethasone, 10 ng/ml of EGF, 1 μg/ml of prolactin, 5 μg/ml of insulin, 10 ng/ml of prostatropin, 100 U/ml of penicillin, and 100 μg/ml of streptomycin. Partially purified sources of prostatropin can be substituted as described in McKeehan et al. [1984] and Chaproniere and McKeehan [1986].
13. Incubate the culture in a humidified atmosphere of 95% air and 5% CO_2 at 37°C. Change the medium at days 3 and 5. The cells should be near confluent by day 7.

Variations

The foregoing procedure can be applied to the culture of human prostate epithelial cells with modifications described in Chaproniere and McKeehan [1986].

The purified growth factor, prostatropin, is identical to heparin-binding fibroblast growth factor type one, previously called acidic FGF or HBGF-1 [Burgess and Maciag, 1989]. The new nomenclature, achieved by consensus, is now FGF-1. Molecular characterization of new members of the FGF ligand family, as well as the FGF receptor family in prostate cells, has revealed that prostate epithelial cells express a specific splice variant (FGF-R2IIIb) of one of the four FGF receptor genes. The specific receptor recognizes FGF-1 and stromal-cell-derived FGF-7 (also called keratinocyte growth factor), but not FGF-2 (previously called bFGF or HBGF-2), described in Yan et al. [1992]. Prostate stromal cells express only the FGF-R2 gene, which recognizes FGF-1 and FGF-2, but not FGF-7. Since stromal cells respond to both FGF-1 and FGF-2, prostatropin/FGF-1 can be substituted for FGF-7 in the above procedure to provide an additional selection for epithelial cells. FGF-2 will not substitute for FGF-1 or FGF-7. If purified FGF-1 is used in the protocol, its activity can be potentiated by the addition of 10–50 μg per ml of heparin.

MESENCHYMAL CELLS

I include under this head those cells which are derived from the embryonic mesoderm, but exclude the hematopoietic system. (See Hematopoietic Cells later in this chapter). Mesenchymal cells include the structural and vascular cells.

Connective Tissue

Connective-tissue cells are generally regarded as the weeds of the tissue culturist's garden. They survive most mechanical and enzymatic explantation techniques and may be cultured in many simple media.

Although cells loosely called fibroblasts have been isolated from many different tissues and assumed to be connective tissue cells, the precise identity of cells in this class remains somewhat obscure. Fibroblast lines (e.g., 3T3 from the mouse) produce types I and III collagen and release it into the medium [Goldberg, 1977]. While collagen production is not restricted to fibroblasts, synthesis of type I in relatively large amounts is characteristic of connective tissue. However, 3T3 cells can also be induced to differentiate into adipose cells [Kuriharcuch and Green, 1978]. It is possible that cells will transfer from one lineage to another under certain conditions, but such transdifferentiation has rarely been confirmed. It is more likely that mouse embryo fibroblastic cell lines are primitive mesodermal cells that may be induced to differentiate in more than one direction.

Human, hamster, and chick fibroblasts are morphologically distinct from mouse fibroblasts, as they assume a spindle-shaped morphology at confluence, producing characteristic parallel assays of cells distinct from the pavementlike appearance of mouse fibroblasts. The spindle-shaped cell may represent a more highly committed precursor and may be more correctly termed a fibroblast. NIH3T3 cells may become spindle shaped if allowed to remain at high cell density.

It has also been suggested that fibroblastic cell lines may be derived from vascular pericytes, connective-tissue-like cells in the blood vessels, but in the absence of the appropriate markers, this is difficult to confirm.

Clearly, cell lines loosely termed fibroblastic can be cultivated from embryonic and adult tissues, but these lines should not be regarded as identical or classed as fibroblasts without confirmation by the appropriate markers. Collagen, type I, is one such marker. Thy I antigen has also been used [Raff et al., 1979], although it may also appear on some hematopoietic cells.

Cultures of fetal or adult fibroblasts can be prepared by primary explant (see Protocol 11.4), warm (see Protocol 11.5) and cold (see Protocol 11.6) trypsinization, and collagenase digestion (see Protocol 11.8).

Adipose Tissue

While it may be difficult to prepare cultures from mature fat cells, differentiation may be induced in cultures of mesenchymal cells (e.g., mouse 3T3-L1) by maintaining the cells at a high density for several days [Kuriharcuch and Green, 1978; Vierick et al., 1996]. An adipogenic factor in serum appears to be responsible for the induction. Primary cultures of fat cells can also be prepared from rat epididymus by collagenase digestion; the following protocol has been abridged from Quon [1998].

Adipose cells (adipocytes) are terminally differentiated, specialized cells whose primary physiological role has classically been described as an energy reservoir for the body. Adipocytes are a storage depot for triglycerides in times of energy excess and a source of energy in the form of free fatty acids released by lipolysis during times of energy need. Recently, the important role of adipose cells as active regulators of carbohydrate and lipid metabolism has received increased attention [Flier, 1995; Reaven, 1995]. It is likely that specific abnormalities in adipose tissue can contribute directly to the pathogenesis of common diseases such as diabetes, hypertension, and obesity [Flier, 1995; Hotamisligil et al., 1995; Reaven, 1995; Walston et al., 1995; Zhang et al., 1994].

PROTOCOL 22.11. PRIMARY CULTURE OF ADIPOSE CELLS

Principle

A modification of the isolation procedure originally described by Rodbell [1964] is optimized to obtain approximately 4 ml of packed adipose cells from the epididymal fat pads of four rats by flotation of disaggregated cells following centrifugation. This protocol can be scaled up or down as needed and has been used to provide insulin-responsive cells suitable for DNA transfer by electroporation.

Materials

Sterile:

DMEM-A: DMEM, pH 7.4, containing 25 mM of glucose, 2 mM of glutamine, 200 nM of (R)-N^6-(1-methyl-2-phenylethyl)adenosine (PIA, Sigma), 100 μg/ml of gentamicin, and 25 mM of HEPES. Prepare in 100 ml aliquots and warm to 37°C before use.

DMEM-B: DMEM-A plus 7% BSA

KRBH buffer: Krebs–Ringer medium, pH 7.4, containing 10 mM NaHCO$_3$, 30 mM HEPES (Sigma), 200 nM adenosine (Boehringer–Mannheim), and 1% (w/v) BSA (Intergen). To make 1 liter of KRBH buffer, add 7 g of NaCl, 0.55 g of KH$_2$PO$_4$, 0.25 g of MgSO$_4$·7H$_2$O, 0.84 g of NaHCO$_3$, 0.11 g of CaCl$_2$, 7.15 g of HEPES, 10 g of BSA, 1 ml of 200-μM adenosine, and UPW to a volume of 1 l. Dispense in aliquots of 100 ml. Warm to 37°C and adjust pH to 7.4 before use

KRBH-A buffer: KRBH with 5% BSA

Collagenase, 20 mg/ml in 2 ml of KRBH buffer, and sterilize by passage through a 0.22-μM filter

Scissors: two pairs of sharp dissecting scissors, 10 cm (4 in); Perry forceps, 12.5 cm (5 in)

Nylon mesh filter, 250 μm, in holder or filter funnel

Polypropylene tubes, 17 mm × 100 mm

Low-density polypropylene vial, 30 ml

Conical centrifuge tube, 50 ml (Corning)

Tissue-culture Petri dish, 60 mm diameter

Wide-bore pipette tips, 200 μl

Nonsterile:

Male Sprague-Dawley rats (145–170 g, CD strain, Charles River Breeding Laboratories)

Plastic box perfused with a gas mixture of 70% CO$_2$, 30% O$_2$

Guillotine

70% ethanol

Humid incubator at 37°C with 5% CO$_2$

Tabletop centrifuge

Pipettor

Protocol

Dissection of Epididymal Fat Pads:

1. Anesthetize rats in a plastic box with a gas mixture of 70% CO$_2$ and 30% O$_2$.
2. Decapitate rats using a guillotine and exsanguinate.
3. Soak the bodies briefly in 70% ethanol.
4. Remove the epididymal fat pads while maintaining the highest level of sterility possible.
   (a) Cut through the skin on the lower abdomen with one pair of scissors to expose the peritoneum.
   (b) Using a second pair of scissors, open the peritoneum and pull up the testes with a pair of forceps.
   (c) Trim fat pads from epididymides, taking care to leave the blood vessels behind.
5. Transport tissue to the culture laboratory.

Collagenase Digestion and Washing of Adipose Cells:

6. Add 4 g of fat pads (approximately equivalent to 8 fat pads) to a 30-ml low-density polypropylene vial containing 4 ml of KRBH buffer at 37°C.
7. Mince fat pads into pieces approximately 2 mm in diameter with scissors.
8. Add 1 ml of collagenase solution to the vial containing the minced fat pads, and incubate the pieces in a shaking water bath at 37°C for approximately 1 h, until the cell mixture takes on a creamy consistency.
9. After collagenase digestion, add 10 ml of KRBH buffer at 37°C to the vial.
10. Mix cells in the vial by swirling, and gently pass the cells through a 250-μm nylon mesh filter into a 50-ml conical tube.
11. Wash the cells by adding 30 ml of KRBH buffer at 37°C to the tube and centrifuging briefly at 200 g in a tabletop centrifuge. Remove infranatant with a pipette. Note that adipose cells will be floating on top of the aqueous buffer.
12. Repeat the washing of cells by adding 40 ml of KRBH buffer, centrifuging, and removing infranatant two additional times.
13. Wash cells twice with 40 ml of DMEM-A at 37°C.
14. After the final wash, resuspend the cells from the surface of the medium in DMEM-A at a cytocrit of approximately 40%.

Primary Culture of Adipose Cells:

15. Transfer 2 ml of the 40% cytocrit DMEM-A suspension, using 200-μl wide-bore pipette tips, into one 60-mm tissue culture dish (Becton Dickinson).
16. Place the cells in a humid incubator at 37°C with 5% CO$_2$ for 1.5 h.

17. Add 5 ml of DMEM-B to each dish.

At this point, one typically performs a glucose up-take assay [Quon 1998] on an aliquot of cells to check their viability.

Muscle

Cardiac muscle can be isolated using the cold-trypsin method (Protocol 11.6) and will show contractions after about 3 d in culture [Deshpande et al., 1993]. Human cardiac myocytes have also been grown [Goldman and Wurzel, 1992]. Smooth muscle cells have been grown from vascular tissue [Subramanian et al., 1991] and cocultured with vascular endothelium [Jinard et al., 1997].

The following protocol for the culture of human adult skeletal muscle cells was contributed by G. Barlovatz-Meimon, S. Bonavaud, and Ph. Thibert, Faculté de Science, Université Paris XII, Val de Marne, Avenue General de Gaulle, F-94010 Creteil, France.

PROTOCOL 22.12. SKELETAL MUSCLE

Principle

It is possible to culture myogenic cells from adult skeletal muscle of several species under conditions in which the cells continue to express at least some of their differentiated traits. Called *satellite cells*, the myogenic cells partially mimic the first steps of skeletal muscle differentiation. They proliferate and migrate randomly on the substratum and then align and finally undergo a fusion process to form multinucleated myotubes [Richler and Yaffe, 1970; Campion, 1984; Hartley and Yablonka-Reuveni, 1990]. Although three to four passages can be performed by means of trypsinization, subculture is no longer possible once differentiation (i.e., fusion) has taken place. For the same reason, proliferation is very difficult to estimate once some cells have started to fuse.

The procedure described here is an enzymatic method of digesting a muscle biopsy. Primary cultures from human healthy muscle biopsies are highly enriched in myogenic cells, as evidenced by at least 85% positivity to desmin by immunostaining at day 10 after seeding.

Outline

Primary cultures can be grown easily in Ham's F12 medium supplemented with 20% fetal calf serum (FCS). Without modifying the culture conditions, these cells proliferate and differentiate by fusing to form multinucleated myotubes, confirming the myogenicity of the cultivated cells.

Materials

Sterile:

Transport medium: Ham's F12 (Seromed 08101) without NaHC0₃ and with 20 mM HEPES (Serva 25245), 75 ml

Growth medium: Ham's F12 with 20% fetal bovine serum (Gibco 011-06290), 200 ml

Pronase solution: 0.15% pronase (Sigma P-6911) and 0.03% EDTA (Merck 8118) with 20 IU/ml of penicillin, and 20 µg/ml of streptomycin (0.4% of Eurobio PES 3000), in Ham's F12 or PBS, 100 ml

PBSA, 100 ml

100-µm nylon mesh

Scalpels

Long, fine scissors

Petri dishes for dissection, 9 cm

Flasks, 25 cm² (four)

Centrifuge tubes, 20 ml, or universal containers (five)

Nonsterile:

Hemocytometer

Protocol

Dissociation:

1. Trim off nonmuscle tissue from the biopsy with a scalpel, and rinse in PBSA.
2. Cut the biopsy into fragments parallel to the fibers and wash in PBSA prior to weighing the biopsy.
3. Place the fragments parallel to each other in the lid of a Petri dish, cut the fragments into thinner cylinders and then, finally, into 1-mm³ pieces, without crushing the tissue. The final cutting can be done in a tube with long scissors, again avoiding crushing.
4. Rinse with PBSA and let the pieces settle; discard the supernatant.
5. Digest the fragments for 1 h at 37°C in pronase solution. (Use 15 ml for a biopsy of 1 to 3 mg.) Shake the tube gently at 10–15-min intervals.
6. Triturate the culture with a pipette after incubation. The medium should become increasingly opaque as more and more cells are released.
7. Let the fragments settle to the bottom by gravity, forming pellet P1 and supernatant S1.
8. Filter S1 through a 100-µm nylon mesh and into a 20-ml centrifuge tube. Shake or pipette the supernatant gently to resuspend the cells.
9. Centrifuge the tube 8–10 min at 350 g. Discard the supernatant by aspiration.
10. Resuspend the pellet very, very gently by means of a rubber-bulb pipette in precisely 10 ml of growth medium, and count the cells with a hemocytometer.

11. Dilute the suspension in growth medium to seed culture flasks with about 1.5×10^4 cells/ml. About $1-2 \times 10^5$ cells/g are obtained from healthy donor biopsies.

12. Add 15 ml of digestion medium to P1, and incubate the fragments for 30 min in a water bath at 37°C, with periodic shaking.

13. Pipette the suspension to disaggregate the cells and then filter the suspension through nylon mesh. Rinse the filter with 20 ml of growth medium.

14. Centrifuge the suspension for 8–10 min at 350 g, count the cells, and seed as before.

15. Transfer the flasks to a 37°C humidified incubator with 5% CO_2.

△ *Safety Note.* The rest of the biopsy and all tubes, pipettes, plates, etc., used in the procedure should be treated with hypochlorite before disposal. (See Disposal in Chapter 6.)

Maintenance of Cultures:

Change the medium very gently 24 h after seeding and then every 3–4 d. The development of these cultures is mainly towards differentiation. The timing of the three phases for human muscle cells is about 4–6 d for peak proliferation; then the cells align at about day 8, and around day 10 to 12 an increase in cell fusion and the formation of myotubes are observed. Nevertheless, one must keep in mind that some cells may differentiate earlier and that others will still proliferate when the majority of the culture is undergoing differentiation.

Subculture:

1. Add a small volume of EDTA gently to the cells and remove it immediately.

2. Add sufficient trypsin solution (0.25%) to form a thin layer over the cells.

3. When cells detach, add 5 to 10 ml of complete growth medium, pass the culture very gently in and out of a pipette, and then centrifuge the cells for 10 min at 350 g.

4. Count an aliquot and seed the cells at the chosen concentration.

Proliferation and Differentiation Indexes

The growth curves of human myogenic cell cultures obtained in Ham's F12 medium supplemented with 20% FCS show the three traditional phases (see Growth Cycle in Chapter 20): the lag phase, the exponential phase, and the plateau, which corresponds to the onset of fusion. The last, evaluated in terms of the number of nuclei incorporated into myotubes or in terms of a fusion index (the percentage of nuclei incorporated into myotubes relative to the total number of nuclei), commences usually around day 8 after plating and rises dramatically around day 10. According to the sample, this chronology can gain or lose one day.

Hence, differentiation, expressed as the number of nuclei per myotube/cm^2, may be observed morphologically. But the differentiation process can also be monitored by the use of biochemical markers (such as the sarcomeric proteins), enzymes involved in differentiation (e.g., creatine phosphokinase and its time-dependent muscle-specific isoform shift), or the appearance of α-actin [Buckingham, 1992].

A family of genes, the best known of which is MyoD, was shown to activate muscle-specific gene expression in myogenic progenitors. [For review, see Buckingham, 1992.] Myogenic cell differentiation involves either the activation of a variety of other genes with concurrent changes in cell surface adhesive properties [Dodson et al., 1990] or the recently shown requirement of cell surface plasminogen activator urokinase and its receptor [Quax et al., 1992].

Variations and Applications

The explant-reexplant method has been developed by the group of Askanas [Askanas et al., 1990; Pegolo et al., 1990]; it has proved reliable for a long time, especially for studies on muscle defects.

When the aim is the study of the mechanism of differentiation, one has the choice between the explant and the enzymatic methods, followed by clonal or nonclonal culture or by three-dimensional cultivation. If the aim of the work is to obtain a large number of cells for biochemical or molecular studies, then the enzymatic method seems to be preferable.

Cartilage

The following protocol for culturing cells from articular cartilage was contributed by François Lemare, Sophie Thenet, Sylvie Demignot, and Monique Adolphe of Laboratoire de Pharmacologie Cellulaire de l'Ecole Pratique des Hautes Etudes, Institut Biomédical des Cordeliers, 15 rue de l'Ecole de Médecine 75006 Paris, France.

Chondrocytes are highly specialized cells of mesenchymal origin that are responsible for synthesis, maintenance, and degradation of the cartilage matrix. A great deal of research in the field of rheumatology has been focused on understanding the mechanisms that induce metabolic changes in articular chondrocytes during osteoarthritis and rheumatoid arthritis. Articular chondrocytes in culture are a very useful tool, but, cultured in a monolayer, they rapidly divide, become fibroblastic, and lose their biochemical characteristics [Benya et al., 1977; von der Mark, 1986]. As early as the first passage, there is a gradual shift from the synthesis of type II collagen to types I and III collagens and from aggrecan to low-molecular-weight proteoglycans. In parallel with the phenotypic modulation, the metabolic response to interleukin-1β is quantitatively affected by subculture [Blanco et al., 1995; Lemare et al., 1998].

PROTOCOL 22.13. CHONDROCYTES IN ALGINATE BEADS

Principle

Among the various methods explored for maintaining the phenotype of chondrocytes [for a review, see Adolphe and Benya, 1992], culture in alginate beads appears the most promising, since it has been shown that this culture system leads to the formation of a matrix similar to that of native articular cartilage [Häuselmann et al., 1996; Petit et al., 1996]. The system maintains the expression of the differentiated phenotype and is also able to restore it in dedifferentiated chondrocytes [Bonaventure et al., 1994; Lemare et al., 1998]. Another advantage over other three-dimensional methods is that cells can easily be recovered after the culture is completed, allowing protein and gene expression studies [Lemare et al., 1998].

Outline

Culture in alginate beads is based on the gelation in calcium chloride of a suspension of chondrocytes in an alginate solution.

Materials

Sterile:

Cartilage dissection medium (CDM): Ham's F12 medium (Gibco), supplemented with netilmycin (20 μg/ml), vancomycin (10 μg/ml), and ceftazidim (30 μg/ml) (all from Sigma), is used for dissection of the joints and removal of cartilage, as well as for enzymatic digestion.

Growth medium: DMEM (glucose, 1 g/l)/Ham's F12 (1:1, v/v), gentamicin, 4 μg/ml, supplemented with 10% FCS.

Note. Different batches of serum have to be compared for their ability to support chondrocyte growth as well as the expression of the differentiated phenotype (assessed, for example, by immunolabeling of type II collagen). When one lot is selected, several months' supply should be purchased to avoid frequently changing batches of serum.

Enzyme solutions for dissociation of cartilage:

(a) Trypsin, 0.05%, porcine pancreatic (Sigma) in Ham's F12

(b) Collagenase, 0.3%, type 1 (Worthington, CLS1) in Ham's F12

(c) Collagenase 0.06% in Ham's F12 + 10% FCS

All solutions are sterilized using 0.22-μm filters.

Note. New batches of collagenase should be tested for efficient cartilage dissociation by comparison with the previous batch.

Trypsin-EDTA solution (for subculture): 0.1% w/v porcine pancreatic trypsin in PBSA with 5 mM (0.02% w/v) of EDTA.

Sodium alginate solution: 1.25% w/v sodium alginate (Fluka) in 20 mM Hepes; 0.15 M NaCl.

Preparation: Using a heated stirrer, first dissolve the Hepes in 0.15 M NaCl and heat to ~60°C. Add the sodium alginate and continue stirring until it dissolves completely (>2 h). Carefully adjust the solution to pH 7.4 with 1 M NaOH; do not exceed pH 7.4, as the addition of acid (HCl or H$_2$SO$_4$) results in irreversible precipitation.

Dispense the solution into aliquots and sterilize it by autoclaving at 121°C for 10 min.

Alternative preparation: Sodium alginate dissolution in HEPES/NaCl can be achieved rapidly (~10 min) in a microwave oven, but care must be taken to avoid caramelization.

Gelation solution: CaCl$_2$, 102 mM, HEPES, 5 mM, pH 7.4

Solubilization solution: EDTA, 50 mM, HEPES, 10 mM, pH 7.4

Sterile magnet

Protocol

Dissection:

1. Prepare cultures from knee, shoulder, and hip joints. Fetal or young donors are preferable to adults, as they provide higher quantities of cells and take longer to senesce. One-month old "Fauve de Bourgogne" or "New Zealand" rabbits are ideal. However, hip joints obtained from surgical resection from human adults are also suitable sources of chondrocytes.

2. Wash tissue pieces thoroughly with CDM before dissection. Dissection should begin without delay. Remove skin, muscle, and tendons from joints. Carefully take cartilage fragments from articulations that are free of connective tissue.

Dissociation:

To prevent the cartilage drying, carry out all dissociation steps in 100-mm non-TC-treated dishes containing CDM. The quantities of enzyme solutions indicated in the rest of the protocol are suitable for tibiofemoral and scapulohumeral articulations from one rabbit but they must be adapted to the quantity of cartilage to be dissociated.

3. Using crossed scalpels (see Protocol 11.4), mince cartilage slices into 1-mm³ pieces.

4. Transfer cartilage fragments into a 30-ml flat-bottomed vial.

5. Add 10 ml of 0.05% trypsin solution, and incubate the fragments with moderate magnetic agitation for 25 min in a sealed vial at room temperature.

6. Remove the trypsin solution after the fragments sediment.

7. Add 10 ml of 0.3% collagenase solution, and incubate the culture with moderate magnetic agitation for 30 min in a sealed vial at room temperature.

8. Remove collagenase solution after the fragments sediment.

9. Add 10 ml of 0.06% collagenase solution.

10. Transfer the suspension into a 100-mm bacterial dish. Rinse the vial with another 10 ml of 0.06% collagenase solution, transfer the rinse into the dish, and incubate the cartilage fragments overnight at 37°C in an incubator with 5% CO_2.

11. Transfer the cell suspension into a 50-ml centrifuge tube and mix on a vortex mixer for a few seconds.

12. Remove residual material left after digestion by passing the digested material through a 70-μm nylon filter (Falcon).

13. Centrifuge the filtrate at 400 g for 10 min.

14. Resuspend the cell pellet in 20 ml of CDM, and count the cells with a hemocytometer.

15. Centrifuge the cells at 400 g for 10 min.

Entrapment:

16. Resuspend the cell pellet in 1 ml of sodium alginate solution, and then dilute the suspension progressively in more sodium alginate solution, until a cellular density of 2 × 10⁶ cells/ml is reached. This progressive dilution is necessary to obtain a homogeneous cell suspension in alginate.

17. Express the cell suspension in drops through a 21G needle into the gelation solution with moderate magnetic stirring, and allow the alginate to polymerize for 10 min to form beads.

18. Wash the beads 3 times in 5 vol NaCl, 0.15 M.

19. Distribute the beads, 5 ml (1 × 10⁷ cells), into 75-cm² flasks containing 20 ml of growth medium.

20. Incubate the culture at 37°C in a humidified atmosphere of 5% CO_2 and 95% air.

Recovery:

21. Discard the medium.

22. Add 2 vol of solubilization solution to the beads.

23. Incubate the culture 15 min at 37°C.

24. Centrifuge the cells at 400 g for 10 min.

25. Resuspend the cell pellet in growth medium containing 0.06% collagenase.

26. Incubate the cells for 30 min at 37°C in a humidified atmosphere of 5% CO_2 and 95% air.

27. Centrifuge the cells at 400 g for 10 min.

28. Resuspend the cell pellet in serum-free DMEM/Ham's F12 medium, and count the cells with a hemocytometer.

29. Centrifuge the cells at 400 g for 10 min.

30. Repeat steps 28, 29, without counting the cells.

Analysis of Chondrocyte Phenotype

Collagen. Type II collagen represents 85–90% of the total collagen synthesized by chondrocytes in cartilage or primary culture. Type II collagen is highly specific to chondrocytes and is considered as their main differentiation marker. The expression of type II collagen can be analyzed by indirect immunofluorescence using specific antibodies. These must be very carefully checked, however, to make sure that there is no cross-reactivity with type I collagen, which often occurs. Polyclonal antibodies against type II collagen can be purchased from Southern Biotechnology Associates, Inc. The immunocytochemical study of collagen expression in chondrocytes cultured in alginate beads necessitates sectioning the beads as described in Lemare et al. [1998].

For a more precise analysis of the collagen phenotype, SDS-PAGE and two-dimensional CNBr peptide maps of purified collagen after [³H]-Proline incorporation have to be performed [Benya, 1981].

Concerning major collagens, SDS-PAGE analysis can reveal the presence of type III collagen and of the $\alpha2(I)$ chain of collagen I. However, it cannot separate the $\alpha1$ chain of type I and type II (i.e., it cannot make a distinction between type I and type II trimer collagen), which is why two-dimensional CNBr peptide mapping is necessary to assert type II collagen expression.

The expression of these genetically distinct collagens can also be analyzed at the transcriptional level (Northern blot) with conventional transfer and hybridization techniques [Lemare et al., 1998]. Because of sequence homology between different collagen genes, hybridization and washing conditions must be tested and need to be stringent enough to avoid cross-hybridizations.

Proteoglycans. Alcian Blue staining at acidic pH is widely used as an indicator of the presence of sulfated aggrecan [Lemare et al., 1998]. As with immunocytochemical studies, this staining necessitates sectioning the beads. A more precise analysis of the different types of proteoglycans synthesized can be performed after $^{35}SO_4$ incorporation [Verbruggen et al., 1990]. cDNA probes for aggrecan [Lemare et al., 1998] and link protein allow an analysis at the transcriptional level of the expression of these specific proteins.

Variations and Applications

Several studies have shown that chondrocytes cultured in alginate beads represent an attractive alternative chondrocyte model [Grandolfo et al., 1993; Häuselmann et al., 1994, 1996; Petit et al., 1996]. The main advantage of this technique over other three-dimensional methods is that cells can easily be recovered after culture, allowing protein and gene expression studies [Lemare et al., 1998].

The number of available cells often becomes a limiting factor in studies of chondrocyte metabolism, as articular cartilage is difficult to obtain in large quantities and has a low cellularity. In this case, the number of chondrocytes available can be amplified by first culturing the cells in a monolayer (2 passages) and then restoring the differentiated properties of chondrocytes by cultivating the cells in alginate beads for two weeks [Bonaventure et al., 1994; Lemare et al., 1998]. In parallel with this phenotypic restoration, the metabolic response to Interleukin-1β of these cells was restored at a quantitative level similar to that of chondrocytes in a monolayer primary culture [Lemare et al., 1998]. Moreover, comparisons of interactions between nitric oxide and prostaglandins in cartilage explants, chondrocytes in a primary monolayer culture, and freshly isolated chondrocytes cultured in alginate beads indicate that there are several subtle differences between the metabolism of the cells in the three models. In particular, the PGE_2/NO production ratios were different in the three models, and chondrocytes cultured in alginate beads present a metabolism closer to that of cartilage explants than do chondrocytes in a monolayer [Lemare et al., manuscript in preparation]. This makes the culture in alginate beads a relevant model for the study of chondrocyte biology.

Bone

Although bone is mechanically difficult to handle, thin slices treated with EDTA and digested in collagenase [Bard et al., 1972] give rise to cultures of os-

teoblasts that have some functional characteristics of the tissue. Antiserum against collagen has been used to prevent fibroblastic overgrowth without inhibiting the osteoblasts [Duksin et al., 1975]. Propagated lines have been obtained from osteosarcoma [Smith et al., 1976; Weichselbaum et al., 1976], but not from normal osteoblasts.

This protocol for the culture of bone cells was contributed by Edith Schwartz, Departments of Orthopedic Surgery and Physiology, Tufts University School of Medicine, 136 Harrison Avenue, Boston, MA 02111.

PROTOCOL 22.14. OSTEOBLASTS

Principle

Bone culture suffers from the inherent problem that the hard nature of the tissue makes manipulation difficult. However, conventional primary explant culture or digestion in collagenase and trypsin releases cells that may be passaged in the usual way.

Materials

Sterile:

Ham's F12 medium with 12% fetal bovine serum, 2.3 mM Mg^{2+}, 100 U/ml of penicillin, and 100 μg/ml of streptomycin SO_4

Trypsin solution to be used for cell passage: Dissolve 125 mg of trypsin (Sigma type XI) in 50 ml of Ca^{2+} Mg^{2+}-free Tyrode's solution. Adjust the resulting solution to pH 7.0 and filter sterilize it. Place aliquots of 5 and 10 ml in Pyrex tubes and store at $-20°C$.

Digestion solution for the isolation of osteoblasts: Solution A: 8.0 g of NaCl, 0.2 g of KCl, 0.05 g of $NaH_2PO_4 \cdot H_2O$ in 100 ml of UPW

Solution B (Collagenase–trypsin solution): Dissolve 137 mg of collagenase (type I, Worthington Biochemicals) and 50 mg of trypsin (Sigma, type III) in 10 ml of solution A. Adjust the combined solution to pH 7.2, and then bring the amount 100 ml with UPW. Filter sterilize and distribute the solution into 10-ml aliquots that are stored at $-20°C$.

Scalpels

Forceps

9-cm Petri dishes

25- or 75-cm² flasks (Corning, Falcon, Nunc)

(a) Explant Cultures

Outline

Small fragments of tissue are allowed to adhere to the culture flask by incubation in a minimal amount of medium. (See Protocol 11.4.) The adherent explants are then flooded and the outgrowth is monitored.

Protocol

1. Obtain bone specimens from the operating room.
2. Rinse the tissue several times at room temperature with sterile saline.
3. If the bone cannot be used immediately, cover the specimen with sterile Ham's F12 medium containing 12% fetal calf serum, penicillin, and streptomycin sulfate. The bone may be stored overnight at 4°C.
4. The next morning, rinse the tissue with Tyrode's solution containing penicillin (100 U/ml) and streptomycin sulfate (100 μg/ml).
5. Place the bone in a Petri dish, and with the use of scalpel and forceps, remove the trabeculae and place them in a second Petri dish.
6. Add 10 ml of Tyrode's solution over the excised trabeculae. Rinse the trabeculae several times with Tyrode's solution until blood and fat cells are removed.
7. To initiate explant cultures, prepare a 25-cm^2 flask by preincubating it with 2 ml of complete medium for 20 min to equilibrate the medium with the gas phase. Adjust the pH as necessary with CO_2, 4.5% $NaHCO_3$, or HCl.
8. Cut the trabeculae into fragments of 1–3 mm.
9. Remove the preincubation medium from the flask, and add 2.5 ml of fresh Ham's F12 medium to the flask. Transfer between 25 and 40 fragments of trabeculae to the flask.
10. With the flask in an upright position, slide the explant pieces along the base of the flask with the aid of an inoculating loop, and distribute the explant pieces evenly.
11. Permit the flask to remain upright for 15 min at 37°C.
12. Slowly restore the flask into a normal horizontal position. The explant pieces will stick to the bottom of the flask.
13. Leave the flask in the horizontal position at 37°C for 5 to 7 d. After this period, check for outgrowth, and replace the medium in the flask with fresh medium. To prevent the explant pieces from becoming detached, lift the flask slowly into the vertical position before carrying it from the incubator to the hood or to the microscope.
14. To maintain cultures, change the medium twice per week.
15. When confluence is reached, the explant pieces are removed, the cell layer is trypsinized, and the

cells are isolated by centrifugation and seeded into flasks or wells.

(b) Monolayer Cultures from Disaggregated Cells

Outline

Trabecular bone is dissected down to 2–5 mm and digested in collagenase and trypsin. Suspended cells are seeded into flasks in F12 medium.

Protocol

1. Wash trabecular bone specimens repeatedly with Tyrode's solution to remove the fat and blood cells. The trabeculae are excised with scalpel and forceps under sterile conditions.
2. After collecting as much bone as possible, wash the remaining blood and fat cells away by rinsing the specimens three times with Tyrode's solution and cut into 2–5 mm fragments.
3. Wash the cut trabeculae with F12 medium containing fetal calf serum.
4. Place the pieces of bone in a small sterile bottle with a magnetic stirrer, and add 4 ml of digestion solution. (This amount should cover the bone specimens.)
5. Stir the solution containing bone fragments at room temperature for 45 min.
6. Remove the suspension of released cells and discard it, since these cells are most likely to contain fibroblasts.
7. Add a second aliquot of 4 ml of digestion solution to the bone fragments, and stir the mixture at room temperature for 30 min.
8. Collect the digestion solution from bone fragments, and centrifuge it for 2 min at 580 g at room temperature.
9. After removing the supernatant, suspend the cells in 4 ml of Ham's F12 medium with 20% fetal calf serum, and count the cells.
10. Centrifuge the suspension at 580 g for 10 min, and resuspend the cells in 4 ml of complete medium. This suspension will become the inoculum.
11. Preincubate 75-cm^2 flasks for 20 min with 8 ml of complete F12 medium to equilibrate with the gas phase.
12. Remove the preincubation medium and add 2 ml of complete F12 medium.
13. Add 4 ml of medium containing the cell suspension. The inoculum should contain 6,000–10,000 cells per cm^2 of surface area.
14. Finally, add another 6 ml of Ham's F12 medium, to give a total volume of 12 ml.

15. In the interim, add an additional 4 ml of digestion solution to the remaining pieces of bone, and repeat the digestion for 30 min. The released cells are harvested, and, if necessary, the digestion step is repeated several more times. With large amounts of bone, the digestion period can be increased to 1–3 h. Cell counts are performed after each digestion period, and the released cells are used to inoculate a different flask.

Passage of Cells in Culture:

16. Remove the pieces of explant.

17. Remove the medium and rinse the cell layer with PBSA, 0.2 ml/cm^2.

18. Add trypsin to the flask, 0.1 ml/cm^2, and incubate at 37°C until the cells have detached and separated from one another. Monitor cell detachment and separation on the microscope. In general, a 10-min incubation is sufficient.

19. Transfer the released cells to a centrifuge tube with an equal volume of Ham's F12 medium with 20% fetal calf serum (FCS).

20. Centrifuge the cells at 600 g for 5 min.

21. Discard the supernatant, and resuspend the cells in complete medium by gentle, repeated pipetting.

22. Set one aliquot aside for the determination of cell concentration and another for DNA determination (see Protocol 20.3).

23. Inoculate the remaining cells into culture flasks or wells that have previously been equilibrated with medium. The cells should reattach within 24 h.

Endothelium

Endothelium has been successfully cultured by collagenase perfusion of bovine aorta (Fig. 15.2e) [Gospodarowicz et al., 1976, 1977; Schwartz, 1978] and human umbilical vein [Gimbrone et al., 1974], trypsinization of white matter from rat cerebral cortex [Phillips et al., 1979], and microdissection of adrenal cortex [Folkman et al., 1979].

Endothelium can be characterized by the presence of factor VIII antigen [Booyse et al., 1975], type IV collagen [Howard et al., 1976], Weibel-Palade bodies [Weibel and Palade, 1964], and, sometimes, the formation of tight junctions, although the last feature is not always demonstrated readily in culture.

Endothelial cultures are good models for contact inhibition and density limitation of growth, as cell proliferation is strongly inhibited after confluence is reached [Haudenschild et al., 1976].

Much interest has been generated in endothelial cell culture because of the potential involvement of endothelial cells in vascular disease, the repair of blood vessels, and angiogenesis in cancer. Folkman and Haudenschild [1980] described the development of three-dimensional structures resembling capillary blood vessels derived from pure endothelial lines *in vitro*. Growth factors, including angiogenesis factor derived from Walker 256 cells *in vitro*, play an important part in maintaining cell proliferation and survival, so that secondary structures can be formed.

The following protocol for the culture of large-vessel endothelial cells was contributed by Bruce Zetter, Department of Surgery, Harvard Medical School, Boston, MA.

PROTOCOL 22.15. VASCULAR ENDOTHELIAL CELLS

Principle

Endothelial cells make up a single cell layer at the inner surface of all blood vessels. The vessels most commonly used to obtain cultured endothelial cells are the bovine aorta [Booyse et al., 1975], bovine adrenal capillaries [Folkman et al., 1979], rat brain capillaries [Bowman et al., 1981], human umbilical veins [Jaffe et al., 1973], and human dermal [Davison et al., 1983] and adipose [Kern et al., 1983] capillaries. Although all endothelia share some properties, significant differences exist between the endothelial cells of large and small blood vessels [Zetter, 1981].

Outline

Endothelial cells released by collagenase incubated within the blood vessel are cultured on a gelatin-coated substrate in medium supplemented with mitogens (human and bovine capillaries) or without mitogens (bovine large vessels).

Materials

Aseptically isolated blood vessels, preferably in 10-cm sections, approximately 5 mm in diameter. If asepsis cannot be guaranteed, clamp both ends of the blood vessel and dip it in 70% alcohol for 30 s.

Collagenase: 0.25% crude collagenase, ~200 U/mg, in PBSA containing 0.5% bovine serum albumin

DMEM supplemented with 10% calf serum and antibiotics

Tissue-culture-grade Petri dishes, 9 cm, or 75-cm^2 flasks coated with 1.5% gelatin in PBSA: Incubate overnight in the gelatin solution, remove the gelatin, add medium with serum, and incubate the mixture until the cells are ready for culture

Two clamps or hemostats, 25 mm

Sharp scissors, 50 mm

Protocol

1. Ligate one end of a 10-cm section of blood vessel 2–10 mm in diameter to a 5-ml plastic syringe.
2. Wash the vessel with 20 ml of PBSA to remove blood.
3. Run the collagenase solution through the blood vessel until it appears at the bottom end, clamp that end with a hemostat, and incubate the vessel containing the collagenase at room temperature for 10 min.
4. Cut the vessel above the clamp with sharp scissors, and collect the collagenase in a 10-cm Petri dish.
5. Rinse the lumen of the vessel with 10 ml of PBSA, and add this to the collagenase from step 4.
6. Centrifuge the pooled digest and wash it at 100 g for 5 min.
7. Wash the digest twice more by resuspending it in a medium and centrifuging it.
8. Resuspend the final pellet in growth medium, and seed the cells into gelatin-coated dishes or flasks, with approximately one 10-cm section of blood vessel, 5 mm in diameter, per 75-cm flask or 10-cm-diameter dish.
9. Subculture by conventional trypsinization. (See Protocol 12.2.)

Identification

Vascular endothelial cells are identified by the production of factor VIII [Hoyer et al., 1973], angiotensin-converting enzyme [del Vecchio and Smith, 1981], the uptake of acetylated low-density lipoprotein [Voyta et al., 1984], the presence of Weibel–Palade bodies [Weibel and Palade, 1964], and the expression of endothelial-specific cell surface antigens [Parks et al., 1985].

Variations

Human endothelial cells are cultured in medium 199 with 10–20% human serum, 25 μg/ml of hypothalamus-derived endothelial mitogen (BTI, Stoughton, MA), and 90 μg/ml of heparin [Thornton et al., 1983]. Capillary endothelial cells also require the direct addition of an endothelial mitogen or a conditioned medium from a tumor culture [Folkman et al., 1979].

NEUROECTODERMAL CELLS

Neurons

Nerve cells appear to be more fastidious in their choice of substrate than most other cells. They will not survive well on untreated glass or plastic, but will demonstrate neurite outgrowth in collagen and poly-

D-lysine. Neurite outgrowth is encouraged by a polypeptide nerve growth factor [NGF; Levi-Montalcini, 1979] and factors, secreted by glial cells [Marchionni et al., 1993], that are immunologically distinct from NGF. Cell proliferation has not been found in cultures of most neurons, even with cells from embryonic stages in which mitosis was apparent *in vivo*; however, recent studies with embryonal stem cells have shown that some neurons can be made to proliferate *in vitro* [Thomson et al., 1998] and recolonize *in vivo* [Snyder et al., 1992].

The following protocol for the monolayer culture of cerebellar neurons was contributed by Bernt Engelsen and Rolf Bjerkvig, Department of Cell Biology and Anatomy, University of Bergen, Žrstadveien 19, N-5009 Bergen, Norway.

PROTOCOL 22.16. CEREBELLAR GRANULE CELLS

Principle

Cerebellar granule cells in culture provide a well-characterized neuronal cell population that is suited for morphological and biochemical studies [Drejer et al., 1983; Kingsbury et al., 1985]. The cells are obtained from the cerebella of 7- or 8-day-old rats, and nonneuronal cells are prevented from growing by the brief addition of cytosine arabinoside to the cultures.

Outline

The cerebella from four to eight neonatal rats are cut into small cubes and trypsinized for 15 min at 37°C in Hanks's BSS. The cell suspension is seeded in poly-L-lysine-coated culture wells or flasks.

Materials

Sterile:
DMEM with
  (a) Glucose ....30 mM
  (b) L-glutamine..2 mM
  (c) KCl..24.5 mM
  (d) Insulin (Sigma I-1882)................... 100 mU/l
  (e) p-aminobenzoic acid (Sigma A-3659)...7 μM
  (f) Gentamycin................................100 μg/ml
  (g) Fetal calf serum, heat inactivated 10%
Hanks's BSS with 3 g/l BSA (HBSS)
Poly-L-lysine, mol. wt. >300,000 (Sigma P-1524)
Cytosine arabinoside (Cytostar, Upjohn; powder)
Trypsin (type II) (0.025% in HBSS)
Silicone (Aquasil, Pierce 42799)
35-mm tissue culture Petri dishes
Scalpels, scissors, and forceps
Pasteur pipettes and 10-ml pipettes

Test tubes, 12 and 50 ml
Nonsterile:
Water bath
Siliconization of Pasteur pipettes:
  (a) Dilute the Aquasil solution 0.1–1% in UPW.
  (b) Dip the pipettes into the solution or flush out the insides of the pipettes.
  (c) Air dry the pipettes for 24 h, or dry for several minutes at 100°C.
  (d) Sterilize the pipettes by dry heat.
Poly-L-lysine Treatment of Culture Dishes:
  (a) Dissolve the poly-L-lysine in UPW (10 mg/l).
  (b) Sterilize the mixture by filtration.
  (c) Add 1 ml of poly-L-lysine solution to each of the 35-mm Petri dishes.
  (d) Remove the poly-L-lysine solution after 10–15 min, and add 1–15 ml of culture medium.
  (e) Place the culture dishes in the incubator (minimum 2 h) until the cells are to be seeded.

Protocol

1. Dissect out the cerebella aseptically and place them in HBSS.
2. Mince the tissue with scalpels into small cubes approximately 0.5 mm³.
3. Transfer the minced tissue to test tubes (12 ml) and wash the tissue three times in HBSS. Allow the tissue to settle to the bottom of the tubes between each washing.
4. Add 10 ml of 0.025% trypsin (in HBSS) to the tissue, and incubate the tube in a water bath for 15 min at 37°C.
5. Transfer the trypsinized tissue to a 50-ml centrifuge tube, and add 20 ml of growth medium to stop the action of the trypsin.
6. Disaggregate the tissue by trituration through a siliconized Pasteur pipette, until a single-cell suspension is obtained.
7. Let the cell suspension stay in the centrifuge tube for 3–5 min, allowing small clumps of tissue to settle to the bottom of the tube. Remove these clumps with a Pasteur pipette.
8. Centrifuge the single-cell suspension at 200 g for 5 min, and aspirate off the supernatant.
9. Resuspend the pellet in growth medium, and seed the cells at a concentration of 2.5–3.0 × 10⁶ cells/dish.
10. After 2–4 d (best results usually are obtained after 2 d), incubate the cultures with 5–10 μM cytosine arabinoside for 24 h.
11. Change to regular culture medium.

Characterization

Neurons can be identified immunologically by using neuron-specific enolase antibodies or by using teta-nus toxin as a neuronal marker. Astrocyte contamination can be quantified by using glial fibrillary acidic protein as a marker.

Variations

A single-cell suspension can be obtained by mechanical sieving through nylon meshes of decreasing diameter (Protocol 11.9) or by sequential trypsinization (i.e., a 3–5-min trypsin treatment). Instead of HBSS, Puck's solution, Krebs, or other buffers with glucose can be used.

Glial Cells

Glial cells have been cultured from avian, rodent, and human brains. Human adult normal brain lines express glial fibrillary acidic protein [GFAP; Burke et al., 1996], high-affinity γ-aminobutyric acid and glutamate uptake, and glutamine synthetase activity [Frame et al., 1980].

Oligodendrocytes do not readily survive subculture, but Schwann cells from the optic nerve have been subcultured using cholera toxin as a mitogen [Raff et al., 1978; Brockes et al., 1979]. Cultures of human glioma can also be prepared by mechanical disaggregation, trypsinization, or collagenase digestion [Pontén, 1975; Freshney, 1980; see also Fig. 15.2c]. The right temporal lobe from human males appears to be marginally better than other regions of the brain [Westermark et al., 1973], but most regions give a good chance of success. The glia–glioma system provides a good model for comparing normal and neoplastic cells under the same conditions.

A number of gliomas have been cultured from rodents, among which the C_6 deserves special mention [Benda et al., 1968]. This cell line expresses the astrocytic marker—glial fibrillary acidic protein—in up to 98% of cells [Freshney et al., 1980a], but still carries the enzymes glycerol phosphate dehydrogenase and 2′3′ cyclic nucleotide phosphorylase [Breen and De Vellis, 1974], both of which are oligodendrocytic markers. The line appears to be an interesting example of a precursor cell tumor that can mature along two distinct phenotype routes simultaneously.

Linser and Moscona [1980] separated the Müller cells of the neural retina from pigmented retina and neurons and demonstrated that full functional development could not be achieved unless the Müller cells (astroglia) were recombined with neurons from the retina. Neurons from other regions of the brain were ineffective.

The following protocol for the purification of rat olfactory nerve ensheathing cells from the olfactory bulb by fluorescence-activated cell sorting was contributed by Susan Barnett, Departments of Neurology and Medical Oncology, University of Glasgow, CRC Beatson Laboratories, Garscube Estate, Bearsden, G61 1BD, Glasgow, Scotland. The work was originally carried out in the laboratory of Dr. Mark Noble at the Ludwig Institute in London, and his assistance is gratefully acknowledged. Dr. Ilse Sommer provided the O4 antibody.

The ability to produce highly enriched populations of individual cell types is vital to the success of biological studies on the diverse elements that control cell growth, differentiation, and function. In this regard, the fluorescence-activated cell sorter (FACS) is a powerful tool for the enrichment of relatively small populations of cells from a heterogeneous cell population. Initially, the cell sorter was devised to characterize and sort cells of the hematopoietic system, for which a large number of cell-specific antibodies were available. With the increasing availability of antibodies that characterize both glia (Table 22.2) and neurons, it has now become possible to transfer this technique to the field of neurobiology [Abney et al., 1983; Williams et al., 1985; Trotter and Schachner, 1988; Barnett et al., 1993; Franceschini and Barnett 1996].

The olfactory bulb is a tissue of considerable interest, since it is the only central nervous system (CNS) mammalian tissue that can support neuronal regeneration throughout life [Moulton, 1974; Graziadei and Monti Graziadei, 1979, 1980]. Once neuronal death occurs, either during the course of normal cell turnover or following injury, newly generated neurons are recruited from a pool of constantly proliferating stem cells. These neurons are able to produce axons that reinnervate the olfactory bulb and reestablish olfaction. It is thought that a population of glial cells present in the olfactory nerve layer of the olfactory bulb, termed olfactory bulb ensheathing cells (OBECs), supports or encourages the growth of olfactory axons within the CNS [Doucette, 1984, 1990]. In fact, recent experiments have demonstrated that OBECs are capable of remyelinating experimentally created demyelinated lesions in rat spinal cord [Franklin et al., 1996]

and that they support the regeneration of cut axons [Li et al., 1997; Ramon-Cueto et al., 1998].

To study the biology of OBECs in detail, it is necessary to separate them from the other major cell types that make up the olfactory bulb: neurons, CNS glial cells, oligodendrocytes and astrocytes, endothelial cells, and meningeal cells.

The rat olfactory bulb fulfills the criteria for successful cell separation by FACS in that (1) it readily forms a single-cell suspension after enzymatic treatment and (2) the mouse monoclonal antibody O4 [IgM; Sommer and Schachner, 1981], in conjunction with antigalactocerebroside antibody [anti-GalC; Ranscht et al., 1982], can be used to distinguish OBECs from other cell types that are present [Barnett et al., 1993, Franceschini and Barnett, 1996]. The O4 antibody recognizes sulphatide, seminolipid, and pro-ligodendroblast antigen [Bansal et al., 1992] and labels many CNS and PNS glial cells, including Schwann cells, oligodendrocytes, and oligodendrocyte-type-2 astrocyte (O-2A) progenitor cells [Gard and Pfeiffer, 1990; Mirskey et al., 1990; Sommer and Schachner, 1981]. The The olfactory bulb contains three cell populations that are labeled by means of the O4 antibody: OBECs, oligodendrocytes, and O-2A progenitors [Barnett et al., 1993]. In contrast to OBECs, oligodendrocytes and O-2A progenitors are also labeled by anti-GalC antibody, which can thereby be used as a negative selection marker. By using these two antibodies in conjunction, it has been found that 8–14% of the O4 positive cells are OBECs and less than 1% of the total O4 positive cells are oligodendrocytes or O-2A progenitors. Moreover, the characteristic highly branched or bipolar morphology of oligodendrocytes and O-2A progenitors, respectively, can easily be distinguished from the flat fibroblastlike morphology of the OBECs.

The O4 positive FACS-purified OBECs can be classified into two types of cell—Schwann-cell-like and astrocytelike, which are distinguished by a panel of neu-

TABLE 22.2. Characterization of Glial Cells by Immunostaining

Cell type	GFAP	Gal-C	O4	A2B5	NGF
O2-A	−	−	−	+	−
Type-1 astrocyte	+	−	−	−	−
Type-2 astrocyte	+	−	−	+	−
Oligodendrocyte	−	+	+[a]	+[a]	−
ONEC	+/−[c]	−	+[b]	−	+

[a]Mature oligodendrocytes lose A2B5 and become GalC$^+$ and O4$^+$ only.

[b]ONECs lose O4 in culture.

[c]GFAP may be fibrous or amorphous. ONECs gain GFAP expression in culture.

ral markers. [See Doucette, 1990; Franceschini and Barnett, 1996.] These two cell types may have different functional roles *in vivo*. The growth factors that regulate their growth and differentiation are currently being investigated. In the presence of conditioned medium from type-1 astrocytes (ACM), OBECs exhibit good viability and growth for at least 14 days. The method for growing astrocytes in culture can be found in Nobel and Murrey [1984] and Wolswijk and Noble [1989]. Recently, an isoform of neuregulin (neudifferentiation factor β) was identified as the factor in ACM [Pollock et al., in press].

PROTOCOL 22.17 OLFACTORY BULB ENSHEATHING CELLS

Principle

The protocol describes a rapid method for purifying a minor population of glial cells from the olfactory bulb by using a cell surface antibody and FACS. These principles however, can be applied to the purification of virtually any glial or neuronal population, provided that two main criteria are fulfilled: the researcher is able to produce a viable single-cell suspension and suitable cell-type-specific antibodies are available for identifying the cells of interest.

Materials

Sterile or Aseptically Prepared:
Poly-L-lysine, stock, 4 mg/ml, dilute 1:300 in sterile UPW, and use at a final concentration of 13.3 μg/ml (Sigma)
Collagenase type I, stock, 2,000 U/ml in L-15 (ICN, Biomedicals Inc. cat no. 195109)
Ca^{2+}- and Mg^{2+}-free Hanks's BSS (HBSS, Gibco)
Leibowitz L-15 medium plus 25 μg/ml of gentamycin (ICN)
DMEM media (Gibco) containing:
(i) 5% fetal calf serum (DMEM-FCS)
(ii) 25 mM (4.5 g/l) glucose and supplemented with 25 μg/ml of gentamicin (Gibco), 0.0286% (v/v) BSA pathocyte (ICN Chemicals), 2 mM-glutamine, 10 mg/ml of bovine pancreas insulin, 100 mg/ml of human transferrin, 0.2 mM progesterone, 0.10 mM putrescine, 0.45 mM-L-thyroxine, 0.224 mM selenium, and 0.49 mM 3,3',5-triiodo-L-thyronine (all from Sigma) [DMEM-BS; Bottenstein and Sato, 1979].
Monoclonal antibodies: O4 (Dr. I. Sommer, Dept. Neurology, INS, Southern General Hospital, Glasgow G51 4BF, Scotland) and antigalactocerebroside [anti-GalC; see Ranscht et al., 1982].

Second antibodies: antimouse IgM-fluorescein and antimouse IgG3-phycoerythrein (Southern Biotechnology Associates, Inc.; antibodies can be purchased from Cambridge Biosciences in the United Kingdom).
Petri dish
Falcon 15 ml polypropylene conical tube with cap
Forceps, curved and straight (size 7)
Tubes, polystyrene, 12 × 75 mm, round bottom with caps [Falcon #2058; these tubes fit both the sample tubing and the sort-collecting port of the FACS].
Scissors, curved, dissecting
Scalpel
Nonsterile:
Sprague-Dawley rats, 6–8 d post partum
Dissecting board and pins
Ethanol, 70%, in a spray bottle
FACS sorter (Becton Dickinson; see Fluorescence-Activated Cell Sorting in Chapter 14).

Protocol

Preparation of Dissociated Cell Suspension of OBECs:
Coat all tissue culture plastics and glass coverslips with diluted poly-L-lysine (final concentration, 13.3 mg/ml; incubate for at least 1 h to a maximum of 24 h), wash the plastics and coverslips once with PBSA, and air dry them prior to use. A greater yield of OBECs is obtained from neonatal rats rather than older animals, although it is possible to make preparations from animals of any age.

1. Kill the rats by decapitation. Use 15–20 animals for a FACS sort of 2–3 h to recover enough OBECs for 10–15 cover slips (10,000 cells per coverslip).
2. Pin the head dorsal side up onto a dissecting board and spray with 70% ethanol.
3. Dip all dissecting instruments in 70% ethanol prior to use, and shake the instruments dry.
4. Remove the skin from the head by using sharp curved scissors, and make a circular cut to remove the top of the skull, revealing the brain and the two olfactory bulbs at the tip of the nose (Fig. 22.1).
5. Using curved forceps, gently release the olfactory bulbs from the brain, and place them in a Petri dish containing a drop of L-15 plus gentamycin.
6. Using a sterile scalpel blade, chop the olfactory bulbs into small pieces.
7. Place the pieces into a small vial containing 1 ml of stock collagenase and 1 ml of L-15.
8. Incubate the pieces of olfactory bulbs at 37°C for 30–45 min.

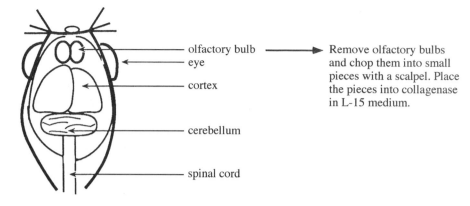

Fig. 22.1. Olfactory Bulb Dissection. Schematic diagram of the isolation of the olfactory bulbs from the brain of newborn rats. (Courtesy of Susan Barnett.)

9. Centrifuge the resultant suspension at 1,000 rpm for 5 min and remove the supernatant.

10. Resuspend the pelleted bulb tissue in 1 ml of Ca^{2+}- and Mg^{2+}-free HBSS.

11. Glial cells are fragile; therefore, to produce a single-cell suspension, the olfactory bulb tissue must be dissociated gently, taking care not to produce air bubbles. Dissociate the tissue through a 19G hypodermic needle followed by a 23G hypodermic needle.

12. Add 5 ml of Ca^{2+}- and Mg^{2+}-free HBSS, and spin the suspension again at 1,000 rpm for 10 min. Remove the supernatant and resuspend the cells in DMEM-FCS at a concentration of 5 $\times$ 10^6 cells/ml.

FACS Sorting of OBECs:

To remove as many of the $O4^+$ oligodendrocytes as possible from the sorted cell population, it is preferable to label them with a second antibody to galactocerebroside [anti-GalC; Ranscht et al., 1982; Barnett et al., 1993], thus allowing discrimination between oligodendrocytes and OBECs. In this way, a dual-color sort can be carried out in which the unwanted cells, now labeled with both antibodies, can be ignored.

1. After tissue dissociation, resuspend 5 $\times$ 10^6 cells in 500 μl of a mixture of O4 and anti-GalC antibody hybridoma supernatants, and incubate at 4°C for 30–45 min in a 15-ml centrifugation tube. There will be more than 5 $\times$ 10^6 cells, so prepare enough tubes for the number of cells. If the hybridoma supernatant does not contain high concentrations of antibody, then an increase in cell number will result in an apparent decrease in the percentage of $O4^+$ cells. A similar consideration applies to the anti-GalC hybridoma supernatant. It is therefore always necessary to titrate antibodies by FACS analysis prior to use.

2. After 30–45 min, wash the cells twice in DMEM-FCS and resuspend in 500 μl of DMEM-FCS containing a 1:100 dilution of both goat antimouse IgM-fluorescein and goat antimouse IgG3-phycoerythrein. Incubate the cells for a further 30 min at 4°C.

3. Wash the cells twice in 15 ml of DMEM-FCS by centrifugation, and resuspend them at a concentration of 2 $\times$ 10^6 cells/ml for sorting in ice-cold DMEM-FCS media in cell-sorting tubes (Becton Dickinson 2085). Keep the cell suspension on ice. The FACS profile for antibody-labeled cells is stable for at least 3–4 h when the cells are kept on ice.

4. This protocol describes the method used for sorting cells on a FACStar cell sorter. However, any type of cell sorter should produce similar results. Align the laser to the 488-nm excitation line. Fluorescence compensation must be set to negate fluorescence spillover into the individual fluorescent channels. A control sample of cells labeled with the second antibodies alone should be analyzed prior to the labeled sample so that any nonspecific cell label can be accounted for and removed by decreasing the gain on the photomultiplier tube.

5. Set a cell sort window around the $O4^+$ $GalC^-$ cells. The average percentage of $O4^+$ olfactory bulb cells should be around 10 $\pm$ 1%.

6. Sort the $O4^+GalC^-$ cells into a polystyrene round-bottom tube containing 4 ml of DMEM-FCS.

7. After sorting, transfer the sorted cells to a 15-ml centrifugation tube, wash out the sort tube with medium, pool the washes, and fill the centrifuge tube with DMEM-FCS. Spin the centrifuge tube at 1,000 rpm for 15 min.

8. Remove the supernatant and resuspend the cells in DMEM-BS. Optimum cell viability and growth for the OBECs may be obtained in a 1:1 dilution of DMEM-BS:ACM [Barnett et al., 1993]. ACM is collected from monolayers of purified cortical astrocytes [Noble and Murrey, 1984] incubated for 48 hours in DMEM-BS.

This protocol produces a highly enriched population of O4$^+$ olfactory bulb glial cells (>97%). These cells may be plated onto poly-L-lysine-coated coverslips, Petri dishes, or flasks and used for further investigations. Current evidence suggests that these cells do correspond to the OBECs described in electron microscope studies [Doucette, 1984; Raisman, 1985], and research continues on their further characterization.

Endocrine Cells

The problems of culturing endocrine cells [O'Hare et al., 1978] are accentuated because the relative number of secretory cells may be quite small. Sato and colleagues [Buonassisi et al., 1962; Sato and Yasumura, 1966] cultured functional adrenal and pituitary cells from rat tumors by mechanical disaggregation of the tumor [Zaroff et al., 1961] and regular monolayer culture. The functional integrity of the cells was retained by intermittent passage of the cells as tumors in rats [Buonassisi et al., 1962; Tashjian et al., 1968]. These lines are now fully adapted to culture and can be maintained without animal passage [Tashjian, 1979] and, in some cases, in fully defined media [Hayashi and Sato, 1976].

Fibroblasts have been reduced in cultures of pancreatic islet cells by treatment with ethylmercurithiosalicylate [Braaten et al., 1974] and have also been purified by density gradient centrifugation [Prince et al., 1978] and by centrifugal elutriation [Bretzel et al., 1990]. The islet cells apparently produce insulin, but not as propagated cell lines.

Pituitary cells that produce pituitary hormones for several subcultures have been isolated from the mouse [de Vitry et al., 1974]. Although normal human pituitary cells do not survive well, and pituitary adenoma cells gradually lose the capacity to synthesize hormones, three-dimensional cultures of pituitary adenomas continue to secrete prolactin [Guiraud et al., 1991].

Melanocytes

The following protocol for the culture of human melanocytes was contributed by Hee-Young Park, Mina Yaar, and Barbara A. Gilchrest, Department of Dermatology, Boston University School of Medicine, 80 East Concord St., Boston, MA 02118-2394.

PROTOCOL 22.18. MELANOCYTES

Principle

The high substrate dependency of cultured keratinocytes has been utilized to obtain preferential melanocyte attachment and growth in a hormone-supplemented medium [Naeyaert et al., 1991]. The system circumvents the problem of keratinocyte contamination, while supporting good melanocyte proliferation with minimal supplementation of serum [Gilchrest et al., 1984; Wilkins et al., 1985] in the absence of chemotherapeutic agents or tumor promoters such as phorbol esters. These substances down-regulate protein kinase C (PKC) and, hence, tyrosinase activity, curtailing the production of pigment [Park et al., 1993]. With conventional serum supplementation (5–20%), melanocyte growth is far better than reported in other systems, and fibroblast overgrowth can be prevented by reducing the Ca^{2+} concentration to 0.03 mM [Naeyert et al., 1991]. The system described here can be further modified by removing cholera toxin (which otherwise promotes melanocyte growth by greatly increasing intracellular levels of cAMP) and substituting dibutyryl cAMP. This medium then provides physiologic levels of both PKC and cAMP that permit studies of cAMP-modulating agents [Park et al., 1999] under conditions that are optimal for melanogenesis studies.

Outline

Epidermis, stripped from small fragments of skin following trypsinization, is dissociated in EDTA and cultured in serum-free medium supplemented with BFGF.

Materials

Sterile:
PBSA
Fetal bovine serum
Hormone-supplemented medium: Medium 199 (Gibco 400–1200), 93.1 ml
EGF (Collaborative Research), 0.1 ml, 10 μg/ml stock concentration, 10 ng/ml final concentration
Transferrin (Sigma), 0.1 ml, 10 mg/ml stock, 10 μg/ml final
Insulin (Sigma), 0.1 ml, 10 mg/ml stock, 10 ug/ml final
Triiodothyronine (Sigma), 0.1 ml, 1 × 10^{-6} M stock, 1 × 10^{-9} M final
Hydrocortisone (Calbiochem), 0.5 ml, 200× stock, 1.4 × 10^{-6} M final
Cholera toxin (Calbiochem), 1.0 ml, 1 × 10^{-7} M stock, 1 × 10^{-9} M final

Cholera toxin can be replaced by bovine pituitary extract (Clonetics), 35 μg/ml, stock 0.7 ng/ml final, and dibutyryl cAMP (Sigma), 1 mM stock, 100 μM final

Basic FGF (Amgen), 5.0 ml, 200 ng/ml stock, 10 ng/ml final

Trypsin/EDTA: Trypsin 0.25% in PBSA with 0.1% (3.5 mM) EDTA (Gibco 610–5050)

Tissue culture dishes (Falcon, Becton Dickinson), 100, 60, and 35 mm in diameter

Pipettes

Centrifuge tubes, 15 and 50 ml

Nonsterile:

Humidified incubator (37°C, 7% CO_2)

Water bath

Electronic cell counter or hemocytometer

Protocol

Establishing the Primary Culture

Day 1:

1. Rinse the skin specimen in 70% ethanol and then twice in PBSA.
2. Transfer the tissue to a sterile 100-mm dish, epidermal side down.
3. With dissecting scissors, excise subcutaneous fat and deep dermis.
4. Cut the remaining tissue into 2 × 2-mm pieces with a scalpel, rolling the blade over the tissue. (Do not use a sawing action.)
5. Transfer the tissue fragments to cold 0.25% trypsin in a 15-ml centrifuge tube.
6. Incubate the tissue fragments 1–1½ h at room temperature. For some donors, incubation at 37°C for 1–2 h followed by incubation at 4°C for 18–24 h may give higher melanocyte yields.

Day 2:

7. Gently tap the centrifuge tube to dislodge fragments that have settled at the bottom, and then rapidly pour the contents of the tube into a 60-mm dish.
8. Using forceps, transfer the tissue fragments individually to a dry 100-mm dish, epidermal side down. Gently roll each tissue fragment against the dish. The epidermal sheet should adhere to the dish and allow a clean separation of the dermis with forceps. Discard the dermal fragments.
9. Transfer all epidermal sheets to a sterile 15-ml centrifuge tube containing 5 ml of 0.02% (0.7 mM) EDTA. Take care to place each epidermal sheet in the EDTA solution, not on the plastic wall of the centrifuge tube.
10. Gently vortex the tube to disaggregate the epidermal sheets into a single-cell suspension.

11. Centrifuge the cells for 5 min at 350 g, and then aspirate the supernatant.
12. Resuspend the pellet in serum-free melanocyte medium.
13. Count the cells with a hemocytometer, and inoculate 1 × 10^6 cells (approximately 2–4 × 10^4 melanocytes) per 35-mm dish in 2 ml of hormone-supplemented medium containing 5% FBS to facilitate attachment of the cells.
14. Refeed the culture with fresh serum-free medium containing growth factors and bovine pituitary extract twice weekly.

Days 3–30:

15. At 24 h, the cultures will contain primary keratinocytes with scattered melanocytes (Fig. 22.2a). Keratinocyte proliferation should cease with several days, and colonies should begin to detach during the second week. By the end of the third week, only melanocytes should remain (Figs. 22.2b,c). In most cases, cultures attain near confluence and are ready to passage within 2–4 weeks.

Subculture

1. Gently rinse the culture dish twice with 0.02% (0.7 mM) EDTA.
2. Add 3 ml of 0.25% trypsin/0.1% (2.5 mM) EDTA, and incubate at 37°C. Examine the dish under phase microscopy every 5 min to detect cell detachment.
3. When most cells have detached, inactivate the trypsin with soybean trypsin inhibitor (Sigma, 10 μg/ml) or 1 ml of medium containing 10% calf serum. (Melanocytes maintained under serum-free conditions greatly benefit from a "serum kick" at the time of subculture.)
4. Pipette the contents of the dish to ensure complete melanocyte detachment.
5. Aspirate and centrifuge the cells for 5 min at 350 g.
6. Aspirate the supernatant, resuspend the cells in a calcium-free melanocyte medium containing 5% FBS without calcium, and replate at 2–4 × 10^4 cells per 100-mm dish. FBS is required for good melanocyte attachment (>75%) to plastic dishes, but can be omitted if the dishes are coated with fibronectin or type I/III collagen [Gilchrest et al., 1985].
7. Refeed the culture twice a week with serum-free or serum-containing melanocyte medium. The serum-containing medium will give a greater melanocyte yield, but encourages the growth of fibroblasts. Reducing the calcium concentration to 0.03 mM has no effect on the proliferation of melanocytes, but inhibits fibroblast overgrowth [Naeyaert et al., 1991].

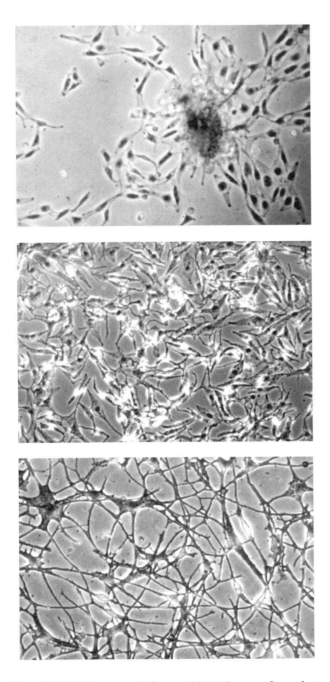

Fig. 22.2. Melanocyte Cultures. (a) Culture of newborn foreskin-derived melanocytes 1 week after inoculation. Note the multiple keratinocyte colonies with central stratification and tightly apposed epithelial cells at the periphery. The melanocytes are the relatively small, dark dendritic cells, most of them in contact with the keratinocyte colonies by means of dendritic projections (seen via phase contrast). (b) Ten-day-old primary cultures of newborn epidermal melanocytes in medium lacking TPA. Many cells display branching dendrites, and other cells display bipolar to polygonal morphology (seen via phase contrast, ×320). (c) Secondary culture of TPA-treated epidermal melanocytes. Note the dendritic morphology of the cells and their slender spindle shape.

Confirmation of Melanocytic Identity

Melanocyte cultures may be contaminated initially with keratinocytes and at any time by dermal fibroblasts. Both forms of contamination are rare in cultures established and maintained by an experienced technician or investigator, but are common problems for the novice. The cultured cells can be confirmed to be melanocytes with moderate certainty by frequent examination of the culture under phase microscopy, assuming that the examiner is familiar with the respective cell morphologies (Figs. 22.2b,c). More definitive identification is provided by electron microscopic examination, DOPA staining, or immunofluorescent staining with Mel 5 antibody, directed against tyrosinase-related protein-1.

HEMATOPOIETIC CELLS

Hematopoietic cells have been grown in colony-forming assays, in long-term cultures from bone marrow, and as continuous cell lines. Under the control of specific growth factors, hematopoietic stem cells, or CFU-A, can be recloned from agar cultures. (See Protocol 22.20.) MIP-1α appears to be the main factor that keeps the stem cells in the regenerative compartment. However, most suspension colonies, which contain cells of only one lineage, survive only as primary cultures that lose their repopulation efficiency, and hence, cannot be subcultured. Granulocytic colonies are the most common; but under the appropriate conditions, other lineages can be produced. (See Protocol 22.20.)

A number of myeloid cell lines have been developed from murine leukemias [Horibata and Harris, 1970] and, like some of the human lymphoblastoid lines [Collins et al., 1977], have been shown to make globulin chains and, in some cases, complete α- and γ-globulins. (See Production of Monoclonal Antibodies in Chapter 27.) Some of these lines can be grown in serum-free medium [Iscove and Melchers, 1978]. T-cell lines require T-cell growth factors [e.g., interleukin IL-2; Gillis and Watson, 1981], and B-cell growth factors have also been described [Howard et al., 1981; Sredni et al., 1981; see also Freshney et al., 1994].

The first human lymphoblastoid cell lines were derived by culturing peripheral lymphocytes from blood at very high cell densities ($\sim 10^6$/ml), usually in deep culture (>10 mm) [Moore et al., 1967]. Subsequently, immortalization was shown to be due to EBV, and reproducible techniques are now available for this procedure [Bolton and Spurr, 1996].

Rossi and Friend [1967] demonstrated that a mouse RNA virus (the "Friend virus") could cause splenomegaly and erythroblastosis in infected mice.

Cell cultures taken from minced spleens of these animals could, in some cases, give rise to continuous cell lines of erythroleukemia cells. All of these lines are transformed by what is now recognized as a complex of defective and helper viruses derived from Moloney sarcoma virus [Ostertag and Pragnell, 1978, 1981]. Some cell lines can produce a virus that is infective *in vivo*, but not *in vitro*, and the cells can also be passaged as solid tumors or ascites tumors in DBA2 or BALB-C mice.

Treating cultures of Friend cells with a number of agents, including DMSO, sodium butyrate, isobutyric acid, and hexamethylene bisacetamide, promotes erythroid differentiation [Friend et al., 1971; Leder and Leder, 1975]. Untreated cells resemble undifferentiated proerythroblasts, while treated cells show nuclear condensation, a reduction in cell size, and an accumulation of hemoglobin (see Soluble Inducers in Chapter 16) to the extent that centrifuged cell pellets are red in color. Evidence for differentiation can be demonstrated by staining the cells for hemoglobin with benzidine, (see Plate 26) and *in situ* hybridization of globin-specific messenger RNA (see Plate 37). The human leukemic cell line K562 can also be induced to differentiate with sodium butyrate and hemin, though not with DMSO [Andersson et al., 1979c].

Macrophages may be isolated from many tissues by collecting the cells that attach during enzymatic disaggregation. The yield is rather low, however, and a number of techniques have been developed to obtain larger numbers of macrophages. Mineral oil or thioglycollate broth [Adams, 1979] may be injected into the peritoneum of a mouse, and 3 days later the peritoneal washings will contain a high proportion of macrophages. If necessary, macrophages may be purified by their ability to attach to the culture substrate in the presence of proteases. (See Selective Adhesion in Chapter 13.) Macrophages can be subcultured only with difficulty, because of their insensitivity to trypsin. Methods have been developed using hydrophobic plastics (e.g., Petriperm dishes, Heraeus).

There are some reports of propagated lines of macrophages, mostly from murine neoplasia. Normal mature macrophages do not proliferate, although it may be possible to culture replicating precursor cells by the method of Dexter. (See Protocol 22.19.)

Long-Term Bone Marrow Cultures

The following protocol for the long-term culture of bone marrow was contributed by E. Spooncer, Department of Biomolecular Sciences, UMIST, Sackville Street, Manchester, U.K. M60 1QD.

PROTOCOL 22.19. LONG-TERM HEMATOPOIETIC CELL CULTURES FROM BONE MARROW

Principle
By culturing whole bone marrow, the relationship between the stroma and stem cells is maintained, and in the presence of the appropriate hematopoietic cell and stromal cell interactions, stem cells and specific progenitor cells can continue proliferating over several weeks [Dexter et al., 1984; Spooncer et al., 1992]. Progenitor cells from fresh marrow or long-term cultures may be assayed by clonogenic growth in soft agar [Heyworth and Spooncer, 1992] or in mice [Till and McCulloch, 1961].

Outline
Marrow is aspirated into growth medium and maintained as an adherent cell multilayer for at least 12, and up to 30, weeks. Stem cells and maturing and mature myeloid cells are released from the adherent layer into the growth medium. Granulocyte/macrophage progenitor cells can be assayed in soft gels.

Materials
All reagents must be pretested to check their ability to support the growth of the cultures.
Sterile:
Fischer's medium (Gibco) supplemented with 50 U/ml of penicillin and 50 μg/ml of streptomycin and containing 16 mM (1.32 g/l) of NaHCO$_3$
Growth medium: Fischer's as above, 100-ml aliquots supplemented with 1×10^{-6} M hydrocortisone, sodium succinate, and 20% horse serum (hydrocortisone sodium succinate made up as 10^{-3} M stock in Fischer's medium and stored at $-20°$C)
Syringes, 1 ml, with 21G needles
Gauze, swabs, scissors, forceps
25-cm^2 tissue culture flasks
Nonsterile:
Five mice: (C57Bl/6 × DBA/2)F$_1$ bone marrow performs well in long-term culture, but marrow from some strains (e.g., CBA) does not [Greenberger, 1980]

Protocol
1. Kill the donor mice by cervical dislocation.
2. Wet the fur with 70% alcohol and remove both femurs. Collect 10 femurs in a Petri dish on ice containing Fischer's medium. One femur contains 1.5–2.0 × 10^7 nucleated cells.
3. In a laminar flow hood:
   (a) Clean off any remaining muscle tissue using gauze swabs.

(b) Hold the femur with forceps and cut off the knee end. The 21G needle should fit snugly into the bone cavity.

(c) Cut off the other end of the femur as close to the end as possible.

(d) Insert the tip of the bone into a 100-ml bottle of growth medium, and aspirate and depress the syringe plunger several times until all the bone marrow is flushed out of the femur.

(e) Repeat steps (a)–(d) with the other nine bones.

4. Disperse the marrow to a suspension by pipetting the large marrow cores through a 10-ml pipette. There is no need to disaggregate small clumps of cells.

5. Dispense 10-ml aliquots of the cell suspension into 25-cm² tissue culture flasks, swirling the suspension often to ensure an even distribution of the cells in the 10 cultures.

6. Gas the flasks with 5% CO_2 in air and tighten the caps.

7. Incubate the cultures horizontally at 33°C.

8. Feed the cultures weekly:

(a) Agitate the flasks gently to suspend the loosely adherent cells.

(b) Remove 5 ml of growth medium, including the suspension cells; take care not to touch the layer of adherent cells with the pipette.

(c) Add 5 ml of fresh growth medium to each flask; to avoid damage, do not dispense the medium directly onto the adherent layer.

(d) Gas the cultures and replace them in the incubator.

Analysis

Cells harvested during feeding can be investigated by a range of methods, including morphology, CFC assays (see Protocol 22.20), and the *in vivo* CFU-S assay for stem cells [Till and McCulloch, 1961].

Variations

Mouse erythroid [Dexter et al., 1981], B-lymphoid [Whitlock et al., 1984], and human long-term cultures [Gartner and Kaplan, 1980; Coutinho et al, 1992] have been grown.

Hematopoietic Colony-Forming Assays

Hematopoietic progenitor cells may be cloned in suspension in semisolid media in the presence of the appropriate growth factor(s) [Heyworth and Spooncer, 1992]. Pure or mixed colonies will be obtained, depending on the potency of the stem cells that are isolated. Assays for the detection of granulocyte and macrophage colony-forming cells (GM-CFC), erythroid burst-forming units (BFU-E), mixed colony-forming cells (CFC-mix), and granulocyte, erythrocyte, macrophage, and megakaryocyte colony-forming cells (CFC-GEMM) are described in Testa and Molineux [1993], and their place in routine hematopoietic cell culture technology is already well established [Metcalf, 1990].The following protocol has been abridged from Freshney [1994].

PROTOCOL 22.20. HEMATOPOIETIC COLONY FORMING ASSAYS

Outline

Suspend bone marrow cells in agar or methocel, and seed the cells into dishes with the appropriate growth factors.

Principle

The efficiency of growth of colonies in these assays is increased by using methocel instead of agar as the semisolid phase. This practice makes for a tighter colony that is easier to evaluate and count. Because methocel is a high-viscosity liquid, and not a gel like agar, cells will sediment through it, albeit slowly, and plate out on the plastic base of the dish. This route places them all in one focal plane for subsequent observation. The colonies will form if grown in an atmosphere of 5% CO_2 in air, but this may be at the expense of adequate hematopoiesis, and ideally, the gas phase should be 10% CO_2 and 5% O_2 in air [Bradley et al., 1978]. If an incubator with this gas mixture is not available, a cylinder of mixed gases can be rented. Place the dishes in a plastic box with a lid with a hole in it. Seal the box with plastic tape, and gas via the hole before sealing it. A dish of water in the box will keep the atmosphere humid.

Materials

Sterile or Aseptically Prepared:
Bone marrow cells. (See Protocol 22.19, Steps 1–4 for preparation.) Count the nucleated cells in a hemocytometer after staining them with methylene blue, or lyse the cells with Zapoglobin (Coulter) and count the nuclei on an electronic cell counter. Each femur will yield 1.0–1.5 × 10⁷ cells.
Methylcellulose, 4,000 cps (Fluka)
Noble agar (Difco)
Alpha MEM stock (Gibco)
FBS
Growth factors (Table 22.3), either recombinant (R&D) or from a conditioned medium
BSA, 10% in PBSA
Petri dishes, 30 mm

TABLE 22.3. Addition of Cells and Growth Factors for Colony-Forming Assays

| | Growth factors/ml | | | | | |
Assay	IL-6	rMurGM†	rIL-3*	Epo	Cells	Incidence/10^5
BFU-E		0.1 ng	1 ng	2 U	5×10^4	40–80
CFC-mix			1 ng	2 U	5×10^4	100–180
CFU-GEMM*	100 ng		1 ng	2 U	5×10^4	92–106

*10% Wehi-CM can be substituted for rIL3 [Bazill et al., 1983].

†10% AF1 19T CM can be substituted for GM-CSF [Pragnell et al., 1988].

Wehi-cell-conditioned medium (Wehi-CM). Wehi 3B is a mouse myelomonocytic cell line that when cultured, produces IL-3 (multi-CSF) into the medium [Bazill et al., 1983]. See Protocol 13.2 for a method for making a conditioned medium.

Methylcellulose
  (a) Weigh out 7.2 g of methocel, and add it to a 500-ml bottle containing a large magnetic stirrer bar.
  (b) Sterilize the methocel by autoclaving.
  (c) Add 400 ml of sterile UPW heated to 90°C to wet the methocel.
  (d) Stir the mixture at 4°C overnight to dissolve the methocel. The solution is now methocel 2×. It is more accurate to use a syringe (without a needle) than a pipette to dispense methocel.

Alpha medium stock solution:
  (a) Alpha medium, powder (Gibco 10 l pack size)
  (b) MEM vitamin stock, 100×, 100 ml
  (c) Gentamicin sulphate, 200 mg
    Stir the medium on a heated stirrer until it dissolves, and make to 3 l with UPW. (Do not allow the temperature to rise above 37°C.) Before final filtration through a 0.22-μm filter, it is advantageous to prefilter the medium through stacked filters of pore sizes 5, 1.2, 0.8 and 0.45. Dispense the medium into 21 ml aliquots and store it in premeasured volumes at −20°C.

Alpha medium, 2×:
  (a) Alpha medium stock solution   21 ml
  (b) Foetal bovine serum (FBS)   25 ml
  (c) Glutamine (200 mM)   1 ml
  (d) NaHCO$_3$, 7.5%   3 ml
    Mix the ingredients in a sterile bottle and equilibrate to 37°C.

Nonsterile:
With agar, use two water baths, one at 37°C and the other at 55°C.
Incubator: gas phase, 10% CO_2, 5% O_2, 85% N_2

(a) BFU-E, CFC-Mix, and CFU-GEMM

Protocol
1. Mix an equal volume of Alpha 2× medium to which 1% BSA has been added, with 2× methocel to make the required amount of medium for the experiment. Keep the mixture cold; methocel is more liquid when it is cold.
2. Set up cultures in triplicate, but make enough mix for 4 dishes, as methocel clings to the side of tubes and some is always lost.
3. Add the required concentration of growth factors to the tube (Table 22.3).
4. Add 5×10^4 cells.
5. Increase the total volume to 4.4 ml with the addition of 1× medium, and mix the tubes on a vortex mixer.
6. Using a syringe, plate out 1 ml of medium into 3-cm non-tissue-culture-grade dishes.
7. Incubate the culture at 37°C in a humid atmosphere of 10% CO_2 and 5% O_2 in air for 8–15 d.

Identification of the Colonies
BFU-E colonies can be either single colonies composed of very small cells or multicentric colonies (bursts), each with tightly packed very small pink or red cells. The incidence of these colonies is 40–80/10^5 bone marrow cells.

CFC-mix colonies can be single, compact colonies, usually with a halo of cells of widely varying size. They may be multicentric, but the cell population is obviously heterogeneous. The incidence of colonies in this assay is between 100 and 180/10^5 bone marrow cells, of which only about 10% will be mixed colonies containing erythroid cells. If the erythroid cells are not red, it can be very difficult for the inexperienced eye to identify the colonies accurately. The colonies should be photographed and then picked out, after which cytospins should be made that can be fixed and stained with 10%

Giemsa and help sought with identification of the cells [Heyworth & Spooncer, 1992].

(b) GM-CFC

1. Use 3 cm of non-tissue-culture-grade Petri dishes, lay out the dishes, and label them.
2. Make a 0.3% agar medium (see Protocol 13.4) and keep it at 37°C.
3. Prepare bone marrow cells. (See Protocol 22.19.) Count the nucleated cells after staining the cells with methylene blue, or lyse the cells with Zapoglobin (Coulter), and count the nuclei with an electronic cell counter. Each femur will yield $1.0-1.5 \times 10^7$ nucleated cells.
4. Add 0.1 ng/ml of rMurGM-CSF to each dish.
5. Add 1 ml of agar medium containing 7.5×10^4 cells and swirl gently to mix the cells and agar. Allow the agar to set at room temperature.
6. Alternatively, 0.8% methocel medium can be used. This generally gives a tighter colony, which is easier to count.
7. Place the dishes in a clean plastic box.
8. Incubate the dishes for 6 d in a humidified incubator at 37°C in an atmosphere of 5% CO_2 in air.
9. Using an inverted microscope, count the colonies that contain more than 50 cells, and express the number as colonies/10^5 cells seeded.
10. The incidence of GM-CFC in normal bone marrow is $100-120/10^5$ cells.

GONADS

Ovarian granulosa cells can be maintained and are apparently functional in primary culture [Orly et al., 1980], but specific functions are lost on subculture. A cell line started from Chinese hamster ovary [CHO-KI; Kao and Puck, 1968] has been in culture for many years, but its lineage still has not been identified. Although epithelioid at some stages of growth, it undergoes a fibroblasticlike modification when it is cultured in dibutyryl cyclic AMP [Ilsie and Puck, 1971].

Cellular fractions from testis have been separated by velocity sedimentation at unit gravity, but no prolonged culture of the fractions has been reported. The TM4 is an epithelial line from mouse testis, although its differentiated features have not been reported. Sertoli cells have also been cultured from testis [Mather, 1979].

Germ Cells

There have been many reports of germ cell cultures, but the cultures are derived from early embryos rather than the gonads. When cells from an embryo are implanted into an adult (e.g., under the kidney capsule), they can give rise to tumors known as teratomas, or embryonal carcinomas. These kinds of tumor also arise spontaneously when groups of embryonic cells or single cells are carried over into the adult, often at an inappropriate site. Artificially derived teratomas have been used extensively to study differentiation [Martin and Evans, 1974; Martin, 1975, 1978], because they may develop into a variety of different cell types (muscle, bone, nerve, etc.). Teratoma cells on feeder layers of, for example, SCI mouse fibroblasts, will proliferate, but not differentiate, whereas, when the cells are grown on gelatin without a feeder layer or in nonadherent plastic dishes, nodules form that eventually differentiate.

Similar germ cell cultures have been derived from human embryos and show differentiation down a number of different pathways, suggesting a possible future role in transplantation [Thomson et al., 1998].

CHAPTER 23

Tumor Cells

The culture of cells from tumors, particularly spontaneous human tumors, presents similar problems to the culture of specialized cells from normal tissue. The tumor cells must be separated from normal connective tissue cells, preferably by provision of a selective medium that will support tumor cells, but not normal cells. While the development of selective media for normal cells has advanced considerably (see Selective Media in Chapter 9 and Chapter 22), progress in tumor culture has been limited by variation both among and within samples of tumor tissue, even from the same tumor type. It is often surprising to find that tumors which grow *in vivo*, largely as a result of their apparent autonomy from normal regulatory controls, fail to grow *in vitro*.

There are many possible reasons for the failure of some tumor cultures to survive. Their nutritional requirements may be different from those of the equivalent normal cells, or perhaps attempts to remove stroma may actually deprive the tumor cells of a matrix or of nutritional or informational stimuli necessary for survival. Alternatively, dilution of tumor cells to provide a sufficient amount of nutrients per cell may also dilute out autocrine growth factors produced by the cells. Strictly speaking, truly autocrine factors should be independent of dilution if they are secreted onto the surface of the cell and are active on the same cell, but it is possible that some so-called autocrine growth factors are in fact often paracrine—i.e., they act on adjacent cells and not only on the cell releasing them. Hence, a closely interacting population is required. Interaction with certain types of stromal cells may substitute for homotypic paracrine interaction if the stroma are able to make the requisite growth factors, either spontaneously or in response to the tumor cells.

It may be incorrect to assume that the growth factor dependence of a tumor cell is similar to that of the normal cells of the tissue from which it was derived. Tumor cells may produce endogenous autocrine growth factors, such as TGF-α, and the provision of exogenous growth factors, such as EGF, may compete for the same receptor. Furthermore, the response of a tumor cell to a growth factor, or hormone, will depend on what other growth factors are present, some of which may be tumor cell derived, and on the status of the cell. A normal cell, capable of expressing growth suppressor and senescence genes, may respond differently from a cell in which one or more of these genes is inactive or mutated, and in which antagonistic, growth-promoting oncogenes are overexpressed.

So there are many possible reasons for a tumor cell population responding differently to the nutritional and mitogenic environment optimized for normal cells of the same lineage. To confirm this difference more information is required on the nutritional requirements of tumor cells, but, given the heterogeneity of tumors, the task is a daunting one. Since the potential therapeutic benefit to be derived from knowledge of the nutritional requirements of tumor cells is not likely to be great, greater emphasis has been placed on the response of tumor cells to growth factors and the differences in signal transduction, areas for which the potential therapeutic benefit may be greater.

Optimization of nutritional conditions has been restricted to certain specific types of tumor for which well-characterized continuous cell lines are available, and the result is the generation of selective media for cells such as HeLa, the small-cell lung cancer cell lines, such as NCI-H69, and basal cell carcinoma of skin. This means that some continuous cell lines may be maintained serum free, but, unfortunately, the selective conditions developed may be specific to continuous cell lines from that tumor, or even to a particular continuous cell line.

Finally, it is probable that many cells in a tumor have a limited life span, due to genetic aberration, terminal differentiation, apoptosis, or natural senescence and only a few cells, analogous to a stem cell population in normal tissue, have the potential for continuous survival. Dilution into culture may reduce the number of these cells, as well as their interaction with other cells, such that survival is impossible. Cells from multicellular animals, unlike prokaryotes, do not survive readily in isolation. Even a tumor is still a multicellular organ and may require continuing cell interaction for survival. The lethality of the tumor to the host lies in its uncontrolled infiltration and colonial growth, but the origin of the bulk of the cell population may reside in a relatively small population of transformed stem cells. This pool of stem cells may be so small that its dilution on explantation deprives it of some of its prerequisites for survival, particularly paracrine growth factors from stromal elements and other tumor subclones.

In sum, the goal is either to create the correct, defined nutritional and hormonal environment or, failing that, to provide a sustaining environment, as yet undefined, but nevertheless able to permit the survival of an appropriate or representative population. There has been a continuing trend to use serum and feeder layers in order to get tumor cells to grow, and only a few tumors have responded to serum-free culture. As many transformed cells are not inhibited by TGF-β, there has not been the same need to eliminate serum, other than to repress fibroblastic growth, which remains a major problem. The adaption of medium designed for equivalent normal cells is still the most logical approach to obtaining cell lines from tumors.

SAMPLING

In addition to preventing the overgrowth of connective tissue or vascular cells, both of which are stimulated to invade and proliferate by many tumors, tumor cell culture has the additional aim of separating the transformed cells from the normal equivalent tissue cells, which may have similar characteristics. Furthermore, while any section of gastric epithelium may be regarded as representative of that particular zone of the gastric mucosa, tumor tissue, dependent as it is on genetic variation and natural selection for its development, is usually heterogeneous and composed of a series of often diverse subclones displaying considerable phenotypic diversity. Ensuring that cultures derived from this heterogeneous population are representative is difficult, and can never be guaranteed unless the whole tumor is used and survival is 100%. Since these conditions are practically impossible to achieve, the average tumor culture is a compromise. Assuming that representative subpopulations have been retained and are able to interact, the corporate identity may be similar to the original tumor.

The problem of selectivity is accentuated when sampling is carried out from secondary metastases, which often grow better, but may not be typical either of the primary tumor or of all other metastases.

In view of these practically overwhelming problems facing tumor culture, it is almost surprising that the field has produced any valid data whatsoever. In fact, it has, and this may result from (1) the aforementioned autonomy of tumor populations, which may have allowed the proliferation of tumor cells under conditions in which normal cells would not multiply; (2) the increased size of the proliferative pool in tumors, which is larger than that of most normal tissues; (3) the ability of tumor cells to give rise to tumors as xenografts in immune-deprived mice; and (4) the propensity of malignantly transformed cells to give rise to continuous immortalized cell lines more frequently than normal cells. This last feature, more than any other, has allowed extensive research to be carried out on tumor cell populations, even on apparently normal differentiation processes, in spite of the uncertainty of their relationship to the tumor from which they were derived.

The uncertainty of the status of continuous cell lines remains, but nevertheless they have provided a valuable source of human cell lines for molecular and virological research. The question of whether they represent advanced stages of progression of a tumor whose development has been accelerated in culture, a cryptic stem cell population, or a purely *in vitro* artefact is still to be resolved. They are certainly distinct from most early-passage tumor cultures and should be regarded as a valuable resource, albeit genetically and phenotypically distinct from early-passage cell lines, but predominantly of the genotype of the parental cell from which they were derived. Their immortality is more likely to be due to the deletion or suppression of genes inducing senescence [Pereira-Smith and Smith, 1988; Goldstein et al., 1989; Holt et al., 1996; Sasaki et al., 1996] and to increased telomerase activity [Bryan and Reddel, 1997; Bodnar et al., 1998] than to overexpression of genes conferring malignancy per se.

DISAGGREGATION

Some tumors, such as human ovarian carcinoma, some gliomas, and many transplantable rodent tumors are readily disaggregated by purely mechanical means, such as pipetting and sieving (see Mechanical Disaggregation in Chapter 11), which may also help to minimize stromal contamination, as stromal cells are often more tightly locked in fibrous connective tissue. Many of the common human carcinomas, however, are hard, or scirrhous, and the tumor cells are contained within large amounts of fibrous stroma, making mechanical disaggregation difficult, although scraping the cut surface of scirrhous tumors has been used in the so-called spillage technique [Lasfargues, 1973; Oie et al., 1996], to release tumor cells from the fibrous stroma.

Enzymatic digestion has proved to be preferable to mechanical disaggregation in most cases. Although trypsin has often been used for this purpose, its effectiveness against fibrous connective tissue is limited, and it can reduce the seeding efficiency [Lounis et al., 1994]; crude collagenase has been found to be more effective with several different types of tumor [Dairkee et al., 1997]. Enzymatic disaggregation also releases many stromal cells, requiring selective culture techniques for their elimination. (See Selective Culture in this chapter.) Collagenase exposure may be carried out over several hours, or even days, in complete growth medium. (See Protocol 11.8.)

Extensive necrosis is also a problem of tumors that is not usually encountered with normal tissue. Usually, the attachment of viable cells allows necrotic material to be removed on subsequent feeding, but if the amount of necrotic material is large and not easily removed at dissection, it may be advisable to use a Ficoll-paque separation (see Protocol 11.10) to remove necrotic cells.

PRIMARY CULTURE

Some cells—e.g., macrophages—attach to the substrate during collagenase digestion, but may be removed by transferring the disaggregated cell suspension to a fresh flask when the collagenase is removed. The adherent cells may be retained and cultured separately. The reseeded cells will contain many stromal cells (principally fibroblasts and endothelium), some of which may be removed by a second transfer to a fresh vessel in 2–4 h, since tumor cells, particularly clusters of malignant epithelium, are often less adhesive and take longer to attach. This method of removal by serial transfer is generally only partially successful, however, and it will usually require selective culture conditions for the complete removal of the stromal cells.

Physical separation techniques have also been used to remove stromal contaminants [Csoka et al., 1995; Oie et al., 1996; see also Chapter 14], but, in general, these methods are suitable only if the cells are to be used immediately, as stromal overgrowth usually follows in the absence of selective conditions.

Cloning is a method that suggests itself, but there are several limitations. First, tumor cells in primary culture often have poor plating efficiencies (<0.1%). Furthermore, because of the heterogeneity of tumor cell populations, several clones must be isolated to be at all representative. However, by the time that a clone has grown to sufficient numbers to be of potential analytical value, it may have changed considerably, and it may even have become heterogeneous itself, due to genetic instability. Cloned isolates from a tumor should be studied collectively, and even in coculture, for a meaningful interpretation.

There has also been some difficulty in propagating cell lines from primary clones, particularly from clones isolated by the suspension method. It may be that although these cells are clonogenic, few of them really are stem cells, or, if they are, they mature spontaneously, due to the suspension mode of growth, and lose their regenerative capacity. Nevertheless, cell strains cloned directly from tumors would be valuable material for studying tumor clonal diversity and interaction, and they represent a key area of study for future investigation.

CHARACTERIZATION

The isolation of cells from tumors may give rise to several different types of cell line. Besides the neoplastic cells, connective-tissue fibroblasts, vascular endothelial, and smooth-muscle cells, infiltrating lymphocytes, granulocytes, and macrophages, as well as elements of the normal tissue in which the neoplasia arose, can all survive explantation. The hematopoietic components seldom form cell lines, although hematopoietic cell lines have been derived from small-cell carcinoma of the lung, causing serious confusion, since this carcinoma also tends to produce suspension cultures that can express myeloid markers [Ruff and Pert, 1984]. Macrophages and granulocytes are so strongly adherent and nonproliferative that they are generally lost at subculture. Smooth muscle does not propagate readily without the appropriate growth factors and selective medium, so the major potential contaminants of tumor cultures are fibroblasts, endothelial cells and the normal equivalents of the neoplastic cells.

Of these contaminants, the major problem lies with the fibroblasts, which grow readily in culture and may also respond to tumor-derived mitogenic factors. Similarly, endothelial cells, particularly in the absence of fibroblasts, may respond to tumor-derived angiogenesis factors and proliferate readily. The role of normal equivalent cells is harder to define, as their similarity to the neoplastic cells has made the appropriate experiments difficult to analyze. Characterization criteria should be chosen to exclude nontumor cells. For example, endothelial cells are factor VIII positive, contact inhibited, and sensitive to density limitation of growth; fibroblasts have a characteristic spindle-shaped morphology, are density limited for growth (though less so than endothelial cells), have a finite life span of 50 generations or so, make type I collagen, and are rigidly diploid.

In general, the normal cell component, phenotypically equivalent to the tumor cells, is harder to identify and eliminate. The cells will be diploid, although some tumor cells may be close to diploid. They are usually anchorage dependent and will have a finite life span [although, again, there are cases of normal epithelial cell lines becoming continuous; Boukamp et al., 1988], and, if the cells are epithelial, they are more likely to be inhibited by serum TGF-β. The tumor cells are more likely to show genetic aberrations, such as oncogene amplification, translocations, and suppressor gene deletions, identifiable by FISH (see Protocol 27.2). The tumor cells are also likely to be angiogenic, and show a higher expression of urokinaselike plasminogen activator (uPA), and will be invasive (see Chapter 17).

From a behavioral aspect, the ability of neoplastic cells to grow on a preformed monolayer of the normal cells of same type is a good criterion for tumor cell identity and a potential model for separation. The normal cells also provide a feeder layer to sustain the tumor cells. Glioma, for example, will grow readily (better than on plastic in some cases) on a preformed monolayer of normal glial cells [MacDonald et al., 1985] (see Fig. 13.13), but their normal counterparts will not, and the same may be true for hepatoma cells and skin carcinomas.

Normal cells tend to have a low growth fraction (i.e., the labeling index with prolonged [³H]-thymidine exposure; see Contact Inhibition in Chapter 17) at saturation density, while neoplastic cells continue to grow faster postconfluence. The maintenance of cultures at high density can sometimes provide conditions for overgrowth of the neoplastic cells.

DEVELOPMENT OF CELL LINES

Primary cultures of carcinoma cells do not always take readily to trypsin passage, and many of the cells in the

primary culture may not be capable of propagation, due to genetic or phenotypic aberrations, terminal differentiation, or nutritional insufficiency. Nevertheless, some primary cultures from tumors can be subcultured opening up major possibilities. Evidence for tumor cells in the subculture implies that they have not been overgrown, may even have a faster growth rate than contaminating normal cells, and may be available for cloning or other selective culture methods. (See Selective Media in Chapter 9 and Selective Inhibitors in Chapter 13.)

One of the major advantages of subculture is amplification. Replicate cultures can be prepared for characterization and assay of specific parameters such as genomic alterations, changes in gene expression, chemosensitivity, and invasiveness. Disadvantages of subculture include evolution away from the phenotype of the tumor, due to the inherent genetic instability of the cells and selective adaptation of the cell line to the culture environment.

Continuous Cell Lines

One major criterion for the neoplastic origin of a culture is its capacity to form a continuous cell line. The constituent cells of this line are normally aneuploid, heteroploid, insensitive to density limitation of growth, anchorage independent, and often tumorigenic. The relationship of this cell line to the primary culture and the parent tumor is still difficult to assess, however, as such cells are not always typical of the tumor population. The cells of a continuous cell line may be either (1) further transformation stimulated by adaptation to culture, made possible by the unstable genotypic characteristics of tumor cells, or (2) a specific subset or stem cell population of the tumor. Currently, the second possibility seems more likely, as the emergence of a continuous cell line is often from colonies within the monolayer, suggesting that the continuous line arises from a minor subset of the tumor cell population, with cell culture merely providing the appropriate conditions for their expansion.

The capacity to form continuous cell lines is a useful criterion for a malignant origin, and some authors maintain that the characteristics of the cell lines (e.g., tumorigenicity, histology, chemosensitivity, etc.) still correlate with the tumor of origin [Tveit and Pihl, 1981; Minna et al., 1983]. In any event, these cell lines provide useful experimental material, although the time required for their evolution makes immediate clinical application difficult.

SELECTIVE CULTURE

Three main approaches have been adopted to select tumor cells in primary culture: selective media, confluent feeder layers, and suspension cloning. (See Selective Media in Chapter 9 and Selective Inhibitors in Chapter 13.)

Selective Media

There are only a few media that have been developed as selective agents for tumor cells, due to their inherent problems of variability and heterogeneity. HITES [Carney et al., 1981; see Table 9.2] is one such medium and may owe its success to the production of peptide growth factors by small-cell lung cancer, for which the medium was developed. A proportion, but not all, of small-cell lung cancer biopsies will grow in pure HITES; others will survive with a low-serum supplement (e.g., 2.5%). HITES medium is modified RPMI 1640 with hydrocortisone, insulin, transferrin, estradiol, and selenium. Of these constituents, selenium, insulin, and transferrin are probably the most important and are found in many serum-free formulations. (See Table 9.2.)

The NCI group has also produced a selective medium for adenocarcinoma; reputedly suitable for lung, colon, and, potentially, many other carcinomas [Brower et al., 1986]. It is also based on RPMI 1640, supplemented with selenium, insulin, and transferrin, with the addition of hydrocortisone, EGF, triiodothyronine, BSA, and sodium pyruvate. (See Table 9.2.)

A simplified version of this medium, RPMI 1640 plus selenium, insulin, and transferrin, and supplemented with 2.5% fetal bovine serum, may be suitable for a number of tumors with minimal stromal overgrowth. A selective, serum-free medium was also reported for bladder carcinoma [Messing et al., 1982].

Other types of selective media depend on the metabolic inhibition of fibroblastic growth and are not specifically optimized for any particular type of tumor. (See Selective Inhibitors in Chapter 13.) However, inhibitors have not been found to be generally effective, with the exception of the use of monoclonal antibodies against fibroblasts by Edwards et al. [1980] and Paraskeva et al. [1985]. These antibodies have proved useful in establishing cultures from laryngeal and colon cancer. Antibodies have also proved useful in either positively selecting epithelial cells or negatively sorting stromal cells from tumor cell suspensions by panning or magnetic sorting. (See Magnetic Sorting in Chapter 14.)

Confluent Feeder Layers

The use of confluent feeder layers (see Figs. 13.3 and 13.12), perhaps more than any other method, has been applied successfully to many types of tumor. Smith and others [e.g., Lan et al., 1981] used confluent feeder layers of fetal human intestine, FHS74Int, to grow epithelial cells from mammary carcinoma, using media conditioned by other cell lines, although later reports suggest that selective culture in MCDB 170 is a more reproducible approach [Hammond et al., 1984]. Feeder layers of mouse 3T3 or STO embryonic fibroblasts were used successfully with breast, colon, and basal cell carcinoma [Rheinwald and Beckett, 1981; Leake et al., 1987].

Feeder layer techniques rely on the prevention of fibroblastic overgrowth by a preformed monolayer of other contact-inhibited cells. They are not selective against normal epithelium, as normal epidermis and normal breast epithelium both form colonies on confluent feeder layers. Results from glioma [MacDonald et al., 1985], however, suggest that selection against equivalent normal cells may be possible on a homologous feeder layer. Glioma grown on normal glial feeder layers should lose any normal glial contaminants. By the same argument, breast carcinoma seeded on confluent cultures of normal breast epithelium— e.g., milk cells (see Breast in Chapter 22)—could become free of any contaminating normal epithelium.

PROTOCOL 23.1 GROWTH ON CONFLUENT FEEDER LAYERS

Outline

Treat feeder cells in the midexponential phase with mitomycin C, and reseed the cells to give a confluent monolayer. Seed tumor cells, dissociated from the biopsy by collagenase digestion, or from a primary culture with trypsin, onto the confluent monolayer. (Fig. 23.1.) Colonies from epithelial tumors may form in 3 weeks to 3 months. Fibrosarcoma and gliomas do not always form colonies, but may infiltrate the feeder layer and gradually overgrow.

Materials

Sterile:
Feeder cells (e.g., 3T3, STO, 10T1/2, or FHS 74 Int)
Mitomycin C (Sigma), 1 mg/ml

Note. It is advisable to do a dose-response curve with mitomycin C when using feeder cells for the first time, to confirm that this dose allows the feeder layer to survive for 2–3 weeks, but does not permit further replication in the feeder layer after about two doublings, at most. In this case, proceed as follows:

(a) Treat the cells overnight (18 h) in 25-cm² flasks with 1–100 μg/ml of mitomycin C.
(b) Trypsinize the cells, and reseed the entire contents of the flasks into a 75-cm² flask in 20 ml of fresh medium.

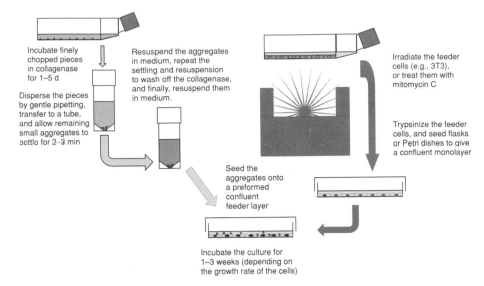

Fig. 23.1. Confluent Feeder Layers. Epithelial clusters from collagenase digestion will form colonies of epithelial cells on confluent feeder layers, such as fetal intestinal epithelium (FHS 74 Int), normal human glia, or irradiated 3T3 or STO cells. Selection is against stromal components, but not normal epithelium.

(c) Grow the cells for 3 weeks, feeding them twice per week.

(d) Stain the culture, and check for surviving colonies.

Growth medium

Collagenase, 2,000 U/ml, CLS grade (Worthington) or equivalent

Trypsin, 0.25%, in PBSA

Tumor biopsy or primary culture

Forceps, fine curved

Scalpels with #22 blades

Petri dishes for dissection, as for primary culture

Protocol

1. Grow up the feeder cells to 80% confluence in six 75-cm² flasks.
2. Add mitomycin C to give the appropriate final concentration usually around 5 µg/ml.
3. Incubate the cells overnight (~18 h) in mitomycin C.
4. Remove the medium with mitomycin C, and wash the monolayer with fresh medium
5. Grow the cells for a further 24–48 h.
6. Trypsinize the cells, and reseed them in 25-cm² flasks at 5×10^5 cells/ml (1×10^5 cells/cm²). Incubate the cultures for 24 h.
7. If you are using biopsy material, the biopsy should be dissected and placed in collagenase during step 2.
8. Seed a cell suspension from the biopsy, approximately 20–100 mg/flask, into two of the 25-cm² flasks, such that each flask holds 6 ml of suspension. Remove 1 ml of the suspension from each flask, and add it to 4 ml of medium in each of two more flasks. The third pair of flasks should be kept as controls to guard against feeder cells surviving the mitomycin C treatment.

If you are using a primary culture, trypsinize or dissociate the cells in collagenase, 200 U/ml final (see Protocol 11.8), and seed onto a feeder layer at 10^5 cells/ml in two flasks and 10^4 cells/ml in two flasks.

If the cells are from a glioma or fibrosarcoma, colonies may not appear, and the surviving tumor will be confirmed only by subculturing the cells without a feeder layer (by which time contaminating normal cells should have been eliminated).

It is essential to confirm the species of origin of any cell line derived by this method, in order to guard against accidental contamination from resistant cells in the feeder layer. The species of origin can be confirmed by chromosomal analysis (see Protocol 15.9) and lactate dehydrogenase isoenzyme electrophoresis (see Protocol 15.11) if the feeder is of a different species from the primary culture. If feeder cells of the same species as the primary culture are used, it is necessary to fingerprint both the feeder layer cells and any culture that is generated and compare the results with a portion of the biopsy or other tissue taken from the donor. (See Protocol 15.10.)

Suspension Cloning

The transformation of cells *in vitro* leads to an increase in their clonogenicity in agar [Macpherson and Montagnier, 1964] (see Protocols 13.4 and 13.5); tumorigenicity has also been shown to correlate with cloning in Methocel [Freedman and Shin, 1974]. Since cells may be cloned in suspension directly from disaggregated tumors [Hamburger and Salmon, 1977], or at least colonies may grow (they may not be clones) in preference to normal stromal cells, suspension cloning would seem to be a potentially selective technique. However, the colony-forming efficiency is very low (often <0.1%), and it is not easy to propagate the colonies isolated from agar. On the whole, this method has not been successful for generating cell lines, and it has had greater exploitation in the assay of primary cultures from tumor biopsies. (See Predictive Testing in Chapter 21.)

Histotypic Culture

Apart from organ culture itself (see Organ Culture in Chapter 24), which is not fundamentally different for tumor tissue than for normal tissue, methods with particular application to tumor culture are spheroid culture (see Protocol 24.2) and filter-well culture (see Protocol 24.4).

Spheroids. Normal stromal cells do not form spheroids or even become incorporated in tumor-derived spheroids. Hence, cultures from tumors allowed to form spheroids on nonadhesive substrates, like agarose (see Protocol 24.2), will tend to overgrow their stromal component.

Some cultures from breast and small-cell lung carcinoma can generate spheroids or nonspheroidal cellular organoids that float off and may be collected from the supernatant medium, leaving the stroma behind. However, the spheroids or organoids do not always appear soon after culture and can sometimes take weeks, or even months, to form, suggesting derivation from a minority cell population in the tumor.

Spheroid generation does not arise in all tumor cultures, but has been described in neuroblastoma, melanoma, and glioma. (See Spheroids in Chapter 24.) Its potential for generating cell lines has not been fully explored, however, as the bulk of attention has been given to forming either attached monolayers or suspension colonies in agar for assay purposes.

Filter Wells. (See Invasiveness in Chapter 17.)

Xenografts

When cultures are derived from human tumors, the scarcity of material and the infrequency of rebiopsy means that it is difficult to make several attempts at culture the same tumor, using different selective techniques. The growth of some tumors in immune-deprived animals [Rofstad, 1994], provides an alternative approach that makes much greater amounts of tumor available. It has sometimes been found that cultures can be initiated more easily from xenografts than from the parent biopsy, but whether this is due to the availability of more tissue, progression of the tumor, or modification by the heterologous host (e.g., by murine retroviruses) is not clear.

Two main types of host are in current use: the genetically athymic nude mouse, which is T-cell deficient [Giovanella et al., 1974], and neonatally thymectomized animals that are subsequently irradiated and treated with cytosine arabinoside [Selby et al., 1980; Fergusson et al., 1980]. The first type of host is expensive to buy and difficult to rear, but maintains the tumor for longer. Thymectomized animals are more trouble to prepare, but cheaper and easier to provide in large numbers. They do, however, regain immune competence and ultimately reject the tumor after a few months. Take rates for tumors can be enhanced by using mice that are asplenic as well as athymic, genetically (e.g., *scid* mice) or by splenectomy, or by sublethally irradiating nude mice. Implantation with fibroblasts or Matrigel has also been reported to improve tumor take [Topley et al., 1993].

If access to a nude mouse colony is available, or facilities exist for neonatal thymectomy and irradiation, xenografting should be considered as in generating a culture. Although only a small proportion of tumors may take, the resulting tumor will probably be easier to culture, and repeated attempts at culture may be made with subsequent passage of the tumor in mice. However, particular care must be taken, as with isolation from mouse feeder layers, to ensure that the cell line ultimately surviving is human, and not mouse, by proper characterization with isoenzyme and chromosome analysis.

Preservation of Tissue by Freezing

It is often difficult to take advantage of a large biopsy and utilize all of the valuable material that it provides. It is possible in these cases to preserve the tissue by freezing.

PROTOCOL 23.2. FREEZING BIOPSIES

Outline
Chop the tumor, expose the pieces to DMSO, and freeze aliquots in liquid nitrogen.

Materials
Sterile:
Biopsy
Plastic ampules, 1.2 ml (Nagle Nunc)
DBSS (see Reagents Appendix)

Growth medium with antibiotics (see Collection Medium in Reagents Appendix)

Dimethyl sulfoxide (self-sterilizing if placed in a sterile container)

Instruments (scalpels, forceps, dishes, etc., as for primary culture)

Protocol

1. After removing necrotic, fatty, and fibrous tissue, chop the tumor into about 3–4 mm pieces, and wash the pieces in DBSS, as for primary culture.

2. Place four or five pieces in each ampule.

3. Add 1 ml of growth medium containing 10% DMSO to the pieces, and leave them for 30 min at room temperature.

4. Freeze the ampules at 1°C/min (see Protocol 19.1), and transfer them to the gas phase of a liquid-nitrogen freezer.

5. To thaw an ampule, place it in 37°C water (with appropriate precautions; see Protocol 19.2).

6. Swab the ampule thoroughly in alcohol, open it, allow the pieces to settle, and remove half of the medium.

7. Replace the medium slowly with fresh, DMSO-free medium. Mix by gentle shaking, and allow to stand for 5 min.

8. Gradually replace all of the medium with DMSO-free medium, transfer the pieces to a Petri dish, and proceed as for regular primary culture, but allowing twice as much material per flask.

SPECIFIC TUMOR TYPES

Reference to the general protocols described in Chapter 11, together with the selective culture techniques cross-referenced previously, will provide a good starting point for culturing most tumor types. In general, a reasonable approach to tumor culture is to combine collagenase digestion (e.g., Protocol 11.8) with the tissue-specific approaches given in the protocols in Chapter 22, with or without the use of a feeder layer. Some specific examples of tumor culture are briefly discussed below.

Breast

Breast carcinoma can be cultured from organoids derived from collagenase digestion [Leake et al., 1987; Dairkee et al., 1995, 1997] and propagated on feeder layers or in MCDB 170. Regulating the microenvironment by generating gradients of oxygen, nutrients, and waste metabolites produced a phenotype similar to the carcinoma *in vivo*.

Lung

Both small-cell lung carcinoma (SCLC) and non-small-cell carcinoma (NSCLC) have been cultured successfully [Oie et al., 1996], using serum-free selective media—HITES for SCLC and ACL4 for NSCLC—mechanical spillage, and density gradient separation on Ficoll for isolating the cells. A substantial panel of these cell lines has been accumulated by the NCI, and some are available through the ATCC. The effects of matrix on oncogene and growth factor expression have also been studied [Pavelic et al., 1992] and have been used to facilitate culture of lung carcinoma cells from bone marrow micrometastases [Pantel et al., 1995].

Colon

Serum-free conditions for the culture of some human colorectal cancer cell lines have been described [Fantini et al., 1987; Murakami and Masui, 1980], but these conditions are generally not suitable for newly isolated carcinoma cultures, which require serum. Some success has also been reported with colorectal carcinoma [Paraskeva and Williams, 1992 (see Protocol 22.5); Park and Gazdar, 1996], from biopsies which have been taken from both primary tumors and metastases [Danielson et al., 1992]. As in lung carcinoma, colorectal tumors occur with neuroendocrine properties, and some success has been reported on the use of HITES medium (see Table 9.2) with them [Lundqvist et al., 1991]. Density centrifugation on Percoll has been used to purify colonic carcinoma cells for primary culture in conventional medium (RPMI 1640 with 10% FB) [Csoka et al., 1995].

Pancreas

Cell lines from pancreatic primary tumors or metastases have been isolated and propagated in RPMI 1640 supplemented with fetal bovine serum. The cell lines were adapted to protein-free medium for the examination of cell products [Yamaguchi et al., 1990].

Ovary

A number of cell lines have been established from ovarian epithelial tumors (e.g., OAW series [Wilson et al., 1996], OVCAR-3 [Hamilton et al., 1983], A2780, [Tsuruo et al., 1986]), some in serum-free medium [Jozan et al., 1992] and others in serum-containing medium (e.g., OSE medium, 50:50 M199:MCDB105, supplemented with 15% FBS, used following collagenase digestion [Lounis et al., 1994]). Density centrifugation on Percoll has also been used to purify ovarian carcinoma cells for primary culture, as for colonic carcinoma [Csoka et al., 1995].

Prostate

The matrix-assisted method, described above for lung carcinoma, has also proved to be successful in isolating cell lines from prostate tumors [Pantel et al., 1995]. Serum-free culture has also been used for indicating cultures from normal prostate and from benign and malignant tumors [Chopra et al., 1996]. (See also Protocol 22.10.)

Skin

Melanoma. Pigment cells from skin do not readily survive without the appropriate growth factors (see Protocol 22.18), although cultures can be obtained from melanomas with a reasonable degree of success [Creasey et al., 1979; Mather and Sato, 1979a,b]. Primary melanomas are often contaminated with fibroblasts, but cloning on confluent feeder layers of normal cells [see Protocol 23.1; Creasey et al., 1979; Freshney et al., 1982b] may be possible. In general, greater success has been obtained with secondary growth from lymph nodes, or from distant metastatic recurrences.

MCDB 153, supplemented with FGF-2, insulin, transferrin, α-tocopherol, bovine pituitary extract, hydrocortisone, and 5% serum, with catalase and PMA added for the first two passages, has been used to grow melanocytes from normal skin, dysplastic nevi, and melanotic metastases [Levin et al., 1995]. Geneticin, 100 μg/ml, has been used to inhibit fibroblast growth. (See also Protocol 22.18.)

Basal Cell Carcinoma. The 3T3 cell feeder-layer technique has proved successful for culturing basal cell carcinoma of skin [see Protocol 22.1; Rheinwald and Beckett, 1981].

Squamous Cell Carcinoma. SCC and erythroplakias have also been cultured on 3T3 feeder layers [Edington et al., (in preparation)].

Cervix

Benign and malignant tumors may be established from cervical biopsies using the 3T3 feeder layer technique described for normal cervix. (See Protocol 22.4.)

Neuroblastoma

Several lines of neuroblastoma (e.g., SK-N-BE(2) [Biedler and Spengler, 1976], Tumilowicz et al., 1970]) have been isolated and are of particular interest, because of their potential for differentiation [Dimitroulakos et al., 1994].

Seminoma

Testicular seminomas have been cultured using STO cells as a feeder layer and then have been supplemented with stem cell factor (SCF), LIF, and FGF-2, as used for embryonal stem cell cultures [Olie et al., 1995]. Although the cultures were heterogeneous, however, they did not give rise to primordial germ cell cultures.

CHAPTER 24

Organotypic Culture

CELL INTERACTION AND PHENOTYPIC EXPRESSION

The historical divergence between maintenance of a fragment of explanted tissue and propagation of the cells that grew out from it led to the development of organ culture and cell culture (see Types of Tissue Culture in Chapter 1), and it is cell culture that has become dominant. Now, although the potential uses of propagated cell lines is far from exhausted, many people are reverting to the notion that nutritional and hormonal supplementation are in themselves inadequate to re-create full structural and functional competence in a given cell population. The vital missing factor is cell interaction and the signalling capacity that it entails.

Reciprocal Interactions

Interacting populations of cells have a mutual effect on their respective phenotypes, and the resultant phenotypic changes lead to new interactions. Cell interaction is therefore not a single event, but, instead, a cascade of events. Similarly, exogenous signals do not initiate a single event, as may be the case with homogeneous populations, but initiate a new cascade, as a result of the exogenously modified phenotype of one or both partners. For example, alveolar cells of the lung synthesize and release surfactant only in response to hormonal stimulation of adjacent fibroblasts [Post et al., 1984]; similarly the response of prostate epithelium to stromal signals is in turn activated by androgen binding to the stroma [Thomson et al., 1997].

Epithelium differentiates in response to matrix constituents that are often determined jointly by the ep-
ithelium on one side and connective tissue on the other, as may be the case with the interaction between epidermis and dermis *in vitro* [Fusenig, 1994a; Limat et al., 1995]. Hence, the whole integrated tissue may easily, and understandably, respond differently to simple ubiquitous signals, not because of the specificity of the signal or the receptor affinity, but because of the quality of the microenvironment encoded in the juxtaposition of one cell type with a specific correspondent. As in human society, the response of one individual to an exogenous stimulus is dictated as much by the spatial and temporal relationship of the individual with other individuals as by the endogenous makeup of the individual. Likewise, a primitive neural crest cell may become a neuron, an endocrine cell, or a melanocyte depending on its ultimate location, its interaction with adjacent cells, and its response, mediated by neighboring cells, to hormonal stimuli.

In essence, this preamble establishes that while some cell functions, such as cell proliferation, glycolysis, respiration, and gene transcription, proceed in isolation, their regulation as related to a functioning multicellular organism ultimately depends on the interaction among cells of the appropriate lineage, the appropriate stage in that lineage, and on the interaction between cells of different lineages occupying the same microenvironment. This concept suggests that if you want to study the biology of isolated cells, or use the cells as a substrate, conventional monolayer or suspension cultures may be adequate, but if you want to learn something of the integrated function, or dysfunction, of whole organs, a histotypic or organotypic model will be required.

Choice of Models

There are two major ways to approach this goal. One is to accept the cellular distribution within the tissue, explant it, and maintain it as an organ culture. The second is to purify and propagate individual cell lineages, study them alone under conditions of homologous cell interaction, recombine them, and study their mutual interactions. These approaches have given rise to three main types of technique: (1) organ culture, in which whole organs, or representative parts, are maintained as small fragments in culture and retain their intrinsic distribution, numerical and spatial, of participating cells; (2) histotypic culture, in which propagated cell lines are grown alone to high density in a three-dimensional matrix; and (3) organotypic culture, in which cells of different lineages are recombined in experimentally determined ratios and spatial relationships to re-create a component of the organ under study.

Organ culture seeks to retain the original structural relationship of cells of the same or different types, and hence their interactive function, in order to study the effect of exogenous stimuli on further development [Lasnitzki, 1992]. This relationship may be preserved by explanting the tissue intact or recreated by separating the constituents and recombining them, as in the now-classical experiments of Grobstein and Auerbach and others in organogenesis [Auerbach and Grobstein, 1958; Cooper, 1965; Wessells, 1977]. (See also Cell Interaction in Chapter 16.) Organotypic culture represents the synthetic approach, whereby a three-dimensional, high-density culture is regenerated from isolated (and, preferably, purified and characterized) lineages of cells that are then recombined, after which their interaction is studied, and, in particular, their response to exogenous stimuli is characterized.

The exogenous stimuli may be regulatory hormones, nutritional conditions, or xenobiotics. In each case, the response is likely to be different from the responses of a pure cell type in isolation, grown at a low cell density.

Several types of system have been described to study isolated, whole, undisaggregated tissue, recombinations of tissues, or purified cell lineages. Since these types of system may provide models for quite distinct kinds of investigation, each is described separately.

ORGAN CULTURE

Gas and Nutrient Exchange

A major deficiency in tissue architecture in organ culture is the absence of a vascular system, limiting the size (by diffusion) and potentially the polarity of the cells within the organ culture. When cells are cultured as a solid mass of tissue, gaseous diffusion and the exchange of nutrients and metabolites becomes limiting. The dimensions of individual cells cultured in suspension or as a monolayer are such that diffusion is rapid, but aggregates of cells beyond about 250 μm in diameter (5,000 cells) start to become limited by diffusion, and at or above 1.0 mm in diameter ($\sim$2.5 $\times$ 10^6 cells) central necrosis is often apparent. To alleviate this problem, organ cultures are usually placed at the interface between the liquid and gaseous phases, to facilitate gas exchange while retaining access to nutrients. Most systems achieve this by positioning the explant on a raft or gel exposed to the air (see Fig. 1.2), but explants anchored to a solid substrate can also be aerated by rocking the culture, exposing it alternately to a liquid medium and a gas phase [Nicosia et al., 1983; Lechner and LaVeck, 1985; see Protocol 22.9], or by using a roller bottle or tube (see Protocol 25.3).

Anchorage to a solid substrate can lead to the development of an outgrowth from the explant and resultant alterations in geometry, although this effect can be minimized by using a nonwettable surface. One of the advantages of culture at the gas–liquid interface is that the explant retains a spherical geometry if the liquid is maintained at the correct level. If the liquid is too deep, gas exchange is impaired; if it is too shallow, surface tension will tend to flatten the explant and promote outgrowth.

Increased permeation of oxygen can also be achieved by using increasing O_2 concentrations up to pure oxygen or by using hyperbaric oxygen. Certain tissues—e.g., thyroid [de Ridder and Mareel, 1978], and prostate, trachea, and skin [Lasnitzki, 1992], particularly from a newborn or an adult—may benefit from elevated O_2 tension, but often, this benefit is at the risk of O_2-induced toxicity. As increasing the O_2 tension will not facilitate CO_2 release or nutrient–metabolite exchange, the benefits of increased oxygen may be overridden by other limiting factors.

Structural Integrity

Structural integrity, above other considerations, was and is the main reason for adopting organ culture as an *in vitro* technique in preference to cell culture. While cell culture utilizes cells dissociated by mechanical or enzymic techniques or spontaneous migration, organ culture deliberately maintains the cellular associations found in the tissue. Initially, organ culture was selected to facilitate histological characterization, but ultimately it was discovered that certain elements of phenotypic expression were found only if cells were maintained in close association.

It is now recognized that associated cells do exchange signals via junctional communications (gap junctions), via paracrine factors, and via cell adhesion molecules. (See Cell Adhesion in Chapter 2 and Cell Interaction in Chapter 16.) Signaling between cells is most striking during organogenesis, but is probably also required for the maintenance of fully mature tissues.

Therefore, maintenance of the structural integrity of the original tissue may preserve the correct homologous and heterologous cellular interactions present in the original tissue and maintain the correct configuration of the extracellular matrix.

Growth and Differentiation

There is a relationship between growth and differentiation such that differentiated cells no longer proliferate. (See Proliferation and Differentiation in Chapter 16.) It is also possible that cessation of growth may in itself contribute to the induction of differentiation, if only by providing a permissive phenotypic state that is receptive to exogenous inducers of differentiation. Because of density limitation of cell proliferation and the physical restrictions imposed by organ culture geometry, most organ cultures do not grow, or, if they do, proliferation is limited to the outer cell layers. Hence, the status of the culture is permissive to differentiation and, given the appropriate cellular interactions and soluble inducers (see Induction of Differentiation in Chapter 16), should provide an ideal environment for differentiation to occur.

Limitations of Organ Culture

Analysis of organ cultures depends largely on histological techniques and they do not lend themselves readily to biochemical and molecular analyses. Biochemical monitoring requires reproducibility between samples, which is less easily achieved in organ culture than in propagated cell lines, due to sampling variation in preparing an organ culture, minor differences in handling and geometry, and variations in the ratios of cell types among cultures. (See Table 1.3.)

Organ cultures are also more difficult to prepare than replicate cultures from a passaged cell line and do not have the advantage of a characterized reference stock to which they may be related. Organ cultures cannot be propagated, and hence each experiment requires recourse to the original donor tissue. Preparation is labor intensive, and as a result, the yield of usable tissue is often too low to be of value in biochemical or molecular assays. Furthermore, as the population of reacting cells may be a minor component of the culture, it is difficult to analyze the biochemical nature of the response and attribute it to the correct cell type, other than by autoradiographic, histochemical, or immunocytochemical techniques, which tend to be more qualitative than quantitative.

Organ culture is essentially a technique for studying the behavior of integrated tissues rather than isolated cells. It is precisely in this area that a future understanding of the control of gene expression (and ultimately of cell behavior) in multicellular organisms may lie, but the limitations imposed by the organ culture system are such that recombinant systems between purified cell types may contribute more information at this particular stage. However, there is no doubt that organ culture has contributed a great deal to our understanding of developmental biology and tissue interactions and that it will continue to do so in the absence of adequate synthetic systems.

Types of Organ Culture

As techniques for organ culture have been dictated largely by the requirement to place the tissue at a location that allows optimal gas and nutrient exchange, most of these techniques put the tissue at the gas–liquid interface on semisolid gel substrates of agar [Wolff and Haffen, 1952] or clotted plasma [Fell and Robison, 1929] or on a raft of microporous filter, lens paper, or rayon supported on a stainless steel grid [Lasnitzki, 1992; see Fig. 1.2] or adherent to a strip of Perspex or Plexiglas. This type of geometry is now most easily attained with filter-well inserts. (See Protocol 24.4.) The following protocol uses organ primordia from chick embryo but is applicable to many other types of tissue.

PROTOCOL 24.1 ORGAN CULTURE

Outline

Dissect out the organ or tissue, reduce it to 1 mm³, or to a thin membrane or rod, and place it on a support at the gas (air)–medium interface. Incubate it in a humid CO_2 incubator, changing the medium as required.

Materials

Sterile or Aseptically Prepared:
Instruments for dissection
Medium (e.g., M199), with or without serum
Filter-well inserts, non-tissue-culture treated (e.g., Costar Transwells polycarbonate #3423)
Multiwell plates, 12 well (Costar)
Nonsterile:
Fertile hen's eggs at 8 d of incubation

Protocol

1. Place the filter-well inserts in the wells of the multiwell plates, and add sufficient medium to reach the level of the bottom of the filter (~1 ml).

2. Place the dishes in a humid CO_2 incubator to equilibrate at 37°C.

3. Prepare the tissue, or dissect out whole embryonic organs (e.g., 8-d femur or tibiotarsus of a chick embryo; see Protocols 11.2 and 11./). The tissue must not be more than 1 mm thick, preferably less, in one dimension. (For example, 8-d embryonic tibiotarsus is perhaps 5 mm long, but only 0.5–0.8 mm in diameter. A fragment of skin might be 10 mm square but only 200 μm thick. Tissue that must be chopped down to size, such as liver or kidney, should be no more than 1 mm³.)

4. For short dissections (<1 h), HBSS is sufficient, but for longer dissections, use 50% serum in HBSS buffered with HEPES to pH 7.4.

5. Take the dishes from the incubator, and transfer the tissue carefully to filters. A Pasteur pipette is usually best for this task and can be used to aspirate any surplus fluid transferred with the explant, although care should be taken not to puncture the filter. Wet the inside of the pipette with medium before aspiration, to prevent fragments of tissue from sticking to the pipette.

6. Check the level of medium, making sure that the tissue is wetted, but not totally submerged, and return the dishes to the incubator.

7. Check after 2–4 h to ensure that a film of medium remains over the filter and explant, but that it is not deep enough for the explant to float.

8. Incubate the dishes for 1–3 weeks, changing the medium every 2 or 3 days and sampling as required.

Analysis

The analysis of an organ culture is usually accomplished by histology, autoradiography, or immunocytochemistry, but assay of total amounts of cellular constituents or enzyme activity is possible, although variation between replicates will be high in such cases.

Variations

Most variations involve the following aspects:

1. **Medium.** M199 or CMRL 1066 may be used with or without serum, and BJG may be used [Biggers et al., 1961] for cartilage or bone.

2. **Type of support.** Organ cultures may be supported by a filter (e.g., polycarbonate) lying on top of a stainless-steel grid in a center-well organ culture dish (Falcon #3037). (Fig. 24.1.) How-

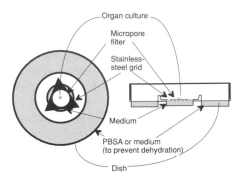

Fig. 24.1. Organ Culture. Small fragment of tissue on a filter laid on top of a stainless steel grid over the central well of an organ culture dish.

ever, the standard technique for filter-well inserts has a number of advantages in terms of handling, range of sizes, materials, and matrix coatings. (See later.) Different types of tissue may be combined on opposite sides of a filter to study their interaction. (See Cell Interaction in Chapter 16.) Furthermore, with small filter-well inserts (e.g., 6.5 mm, as in Corning Costar #3423), the well formed on the top side of the filter assembly generates a meniscus of medium with a large surface area available for gas exchange. It is also possible to alter the configuration of the tissue by raising or lowering the level of medium in the dish, and therefore also in the well; deeper medium gives a spherical explant, and shallower medium flattens the explant.

3. **O_2 tension.** Embryonic cultures are usually best kept in air, but late-stage embryo, newborn, and adult tissue are better kept in elevated oxygen [Trowell, 1959; de Ridder and Mareel, 1978; Zeltinger and Holbrook, 1997].

4. **Stirred or static cultures.** Stirred cultures of small tissue fragments have been used for confrontational cultures for assay of invasion [Mareel et al., 1979; Bjerkvig et al., 1986a,b]. (See Protocol 24.5.)

4. **Rocking or rotated cultures.** The tissue is anchored to a substrate and subjected alternately to liquid culture medium and the gas phase by placing the culture vessel on a rocking platform [see Protocol 22.9; Nicosia et al., 1983], or by anchoring the tissue to the wall of a rotating flask or tube. (See Roller Rocks in Chapter 4, and Protocol 25.3.)

Organ cultures are useful in the demonstration of processes such as embryonic induction, where it is important to maintain the integrity of whole tissue and the opportunity for separated coculture. However, they are slow to prepare and present problems of re-

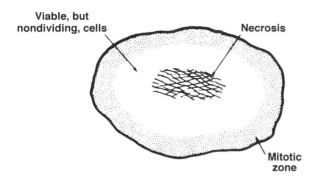

Fig. 24.2. Section of Organ Culture. Diagrammatic representation of the expected distribution of mitoses (stippled area) and necrosis (shaded central area) in an organ culture explant.

producibility between samples. Growth is limited by diffusion (although growth is perhaps not necessary and may even be undesirable), and mitosis is nonrandomly distributed throughout the explant. Mitosis occurs only around the periphery, while the centers of explants frequently become necrotic. (Fig. 24.2.) It has been argued that this type of geometry makes organ cultures good models of tumor growth, for which peripheral cell division is often accompanied by central necrosis.

HISTOTYPIC CULTURE

Various attempts have been made to regenerate tissuelike architecture from dispersed monolayer cultures. Green and Thomas [1978] showed that human epidermal keratinocytes will form dematoglyphs (i.e., friction ridges) if they are kept for several weeks without transfer, and Folkman and Haudenschild [1980] were able to demonstrate the formation of capillary tubules in cultures of vascular endothelial cells cultured in the presence of endothelial growth factor and medium conditioned by tumor cells. However, the most significant development has been the introduction of filter-well inserts, which give the opportunity for the formation of both high-density polarized cultures and heterotypic combinations of cell types to create organotypic cultures. (See Filter Well Inserts in this chapter.)

Gel and Sponge Techniques

Leighton first demonstrated that cells penetrate cellulose sponge [Leighton et al., 1968]. Both normal and malignant cells can do this, and this ability does not seem to reflect malignant behavior. Collagen coating of the sponge may facilitate occupation, and Gelfoam (a gelatin sponge matrix used in reconstructive sur-

gery) may be used in place of cellulose [Sorour et al., 1975]. These systems require histological analysis and are limited in dimensions, like organ cultures, by gaseous and nutrient diffusion.

Collagen gel (native collagen, as distinct from denatured collagen coating) provides a matrix for the morphogenesis of primitive epithelial structures. Mammary epithelium forms rudimentary tubular and glandular structures when grown in collagen [Gomm et al., 1997].

Hollow Fibers

Since medium supply and gas exchange become limiting at high cell densities, Knazek et al. [1972; Gullino and Knazek, 1979] developed a perfusion chamber from a bed of plastic capillary fibers, now available commercially (Endotronics, Cellco). The fibers are gas and nutrient permeable and support cell growth on their outer surfaces. Medium, saturated with 5% CO_2 in air, is pumped through the centers of the capillaries, and cells are added to the outer chamber surrounding the bundle of fibers. (See Figs. 25.6 and 7.11.) The cells attach and grow on the outside of the capillary fibers, fed by diffusion from the perfusate, and can reach tissue-like cell densities. Different plastics and ultrafiltration properties give molecular weight cut-off points at 10, 50, or 100 kDa, regulating the diffusion of macromolecules.

It is claimed that cells in this type of high-density culture behave as they would *in vivo*. For example, in such cultures, choriocarcinoma cells release more human chorionic gonadotrophin [Knazek et al., 1974] than they would in conventional monolayer culture and colonic carcinoma cells produce elevated levels of CEA [Rutzky et al., 1979; Quarles et al., 1980]. However there are considerable technical difficulties in setting up the chambers, and they are costly. Furthermore, sampling cells from these chambers and determining the cell concentration are difficult. Overall, however, hollow fibers appear to present an ideal system for studying the synthesis and release of biopharmaceuticals and are now being exploited on a semiindustrial scale. (See Perfusion Systems in Chapter 25.)

Spheroids

When dissociated cells are cultured in a gyratory shaker, they may reassociate into clusters. Dispersed cells from embryonic tissues will sort during reaggregation in a highly specific fashion [Linser and Moscona, 1980]; for example, Muller cells of chick embryo retina reaggregated with neuronal cells from the retina were inducible for glutamine synthetase, but those reaggregated with neurons from other parts of the brain were not. Cells in these heterotypic aggregates appear to be capable of sorting themselves into groups and forming tissue-like structures.

Homotypic reaggregation also occurs fairly readily, and spheroids generated in gyratory shakers or by growth on agar have been used as models for chemotherapy *in vitro* [Twentyman, 1980] and for the characterization of malignant invasion [Mareel et al., 1980]. As with organ cultures, the growth of spheroids is limited by diffusion, and a steady state may be reached in which cell proliferation in the outer layers is balanced by central necrosis.

The following protocol for preparing multicellular tumor spheroids has been contributed by M. Boyd and T. E. Wheldon, Department of Radiation Oncology, Glasgow University, CRC Beatson Labs, Garscube Estate, Bearsden, Glasgow G61 1BD, Scotland.

PROTOCOL 24.2 SPHEROIDS

Principle

Multicellular tumor spheroids provide a proliferating model for avascular micrometastases. The three-dimensional structure of spheroids allows the experimental study of aspects of drug penetration and resistance to radiation or chemotherapy that are dependent on intercellular contact. Spheroids are also well suited to the study of "bystander effects" in experimental targeted or gene therapy. Human tumor spheroids are more easily developed from established cell lines or from xenografts than from primary tumors [Sutherland, 1988].

From a single-cell suspension (trypsinized monolayer or disaggregated tumor), cells can be inoculated into magnetic stirrer vessels (Techne) and incubated to allow the formation of small aggregates over 3–5 d. Alternatively, aggregates may be formed from cell suspensions in stationary flasks, previously base coated with agar. Aggregates may be left in the original flasks or transferred individually (by pipette) to multiwell plates, where continued growth over weeks will yield spheroids of maximum size, about 1,000 μm [Yuhas et al., 1977; Sutherland, 1988].

Outline

Trypsinize the monolayer, or disaggregate the primary tissue, and seed the cells onto an agar-coated substrate. Transfer the aggregates to 24-well plates for analysis.

Materials

Sterile:
Noble agar (Difco)

Growth medium
Ultrapure water (UPW)
Trypsin, 0.25%, in PBSA
Flasks, 25-cm², or multiwell plates, 24 well
Petri dishes, 9 cm

Note. When agar coating is used, all flasks, plates, and dishes should be sterile, but not necessarily tissue culture grade.

Protocol

Agar Coating In 25-cm² Flasks:
1. Add 1 g of Noble agar (Difco) to 20 ml of UPW in a 100-ml borosilicate glass bottle with a loosely screwed-on cap.
2. Heat the agar in a water bath at 100°C for 10 min or until the agar has completely dissolved.
3. Add the contents of the bottle immediately to 60 ml of growth medium, previously heated to 37°C, and put 5-ml aliquots into each flask. Ensure that the agar is free from bubbles.
4. The agar will set at room temperature in ~5 min giving a 1.25%-agar-coated flask.

Agar Coating In Multiwell Plates:
1. Add 0.5 g of agar to 10 ml of UPW, heat as in step 2 for 25-cm² flasks, and then add 40 ml of UPW.
2. Place 0.5 ml of the resulting solution in each well of a 24-well plate to give a base coat of 1% agar. Accuracy and careful placement is important to ensure easy well-to-well focus of the microscope in subsequent viewing of spheroids.

Spheroid Initiation:
1. Trypsinize the confluent monolayer (for established lines; see Protocol 12.2) or disaggregate (for solid tumors; see Protocols 11.5, 11.6, and 11.8) to give a single-cell suspension.
2. Neutralize the trypsin with medium containing serum (if necessary).
3. Count the number of cells, using an electronic cell counter or a hemocytometer.
4. Place 5×10^5 cells in 5 ml of growth medium in each agar-coated 25-cm² flask, and incubate the cultures. If the cells are capable of spheroid formation, small aggregated clumps (about 100–300 μm in diameter) will form spontaneously in 3–5 d.

For subsequent growth, spheroids should be transferred to new 25-cm² flasks or 24-well plates.

Transfer to 25-cm² Flasks:

1. Transfer the contents of the original flasks to conical centrifuge tubes or universal containers.
2. Allow the spheroids to settle, and remove single cells with the supernatant.
3. Resuspend the spheroids in fresh medium, and transfer the suspensions to new agar-coated flasks, where growth will proceed by division of cells in the outer layer.

Transfer to 24-Well Plates:

1. Transfer the contents of each 25-cm² flask into a 6-cm Petri dish.
2. Add 0.5 ml of medium to each agar-coated well of a 24-well plate.
3. Select individual spheroids of chosen dimensions under low-power magnification (×40), and, using a Pasteur pipette and a Pi-Pump (Shuco International) or another pipetting aid with suitably fine control, transfer selected spheroids of similar diameter individually to the agar-coated wells of the 24-well plate.
4. Place the plate in a CO_2 incubator.
5. Replace the medium in the plate once or twice weekly (exchanging 0.5 ml each time), or add 0.5 ml of medium (without removing any medium) once or twice weekly, giving 2 ml/well after 2–4 weeks.

Analysis

Spheroid growth in wells or flasks may be quantified by regular (e.g., 2–3 times/week) measurement of the diameters of the spheroids, using a microscope eyepiece micrometer or graticule, or, preferably, by measurement of the cross-sectional area using an image analysis scanner. The most accurate growth curves are obtained when spheroids are grown in wells and are individually monitored.

Variations

Transfectant mosaic spheroids. Spheroids can be grown from populations of cells that have been transfected with different genes. The cells are first grown in monolayer and then are transfected with the transgene and subjected to selection for transgene-expressing cells. Mosaic spheroids are formed by the addition of both transfected and non-transfected monolayer cells in any desired proportions [Boyd et al., 1999]. The different cell populations are distributed throughout the resultant spheroids in approximately uniform mosaic patches, maintaining the same proportions of transfected to nontransfected cells as were added at the formation stage. (Fig. 24.3, see also Plate 33.)

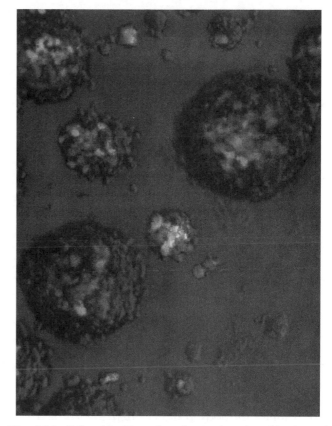

Fig. 24.3. Spheroids. Transfected mosaic spheroids derived from the human glioma cell line MOG-G-UVW. The spherhoids, ranging in size from 100 to 500 μm in diameter, are composed of mixtures of cells transfected with the GFP gene and cells transfected with the NAT gene. The relative proportions of the cell lines present in the spheroids reflect the proportions of the cells added to the spinner flask. (See also Plate 33.)

Applications

Spheroids have wide applications in the modeling of avascular tumor growth [Ward and King, 1997], the role of three-dimensional spatial configurations in gene expression in cell populations [Waleh et al., 1995; Dangles et al., 1997], and in the assessment of cytotoxic treatment. Treatment end points include growth delay, determination of the proportion of spheroids sterilized ("cured") by treatment, and colony formation in the monolayer following the disaggregation of treated spheroids [Freyer and Sutherland, 1980]. An important area is the use of spheroids to study the penetration of cytotoxic drugs, antibodies, or other molecules used in targeted therapy [Sutherland, 1988; Carlsson and Nederman, 1989]. This category represents a special application that is not possible in single-cell suspensions or monolayer cultures. Spheroids have also proved useful in the study of cell killing by biologically targeted radionuclides [Mairs and Wheldon, 1996]. Spheroid cultures have also been used in "confrontation experiments" to assess the invasiveness of spheroids derived from malignant cell populations that are grown in close proximity to normal cell cultures [de Ridder, 1997]; such techniques have also been used in nononcological studies of disease processes, such as studies of rheumatoid arthritis [Ermis et al., 1998] Mosaic spheroids are a new variant form that has special applications in the assessment of bystander effects. For example, a current difficulty of gene therapy for cancer is the inefficiency of gene transfer procedures, leading to the requirement for bystander effects to eliminate cells in a tumor population that have not been transfected successfully. Mosaic spheroids mirror this situation *in vitro* and allow evaluation of different forms of the bystander effect, such as radiation cross fire when transfected cells are targeted with a radioactive agent [Boyd et al., 1999].

Immobilization of Living Cells in Alginate

The technique of encapsulating living cells within alginate beads has been widely used in experimental research—e.g., encapsulation of hybridoma cells for monoclonal antibody production [Lang et al., 1995] and of hormone-producing cells used in animal models for the treatment of diabetes mellitus [Soon-Shiong et al., 1992]. (See also Protocol 22.13, Fig. 24.4, and Plate 34.)

Alginate is found primarily in brown seaweel *Laminaria, Macroaystis*, and *Ascophyllum* and consists mainly of two types of monosaccarides: L-guluronic acid (G) and D-mannuronic acid (M). It is composed of alternating molecules of M and G, and divalent cations bind strongly between separate G blocks and initiate the formation of an extended alginate gel network. Alginate gels can be formed into beads by dripping the alginate solution into a buffer containing divalent cations, such as Ca^{2+}. Mechanical strength, volume, stability, and porosity correlate with the G content such that alginate beads with a high G content have

the largest pore sizes, ranging between 5 and 200 nm [Martinsen et al., 1989; Miura et al., 1986]. Pores of such sizes allow free diffusion of macromolecules out of, as well as into, the alginate. Host immune reactions to the alginate can be reduced substantially by using alginate with a high concentration of G and a low concentration of M [Otterlei et al., 1991].

At present, numerous cell types can be genetically engineered to produce specific proteins of choice. By encapsulating such cells in alginate, a valuable vehicle is obtained for delivering specific recombinant proteins to the organism. Thus, such alginate "bioreactors" may have an important therapeutic potential for the treatment of a number of diseases, in which the alginate may prevent the encapsulated cells from being destroyed by the immune system.

The following protocol and preceding introduction to alginate encapsulation have been contributed by Tracy-Ann Read and Rolf Bjerkvig, Department of Anatomy and Cell Biology, University of Bergen, Norway.

PROTOCOL 24.3. ALGINATE ENCAPSULATION

Outline

Culture potential producer cells in 75-cm² flasks containing growth medium prescribed for the particular cell line of choice.

Materials

Sterile:
Growth medium for selected cells
PBSA
Trypsin (concentration appropriate to subculture regime for cells)
Saline solution containing:
   (a) NaCl, 8.0g
   (b) D-glucose, 1.0 g
   (c) Make up to 1 l with UPW
   (d) Adjust to pH 7.2–7.4 with HCl or NaOH
   (e) Sterilize by autoclaving
$CaCl_2$, 0.1 M, containing
   (a) Saline solution, 500 ml
   (b) $CaCl_2 \cdot 2H_2O$, 7.35 g
   (c) Adjust to pH 7.2–7.4 with HCl or NaOH
   (d) Sterilize by autoclaving
Sodium alginate (PRONOVA (TM) UP LVG): ultrapure, low viscosity, high guluronic acid content
Sterilizing filters, 0.45 μm, nonpyrogenic (millipore)

Protocol

1. Dissolve the alginate in saline solution to a concentration of 1.5%, and shake the resulting so-

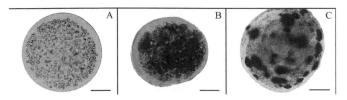

Fig. 24.4. Alginate Encapsulation. Light-microscopic images of cells encapsulated in alginate after 2 hours, 3 weeks, and 4 months, respectively, *in vitro*. Within the alginate beads, both cell death and cell proliferation will occur, and for many cell lines, multicellular spheroids will form inside the beads. (Magnification is 140×; bar = 70 μm.)

lution for a minimum 4 h at room temperature until the alginate has fully dissolved. Alginate dissolved in saline solution can be stored at 4°C for up to 3 d.

2. Filter the alginate 3 times using sterile nonpyrogenic 0.45-μm filters. The final filtration should be carried out immediately before the alginate is used.

3. Grow the cells to confluence. Trypsinize the cells with 3 ml of trypsin, and count them. Centrifuge the cells at 900 rpm for 4 min and remove the supernatant completely.

4. Mix the cells with the alginate to a concentration of 2×10^6 cells/ml by resuspending the cells gently in the alginate. At this point, it is crucial not to generate air bubbles in the alginate, as doing so will result in holes in the beads.

5. Transfer the alginate-suspended cells to a sterile syringe capped with a 27G needle.

6. Drip the cell–alginate suspension, applying a circular movement, into a beaker containing 0.1 M CaCl$_2$, which initiates the formation of alginate beads. Allow the beads to gel for 10 min, and then wash them three times in PBSA. Finally, wash them in growth medium.

7. Culture the beads in 175-cm^2 culture flasks containing growth medium at 37°C, 100% relative humidity, and 5% CO$_2$ in air.

FILTER-WELL INSERTS

The opportunities provided by filter-well inserts for cell interaction, stratification, and polarization have made them a popular culture system in many areas. Mauchamp demonstrated the development of polarity and functional integrity in thyroid epithelium explanted on a collagen-coated filter in a specially constructed mount [Chambard et al., 1983]. (See Polarity and Cell Shape in Chapter 16.) Filter-well inserts have been used to generate stratified epidermis [e.g., Kondo et al., 1997] and polarized intestinal [Halleux and Schneider, 1994] and kidney [Mullen et al., 1997] epithelium. Others have used them to study invasion by granulocytes or malignant cells [McCall et al., 1981; Elvin et al., 1985; Repesh, 1989; Schlechte et al., 1990; Brunton et al., 1997].

One of the major advantages of filter-well inserts is that they allow the recombination of cells at very high, tissue-like densities, with ready access to medium and gas exchange, but in a multireplicate form. Filter-well inserts are now available from several suppliers (see Sources of Materials—Filter-Well Inserts in Trade Index; Figs. 24.5 and 24.6; Plate 35), in a variety of translucent or transparent materials, including poly-

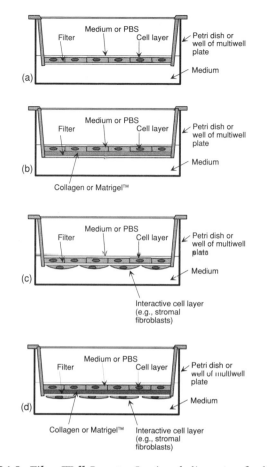

Fig. 24.5. Filter-Well Inserts. Sectional diagram of a hypothetical filter-well insert. (a) Monolayer grown on top of the filter. (b) Monolayer grown on matrix on top of a filter. (c) Interactive cell layer added to the underside of a filter. (d) Interactive cell layer added to the underside of the filter with matrix coating.

Fig. 24.6. Transwells. Costar "transwells" in a twelve-well plate, with a filter-well insert alongside.

TABLE 24.1. Types of Filter-Well Inserts

Make	Name	Material	Qualities	Transparency	Porosity (μm)
Millipore	Millicel	Nitrocellulose	Mesh	Opaque	0.45
		Polyolefin	Mesh	Transparent	0.45
Corning Costar	Transwells	Polycarbonate	Absolute	Transparent	5–8
				Translucent[a]	0.45–1.0
Falcon	Inserts	Polyethylene teraphthalate	Absolute	Transparent	0.45–3
Nunclon	Anocel	Ceramic	Sieve	Transparent	0.01
Earl-Clay	Ultraclone	Collagen	Mesh	Transparent	

[a]The higher the pore frequency, the lower the transparency. Low-porosity filters have a high pore frequency and are, consequently, less transparent.

carbonate, PTFE, and polyethylene teraphthalate, and ranging in size from 6.5 mm–9 cm, suitable for 24-well, 12-well, and 6-well plates, or larger dishes. (Table 24.1.) Filters can be obtained precoated with collagen, laminin, fibronectin, or Matrigel.

PROTOCOL 24.4 FILTER-WELL INSERTS

Outline
Seed cells into filter-well inserts, and culture the cells in excess medium in multiwell plates.

Materials
Sterile:
Approximately 0.5×10^6 cells per cm² of filter
Growth medium, 1–20 ml per filter (depending on the vessel the housing filter)
Filter-well inserts
Multiwell plates for filter inserts: 6, 12, or 24 well
Forceps, curved
Nonsterile:
Pipettor

Protocol
1. Place the filter wells in the plate or dish.
2. Add medium, tilting the dish to allow the medium to occupy the space below filter and to displace the air with minimum entrapment. Add medium until it is level with the filter (2.5 ml for a 6-well plate; 1.0 ml for a 24-well plate).
3. Level the dish, and add 2×10^6 cells in 2 ml of medium to the top of the filter for a 25-mm-diameter filter, or 5×10^5 in 200 μl of medium for a 6.5-mm-diameter filter, taking care not to perforate the filter.
4. Place the dish in a humid CO_2 incubator in a protective box. (See Boxed Cultures in Chapter 5.) It is critical to avoid shaking the box, and the

cultures should not be moved in the incubator, in order to avoid spillage and resultant contamination.
5. Monolayers should become established in 3–5 d, although 5–10 d or longer may be required for histotypic differentiation (e.g., polarized transport) to become established.
6. Cultures may be maintained indefinitely, replacing the medium or transferring the insert to a fresh well or dish every 3–5 d.

Analysis
1. *Permeability.* Some epithelial cells (e.g., MDCK and Caco-2) and endothelial cells (e.g., from umbilical vein) form tight junctions several days after reaching confluence. This process is accelerated by precoating the membranes with collagen. Transepithelial permeability then becomes restricted to physiologically regulated transport through the cells, and pericellular transport falls to near zero. The process can be monitored by looking at dye (e.g., lucifer yellow), [14C]-methylcellulose, or [14C]-inulin transfer across the membrane, or by increasing the transepithelial electrical resistance (TEER).
2. *Polarized transport.* The addition of labeled glucose or amino acid to the upper compartment of the filter well insert will show transport to the lower compartment, while the converse does not occur. If the cells possess P-glycoprotein or some other efflux transporter, cytotoxins (e.g., vinblastine) added to the lower compartment will be transported to the upper compartment, but not vice versa.
3. *Penetration of cells through the filter.* Trypsinize and count each side of the filter in turn (trypsinized cells will not pass through even an 8-μm filter, as their spherical diameter in suspension exceeds this size), or fix the filter, embed, and section and examine by electron microscope or conventional

histology. Visualization is possible in whole mounts by mounting the fixed, stained (Giemsa) filter on a slide in DPX under a coverslip under pressure, to flatten the filter. Differential counting can then be performed by alternately focusing on each plane.

4. *Detachment of cells from the filter to the bottom of the dish.* Count the cells by trypsinization or scanning.

5. *Partition above and below the filter.* Either count the cells as in (3), or prelabel the cells with rhodamine or fluorescein isothiocyanate (5 μg/ml for 30 min in a trypsinized suspension) and measure the fluorescence of solubilized cells (0.1% SDS in 0.3 N NaOH for 30 min) trypsinized from either side of the filter.

6. *Cellular invasion.* Precoat the filter with a cell layer (normal fibroblasts, MDCK, etc.), use microscopic examination to ensure that confluence is achieved, and then seed EDTA-dissociated test cells on top of a preformed layer (10^5–10^6 cells per filter). If the test cells are RITC- or FITC-labeled, then fluorescent measurements will reveal the appearance of the cells below filter.

7. *Matrix invasion.* Coat the filter with Matrigel, apply the cells above the Matrigel, and monitor the appearance of the cells below the filter [Repesh, 1989; Schlechte et al., 1990]. Alternatively, seed the cells onto the lower surface of the filter, coat the upper surface of the filter with Matrigel, and monitor the invasion of the cells into the Matrigel by confocal microscopy [Brunton et al., 1997].

Variations

1. *Depth of medium.* The depth of medium above the filter will regulate oxygen tension at the level of the cells. Keratinocytes or Type II pneumocytes from lung alveoli will require little medium and a high oxygen tension, while enterocytes, such as Caco-2, may be better off submerged and with a lower oxygen tension.

2. *Filter porosity.* 1-μm filters allow cell interaction and contact without transit across the filter. 8-μm filters allow live cells to cross the filter. 0.2-μm filters probably do not allow cell contact. Low-porosity filters may be used to study cell interaction without permitting the cells to intermingle.

3. *Transfilter combinations.* Invert the filter, and load the underside first with 0.5 ml, 2×10^6 cells/ml, of cell suspension: Place the filter upside down in a Petri dish, add the cell suspension, and place the lid on the top of the dish and touching the drop of cell suspension, before all of the medium drains through the filter. Incubate the filter for

18 h. Capillarity will hold the medium and cells until the cells sediment onto the filter and attach [Brunton et al., 1997]. The next day, invert the filter and load the well with interacting cells or Matrigel, as described in Protocol 24.4.

CULTURES OF NEURONAL AGGREGATES

The following protocol for aggregating cultures of brain cells has been contributed by Rolf Bjerkvig, Department of Cell Biology and Anatomy, University of Bergen, Žrstadveien 19, N-5009 Bergen, Norway.

PROTOCOL 24.5. NEURONAL AGGREGATES

Principle

Aggregating cultures of fetal brain cells have been extensively used to study neural cell differentiation [Seeds, 1971; Trapp et al., 1981; Bjerkvig et al., 1986a]. The aggregating cells follow the same developmental sequence as observed *in vivo*, leading to an organoid structure consisting of mature neurons, astrocytes, and oligodendrocytes. A prominent neuropil is also formed. In tumor biology, the aggregates can be used to study brain tumor cell invasion *in vitro* [Bjerkvig et al., 1986b].

Outline

Remove brains from fetal rats at day 17 or 18 of gestation, and prepare the brains as a single-cell suspension. Form brain aggregates by overlay cultures in agar-coated multiwells. The cells in the aggregates form a mature organoid brain structure during a 20-day culture period.

Materials

Sterile:

Dulbecco's modification of Eagle's medium, containing 10% heat-inactivated newborn calf serum; four times the prescribed concentration of nonessential amino acids; L-glutamine, 2 mM; penicillin, 100 U/ml; streptomycin, 100 μg/ml

Phosphate-buffered saline (PBS) with Ca^{2+} and Mg^{2+}

Trypsin type II (0.025% in PBSA)

Agar (Difco)

Multiwell tissue culture dishes (24 wells; Nalge Nunc)

Petri dishes, 10 cm

Test tubes, 12 ml

Scalpels, scissors, and surgical tweezers

Erlenmeyer flasks, 2 at 100 ml

To coat the microwells with agar-medium, use the following procedure:

(a) Prepare a 3% stock solution (3 g of agar in 100 ml of PBSA) in an Erlenmeyer flask.

(b) Heat the flask in boiling water until the agar is dissolved. Place an empty Erlenmeyer flask in boiling water, and add 10 ml of hot agar solution to it.

(c) Slowly add warm complete growth medium to the flask until a medium-agar concentration of 0.75% is reached.

(d) Add 0.5 ml of warm medium-agar solution to each well in the multiwell dish.

(e) Allow agar to cool and gel.

The multiwell dishes can be stored in a refrigerator for 1 week.

Nonsterile:

Water bath

Protocol

1. Dissect out, aseptically, the whole brains from a litter of fetal rats at day 17 or 18 gestation, and place the tissue in a 10-cm Petri dish containing PBSA.
2. Using scalpels, mince the tissue into small cubes, ~0.5 cm^3.
3. Transfer the tissue to a test tube, and wash it three times in PBSA. Allow the tissue to settle to the bottom of the tube between each washing.
4. Add 5 ml of trypsin solution to the tissue, and incubate in a water bath for 5 min at 37°C.
5. Shear the tissue by trituration through a Pasteur pipette approximately 20 times.
6. Allow the tissue to settle for 3 min and transfer the clump-free milky cell suspension to a test tube containing 5 ml of growth medium.
7. Add 5 ml of fresh trypsin to the undissociated tissue, and repeat the trypsinization and dissociation procedure twice more.
8. Spin the cell suspension at 200 g for 5 min.
9. Aspirate the supernatant, resuspend the cells, and pool them in 10 ml of growth medium.
10. Count the cells, and add 3 × 10^6 cells to each agar-coated well. The volume of the overlay suspension should be 1 ml.
11. Place the multiwell dish in a CO_2 incubator for 48 h.
12. Remove the aggregates to a sterile 10-cm Petri

dish, and add 10 ml growth medium to the dish.

13. Transfer larger aggregates individually to new agar-coated multiwells by using a Pasteur pipette.
14. Change the medium every third day by carefully removing and adding new overlay medium.

During 20 days in culture, the aggregates will become spherical and develop into an organoid structure.

Analysis

The next step is fixation and embedding in paraffin or epon for histological or electron microscopic evaluation. Oligodendrocytes, astrocytes, and neurons are identifiable by transmission electron microscopy or by immunohistochemical localization of myelin basic protein, glial fibrillary acidic protein, and neuron-specific enolase, respectively.

Variations

A single-cell suspension can be obtained by mechanical sieving through steel or nylon meshes [Trapp et al., 1981]. Reaggregation cultures can also be obtained using a gyratory shaker. Select a speed (about 70 rpm) such that the cells are brought into vortex, thereby greatly increasing the number of collisions between cells. This movement also prevents cell attachment to the culture flasks.

Organotypic Culture

The advent of filter-well technology, boosted by its commercial availability, has produced a rapid expansion in the study of organotypic culture methods. Skin equivalents have been generated by coculturing dermis with epidermis (Mattek Epiderm, Episkin), with an intervening layer of collagen, or with dermal fibroblasts incorporated into the collagen [Limat et al., 1995], and models for paracrine control of growth and differentiation have been developed with cells from lung [Speirs et al., 1991], prostate [Thomson et al., 1997], and breast [Van Roozendahl et al., 1992].

The opportunity for heterotypic cell interaction has also opened up numerous opportunities for studying inflammation and irritation *in vitro* (see Inflammation in Chapter 21) and for creating other models for tissue interaction with increased *in vivo* relevance [Liu et al., 1995; Emura et al., 1997; Gomm et al., 1997].

CHAPTER 25

Scale-Up

The threshold between normal laboratory scale usage and large scale or bulk culture is purely arbitrary, but is generally regarded as the level of cell production, above which specialized apparatus and procedures will be required. While $1 \times 10^9 - 1 \times 10^{10}$ cells can be produced in simple stirrer cultures of around 1–10 l capacity (see Propagation in Suspension in Chapter 12) larger scale cultures of $1 \times 10^{11} - 1 \times 10^{12}$ cells will require apparatus ranging from a 100 l laboratory-scale fermentor to semi-industrial pilot plant with capacities of from 100–1,000 l. Full scale industrial production uses 5,000–20,000 l bioreactors but these are beyond the scope of this book. The terms *stirrer culture* or *spinner culture* are synonymous and tend to be used for simple culture systems at the low end of the laboratory range of equipment. The terms *fermentor* and *bioreactor*, while not synonymous, have considerable overlap. The name *fermentor* derives from microbiological culture systems, designed originally for bacteria and yeast, and was used initially for laboratory equipment of around 50–100 l capacity, but the increase in scale in line with developments in the biotechnology industry, coupled with a greater diversity in design to cope with monolayer cultures as well as suspension, has led to the introduction of the name *bioreactor*.

The method employed to increase the scale of a culture depends on whether the cells proliferate in suspension or must be anchored to the substrate. Methods for suspension cells are generally simpler and will be dealt with first. Monolayer techniques, being dependent on increased provision of growth surface, tend to be more complex and will be dealt with later. Many of the monolayer techniques are also applicable to suspension cells.

SCALE-UP IN SUSPENSION

Scale-up of suspension cultures involves, primarily an increase in the volume of the cultures. Agitation of the medium is necessary when the depth exceeds 5 mm, and above 5–10 cm (depending on the ratio of surface area to volume), sparging with CO_2 and air is required to maintain adequate gas exchange. (Figs. 25.1 and 25.2.) Stirring of such cultures is best done slowly with a magnet encased in a glass pendulum (Techne, Integra) or with a large-surface-area paddle (Bellco). The stirring speed should be between 30 and 100 rpm, sufficient to prevent cell sedimentation, but not so fast as to create shear forces that would damage the cells. Antifoam (Dow Chemical Co.) or Pluronic F68 (Sigma), 0.01–0.1%, should be included when the serum concentration is above 2%, particularly if the medium is sparged. In the absence of serum, it may be necessary to increase the viscosity of the medium with (1–2%) carboxymethyl cellulose (molecular weight $\sim 10^5$).

PROTOCOL 25.1. STIRRED 4-LITER BATCH SUSPENSION CULTURE

The procedure for setting up a 4-l culture of suspended cells is as follows:

Outline
Grow a pilot culture of cells, and add it to a prewarmed, pregassed aspirator of medium. Stir the cell suspension slowly, with sparging, until the required

Fig. 25.1. Large Stirrer Flask. Techne 5-l stirrer flask on a magnetic stirrer. Note the offset pendulum that makes an excursion in the annular depression in the base of the flask. The side arms are for sampling or perfusion with CO_2 in air, which would be required with this volume of medium. (See Figure 25.2.)

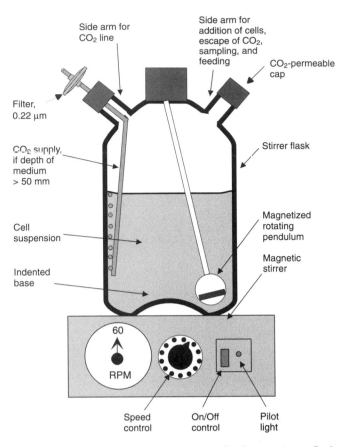

Fig. 25.2. Stirrer Culture. Diagram of a large stirrer flask suitable for volumes up to 8 l.

cell concentration is reached, and then harvest the cells.

Materials

Sterile or Aseptically Prepared:

Starter culture vessel (small stirrer flask; see Figs. 7.6 and 12.7)

Growth medium

Antifoam (silicone: Dow Corning, Merck; Pluronic F68: Sigma)

Prepared large stirrer flask (Bellco, Techne; see Figs. 25.1 and 25.2) with:

(a) A magnet enclosed in a glass pendulum, or a suitable alternative stirrer (see Mixing and Aeration in this chapter) suspended from the top cap (e.g., Techne).

(b) One inlet port with a removable CO_2-permeable cap (i.e., a cap with an integral microporous membrane; see Venting in Chapter 7), or a cap with a silicone diaphragm pierced by a wide-bore needle with a Luer connection fixed to a sterile inline filter (e.g., Millex).

(c) A second inlet port carrying a tube for the gas

line (e.g., Bellco #1965) with a 2–3-mm internal diameter, which reaches almost to the bottom of the vessel, but remains clear of the pendulum when it is stirring, with the external entry terminating in a female Luer fitting guarded by aluminum foil.

(d) Sterilize the stirrer culture vessel by autoclaving, at 100 kPa (15 lb/in²) for 20 min, fully assembled, but with the CO_2-permeable cap removed and sterilized separately.

An in-line sterile Luer-fitting micropore filter, 25 mm diameter, 0.2 μ porosity, (e.g., Millex), to attach to the gas entry line at the time of use

Nonsterile:

Magnetic stirrer

Supply of 5% CO_2, preferably from a metered supply at ~10–30 kPa (2–5 lb/in²)

PBSA for counting cells

Electronic cell counter or hemocytometer

Protocol

1. Set up a starter culture (see Protocol 12.3 and Fig. 12.7) with 200 ml medium, and seed it with cells at 5×10^4–1×10^5 cells/ml. Place the

culture on the magnetic stirrer, rotating at 60 rpm, and incubate the culture until a concentration of $5 \times 10^5 - 1 \times 10^6$ cells/ml is reached. (*Note*: do not exceed 1×10^6 cells/ml as the cells may enter apoptosis)

2. Set up the large stirrer flask as in Fig. 25.2.
3. Add 4 l of medium to the flask.
4. Add 0.4 ml of antifoam to the flask, using a disposable pipette or syringe.
5. Close the side arm with the CO_2-permeable cap.
6. Place the flask on the magnetic stirrer in a 37°C room or incubator.
7. Connect the 5%-CO_2 air line via the sterile, micropore in-line filter to the gas inlet Luer fitting.
8. Turn on the gas at a flow rate of approximately 10–15 ml/min.
9. Stir the culture at 60 rpm.
10. Incubate the large stirrer flask for about 2 h to allow the temperature and CO_2 tension to equilibrate.
11. Turn off the CO_2, and disconnect the flask from the CO_2 line.
12. Bring the large stirrer flask and the starter culture back to the laminar-flow hood.
13. Transfer the starter culture to the large stirrer flask by pouring in one single smooth action.
14. Return the large stirrer flask to the 37°C incubator or hot room.
15. Restart the stirrer at 60 rpm.
16. Reconnect the 5%-CO_2 line, and adjust the gas flow to 10–15 ml/min. (Estimate the rate of flow by counting the bubbles if there is no flow meter on the line.)
17. Incubate the culture for 4–7 d, sampling every day to check the cell proliferation rate:
    (a) Bring the stirrer flask back to the laminar-flow hood;
    (b) Remove the side-arm cap;
    (c) Withdraw 5–10 ml of the cell suspension;
    (d) Count the cells, and check their viability by dye exclusion.
18. When the cell concentration reaches the desired level,
    (a) Turn off the gas and the magnetic stirrer;
    (b) Disconnect the stirrer flask from the gas supply;
    (c) Take the culture to the laboratory, and pour off the cells into centrifuge bottles.
    (d) Centrifuge the cells at 100 *g* for 10 min, and collect the cells (from the pellet) or the supernatant medium, as appropriate.

Analysis

Monitor the growth rate of the cells daily. For the best results, the cells should not show a lag period of more than 24 h and should still be in exponential growth when harvested. Plot the cell counts daily, and harvest the cells at approximately 1×10^6 cells/ml. Suspension cultures tend to enter apoptosis if this concentration is exceeded.

Variations

Adding Medium. The following procedures are alternatives for adding medium to vessels larger than 5 l:

1. Buy medium in media bags that can be hooked up alongside the stirrer flask and allowed to run in unattended.
2. Sterilize the stirrer flask with a premeasured volume of ultrapure water, and make up the medium *in situ*. Mark the side of the flask to indicate the level of water, so that any water lost by evaporation during autoclaving can be made up.
3. Use an autoclavable medium, and autoclave it *in situ*.

Harvesting. Pouring from a large stirrer flask can be difficult, so the cell suspension can be collected either by removing the in-line filter from the gas line and attaching a peristaltic pump, tilting the vessel when the fluid is near to the bottom, or by attaching the gas line to the other port and blowing the cells out through the gas line (after replacing the micropore filter with a flexible tube).

Continuous Culture

If it is required that the cells be kept at a set concentration, the culture can be maintained in a *chemostat* or *biostat*. In this type of culture system, the cells are grown to the mid-log phase (monitored by daily cell counts), a measured volume of cells is removed each day, and replaced with an equal volume of medium. Alternatively, the cells may be run off continuously, at a constant rate, at mid-log phase, and medium added at the same rate. The latter will require a stirrer vessel with four ports, two for CO_2 inlet and outlet, one for medium inlet and one for spent medium outlet, collecting into a reservoir (Fig. 25.3). The flow rate of medium may be calculated from the growth rate of the culture [Griffiths, 1992] but is better determined experimentally by serial cell counting at different flow rates of the medium. The flow rate may be regulated by a variable peristaltic pump on the inlet line.

Production of cells in bulk, in the 1–20-l range, is best done by the batch method described in Protocol 25.1. The steady-state method is required for monitoring metabolic changes related to cell density, but is more expensive in medium and is more likely to lead to contamination. However, if the operation is in the 50–1,000-l range, then more investment and time is spent in generating the culture, and the batch method

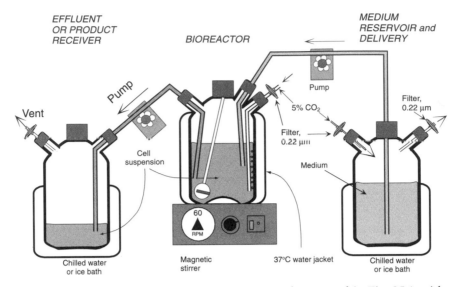

Fig. 25.3. Biostat. A modification of the suspension culture vessel in Fig. 25.1, with continuous matched input of fresh medium and output of cell suspension. The objective is to keep the culture conditions constant rather than to produce large numbers of cells. (See also Figs. 25.17 and 25.18.) Bulk culture, per se, is best performed in batches in the apparatus in Fig. 25.1 or Fig. 25.17.

becomes more costly in time, materials, and down time; thus, continuous culture may be better for such cases.

Suspension cultures can also be grown in bottles rotating on a special rack, as for monolayer cultures. (See Protocol 25.3.)

Adherent Cells. Anchorage-dependent cells cannot be grown in liquid suspension, except on microcarriers (see Protocol 25.4), but transformed cells (e.g., virally transformed or spontaneously transformed continuous cell lines, such as HeLa-S$_3$) can. Because these cells are still capable of attachment, the culture vessels will need to be coated with a water-repellent silicone (e.g., Repelcote), and the calcium concentration may need to be reduced. S-MEM medium (Gibco, Sigma) is a variation of Eagle's MEM with no calcium in the formulation and has been used for the culture of HeLa-S$_3$ and other cells in suspension.

Scale and Complexity

Standard bench-top stirrer cultures operate satisfactorily up to around 10 l for the bulk production of cells or medium. If, however, more attention must be paid to process control, then a controlled fermentor or bioreactor should be used. These culture vessels have regulated input of medium and gas and provide the capability for data collection from oxygen, CO$_2$, and glucose electrodes in the culture vessel. They are regulated by a programmable control unit that re- cords and outputs data and can be used to regulate gas and liquid input and output, stirring speed, temperature, and so on. (See Sources of Materials— Fermentors in Trade Index.)

Mixing and Aeration

Problems of increased scale in suspension cultures revolve around mixing and gas exchange. In contrast to simple stirrer cultures, most laboratory-scale fermentors agitate the medium with a rotating turbine or paddle. Designs of these fermentors vary, mostly to try to achieve maximum movement of liquid with minimum shear. Successful designs usually employ a slowly rotating large-bladed paddle with a relatively high surface area [Griffiths, 1992] and some additive designed to minimize the harmful effects of shear— e.g., Pluronic F68, carboxymethylcellulose (CMC), and polyvinylpyrrolidone (PVP) [Cherry and Papoutsakis, 1990].

Culture Bags. Plastic bags (e.g., Du Pont, ICN) can be used to culture suspension cells. They are gas permeable and can be agitated by rocking on trays on a flat rocking platform.

Air-Lift Fermentors. Large-scale fermentors tend to use air lift (Fig. 25.4): 5% CO$_2$ in air is pumped into a porous steel ring at the base of the central cylinder, and bubbles stream up the center, carrying a flow of liquid with them, and are released at the top, while

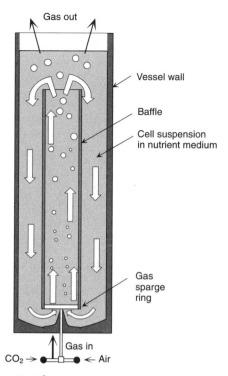

Fig. 25.4. Air-Lift Fermentor. In an air-lift fermentor, the bioreactor consists of two concentric cylinders, with the inner cylinder being shorter at both ends than the outer, thereby creating an outer and inner chamber. The bottom of the inner chamber carries a sintered steel ring through which 5% CO_2 in air is bubbled. The bubbles rise, carrying the cell suspension with them. CO_2 is vented from the top, and displacement ensures the return of the cell suspension down the outer chamber. (Modified from Griffiths [1992].)

the medium is recycled to the bottom of the cylinder. This fairly simple type of fermentor is used extensively in the biotechnology industry, up to capacities of 20,000 l.

Rotating Chambers. Mixing and aeration can also be achieved by rolling the culture vessel, either in a conventional roller bottle (see Protocol 25.3), or in two- or three-compartment chambers (Fig. 25.5). If the cell suspension is limited to one small compartment, then the cell concentration can be quite high, as the cells are not diluted by the bulk of the medium. The product concentration (e.g., antibody) accumulates in the cellular compartment, while nutrients and waste products diffuse across the semi-permeable membrane to and from the medium compartment. Two examples of this kind of design are, (1) a two compartment cylinder, with cells in the smaller compartment and medium in the larger, separated by a semi-permeable membrane (Heraeus, see Fig. 25.5a), and (2) two semi-permeable tubes carrying the cells within an outer cylinder containing the medium (Techne, see Fig. 25.5b). Both rotate to ensure mixing and in each case the medium can be sparged or replaced without disturbing the cells or product.

Perfused Suspension Culture. Hollow-fiber and membrane-perfusion systems also operate on the principle of compartmentalization. The cells are retained in a low-volume compartment at a very high concentration, and medium is perfused through hollow fibers within the cell compartment (Fig. 25.6), or through an adjacent membrane compartment (Membroferm, Fig. 25.7). Regulation of gas exchange is external to the culture chamber. One version of the Membroferm allows for the product to be collected in a third membrane-bound compartment. Like compartments can also be linked for serial perfusion.

Fluidized Bed Reactors for Suspension Cultures. Although microcarriers were originally conceived for monolayer cultures (see Microcarriers in

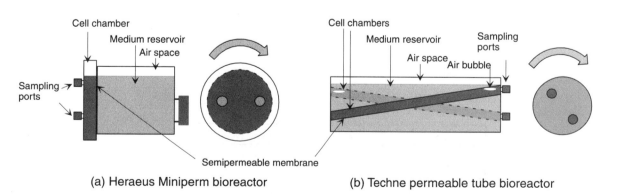

(a) Heraeus Miniperm bioreactor   (b) Techne permeable tube bioreactor

Fig. 25.5. Rotating Chamber Systems. (a) Heraeus Miniperm. The concentrated cell suspension in the left chamber is separated from the medium chamber by a semipermeable membrane. High-molecular-weight products remain with cells and can be harvested from the sampling ports, while replenishment of the medium is carried out via the right-hand ports. Mixing is achieved by rotating the chamber on a roller rack. (b) The Techne system relies on enclosing the cells in angled, permeable tubes, with an air bubble located in each tube. As the culture rotates, the bubble moves up and down the tubes, ensuring mixing.

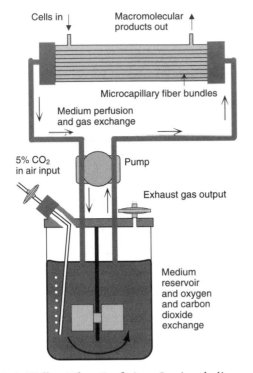

Fig. 25.6. Hollow-Fiber Perfusion. Sectional diagram of a hollow-fiber perfusion system. Medium is circulated from a reservoir to the culture chamber by a peristaltic pump. Aeration and CO_2 exchange are regulated in the reservoir of medium.

this chapter), their principle has been extended to high-density, porous microcarriers, in which suspension cells can become trapped in the interstices of the bead matrix. Because of the higher density of the microcarriers they can be perfused slowly from below, at such a rate that their sedimentation rate matches the flow rate. The beads therefore remain in stationary suspension, while the medium perfuses past, constantly replenishing nutrients and collecting the product into a downstream reservoir. Gas exchange is ex-

ternal to the reactor, and no mechanical mixing is required.

NASA Bioreactor. Intrigued by the concept of growing cells in zero gravity, NASA constructed a rotating chamber in which cells, growing in suspension, achieved simulated zero gravity by slowly rotating the chamber, altering the sedimentation vector continuously. (Fig. 25.8.) The cells remain stationary, are subject to zero shear force, and tend to form three-dimensional aggregates, which, reputedly, enhances product formation. When the rotation stops, the aggregates sediment, and the medium can be replaced.

SCALE-UP IN MONOLAYER

For anchorage-dependent monolayer cultures, it is necessary to increase the surface area of the substrate in proportion to the number of cells and the volume of medium. This requirement has prompted a variety of different strategies, some simple and others complex.

Multisurface Propagators

Nunc Cell Factory. The simplest system for scaling up monolayer cultures is the Nunclon Cell Factory. (Fig. 25.9; also Table 7.1.) This system is made up of rectangular Petri-dish-like units, with a total surface area of 600–24,000 cm^2, interconnected at two adjacent corners by vertical tubes. Because of the positions of the apertures in the vertical tubes, medium can flow between compartments only when the unit is placed on end. When the unit is rotated and laid flat, the liquid in each compartment is isolated, although the apertures in the interconnecting tubes still allow connection of the gas phase. The cell factory has the advantage that it is not different in the geometry or the nature of its substrate from a conventional flask

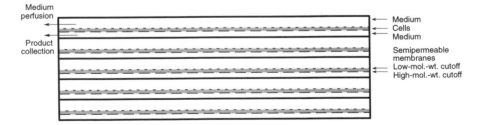

Fig. 25.7. Membroferm. A lamellated structure of semipermeable membranes, based on a triple unit wherein cells within a double membrane are fed from an upper compartment via a membrane with a low-molecular-weight cutoff. The cells secrete product into the lower compartment via a membrane that is microporous with a high-molecular-weight cutoff. Each base unit of three elements can be connected to create a perfusion system.

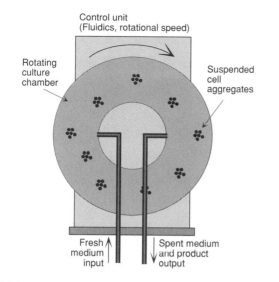

Fig. 25.8. NASA Bioreactor. Cells are maintained in the central chamber, rotating at a speed such that they do not sediment.

or Petri dish. The recommended method of using the cell factory is as follows (see Fig. 25.10):

PROTOCOL 25.2. NUNC CELL FACTORY

Outline

Prepare a cell suspension in medium, and run the suspension into the chambers of the unit. Lay the unit flat, and gas it with CO_2. Seal and incubate the unit.

Fig. 25.9. Nunc Cell Factory. Both smaller ($\sim$1,500 cm²) and larger ($\sim$100,000 cm²) versions of this type of culture vessel are available, working on the same principle.

Materials

Sterile:
Monolayer cells
Growth medium
Trypsin, 0.25% crude
PBSA
Nunc Cell Factory multi-compartmented culture chamber
Silicone tubing and connectors
Nonsterile:
Hemocytometer or electronic cell counter and counting fluid

Protocol

1. Trypsinize the cells (see Protocol 12.2), resuspend them, and dilute the suspension to 2×10^4 cells/ml in 1,500 ml of medium.
2. Place the chamber on a long edge, with a supply tube to the bottom (see Fig. 25.10), and run the cells and medium in through the supply tube. Medium in all chambers will reach the same level.
3. Clamp off the supply tube, and disconnect it from the medium reservoir.
4. Rotate the chamber through 90° in the plane of the monolayer, so that the unit lies on a short edge, with the supply tube at the top. (See Fig. 25.10.)
5. Rotate the chamber through 90° perpendicular to the plane of the monolayer, so that it lies flat on its base, with the culture surfaces horizontal. To transport it to the incubator, tip the medium away from the supply port.
6. If it is necessary to gas the culture, loosen the clamp on the supply line, purge the unit with 5% CO_2 in air for 5 min, and then clamp off both the supply and the outlet. The chamber may be gassed continuously if desired.
7. To change the medium (or collect the medium), follow step 5 in reverse and then step 4 in reverse. Flame the clamped line, open the clamp, and drain off the medium.
8. Replace the medium as in steps 2–6.
9. To collect the cells,
   (a) Remove the medium as in step 7;
   (b) Add 500 ml of PBSA and then remove it;
   (c) Add 500 ml of trypsin at 4°C, and remove it after 30 s.
   (d) Incubate the cells with the residual trypsin for 15 min;
   (e) Add medium to the cells, and rock the chamber to resuspend the cells;
   (f) Run the medium off the cells as in step 7.
10. The residue may be used to seed the next culture, although this method makes it difficult to

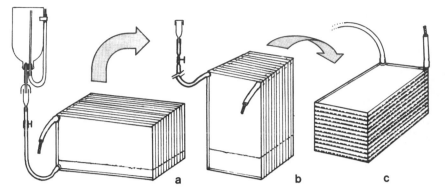

Fig. 25.10. Filling Nunc Cell Factory. (a) Run the medium in. (b) Rotate the system onto its short side away from the inlet. (c) Lay the system down flat and seal the inlet, or connect the inlet to a 5%-CO_2 line.

control the seeding density. It is better to discard the chamber and start fresh.

Analysis

Monitoring cell growth in these chambers is difficult, so a single tray is supplied to act as a pilot culture. It is assumed that the single tray will behave as the multichamber unit.

The supernatant medium can be collected repeatedly for virus or cell product purification. The collection of cells for analysis depends on the efficiency of trypsinization.

This technique has the advantage of simplicity, but can be expensive if the unit is discarded each time cells are collected. It was designed primarily for harvesting supernatant medium, but is a good method for producing large numbers of cells (3×10^8–3×10^9) for a pilot run or on an intermittent basis.

Multiarray Disks, Spirals, and Tubes

Disks, spirals, and tubes have all been used to increase the surface area for monolayer growth, but few of these systems are now available on a commercial basis. Most matrix or multisurface propagators have now gone towards perfusion (see Perfusion Systems in this chapter), with the emphasis on product recovery. Corning Costar markets a multisurface perfusion system, called CellCube, that is a hollow polystyrene cube with multiple inner lamellae, perfused with oxygenated, heated medium. The inner lamellae are capable of supporting monolayer growth on both surfaces.

Roller Culture

If cells are seeded into a round bottle or tube that is then rolled around its long axis, the medium carrying the cells runs around the inside of the bottle. (Figs. 25.11 and 25.12.) If the cells are nonadherent, they will be agitated by the rolling action, but will remain in the medium. If the cells are adhesive, they will gradually attach to the inner surface of the bottle and grow to form a monolayer. This system has three major advantages over static monolayer culture: (1) the increase in surface area; (2) the constant, but gentle, agitation of the medium; and (3) the increased ratio of the medium's surface area to its volume, which allows gas exchange to take place at an increased rate through the thin film of medium over cells not actually submerged in the deep part of the medium.

PROTOCOL 25.3. ROLLER BOTTLE CULTURE

Outline

Seed a cell suspension in medium into a round bottle, and rotate the bottle slowly on a roller rack.

Materials

Sterile or Aseptically Prepared:
Medium and medium dispenser
PBSA
0.25% crude trypsin
Monolayer culture
Roller bottles
Nonsterile:
Hemocytometer, or electronic cell counter and counting fluid
Supply of 5% CO_2
Roller rack (see Fig. 25.11)

Protocol

1. Trypsinize the cells, and seed them at the usual density for these cells.

Note. The gas phase is large in a roller bottle, so, with media based on Hanks' salts and a gas phase of

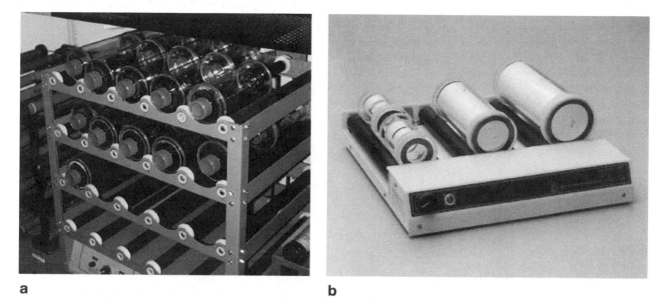

a **b**

Fig. 25.11. Roller Culture Bottles on Racks. (a) Large benchtop or free-standing extendable rack (Bellco). (b) Small benchtop rack (Courtesy of New Brunswick Scientific).

air (see Table 8.1), it may be necessary to blow a little 5% CO_2 into the bottle (e.g., for 2 s at 10 l/min). If the medium is CO_2/HCO_3-buffered, then the gas phase should be purged with 5% CO_2 (for 30 s–1 min at 20 l/min, depending on the size of the bottle; see Protocol 12.2).

2. Place the bottle on the roller rack, and rotate it at 20 rev/h until the cells attach (24–48 h).
3. Increase the rotational speed to 60–80 rev/h as the cell density increases.

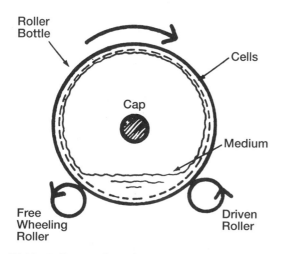

Fig. 25.12. Roller Bottle Culture. The cell monolayer (dotted line) is constantly bathed in liquid, but is submerged only for about one-fourth of the cycle, enabling free gas exchange.

4. To feed the cells or harvest the medium, take the bottles to a sterile work area, draw off the medium as usual, and replace it with fresh medium. (See Protocol 12.1.) A transfusion device (see Fig. 4.14) or media bag (e.g., Sigma) is useful for adding fresh medium, provided that the volume is not critical. If the volume of medium is critical, it may be dispensed by pipette or metered by a peristaltic pump (Camlab, Jencons). (See Fluid Handling in Chapter 4.)
5. To harvest the cells, remove the medium, rinse the cells with 50–100 ml of PBSA, and discard the PBSA. Then add 50–100 ml of trypsin at 4°C, and roll the bottle for 15 s by hand or on a rack at 20 rpm. Draw off the trypsin, incubate the bottle for 5–15 min, and add medium to the bottle. Finally, rock and rotate the bottle and/or wash off the cells by pipetting.

Analysis

Monitoring cells in roller bottles can be difficult, but it is usually possible to see the cells on an inverted microscope. However, with some microscopes, the condenser needs to be removed, and with others, the bottle may not fit on the stage. Therefore, choose a microscope with sufficient stage accommodation.

For repeated harvesting of large numbers of cells or for collecting supernatant medium, the roller-bottle system is probably the most economical, although it is labor intensive and requires investment in a roller rack. (See Fig. 25.11.)

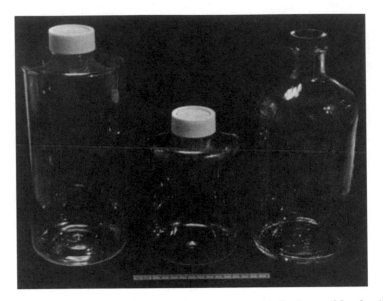

Fig. 25.13. Examples of Roller Culture Bottles. Center and left, disposable plastic (Falcon, Corning); right, glass.

Variations

Aggregation. Some cells may tend to aggregate before they attach. This behavior is difficult to overcome, but may be reduced by lowering the initial rotational speed to 5, or even 2, rev/h; trying a different type or batch of serum; or precoating the surface of the roller bottle with fibronectin or polylysine. (See Treated Surfaces in Chapter 7.)

Size. A range of bottles (around 500–1800 cm²), both disposable and reusable, is available. (See Table 7.1; Fig. 25.13; and Sources of Materials—Roller Bottles in Trade Index.) Some bottles are provided with a ribbed inner surface to increase the surface area available for cell growth.

Volume. The volume of the medium may be varied. A low volume will give better gas exchange and may be better for untransformed cells. Transformed cells, which are more anaerobic, grow faster and produce more lactic acid and thus may be better in a larger volume of medium. The volumes given in Table 7.1 are mean values and may be halved or doubled as appropriate.

Mechanics. Roller racks are preferable for bottles (see Fig. 25.11b), as they are economical in space and allow easy observation of the bottles. Roller drums (see Fig. 25.14), on the other hand, require more space for a given volume of culture bottle, but can be used for large numbers of smaller bottles or tubes. Small bottles and tubes can also be rotated on a roller rack by enclosing them in a cylindrical holder. See Fig. 25.11a.)

Fig. 25.14. Roller Drum Apparatus. Roller drums are used for roller culture of large numbers of small bottles or tubes. (Courtesy of New Brunswick Scientific.)

Microcarriers

Monolayer cells may be grown on microbeads ~150 μm (i.e., 90–300 μm) in diameter and made of plastic, glass, gelatin, or collagen [Griffiths, 1992]. (Table 25.1.) Culturing monolayer cells on microbeads gives a maximum ratio of the surface area of the culture to volume of the medium, up to 90,000 cm²/l, depending on the size and density of the beads, and has the additional advantage that the cells may be treated as a suspension. While the Nunc Cell Factory gives an increase in scale with conventional geometry, microcar-

TABLE 25.1. Microcarriers

Name	Supplier	Composition	Specific gravity	Diameter, μm
DEAE Dextran				
Cytodex 1	Amersham Pharmacia	DEAE-dextran	1.03	160–230
Cytodex 2	Amersham Pharmacia	DEAE-dextran	1.04	115–200
Microdex	Dextran Products	DEAE-dextran	1.03	150
Dormacell	JRH Biosciences	DEAE-dextran	1.05	140–240
Plastic				
Acrobeads	Galil	Coated polyacrolein	1.04	150
Biosilon	Nalge Nunc	TC-treated polystyrene	1.05	160–300
Biocarriers	Biorad	Polyacrylamide/DMAP	1.04	120–180
Bioplas	Whatman, Cellon	Polystyrene	1.04	150–210
Cytospheres	Lux	TC-treated polystyrene	1.04	160–230
Bio-spex PlastiSPEX	JRH Biosciences	Polystyrene	1.02, 1.03, 1.04	90–210
Rapidcell P	ICN	Polystyrene	1.02, 1.03	90–210
Gelatin				
Ventregel	Ventrex (Bayer)	Gelatin	1.03	150–250
Cytodex 3	Amersham Pharmacia	Gelatin-coated dextran	1.04	130–210
Cultisphere-G	Sigma	Gelatin	N/A	120–180
Cultisphere-GL	Sigma	Gelatin	N/A	150–330
Cultisphere-S	Sigma	Gelatin	N/A	120–180
Glass				
Bioglas	Whatman, Cellon	Glass-coated plastic	1.03	150–210
Bio-soex GlasSPEX	JRH Biosciences	Glass-coated polystyrene	1.02, 1.03, 1.04	90–210
Rapidcell G	ICN	Glass	1.02, 1.03	90–210
Ventreglas	Ventrex (Bayer)	Glass	1.03	90–210
Glass	Sigma	Glass, reusable	1.03, 1.04	95–210
Cellulose				
DE-52/53	Whatman	DEAE-Cellulose	1.03	Fibers
Collagen or collagen coated				
Bio-spex	JRH Biosciences	Collagen-coated polystyrene	1.2, 1.03, 1.04	90–210
Cytodex 3	Amersham Pharmacia	Collagen	1.04	130–210
Rapidcell C	ICN	Collagen coated	1.02, 1.03	90–210
Biospheres	Whatman, Cellon	Collagen-coated	1.02	150–210

riers require a significant departure from the usual substrate design. However, this difference has relatively little effect at the microscopic level (see Fig. 25.15; Plate 36), as the cells are still growing on a smooth surface at the solid–liquid interface, although some cells are influenced by the radius of curvature and may prefer larger bead diameters.

The major difference created by microcarrier systems is in the mechanics of handling [Griffiths, 1992]. Efficient stirring without grinding the beads is essential and is achieved with a suspended rotating pendulum (Techne) (see Figs. 25.1 and 25.2) or paddle (Bellco), as for suspension cell culture, rotating at 30 rpm. Technical literature is available from microcarrier suppliers to assist in setting up satisfactory cultures, and a number of protocols for microcarrier systems have been published [e.g., Griffiths, 1992].

PROTOCOL 25.4. MICROCARRIERS

Outline
Seed the cells at a high cell and bead concentration, dilute, stir, and sample the cells as required.

Materials
Sterile:
Growth medium
Microcarriers (e.g., Cytodex 1, Amersham Pharmacia)
Starter culture

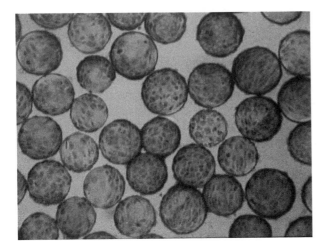

Fig. 25.15. Microcarrier Culture. Vero cells growing on microcarriers. (Courtesy of Flow Laboratories [now ICN].) (See also Plate 36.)

Culture vessel (Techne, Bellco)
Nonsterile:
Magnetic stirrer (Techne, Bellco)

Protocol

1. Suspend the beads at 2–3 g/l, in one-third of the final volume of medium required.
2. Trypsinize and count the cells. Seed the cells at three to five times the normal seeding concentration into the bead suspension.
3. Stir the culture at 10–25 rpm for 8 h.
4. Add medium to reach the final volume, dictated by a bead concentration of 0.7–1 g/l.
5. Increase the stirring speed to approximately 60 rpm.
6. If the pH falls, feed the culture by switching off the stirrer for 5 min, allowing the beads to settle, and then replacing one-half to two-thirds of the medium.
7. To harvest the cells, remove the medium, wash the cells by settling, digest the beads with trypsin EDTA, spin down the culture, and wash the cells.

Analysis

Cell counting on beads can be difficult, so the growth rate of the cells should be checked by determination of DNA (see Protocol 20.3); protein (see Protocol 20.4), if nonproteinaceous beads are used, or dehydrogenase activity, using the MTT assay (see Protocol 21.4) on a sample of the beads.

Variation

Most variations on this method arise from the choice of bead or design of the culture vessel and stirrer [Griffiths, 1992]. Bead density varies from

1.03–1.05 g/cc and influences the stirring speed, since a higher speed is required for higher density beads. Composition, or coating, of the beads will also influence attachment, and more fastidious cells may prefer gelatin or collagen beads, which are also soluble and can be removed by protease activity. Glass beads are the easiest to recycle.

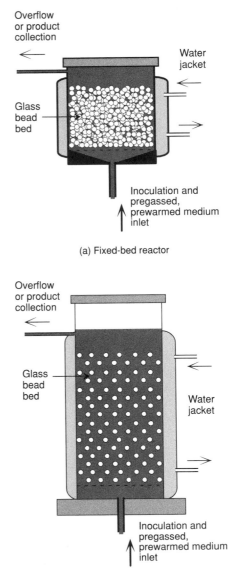

(a) Fixed-bed reactor

(b) Fluidized-bed reactor

Fig. 25.16. Fixed- and Fluidized-Bed Reactors. Cells grown on the surface of beads are perfused with medium. In the fixed bed reactor (a), the beads are usually glass and are settled in a dense bed resting on a perforated base at the bottom of the culture vessel. Once the culture is established, the beads do not move, and medium percolates around the them. In the fluidized bed (b), beads of a lower density are used and float on the low-velocity perfusate.

Perfused Monolayer Culture

Perfusion is frequently used to facilitate medium replacement and product recovery. The Cellcube, (Corning-Costar) is a perfused, multisurface, single-use propagator with growth-surface areas from 21,250 to 85,000 cm², with associated pumps, oxygenator, and system controller.

Membrane Perfusion. Many systems depend on filter membrane technology as exploited by Millipore (MCCS), in which the culture bed is a flat, permeable sheet that is folded over into many layers, and Membroferm, which is compartmentalized in such a way that the cells, medium supply, and product occupy different membrane compartments. (See Perfused Suspension Culture in this chapter and Fig. 25.7.)

Hollow-Fiber Perfusion. There are a number of systems based on hollow-fiber perfusion (Cellco, En-

dotronics, Integra, CD Medical). In these systems, adherent cells grow on the outer surface of the perfused microcapillary bundles. High-molecular-weight products concentrate in the outer space with the cells, while nutrients are supplied and metabolites removed via the inner space. (See Fig. 25.6; Hollow Fibers in Chapter 24.) The potential of these systems for adherent cells lies in the re-creation of high, tissue-like cell densities, matrix interactions, and the establishment of cell polarity, all of which may be important for post-translational processing of proteins and for exocytosis. Although originally designed for attached cells, hollow-fiber systems are also used extensively for suspension cells, such as hybridomas. (See Production of Monoclonal Antibodies in Chapter 27.)

Matrix Perfusion. A porous ceramic matrix has been used in the Opticell system, which is available at 4,250 cm² and units of 4.8 m², and provided with a

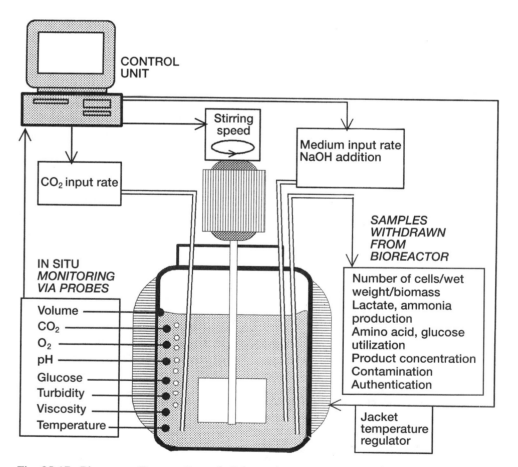

Fig. 25.17. Bioreactor Process Control. Schematic representation of a paddle-stirred bioreactor with direct-reading probes on the left, feeding to a control unit (top left) that stores the data and also regulates conditions within the bioreactor. A sampling port on the right withdraws the cell suspension from the bioreactor for analysis.

perfusion control system. This system is a rigid matrix and is perfused via fine parallel channels running the length of the matrix.

Fixed-Bed Reactors. Systems have also been developed with beds of glass beads (Fig. 25.16a), with the medium being perfused upward through the bed or percolating downward by gravity. The product is collected with the spent medium in a reservoir. These systems are described in greater detail by Speir and Griffiths [1985–1990] and Griffiths [1992]. Fixed bed reactors can also use a porous ceramic matrix, perfused through microchannels (Opticell).

Fluidized-Bed Reactors for Monolayer Cultures. In fluidized bed reactors, porous beads of a relatively low density—made of ceramics, or a mixture of ceramics and natural products, such as collagen (Verax)—are suspended in an upward stream of medium when the flow rate of the medium matches the sedimentation rate of the beads. (See Fig. 25.16b.) While suspension cells lodge in the beads by entrapment, monolayer cells attach to the outer surfaces as well as the interstices of the porous bead.

Microencapsulation. Sodium alginate behaves as a sol or gel, depending on the concentration of divalent cations. It will gel as a hollow sphere around cells in suspension in a high concentration of divalent cations. (See Protocols 24.3 and 22.13; Fig. 24.4.) Because the alginate acts as a barrier to high-molecular-weight molecules, macromolecules secreted by the cells are trapped within the vesicle, while nutrients, metabolites, and gas freely permeate the gel. The product and cells are recovered by reducing the concentration of divalent cations. These gels have a low immunoreactivity and can be implanted *in vivo*.

Monitoring Growth

The progress of suspension cultures is monitored via pH, oxygen, CO_2, and glucose electrodes that read from the culture *in situ*, and by assaying the utilization of nutrients, such as glucose and amino acids, or the buildup of metabolites, such as lactate and ammonia, and products, such as immunoglobulins from hybridomas. (Fig. 25.17.) The number of cells and other parameters, such as ATP, DNA, and protein, are determined in samples drawn from the culture and are used to calculate the total biomass. The temperature of the medium is regulated by preheating the input medium and by heating the surrounding water jacket regulated by feedback from the temperature probe. The flow rate of the medium is controlled and can match the output to the sample line, if the suspension is running as a biostat (see Continuous Culture in this chapter), and the stirring speed can be regulated, along with the viscosity of the medium, to reduce the shear stress.

There is a recurrent problem when monolayer cell cultures are scaled up, particularly in a fixed-bed or hollow-fiber bioreactor: it is no longer possible to observe the cells directly, and both monitoring the progress of a culture by cell counting and determining the biomass become difficult. One approach to this problem [Brindle, 1998] uses nuclear magnetic resonance (NMR) to assay the contents of the culture chamber. By placing the cells in a perfused hollow-fiber chamber within the magnetic field of an NMR spectroscope, a characteristic NMR spectrum is generated, enabling the identification and quantitation of specific metabolites. (Fig. 25.18.) NMR spectroscopes can also be used as imaging devices, producing a quasi-optical section through the chamber to reveal the distribution of the cells, and even distinguishing between proliferating and nonproliferating zones of the culture.

Fig. 25.18. Analysis by NMR. Cell growth in hollow fibers, analyzed by NMR. (a) NMR detectors interfaced to a high-field NMR spectrometer. The environmental cabinet in the background contains buffer reservoirs, pumps, and gas-exchange equipment, and maintains the bulk of the apparatus at 37°C. Oxygenated growth medium is pumped at a flow rate of 50 ml/min through insulated tubing to the hollow-fiber cartridge, which is situated in the bore of the high-field magnet (foreground). (b) [^{31}P] NMR spectrum of CHO cells growing in a hollow-fiber reactor. The cells had been grown on macroporous beads in the extracapillary space of a specially constructed hollow-fiber cartridge that could be accommodated within a 25-mm-diameter NMR probe. The cells were present at a density of approximately 7×10^7 cells/ml. *Abbreviations:* PME, phosphomonoesters, including phosphocholine and phosphoethanolamine; Pi, extracellular inorganic phosphorus; Pi$^{(acid)}$, an acidic (pH 6.7) extracellular Pi pool within the bioreactor; PCr, phosphocreatine; γ-ATP, γ-phosphate of ATP; α-ATP, α-phosphate of ATP; DPDE, diphosphodiesters, including UDP-glucose; β-ATP, β-phosphate of ATP. The chemical shift scale is referenced to phosphocreatine at 0.0 ppm. (Courtesy of Dr. Kevin Brindle.)

a

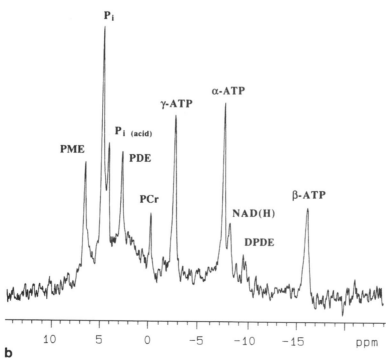

b

Regulations regarding the use of transformed cells for the production of biopharmaceuticals are now being relaxed, but a number of arguments for anchorage-dependent cells remain. While cells that grow in suspension are easier to manage, anchored cells, with their resultant potential for developing polarity, may yet represent a better model for post-translational modification of proteins and the appropriate membrane flux, leading to their secretion from the cells in a bioactive form.

CHAPTER 26

Specialized Techniques

LYMPHOCYTE PREPARATION

There is a variety of methods for the preparation of lymphocytes, but flotation on a combination of Ficoll and sodium metrizoate (e.g., Hypaque) is the most widely used technique [Boyum, 1968a,b]. Products similar to Hypaque are available from Nygaard, Pharmacia, and ICN and come with instructions for use. The generic term used throughout this book for these products is Ficoll-paque.

PROTOCOL 26.1. PREPARATION OF LYMPHOCYTES

Outline
Layer whole citrated blood or plasma, depleted in red cells by dextran-accelerated sedimentation, on top of a dense layer of Ficoll-paque. After centrifugation, most of the lymphocytes are found at the interface between the Ficoll/metrizoate and the plasma.

Materials
Sterile:
Blood sample in heparin or citrate (concentration determined by collection container which will already contain citrate or heparin)
Clear centrifuge tubes or universal containers
PBSA
Ficoll-paque, adjusted to 1.077 g/cc (ICN, Nygaard, Amersham Pharmacia)
Serum-free medium

Syringe, Pasteur pipettes, or pipettor
Nonsterile:
Hemocytometer or electronic cell counter
Centrifuge

Protocol
1. Collect the blood sample in citrated or heparinized container and transport to laboratory.

△ *Safety Note.* Human blood may be infected with agents such as AIDS virus or hepatitis and should be handled with great care. (See Human Biopsy Material in Chapter 6.)

2. Dilute 1:1 with PBSA, and layer 9 ml onto 6 ml Ficoll-paque. This should be done in a wide, transparent centrifuge tube with a cap such as the 25-ml Sterilin or Nunc universal container, or with double these volumes in a clear plastic Corning 50-ml tube.
3. Centrifuge the suspension for 15 min at 400 g (measured at the center of the interface).
4. Carefully remove the plasma/PBSA without disturbing the interface.
5. Collect the interface with a syringe or Pasteur pipette, and dilute it to 20 ml in serum-free medium (e.g., RPMI 1640).

△ *Safety Note.* Pasteur pipettes and syringes with sharp needles should not be used with human blood. Instead, use a 1-ml pipettor or a syringe with a blunt canula.

6. Centrifuge the diluted cell suspension from the interface at 70 g for 10 min.
7. Discard the supernatant fluid, and resuspend the pellet in 2 ml of serum-free medium. If several washes are required—e.g., to remove serum factors—resuspend the cells in 20-ml of serum-free medium, and centrifuge two or three times more, and finally resuspend the pellet in 2 ml of serum-free medium.
8. Stain a sample of the cells with methylene blue, and count the nucleated cells on a hemocytometer. Alternatively, lyse a sample of the cells with Zapoglobin (Beckman Coulter), and count the nuclei on an electronic cell counter with a 70–100 μm orifice tube.

Lymphocytes will be concentrated at the interface, along with some platelets and monocytes. Some granulocytes may be found in the interface although most will be found mostly in the Ficoll/metrizoate, in the pellet created in step 3. Monocytes and residual granulocytes can be removed from the interface fraction by taking advantage of their adherence to glass (beads or the surface of a flask) or to nylon mesh. Use a positive sort by MACS (see Magnetic Sorting in Chapter 14) or flow cytometry (see Fluorescence-Activated Cell Sorting in Chapter 14) with specific lymphocyte surface markers if purer preparations are required.

Blast Transformation

Lymphocytes in purified preparations or in whole blood may be stimulated with mitogens such as phytohemagglutinin (PHA), pokeweed mitogen (PWM), and antigen [Berger, 1979; Hume and Weidemann, 1980]. The resultant response may be used to quantify the immunocompetence of the cells. PHA stimulation is also used to produce mitosis for chromosomal analysis of peripheral blood [Rooney and Czepulkowski, 1986; Watt and Stephen, 1986]. (See Protocol 15.9.)

Protocol

1. Using the washed interface fraction from step 7 of Protocol 26.1, incubate 2×10^6 cells/ml in medium, 1.5–2.0 cm deep, in HEPES or CO_2-buffered DMEM, CMRL 1066, or RPMI 1640 supplemented with 10% autologous serum or fetal bovine serum.
2. Add PHA, 5 μg/ml (final concentration), to stimulate mitosis from 24 to 72 h later.
3. Collect samples at 24, 36, 48, 60, and 72 h, and prepare smears or cytocentrifuge slides of the samples to determine the optimum incubation time (i.e., the peak mitotic index).
4. Add 0.001 μg/ml (final concentration) of Colcemid for the 2 h during which the peak of mitosis is anticipated from observations made in step 3 [Berger, 1979].
5. Centrifuge the cells after the Colcemid treatment, resuspend the pellet in 0.075 M KCl for hypotonic swelling, and proceed as for chromosome preparation (see Protocol 15.9).

AUTORADIOGRAPHY

This section is intended to cover microautoradiography of any small molecular precursor into a cold acid-insoluble macromolecule, such as DNA, RNA, or protein. Other variations may be derived from this text or found in the literature [Rogers, 1979; Stein and Yanishevsky, 1979]. (See also Protocol 27.1.) Since autoradiography is used extensively to localize material in blots from electrophoresis as well as in microscope preparations, the two methods may be distinguished as macroautoradiography and microautoradiography, respectively. (See Glossary.)

Isotopes suitable for microautoradiography are listed in Table 26.1. A low-energy emitter (e.g., ^3H or ^{55}Fe) in combination with a thin emulsion gives high

---

PROTOCOL 26.2. PHA STIMULATION OF LYMPHOCYTES

Materials

Sterile:
Medium + 10% FBS or autologous serum
Phytohemagglutinin (PHA), 50 μg/ml
Test tubes or universal containers
Microscope slides
Colcemid, 0.01 μg/ml in BSS
0.075 M KCl

TABLE 26.1. Isotopes Suitable for Autoradiography

Isotope	Emission	Energy (mV) (mean)	$T_{1/2}$
^3H	β^-	0.018	12.3 yr
^{55}Fe	X-rays	0.0065	2.6 yr
^{125}I	X-rays	0.035	60 days
		0.033	
^{14}C	β^-	0.155	5,570 yr
^{35}S	β^-	0.167	87 days
^{45}Ca	β^-	0.254	164 days

intracellular resolution. Slightly higher energy emitters (e.g., ^{14}C and ^{35}S) give localization at the cellular level. Still higher energy isotopes (e.g., ^{131}I, ^{59}Fe, and ^{32}P) give poor resolution at the microscopic level, but are used for macroautoradiographs of chromatograms and blots from DNA, RNA, and protein electrophoresis, for which the absorption of low-energy emitters would limit the detection of incorporation. Low concentrations of higher energy isotopes (^{14}C and above), used in conjunction with thick nuclear emulsions, produce tracks that are useful in locating a few highly labeled particles (e.g., virus particles infecting a cell population or tissue).

Tritium is used most frequently for autoradiography at the cellular level, because the β particles released have a mean range of about 1 μm in aqueous media, giving very good resolution. Tritium-labeled compounds are usually less expensive than the ^{14}C- or ^{35}S-labeled equivalents and have a longer half-life. Because of their low energy of emission, however, it is important that the radiosensitive emulsion be positioned in close proximity to the specimen, with nothing between the cell and the emulsion. Even in this situation, only the incorporation in the top 1 μm of the specimen will be detected.

β-particles entering the emulsion produce a latent image in the silver halide crystal lattice within the emulsion at the point where they stop and release their energy. The image may be visualized as metallic silver grains by treatment with an alkaline reducing agent (developer) and subsequent removal of the remaining unexposed silver halide by an acid fixer.

The latent image is more stable at low temperatures and in anhydrous conditions, so its sensitivity (signal versus background) may be improved by exposure in a refrigerator over desiccant. This reduces the background silver-grain formation by thermal activity.

PROTOCOL 26.3. MICROAUTORADIOGRAPHY

Outline
Incubate cultured cells with the appropriate isotopically labeled precursor (e.g., [^3H]-thymidine to label DNA), wash, fix, and dry the cells. (Fig. 26.1.) Perform any necessary extractions (e.g., to remove unincorporated precursors). Coat the specimen with emulsion in the dark and leave it to expose. When the specimen is subsequently developed in a photographic developer, silver grains can be seen in overlying areas where radioisotope was incorporated. (Fig. 26.2; see Plate 37.)

Materials

Setting up the culture:
Sterile:
Cells
PBSA
Trypsin
Growth medium
Coverslips or slides, and Petri dishes (may be non-tissue-culture grade if coverslips or slides are used) or plastic bottles
Nonsterile:
Hemocytometer, or electronic cell counter and counting fluid

Labeling with the Isotope:
Sterile:
Isotope
HBSS
Nonsterile:
Protective gloves
Containers for disposal of radioactive pipettes
Container for radioactive liquid waste

Fixing and Processing the Cells:
Nonsterile:
Acetic methanol (1:3, ice-cold, freshly prepared)
DPX
TCA, 0.6 N
Deionized water
Gelatin: 0.2 g in 200 ml of UPW; microwave for 2 min, cool, and filter

Setting up Autoradiographs:
Nonsterile:
Safelight: Kodak Wratten II or equivalent
Emulsion (Amersham Hypercoat Emulsion LM-1, RPN 40)

Note. It is convenient to melt the emulsion and disperse it into aliquots in dipping vessels (Amersham) suitable for the number of slides to be handled at one time. If the slides are sealed in a dark box, they may be stored at 4°C until required.

Light-tight microscope slide boxes (Clay Adams, Raven)
Silica gel (Fisher)
Dark vinyl tape
Black paper or polyethylene

Note. All glassware must be carefully washed and free of isotopic contamination. Plastic coverslips should be used in preference to glass, to minimize

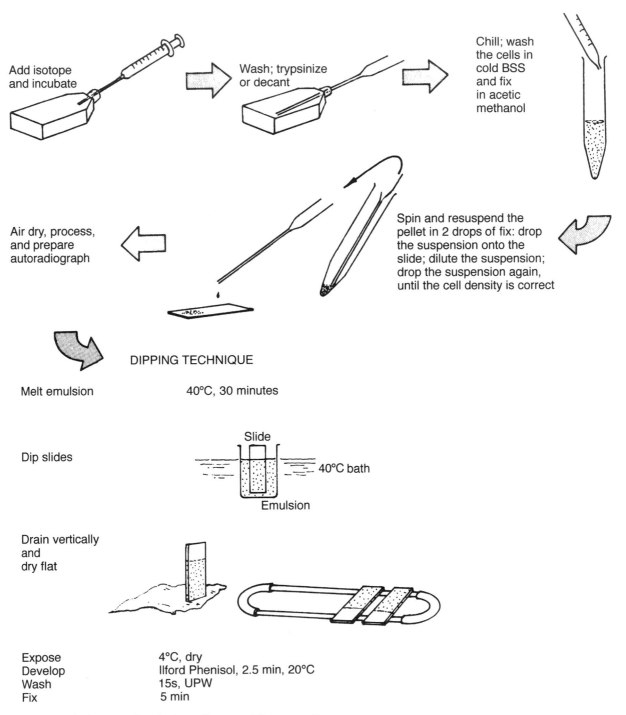

Add isotope
and incubate

Wash; trypsinize
or decant

Chill; wash
the cells in
cold BSS
and fix
in acetic
methanol

Air dry, process,
and prepare
autoradiograph

Spin and resuspend the
pellet in 2 drops of fix: drop
the suspension onto the
slide; dilute the suspension;
drop the suspension again,
until the cell density is correct

DIPPING TECHNIQUE

Melt emulsion   40°C, 30 minutes

Dip slides

Slide

40°C bath

Emulsion

Drain vertically
and
dry flat

Expose   4°C, dry
Develop   Ilford Phenisol, 2.5 min, 20°C
Wash   15s, UPW
Fix   5 min

Wash, dehydrate, and stain the cells; mount the coverslip

Fig. 26.1. Microautoradiography. Steps for preparing a microautoradiograph from a cell culture.

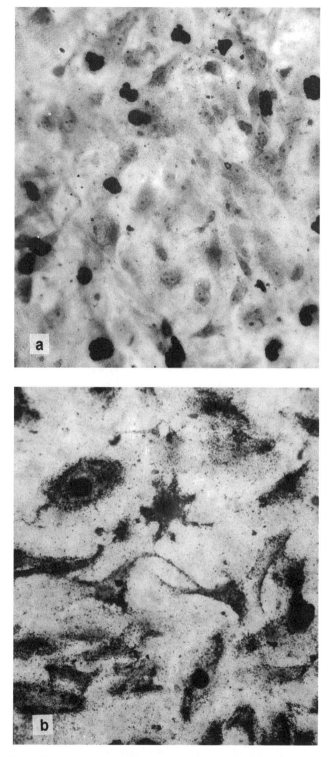

Fig. 26.2. Microautoradiograph. This pair of pictures is an example of [³H]-thymidine incorporation into a cell monolayer. Normal glial cells were incubated with 0.1 μCi/ml (3.7 KBq/ml), 200 Ci/mMol (7.4 GBq/μmol), of [³H]-thymidine for 24 h, washed, and processed as described in Protocol 26.3. (a) Typical densely labeled nuclei, suitable for determining the labeling index. (See Protocol 20.10.) (b) A similar culture, infected with mycoplasma; cytoplasm is also labeled now.

radioactive background. Be particularly careful to prevent spillages, and immediately mop up any that do occur. Wear gloves, and change them regularly—e.g., when you move from incubation (a high level of isotope) to handling washed, fixed slides (a low level of isotope).

Processing Autoradiographs:
Nonsterile:
Developer: Phenisol (Ilford), 20%
Stop bath: 1% acetic acid
Photographic fixer: 30% sodium thiosulphate
Coverslips (#00)

Staining:
Nonsterile:
Hematoxylin (filter before use) or Giemsa stain (see
  Protocol 15.2; Gurr, Merck)
Phosphate buffer, 0.01 M, pH 6.5
Ethanol, 50, 70, and 100%
Histoclear (Fisher)
DPX (Merck)

Protocol

Setting up the Culture:
1. Set up replicate monolayer cultures on coverslips, slides (Nunc, Bellco), or Petri dishes.
2. Incubate the cultures at 37°C until the cells reach the appropriate stage for labeling.

Adding the Isotope:
3. Add the isotope, usually in the range of 0.1–10 μCi/ml (~4.0 KBq–0.4 MBq/ml), 100 Ci/mmol (~4 GBq/μmol), for 0.5–48 h as appropriate.

△ *Safety Note.* Follow local rules for handling radioisotopes. Since such rules vary, no special recommendations are made here, other than the following: Wear gloves; do not work in horizontal laminar flow; use a shallow tray with an absorbent liner to contain any accidental spillage; incubate the cultures in a box or tray labeled for use with radioactivity; and regulate the disposal of radioactive waste according to local limits. (See Radiation in Chapter 6.)

4. Remove all medium containing isotope, and wash the cells carefully in BSS, discarding the medium and washes.

△ *Safety Note.* ³H-nucleosides are highly toxic, due to their ultimate localization in DNA. (See Ingestion in Chapter 6.)

Fixing and Processing the Cells:

5. Fix the cells in ice-cold acetic methanol for 10 min.

6. Prepare the slides:

   (a) Coverslips: Mount the coverslips on a slide with DPX or Permount, cells uppermost.

   (b) Cell suspensions (from growth on suspension or trypsinized): Centrifuge the cells onto a slide (see Protocol 15.5) or make drop preparations (see Protocol 15.9, steps 4–12).

   (c) Prepare several extra control slides for later use, to determine the correct duration of exposure. All preparations are referred to as "slides" from now on.

7. Extract acid-soluble precursors (when labeling DNA, RNA, or protein) with ice-cold 0.6 N TCA (3 × 10 min), and perform any other control extractions (e.g., with lipid solvent or enzymatic digestion).

8. Dip the slides in the gelatin solution for 2 min.

9. Drain the slides vertically, and let them dry. (The slides may be stored dry at 4°C.)

Setting up the Autoradiographs:

10. Take the slides to the darkroom.

11. Under a dark-red safelight, melt the emulsion in a water bath at 46°C.

12. Mix the emulsion gently with a clean slide, taking care not to create bubbles.

13. Dip the slides:

   (a) Dip the slides in the emulsion for 5 s at 46°C, making sure that the cells are completely immersed. Note that the temperature is critical and will determine the thickness of the emulsion and, consequently, the resolution.

   (b) Withdraw the slides, and drain them for 5 s.

   (c) Dip the slides again in the emusion for 5 s.

   (d) Withdraw the slides, and drain them vertically for 5 s.

   (e) Wipe the back of each slide with a paper tissue.

   (f) Allow the slides to dry flat on a tissue or piece of filter paper for 10 min.

   (g) Place the slides on a rack in a light-tight box, and allow them to dry completely. Do not force the slides to dry; dry them slowly in humid conditions to avoid crocking the emulsion.

14. When the slides are dry (2–3 h), transfer them to light-tight microscope slide boxes with a desiccant, such as silica gel. Make sure that you do not touch the slides and that they do not touch each other or the desiccant.

15. Seal the boxes with dark vinyl tape (e.g., electrical insulation tape), wrap them in black paper or polythene, and place them in a refrigerator. Make sure that this refrigerator is not used for the storage of isotopes.

16. Leave the boxes at 4°C for 1–2 weeks. The exact time required will depend on the activity of the specimen and can be determined by processing one of the extra slides at intervals. Slides with prolonged exposure times have an increased background and are prone to latent image fade. It is better to increase the activity of the label than to expose a slide for too long.

Processing the Autoradiographs:

17. Return the boxes to the darkroom (under a dark-red safelight), unseal the boxes, and allow the slides to come to atmospheric temperature and humidity (~2 min).

18. Prepare the solutions for development at 20°C.

19. Place the slides in the developer for 2.5 min with gentle intermittent agitation.

20. Wash the slides briefly in UPW.

21. Transfer the slides to a photographic fixer for 5 min.

22. Rinse the slides in deionized water, and then place them in hypo clearing agent (Kodak) for 1 min.

23. Wash the slides in cold running water, or with five changes of water over 5 min.

24. Dry the slides, and examine them on the microscope. Phase contrast may be used by mounting a thin glass (#00) coverslip in water. Remove the coverslip when you are finished examining the slides and before the water dries out, or else the coverslip will stick to the emulsion.

Staining:

25. To stain with hematoxylin,

   (a) Stain the slides in freshly filtered hematoxylin for 45 s;

   (b) Rinse the slides in running water for 2 min.

   (c) Dehydrate the slides in a succession of 50, 70, and 100% EtOH;

   (d) Clear the slides in Histoclear;

   (e) Mount coverslips in DPX.

26. To stain with Giemsa:

   (a) Immerse the dry slides in neat Giemsa stain for 1 min;

   (b) Dilute the stain *in situ* 1:10 in 0.01 M phosphate buffer (pH 6.5) for 10 min;

   (c) Remove the staining solution by upward displacement with water. The slides should not be withdrawn or the stain poured off, or else the scum that forms on top of the stain will adhere to the specimen;

(d) Rinse the slides thoroughly under running tap water until the color is removed from emulsion, but not from the cells.

If a coverslip is used, it must be #00 with a minimum of mountant, to allow sufficient working distance for a 100× objective.

Analysis

Qualitative

Determine the specific localization of grains—e.g., over nuclei only, or over one cell type rather than another.

Quantitative

1. *Grain counting.* Count the number of grains per cell, per nucleus, or localized elsewhere in the cells. This requires a low grain density—about 5–20 grains per nucleus, 10–50 grains per cell—no overlapping grains, and a low uniform background.

2. *Labeling index.* Count the number of labeled cells as a proportion of the total number of cells. (See also Protocol 20.10.) The grain density should be higher for this assessment than for grain counting, to ease the recognition of labeled cells. If the grain density is high (e.g., ~100 grains per nucleus), set the lower threshold at, say, 10 grains per nucleus or per cell; but remember that low levels of labeling, significantly over background, may yet contain useful information, e.g., regarding DNA repair.

Microautoradiography is a useful tool for determining the distribution of isotope incorporation within a population, but it is less suited to total quantitation of isotope uptake or incorporation, for which scintillation counting is preferable.

Variations

Isotopes of two different energies—e.g., ^3H and ^{14}C —may be localized in one preparation by coating the slide first with a thin layer of emulsion, then coating that layer with gelatin alone, and finally coating the gelatin with a second layer of emulsion [Rogers, 1979]. The weaker β emission from the ^3H is stopped by the first emulsion and the gelatin overlay, while the higher energy β emission from the ^{14}C, having a longer mean path length, of around 20 μm, will penetrate the upper emulsion.

Adams [1980] described a method for autoradiographic preparations from Petri dishes or flasks such that liquid emulsion is poured directly onto fixed preparations without the need for trypsinization.

Soft β-emitters may also be detected in electron microscope preparations, using very thin films of emulsion or silver halide sublimed directly onto the section [Rogers, 1979].

Fluorescent and luminescent probes (Amersham International) are now being used in place of radioisotopically labeled probes. (See Protocol 27.2.) The resolution with this method is often superior, quantitation is possible by confocal microscopy (Biorad) or a CCD camera (Dage-MTI), and disposal of reagents is environmentally friendly. However, the equipment for microscopic evaluation is expensive (~$50,000–200,000).

TIME-LAPSE RECORDING

Time-lapse recording was developed primarily as a cinematography technique by which naturally slow processes can be observed at a greatly accelerated rate. At its inception, the technique required a camera operated automatically by a signal from a timing device. Scientific cinematography started with the first cine camera itself when Marey used it to record animal movement in 1888. In the 1950s, Michael Abercrombie was the first to use the technique for behavior studies of tissue culture cells in a rigorously quantitative way [Abercrombie and Heaysman, 1954]. Video systems gradually replaced cine film, and "video microscopy" optimized their use in combination with light microscopes [Inoué and Spring, 1997].

The quality of video imaging progressively improved, and its digital form eventually superseded 16-mm film. The microscopy technique itself, however, remains the critical link in the whole process. Commercially available microscopy techniques usually require laborious manual, or at least semi-interactive, frame-by-frame analysis of the recordings. Automated analysis of cell behavior by digital image processing was achieved with highly refractile organisms [Wessels et al., 1996]. Fully automatic analysis with unstained vertebrate tissue culture cells requires special microscopy: digitally recorded interference microscopy with automated phase shifting [Dunn and Zicha, 1997]. Automatic analysis of cell behavior is an important aspect of time-lapse recording, because large numbers of data usually are required to investigate phenomena in this area with high intrinsic variability.

This background information and the following protocol for time-lapse recording has been contributed by Daniel Zicha, Imperial Cancer Research Fund, Light Microscopy Laboratory, 44 Lincoln's Inn Fields, London, WC2A 3PX, England, United Kingdom.

PROTOCOL 26.4. TIME-LAPSE VIDEO

Outline

Prepare cells in a culture chamber with suitable optical properties for the microscopy technique. The microscope is equipped with a video camera connected to a recording device, and additional equipment provides the desired temperature for the cell culture. During recording, individual exposures are triggered by a controller that also operates a shutter, eliminating unnecessary illumination of cells.

Materials

Medium: MEM with Hanks' salts, 12 mM bicarbonate, 10% calf serum (SML–Serum Medium Laborbedarf, Lauphelm, Germany), and 4 mM L-glutamine

Trypsin/EDTA: trypsin (Difco), 0.25% in EDTA, 0.5 mM

Coverslips, 18 × 18 mm, No. 1.5 (Merck), washed with concentrated sulphuric acid and repeated boiling in UPW using Teflon coverslip holders (Eppendorf), and sterilized by autoclaving

Petri dishes, plastic, 35 mm

Filming chamber:
  (a) 1-mm thick glass slide with a 15-mm-diameter hole in the center covered with a #3 coverslip, cemented to one face of the slide by UV-curing glass cement (Southern Watch and Clock Supplies).
  (b) Sterilize the chamber with 70% alcohol, followed by a blow with clean compressed air
  (c) Wax mixture: beeswax (Fisher), soft yellow paraffin (Fisher), and paraffin wax (melting point of 46°C; Fisher), in a ratio of 1:1:1

Microscope: Axiovert 135 TV (Zeiss)

Perspex box, designed to encapsulate the stage of the microscope and is equipped with a controlled heating system (custom made from controller 208 2739, T probe 219 4674, heater element 224 565, low noise a.c. fan 583-325; RS Components, UK)

Computer-controlled shutter and filter wheel (Ludl Electronic)

Camera: RTE/CCD-1300-Y/HS (Princeton Instruments)

Computer: Macintosh G3

Imaging software IPlab (Scanalytics), which records microscope images in time-lapse mode

Protocol

1. Switch on all the equipment, especially the heater, well in advance, to allow the temperature within the cabinet to stabilize.
2. Release adherent culture cells into suspension, using trypsin/EDTA.
3. Seed around 20,000 cells on a coverslip in a Petri dish, and allow the cells to settle at 37°C in a humidified atmosphere with 3% CO_2 overnight.
4. Mount the coverslip on the filming chamber, which should be filled with culture medium, leaving a small bubble to ensure CO_2 equilibration with the Hanks' salts. Seal the coverslip in place with the hot wax mixture.
5. Use the appropriate microscopy technique—for example, phase contrast and fluorescence with LD achroplan 32×/0.4 Ph2 objective. A higher power objective with a short working distance will require an upright microscope to observe adherent cells moving over the ceiling of the chamber.
6. Prepare the computer for image acquisition:
   (a) Calibrate the image size using the stage micrometer.
   (b) Choose the exposure time; 0.1 s is usually sufficient for phase contrast, whereas weak fluorescence requires a longer exposure of 0.4 s or more.
   (c) Select the time-lapse interval, the number of frames, and the file name. Analysis of cell translocation in slowly moving, fibroblastlike cells will give satisfactory results at 5-min lapse intervals. Observation of cell morphology that changes much faster will require lapse intervals of 1 min or even less. Also, translocation of fast cells, such as neutrophil leukocytes, will need 1 min or shorter lapse intervals.
   (d) Start the following script to perform the alternating acquisition of phase-contrast and fluorescence images:
      (i) Set indexed file name for save;
      (ii) Insert START label;
      (iii) Set the filter wheel for fluorescence, and open the arc lamp shutter;
      (iv) Full acquire for fluorescence image using the appropriate exposure time;
      (v) Set the filter wheel for phase contrast, and close the arc lamp shutter;
      (vi) Save the fluorescence image;
      (vii) Open the halogen lamp shutter;
      (viii) Full acquire for phase-contrast image using the appropriate exposure time;
      (ix) Close the halogen map shutter;
      (x) Save the phase-contrast image;
      (xi) Pause for the time-lapse interval;
      (xii) Close all windows;
      (xiii) Loop to START label unless the specified number of frames have been recorded;

(xiv) Quit the program.

Separate phase-contrast and fluorescence movie sequences can be viewed either independently or in combination using iPLab animation.

7. Store the image sequences, using a CD writer.

△ *Safety Notes.* (1) Take care when washing coverslips with concentrated acid. (2) The controlled electrical heater, built from Radio Spares (RS) components, has to be properly insulated and earthed, and the fan has to be mounted in a position that prevents the rotating blades from being touched accidentally.

Analysis

The imaging software IPLab can be used for the interactive measurement of positions, areas, and intensities in the images. Theory of moments provides a good basis for automatic analysis of cell behavior [Dunn and Brown, 1990]. It can be applied to fluorescence images for the evaluation of cell translocation using centroid positions, of cell spreading using cell area, of the intensity of the fluorescence signal, or of the shape factors in morphological studies.

Variations

The choice of medium depends on the requirements of the cells under observation. Medium with Hanks' salts has the advantage that its pH equilibrates automatically in a sealed culture chamber. An open chamber can be designed that contains a Petri dish positioned over a hole in the bottom of the chamber. This arrangement requires humidification and a supply of a mixture of CO_2 and air into the chamber. Hanks-based medium with 12 mM bicarbonate will need around 3% CO_2, and Dulbecco's MEM will need 10% CO_2. An open chamber requires an inverted microscope, and high-power magnification can be achieved when glass-bottom, No. 1.5, uncoated, γ-irradiated 35-mm microwell dishes (MatTek) are used. Specialized chambers with local heating and a through-flow option are available (see Fig. 4.21b; Intracell, Royston). It is no longer worth giving serious consideration to film-based methods, or even disc or tape time-lapse video recorders, because digital recording on a computer now provides greater convenience as well as other features—namely, image processing, including contrast enhancements, background-noise subtraction, and support for analysis. An alternative time-lapse recording software for a PC is MetaMorph Imaging Software (Universal Imaging Corporation, PA, USA). Differential interference contrast (DIC) is a common alternative to phase contrast that improves the contrast of details at high resolution.

CONFOCAL MICROSCOPY

Time-lapse records, or real-time records made on videotape, can be made via the confocal microscope with a conventional high-resolution video camera or CCD. These systems allow the recording of events within the cell depicted by the distribution and relocation of, and changes in the staining intensity of fluorescent probes. Fluorescent imaging can be used to localize cell organelles, such as the nucleus or Golgi complex [Lippincott-Schwartz et al., 1990]; measure fluctuations in intracellular calcium [Cobbold and Rink, 1987]; and follow the penetration and movement within the cell of a drug [Neyfakh, 1987; Bucana et al., 1990]. Measurements can be made in three dimensions, as the excitation and detection system is capable of visualizing optical sections through the cell, and over very short periods of time.

CELL SYNCHRONY

In order to follow the progression of cells through the cell cycle, a number of techniques have been developed whereby a cell population may be fractionated or blocked metabolically so that on return to regular culture, the cells will all be at the same phase.

Cell Separation

Techniques for cell separation have been described in Chapter 14. Sedimentation at unit gravity can be used [Shall and McClelland, 1971; Shall, 1973], but centrifugal elutriation is preferable if a large number of cells ($>5 \times 10^7$) is required [Breder et al., 1996; Mikulits et al., 1997]. (See Figs. 14.9 and 14.10.) Fluorescence-activated cell sorting (see Figs. 14.12 and 14.13) can also be used, in conjunction with a nontoxic, reversible DNA stain, such as Hoechst 33342. The yield is lower for this method than for centrifugal elutriation ($\sim 10^7$ cells or less), but the purity of the fractions is higher.

One of the simplest techniques for separating synchronized cells is mitotic shakeoff: Monolayer cells tend to round up at metaphase and detach when the flask in which they are growing is shaken. This method works well with CHO cells [Tobey et al., 1967; Petersen et al., 1968] and some sublines of HeLa-S_3. Placing the cells at 4°C for 30 min to 1 h, then incubating until metaphases appear, enhances the yield at shake-off [Miller et al., 1972].

Blockade

Two types of blocking have been used:

(1) DNA synthesis inhibition (S phase). The agents used to inhibit DNA synthesis include thymidine,

hydroxyurea, cytosine arabinoside, and aminopterin, [Stubblefield, 1968]. The effects of these agents are variable, because many are toxic, since blocking cells in cycle at phases other than G1 tends to lead to deterioration of the cells. Hence, the culture will contain nonviable cells, cells that have been blocked in the S phase but are viable, and cells that have escaped the block [Yoshida and Beppu, 1990].

(2) Nutritional deprivation (G1 phase). In these cases, serum [Chang and Baserga, 1977] or isoleucine [Ley and Tobey, 1970] is removed from the medium for 24 h and then restored, whereupon transit through the cycle is resumed in synchrony [Yoshida and Beppu, 1990].

A high degree of synchrony (e.g., >80%) is achieved only in the first cycle; by the second cycle, the degree of synchrony may be <60%, and by the third cycle, cell cycle distribution may be close to random. A chemical blockade is often toxic to the cells, and nutritional deprivation does not work well in many transformed cells. Physical fractionation techniques are probably most effective and do less harm to the cells. (See Chapter 14.)

CULTURE OF AMNIOCYTES

The human fetal karyotype can be determined by culturing amniotic fluid cells obtained by amniocentesis. Amniocentesis can now be performed from 11 weeks of gestation onward, although most amniocenteses are still performed from 15 to 18 weeks of gestation.

Inborn errors of metabolism and other sex-linked or autosomal recessive and dominant conditions are diagnosed mainly on placental tissues obtained by chorionic villus sampling and using direct techniques (i.e., using uncultured material). In addition, methods are now available, using commercial probes (Aneu-Vysion from Vysis), to detect the main trisomies and sex chromosome aneuploidy using uncultured amniocytes. Other approaches to detecting trisomies, using molecular polymorphisms, are also available. However, most prenatal diagnoses for Down's syndrome are still based on a complete chromosome analysis, and a full chromosome analysis from amniotic fluid requires the culture of cells.

Protocol 26.5, particularly the coverslip method, was originally submitted by Marie Ferguson-Smith, East Anglian Regional Genetics Service, Addenbrookes Hospital, Hills Road, Cambridge, CB2 2QQ, England, United Kingdom. It has since been updated, and the alternative closed-tube method has been described by Mike Griffiths, Regional Genetics Laboratory, Birmingham Women's Hospital, Edgbaston, Birmingham, B15, 2TG, United Kingdom.

PROTOCOL 26.5. CULTURE OF AMNIOCYTES

Principle

The principle of culturing amniotic fluid cells is based on separating the cells by centrifugation and setting up the cell suspension in a suitable culture vessel. A variety of approaches exist that use either closed or open systems to grow the cells in tubes, flasks, slide chambers, or on coverslips in Petri dishes. Good-quality, rapid results can be obtained with any of these techniques, but a closed-tube system has the advantage of robustness, simplicity, and minimal use of resources.

Closed-Tube Culture System:

Outline

Incubated cell cultures in a plastic Leighton tube in a standard incubator at 37°C, and harvest the cells using a trypsin suspension method.

Materials

Sterile or Aseptically Prepared:
Amniotic fluid sample, 10–15 ml
Complete medium: 100 ml of Ham's F10 medium with 20 mM Hepes buffer, 2 mM glutamine, 100 U/ml penicillin, and 100 μg/ml streptomycin with 10% fetal bovine serum or 2% Ultroser G (Gibco) PBSA
Colcemid, 10 μg/ml: Add 1 ml to 9 ml of sterile PBSA for a working solution, and then add 0.1 ml of this for each 1 ml of medium in the culture, to reach a final concentration of 0.1 μg/ml
Thymidine (Sigma): Make a stock solution of 15 mg/ml in PBSA, and filter sterilize it. Add 0.1 ml of the stock thymidine solution to 10 ml of medium. Change to use the thymidine medium before harvesting cells. The final concentration in the medium should be 0.15 mg/ml.
Bromodeoxyuridine, BUdR (Sigma): A vial contains 250 mg; dissolve all of it in 25 ml of sterile PBSA, and filter sterilize the solution. Add 0.5 ml of the solution to 11.5 ml sterile of PBSA and 8 ml of diluted colcemid solution. Dispense the resultant solution into aliquots, and store them frozen. Add 0.1 ml of the BUdR-colcemid mix to each 1 ml of culture. The final concentrations should be 25 μg/ml of BUdR and 0.04 μg/ml of colcemid.
Trypsin (Bacto, Difco): Reconstitute the contents of the vial with 10 ml of UPW, following manufacturer's instructions to give a 5% w/v solution.

Versene (EDTA), 0.5 mM in isotonic saline (Gibco)

Trypsin/versene (TV) solution for subculture and harvest: 2 ml of reconstituted Bacto trypsin in 100 ml of versene giving 0.1% w/v trypsin

Universal containers for collecting samples of amniotic fluid

Plastic, disposable, 10-ml pipettes, and a pipetting aid for setting up cultures

Leighton tubes (flat sided, plastic) and caps (Nunc)

Syringes, 20 ml; plastic and plastic mixing needles (Henley Medical) for feeding cultures

Transfer pipettes/pastettes, 1 ml and 3 ml, plastic, disposable

Histopaque-1077 (Sigma); adapt the manufacturer's instructions

Nonsterile:

Disinfectant (e.g., Virkon (Merck)); follow the manufacturer's instructions

Potassium chloride hypotonic solution, 0.075 M KCl in UPW; add 5.574 g of analar potassium chloride to 1 l of UPW

Fixative: Freshly mixed, 3 parts analar methanol to 1 part analar glacial acetic acid

Transfer pipettes/pastettes, 1 ml, plastic, disposable

Glass microscope slides, precleaned (Berliner Glas KG, from Skan, through Richardsons of Leicester)

Coplin jars

Hydrogen peroxide solution, 5% v/v: Dilute 1 part 30% hydrogen peroxide with 5 parts water

Saline solution, 0.9% w/v NaCl in UPW

Trypsin/saline solution for banding: Add 1.4 ml of reconstituted Bacto trypsin (Difco) to 50 ml of saline giving 0.14% w/v trypsin

Buffer, pH 6.8 (Gurr, Merck): Add one buffer tablet to 1 l of UPW

Leishman stain (Gurr, Merck), usually diluted 1 part to 4 parts pH-6.8 buffer

DPX slide mountant and coverslips

Equipment:

Oven at 60°C

Hot plate at 75°C

Bright-field and phase-contrast upright microscope, to assess slide making, banding, and for chromosome analysis

Protocol

△ *Safety Notes.*

(a) Wear gloves and a laboratory coat. All samples should be handled under sterile conditions in a class II microbiological safety cabinet. (Laminar-flow hood; see Human Biopsy Material in Chapter 6)

(b) Discarded media and hypotonic supernatants (but not fixative) should be poured off into a disinfectant solution—e.g., Virkon or hypochlorite. Fixative should be discarded into sodium bicarbonate solution to neutralize the acid. After standing 2 h both types of waste may be discarded into normal drainage, with plenty of running water.

(c) Bromodeoxyuridine (BUdR) is a known mutagen, and its use is optional. An appropriate local safety assessment (e.g., COSHH) should be undertaken before proceeding with any work involving this chemical. Solutions and supernatants containing significant amounts of the chemical may be discarded into sealable vessels containing vermiculite (or a similar absorbent material) and destroyed by incineration.

(d) Protocols involving the use of methanol and acetic acid for fixing and preparing slides should be carried out in a fume hood.

Setting up and Monitoring Tube Cultures:

1. Expect approximately 10–15 ml of amniotic fluid to be delivered to the laboratory in a plastic sterile universal container.

2. Use a sterile 20-ml syringe fitted with a sterile plastic mixing needle (or a 10-ml pipette) to divide the sample between three Leighton tubes. Label the tubes with patient and culture identification details. If the sample is small (a volume of 5 ml or less), set up only two cultures.

3. Centrifuge the tubes for 5 min at 150 g.

4. Pour the supernatants back into their original container (for biochemical assay or immunoassay—e.g., α-fetoprotein (AFP) estimation), or discard them into disinfectant solution (Virkon).

5. Resuspend the cell pellets in 1 ml of culture medium per Leighton tube.

6. Incubate the tubes at 37°C.

7. Leave the cultures undisturbed for 5–7 d, to allow the cells to settle and establish colonies.

8. Assess the cultures. Depending on the degree of cell growth, either add 0.5 ml of fresh medium, or remove and discard the old medium and add approximately 1 ml of fresh medium.

9. Thereafter, reassess cell growth as necessary (every 2–4 d), changing the medium as appropriate until there is sufficient growth for a harvest, usually 7 to 12 d after the cultures were initiated. Change the medium on the day prior to harvesting, if possible.

Notes.

(a) Rapid cell growth can be encouraged by spreading colonies using trypsin. (See Dispensing and Subculturing below). Overgrown cultures can be recovered by subculturing into several additional Leighton tubes or flasks if large quantities of cells are required.

(b) Bloodstained amniotic fluid samples may not grow as well as clear samples. One approach to working with such samples is to separate some of the amniocytes from the contaminating red blood cells, using density gradient centrifugation (e.g., Histopaque; see Protocol 11.10).

(c) Assessment of heavily bloodstained samples at 6 to 7 d is usually not possible without removing the bloodstained medium first. The medium can be collected into supplementary tubes, which may be discarded if the original tubes show growth, or can be incubated further if necessary.

(d) Supplementary tubes may be established at the first change of medium in any case when discarding the original suspension is undesirable.

Dispersing and Subculturing:

1. Prewarm approximately 1.5 ml of sterile TV solution for each tube to be processed to 37°C.
2. Remove the medium from the culture tube, and rinse it once with 1 ml of sterile TV solution.
3. To disperse the culture,
   (a) Add 0.2 ml of TV to the culture, and incubate the culture at 37°C for 1–2 min to detach the cells.
   (b) Check the cells on an inverted microscope. When the cells are in suspension, add 1 ml of culture medium, and return the tube to the incubator. (The serum in the medium will inactivate the TV solution.)
4. To subculture,
   (a) Add 0.5 ml of TV to the culture, and incubate it at 37°C for 1–2 min to detach the cells.
   (b) Check the culture on an inverted microscope. When the cells are in suspension, add 1–2 ml of culture medium, and divide the suspension between an appropriate number of subculture tubes. Two to eight subcultures may be seeded, depending on the initial cell density. It is usually worth varying the concentration of cells in each subculture tube, as doing so improves the chances of being able to harvest a tube at an optimal cell density.
   (c) Top up each subculture with fresh medium to a final volume of 1 ml, and then incubate.
5. Check that the cells have resettled the next day, and change the medium.

Routine Tube Harvesting:

Harvests can be routinely carried out on primary cultures so long as other cultures are available as a backup. If only one culture remains, subculture it

prior to harvesting the cells, or salvage it after the harvest.

1. When the cultures are ready to be harvested, add 0.1 ml of diluted colcemid to each culture.
2. Incubate the cultures for as long as is necessary to accumulate enough rounded-up mitotic cells. This is typically 2 to 3 h, but may be as little as half an hour for very active cultures, or more than 4 h for slow-growing cultures.
3. Remove the medium into Virkon solution, and drain the tube briefly onto a paper towel.
4. Add 2 ml of TV, and incubate the cultures at 37°C for 3 min to detach the cells.
5. When the cells are in suspension, add 7 ml of KCl hypotonic solution, and leave the cultures at 37°C for 5 min.
6. Centrifuge the tubes at 150 g for 5 min.
7. Carefully remove the supernatant by pouring it into Virkon solution. Drain the tube briefly onto a paper towel. Flick the tube gently to resuspend the cells in the small amount of remaining liquid.
8. Slowly fix the cells using fresh fixative:
   (a) Flick the tube, and add the first 1 to 2 ml of fixative drop by drop, continually agitating the cells to avoid cell clumping.
   (b) Add a further 3 to 4 ml of fixative.
9. Centrifuge the tubes at 150 g for 5 min.
10. Remove the supernatant fixative by pouring it into sodium bicarbonate solution.
11. Gently resuspend the cell pellet, and add 5 ml of fixative to it.
12. Change the fixative by repeating the centrifugation, pour-off, and refixation steps (steps 9–11).
13. Centrifuge the cells again, pour off the supernatant, resuspend the cell pellet in residual fixative, and make slides of the cells. (See later steps for slide preparation.)

Notes.

(a) Fixed-cell suspensions can be stored at −20°C at any of the fixed stages. Change the fix twice before making slides.

(b) Salvage harvests may be used as an alternative to subculture or when only a single culture remains. This method requires the addition of sterile colcemid and sterile TV in the initial stages of the harvest, followed by transfer of the cells to a separate centrifuge tube after the cells have detached, but before the addition of hypotonic solution. Fresh medium can be added to the original culture tube; the medium should then be changed the next day, after the cells have resettled, to remove residual traces of trypsin and colcemid.

(c) If a harvest produces an unacceptably low yield

of metaphases, then alternative strategies such as the following can be used:

(i) Subculture the cells into a flask that is supported at a slight angle to the horizontal while the cells settle. Tilting the flask in this manner ensures a leading edge of the cells that are always growing at an optimal rate across the whole width of the flask.

(ii) Expose the cells to a reduced concentration of colcemid overnight.

(d) Longer chromosomes for higher resolution analysis may be produced by using either thymidine synchronization or overnight exposure to colcemid in the presence of bromodeoxyuridine. However, both these approaches work best with cells in an exponential growth phase, requiring careful assessment of the growth.

Thymidine Synchronized Harvesting:

1. On the morning of the day before the cultures are to be harvested, change the medium in the cultures, using the thymidine-supplemented medium (to a final thymidine concentration of 0.15 mg/ml).

2. Early on the day of harvest, rinse out the thymidine, using prewarmed medium. Pour off the medium, and rinse the cells twice. Add 1 ml of medium (without thymidine), which releases the thymidine block. The time of release depends on the time you intend to harvest the culture.

3. Incubate the culture for 4 h then add 0.1 ml of diluted colcemid.

4. Incubate the culture for a further 2 h, and then harvest as for routine tube harvests. (Start with step 2 of Routine tube harvesting, earlier in this protocol.)

Bromodeoxyuridine Overnight Colcemid Harvests:

1. Change the medium in the cultures to be harvested during the morning of the day before the harvest.

2. In the afternoon of the same day, add 0.1 ml of BUdR–colcemid solution for each 1 ml of culture medium (final concentrations in culture: BUdR, 25 μg/ml; colcemid, 0.04 μg/ml).

3. The next morning (after about 20 h of exposure), pour off the BUdR–colcemid medium into a sealable container filled with vermiculite. Add 2 ml of prewarmed TV to the culture tube, and incubate the tube for 3 min to detach the cells.

4. Check that the cells are in suspension, and then add 7 ml of 0.075 M KCl hypotonic solution. Incubate the culture for 5 min.

5. Centrifuge the culture at 150 g for 5 min.

6. Pour off the hypotonic supernatant into the vermiculite container.

7. Gently resuspend the cell pellet, slowly fix the cells and continue the harvest, starting with step 6 of Routine tube harvesting, earlier in this protocol.

Slide Preparation:

1. Use cleaned slides. The slides may be purchased as precleaned slides; however, it may still be beneficial to add 1 drop of fresh fix to each slide and wipe the slide with a paper towel immediately prior to use. If it is possible to control the surrounding environment, make the slides at 20–25°C and a relative humidity of 40–50%.

2. Place one drop of cell suspension onto each slide by dropping the suspension from a height of 1–3 cm. Allow the drop to dry naturally.

3. Assess the quality of spreading using a phase-contrast microscope. If the spreading is acceptable, make the rest of the slides, adjusting the cell density by adding extra drops of fix to the suspension if necessary.

4. In some circumstances, the spreading may need to be improved, and the following suggestions may be helpful:

   (a) Breathe on the slide first, and then place one drop of cell suspension onto the slide from a height of 1–3 cm. Allow the slide to dry naturally.

   (b) Place one drop of cell suspension onto the slide, either with or without breathing first. Leave the slide for a few seconds, and then place a drop of fresh fix on top before the first drop has dried. Allow the slide to dry.

   (c) Place one drop of suspension onto the slide, either with or without breathing on the slide first. Leave the slide for several seconds, and then, as the drying surface becomes dimpled or Newton rings become visible, place a drop of fresh fix on top. Allow the slide to dry.

   (d) If the quality of spreading is still unacceptable, place the fixed cell suspension in the freezer in fix overnight, and remake the slides the next day.

G-Banding with Trypsin:

1. Pretreatment of the slide may be carried out prior to trypsin exposure, but, depending on the age of the slide, may not be essential. Hydrogen peroxide pretreatment works well with fresh slides. Concentrated Hanks' salt solution works well with older slides. Coplin jars are suitable staining vessels. Slides that are not pretreated may also be used.

2. Any of the following pretreatment methods may be used:

   (a) Immerse the fresh slides (after drying for 1 h) in 5% hydrogen peroxide for 20 s to 2 min.

(b) Incubate the fresh slides in an oven at 60°C for a few hours or overnight, to age the slides. Immerse the slides in 5× Hanks' BSS for 5 min.

(c) Immerse 1 d or older slides in pH-6.8 buffer for 60 s.

3. Rinse the slides well in pH-6.8 buffer.

4. Immerse the slides in 0.14% trypsin in saline solution for 3 s to 2 min.

5. Rinse the slides well in pH-6.8 buffer.

6. Stain the slides with Leishman stain, freshly diluted 1:4 with pH-6.8 buffer, for 4 min.

7. Wash the slides with tap water, and drain them to dry or blot them dry with care.

8. Assess banding under a bright-field microscope at 400× magnification. If the slides are underbanded, they may be destained in pH-6.8 buffer or methanol and the procedure repeated. If the slides are overbanded, start the procedure again with another slide and vary the exposure time to trypsin or change the pretreatment.

9. Leave the slides on a hot plate (75°C) for a few minutes to ensure that they are completely dry, and then mount them with a glass coverslip, using DPX.

Chromosome Painting: See Protocol 27.2.

Open-Coverslip Culture System:

Outline
Culture the cells on coverslips in a Petri dish in a 5%-CO_2-in-air, 95%-humidity incubator.

Materials
The materials for this procedure are mostly the same as for the closed-tube culture system, with the following variations:

Sterile:
Sodium-bicarbonate-buffered Hams F10 medium with L-glutamine, penicillin, streptomycin, fetal bovine serum, and Ultroser G, as for the closed-tube system
Plastic Petri dishes, 35 mm, with vented lids
Glass coverslips, 22 mm, dry heat sterilized
Forceps

Nonsterile:
Hypotonic solution, 0.05 M KCl in UPW
Trypsin, 0.8% in saline solution, for banding (0.4 ml of reconstituted Difco trypsin in 50 ml of saline)

Protocol
Setting Up and Cell Culture:

1. Centrifuge the sample in two universal containers at 150 g for 10 min. If the sample arrives in only one universal container, split the sample into a second universal container before centrifuging, to avoid the risk of losing a sample in the centrifuge, due to breakage.

2. Place sterile coverslips in 2 Petri dishes, and label the lids and dishes with the sample designation.

3. Remove the supernatant with a pipette, leaving 1 ml of fluid above the cell pellet.

4. Resuspend the cell pellet(s), and transfer the suspension to the Petri dishes.

5. Add 3 ml of medium to each dish, and incubate the cultures at 38°C in a 5%-CO_2-in-air atmosphere in a humid incubator.

6. After 5-7 d inspect the cultures, for cell growth using an inverted microscope.

7. Partially change the medium by removing approximately 2 ml of the old medium with a sterile pastette and repacing it with an equal volume of fresh medium.

8. Inspect the cultures and change the medium twice a week. Cells are ready to be harvested when a sufficient number of actively growing colonies of suitable size have developed, usually 10 to 14 d after the culture was initiated.

In situ Harvesting of Coverslips:

1. Transfer the coverslip to a new Petri dish (appropriately labeled) with medium up to 24 h before the intended harvest. The original dish should be kept as a reserve source of extra cells.

2. Add 0.3 ml of diluted colcemid to the Petri dish to give a final concentration of 0.1 µg/ml, and incubate the dish for 2-3 h.

3. Gently remove the medium with a pastette, and replace it with 3 ml of 0.05 M KCl hypotonic solution, prewarmed to 37°C. Incubate the culture for 10 min.

4. Remove the Petri dish from the incubator, and place it on a suitable tray, with the labeled lid under the dish.

5. Add 6 drops of fixative gently down the side of the Petri dish. Allow the fixative to disperse in the hypotonic solution. Leave the dish for 1 min.

6. Add another 6 drops of fixative to the Petri dish and leave for 1 min.

7. Remove 2 ml of supernatant from the Petri dish, and replace it with 2 ml of fresh fixative, adding the fixative gently down the side of the dish. Leave the dish for 2 min.

8. Remove all supernatant from the Petri dish, and replace it with 3 ml of fresh fixative, adding the fixative gently down the side of the dish. Leave the dish for 10 min.

9. Remove all of the fixative, and allow the coverslip to air dry within the dish. After removal of the fixative, the tray can be raised along one side to encourage the residual fixative to drain off the coverslips.

G-Banding with Trypsin:

1. Age the coverslip (still in the Petri dish for identification) in an oven at 60°C overnight or on a hot plate at 75°C for 2 h.
2. Make up 0.04% trypsin in saline solution, and, immediately prior to banding, mixt 1 part Leishman stain with 4 parts pH-6.8 buffer.
3. Using forceps, place the coverslip in the trypsin solution, and agitate it gently for a few seconds (the exact amount of time varies according to the age of the material).
4. Rinse the coverslip in buffer, and return it to the Petri dish.
5. Immediately flood the Petri dish with diluted Leishman stain for 2–4 min (depending on the batch of stain).
6. Rinse the coverslip in buffer, and stand it against the side of the Petri dish, on absorbent paper, to drain, and allow it to dry.
7. Coverslips may be examined for the quality of banding before being mounted.
8. Mount the coverslip on a labeled slide, using DPX.

CULTURE OF CELLS FROM POIKILOTHERMS

The approach to the culture of cells from cold-blooded animals (poikilotherms) has been similar to that employed for warm-blooded animals, largely because the bulk of present-day experience has been derived from culturing cells from birds and mammals. Thus, the dissociation techniques for primary culture employ proteolytic enzymes, such as trypsin with EDTA as a chelating agent. Fetal bovine serum appears to substitute well for homologous serum or hemolymph (and is more readily available), but modified media formulations may improve cell growth. A number of these media are available through commercial suppliers (see Sources of Materials—Media in Trade Index), and the procedure for using those media is much the same as for mammalian cells: Try those media and sera that are currently available, assaying for growth, plating efficiency, and specialized functions. (See Culture Testing in Chapter 10.) Since the development of media for many invertebrate cell lines is in its infancy, it may prove necessary to develop new formulations if an untried class of invertebrates or type of tissue is examined. Most of the accumulated experience so far relates to insects and molluscs.

Two reviews cover some of the early developments of the field [Vago, 1971, 1972; Maramorosch, 1976], and some recent exploitation in biotechnology is described in a publication of the European Society for Animal Cell Technology [Jain et al., 1991; Klöppinger et al., 1991; Speir et al., 1991]. The latter relates to the use of the baculovirus vector in insect cell lines, which has many of the advantages of post-translational modification found in mammalian cells, without the regulatory problems related to the isolation of biopharmaceuticals from human and mammalian cells.

Culture of vertebrate cells other than birds and mammals has also followed procedures for warm-blooded vertebrates, and so far there has not been a major divergence in technique. Since this is a developmental area, certain basic parameters will still need to be considered to render culture conditions optimal, and if a new species is being investigated, optimal conditions for growth may need to be established—e.g., pH, osmolality (which will vary from species to species), nutrients, and mineral concentration. Temperature may be less vital, but it should be fixed within the appropriate environmental range and regulated within ±0.5°C; overheating is particularly damaging.

Fish Cells

Fish cell culture has become increasingly popular, due to the growing commercial interest in fish farming. Protocols 26.6–26.8 for culturing cells from zebrafish, have been abridged from Collodi [1998].

The zebrafish possesses many favorable characteristics that make it a popular nonmammalian model for studies of vertebrate development and toxicology [Powers, 1989; Driever et al., 1994]. Zebrafish reach sexual maturity in approximately 3 months, and females produce 100 to 200 eggs each week throughout the year. Embryogenesis is completed outside of the mother in 3 to 4 d, and the large, transparent embryos are amenable to experimental manipulations involving cell labeling or ablation techniques [Westerfield, 1993]. *In vitro* approaches utilizing embryo cell cultures have also been employed for the study of zebrafish development. Cell lines and long-term primary cultures, initiated from blastula, gastrula, and late-stage embryos, have been established [Collodi et al., 1992; Ghosh and Collodi, 1994; Sun et al., 1995a,b; Peppelenbosch et al., 1995].

Cell Lines Derived from Early-Stage Embryos. Since differentiation occurs during zebrafish gastrulation, the cells in earlier stage embryos, such as the blastula, are pluripotent [Kane et al., 1992], and methods have been developed for the culture of cells from these early-stage embryos. The ZEM-2 cell line, initiated from mid-blastula-stage embryos, has been growing in culture for more than 300 generations in medium containing low concentrations of fetal bovine and trout sera, insulin, trout embryo extract, and medium conditioned by buffalo rat liver cells [Ghosh and Collodi, 1994].

A fibroblastic cell line, ZEF, has also been derived from early-stage embryos in medium supplemented with FBS and FGF. Once established, the line has been maintained in LDF medium (see Reagents and Media below) containing 10% FBS [Sun et al., 1995a], and ZEF cells have been utilized as a feeder layer for primary cultures of zebrafish embryo cells [Sun et al., 1995a, Bradford et al., 1994a].

Primary Cultures Derived from Early-Stage Embryos. In addition to the continuously growing embryo cell lines that are available, methods have been developed for the initiation of primary cultures derived from early zebrafish embryos [Sun et al., 1995b, Bradford et al., 1994a,b]. Primary cultures, derived from early gastrula-stage embryos, maintained a diploid chromosome number and exhibited a morphology characteristic of pluripotent ES cells when derived on a feeder layer of ZEF fibroblasts in medium containing FBS, trout serum, fish embryo extract, insulin, and leukemia inhibitory factor [Sun et al., 1995b; Bradford et al., 1994a].

Cell Lines Derived From Late-Stage Embryos. Late-stage zebrafish embryos (20–24 h postfertilization) have been used for the derivation of three fibroblastic cell lines: ZF29, ZF13 [Peppelenbosch et al., 1995], and ZF4 [Driever and Rangini, 1993]. The lines were derived in Leibowitz's L-15 (ZF29 and ZF13) or a mixture of Ham's F12 and Dulbecco's modified Eagle's media (ZF4) supplemented with FBS.

Reagents and Media for Protocols 26.6–26.8
Sterile:
LDF basal medium:
    Leibowitz's L-15 medium, 100 ml
    DMEM, 70 ml
    Ham's F-12, 30 ml
    Sodium selenite 6 μM, 200 μl
    Store refrigerated at 4°C.
LDF primary medium: LDF basal medium, plus
    FBS, 1%
    Trout serum, 0.5%
    Trout embryo extract, 40 μg of protein/ml
    Insulin, 10 μg/ml
    Leukemia inhibitory factor, human, recombinant, 10 ng/ml
LDF maintenance medium: LDF basal medium, plus
    FBS, 1%
    Trout serum (Sea Grow, East Coast Biologicals), 0.5%
    Trout embryo extract, 40 μg of protein/ml
    Insulin, 10 μg/ml

BRL-conditioned medium, 50%
D medium: DMEM/F12/10FB: 50/50 DMEM/Ham's F12 with 10% fetal bovine serum (FBS)
Trout embryo extract:
(a) Collect the embryos (Shasta Rainbow or other strains of trout, 28 d postfertilization, reared at 10°C; or Zebrafish, three days postfertilization, reared at 28°C), and store them frozen at −80°C.
(b) To prepare the extract, thaw the embryos (approximately 150 g) and homogenize them in 10 ml of LDF for 2 min on ice, using a Tissuemizer homogenizer (Tekmar).
(c) Pass the homogenate through several layers of cheesecloth to remove the chorions, and then centrifuge the homogenate at (20,000 g for 30 min at 4°C).
(d) After centrifugation, collect the supernatant, leaving behind the bright-orange lipid layer present on the surface.
(e) Transfer the supernatant to a new tube, and centrifuge it as before (in step (c)).
(f) Collect the supernatant, leaving behind any remaining lipid, and then ultracentrifuge it at 100,000 g for 60 min at 4°C.
(g) After ultracentrifugation, collect the supernatant, leaving behind the lipid layer. Dilute the supernatant with LDF (1:10), and filter sterilize.
(h) The extract must be passed through a series of filters (1.2 μm, 0.45 μm, and 0.2 μm).
(i) Store the extract frozen at −80°C in 0.5-ml aliquots.
(j) To use the extract for cell culture, measure the concentration of protein (see Protocol 20.4) and then dilute the extract with LDF to the desired working concentration.
(k) Store the diluted extract refrigerated at 4°C for a maximum of two months.
BRL cell-conditioned medium:
(a) Culture BRL cells (ATCC) at 37°C in 75-cm² flasks in DMEM/F12/10FB.
(b) When the cultures become confluent, replace the FD medium with LDF supplemented with 2% FBS, and incubate the cells at 37°C.
(c) After 5 d, remove the LDF, filter it, and store it frozen at −20°C.
(d) Add fresh LDF to the BRL cultures, and repeat the process 5 d later. Conditioned LDF medium can be collected up to three times from the same flask before the cells must be split and allowed to grow again to confluence.
Holtfreter's buffer:
    NaCl, 70 g
    KCl, 1.0 g

NaHCO$_3$, 4.0 g
CaCl$_2$, 2.0 g
UPW, to 1000 ml
Store at 4°C. Prepare a working solution by diluting 1:20 with UPW.
Pronase E, 0.5 mg/ml in Holtfreter's buffer
Trypsin, 1%, EDTA, 1 mM, in PBSA

PROTOCOL 26.6. FIBROBLAST FEEDER LAYERS FROM ZEBRAFISH EMBRYOS

Outline
Prepare feeder layers of embryonic fibroblasts from gastrula-stage zebrafish [Sun et al., 1995a] by removing the chorion in pronase, culturing the cells in FGF-supplemented medium, and selecting the fibroblasts by differential trypsinization.

Materials (See above for preparation of reagents)
Sterile:
Embryos, eight hours postfertilization
Pronase
Trypsin
FBS
LDF primary medium with 10 ng/ml bovine FGF (*a* + *b* mixture)
LDF basal medium with 10% FBS
Flasks, 25 cm²

Protocol
1. Collect approximately 30 embryos (eight hours postfertilization).
2. Remove the chorion by pronase treatment [Sun et al., 1995a]
3. Incubate the embryos in trypsin (1 min) while gently pipetting to dissociate the cells.
4. Add FBS (10% final concentration) to stop the action of the trypsin.
5. Collect the cells into a pellet by centrifugation (at 500 *g* for 10 min).
6. Resuspend the cell pellet in 5 ml of LDF primary medium containing FGF, and transfer the cells to a 25-cm² flask.
7. Allow the cells to attach and grow to confluency at 26°C.
8. When the cells are confluent, passage the culture in the same medium.
9. After 2–3 passages, a mixed population of epithelial and fibroblastic cells will be present in the culture, and the cells can be maintained in LDF basal with 10% FBS. Select the fibroblasts for further culture, by differential trypsinization:
    (a) Treat the culture with trypsin for 1 min to remove most of the fibroblasts, and leave the epithelial cells attached to the plastic.
    (b) Transfer the fibroblasts to another flask, and repeat this process when the culture becomes confluent.
    (c) After two or three passages, the culture will consist of a homogeneous population of fibroblasts.
10. To prepare feeder layers of growth-arrested fibroblasts,
    (a) Add the cells to the appropriate culture dish or flask, and allow the fibroblasts to grow into a confluent monolayer.
    (b) Add mitomycin C, 10 μg/ml, to the cultures, and incubate the cultures for 3 h, at 26°C.
    (c) After being rinsed three times with LDF, the growth-arrested fibroblasts can be used as feeder layers for zebrafish embryo cell cultures.

PROTOCOL 26.7. PRIMARY CULTURES FROM ZEBRAFISH EMBRYOS

Outline
Collect embryos, remove the chorion, disaggregate the embryos in trypsin, and grow primary cultures derived from zebrafish blastula- and early gastrula-stage embryos on feeder layers of embryonic fibroblasts.

Materials (See Reagents and Media, above
Sterile:
LDF primary medium
Holtfreter's buffer
Dilute bleach, 0.1% in UPW
Pronase E solution
Feeder layers of embryonic fibroblasts
Human recombinant leukemia inhibitory factor (LIF),1 μg/ml

Protocol
1. Harvest embryos at the midblastula or early gastrula stage, and rinse them several times with clean water.
2. After rinsing, transfer the embryos into 60-mm Petri dishes (50–100 embryos/dish), take them to a laminar flow hood, and maintain them under aseptic conditions.
3. Soak the embryos for 2 min in dilute bleach, and rinse them several times in sterile Holtfreter's buffer.
4. Dechorionate the embryos by incubating them in 2 ml of Pronase E solution for ~10 min, and then gently swirl the embryos in the Petri dish

to separate them from the partially digested chorion.

5. Tilt the dish to collect the embryos on one side, and gently remove 1.5 ml of the Pronase solution with a Pasteur pipette. To prevent the dechorionated embryos from adhering to the dish and rupturing, keep the dish tilted so that the embryos remain suspended in the remaining Pronase solution.

6. Gently rinse the embryos by adding 2 ml of Holtfreter's buffer and swirling gently.

7. Tilt the dish and remove most of the Holtfreter's buffer, leaving the embryos suspended in ~0.5 ml.

8. Repeat the rinse procedure two more times.

9. After the final rinse, leave the embryos suspended in 0.5 ml of Holtfreter's buffer, and add 2 ml of trypsin solution.

10. Incubate the embryos in the trypsin for 1 min and then dissociate the cells by gently pipetting 3–4 times.

11. Immediately transfer the cell suspension into a sterile polypropylene centrifuge tube, and add to the tube 200 μl of FBS to stop the trypsin.

12. Collect the cells by centrifugation (at 500 g, 5 min) and resuspend the pellet in LDF primary medium (without FBS or trout serum).

13. Seed the cells at 1×10^4 cells/cm² onto feeder layers of growth-arrested embryonic fibroblasts, contained in multiwell dishes or flasks.

14. Allow the cells to attach to the feeder layers (~15 min) before adding FBS and trout serum. Human recombinant leukemia inhibitory factor (10 ng/ml) is used in the medium in preference to BRL-conditioned medium [Sun et al., 1995a,b].

PROTOCOL 26.8. CELL LINES FROM ZEBRAFISH EMBRYOS

Materials (See Reagents and Media, above)
Sterile or Aseptically Prepared:
ZEM-2 cells (or equivalent)
LDF maintenance medium

Protocol
1. Grow cultures derived from early zebrafish embryos, such as ZEM-2, in LDF maintenance medium to ~70% confluency.
2. Incubate the cultures at 26°C in ambient air.
3. Change the medium approximately every 5 d.
4. Subculture by trypsinization. (See Protocol 12.2.)

Trout serum and embryo extract have also been shown to stimulate the growth of embryo cells from other fish species [Collodi and Barnes, 1990].

Insect Cells
There has been considerable interest for some time in the culture of insect cells for studies of pest control and environmental toxicology. However, the greatest increase in the usage of insect cell culture has resulted from the use of baculovirus for gene cloning [Midgley et al., 1998]. Baculovirus is often grown in Sf9 cells, a continuous cell line from the Fall army worm *Spodoptera fungiperda*. Protocol 26.9, for the culture of Sf9 cells, has been abridged from Midgley et al. [1998]. In this method, Sf9 cells are kept growing continuously in a magnetic spinner culture flask (see Protocol 12.3) or a flat-bottom flask with a magnetic stirrer bar mixing at about 80 rpm, ensuring that the stirrer is not a source of heat. Ideally, the cells should be maintained at 27°C, but it is possible to grow them without an incubator in a room with constant temperature between 20–28°C. CO_2 is not required for these media. Cells can be maintained in standard plastic tissue culture flasks, but, since the cells attach to the surface of the flask, they must be detached for subculture by scraping or dislodging the cells with a jet of medium. However, this method will result in a lot of cell death, since the cells attach quite tightly to plastic when grown in the presence of serum. The cells should have a population-doubling time of <24 h.

PROTOCOL 26.9. PROPAGATION OF INSECT CELLS

Outline
Disperse the cells mechanically from the monolayer, and propagate them in suspension at 27°C.

Materials
Sterile or Aseptically Prepared:
Cells: Sf9 [Smith et al., 1983] (ATCC #CRL-1711) or lines derived from the cabbage looper *Trichoplusia ni* (Tn368, or BTI-TN-5B1-4; also known as "High Five," available from Invitrogen)
Growth medium: EX-CELL 400 (JRH Biosciences) containing 2 mM L-glutamine, supplemented with 5% FBS, and 5 ml of penicillin/streptomycin solution (penicillin, 50 U/ml; streptomycin, 50 μg/ml). Store the medium at 4°C in the dark, and always warm it to room temperature before use. Sf9 cells are very sensitive to changes in growth medium, so for any change (e.g., to use serum-free EX-CELL 400), acclimatize the cells by gradually adding the new medium over a number of days.

Dimethylsulphoxide (DMSO),10% in FBS
Pluronic F68
Culture flasks
Spinner flask and magnetic stirrer
Incubator at 27°C (CO$_2$ not required)

Protocol

Routine Maintenance:

1. Detach the cells from the flask culture by scraping or dispersing them with a jet of medium, or use cells grown in suspension in a spinner flask.
2. Count the cells by hemocytometer, and determine their viability by dye exclusion with trypan blue or naphthalene black.
3. Seed the spinner flask at 0.5–1 × 10^6 viable cells per ml (20–100 ml in a 500-ml spinner flask).
4. Incubate the cells at 20–28°C. (27°C is optimal.)
5. Dilute the cells to 0.5–1 × 10^6 cells per ml every 48–72 h, or when there are about 4–5 × 10^6 cells/ml.
6. Transfer the cells to a clean flask every 3–4 weeks.
7. If the cells clump, try stirring them slightly faster, and add the surfactant Pluronic F-68 (0.5–1.0% v/v) to reduce shearing.

Freezing Cells for Storage:

1. Count the cells, and centrifuge at 1000 rpm (~200 g) for 5 min.
2. Resuspend the pellet at 1 × 10^7 cells/ml in 10% DMSO in FBS.
3. Dispense the cells into aliquots, place the aliquots into ampules, and chill the ampules on ice for an hour.
4. Pack the tubes into a styrofoam container.
5. Freeze the tubes slowly (~1°C/min) overnight at −70°C.
6. Transfer the tubes to a liquid-nitrogen freezer.
7. To recover the frozen cells, thaw the ampules rapidly at 37°C. (If the cells are stored in the liquid phase, take care to thaw them in a covered vessel, to avoid risk of injury from explosion of the ampule.)
8. Transfer the cells into a 25-cm^2 flask containing 5 ml of medium. Tip the flask to spread out the cells evenly.
9. Remove the medium after 2–3 h, when most of the cells should have attached, and add 5 ml of fresh medium.
10. Leave the cells 2–3 d to recover before detaching, and then transferring them to a stirrer flask or a larger plastic flask as described previously.

Molecular Techniques

MOLECULAR BIOLOGY IN CELL CULTURE

A wide range of molecular techniques is used in association with cultured cells, both in analytical and preparative procedures at the laboratory scale, and at the developmental and production level in the biotechnology industry. Although the focus of this book is on basic procedures in tissue culture, it is appropriate to present some of the molecular techniques that are used most frequently. However, there is neither space nor justification to present basic molecular techniques, such as DNA purification, molecular cloning, and hybridization, so the techniques presented throughout this chapter are those which impinge most directly on current usage of cell culture. It is assumed that those who use them will have a knowledge of basic molecular methods, or, if not, will have access to descriptions of these methods in one of the number of basic molecular texts that are available [e.g., Sambrook et al., 1989; Ausubel et al., 1996].

IN SITU MOLECULAR HYBRIDIZATION

Nucleic acid hybridization is used routinely for the detection of specific nucleotide sequences in DNA (Southern blotting) or RNA (Northern blotting). This technique can be applied as an *in situ* cytological technique with fixed cells to detect nucleotide sequences *in situ*. (See Plate 37.) Protocol 27.1, for *in situ* hybridization, has been contributed by W. Nicol Keith, CRC Department of Medical Oncology, University of Glas-

gow, CRC Beatson Laboratories, Garscube Estate, Switchback Rd., Glasgow G61 1BD, United Kingdom.

Analysis of RNA Gene Expression by *in Situ* Hybridization

Molecular techniques can be roughly broken down into two groups: lysate analysis and *in situ* analysis. With lysate methods (Southern blot analysis and PCR), tissue biopsies are homogenized and the spatial relationships between the cells of the tissue are destroyed [Murphy et al., 1995]. This process leads to a loss of information on heterogeneity and small subpopulations, particularly in tumor biopsies, and presents an averaging of changes. However, quantitation can be simpler and more accurate than *in situ* approaches. In comparison, *in situ* techniques, such as RNA *in situ* hybridization, allow the visualization of gene expression in individual cells within their histological context [Soder et al., 1997, 1998].

PROTOCOL 27.1. AUTORADIOGRAPHIC *IN SITU* HYBRIDIZATION

Principle

The principle of *in situ* hybridization is based on the specific binding of a labeled nucleic acid probe to a complementary sequence in a tissue sample, followed by visualization of the probe. This process enables both detection and localization of the target sequence. A number of prerequisites for the success of this procedure include retention of the nucleic acid sequences in the sample and accessibility of these sequences to the probe.

Specimens suitable for *in situ* hybridization (ISH) include cells from culture and tissue from samples of whole or biopsied organs.

Outline

Hybridize radiolabeled probes to fixed cells on microscope slides. Then visualize sites of hybridization by autoradiography.

Materials

Microcentrifuge tubes (e.g., Eppendorf)
Microcentrifuge (Eppendorf)
Linearization of Plasmid DNA:
RNA labeling kit (Amersham, RPN 3100)
Diethylpyrocarbonate (DEPC; Sigma)
DEPC-water: 1% DEPC in distilled water (dH$_2$O), autoclave
Phenol-chloroform iso-amyl alcohol, pH 8 (Sigma)
NaAc, 3 M, pH 8.0
Absolute alcohol, analytical grade
Agarose, 1%, electrophoresis grade (Gibco)
Glycogen, 20 mg/ml (Boehringer Mannheim)
Probe-Labeling Reagents and Solutions:
Dithiothreitol (DTT), 0.2 M (Sigma)
DTT, 50 mM: divided into aliquots and stored at −20°C
[^{35}S]-UTP (Amersham SJ 603)
G50 Sephadex columns (Pharmacia Biotech)
Column buffer: 0.3 M Na Acetate, 1 mM EDTA, 1% SDS, autoclaved
Phenol, pH 5.0 (Sigma)
Chloroform iso-amyl alcohol (Biogene)
Scintillation fluid: Ecoscint A (National Diagnostics)
In Situ Hybridization Reagents and Solutions:
Histoclear (Fisher)
NaCl-DEPC: 0.85% NaCl, 1% diethyl pyrocarbonate (DEPC), autoclaved
PBSA, 1% DEPC, autoclaved
EDTA, 0.5 M: Dissolve in 1% DEPC in dH$_2$O; pH 7.5, autoclaved
Proteinase K Buffer: 1 M Tris HCl, 0.5 M EDTA, 1% DEPC, pH 7.5, autoclaved
Proteinase K stock solution (Sigma): 20 mg/ml in DEPC H$_2$O; divide into aliquots and store at −20°C
Formalin
Triethanolamine, 0.1 M, with 1% DEPC, autoclaved
Acetic anhydride (Sigma)
DTT, 1 M: divided into aliquots and stored at −20°C
Hybridmix, 60%:
    Formamide (Fluka), 6 ml
    Dextran sulphate, 50%, in 1% DEPC, 2 ml

SSC, 20×, 1 ml
Tris HCl, 1 M, 100 μl
Denhardts solution, 50×, 200 μl
SDS, 10%, 100 μl
tRNA (10 mg/ml; Sigma), 400 μl
Salmon DNA (10 mg/ml; Sigma), 200 μl
Store at −20°C in 400-μl aliquots.
Washing Reagents and Solutions:
SSC, 20× (see Reagents Appendix; also from Gibco); dilute to 5× SSC, 2× SSC, and 0.1× SSC
Formamide, 50% in 2× SSC
β-mercaptoethanol (Sigma)
RNase buffer: 0.5 M NaCl, 0.5 M EDTA, 1 M Tris, pH 7.5
RNaseA stock solution (Sigma): 10 mg/ml in DEPC H$_2$O; store at −20°C in 400-μl aliquots
Gelatin: 0.2 g in 200 ml of dH$_2$O; microwave for 2 min, cool, and filter
Autoradiography Reagents and Solutions:
See Protocol 26.3.

Protocol

Handling RNA:
All solutions involved in the preparation of the probe and up to the posthybridization wash steps must be free from RNase. Solutions should be treated with DEPC and autoclaved for 20 min at 121°C. This procedure removes the majority of RNases, but, due to the ubiquitous nature of RNases, it is not a substitute for care in handling the solutions, glassware, and pipettes. A set of pipettes dedicated for use only with RNA is worthwhile, and regular treatment of the pipettes with DEPC–water overnight or with a proprietary anti-RNase solution, such as RNase-Zap (Ambion), may be useful as well. All glassware should be wrapped in aluminium foil and autoclaved prior to use. Plastic Eppendorfs may be treated with DEPC–water or RNase-Zap prior to autoclaving.

Probe Preparation:
RNA probes (riboprobes)
Preparing RNA probes requires use of a DNA template of the target sequence, and generation of sense and antisense RNA probes, with radioactive nucleotides incorporated, is possible.

Single-stranded RNA probes are ideal if high sensitivity is required; probes of 200–1000 base pairs (bp) have been used, but probes of 150–200 bp are probably optimal, as tissue penetration can become reduced with longer probe size. Limited alkaline hydrolysis can be used to reduce probe size as required. The RNA–RNA or RNA–DNA hybrids are more stable than their oligonucleotide or DNA counterparts, rendering them the most popular probes.

Commercial probes and control probes

Commercially available (Ambion) DNA templates for Actin and GAPDH can be used to generate RNA probes for use as positive controls, since they are housekeeping genes and are ubiquitously expressed. Sense probes are commonly used as negative controls and are superior to the omission of a probe as a control. Well-characterized tumor samples with a range of RNA expression can be used as positive specimens during each run of slides. Tumor samples of a variety of tissues are also available commercially.

Linearization of Plasmid DNA:

1. Use 10–20 μg of DNA in a microcentrifuge tube.
2. Add 10 units of restriction enzyme per μg of DNA, and set up digestion as recommended by suppliers of the enzyme.
3. Leave the reaction at 37°C for 3 h or overnight.

Phenol Chloroform Extraction:

4. Add 400 μl of phenol-chloroform-isoamyl alcohol (pH 8.0), and vortex the mixture.
5. Spin the mixture for 3 min at 13,000 rpm at room temperature.
6. Collect the supernatant, transfer it to a fresh tube, and add 10 μl of 3 M NaAc (pH 8) to it.
7. Add 250 μl of 100% ethanol (stored at −20°C).
8. Add 1 μl of glycogen to help precipitate the DNA.
9. Place the tube on dry ice for 1 h.
10. Spin for 15 min at 13,000 rpm.
11. Discard the supernatant, and keep the pellet.
12. Add 400 μl of 70% ethanol (stored at −20°C) to wash the pellet.
13. Spin for 10 min at 13,000 rpm.
14. Remove the 70% ethanol, and air dry the pellet.
15. Resuspend the pellet in 10–20 μl of DEPC H$_2$O, depending on the amount of DNA used (see step 1), aiming for a final concentration of 1 μg/μl.
16. Run 0.5 μl of this suspension on a 1% agarose gel.

RNA Labeling:

This part of the procedure involves the incorporation of radioactive nucleotides. Use the Amersham kit (RPN3100) according to the pack insert, with reference to the following method:

1. Combine the following in a microcentrifuge tube:
   (a) 4 μl of 5× transcription buffer,
   (b) 1 μl of 0.2 M DTT,
   (c) 1 μl of HPR1,
   (d) 0.5 μl of ATP, CTP, and GTP,
   (e) 1 μl of linearized DNA template (1 μg/ml),
   (f) 9.5 μl of [^{35}S]-UTP, and
   (g) 2 μl of RNA polymerase.

2. Mix the components, and place the solution at 37°C for 1.5 hours.

DNase Extraction of the DNA Template:

3. Add 10 U of DNase I.
4. Add 1 μl of RNase inhibitor.
5. Mix the solution, and place it at 37°C for 10 min.

Removal of Unincorporated Nucleotides:

6. Equilibrate a G50 Sephadex column with 2 ml of column buffer.
7. Add the probe to the column.
8. Add 400 μl of column buffer to the column, and allow the buffer to run through.
9. Add a further 400 μl of column buffer and collect the eluate in an Eppendorf tube.

Phenol-Chloroform Extraction:

10. Add 400 μl of phenol (pH 5.0) to the tube, vortex, and spin for 3 min at 13,000 rpm.
11. Collect the supernatant, transfer it to a fresh microtube and add 400 μl chloroform-isoamyl alcohol to it. Vortex, and spin for 3 min at 13,000 rpm.
12. Collect the supernatant and remove 1 μl of it for counting the incorporation.
13. Add 2.5 vol. of 100% ethanol (stored at −20°C) to the remaining supernatant.
14. Add 1 μl of yeast glycogen to facilitate precipitation.
15. Place the tube on dry ice for 30 min.
16. Spin for 15 min at 13,000 rpm.
17. Remove the alcohol and leave the pellet undisturbed.
18. Wash the pellet with 70% ethanol, spin at 13,000 rpm for 10 min, and remove the ethanol.
19. Air dry the pellet and resuspend it in 50 mM DTT, calculating the volume of the 50 mM DTT as follows:
    (a) Add the 1 μl of supernatant from step 12 to 2–3 ml of scintillation fluid and determine counts per minute (CPM) on scintillation counter.
    (b) Volume DTT = $\dfrac{\text{CPM} \times 400}{6 \times 10^5}$
    where 400 is the volume after phenol/chloroform extraction and 6×10^5 is the required total CPM in the DTT solution.
    (c) This is the volume of 50 mM DTT that the probe should be re-suspended in.
20. Remove 1 μl and count again to confirm the activity of the probe.

In Situ Hybridization:

Preparation of the specimen:

The objective of this part of the procedure is to preserve the architecture and morphology of the tissue and to retain the RNA products. Rapid processing of the tissue sample, either by freezing or fixing in formalin, enables the RNA to be preserved. Cross-linking fixatives, such as 4% paraformaldehyde and 4% formaldehyde, are the fixatives of choice for the retention and/or accessibility of cellular RNA. The length of fixation will depend on the size of the specimen. Longer fixation times result in better tissue morphology, but may reduce access to the probe. Paraffin wax is the embedding medium of choice. It allows sectioning down to 1 μm in thickness and is easily removed prior to hybridization. As the sections will be processed through a number of solutions, coated slides are recommended to hold the specimen on the slide. Frozen samples should be chilled to −70°C, and, following cryosectioning, they should be placed on a coated slide and fixed.

Tissue preparation prehybridization:

This treatment of the tissue prior to hybridization attempts to increase the access of the probe to the target RNA sequence and to reduce nonspecific background binding. The specimen is subjected to protease treatment to increase the accessibility of the target nucleic acid to the probe, especially if the probe is greater than 100 bp. It is important to post-fix the specimen in formaldehyde, to prevent disintegration of the tissue. Nonspecific binding to amino groups is reduced by acetylation with acetic anhydride. During tissue preparation, great care must be taken to protect the specimen from RNase. All glassware must be treated to remove any contamination, all solutions must be treated with DEPC, and gloves must be worn throughout the procedure. Handling of the tissue sections should be kept to a minimum.

Hybridization:

The hybridization temperature can be critical for some probe/target sequences. Formamide in the hybridization buffer, as a helix destabilizer, reduces the melting point of the hybrids and enables reduction of the temperature of hybridization. The lower temperature helps to preserve tissue architecture. A temperature of 52°C has been found to be the optimal. Dextran sulphate in the hybridization buffer, by volume exclusion, increases the concentration of the probe and reduces hybridization times. Although the hybridization reaction is almost complete after 5–6 h, it is convenient to leave the reaction overnight. The sodium ion concentration in the buffer serves to stabilize the hybrids.

Posthybridization washing:

The main objective of posthybridization washing is to remove unbound and nonspecifically bound probes by selection of the temperature, salt concentration, and formamide concentration. The use of

RNase enables the digestion of single-stranded RNA, unbound to the target, but does not affect the bound RNA–RNA complexes.

Pretreatment of Paraffin Sections:

1. Dewax the paraffin sections with Histoclear, twice for 10 min.
2. Rehydrate through an ethanol series: 100%, 90%, 70%, 50%, and 30%, for 10 s each.
3. Rinse in 0.85% NaCl and PBSA solutions for 5 min each
4. Digest the section in Proteinase K, 400 μl of Proteinase K stock in 200 ml of Proteinase K buffer for 7.5 min
5. Rinse in PBSA; 3 min.
6. Postfix in 4% formalin or 4% paraformaldehyde.
7. Rinse in DEPC-water for 1 min.
8. Acetylate in 200 ml of 0.1 M triethanolamine with 500 μl of acetic anhydride for 10 min, stirring throughout in a fume hood.
9. Rinse in PBSA and 0.85% NaCl, for 5 min each.
10. Dehydrate through the ethanol series: 30%, 50%, 70%, 90%, and 100%, for 10 s each.
11. Air dry the section.

Preparation of the Probe and Hybridization:

12. For 20 paraffin sections, combine 16 μl of 1 M DTT, 344 μl of 60% Hybridmix, and 40 μl of the probe. Vortex and spin the solution briefly.
13. Denature the probe at 80°C for 3 min. Cool it on ice.
14. Apply 20 μl of the probe mix from step 12 to each tissue section, and cover it with a glass coverslip.
15. Hybridize the solution at 52°C overnight in a humidified chamber.

Posthybridization Wash:

16. Preheat the solutions to the required temperature.
17. Wash the sections in 200 ml of 5× SSC with 250 μl of β-mercaptoethanol for 30 min at 50°C.
18. Wash in 200 ml of 50% formamide and 2× SSC with 1.4 ml of β-mercaptoethanol for 20 min at 65°C.
19. Wash in 200 ml of RNase buffer twice for 10 min each time at 37°C.
20. Wash in 200 ml of RNase buffer with 400 μl of RNase A solution for 30 min at 37°C.
21. Repeat step 19 but for 15 min each wash.
22. Repeat step 18.
23. Wash in 200 ml 5× SSC and 200 ml of 0.1× SCC, for 15 min each at 50°C.
24. Dehydrate in an ethanol series: 50%, 70%, and 100% for 1 min each.
25. Air dry the sections.

26. Dip the slides in the gelatin solution for 1 min, and then air dry them.

Autoradiography:

See Setting up autoradiographs in Protocol 26.3.

Analysis

(1) Examine the sections using light microscopy under bright- and dark-field illumination.

(2) Score the sections with reference to positive and negative controls.

Fluorescence *in Situ* Hybridization in the Analysis of Genes and Chromosomes

Protocol 27.2, for fluorescence *in situ* hybridization (FISH), has been provided by Nicol Keith, CRC Department of Medical Oncology, University of Glasgow, Garscube Estate, Bearsden, Glasgow G61 1BD, United Kingdom.

In situ hybridization is the most direct way of determining the linear order of genes on chromosomes. By using chromosome- and gene-specific probes, numerical and structural aberrations can also be analyzed within individual cells. These techniques have a wide variety of applications in the diagnosis of genetic disease and the identification of gene deletions, translocations, and amplification during cancer development [Bar-Am et al., 1992; Kallioniemi et al., 1992; Matsumura et al., 1992; Ried et al., 1997; Telenius et al., 1992].

PROTOCOL 27.2. FISH USING SINGLE-COPY GENOMIC PROBES AND CHROMOSOME PAINTING

Principle

Nucleic acid probes are labeled nonisotopically by the incorporation of nucleotides modified with molecules such as biotin or digoxigenin. After hybridization of the labeled probes to the chromosomes, detection of the hybridized sequences is achieved by forming antibody complexes that recognize the biotin or digoxigenin within the probe. The hybridization is visualized by using antibodies conjugated to fluorochromes. The fluorescent signal can be detected in a number of ways. If the signal is strong enough, standard fluorescence microscopy can be used. However, data analysis and storage can be improved considerably by the use of digital imaging systems such as confocal laser scanning microscopy or cooled CCD camera. The major advantages of fluorescence *in situ* hybridization (FISH) are that it is nonisotopic, rapid, good for data storage and manipulation, and sensitive. It also shows accurate sig-

nal localization, allows simultaneous analysis of two or more fluorochromes, and provides a quantitative and spatial distribution of the signal.

Outline

Hybridize biotinylated or digoxigenin-labeled probes to denatured chromosomes and detect the probes by double-antibody fluorescent staining.

Materials

SSC, 2× (see Reagents Appendix for SSC, 20×)
SSC, 1×
PBSA
Glycogen, 20 mg/ml (Boehringer Mannheim)
EtOH, 70%
EtOH, 100%
Fixative: Methanol:acetic acid, 3:1
RNase, 100 μg/ml in 2× SSC
Paraformaldehyde, 1%, in PBSA
Formamide, 70%, in 2× SSC
Hybridization buffer: formamide (50%), dextran sulfate (5%) 2× SSC, salmon sperm DNA (500 μg/ml)
Labeled probe:
Large cosmid clones are most suitable for probes to detect single-copy genes. However, cDNA probes can be used as well. DNA is labeled by nick translation, using commercial kits. The nick translation kit marketed by Boehringer Mannheim can be used to incorporate either biotin or digoxigenin; follow the manufacturer's instructions.
Precipitation of a probe containing repetitive sequences:
Large cosmid probes often contain repetitive sequences that, if not suppressed prior to hybridization, result in high levels of nonspecific hybridization. The repeat sequences can be suppressed by competition with unlabeled human Cot1 DNA sequences that are enriched for repeat sequences. The Cot1 DNA can be included at the precipitation step (step (b) in the following procedure).

(a) For a 20-μl nick translation reaction, add to the probe 1 μl of 0.5 M EDTA, 2.5 μl of 4.0 M LiCl, 1 μl of glycogen, 100- to 1,000-fold excess human Cot1 DNA (Life Technologies), and 100 μl of ethanol.

(b) Place on dry ice for 30 min or at −20°C overnight to allow precipitate to form.

(c) Spin in the microfuge for 15 min to pellet the DNA.

(d) Wash the pellet in 70% ethanol.

(e) Spin to repellet DNA and dry the pellet.

(f) Resuspend DNA at 2–10 ng/μl in hybridization buffer.

Precipitation of probes without repetitive sequences:

Follow the foregoing protocol, but leave out Cot1 DNA from the precipitation.

Probe denaturation:

(a) For probes containing repetitive sequences that need to be suppressed using Cot1 DNA, heat the probe mix to 70°C for 10 min. Place the probe mix at 37°C for 1 h prior to application to a slide.

(b) For probes without repetitive sequences, heat the probe mix to 70°C for 10 min. Chill the probe mix on ice for 10 min.

Probe detection buffer, TBST: 0.05% Tween 20 in 0.1 M Tris, 0.15 M NaCl; pH 7.5

Formamide, 50%, in 1× SSC

Antibodies:

(a) First antibody (e.g., sheep polyclonal antiserum to digoxigenin or biotin; titration recommended by supplier [Boehringer Mannheim]);

(b) Second antibody (e.g., FITC-conjugated donkey antisheep IgG [Jackson Immunoresearch Inc.]);

(c) Dilute the antibodies in 3% BSA in TBST.

Rubber latex adhesive

Mountant: Vectashield; contains inhibitor of photobleaching (Vector Labs)

Protocol

Chromosome Preparation and Denaturation

1. Prepare metaphase-arrested cells by the standard technique (see Protocols 15.9 and Protocol 26.5), and drop the fixed cells onto slides. Mark the areas of spread with a diamond pencil.

2. Refix the cells for one hour in fresh methanol: acetic acid, 3:1, and then air-dry the slides.

3. Rinse the slides in 2× SSC for 2 min at room temperature.

4. Incubate in 100 μg/ml of RNase in 2× SSC at 37°C for 1 h.

5. Rinse in PBSA.

6. Refix in freshly prepared 1% paraformaldehyde in PBSA for 10 min at room temperature. (This step is optional.)

7. Rinse in PBSA.

8. Dehydrate for 2 min in 2 lots each of 70% ethanol, and 100% ethanol. Air dry the slides.

9. Denature the chromosomes in 70% formamide in 2× SSC at 70°C for 2–4 min (determine the appropriate amount of time experimentally, starting at 2 min). Make sure that the 70% formamide is at 70°C before using it.

10. Wash the slides in several changes of ice-cold 70% ethanol.

11. Dehydrate as in step 8, and air dry the slides.

12. The chromosomes are now ready for hybridization.

Hybridization

13. Apply 10 μl of the denatured probe over the areas of spread, and cover the spreads with 22 × 22-mm coverslips.

14. Seal the coverslips around the edges with rubber latex adhesive.

15. Place the slides in a humidified box at 37°C overnight.

Probe Detection

16. Remove the coverslips by immersing the slides in 2× SSC (at room temperature) and peeling off the adhesive. Place the slides in a Coplin jar.

17. Soak the slides, two times for 10 min each, in 50% formamide, 1× SSC, at 42°C.

18. Wash the slides, two times for 10 min each, in 2× SSC at 42°C.

19. Rinse in TBST.

20. Block nonspecific binding by incubating the slides with 3% bovine serum albumin (BSA) in TBST for 30–60 min at 37°C.

21. Add the first antibody to the slides. Use 100 μl of antibody per slide, and cover each slide with a Parafilm coverslip.

22. Incubate the slides for 1 h at 37°C.

23. Wash the slides in 500 ml of TBST for 10 min at room temperature.

24. Add the second, fluorochrome-conjugated, antibody, in 3% BSA/TBST, to the slides at a titration recommended by the supplier or determined by experiment.

Note. Be sure to use the correct antibody combinations—for example, sheep polyclonal antiserum to digoxigenin as a first antibody, followed by FITC-conjugated donkey antisheep IgG.

25. Incubate for 30 min at 37°C.

26. Wash for 30 min in 500 ml of TBST at room temperature.

27. Counterstain with 0.8 μg/ml of DAPI and/or 0.4 μg/ml of propidium iodide in TBST for 10 min. Mount coverslips in an antifade mountant.

28. View the slides using fluorescence microscopy, with appropriate filter combinations.

Variations

Chromosome Painting. Chromosome paints are available commercially from a number of sources, including Life Technologies, Cambio, and Oncor. It is therefore no longer necessary to prepare your own paints. The hybridization and detection protocols vary with each commercial source. However, in general, section (a) (*Chromosome preparation and denaturation*) of the foregoing protocol can be used prior to hybridization. Hybridization and detection can then be carried out according to the supplier's instructions. If the paint is labeled with biotin, such as is the case for the Cambio paints, section (c) (*Probe detection*) can be followed, using the appropriate antibody combinations.

Recent Advances

Classically, karyotypic analysis is carried out by chromosome banding, using dyes that differentially stain the chromosomes. (See Protocol 26.5.) Thus, each chromosome is identified by its banding pattern. However, traditional banding techniques cannot characterize many complex chromosomal aberrations. Recently, new karyotyping methods based on chromosome painting techniques—namely, spectral karyotyping (SKY) and multicolor fluorescence *in situ* hybridization (M-FISH)—have been developed. These techniques allow the simultaneous visualization of all 24 human chromosomes in different colors. Furthermore, visualization of the resulting fluorescence patterns by computer programs makes these techniques more sensitive than the human eye. These techniques are proving to be highly successful in the identification of new chromosomal alterations that were previously unresolved by traditional approaches [Wienberg and Stanyon, 1997; Ried et al., 1997; Macville et al., 1997].

SOMATIC CELL FUSION

Somatic cells fuse if cultured with inactivated Sendai virus or with polyethylene glycol (PEG) [Pontecorvo, 1975]. A proportion of the cells that fuse progress to nuclear fusion, and a proportion of these cells progress through mitosis, such that both sets of chromosomes replicate together and a hybrid is formed. In some interspecific hybrids—e.g., human–mouse—one set of chromosomes (the human) is gradually lost [Weiss and Green, 1967]. Thus, genetic recombination is possible *in vitro*, and, in some cases, segregation is possible as well.

Since the proportion of viable hybrids is low, selective media are required to favor the survival of the hybrids at the expense of the parental cells. TK⁻ and HGPRT⁻ mutants (see Selection of Hybrid Clones in this chapter) of the two parental cell types are used,

and the selection is carried out in HAT medium (hypoxanthine, aminopterin, and thymidine) (Fig. 27.1) [Littlefield, 1964a]. Only cells formed by the fusion of two different parental cells (heterokaryons) survive, since the parental cells and fusion products of the same parental cell type (homokaryons) are deficient in either thymidine kinase or hypoxanthine guanine phosphoribosyl transferase. The parental cells and homokaryons cannot, therefore, utilize thymidine or hypoxanthine from the medium, and since aminopterin blocks endogenous synthesis of purines and pyrimidines, they are unable to synthesize DNA.

The following protocol for somatic cell fusion has been contributed by Ivor Hickey, Division of Genetic Engineering, School of Biology and Biochemistry, Queen's University, Medical Biology Centre, 97 Lisburn Road, Belfast BT9 7BL, United Kingdom.

PROTOCOL 27.3. CELL HYBRIDIZATION

Principle

Although many cell lines undergo spontaneous fusion, the frequency of such events is very low. In order to produce hybrids in significant numbers, cells are treated with either inactivated Sendai virus [Harris and Watkins, 1965] or, more commonly, the chemical fusogen polyethylene glycol (PEG) [Pontecorvo, 1975]. Selection systems that kill parental cells but not hybrids are then used to isolate clones of hybrid cells.

Outline

Bring the cells to be fused into close contact, either in suspension or in monolayers. Treat the cells with PEG, briefly to minimize cell killing. Usually, the cells are given a 24-h period to recover before selection for hybrids.

Materials

Sterile:
PEG 1,000 (Merck):
  (a) Autoclave the PEG to liquefy and sterilize it.
  (b) Allow it to cool to 37°C, and then mix it with an equal volume of serum-free medium, prewarmed to 37°C.
  (c) Adjust the pH to approximately 7.6–7.9, using 1.0 M NaOH.
  (d) Store the solution at 4°C for up to 2 weeks.
Complete growth medium
Serum-free growth medium
NaOH, 1.0 M
Petri dishes, 50 mm
Universal containers

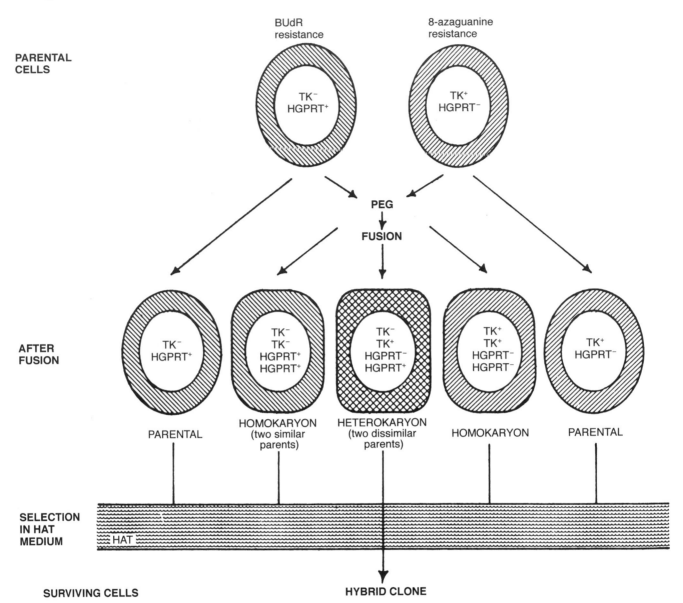

Fig. 27.1. Somatic Cell Hybridization. Selection of hybrid cells after fusion. (See text.)

Protocol

Monolayer Fusion:

1. Inoculate equal numbers of the two types of cells to be fused into 50-mm tissue culture dishes. Between 2.5×10^5 and 2.5×10^6 of each parental cell line per dish is usually sufficient. Incubate the mixed culture overnight.
2. Warm the PEG solution to 37°C. It may be necessary at this point to readjust its pH, using NaOH.
3. Remove the medium thoroughly from the cultures and wash them once with serum-free medium. Add 3.0 ml of the PEG solution and spread it over the monolayer of cells.
4. Remove the PEG solution after exactly 1.0 min, and rinse the monolayer three times with 10-ml

of serum-free medium before returning the cells to complete medium.

5. Culture the cells overnight before adding selection medium.

Suspension Fusion:

1. Centrifuge a mixture of 4×10^6 cells of each of the two parental cell lines at 150 g for 5 min at room temperature. Carry out centrifugation and subsequent fusion in 30-ml plastic universal containers or centrifuge tubes.
2. Resuspend the pellet in 15 ml of serum-free medium, and centrifuge again.
3. Aspirate off all of the medium, and resuspend the cells in 1 ml of PEG solution by gently pipetting.
4. After 1.0 min, dilute the suspension with 9 ml of serum-free medium, and transfer half of the sus-

pension to each of two universal containers or centrifuge tubes containing a further 15 ml of serum-free medium.

5. Centrifuge the suspensions at 150 g for 5 min. Remove the supernatant, and resuspend the cells in complete medium.

6. After overnight incubation, clone the cells in selection medium.

Variations

A large number of variations of the PEG fusion technique have been reported. While the procedure described here works well with a range of mouse, hamster, and human cells in interspecific and intraspecific fusions, it is unlikely to be optimal for all cell lines. Inclusion of 10% DMSO in the PEG solution has the advantage of reducing its viscosity and has been reported to improve fusion [Norwood et al., 1976]. Also, the molecular weight of the PEG used need not be 1,000 D. Preparations with molecular weights from 400 to 6,000 D have been successfully used to produce hybrids. Although now largely superseded by PEG as a fusogen, for reasons of convenience, Sendai virus fusion remains a reliable method. If a source of virus is available, the method of Harris and Watkins [1965] can be used.

Selection of Hybrid Clones

The method of selection used in any particular instance depends on the species of origin of the two parental cell lines, the growth properties of the cell lines, and whether selectable genetic markers are present in either or both cell lines. Hybrids are most frequently selected using the HAT system: 10^{-4} M hypoxanthine, 6×10^{-7} M aminopterin, and 1.6×10^{-6} M thymidine [Littlefield, 1964a]. This system can be used to isolate hybrids made between pairs of mutant cell lines deficient in the enzymes thymidine kinase (TK$^-$) and hypoxanthine guanosine phosphoribosyl transferase (HGPRT$^-$), respectively. TK$^-$ cells are selected by exposure to BUdR and HGPRT$^-$ cells by exposure to thioguanine, following the procedures described by Biedler in Chapter 13. (See Protocol 13.9.) When only one parent cell line carries such a mutation, HAT selection can still be applied if the other cell line does not grow, or grows poorly in culture (e.g., lymphocytes, senescing primary cultures).

Differential sensitivity to the cardiac glycoside ouabain is an important factor in the selection of hybrids between rodent cells and cells from a number of other species, including human. Rodent cells are resistant to concentrations of this antimetabolite up to 2.0 mM, while human cells are killed at 10^{-5} M ouabain. The hybrids are much more resistant to ouabain than the human parental cells. If a rodent cell line that is

HGPRT deficient is fused to unmarked human cells, then the hybrids can be selected in medium containing HAT and low concentrations of ouabain.

Although many other selection systems have been reported, only complementation of auxotrophy [Kao et al., 1969] has been widely used.

It must be stressed that, whichever method is used to isolate clones of putative hybrid cells, confirmation of the hybrid nature of the cells must be obtained. This is usually done using cytogenetic (see Protocol 15.9) or biochemical techniques. In certain cases, comparing the number of hybrids with the frequency of revertants may be the only way of making this confirmation.

Nuclear Transfer

Genetic recombination experiments can also be carried out with isolated nuclei, but the major interest in this technique is related to cloning individual animals [Kono, 1997; Wolf et al., 1998]. Nuclei can be isolated by centrifuging cytochalasin B–treated cells and fusing the extracted nuclei to recipient whole cells or enucleated cytoplasts in the presence of PEG. (Fig. 27.2.) However, in animal cloning experiments micromanipulation techniques are used to remove the nucleus from one cell and inject it into a fertilized, preimplantation egg from which the existing nucleus has been removed.

Monochromosomal Transfer

Chromosomes may be isolated from metaphase cells as micronuclei after prolonged colcemid treatment, followed by cytochalasin B treatment and centrifugation. (Fig. 27.3.) Incubation of these micronuclei with

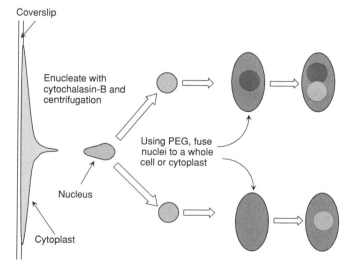

Fig. 27.2. Nuclear Transfer. Whole nuclei extruded by treatment with cytochalasin B hybridized with whole cells or enucleated cytoplast.

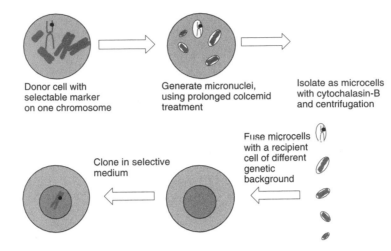

Fig. 27.3. Monochromosomal Transfer. Chromosomes, carrying a selectable marker, are isolated by centrifugation, as in Fig. 27.2, from cells in mitotic arrest which was induced by prolonged colcemid treatment. The microcells, containing individual chromosomes are then fused to whole cells.

whole host cells in the presence of PEG and PHA results in fusion with the host cells and, ultimately, their incorporation into the nucleus. Incorporation of a selectable marker (e.g. HyTK) into individual chromosomes of the donor cell allows the selection of resistant clones containing the marked chromosome with hygromycin after incorporation into the host cell [Newbold and Cuthbert, 1998].

PRODUCTION OF MONOCLONAL ANTIBODIES

Monoclonal antibodies have become indispensable tools in research, diagnostics, and therapeutics. They have gradually replaced polyclonal antibodies since hybridoma technology was first introduced by Kohler and Milstein in 1975. The following protocol was contributed by Janice Payne and Tina Kuus-Reichel of Hybritech Inc., a subsidiary of Beckman Coulter Inc., P.O. Box 296006, San Diego, CA 92166.

PROTOCOL 27.4. PRODUCTION OF MONOCLONAL ANTIBODIES

Principle

Hybridomas are produced by fusing a nonsecreting myeloma cell with an antibody-producing B-lymphocyte in the presence of polyethylene glycol. (Fig. 27.4.) The myeloma cell is deficient in an enzyme hypoxanthine-guanine phosphoribosyl transferase (HGPRT) or thymidine kinase (TK) necessary for

DNA synthesis and cannot survive in selection medium containing hypoxanthine, aminopterin, and thymidine. Any unfused B-lymphocytes from the spleen cannot survive in culture for more than a few days. Any B-cell–myeloma hybrids should contain

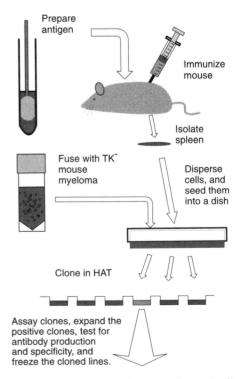

Fig. 27.4. Production of Hybridomas. Schematic diagram of the production of hybridoma clones capable of secreting monoclonal antibodies.

the genetic information from both parent cells and are thus able to survive in the HAT selection medium. They can be cultured indefinitely and will produce unlimited quantities of antibody. Supernatants from surviving hybridomas are screened for antibody by ELISA. Those hybridomas selected are then subcloned to ensure that they are producing antibody that is specific for a single epitope. Antibody production can be scaled up *in vivo* as ascites in mice or *in vitro* as a suspension culture. Hybridomas also grow very well in various hollow-fiber and fermentation culture systems.

Outline

Using polyethylene glycol (PEG), fuse spleen cells from an immunized mouse with myeloma cells (P3.653). Select hybrid colonies (hybridomas) in HAT medium. Screen the supernatants by ELISA 10–14 d after fusion (the ELISA screening protocol should be developed prior to the fusion), and expand, freeze, and subclone the desired hybridomas, to ensure monoclonality.

Materials

Sterile or Aseptically Prepared:

Mice (Balb/c or A/J), 6–10 weeks old

P3.653 myeloma: This myeloma cell line from ATCC does not secrete immunoglobulin and performs well in fusions

Antigen (125 μg per mouse is optimal; small antigens can be conjugated to keyhole limpet hemocyanin (KLH, Sigma) to increase antigenicity)

Adjuvants (Alum or Freund's adjuvant; complete and incomplete, Sigma): A detailed description of these and other adjuvants can be found in Vogel and Powell [1995]

TCD (T-Cell depletion) buffer: Hanks' balanced salt solution + 10 mM HEPES + 0.3% BSA

NH$_4$Cl, 0.16 M

Antimouse Thy 1.2 antibody (Accurate)

Rabbit complement, Low Tox M (Accurate)

PBSA

SFM (serum-free medium): MEM that has been stored at room temperature, with the cap of the container loosened to allow the release of CO$_2$; the pH must be alkaline

PEG (polyethylene glycol): Melt 10.5 ml of PEG 1450 (Sigma) in a 56°C water bath; add 19.5 ml of warm sterile MEM (pH 8.3 to 8.7); mix well; and allow the solution to equilibrate, with the cap of the container loosened, for 5–7 d before fusion.

HAT stock (100×) (10 mM hypoxanthine, 40 μM aminopterin, 1.6 mM thymidine): 1.36 g of hypoxanthine, 729 μl of aminopterin (from 25-mg/ml stock), 0.387 g of thymidine, 0.022 g of glycine

dissolved in 4 ml of 5 M NaOH + 26 ml UPW. Make up to 1 L with UPW. Filter sterilize the solution.

HAT medium: Basal medium, such as MEM or RPMI, + 10% FBS + 20% spleen-conditioned medium plus HAT stock (final dilution 1:100)

SCM (spleen-conditioned medium): Tease 3–5 naive mouse spleens in PBSA. Count the cells, and resuspend them at 1 × 10^6 cells/ml in MEM + 10% horse serum. Transfer the suspension to a 500-ml spinner flask, and incubate it at 37°C in 5% CO$_2$ for 48 h. Remove the cells by centrifugation, and store the supernatant frozen.

8-Azaguanine stock, 10 mM (100×): 1.52 g of 8-azaguanine dissolved in 4 ml of 5 M NaOH + 21 ml of UPW. Make the solution up to 1 L with UPW. Filter sterilize the stock.

MEM, 10% fetal calf serum, 0.1 mM 8-azaguanine (for maintenance of P3.653 myeloma)

HT stock (100×): 0.408 g of hypoxanthine, 0.1161 g of thymidine, 0.0067 g of glycine, dissolved in 2 ml of 5 M NaOH + 8 ml of UPW. Make the solution up to 300 ml with UPW. Filter sterilize the stock.

HT medium: HT stock diluted 100× in basal medium (as for HAT medium, above)

Syringes, 1 ml with 25G needles

Syringes, 1 ml, with 23G needles, ×2

Dissecting instruments (scissors and forceps)

Petri dishes, 60 × 15 mm

Centrifuge tubes, 15 ml and 50 ml

Multiwell plates, 24 well and 96 well

Culture flasks

100-ml Nalgene bottle

Pipette tips

Reservoir for multipipettor (100 ml, Matrix Technologies; Corning Costar)

Nonsterile:

Trypan blue (Sigma)

Multipipettor, 12 channel (Matrix Technologies)

Inverted phase-contrast microscope

Unopette microcollection system (Becton Dickinson)

Hemocytometer

Protocol

Immunization:

1. Bleed the mice on day 0 prior to the initial injection, and check the serum for background antigen reactivity.

2. Immunize the mice (A/J or Balb/c) with antigen emulsified in Freund's adjuvant or mixed with a 1/10 volume of alum and vortexed. Give three injections of antigen intraperitoneally, according to the following schedule:

Day	Amount of antigen	Adjuvant
0	50 μg	Alum or complete Freund's adjuvant
14	25 μg	Alum or incomplete Freund's adjuvant
28	25 μg	Alum or PBSA

3. Bleed the mice on day 35 and measure the serum titer of antibody by an ELISA assay.

4. Dilute the serum serially 1:4 after a 1:30 dilution, and up to 1:30720.

5. Select mice with the highest ratio of serum titers to antigen for fusion.

6. Give the selected mice a final boost of 10 μg of antigen i.v. or 25 μg of antigen i.p. 3 d prior to fusion.

Myeloma:

1. It is convenient to perform fusions on a Thursday, with the mice receiving a final boost of antigen on a Monday.

2. Maintain the P3.653 myeloma cell line in MEM + 10% fetal calf serum + 8-azaguanine.

3. Dilute the P3.653 cells to 3.5 × 10^5 cells/ml each day for the three days prior to fusion.

T-Cell Depletion:

1. Bleed mice with appropriate serum titers, and sacrifice them by cervical dislocation. Aseptically remove the spleens, and place them in a sterile Petri dish with 5 ml of sterile PBSA. Gently tease the spleens with two 23G needles on 1-ml syringes. Teasing spleens roughly will result in a high concentration of fibroblasts.

2. Transfer the cells to a 15-ml conical tube, and allow clumps to settle. Transfer spleen cells (without clumps) to a 50-ml conical tube, and, following a 1:100 dilution in a Unopette, count the cells with a hemocytometer.

3. Spin the cells at 1,000 rpm for 8 min. To lyse the red blood cells, resuspend the resultant pellet in 0.84% NH$_4$Cl (10 ml/spleen), and incubate the suspension at 4°C for 15 min. Underlayer the cell suspension with 14 ml of horse serum, and spin the solution at 1,500 rpm for 8 min.

4. Resuspend the resultant pellet in 50 ml of TCD buffer, and spin the suspension at 1,000 rpm for 8 min.

5. For T-cell depletion, resuspend the resultant cell pellet in anti-Thy 1.2 at a final concentration of 1 × 10^7 cells/ml. (Make a 1:500 dilution of anti-Thy 1.2 in TCD buffer, and filter sterilize it.) Incubate the suspension at 4°C for 45 min, and then spin it at 1,000 rpm for 8 min. Resuspend the resultant pellet in rabbit complement (recon-

stituted in 1 ml of cold UPW, diluted 1:12 in TCD buffer, and filter sterilized). Incubate the suspension at 37°C for 45 min, and then spin it at 1,000 rpm for 8 min. Count the cells by trypan blue exclusion on a hemocytometer. B-cell recovery should be 30–50%.

Fusion:

1. Mix the myeloma and B-cells in a 50-ml centrifuge tube. One fusion can be done on a maximum of 1.2 × 10^8 spleen cells. Mix the spleen cells with P3.653 myelomas at a ratio of 4:1; thus, the maximum number of P3.653 cells per fusion is 3 × 10^7 cells.

2. Centrifuge the suspension at 1,000 rpm for 8 min.

3. Break up the resultant pellet by tapping, and add 1 ml of PEG to the tube over 15 s.

4. Mix the suspension by gently swirling the tube for 75 s.

5. Add 1 ml of SFM over 15 s, and gently swirl the tube for 45 s.

6. Add 2 ml of SFM over 30 s, and swirl the tube for 90 s.

7. Add 4 ml of HAT medium over 30 s, and swirl the tube for 90 s.

8. Finally, add 8 ml of HAT medium over 30 s, and swirl tube for 90 s.

9. Add this volume (16 ml) to a sterile Nalgene bottle containing the calculated amount of HAT medium (125 ml if the maximum cell concentration has been used). 16 ml, containing 1.5 × 10^8 cells, from step 1 in this section of the protocol plus 125 ml of HAT medium in the bottle makes 141 ml. With the wash in the next step (step 10), the total volume is 150 ml and will result in a final concentration of 1 × 10^6 cells/ml.

10. Wash the 50-ml conical tube with 9 ml of HAT medium, and add this volume to the bottle.

11. Mix the contents of the bottle well, and transfer the cells to the sterile reservoir.

12. Using a 12-channel multipipettor, plate the cells at 200 μl/well into a sterile 96-well plate. The final concentration is then 2 × 10^5 cells/well.

Selection of Hybridomas:

1. Feed the fusion plates 5 d after fusion, by aspirating most of the culture media from the wells and replacing it with 150–200 μl/well of fresh HAT medium.

2. Feed the plates twice per week.

3. Screen the clones for selection of positive hybridomas, usually two weeks after fusion, using ELISA.

4. After a further 48 h, retest those clones that tested positive in the previous step.

5. Expand the most productive hybridomas by culturing them in two wells of a 96-well plate in media containing 10% FBS and HT.

6. Retest the clones, expand the positive hybridomas to a 24-well plate, and wean them off HT medium, at which time 2 ml of culture supernatant should be harvested for screening. At this step, enough volume is harvested to perform several selection assays to ensure that the antibody is directed only at the antigen of interest.

7. Expand the hybridomas to be kept to 4 wells of a 24-well plate, and cryopreserve them. (See Protocol 19.1.)

8. Perform a second cryopreservation after expanding the hybridoma to a 75-cm² flask.

Screening

Take care in developing the screening strategy to obtain a monoclonal antibody with the characteristics that you want. Hybridoma culture supernatants should be screened as early as feasible for desired reactivity patterns. After initial selection by ELISA for reactivity to the immunogen, the expanded culture supernatant should be tested in the application for which it was developed (e.g., Western blot, competitive immunoassay, flow cytometry, etc.). A more detailed discussion of ELISA and other immunoassays can be found in Knott et al. [1997].

Subcloning

To ensure monoclonality, subclone hybridomas of interest. This can be done by serially diluting cells and plating the equivalent of 1 cell per 3 wells in a 96-well plate or by sorting with an automated cell deposition unit (ACDU) on a FACStarplus (Becton Dickinson) and plating at one cell per well. Subcloning can be done on top of a mouse spleen feeder layer plate. Feeder layers are prepared by teasing a naive mouse spleen and resuspending the cells at 1×10^6 cells/ml. The cells are then plated in a 96-well microtitration plate at a final concentration of 2×10^5 cells/well. After subcloning, colonies can usually be seen at day 5 and must be checked visually for monoclonality. Plates are fed with fresh medium beginning on day 7. Screening for positive hybridomas is usually done between days 10–14. Those clones selected are then expanded and frozen in the same way as the parental hybridoma.

Antibody Production

Concentrated antibody from clones of interest can be produced *in vivo* as ascites in IFA primed mice (Balb/c or nu/nu [Gillette, 1987]) or *in vitro* as a suspension culture. Several hollow-fiber cell culture systems are also available (Unisyn, Cellex, Cellco, Integra). (See Hollow Fibers in Chapter 24 and Perfusion Systems in Chapter 25.) When hybridomas are inoculated into a hollow-fiber system, the cells are maintained in a compartment of the bioreactor, while fresh media and waste from the cells are recirculated. High concentrations of antibody are produced in the cell compartment, and culture supernatant containing antibody can be harvested at multiple time-points.

DNA TRANSFER

In order to study the function of individual genes, the sequence of interest can be cloned and then transferred into host cells by a variety of techniques, such as transfection, lipofection, and retroviral infection. (Fig. 27.5.) Cloned DNA is often conveniently maintained as part of a bacterial plasmid. Many plasmids can attain a high copy number during bacterial growth, thus ensuring a plentiful stock of DNA for experimentation. Plasmid DNA is purified from the bacteria prior to use. Once the sequence of interest has been cloned, it can be manipulated further to isolate subclones containing, for example, promoter sequences. By genetic manipulation, promoter sequences can be linked to a reporter gene (e.g., β-gal or CAT) whose products (e.g., β-galactosidase, chloramphenicol acetyl transferase) can be readily assayed subsequent to transfection. In this way, tissue-specific gene expression can be analyzed in detail. Oncogenes and tumor suppressor gene function can also be analyzed by similar manipulations.

Transfections may be *transient* or *stable*. Transient transfections are short term and used shortly after transfection, and the efficiency of transfection is de-

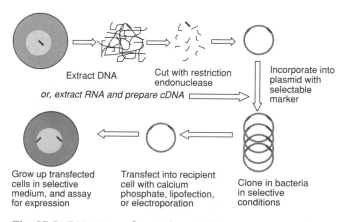

Extract DNA Cut with restriction endonuclease Incorporate into plasmid with selectable marker

or, extract RNA and prepare cDNA

Grow up transfected cells in selective medium, and assay for expression Transfect into recipient cell with calcium phosphate, lipofection, or electroporation Clone in bacteria in selective conditions

Fig. 27.5. DNA Transfer. Isolated DNA endonuclease fragments amplified by gene cloning techniques, added to whole cells, and incorporated by treatment with lipofection, electroporation, or coprecipitation with $Ca_3(PO_4)_2$.

termined by reporter gene assays. DNA used for stable transfection contains a selectable marker, such as *neo* or *hyg B*, that confers resistance to G418 (geneticin, an analogue of neomycin) or hygromycin, respectively. Transfected cells are then selected by continued exposure to the selection agent, and resistant clones can be isolated. (See Protocols 13.6 and 13.8.)

Three protocols are presented below, two for stable transfection (using calcium phosphate and electroporation) and one for transient (using lipofection). Selection protocols for stable transfection can be added to the lipofection protocol, provided that the construct used for transfection contains the appropriate selectable marker.

Coprecipitation with Calcium Phosphate

The following protocol for gene transfer into mammalian cells by the calcium phosphate technique has been contributed by Demetrios A. Spandidos, Medical School, University of Crete, Heraklion, Crete, Greece. (See also Protocol 17.1.)

PROTOCOL 27.5. STABLE DNA TRANSFECTION WITH CALCIUM PHOSPHATE

Principle

The calcium phosphate technique for introducing genes into mammalian cells was first described by Graham and Van der Eb [1973] and is still widely used. In this method, exogenous DNA is mixed with calcium chloride and is then added to a solution containing phosphate ions. A calcium-phosphate–DNA coprecipitate is formed, which is taken up by mammalian cells in culture, resulting in expression of the exogenous gene. This method can be used to introduce any DNA into mammalian cells for transient expression assays or long-term transformation.

Outline

Transfect cells with the appropriate DNA carrying a selectable marker (e.g., the aminoglycoside phosphotransferase (*aph*) gene), and apply selection to eliminate the cells that have not taken up and expressed the exogenous gene.

Materials

Geneticin (G418, Gibco)

Methocel MC4000 CP (Fluka)

Plasmids carrying the *aph* gene (see Spandidos and Wilkie [1984a,b] for a variety of *aph* recombinant plasmics), 100 μg/ml in TEB

Carrier salmon sperm DNA (Sigma), 100 μg/ml in TEB

Growth medium, SF12: Ham's F12, supplemented with Eagle's MEM amino acids (ICN), with 15% fetal bovine serum (Sterile Systems, Inc.)

Liquid-Medium Selection Procedure:

HEPES-buffered saline, 2× (2 × HBS):

NaCl, 1.63 g

HEPES, 1.19 g

Na$_2$HPO$_4 \cdot$2H$_2$O, 0.023 g

Distilled water, to 100 ml

Adjust the pH to 7.1 with 0.5 M NaOH. Filter sterilize the solution, and store it at 4°C. The final composition of the 2 × HBS is 0.28 M NaCl, 1.5 mM Na$_2$HPO$_4$, and 50 mM HEPES, with pH 7.1.

CaCl$_2$, 2.5 M: Dissolve 10.8 g of CaCl$_2 \cdot$6H$_2$O in 20 ml (final volume) of distilled water. Filter sterilize the solution, and store it at 4°C.

Tris-EDTA buffer (TEB): 0.1 mM EDTA, 1.0 mM Tris-HCl; pH 8.0. Mix 50 μl og 0.2 M EDTA (pH 8.0) and 100 μl of 1.0 M Tris-HCl (pH 8.0) with distilled water to a final volume of 100 ml. Filter sterilize the solution, and store it at 4°C.

Methocel Selection Procedure:

Methocel medium: Mix 3 g of Methocel with 200 ml of distilled water, and autoclave the solution. The Methocel dissolves to yield a clear solution that can be stored at 4°C for at least 6 months. (See Reagent Appendix.) Just before use, warm the medium to 37°C and add

Ham's F12 medium, 10×, 22.0 ml

MEM essential amino acids, 50×, 4.0 ml

Sodium pyruvate, 0.1 M, 4.0 ml

Glutamine, 0.2 M, 2.5 ml

Sodium bicarbonate, 7.5%, 5.0 ml

Next, add 100 ml of serum and the appropriate concentration of the drugs to be used for selection. Thus, for Methocel medium containing geneticin (200 μg/ml), add 3.4 ml of 20-mg-of-geneticin/ml water (stock solution, filter sterilized). The final concentrations of Methocel and serum are 0.9% and 30%, respectively. Note that the composition of the Methocel medium in step 2 of the second part of the protocol depends upon the cell line being used, while the nature of the components in step 3 depends upon the selection marker being used.

Protocol

Liquid-Medium Selection Procedure for Attached Cells:

1. Harvest exponentially growing cells (i.e., mouse LATK⁻ cells) by trypsinization. (See Protocol 12.2.)

2. Replate the cells at a density of 5 × 10⁵ cells per flask (25-cm² growth area) in 5 ml of SF12 medium containing 15% fetal bovine serum.

3. Incubate the culture at 37°C for 24 h.

4. Add 1 ml of 2 × HBS to a bijou vial.

5. Into a second, plastic bijou vial, place donor DNA plus carrier DNA to give a final concentration of 80 $\mu g/ml$ in 0.5 ml of TEB (e.g., 0.2 ml of plasmid DNA @ 100 $\mu g/ml$ + 0.2 ml of carrier DNA @ 100 $\mu g/ml$ + 0.1 ml TEB).

6. Add 0.4 ml of TEB and 0.1 ml of 2.5 M $CaCl_2$ to the vial, and mix the solution.

7. Add this DNA solution slowly (over about 30 s), with continuous mixing, to the 1 ml of 2 × HBS in the first bijou vial.

8. Mix the contents of the vial immediately by vortexing, and leave the solution at room temperature for 30 min. The DNA concentration at this stage is 20 $\mu g/ml$.

9. After the incubation, a fine precipitate will have formed.

10. Add 0.5 ml of this DNA–calcium phosphate suspension to each flask containing cells in 5 ml of growth medium.

11. Incubate the flasks at 37°C for 24 h, to allow absorption of the DNA–calcium phosphate co-precipitate by the cells.

12. Preselection expression stage:

    (a) Replace the medium in the flasks with fresh, prewarmed medium.

    (b) Incubate the flasks at 37°C for a further 24 h, to allow expression of the transferred gene(s) to occur.

13. Selection stage:

    (a) Replace the medium with an appropriate selection medium—in this case, SF12 medium containing 15% serum and 200 $\mu g/ml$ geneticin. (Cultured cell lines differ in their sensitivity to geneticin, and the most suitable concentration of geneticin to use must be determined empirically; see Protocols 21.3 and 21.4.)

    (b) Renew the selection medium every 2–3 d for up to 2–3 weeks when the colonies are routinely counted.

    (c) Pick colonies. (See Protocol 13.6.)

Methocel Selection Procedure for Anchorage-Independent Cells:

1. Start the transformation as described in the protocol for liquid-medium selection (steps 1–11). After allowing for preselection expression (step 12), trypsinize and then count the cells.

2. Mix 0.2 ml of cells with 20 ml of Methocel medium in a plastic universal container, and plate the cells on bacteriological plates (9-cm diameter). Up to 2 × 10⁶ cells per plate can be plated; the choice of plating density depending on the transformation frequency expected.

3. Incubate the plates at 37°C for 7–10 d depending on the population-doubling time of the recipient cell line.

4. If required, pick individual colonies. (See Protocol 13.8.)

Analysis

Count the colonies using an inverted microscope.

Lipofection

The original method for cationic lipid-mediated DNA transfection into cultured cells [Felgner et al., 1987] was improved in 1993 by replacement of the mono-cationic lipid reagent with a polycationic one, Lipofectamine [Ciccarone et al, 1993; Hawley-Nelson et al., 1993]. The method is based on an ionic interaction of DNA and liposomes to form a complex, which can deliver functional DNA into cultured cells. Plasmid DNA is complexed, but not encapsulated, within uni-lamellar liposomes 600–1200 nm in size [Mahato et al., 1995a], formed by cationic lipids in water.

The advantages of cationic liposome-mediated transfection over other methods include generally higher efficiency; the ability to transfect successfully a wide variety (over 300 reported) of eucaryotic cell lines, many of which are refractive to other transfection procedures; and relatively low cell toxicity. Another advantage is that the basic procedure of DNA transfection can be adapted for transfection with RNA, synthetic oligonucleotides, proteins, and viruses. Finally, cationic liposomes can be used for the successful delivery of functional genes or viral genomes *in vivo* [Mahato et al., 1995b; Tagawa et al., 1996; Thorsell et al., 1996]. Its disadvantage is the relatively high cost of reagents, which practically precludes large-scale use.

The following protocol has been abridged from Bichko [1998].

PROTOCOL 27.6. TRANSIENT TRANSFECTION BY LIPOFECTION

Materials

Sterile or Aseptically Prepared:

Cationic lipids

Lipofectamine: 3:1 (w/w) liposome formulation of the polycationic lipid DOSPA (2,3-dioleyloxy-N[2(sperminecarboxamido)ethyl]-N,N-dimethyl-1-propanaminium tryfluoroacetate) and the neutral lipid DOPE (dioleoylphosphatidylethanol-amine) in water (Gibco)

Lipofectin: 1:1 (w/w) liposome formulation of the cationic lipid DOTMA (N-[1-(2,3 dioleyloxy)-propyl]-n,n,n-trimethylammonium-chloride) and the neutral lipid DOPE in water (Gibco)

LipofectACE: 1:2.5 (w/w) liposome formulation of the cationic lipid DDAB (dimethyl dioctadecylammonium bromide) and the neutral lipid DOPE in water (Gibco)

Cells for transfection

DNA for transfection

Growth medium: DMEM (1×) with 10% FBS, penicillin (100 U/ml) and streptomycin (100 μg/ml, or as appropriate to the cells being used)

Reduced serum medium: OPTI-MEM 1 (contains 2% FBS; Gibco)

Serum-free medium: DMEM without serum or antibiotics

PBSA

Buffered saline:
  NaCl, 0.15 M
  K_2HPO_4, 0.006 M
  KH_2PO_4, 0.002 M

Trypsin, 0.25%, in PBSA

Multiwell plates, 6 well, or 35-mm Petri dishes

Protocol

(a) Transfection of Adherent Cells

1. Seed approximately 1×10^5 to 3×10^5 cells per well in 6-well plates in 3 ml of growth medium.
2. Incubate the cells at 37°C in a CO_2 incubator until the cells are 50% to 90% confluent. This step usually takes 18–24 h and should not take less then 16 h.
3. Before transfection, prepare the DNA and lipid solutions in sterile tubes:
   (a) For the DNA solution, dilute 1–2 μg of DNA into 0.5 ml of reduced-serum medium.
   (b) For the lipid solution, dilute 2 to 25 μl of cationic lipid reagent into 0.5 ml of serum-free medium.
   (c) Combine the two solutions, mix the resultant solution gently, and incubate it at room temperature for 15–45 min, to allow the formation of DNA–lipid complexes.
4. Rinse the cells once with 2 ml of serum-free medium.
5. Overlay the DNA–lipid complex onto the cells. Antibacterial agents should be omitted during transfection.
6. Incubate the cells with the complexes at 37°C in a CO_2 incubator for 2–24 h; five or 6 hours are usually enough.
7. Following incubation, remove the transfection mixture from the cells, and replace it with 3 ml of complete growth medium.
8. Replace the growth medium with fresh medium 18 to 24 h after the start of transfection.

9. Assay the growth medium or cells for transient gene activity as appropriate. (See Protocols 27.8 and 27.9.)

(b) Transfection of Suspension Cells

1. Prepare the transfection mixture in sterile tubes as follows:
   (a) Dilute 2–5 μg of DNA in 0.5 ml of reduced-serum medium.
   (b) Dilute 2–20 μl of cationic lipid reagent in 0.5 ml of serum-free medium.
   (c) Combine the two solutions, mix the resultant solution gently, and incubate it at room temperature for 15–45 min, to allow the formation of DNA–lipid complexes.
2. Centrifuge a cell suspension containing approximately 1×10^6 to 2×10^6 cells, and aspirate the medium.
3. Resuspend the cells in the transfection mixture, and transfer the suspension to a 35-mm dish.
4. Incubate the cells in a CO_2 incubator for 4 to 6 h.
5. To each dish, add 0.5 ml of growth medium, supplemented with 30% serum. (A large amount of serum is necessary to protect cells in suspension, which are more sensitive to the toxic effects of liposome reagents.)
6. Incubate the dishes in a CO_2 incubator overnight.
7. Add 2 ml of complete growth medium to each dish, and incubate the dishes in a CO_2 incubator.
8. At 24 to 72 h after the start of transfection, assay the cells or medium for gene activity as appropriate. (See Protocols 27.8 and 27.9.)

Electroporation

DNA can be introduced into cells by electroporation, when a high cell concentration is briefly exposed to a high-voltage electric field in the presence of the DNA to be transfected [Chu et al., 1987]. Small holes are generated transiently in the cell membrane [Zimmerman and Vienken, 1982], and the DNA is allowed to enter the cell and, in some of the cells, becomes incorporated into the genome. Equipment for electroporation is available commercially (BioRad). Most cells refractory to chemical methods of gene transfer are successfully transfected by electroporation [Andreason and Evans, 1988; Chu et al., 1987].

Electroporation is usually performed at a constant capacitance setting (and therefore a constant pulse duration), with various field strengths (500–1500 kV/cm) for pilot investigations. For most cells, the settings at which approximately 20–50% of the cells remain viable after electroporation are sufficient for DNA transfer [Chu et al., 1987; Andreason and Evans, 1988]. Electroporation is usually performed at room temperature, and the cells are subsequently kept on ice, to extend the period of time that the membrane pores remain open [Andreason and Evans, 1989].

There is a linear relationship between DNA concentration, DNA uptake, and reporter gene expression [Chu et al., 1987]. It is believed that linearized DNA is more efficient for the production of stable transfectants than is supercoiled DNA, presumably due to the increased efficiency with which linear DNA integrates into the genome DNA [Potter et al., 1984; Chu et al., 1987]. Electroporation results in the integration of DNA in low copy number [Boggs et al., 1986; Toneguzzo et al., 1988], although the copy number introduced can be adjusted by altering the concentration of DNA in the cell suspension. Chemical methods of transfection usually result in the integration of large concatamers, which may inherently interfere with cell function and obscure investigations involving specific gene overexpression [Robins et al., 1981; Kucherlapati and Skoultchi, 1984].

Suspension cells are more easily transfected by electroporation than are adherent cells, since adherent cells must be detached from the culture vessel. The drawbacks of electroporation include its requirement for more cells and DNA than chemical methods of gene transfer and its variability in optimal parameters between cell types.

The above introduction and the following protocol, have been abridged from Cataldo et al. [1998]. The protocol describes electroporation conditions for the stable transfection of a clone (Y10) of the murine megakaryocytic cell line L8057 [Ishida et al., 1993]. L8057-Y10 cells grow in suspension and are maintained in Ham's F12 medium supplemented with 10% heat-inactivated FBS, penicillin (50 U/ml), and streptomycin (50 μg/ml). The overall scheme presented here is for the stable transfection of L8057-Y10 cells.

PROTOCOL 27.7. STABLE TRANSFECTION BY ELECTROPORATION

Materials

Sterile or Aseptically Prepared:

Expression vectors and preparation of DNA:

pSVTKGH [Selden et al., 1986], linearized

This vector contains the SV40 enhancer and the herpes virus thymidine kinase promoter sequences that drive the expression of human growth hormone (hGH). hGH is secreted directly into the tissue culture medium and is detected by radioimmunoassay [Ravid et al., 1991].

pcDNA3 (Invitrogen, San Diego, CA), linearized

This expression vector contains the human cytomegalovirus enhancer–promoter sequences upstream from its multicloning site, and the polyadenylation

signal and transcription–termination sequences of the bovine growth hormone gene (bGH) downstream from the multicloning site. The pcDNA3 vector also contains the gene for neomycin resistance, alleviating the need to cotransfect an antibiotic resistance gene for the selection of stable transformants overexpressing a particular gene of interest. pcDNA3 may also be used as a selectable marker when cotransfected with reporter constructs, as presented in the protocol.

Tissue culture:

L8057-Y10 cells

Flasks, 75 cm²

Plates, six well

PBSA

Medium F12/FB: Ham's F12 with glutamine (2 mM), penicillin (50 U/ml), and streptomycin (5 μg/ml), supplemented with 10% FBS (heat inactivated; Gibco #16140-014)

Geneticin (Gibco), 50 mg/ml in PBSA (adjust for potency—e.g., if the potency of a batch is 731 μg/mg, dissolve 64.4 mg of G418 per ml of PBSA)

Selective medium: F12/FB containing 600 μg/ml of G418

Electroporation:

Electroporation buffer [Ravid et al., 1991]:

NaCl, 30.8 mM

KCl, 120.7 mM

Na_2HPO_4, 8.1 mM

KH_2PO_4, 1.46 mM

$MgCl_2$, 5 mM

Gene Pulser Cuvettes

Nonsterile:

Gene Pulser II Apparatus (BioRad)

Gene Pulser II Capacitance Extender Plus

Protocol

1. Seed L8057-Y10 cells at 5×10^5 cells/ml in a T-75 culture flask, and incubate the flask at 37°C in 5% CO_2 in F12/FB.

2. During the late log phase, collect the cells by centrifugation at 4°C at 380 g for 5 min.

3. Wash the resultant cell pellet in 10 ml of PBSA, and centrifuge at 4°C at 380 g for 5 min.

4. Wash the resultant cell pellet in 5 ml of electroporation buffer, and count the cells, using a hemocytometer. Collect the cells by centrifugation at 4°C at 380 g for 5 min.

5. Resuspend the cells at 1×10^6 cells/0.8 ml of electroporation buffer.

6. Transfer 0.8 ml of cells into prechilled electroporation cuvettes.

7. Add 50 μg of linearized pSVTKGH, along with 5 μg of linearized pcDNA3, to the cell suspension. (When analyzing a specific gene pro-

moter–reporter gene expression, a suitable negative control is to electroporate the cells in the absence of plasmid DNA.)

8. Mix the DNA/cell suspension by holding the sides of the cuvette and flicking the bottom. Incubate the suspension on ice for 10 min.

9. Electroporate the suspension at 400 volts/500 μF, and record the duration of the shock. Remove the cuvette from the shocking chamber, and incubate it on ice for 10 min.

10. Transfer the electroporated cells into 10 ml of F12/FB, rinse the cuvette with medium to remove all of the cells, and collect the cells by centrifugation.

11. Resuspend the cells in 20 ml of F12/FB, and culture the suspension in a T-75 flask at 37°C in 5% CO_2 to allow expression of the neomycin selectable marker gene.

12. After 24–48 h, collect the cells by centrifugation, and then resuspend the cells in 20 ml of selective medium.

13. Change the selective medium every 2–4 d for at least 2 weeks, to remove the debris of dead cells and to permit resistant cells to grow.

14. Assay for hGH expression in the cell supernatant. (See Reported Gene Expression below.)

Variations

Geneticin. Cultured cell lines differ in their sensitivity to geneticin, and the most suitable concentration to use must be empirically determined by doing a kill curve on the cell line that is being transfected. The cells should remain in selection medium throughout their growth period.

Picking Colonies. See Protocols 13.6 and 13.8.

Transient Transfection. Use circular plasmids, such as pCMVβ-gal, to replace pcDNA3. After step 10, resuspend the cells in 2 ml of culture medium without selection agent, and culture them in a six-well plate for three to four days. Assay for transient hGH expression in the cell supernatant, and prepare a cell lysate to determine β-galactosidase activity.

Reporter Gene Expression. In the case of transient transfection, a plasmid containing the β-galactosidase reporter gene is cotransfected, in order to normalize hGH gene expression for overall electroporation efficiency. There are commercial kits available for the detection of both hGH secretion (Nichols Institute Diagnostics, #40-2205) and β-galactosidase activity (Promega, #E2000).

Other DNA Transfer Methods

Retroviral Infection. Retroviruses have a high efficiency of gene transfer, are able to incorporate larger DNA fragments than plasmids, and infect host cells spontaneously [Ausubel et al., 1996; Hicks et al., 1998]. The introduced gene becomes permanent, and the process of insertion is achieved by normal cellular processes, is not harmful to the host cell, and does not cause any other genetic alterations.

Baculovirus. Inserting genomic sequences into baculoviruses, which are then propagated in insect cells (such as Sf9 cells), also allows large sequences (>100 kbp) to be cloned. The proteins produced have post-translational modifications that are not available in prokaryotic systems, although there are differences in processing from mammalian cells. Baculoviruses are not transmissible to mammalian cells, and Sf9 cells are unlikely to carry any risk of contamination with mammalian viruses, provided that any mammalian-derived supplements (e.g., FBS) are thoroughly screened before use. Sf9 can be subcultured without trypsin [see Protocol 26.9; Midgley et al., 1998].

Yeast Artificial Chromosomes. Yeast artificial chromosomes (YACs) [Strauss, 1998] also provide a genome that is capable of packaging larger sequences of DNA than bacterial plasmids, with downstream post-translational processing such as found in eukaryotic cells, although this latter aspect is likely to have significant differences from that in mammalian cells. Propagation in yeast also gives a high-yield and stable culture system and is less difficult to maintain than large-scale insect or mammalian cell cultures.

Mammalian Artificial Chromosomes. There has been considerable success in applying the principle of YACs to mammalian systems [Ikeno et al., 1998; Ascenzioni et al., 1997], in order to incorporate large mammalian sequences containing one or more structural and multiple regulatory genes into one construct. These constructs are known as mammalian artificial chromosomes (MACs) and are introduced into mammalian cells by monochromosomal transfer techniques (see Monochromosomal Transfer in this chapter). The technology has now progressed to the construction of human artificial chromosomes opening up a whole new era for genetic therapy [Willard, 1998; Warburton and Kipling, 1997].

Reporter Genes

Assays of transfected cells for reporter genes, such as β-gal or CAT, confirm that the DNA construct has been incorporated and is expressed by the transfected population. β-gal staining can also be used to track cells in cell interaction studies [Bradley and Pitts,

1994]. The following protocol is abridged from Bichko [1998], based on Sanes et al. [1986], and modified by Life Technologies.

PROTOCOL 27.8. *IN SITU* STAINING FOR β-GALACTOSIDASE

Materials
Nonsterile:
PBSA

Substrate: X-gal (Gibco), 20 mg/ml in dimethylformamide. Store at −20°C in the dark, in a polypropylene tube, for up to 6 months.

Fixative: Formaldehyde, 1.8%; glutaraldehyde, 0.05% in PBSA. Prepare the fixative by combining 85 ml of water, 10 ml of 10× PBSA, 5 ml of formalin (37% formaldehyde solution), and 0.2 ml of glutaraldehyde (25% solution). Store at 4°C.

Stain solution: 5 mM potassium ferricyanide, 5 mM potassium ferrocyanide, 2 mM $MgCl_2$ in PBSA. Store at 4°C.

Substrate/stain solution: 1 mg/ml of X-gal in stain solution. Prepare immediately before using.

Formalin, 10%, in PBSA.

In this case, the expression plasmid for β-gal would be used as a reporter gene for transfections (for example, pCMVβgal [MacGregor and Caskey, 1989]).

Protocol
1. Wash the cells (e.g., in 6-well plates) once with 2 ml of PBSA.
2. Fix the cells with 1 ml of fixative for 5 min at room temperature.
3. Wash the cells twice with 2 ml of buffered saline.
4. Add 1 ml per well of substrate/stain solution to the cells, and incubate them 2 h to overnight at 37°C.
5. Rinse the cells in each well with 2 ml of buffered saline. Observe the cells on an inverted microscope, and count the blue (β-gal positive) cells.
6. To store the plates, fix the cells in each well with 1 ml of 10% formalin in buffered saline for 10 min at room temperature, rinse the cells with buffered saline, and store in buffered saline at 4°C.

An expression plasmid for CAT (for example, pCMVCAT [Boshart et al., 1985]) can also be used as a reporter gene for transfection.

The following protocol is abridged from Bichko [1998], based on Neumann et al. [1987], and modified by Life Technologies.

PROTOCOL 27.9. CHLORAMPHENICOL ACETYLTRANSFERASE (CAT) ASSAY

Materials
Sterile:
PBSA
Multiwell plates: 6 well, 35 mm
Nonsterile:
Tris buffer: Tris-HCl, 0.1 M, pH 8.0
Tris/Triton: Tris-HCl, 0.1 M, pH 8.0, 0.1% Triton X-100; store at 4°C
CAT dilution buffer: Tris-HCl, 0.1 M, pH 8.0, 50% glycerol, 0.2% BSA
Substrate: 250 mM chloramphenicol (Gibco) in 100% ethanol; store in aliquots at −70°C.
[14C]-CoA: [14C]-butyryl Coenzyme A (10 μCi/ml; Amersham Pharmacia)
CAT enzyme standards (Gibco): Prepare standard solutions of 0.2, 1, 2, 4, and 10 U/ml in CAT dilution buffer. Store at 4°C.
Liquid-scintillation cocktail: Econofluor (Gibco)
Deionized, distilled water (DDW)
Polypropylene scintillation vials, 3.5 ml
Microcentrifuge tubes
Microcentrifuge
Ice bath

Protocol
Cell Harvesting for 6-Well, 35-mm Plates:
1. At 24–72 h after transfection, wash the cells once with buffered saline.
2. Put the plates on ice, and add 1 ml of Tris/Triton per well.
3. Freeze the plates for 2 h at −70°C.
4. Thaw the plates at 37°C, and then put them on ice.
5. Transfer the cell lysates to microcentrifuge tubes, and spin the tubes for 5 min at maximum speed.
6. Collect the supernatants, and heat them at 65°C for 10 min to inactivate the inhibitors of CAT.
7. Centrifuge at maximum speed for 3 min and collect the supernatant ("cell extract"). Keep the supernatant at −70°C.
CAT Assay:
8. Put 5–150 μl of cell extract from each sample into a 3.5-ml polypropylene scintillation vial, and add sufficient 0.1 M Tris to it to reach a final volume of 150 μl.
9. For a negative control, use 150 μl of 0.1 M Tris.
10. For a positive control,
   (a) Add 150 μl of 0.1 M Tris to each of 5 vials.
   (b) Add 5 μl of each CAT standard solution to

the vials, to give a standard curve of 1, 5, 10, 20, and 50 mU of CAT.

11. To each sample (including the controls), add 100 μl of the following mixture:
    (a) 84 μl of UPW,
    (b) 10 μl of 0.1 M Tris,
    (c) 1 μl of chloramphenicol, and
    (d) 5 μl (50 nCi) of [^{14}C]-CoA.

12. Cap the samples, and incubate them at 37°C for 2 h.

13. Add 3 ml of Econofluor to all tubes, and then recap the tubes.

14. Mix the contents by inverting the tubes.

15. Incubate the tubes at room temperature for 2 h.

16. Count the samples for 30 s in a liquid-scintillation counter.

CHAPTER 28

Problem Solving

No matter how well a laboratory is run, problems arise when new staff, new techniques, or any other new development destabilizes its normal flow of activity. One solution to such problems is to make sure that procedures do not change—i.e., to define standard operating procedures (SOPs) and ensure that deviations from these procedures are made only following exhaustive testing of the possible repercussions. However, this is often difficult, particularly in a research environment, where progress demands change and new procedures are introduced continually.

The advice given in the preceding chapters has concentrated on practical, "how to do it" instructions, sometimes with indications of what might go wrong. This chapter attempts to summarize these potential problems under topic headings, and adds a few more potential difficulties, queries, and, hopefully, solutions. The bold headings provide the subject area and potential problems; the bulleted text suggests remedies.

SLOW CELL GROWTH

Is the problem restricted to your own stocks, or are other researchers having similar problems?

If the problems are restricted to your own stock,
Check your cells for contamination. (See Contamination in this chapter, and Monitoring for Contamination in Chapter 18.)

Check the growth of your cells with other media and sera (i.e., a different batch or supplier). If you normally use powder or 10×, buy in an alternative stock of 1×. If this change indicates that the problem is with the medium, see the subsection on medium later in this section.
If the problem is not with the medium, it may be with your cells.

If the problem is with your cells,
Thaw out another ampule from the freezer, and compare it with your current stock. Handle the two cultures separately, in case you have a contamination in the original stock. Do this before checking (or while checking) the following:

Maintenance regime. Count your cells at subculture, and set up a growth curve (see Protocols 20.7 and 20.8) to compare with your previous records for these cells. Check if the following factors apply to your cells:

The seeding density was too low at transfer.
The cells were subcultured too frequently.
The cells were allowed to remain for too long in the plateau phase before subculture.

Subculture routine. Check if the following factors apply to your procedures:

Was there a change in the batch of trypsin or another dissociation agent? Check the batch numbers and suppliers.

To assess the severity of dissociation, check to determine if:
   (a) The duration of exposure to trypsin (or other agents) was too long.
   (b) The agent was too concentrated or had specific activity that was too high.
   (c) The incubator used for trypsinization was too warm.
Pipetting during dissociation was too vigorous.
The cells are sensitive to EDTA (if EDTA was used).
The wrong diluent was used for trypsin.
A bad batch of trypsin diluent was used.

Contamination of the cells.
If you are working without antibiotics (and you should be!), most contaminations will be obvious.
Mycoplasma will not be obvious, and your cells should be checked regularly for it. (See Protocol 18.2.)
See Microbial Contamination in this chapter and Sources of Contamination in Chapter 18 for possible routes and causes.

If the problem is more general and other people are having difficulty as well, check shared facilities and reagents

Equipment. Check the following:

Hot room and incubators
The temperature and stability are inadequately controlled
   (a) The thermostats are faulty.
   (b) The access to the incubator is too frequent.
   (c) The hot room door is being left open.
   (d) The circulating fan(s) has failed or is overheating.
Check with portable recording thermometer (see Temperature recording in Chapter 4).

Humidity of CO_2 incubators
The water tray is not filled.
There is leakage around the doors.
The access to the incubator is too frequent.
Check evaporation rate by weighing Petri dish of PBSA daily.

CO_2 concentration in CO_2 incubators
Check the pH *in situ* with a Petri dish of pretested medium.
The access to the incubator is too frequent.
There are problems with the CO_2 controller.
   (a) Check with CO_2 tester (Carborite).
   (b) Recalibrate zero settings and CO_2 concentration on read-out with standard gas mixtures.

Have any changes occurred in the laboratory?
Check if the following factors have undergone any changes:

Staff
Culture staff
Preparation staff
Preparation procedures and training

Materials
Chemical contamination (see Chemical Contamination in this chapter)

Media (see later)
Supplier
Batch
Type
Storage

Procedures
Incubation times
Speed of operations
Sequence of operations
Location
Scale

Equipment
Replacements: has any new equipment been added?
Operational status: has any equipment become faulty?
Location: has position or location of equipment being used, or adjacent equipment, been changed?
Operating temperature: is the equipment working within its optimal temperature range and that of your samples within it?

Medium
Are there problems with either your own stocks or with general stocks?

Age
Check the batch number and purchase date of the medium. It should not be more than 1 year old.
If the medium contains glutamine, then it is stable only for 1 month at 4°C.

Storage conditions
Is the temperature of the cold room or refrigerator at or below 4°C?
Is the medium stored in the dark, or at least in tungsten light? If not, it will degrade in fluorescent light, unless the light has a low UV output.

Adequacy of the medium
Check the medium against other media. (See Selection of Medium and Serum in Chapter 8.)
Buy in 1× medium, and compare it with your own.

If you already use 1× medium, try another supplier, and check individual additions to the medium (e.g., serum, growth factors, hormones, etc.).

Frequency of changing the medium
Check the cell concentration and pH. If the pH falls below 7.0 in under 48 h, either the pH is too low at the start of the growth period, the cell concentration is too high, or there is a contamination.

pH
Check that the pH is between 7.0 and 7.4 throughout the period of culture.

pH fluctuations
Check the CO_2 supply to the incubator. (See *Equipment*, above)
Check the CO_2 regulation of the incubator. (See *Equipment*, above)

High pH
The CO_2 concentration in the incubator is too low.
The HCO_3^- concentration of the medium is too high.
If you are using DMEM, it should be used with a gas phase of 10% CO_2 if it has been made according to the original recipe.

Osmolality

If the stock medium, one of the components, or the dilution procedure is wrong, the medium may be hyper- or hypotonic.
Check osmolality on an osmometer; it should be between 270 and 340 mOsmol/kg.
Check the preparation procedures.
  (a) Is the amount of water used to dilute a 10× stock, or dissolve powder, correct?
  (b) Has a reagent been added that would alter osmolality?

Accidental omission of a component
Make up a fresh batch of medium (which is quicker than trying to decide which component is missing).

If a constituent of the medium has poor solubility or precipitates before filtration,
Check to ensure that all constituents are dissolved before filtering.

If there is precipitation on storage,
Check that the precipitate redissolves on diluting and/or heating to 37°C.
If the precipitate does not redissolve, discard the batch.

If precipitation recurs, complain to the supplier, or change your supplier.

Defective component
Replace the components one at a time from an alternative source.
Keep a record of the batch numbers.

New batch of stock medium that appears to be faulty
Compare the batch with a previous batch, if one is still available.
Compare the batch with a batch of another supplier if no previous batches are available.

If medium is BSS-based, is BSS satisfactory?
Check with other users.
Try alternative sources of BSS or a different formulation.

If medium is water based, is ultrapure water satisfactory?
Check with other users
Check against fresh 1× medium, bought in complete.

CHOICE OF MEDIUM

Selection of the correct medium can be critical

Has the medium been changed since the cells were acquired?
Revert to the previous type of medium.
Obtain the medium from the same source as that of the supplier or originator of the cell line.
Compare the constituents of the medium with those of the supplier or originator of the cell line.

If you are starting with a new culture, then you may need to screen several media. Screen the media for the following parameters:

Type of medium (see Selection of Medium and Serum in Chapter 8).
Type of serum (see Selection of Medium and Serum in Chapter 8).
Concentration of serum (see Protocols 20.7, 20.8, 21.4, and 21.3)
If serum-free is required, see Selection of Serum-Free Medium in Chapter 9.

UNSTABLE REAGENTS

Glutamine

Store it frozen at −20°C.

As it is reduced to 50% in 3–5 d at 37°C, replace medium after 3 d.

Use a stable alternative—e.g., dipeptide Glutamax—but test it first.

Serum

Store it frozen.

If it is partially thawed, thaw it completely, mix it, and refreeze it.

Test batches of serum before use.

Overlap batches, in case a deficiency is slow to appear. (It may take 2–3 subcultures.)

Other Constituents

Store at 4°C.

Most constituents of media are stable for 1–2 weeks at 37°C.

Trypsin

Store concentrated stock frozen.

Make using stock up from concentrate, and store the diluted solution at 4°C for a maximum of 2 weeks.

Trypsin will degrade unstable if left at room temperature for over 30 min.

PURITY OF CONSTITUENTS

Is the water purifier working correctly?

Monitor the output
Test the conductivity of the water.

Test the total organic content (TOC; Millipore instrument) of the water.

Check medium made with the water against 1× medium.

Check the maintenance regime
When was the deionizer last changed?

When was the reverse osmosis cartridge changed?

Is all connecting tubing clean? Are there any signs of algae, fungi, or bacteria?

Check the storage vessel for algal or fungal contamination.

Check for release of deionizer resin into output water.

Chemical contamination or residue in the glass boiler of the still

Dismantle the glass boiler, and clean it in N HCl.

Chemical traces in the plastic tubing

Change the tubing to new, inert, washed, sterilized, plastic tubing (see Miscellaneous Equipment in Chapter 10).

Bicarbonate

Is the concentration correct?
Check the conductivity or osmolality against a reference standard solution.

Try another batch. (Make it up or buy it in.)

Check for signs of precipitate.

Check if you are using the right amount for the medium. (See Tables 8.1, 8.3, 9.1, and 9.2.)

Antibiotics

Evaluate the following factors:
Frequency of use.

Concentration.

Batch number.

Combinations.

Fungicide (e.g., amphotericin B can be toxic).

Serum

Are you using a new batch?
Check the supplier's quality control.

Compare the batch with a previous batch or other batches.

Concentration
Too low?

Too high?

Reconfirm the lack of toxicity, growth promotion, and plating efficiency
Create a growth curve. (See Protocols 20.7 and 20.8.)

Run a clonal growth assay. (See Protocol 20.9.)

PLASTICS

Are you using a new make, type, or batch?

Check your batch against a previous batch.

Try an alternative supplier.

GLASSWARE

Wash-up

Are other cells showing symptoms of deterioration or impaired growth?

Are other users having trouble?

Is there trace contamination on the glass of storage bottles or pipettes? (See Chemical Contamination in this chapter.)

Have the caps not been properly rinsed? (Adding a few ml of BSS will show by a pH change if detergent has been left in the cap.)

Has any chemical glassware been mixed in with tissue culture glassware?

MICROBIAL CONTAMINATION

See also Table 18.1.

Single User

Sporadic

Single species
Check the aseptic technique of the operator. (See Chapter 5.)
Are there any media or reagents that are unique to that user?
Check personal cleanliness of the operator.
  (a) Are hands washed before and after culture work?
  (b) Is the lab-coat changed?
  (c) Is long hair tied back?
Get the operator to wear gloves.

Mycoplasma
Check with other users for evidence of mycoplasma contamination.
Ensure that the screening program is operative.
Imported cell lines are the most common source of infection. Quarantine them before screening for use in the main tissue culture area. Keep records on importation.
Check the natural-product reagents (serum, trypsin) with an indicator cell line. (See Protocol 18.2.)
Restrict the use of antibiotics to primary culture and critical experiments.

Multispecific
Is technique bad?
Check use of laboratory coats:
  (a) Is the lab-coat changed before commencing culture work?
  (b) Is the lab-coat buttoned? (Even people passing through with a flapping coat can disrupt the laminar air flow.)
Check with other users of the same hood; the hood may be faulty and need service.

Repeated

Single species
The problem is usually a reagent or cell line; is there a unique reagent or cell line that no one else uses?

Multispecific
Is the technique bad. (See Sterile Handling; Standard Procedure in Chapter 5.)
Is the procedure nonstandard?
Is the location bad, or is there too much equipment or traffic?
Is the hood overcrowded?
Are nonsterile reagents being used?
Check use of laboratory coats (see Sporadic multispecific, above).

Continuous

Single species

Contaminated solution
Mix medium (antibiotic free) 1:1 with nutrient broth (e.g., L-Broth), and incubate the solution.
Plate the medium out on blood agar, and incubate the solution, upside down, with blank controls.

Contaminated cell line
Check the cells by Hoechst staining. (See Protocol 18.2.)
Mix the cells with broth, and incubate.

Multispecific
Is technique bad? (See Sterile Handling, Standard Procedure in Chapter 5.)
Is the procedure nonstandard?
Is the location bad, or is there too much equipment or traffic?
Is the hood overcrowded?
Are nonsterile reagents being used?

Widespread

Sporadic

Single species
Is an infrequently used reagent or medium contaminated?
Is the incubator contaminated. If so, clean it out. (See Protocol 18.1.)
Is there an increased spore count in the atmosphere? (Check by exposing bacteriological plates when room is quiet.)

Multispecific

Sterilization failure
Check sterilizing ovens for:
 (a) Overcrowding of contents preventing adequate air circulation.
 (b) Integrity of door seals and any other apertures.
 (c) Temperature and duration of sterilization cycle (160°C for 1 h).
Check autoclaves:
 (a) For overcrowding of contents preventing adequate steam circulation.
 (b) For temperature and duration of sterilization cycle (121°C for 15–20 min) with probe in equivalent sample in center of load.
 (c) To ensure all empty vessels are left open for steam circulation. (See Fig. 10.3.)

New staff is not following standard procedure
Check the training of the new staff with supervisor.

Contaminated storage (e.g., cold room or refrigerator)
Clean out the storage facility.
Make sure that sterilized items are not stored unsealed.
Check the turnover of sterile stocks.

Repeated, single species

Contaminated reagent or medium
Check the frequency of use of reagents among users to narrow the problem down to common reagents.
Test likely candidates by incubating them 1:1 in broth.

Contaminated incubator
Clean out the incubator. (See Protocol 18.1.)

Contaminated hood
Check the cleaning schedules.
Clean out the hood below the work surface.
Check the maintenance schedules.
Expose bacteriological plates in the hood, placed at several locations on the work surface.

Repeated, multispecific

Sterilization failure
Check the autoclaves and sterilizing ovens for overcrowding.
Check the autoclaves and sterilizing ovens for electrical or mechanical failure.
Check the printouts and records of the sterilization cycles.

Check the integrity of the sterilization chamber of ovens.

New member of staff is not following standard procedure
Restrict the use of antibiotics.
Check the training of new staff.
Check procedures with the line manager.
Check for changes in procedures and/or requirements.
Redraft SOPs to suit any changed circumstances.

Contaminated storage (e.g., cold room or refrigerator)
Clean out the storage facility.
Make sure that sterilized items are not stored unsealed.

Contaminated room air
Check for contamination of the room's air by leaving blood agar plates open when the room is not in use, but when the ventilation is on.
Check the aseptic technique of operators.
Check the integrity of the hoods.
Check the air supply and air conditioners.
Check the quality of cleaning of incoming equipment.
Is there building work or other disturbance nearby? If so, try to improve isolation of culture laboratory (e.g., close doors, erect screens, exclude nontissue culture staff, clean all equipment and materials entering the room).

Identification of Contamination

Bacterial, fungal
Check for bacterial or fungal contamination of cells or media by using high-power phase contrast.
Incubate the cells in broth.
Plate out the cells on blood agar, and incubate the culture.
Gram stain the cells, or consult a microbiologist.

Mycoplasma
Stain the culture with Hoechst 33258. (See Protocol 18.2.)
Check for cytoplasmic DNA synthesis (incorporation of [^{3}H]-thymidine), using autoradiography. (See Protocol 26.3.)
Get a commercial test done. (See Sources of Materials—Mycoplasma Testing in Trade Index.)

Viral

To test for viral contamination, use:

Transmission or scanning electron microscopy.

Immunostaining with a fluorescent or peroxidase-conjugated antibody. (See Protocol 15.12.)

Elisa assays.

PCR.

Decontamination

Decontaminate cells only if they are irreplaceable. (See Protocol 18.3.)

CHEMICAL CONTAMINATION

Glassware

Residue after cleaning

Keep tissue culture glassware separate from chemical.

Ensure that there is no carry-over from the last rinse of a previous chemical wash in a washing machine.

Carry out spot checks by

(a) visual examination

(b) adding a small volume of BSS with phenol red and looking for pH change

(c) adding a small volume of UPW and checking conductivity.

(d) cloning cells on the glass after sterilization by dry heat

Select detergent carefully. (See Selection of Detergent in Chapter 10.)

Storage

Foil cap all open vessels.

Store in dust-free area.

Pipettes

Residue or blockage after cleaning

Ensure pipettes are washed and dried tip uppermost.

Collect pipettes into detergent but rinse in water only.

Do not allow agar to be used in glass pipettes.

Check pipettes after washing and before sterilization by

(a) visual examination

(b) adding a small volume of BSS with phenol red and looking for pH change

(c) adding a small volume of UPW and checking conductivity.

Make sure cotton plugs are removed before washing.

Water Purification

See Purity of Constituents in this chapter.

Other Reagents

DMSO can be contaminated by dissolving plastic or rubber from container.

Store DMSO in glass or polypropylene with glass or polypropylene cap.

Glycerol deteriorates on longterm storage

Buy in small amounts which will be used within 3–6 months.

Store in dark bottle.

Powders and Aerosols

Handle toxic chemicals which produce powders or aerosols in a fume hood.

Avoid drafts when weighing powders or dispensing liquids.

Control traffic of people and equipment into tissue culture laboratory and preparation areas.

Laboratory coats carry chemical contamination if used in general laboratory.

Change to a clean laboratory coat before entering the tissue culture laboratory.

PRIMARY CULTURE

Poor take in primary culture

Primary explants do not attach

Scratch the substrate through the explant. (See Protocols 11.4 and 22.9.)

Trap explant under a coverslip. (See Protocol 11.4.)

Embed in a plasma clot. (See Protocol 11.4.)

Disaggregation

Incomplete

Incubate the cells in protease for a longer amount of time.

Try cold pretreatment before incubation. (See Protocol 11.6.)

Use an alternative, or additional, protease. (See Table 12.3; Other Enzymatic Procedures in Chapter 11.)

Complete disaggregation but poor attachment

Floating cells are viable

Treat the substrate

(a) Coat the plastic with collagen, fibronectin, laminin, poly-D-lysine, or poly-L-lysine. (See Matrix Coating in Chapter 7.)

(b) Use a feeder layer. (See Feeder Layers in Chapter 7; Protocols 13.3, 22.4, and 23.1.)
The culture may be nonadherent.

Floating cells are mostly nonviable
Adjust the concentration to a viable cell count.
Remove the nonviable cells. (See Protocol 11.10.)

Floating cells are nonviable and few cells have attached

Cell density
The cell density is too low; increase it to up to 1×10^6 cells/ml.

Cell adhesion is poor
Treat the substrate (see under Floating cells, above, in this section)

Enzymes used are too toxic
Change to a different protease. (See Table 12.3; Other Enzymatic Procedures in Chapter 11.)
Reduce the exposure time.
Try cold pretreatment with protease before incubation. (See Protocol 11.6.)

Medium is very acidic
Check the medium for contamination. (See Monitoring for Contamination in Chapter 18.)
Reduce the cell concentration at seeding or 24 h later.
Add HEPES buffer, and vent the flask.

Wrong medium
Try a range of media. (See Selection of Medium and Serum in Chapter 8; Tables 8.3, 8.6, 9.1, and 9.2; and Selection of Serum-Free Medium in Chapter 9.)
Replace the serum with serum-free medium.
Check the literature for media used with your cells (if you have not already done so).

Supplementation of medium
Use different types or batches of serum. (See Testing Serum in Chapter 8.)
Use different growth factors. (See Growth Factors in Chapter 9 and Table 9.3.)
Use other mitogens (e.g., PMA; prostaglandins; hydrocortisone or other steroids; peptide hormones, such as insulin and transferrin; see Table 9.2).
Use conditioned medium. (See Conditioned Medium in chapter 8; Protocol 13.2.)

Wrong Cells Have Been Selected

Overgrowth by fibroblasts or endothelium
Use selective media. (See Selective Media in Chapter 9 and Selective Inhibitors in Chapter 13.)

Use selective substrates. (See Matrix Coating in Chapter 7 and Interaction with Substrate in Chapter 13.)
Use a selective feeder layer. (See Protocol 23.21.)
Check for cross-contamination from a feeder layer or a xenograft host. (See Cross-Contamination in Chapter 18; Protocols 15.9, 15.10, and 15.11.)

Cross-contamination with another cell line
Check for cross-contamination with other cell lines currently in culture. (See Cross-Contamination in Chapter 18.)

Contamination
Check for contamination. (See Monitoring for Contamination in Chapter 18.)
Pretreat the tissue with antibiotics (see Collection Medium DBSS in Reagents Appendix) or 70% ethanol (see step 7 in Protocol 11.3).
Eradicate the contamination, but only if the material is irreplaceable. (See Eradication of Contamination in Chapter 18.)

CLONING

See also Slow Cell Growth in this chapter.

Poor Plating Efficiency

Too few colonies per dish
Increase the seeding concentration.
Use a feeder layer.
Improve the plating efficiency. (See Stimulation of Plating Efficiency in Chapter 13.)

Colonies are too diffuse
Select a bigger Petri dish, to give more space for the colonies to spread out.
Use glucocorticoid (e.g., dexamethasone, 1×10^{-5}– 1×10^{-6} M).

Mycoplasma contamination
Screen for mycoplasma. (See Protocol 18.2.)
Eradicate the mycoplasma, but only as a last resort, if the cell line is irreplaceable. (See Eradication of Mycoplasma in Chapter 18; Protocol 18.3.)

Poor handling
The cells have been out of the incubator too long.
A prolonged amount of time has been spent in dilution.

Medium

Type
Choose a rich medium (e.g., Ham's F12).
If the medium is serum-free (e.g., MCDB 153 for

keratinocytes; see Tables 9.1 and 9.2), try different serum-free media.

CO_2 is essential.

If serum is essential

Fetal bovine serum is usually better than calf or horse serum.

If fetal bovine serum is already being used, increase the concentration.

Select the batch based on the plating efficiency of the cells that you use. (See Protocol 20.9.)

Poor substrate

Has there been a change in the culture wave supplier?

Coat the matrix. (See Matrix Coating in Chapter 7; Protocol 7.1; Cell–Matrix Interactions in Chapter 16; Protocol 22.9: FN/V/BSA.)

Diffuse Colonies

Coat the matrix. (See Matrix Coating in Chapter 7; Protocol 7.1; Cell–Matrix Interactions in Chapter 16; FN/V/BSA in Protocol 22.9.)

Use glucocorticoid (e.g., 1×10^{-7}–1×10^{-5} M dexamethasone) for the first 48–72 h.

Use a feeder layer. (See Protocol 13.3.)

Too Many Colonies per Dish

Reduce the seeding concentration.

Seed the same number of cells into a larger dish.

Overlapping colonies

Reduce the seeding density.

Grow the colonies for a shorter time (fewer cells/colony).

Seed the same number of cells on a larger dish.

Nonrandom Distribution

Add the cells to medium in a bottle, mix the cell suspension, and then seed the dishes.

Do not swirl the dishes to mix cells or when pipetting.

Ensure that the dishes are level.

Make sure that the medium covers all of the bottom of the dish evenly.

Ensure that the incubator is free from vibration.

Restrict access of other users to the incubator.

Label the box or tray containing the dishes with the phrase, "CLONING, DO NOT MOVE."

Place the box or tray at the back of the incubator.

Place the box or tray containing the dishes on a foam pad. (Wash the pad regularly.)

Nonadherent Cells

Anchorage-independent cells

Clone the cells in agar (see Protocol 13.4), agarose, or Methocel (see Protocol 13.5) over an agar underlay or in a non-tissue-culture-grade dish.

Add growth factors or supplements to underlay.

Use tissue culture grade dish and seed feeder layer before pouring underlay.

Poorly attached cells

Clone the cells in Methocel in a tissue-culture-grade dish. (See Protocol 13.5.)

DIFFERENTIATION

Cells Do Not Differentiate

Apply differentiation-inducing conditions. (See Induction of Differentiation in Chapter 16.)

Use a selective medium (see Tables 9.1 and 9.2) at isolation and during propagation.

Loss of Product Formation

Check for the expression of the relevant gene (using PCR or Northern blot [Ausubel et al., 1996]).

Check the reporter gene expression (using PCR or Northern blot [Ausubel et al., 1996]; see Protocols 27.8 and 27.9).

Apply differentiation-inducing conditions. (See Induction of Differentiation in Chapter 16.)

Use the appropriate selective medium for the cell type (see Tables 9.1 and 9.2) at isolation and during propagation.

FEEDING

Regular Monolayers

Use standard procedure. (See Protocol 12.1.)

Clones

pH too high or too low. (See Slow Cell Growth—Medium, pH in this chapter.)

Monolayer

Use standard procedure. (See Protocol 12.1.)

Suspension

Add medium to agar cultures.

Add medium plus Methocel to Methocel clones; the nutrients will diffuse through the Methocel.

SUBCULTURE

This section discusses the problem of poor take or slow growth after subculture; see also Slow Cell Growth in this chapter.

Cell Cycle Phase at Subculture

Cells should be subcultured from the exponential or late-exponential phase. (See Protocol 12.2.)

Plateau-phase cells tend to have a long lag phase.

Some cultures—e.g., hybridomas, mouse leukemias —will deteriorate rapidly in the plateau phase.

Senescence

Check the generation number; the cells may be approaching the end of a finite life span. (See Immortalization of Cell Lines in Chapter 12.)

Medium

See Slow Cell Growth—Medium in this chapter.

Uneven Growth

Nonrandom distribution of cells
See also Cloning—Nonrandom distribution in this chapter.

Incorrect seeding
Have you been pipetting incorrectly, causing the cell suspension to swirl?
Have you been pipetting cells into the middle of dish without mixing?
Have you been swirling the dish to mix the cells?

Incubator vibration
Restrict the frequency of access to the incubator.
Check that the incubator is firmly seated.
Make sure that the shelves are level.
Place a cushion under the flasks, dishes, or plates.

Uneven heating. (See Fig. 7.10.)
Place flasks or dishes on an insulating tile or metal plate.
Check the air and temperature distribution.

CROSS-CONTAMINATION

This section discusses the problem of cross-contamination with another cell line.

Symptoms

Change in appearance
Morphology of cells changes
Cells pile up at high density in plateau.

Change in growth characteristics
Growth rate faster (i.e., shorter PDT).
Cells grow to a higher saturation density

Cell line characteristics
See Chapter 15.

Prophylaxis
Do not share your reagents with other users.
Do not share reagents among cell lines.

Do not handle more than one cell line at a time.
Do not return a pipette to medium after using it with cells.
Do not use pipettors for serial propagation unless filter tips are used.
Handle HeLa and other rapidly growing, high-plating-efficiency cells last.
Authenticate cell lines before freezing them. (See Authentication in Chapter 15.)

Cure
Discard!
If detected early (i.e., there is evidence of a mixed culture), and no other stocks are available, clone the cells. (See Protocol 13.1.)
Isolate likely colonies (see Protocols 13.6 and 13.7).

CRYOPRESERVATION

Poor Recovery

Freezing rate
Change the cooling rate. (See Cooling Rate in Chapter 19) although 1°C/min usually is optimal.
  (a) Change in wall thickness if using insulated container to freeze.
  (b) Use programmable freezer to change rate and shape of cooling curve.

Cell concentration

At freezing
Increase the cell concentration at freezing (optimum is usually from 1×10^6–1×10^7/cells ml. (See Protocol 19.1.)

Thawing and reseeding
Dilute the culture more slowly after thawing (see Protocol 19.2) by gradually adding medium to the cells.
Reseed the cells at 5× normal seeding density. (See Protocol 19.2.)
Pool several ampules (you will need to centrifuge to remove the preservative, see Protocol 19.2)
Remove the nonviable cells. (See Protocol 11.10.)

Preservative

DMSO
Check for chemical contamination in DMSO (from plastic or rubber).
Check for induction of differentiation by DMSO. (See Soluble Inducers in Chapter 16.)
Centrifuge suspension-grown cells to remove the preservative.

Glycerol
Glycerol is toxic if stored for several months in light. (It gets converted to acrolein.)
Dispense glycerol in small volumes.
Store glycerol in the dark.

Changed Appearance after Cryopreservation

Mistaken identity
Check the labeling of the ampule. If label illegible, discard the ampule.
Check the records. (See Freezer Records in Chapter 19.)
Check the authentication. (See Authentication in Chapter 15.)

Changed conditions since the cells were last grown
Are there new members of staff?
Are there new procedures?
Have the suppliers of media changed?
Is a new serum batch being used?

Contamination

Leakage of ampule

Are ampules capped correctly?
Check the seal: The cap should be tight, but the seal should not be distorted.

Water bath on thawing
Swab the ampules carefully after immersing them in warm water.
Thaw the ampules in a heating block.

Loss of Stock

User stock
Restock from the distribution stock.

Distribution stock
Restock from the seed stock.

Loss of seed stock
Check the security of the inventory control; it should be restricted to the curator only.
Restock from a reputable cell bank. (See Sources of Materials—Cell Banks in Trade Index.)

GRANULARITY OF CELLS

Intracellular

Are the cells unhealthy?
Check growth curve. (Protocols 20.7 and 20.8.)

Check plating efficiency. (Protocol 20.9.)
Is there a high rate of phagocytosis? Check on high power phase microscopy for uptake of neutral red or fluorescent dextran (FITC-dextran Sigma).

Extracellular

Uniform particle size
Check for contamination. (See Monitoring for Contamination in Chapter 18.)

Variable particle size
Is there precipitation from the medium. (See Medium—Precipitation on Storage in this chapter.)
Is there precipitation from serum (not usually harmful)?

CELL COUNTING

Hemocytometer

Variable counts

Sampling error
Mix the cell suspension thoroughly before sampling.
The cells should be singly suspended and not clumped. (See Protocol 20.1.)

Use of hemocytometer
Ensure that the coverslip is correctly attached. (Interference colors should be visible; see Hemocytometer in Chapter 20.)
Make sure that the counting chamber is not overfilled or underfilled.
A sufficient number of cells should be counted (>200).

Visibility of cells
The silvering on the counting chamber should be intact.
Use phase-contrast optics.
Use a noncentered light path if phase-contrast not available.
Try staining the cells before counting them.

Electronic

Variable counts

Sampling error
Mix the cell suspension thoroughly before sampling.
The cells should be singly suspended and not clumped. (See Protocol 20.2.)

The count stops, will not start, or counts slowly (i.e., it takes longer than 25 s)

Orifice clogged
Run the wash cycle.
Run the unblock cycle.
Rub the orifice with the tip of your finger or a fine brush.
Soak in detergent for 1–18 h and repeat unblock; repeat wash cycle three times.

*** The count is lower than expected, or the orifice blocks frequently**

Cell suspension aggregated
Disperse the cells by pipetting the original sample vigorously, redilute the sample, and proceed.
Use different disaggregation technique (see Table 12.3).

Background, is high but will not count

Electrode out of beaker or disconnected
Replace the electrode in the beaker.
Check that the electrode is secure.
Replace the electrode if the terminal plate is missing.
There is precipitate or contaminant in the electrolyte (counting fluid or PBSA).

Count sequence will not start

Blocked orifice
Run the wash cycle.
Run the unblock cycles.
Rub the orifice with the tip of your finger or a fine brush.
Soak in detergent for 1–18 h, repeat unblock and repeat wash cycle three times.

Insufficient negative pressure
Check to ensure that the waste reservoir is not full.
Check the pump (see the manual for a diagnostic test) and connections.
Pump failure has occurred; call engineer.

Background is high

Line or radio interference
There is a problem with electrical equipment (motors, fluorescent lights, incubators). Check and eliminate potential problems by fitting suppressors to the equipment. Line filters are available, but are not always effective. Check the grounding (earthing) of the counter, particularly the casing.

Particulate matter in counting fluid
Filter the counting fluid through a disposable Millex or equivalent filter.

VIABILITY

Morphological Appearance

Granularity and vacuolation
Intracellular granularity usually indicates that the cells are unhealthy.
Vacuolation usually indicates that the cells are unhealthy. (See Fig. 12.1.)

Loss of birefringence

Monolayer cells
If the cells are normally birefringent (edge of cell has a halo on phase contrast) and then lose birefringence, then they usually have lost viability —e.g., by drying out during feeding.

Suspension cells
Suspension cells are normally clear and hyaline; granularity or opacity indicates dead or unhealthy cells.

Removing nonviable cells
Monolayer cells will float off if they are nonviable and thus do not need a special procedure for removal.
For suspension cells or primary disaggregates, enrich the viable cells by spinning them through Ficoll-paque. (See Protocol 11.10.)

Testing Viability

Immediate loss

Dye exclusion. See Protocol 21.1.
Primary cultures are usually 50–90% viable.
Cell lines are usually 90–100% viable.
Thawed cells are usually 50–80% viable.

Cytotoxicity

Long-term survival
Use clonogenic assay. (See Protocol 21.3.)
Use microtitration. (See Protocol 21.4.)

In Conclusion

It has been my intention in the foregoing pages to describe the fundamentals of cell culture in sufficient detail that recourse to the literature is required only to extend your work beyond the basic procedures or to acquire some background detail. It is customary, when giving a lecture, to conclude with a summary that highlights the major points raised in the lecture, and that is how I would like to conclude this text.

There are certain requirements that are crucial to successful and reproducible cell culture, and they may be highlighted as follows:

- Work in a clean, uncrowded, aseptic environment, reserved for culture, and clear up when you have finished.
- To avoid the transfer of contamination, including cross-contamination, do not share media, reagents, cultures, or materials with others.
- Do not assume that your work is immune to mycoplasma because you have never seen it; test your cells regularly.
- Keep adequate records, particularly of changes in procedures, media, or reagents.
- Become familiar with the cell lines that you use—their appearance, growth rate, and special characteristics—so that you can respond immediately to any change.

- Ensure that your work does not compromise your own safety or that of others working around you.
- Preserve cell line stocks in liquid nitrogen, and replace working stocks regularly.
- Protect seed stocks of valuable cell lines, and use other stocks for distribution.
- Try to work under conditions that are precisely defined, including minimal use of undefined media supplements, and do not change procedures for trivial reasons.
- Do not mix cell culture with other microbiological work.

To cover all of the fascinating aspects of cell and tissue culture would take many volumes and defeat the objective of this book. It has been more my intention to provide sufficient information to set up a laboratory and prepare the necessary materials with which to perform basic tissue culture, and to develop some of the more important techniques required for the characterization and understanding of your cell lines. This book may not be sufficient on its own, but with help and advice from colleagues and other laboratories, it may make your introduction to tissue culture easier and more satisfying and enjoyable than it otherwise might have been.

Reagent Appendix

NOTE: Dilutions quoted as, for example, 1:10 or 1:100, are v/v and imply that the final volume is 10 or 100 parts, respectively.

Acetic/Methanol
Add 1 part galacial acetic acid to 3 parts methanol
Make up this reagent fresh each time it is used, and keep it on ice

Agar 2.5%
2.5 g agar
100 ml UPW
Boil to dissolve.
Sterilize by autoclaving or boiling for 2 min.
Store at room temperature.

Amido Black
See naphthalene black.

Amino Acids—Essential
See Eagle's MEM: "Amino Acids," Chapter 8.
(Available as 50× concentrate in 0.1 *N* HCl from commercial suppliers such as ICN and Gibco.)
Make up tyrosine and tryptophan together at 50× in 0.1 *N* HCl and remaining amino acids at 100× in ultrapure water. (Working-strength concentrations are given in Chapter 8.)
Dilute for use as in Protocol 10.8.
Sterilize by filtration.
Store in the dark at 4°C.

Amino Acids—Nonessential
Ingredient	g/l (*100×*)
L-alanine	0.89
L-asparagine H₂O	1.50
L-aspartic acid	1.33
Glycine	0.75
L-glutamic acid	1.47
L-proline	1.15
L-serine	1.05
Water	1,000 ml

Sterilize by filtration.
Store at 4°C.
Use at a concentration of 1:100.

Antibiotics
See under specific headings (*e.g.,* penicillin, streptomycin sulfate, kanamycin sulfate, gentamycin, mycostatin).

Antifoam
(*e.g.,* RD emulsion 9964.40; *see* Sources of Materials—Antifoam in Trade Index
Dispense into aliquots and autoclave to sterilize.
Store at room temperature.
Dilute 0.1 ml/liter (i.e., 1:10,000).

Bactopeptone, 5%
5 g Difco bactopeptone dissolved in 100 ml of Hanks' BSS
Stir to dissolve.
Dispense in aliquots appropriate to a 1:50 dilution, and autoclave.
Store at room temperature.
Dilute 1/10 for use.

Balanced Salt Solutions (BSS)
See Table 8.2.
Dissolve each constituent separately, adding CaCl₂ last, and make up to 1 l. Adjust pH to 6.5.

Sterilize the solution by autoclaving or filtration. With autoclaving, the pH must be kept below 6.5 to prevent phosphate from precipitating; alternatively, calcium may be omitted and added later. If glucose is included, the solution should be filtered to avoid caramelization of the glucose, or the glucose may be autoclaved separately (see Glucose, 20%, in this appendix) at a higher concentration (e.g., 20%) and added later.

With autoclaving, mark the level of the liquid before autoclaving. Store the solution at room temperature, and if evaporation has occurred, make up to mark with sterile ultrapure water before use. If borosilicate glass is used, the bottle may be sealed before autoclaving and no evaporation will occur.

Hanks' BSS without phenol red:
Follow the preceding instructions, but omit phenol red.

Broths

See manufacturers' instructions (Difco, MA Bioservices, Gibco) for preparation. *See also* bactopeptone and tryptose phosphate broth.
Sterilize by autoclaving.

Carboxymethylcellulose (CMC)

1. Weigh out 4 g of CMC and place it in a beaker.
2. Add 90 ml of Hanks' BSS, and bring the mixture to boil in order to wet the CMC.
3. Allow the solution to stand overnight at 4°C to clarify.
4. Make volume up to 100 ml with Hanks' BSS.
5. Sterilize the solution by autoclaving. The CMC will solidify again, but will redissolve at 4°C.

For use (e.g., to increase the viscosity of the medium in suspension cultures), use 3 ml per 100 ml of growth medium.

Chick Embryo Extract [Paul, 1975]

1. Remove embryos from eggs as described in Protocol 11.2, and place the embryos in 9-cm Petri dishes.
2. Take out the eyes using two pairs of sterile forceps.
3. Transfer the embryos to flat- or round-bottomed 50-ml containers, two embryos to each container.
4. Add an equal volume of Hanks' BSS to each container.
5. Using a sterile glass rod that has been previously heated and flattened at one end, mash the embryos in the BSS until they have broken up.
6. Let the mixture stand for 30 min at room temperature.
7. Centrifuge the mixture for 15 min at 2,000 *g*.
8. Remove the supernatant, and, after keeping a sample to check its sterility (see Chapter 10), dispense the solution into aliquots and store at −20°C.

Extracts of chick and other tissues may also be prepared by homogenization in a Potter homogenizer or Waring blender [Coon and Cahn, 1966].

1. Homogenize chopped embryos with an equal volume of Hanks' BSS.
2. Transfer the homogenate to centrifuge tubes, and spin at 1,000 *g* for 10 min.
3. Transfer the supernatant to fresh tubes, and centrifuge for a further 20 min at 10,000 *g*.

4. Check the sample for sterility (see Chapter 10), dispense the remainder into aliquots, and store at −20°C.

0.1 M Citric Acid/0.1% Crystal Violet

21.0 g of citric acid
1.0 g of crystal violet
Make up to 1,000 ml with deionized water.
Stir to dissolve. To clarify, filter the solution through Whatman No. 1 filter paper.

CMC

See carboxymethylcellulose.

Colcemid, 100× Concentrate

100 mg of colcemid
100 ml of Hanks' BSS
Stir to dissolve.
Sterilize by filtration.
Dispense into aliquots and store at −20°C

△ *Safety Note.* Colcemid is toxic; handle it with care by weighing in a fume cupboard and wearing gloves.

Collagenase

2,000 U/ml in Hanks' BSS
100,000 units of Worthington CLS-grade collagenase or the equivalent (specific activity 1,500–2,000 U/mg)
50 ml of Hanks' BSS
To dissolve the mixture, stir at 37°C for 2 h or at 4°C overnight.
Sterilize the solution by filtration, as with serum. (See Protocol 10.13.)
Divide into aliquots, each suitable for 1–2 weeks of use.
Store at −20°.

Collagenase–Trypsin–Chicken Serum (CTC) [Coon and Cahn, 1966]

	Volume	Final Concentration
Calcium- and magnesium-free saline [Moscona, 1952], or PBSA, sterile	85 ml	
Trypsin stock, 2.5%, sterile	4 ml	0.1%
Collagenase stock, 1%, sterile	10 ml	0.1%
Chick serum	1 ml	1.0%

Dispense into aliquots and store at −20°C.

Collection Medium (for Tissue Biopsies)

Growth medium	500 ml
Penicillin	125,000 units
Streptomycin	125 mg
Kanamycin	50 mg
or	
Gentamycin	25 mg
Amphotericin	1.25 mg

Store at 4°C for up to 3 weeks or at −20°C for longer periods.

Crystal Violet 0.1% in Water

Crystal violet ....100 mg
Water...................................100 ml
Filter through Whatman No. 1 paper before use.
Available ready made from Merck.

Dexamethasone (Merck)

1 mg/ml (100×)
This reagent comes already sterile in glass vials. To dissolve it, add 5 ml water by syringe to the vial, remove the resulting solution, and dilute to give a concentration of 1 mg/ml. Divide the solution into aliquots and store at −20°C.
1 mg/ml is approximately 2.5 mM. For use, dilute the solution to give 10–50 nM (physiological concentration range), 0.1–1.0 μM (pharmacological dose range), or 25–100 μM (high dose range).
β-methasone (Glaxo) and methylprednisolone (Sigma) may be prepared in the same way.

Dissection BSS (DBSS)

To Hanks' BSS without bicarbonate, previously sterilized by autoclaving, add the following:
Penicillin ....250 U/ml
Streptomycin...................................250 μg/ml
Kanamycin...................................100 μg/ml
or
Gentamycin...................................50 μg/ml
Amphotericin B ....2.5 μg/ml
(All constituents must be sterile.)
Store at −20°C.

EDTA (Versene)

Prepare as a 10-mM concentrate, 0.374 g/l in PBSA.
Sterilize by autoclaving or filtration.
Dilute 1:10, or 1:5 for use at 1.0–2.0 mM, or, exceptionally, 1:2 for use at 5 mM, diluted in PBSA or trypsin in PBSA.

EGTA

As for EDTA, but EGTA may be used at higher concentrations due to its lower toxicity.

Ficoll, 20%

Sprinkle 20 g of Ficoll (Pharmacia) on the surface of 80-ml UPW, and leave it overnight for the Ficoll to settle and dissolve.
Make up to 100 ml in UPW.
Sterilize by autoclaving.
Store at room temperature.

Fixative for Tissue Culture

See acetic/methanol.
Alternatively, use pure anhydrous ethanol or methanol (see Protocol 15.2), 10% formalin, 1% glutaraldehyde, or 5% paraformaldehyde.

Gentamycin (ICN, Gibco, Schering)

Dilute to 50 μg/ml for use.

Gey's Balanced Salt Solution

	g
NaCl	7.00
KCl	0.37
CaCl$_2$	0.17
MgCl$_2$·6H$_2$O	0.21
MgSO$_4$·7H$_2$O	0.07
Na$_2$HPO$_4$·12H$_2$O	0.30
KH$_2$PO$_4$	0.03
NaHCO$_3$	2.27
Glucose	1.00
Water, up to	1,000 ml
CO$_2$	5%

Giemsa Stain

Buffer:

NaH$_2$PO$_4$·2H$_2$O	0.01 M	1.38 g/l
Na$_2$HPO$_4$·7H$_2$O	0.01 M	2.68 g/l

Combine to give pH 6.5.
Dilute prepared Giemsa concentrate (Gurr, BDH, Fisher) 1:10 in 100 ml of buffer.
Filter the solution through Whatman No. 1 filter paper to clarify. Make up a fresh solution each time, because the concentrate precipitates on storage.
Alternatively, apply undiluted Giemsa stain to an anhydrous, fixed preparation for 1–2 min, dilute 10-fold with water, leave 5–10 min, and rinse by upward displacement. (See Protocol 15.2.)

Glucose, 20%

Glucose 20 g
Dissolve in Hanks' BSS and make up to 100 ml.
Sterilize by autoclaving.
Store at room temperature.

Glutamine, 200 m*M*

L-glutamine, 29.2 g
Hanks' BSS, 1,000 ml
Dissolve the glutamine in BSS and sterilize by filtration. (See Protocol 10.10.)
Dispense the solution into aliquots and store at −20°C.

Glutathione

Make 100× stock (i.e., 0.10 M in BSS or PBSA), and dilute to 1 mM for use.
Sterilize by filtration.
Dispense into aliquots and store at −20°C.

Ham's F12

See Chapter 8.

Hanks' BSS

See "Balanced Salt Solutions," in this appendix and Chapter 8.

HAT Medium

Drug	Concentration	Dissolve in	Molarity (100× final)
Hypoxanthine (H)	136 mg/100 ml	0.05 N HCl	1×10^{-2} M
Aminopterin (A)	1.76 mg/100 ml	0.1 N NaOH	4×10^{-5} M
Thymidine (T)	38.7 mg/100 ml	BSS	1.6×10^{-3} M

For use in the HAT selective medium, mix equal volumes of each, sterilize by filtration, and add the mixture to medium at 3% V/V.

Store H and T at 4°C, A at −20°C.

HB Medium
Add the following to CMRL 1066 medium:

Insulin ....5 μg/ml
Hydrocortisone.............................0.36 μg/ml
β-retinyl acetate ....0.1 μg/ml
Glutamine ....1.17 mM
Penicillin ....50 U/ml
Streptomycin................................50 μg/ml
Gentamycin....................................50 μg/ml
Fungisone.......................................1.0 μg/ml
Fetal bovine serum.........................1%

HBSS
See "Hanks's BSS," Chapter 8.

Hoechst 33258 [Chen, 1997]
2-2(4-hydroxyphenol)-6-benzimidazolyl-6-(1-methyl-4-pierpazyl)-benzimidalol-trihydrochloride

Make up 1 mg/ml stock in PBSA or BSS without phenol red, and store the solution at −20°C. For use, dilute 1:20,000 (1.0 μl → 20 ml) in PBSA or BSS without phenol red at pH 7.0.

△ *Safety Note.* Because this substance may be carcinogenic, handle it with extreme care. Weigh in a fume cupboard and wear gloves.

Kanamycin Sulfate ("Kannasyn"), 10 mg/ml
4 1-g vials of kanamycin
Hanks' BSS 400 ml
1. Add 5 ml of BSS from a 400-ml bottle of BSS to each vial.
2. Leave for a few minutes to dissolve.
3. Remove the BSS and kanamycin from the vials, and add them back to the BSS bottle.
4. Add another 5 ml of BSS to each vial to rinse and return to the BSS bottle. Mix well.
5. Dispense 20 ml aliquots of the solution into sterile containers and store at −20°C.
6. Test for sterility: Add 2 ml of reagent to 10 ml of sterile medium, free of all other antibiotics, and incubate the solution at 37°C for 72 h.
7. Use at 100 μg/ml.

Lactalbumin Hydrolysate 5% (10×)
Lactalbumin hydrolysate...................5 g
Hanks' BSS ....100 ml
Heat to dissolve.
Sterilize by autoclaving.
Use at 0.5%.

McIlvaines Buffer, pH 5.5

	To make 20 ml	*To make 100 ml*
0.2 M Na$_2$HPO$_4$ (28.4 g/l)	11.37 ml	56.85 ml
0.1 M citric acid (21.0 g/l)	8.63 ml	43.15 ml

Media
The constituents of some media in common use are listed in Chapters 8 and 9, together with the recommended procedure for their preparation. For those media not described, see Morton [1970], manufacturers' catalogues (under Source of Materials—Media), in the Trade Index or the original references in the literature.

MEM
See "Eagle's MEM," Chapter 8.

2-Mercaptoethanol (M.W. 78)
Stock solution, 5×10^{-3} M (100×)
4 μl in 10 ml HBSS
Sterilize by filtration in fume cupboard.
Store at −20°C or make up a fresh solution each time.

Methocel *See* "Methylcellulose"

Methylcellulose (1.8%)
1. Weigh out 7.2 g of methocel, and add it to a 500-ml bottle containing a large magnetic stirrer bar.
2. Sterilize by autoclaving with the cap loose for penetration of steam.
3. Add 400 ml of sterile UPW heated to 90°C to wet the methocel.
4. Stir at 4°C overnight to dissolve. (The Methocel will form a solid gel if the magnet does not keep stirring.)

The resulting solution is now methocel 2×, and for use, it should be diluted with an equal volume of 2× medium of your choice.

It is more accurate to use a syringe (without a needle) than a pipette to dispense methocel.

For use, add a cell suspension in a small volume of growth medium. (See Chapters 13 and 27.)

Mitomycin C (Stock Solution, 10 μg/ml [50×])
2-mg vial of mitomycin
1. Measure 20 ml of HBSS into a sterile container.
2. Remove 2 ml of HBSS by syringe and add it to a vial of mitomycin.
3. Allow the mixture to dissolve, withdraw the resulting solution, and add it back to the container.
4. Store for 1 week only at 4°C in the dark. (Cover the container with aluminum foil.)
5. For longer periods, store at −20°C.
6. Dilute to 2 μg/10^6 cells, 0.25 μg/ml, for use, but check for effective concentration when using each new cell line as a feeder layer (see Protocols 13.3, 23.1)

△ *Safety Note.* Because mitomycin is toxic, reconstitute it in the vial. Work in a fume hood when handling the substance in powder form.

MTT
3-(4,5-dimethylthiazol-2-yl)-2,5-diphenyltetrazolium bromide (MTT, Sigma)
50 mg/ml in PBSA
Sterilize by filtration.

△ *Safety Note.* MTT is toxic; weigh it in a fume cupboard and wear gloves.

Mycoplasma

Stain (*See* Hoechst 33258)
Mountant: Glycerol in McIlvaines Buffer pH 5.5

To make 40 ml

0.4 M Na_2HPO_4 (56.8 g/l)	11.37 ml
0.2 M Citric acid (42.0 g/l)	8.63 ml
Glycerol	20.00 ml

Add Vectashield (Vector) to reduce fluorescence fade (see manufacturer's instructions)
Check the pH and adjust to 5.5.

Mycostatin (Nystatin) (2 mg/ml, 100×)

Mycostatin	200 mg
Hanks' BSS	100 ml

Make up by same method as kanamycin.
Final concentration, 20 μg/ml.

Naphthalene Black, 1% in Hanks' BSS

Naphthalene black, 1 g
Hanks' BSS, 100 ml
Dissolve as much as possible of the stain in the BSS, and then filter the resulting saturated solution through Whatman No. 1 filter paper.

PBS

See phosphate-buffered saline.

PBSA

See phosphate-buffered saline.

PE

See phosphate-buffered saline/EDTA.

Penicillin

(e.g., Crystapen benzylpenicillin [sodium]) 1,000,000 units per vial
Use 4 vials and 400 ml of Hanks' BSS and store frozen at −20°C in aliquots of 5–10 ml.
Make up as for kanamycin, stock concentration 10,000 U/ml.
Use at 50–100 U/ml.

Percoll (Pharmacia)

Ready made and sterile as purchased, Percoll should be diluted with medium or HBSS until the correct density is achieved.
Check the osmolality. Adjusting it to 290 mOsm/Kg will require the diluent to be hypo- or hypertonic so it is better to dilute a small sample first and check its osmolality, and then scale up.

Phosphate-Buffered Saline (Dulbecco "A" PBSA)

(See Table 8.2)
Dispense and then autoclave oxoid tablets, code BR 14a, 1 tablet per 100 ml of distilled water. Store at room temperature, pH 7.3, and osmolality 280 mOsm/Kg.
PBSB contains calcium and magnesium and should be made up and sterilized separately. Mix with PBSA, if required, immediately before use.
PBS may also be made up from pure single reagents (see Chapter 8) or a premixed powder. (See Sources of Materials

in Trade Index.) In each case, the Ca^{2+} and Mg^{2+} salts should be made up separately and added just before use if they are required. In general, dissociation solutions and simple isotonic salt solutions for rinsing should be made up without Ca^{2+} and Mg^{2+} salts, while incubation solutions should be complete and generally will contain glucose. (See Chapter 8.)

Phosphate-Buffered Saline/EDTA (PBS/EDTA), 10 m*M* (PE)

Make up PBSA.
Add EDTA, 3.72 g/l, and stir.
Dispense, autoclave, and store at room temperature.
Dilute PBSA/EDTA 1:10 to give 1 m*M* for most applications or 1:2 (5 m*M*) for high chelating conditions (e.g., trypsinization of $CaCo_2$ cells).

Phytohemagglutinin

Stock 500 μg/ml (100×).
Lyophilized (Sigma)
Dissolve the powder by adding HBSS by syringe to an ampule. Dispense into aliquots and store at −20°C.
Dilute 1:100 for use.

SF12

Ham's F12 (see Chapter 8) with 2× Eagle's MEM essential amino acids, 1× nonessential amino acids (ICN, Gibco), lacking thymidine, and with 10× folic acid concentration.

Sodium Citrate/Sodium Chloride *See* SSC

SSC 20× (Sodium Citrate/Sodium Chloride)

Trisodium citrate (dihydrate)	0.3 M	88.2 g
NaCl	3.0 M	175.3 g
Water		1000 ml

Dilute to 1× or 2× as appropriate

Streptomycin Sulfate

1 g streptomycin sulphate per vial
(Similar method to that for preparing kanamycin)
1. Take 2 ml from a bottle containing 100 ml of sterile Hanks' BSS, and add the 2 ml to a 1-g vial of streptomycin.
2. When streptomycin has dissolved, return the 2 ml to the 98 ml of Hanks' BSS.
3. Dilute 1:200 for use.
4. The final concentration should be 50 μg/ml.

Trypsin Stock

2.5% W/V in 0.85% (0.14 M) NaCl
Trypsin solutions can be bought commercially. Alternatively, to make up a 2.5% solution in 0.85% NaCl, stir trypsin for 1 h at room temperature or 10 h at 4°C. If the trypsin does not dissolve completely, clarify it by filtration through Whatman No. 1 filter paper.
Sterilize by filtration, dispense into 10–20 ml aliquots and store at −20°C.
Thaw and dilute 1:10 in PBSA or PE for use. Store diluted trypsin at 4°C for a maximum of 3 weeks.

Note. Trypsin is available as a crude (e.g., Difco 1:250) or purified (e.g., Worthington or Sigma 3× recrystallized) preparation. Crude preparations contain several other proteases that may be important in cell dissociation, but may also be harmful to more sensitive cells. The usual practice is to use crude trypsin, unless the viability of the cells is diminished or reduced growth is observed, in which case purified trypsin may be used. Pure trypsin has a higher specific activity and should therefore be used at a proportionally lower concentration (e.g., 0.01 or 0.05%). Check for mycoplasma when preparing from raw trypsin.

Trypsin, Versene, Phosphate (TVP)

Trypsin (Difco 1:250)	25 mg (or 1 ml 2.5%)
Phosphate-buffered saline, PBSA	98 ml
Disodium EDTA (2H$_2$O)	37 mg
Chick serum (ICN)	1 ml

Mix PBSA and EDTA, autoclave the mixture, and store it at room temperature.

Add chick serum and trypsin before use. If powdered trypsin is used, sterilize it by filtration before adding the serum. Dispense the solution into aliquots and store at −20°C.

Tryptose Phosphate Broth

10% in Hanks' BSS

Tryptose phosphate (Difco)	100 g
Hanks' BSS	1,000 ml

Stir until dissolved.

Dispense into aliquots of 100 ml and sterilize in the autoclave.

Store at room temperature

Dilute 1:100 (final concentration, 0.1%) for use.

Tyrodes's Solution

	g
NaCl	8.00
KCl	0.20
CaCl$_2$	0.20
Mg$_2$Cl$_2$ · 6H$_2$O	0.10
NaH$_2$PO$_4$ · H$_2$O	0.05
Glucose	1.00
UPW, up to	1,000 ml
Gas phase	Air

Versene

See EDTA.

Viability Stain

See naphthalene black.

Vitamins

See media recipes (Chapters 8 and 9).

Make up 1000–10,000× concentrates and combine as required to make up a 100× concentrate.

Sterilize by filtration.

Store at −20°C in the dark.

Trade Index

SOURCES OF MATERIALS

This list is not intended to be comprehensive, but merely provides examples of suppliers for each product. The suppliers' addresses are to be found in the next section Suppliers and Other Resources. Additional suppliers are listed in the BiosupplyNet Source Book, Cold Spring Harbor Laboratory Press, and at www.biosupplynet.com, www.uni-rostock.de/, and www.cato.com/biotech/bio-prod.html, www.ispex.ca/naccbiologicals.html, www.biospace.com/service_and_supplier.cfm., www.ucl.ac.uk/~dmcbaly/scioab.html, www.sciquest.com

Agar	Gibco, Difco	**Benchcote**	Johnson & Johnson
Amino acids	Sigma	**Biochemicals**	Boehringer, Merck, Calbiochem,
Aminopterin	Sigma		Fluka, Pierce, Sigma, U.S.
Amphotericin B	BioWhittaker, ICN, Sigma		Biochemical
Ampules, glass	Wheaton	**Bioreactors**	*See* fermentors
Ampules, plastic	Nalge Nunc (Gibco in U.K.),	**Bottles**	Bellco, Corning, Fisher, Schott
	Corning Costar	**Bovine serum**	Amersham, Bayer, Biofluids,
Anemometers	*See* laminar-flow hoods	**albumin**	Becton Dickinson, Calbiochem,
Antibiotics	*See* each individual antibiotic		ICN, Gibco, Intergen, Sigma
Antibodies	Amersham, Biopool, Boehringer,	**BSS, Hanks's,**	*See* media
	DAKO, Gibco, ICN, R&D	**Earle's, etc.**	
	Serotec, Sigma, Upstate	**Camera lucida**	Leica, Olympus, Nikon, Zeiss
	Biotechnology, Vector	**Carboxymethyl-**	Fisher, Merck
Antifoam RD	Dow-Corning	**cellulose (CMC)**	
emulsion 9964.40		**CCD cameras**	BioWorld, Dage-MTI, Hamamatsu,
Antifoams	Merck, Miles, Sigma		Leica, Scanalytics
Automatic pipettes	Alpha, Becton Dickinson,	**Cell banks**	ATCC, Coriell, ECACC, DSMZ,
	Boehringer, Costar, Gilson, ICN,		JCRB, Riken
	Jencons, Labsystems, Rainin	**Cell counters**	Beckman Coulter, BioWorld,
Automatic pipette	Bellco, Volac (*see* Camlab		Schärfe
plugger	in "Suppliers and Other	**Cell scraper**	Becton Dickinson, Corning Costar,
	Resources")		Nalge Nunc
Autoradiographic	*See* emulsion for autoradiography	**Cell sizing**	Beckman Coulter, Schärfe
emulsion		**Cell strainers**	Becton Dickinson, Dynal, Miltenyi
Bactopeptone	Difco, Gibco		Biotec

Centrifugal elutriator	Beckman
Centrifuges	Beckman, Fisher, Heraeus, IEC, Life Sciences International, Sorval (DuPont)
Centrifuge tubes	Alpha, Corning, Falcon (Becton Dickinson), Nalge Nunc
Chamber slides	Applied Scientific, Bayer, Becton Dickinson, Bibby Sterilin, Heraeus, Metachem, Nalge Nunc, Stem Cell Technologies
Chemicals	Merck, Boehringer, Calbiochem, Fisher, Sigma
Chick plasma	Gibco
Chicken serum	Gibco, ICN
Chloros	Hays Chemical Distribution
Chromosome paints	Applied Imaging, Cambio
Cloning rings	Bellco, Fisher, Scientific Laboratory Supplies
Clorox	Polyscience
Closed-circuit TV	Dage-MTI, Hamamatsu, Leica, *see also* microscopes
CMC	*See* carboxymethylcellulose
CO$_2$ automatic change-over unit for cylinders	Air Products and Chemicals Inc., Gow-mac, Lab Impex, Shandon
CO$_2$ controllers	Air Products, Forma, IEC (Hotpak), Lab-Line, Lab Impex
CO$_2$ incubators	Assab, Forma, Heinecke, Heraeus, ICN, IEC (Hotpak), Lab-Line, LEEC, Napco, New Brunswick Scientific, NuAire, Precision Scientific, Sanyo Gallenkamp
Colcemid	Sigma
Collagen	Boehringer, Becton Dickinson, Cambridge Biosciences, Collagen Corporation, Gibco, ICN, Sigma, Universal Biologicals
Collagenase	Worthington, Sigma, Boehringer
Colony counter	BioWorld, NEN Life Sciences, Optomax, Perceptive Instruments
Confocal microscope	Bio-Rad, Biotech Instruments, Leica, Molecular Dynamics, Nikon, Zeiss
Controlled-rate coolers for liquid N$_2$ freezing	Planer, Taylor Wharton
Coverslips:	
glass	Bayer, Fisher, Gibco, Wheaton
plastic	Bayer, ICN, Lux, Gibco
Crystal violet	Merck, Fisher
Cytobuckets	IEC
Cytocentrifuge	CSP, IEC, Miles, Shandon, Wescor
Cytokines	*See* growth factors
Cytometer	*See* flow cytometer, scanning cytometer
Culture flasks, dishes, and plates	*See* tissue culture flasks, etc.
DEAE dextran	Amersham Pharmacia, Bio-Rad
Deionizers	Bellco, Corning, Elga, Millipore
Densitometers	Beckman Instruments, Pall Gelman Sciences, Gilford Instruments, Helena, Joyce-Lobel, Molecular Dynamics
Density meter	Mettler Toledo, Parr
DPX, Permount	*See* stains
Detergents	Alconox, Calbiochem, Decon, ICN, Pierce
Dexamethasone	Sigma, Merck
Dextran	Calbiochem, Fisher, Sigma
Diacetyl fluorescein	Fisher, Sigma
Dialysis tubing	Laboratory suppliers
Disc filter assembly for sterilization	*See* filters
Dishes	*See* tissue culture flasks, etc
Disinfectants	BioMedical Products, Guest Medical, ICN, Johnson & Johnson, Lab Safety Supply, Polysciences, Sigma, Thomas, VWR Scientific
Dispase	Boehringer Mannheim
Dispensers, liquid	Accuramatic, Barnstead, Corning, Gilson, Jencons, Zinsser
DMSO	Merck
DNase	Sigma, Worthington
DNA fingerprinting	Cellmark Diagnostics, ECACC, Laboratory of the Government Chemist
ECM	*See* matrix
EDTA	*See* chemicals
Ehrlenmeyer flasks	*See* glassware
Electronic cell counter	*See* cell counters
Electronic thermometer	Comark, Fisher, Grant, Omega, Rustrak
Electrophoresis	Amersham Pharmacia, Anachem, Bio-Rad, Gibco/BRL, Haake, Innovative Chemistry, Life Sciences International, Shandon
Emulsion for autoradiography	Amersham Pharmacia
Epidermal growth factor	*See* growth factors
Ethidium bromide	Sigma
Extracellular matrix	*See* matrix
Fermentors	Alfa Laval, Bellco, Biotech Instruments, Braun, Cellon, Genetic Research Instrumentation, New Brunswick Scientific
Fibroblast growth factor	*See* growth factors
Fibronectin	*See* growth factors, matrix
Ficoll	Amersham Pharmacia, ICN
Ficoll-paque	Amersham Pharmacia, BioWhittaker, ICN, Nycomed

Micropipettes	Alpha, Anachem, Applied Scientific, Bibby, Boehringer, Costar, Eppendorf, Gilson
Microscopes	Leica, Olympus, Nikon, Zeiss
Microscopes, fluorescence	*See* microscopes
Microscope slides	Lab-Tek, Bellco, general laboratory suppliers, Nalge Nunc
Microtitration plates	*See* tissue culture flasks, etc.
Microtitration plates with removable wells	Gibco
Microtitration plate sealers	*See* plate sealers
Mitomycin C	Sigma
Multipoint pipettors	*See* micropipettes and pipettors
Multiwell plates	*See* tissue culture flasks, etc.
Mycostatin, Nystatin	BioWhittaker, ICN, Gibco, Sigma
Naphthalene black	*See* stains
Needles (for syringes)	*See* syringes
Nucleosides	Sigma
Nutrient broths	Difco, Gibco
Nylon film	Buck Scientific, Portex, Portland Plastics
Nylon mesh	Stanier or *see* cell strainers
Oncostatin M	*See* growth factors
Organ culture grids	Becton Dickinson
Osmometer	Advanced Instruments, Clandon Scientific
Packaging: cartridge paper, semipermeable nylon film	Hospital suppliers, Buck Scientific, Portex, Portland Plastics
PBS A&B	Gibco, ICN, Oxoid, Sigma
Penicillin	BioWhittaker, Gibco, ICN, Sigma
Percoll	Amersham Pharmacia
Perfusion culture	Amicon, Cellco, Endotronics, Microgon
Petri dishes	*See* tissue culture flasks, etc.
Petriperm dishes	Heraeus
Photographic chemicals, films, and papers	Agfa, Eastman Kodak, Fuji, Ilford (and local camera shops)
Phytohemag-glutinin	Amersham Pharmacia, Sigma
Pipettes	Bellco, A. R. Horwell, Corning, Fisher, Volac, Sterilin, Becton Dickinson
Pipetting aids	Alpha, Bellco, Costar, Horwell
Pipette cans	Bellco, Life Sciences International
Pipette cylinders	Bel Art, Fisher, Nalge Nunc
Pipette plugger	Bellco, Volac (John Poulton)
Pipette washer, drier	Bel Art, Shandon, Radleys

Pipettors	Alpha, Anachem, Applied Scientific, Barnstead, Bellco, Bibby, Boehringer, Corning Costar, Eppendorf, Fisher, Gilson, Integra
Plasma, equine	ICN
Plate readers	Bio-Tek, Bio-Rad, Canberra-Packard, Dynex, Fisher, Molecular Dynamics
Plates	*See* tissue culture flasks, etc.
Plate sealers	Anachem, Elkay, ICN
Pleated cartridge filter	Pall Gelman
Pluronic F-68	Gibco, ICN, Serva, Sigma
Polaroid camera	Polaroid
Poly-D-lysine	Sigma
Polystyrene flasks	*See* tissue culture flasks, etc.
Polyvinyl pyrrolidone	Calbiochem, Sigma
Pressure cooker, bench-top autoclave	Astell Scientific, Harvard Apparatus, Inc., LTE, Napco, Valley Forge
Pronase	Sigma
Propidium iodide	Sigma
Protective clothing	Lab Safety Supply, Sigma, Alexandria Workwear
Pumps, peristaltic	Amersham Pharmacia, Cole Parmer, Gilson, Horwell, Millipore, Watson–Marlowe, Zinsser
Pumps, vacuum	Millipore, Pall Gelman, BOC
Quinacrine dihydrochloride	Sigma
Radioisotopes	Amersham International, ICN, New England Nuclear (DuPont), Sigma
Recording thermometer	ABB Kent-Taylor, Comark, Cole Parmer, Fisher, Harvard, Grant, Omega, Pierce, Rustran
Refractometer	Beckman, Cole Parmer, Fisher
Refrigerators	Local discount warehouses
Reusable in-line filter assembly	*See* filters
Reverse osmosis	Barnstead, Elga, Millipore, U.S. Filter
Roccall	Henry Schein Rexodent
Roller bottles, glass	Bellco
Roller bottles, plastic	*See* tissue culture flasks, etc.
Roller bottle rack	Bellco, Integra, New Brunswick Scientific
Rubber pipette bulb	Bel Art, Bellco, Bibby, Fisher, Jencons Scientific, Ltd.
Safety cabinets	*See* laminar-flow hoods
Scalpels	*See* instruments, dissecting
Scanning Cytometer	Compucyte

Scintilation fluid	Amersham Pharmacia, BS & S, National Diagnostics, New England Nuclear (DuPont), Packard
Scintillation vials, minivials	Canberra Packard
Semipermeable nylon film	*See* packaging
Serum	BioWhittaker, Gibco, Globepharm, Hyclone, ICN, JRH Biosciences, Sigma
Serum substitutes:	
Biotain-MPS	BioWhittaker
CPSR	Sigma
Ex-cyte	Bayer
Excell-900	JRH Biosciences
ITS Premix	ICN
Nutridoma	Boehringer
Serxtend	NEN (DuPont)
SIT	Sigma
TCM, TCH	ICN
Ultroser	Gibco
Ventrex	JRH Biosciences
Serum-free media	BioWhittaker, Boehringer, Biofluids, Clonetics, Gibco, Hyclone, ICN, Irvine Scientific, JRH Biosciences, Mediacult, Promacell, Sigma
Sieves	Becton Dickinson, Sigma
Silica gel	Fisher, Merck
Silicones	Fisher, Merck, Sigma
Silicone tubing	Dow Corning, ESCO, Bibby Sterilin
Slide boxes, light, tight	Becton Dickinson, Raven Scientific
Slide containers for emulsion (Cyto-Mailer)	Lab-Tek, Fisher
Specialized cell cultures	Clonetics, BioWhittaker, Cell Systems, Promocell
Stains	Merck, Fisher
Stainless steel mesh	Becton Dickinson, Sigma
Sterility indicators	Applied Scientific, Bennett, Jencons, Popper, Surgicon
Sterilization bags	*See* packaging
Sterilization film	*See* packaging
Sterilizing and drying oven	Astell Hearson, Cole Parmer, Camlab, Fisher, Forma, Harvard, Horwell, LEEC, LTE, New Brunswick Scientific, Precision Scientific
Sterilizing tape (indicator)	Applied Scientific, Barnstead, Fisher, Roboz-Surgical, Shamrock, 3M (laboratory suppliers)
Stills	Corning, Jencons, Steris
Stirrers	*See* magnetic stirrers
Streptomycin	Gibco, Sigma
Syringes	Becton Dickinson, Popper, (and local laboratory suppliers)
Temperature controllers	*See* thermostats, proportional controllers
Temperature recorders	*See* recording thermometer
Thermalog	Bennet, Popper
Thermanox	*See* coverslips, plastic
Thermostats, proportional controllers	Controls & Automation, Fisher, Napco
α-Thioglycerol	Sigma
Thymidine	Sigma
Time-lapse video	*See* CCD cameras, video camera, video recorder
Tissue culture flasks, etc.	Nalge Nunc (Gibco), Becton Dickinson, Bibby, Corning Costar, Linbro, Lux (ICN)
Tissue culture media	*See* media
Trypan blue	*See* stains
Trypsin	ICN, Gibco, Sigma, Worthington
Trypsin inhibitor (soya bean)	Sigma
Tryptophan	*See* amino acids
Tryptose phosphate broth	Difco, Gibco, Oxoid
Tyrosine	*See* amino acids
Universal containers	Sterilin, Nalge Nunc (Gibco)
Vacuum pump	*See* pumps
Video camera	*See* CCD cameras, microscopes
Video recorder	Hamamatsu
Vinblastin	Sigma
Vinyl tape	3M (laboratory suppliers)
Vitamins, solid	Sigma
Vitamins, solution	*See* media
Vitrogen 100	Collagen Corporation (*see* collagen)
Vortex mixer	Gallenkamp, Fisher, general laboratory suppliers
Water purification	Barnstead, Elga, Millipore, U.S. Filter
X-ray film	Eastman Kodak Company, Fuji

SUPPLIERS AND OTHER RESOURCES

ABB Kent–Taylor, Ltd.
Address: Howard Road, Eaton, Socon, Huntingdon,
Cambridgeshire PE19 3EU, England
Tel.: 01480 475321
Fax: 01480 217948
Products & services: Circular chart temperature recorder,
digital recorders

Abbot Laboratories
Address: 5440 Patrick Henry Dr., Santa Clara, CA 95054
Tel.: 408-982-4800; 800-323-9100
Internet address: www.abbot.com
Products & services: Nembutal

Accuramatic
Address: 40–42 Windsor Road, Kings Lynn, Norfolk PE30
5PL, England
Tel.: 01553 777253
Fax: 01553 777253
Products & services: dispensers, liquids, automatic dispenser

Accurate Chemical & Scientific Corp.
Address: 300 Shames Dr., Westbury, NY 11590
Tel.: 516-333-2221; 800-645-6264
Fax: 516-997-4948
E-mail address: info@accuratechemical.com
Internet address: www.accurate_assi_leeches.com
Products & services: Antibodies, cell separation,
microcentrifuges

Advanced Instruments, Inc.
Address: Two Technology Way, Norwood, MA 02062
Tel.: 781-320-9000; 800-225-4034
Fax: 781-320-8181
E-mail address: mail@aitests.com
Internet address: www.aitests.com
Products & services: Osmometer

Advanced Tissue Sciences
Address: 10933 North Torrey Pines Rd., La Jolla, CA 92037-
1005
Tel.: 619-450-5730
Fax: 619-450-5703
Internet address: www.sddt.com/files/library/
lcorporateprofiles/corpstories/ADVTISSU.html
Products & services: Dermagraft-TC(TM), temporary skin
covering, cultured skin grafts, skin equivalent,
keratinocytes

Aire Liquide (UK)
Address: 5 Langley Close, Epsom Downs, Surrey KT18 6HG,
England
Tel.: 01732 279700
Fax: 01732 279707
Products & services: Liquid-nitrogen freezers, liquid-nitrogen
refrigerators, liquid-nitrogen storage vessels

Aire Liquide
Address; Parc Gustave Eiffel, 8 Rue Gutenberg, Bussy, Saint
Georges 77607, Marme-la-Valle, Cedex 3, France
Tel.: 1 64 76 15 00
Fax: 1 64 76 16 99
Products & services: Liquid-nitrogen freezers, liquid-nitrogen
refrigerators, liquid-nitrogen storage vessels

Air Products & Chemicals, Inc.
Address: 7201 Hamilton Blvd, Allentown, PA 18195
Tel.: 610-481-4911; 800-654-4567
Fax: 800-880-5204
E-mail address: info@apci.com
Internet address: www.airproducts.com/
Products & services: CO_2, automatic change-over unit for
cylinders, CO_2 controllers

Air Products Europe
Address: Hersham Place, Molesky Road, Walton on Thames,
Surrey KT12 4RZ England
Tel.: 01932 249 200
Fax: 01932 249 565
Internet address: www.airproducts.com/
Products & services: CO_2, automatic change-over unit for
cylinders, CO_2 controllers

Alconox
Address: 9 E 40th St., No. 200, New York, NY 10016-042
Tel.: 212-532-4040
Fax: 212-532-4301
E-mail address: cleaning@alconox.com
Internet address: www.alconox.com
Products & services: Detergents

Aldrich Chemical Co., Inc.
Address: 1001 W. St. Paul Ave., Milwaukee, WI 53233
Tel.: 414-273-3850; 800-558-9160
Fax: 414-273-4979
Internet address: www.sigma_aldrich.com
Products & services: Chemicals, biochemicals, EDTA

Alexandria Workwear, Plc.
Address: Thornbury, Bristol BS35 2NT, U.K.
Tel.: 01454 416600
Fax: 01454 411100
E-mail address: service@alexandria.co.uk
Products & services: Laboratory coats

Alpha Laboratories
Address: 40 Parham Drive, Eastleigh, Hants S05 4NU,
England
Tel.: 01703 610911
Fax: 01703 643701
E-mail address: alphalabs@ukbusiness.com
Internet address: www.ukbusiness.com/alphalabs
Products & services: Automatic pipettes, micropipettes,
pipettes, pipette aids, Pasteur pipettes, pipette tips,
centrifuge tubes

Alfa Laval AB
Address: S-221 86 Lund, Sweden
Tel.: 46 36 70 00
Fax: 46 39 49 58
E-mail address: group.info@alfalaval.com
Internet address: www.alfalaval.com
Products & services: Fermentors, bioreactors, steam
sterilization

American Society for Cell Biology
Address: Executive Officer; 9650 Rockville Pike, Bethesda,
MD 20814-3992
Tel.: 301-530-7153

Fax: 301-530-7139
E-mail address: ascbinfo@ascb.org
Internet address: www.ascb.org/ascb/

Amersham Pharmacia Biotech, Ltd
Address: Amersham Place, Little Chalfont, Amersham, Bucks HP7 9NA, England
Tel.: 01870 606 1921; 0800 513 313
Fax: 01494 544 350; 0800 616 927
Internet address: www.apbiotech.com
Products & services: Antibodies, cytokines, growth factors, ELISA, fibronectin, radioisotopes, electrophoresis, Ficoll-Hypaque, microcarriers, chromotography, marker beads, Percoll, phytohemagglutinin, peristaltic pumps, DNA standards, Hybond N, Ultraser-G, lymphocyte preparation media, sodium metrizoate, power supplies, Verax

Amersham Pharmacia Biotech, Inc.
Address: 800 Centennial Ave., P.O. Box 1327, Piscataway, NJ 08855-1327
Tel.: Life Sciences: Academic 800-323-9750
  Industrial 800-323-0399
  Biotech Products: 800-526-3593
Fax: Life Sciences: 800-228-8735
  Biotech Products: 800-329-3593
E-mail address: techserv@amersham.com
Internet address: www.apbiotech.com
Products & services: Antibodies, cytokines, growth factors, ELISA, fibronectin, radioisotopes, electrophoresis, Ficoll-Hypaque, microcarriers, chromotography, marker beads, Percoll, phytohemagglutinin, peristaltic pumps, DNA standards, Hybond N, Ultraser-G, lymphocyte preparation media, sodium metrizoate, power supplies, Verax

Amgen, Inc.
Address: 1840 DeHavilland Dr., Thousand Oaks, CA 91320-1789
Internet address: www.ext.Amgen.com
Products & services: Growth factors, FGF

Amicon, Inc. (see also Millipore Corp.)
Address: 72 Cherry Hill Dr., Beverly, MA
Tel.: 508-777-3622; 800-426-4266
Fax: 508-777-6204

Amphioxus Cell Technologies, Inc.
Address: 11222 Richmond Ave, Suite 180, Houston, TX 77082-2646
Tel.: 800-777-2707
Fax: 281-679-7910
E-mail address: activox@amphioxus.com
Products & services: In vitro toxicology assays, human serum proteins, human liver cell line, hepatocytes, hollow-fiber culture

Anachem, Ltd.
Address: Anachem House, 20 Charles Street, Luton, Beds LU2 0EB, England
Tel.: 01582 745000
Fax: 01582 488815
E-mail address: sales@anachem.co.uk
Products & services: Disinfectant, decontamination, swabs, pipettes, electrophoresis, tips, tubes, racks, 96-well microplates, pipettors, nucleic acid purification kits, media, digital imaging, microtitration plates, stirrer flasks, freezer storage, plate sealers, pipette reservoir, Gilson, micropipettes

Ansell Medical
Address: Ansell Medical, 119 Ewell Road, Surbiton, Surrey KT6 6AL
Tel.: 0181 481 1804
Fax: 0181 481 1828
Internet address: www.ansell.com
Products & services: Nitrile gloves

Anton PAAR U.S.A., Inc.
Address: 13 Maple Leaf Ct., Ashland, VA 23005
Tel.: 800-722-7556
Fax: 804-550-1057
Internet address: www.anton-paar.com/ap/
Products & services: Density meter, physicochemical measurement

Appleton Woods, Ltd.
Address: Lindon House, Heeley Rd., Selly Oak, Birmingham B29 6EN, England
Products & services: Temperature indicator strips

Applied Imaging International, Ltd.
Address: BioScience Centre, Times Square, Newcastle upon Tyne, NE1 4EP, England
Tel.: 0191 202 3100
Fax: 0191 202 3101
Products & services: Chromosome paints

Applied Immune Sciences
Address: 5301 Patrick Henry Dr., Santa Clara, CA 95054
Tel.: 408-980-5812
Products & services: Cell separation, immune panning

Applied Scientific
Address: 154 W. Harris Ave., South San Francisco, CA 94080
Tel.: 650-244-9851
Fax: 650-244-9866
E-mail address: custserv@appliedsci.com
Internet address: www.appliedsci.com/
Products & services: Autoclave bags, tape, biohazard bags, culture screening, chamber slides, cell strainer, cell scraper, cryovials, micropipettes, gloves, medium, multiwell plates, pipettes, Pi-pump, pipettors, syringes, needles, flasks, roller bottles, antibiotics

Appligene Oncor Lifescreen
Address: Unit 15, The Metro Centre, Dwight Road, off Tolpits Lane, Watford, WD1 8SS
Tel.: 01923 241 515
Fax: 01923 242 215
E-mail address: custsvcuk@appligene.tpgnet.net
Internet address: http://www.oncorine.com/home
Products & services: Apotag kits

Arnold R. Horwell, Limited
Address: Laboratory & Clinical Supplies, 73 Maygrobe Road, West Hampstead, London NW6 2BP, England
Tel.: 01328 1551
Fax: 01328 5259
Products & services: Automatic pipette plugger, pipette-filling devices, Pi-pump, bulb pipette, pipette aids, pumps, peristaltic, sterilizing and drying oven, vacuum pump, pipetting aids, CO_2 incubators, lab supplies

Assab, Ltd.
Address: 54 Helen Street, Glasgow G51 3HQ, Scotland
Tel.: 0141 425 1133
Fax: 0141 425 1155

Products & services: CO_2 incubators, incubators and ovens, lab equipment and furniture

Astell Scientific, Ltd.
Address: Powerscroft Road, Sidcup, Kent DA14 5DT, England
Tel.: 0181 300 4311
Fax: 0181 300 2247
E-mail address: sales@astell.com
Internet address: www.astell.com
Products & services: Autoclaves, bench-top autoclaves, data loggers, sterilizers, ovens

ATCC (American Type Culture Collection)
Address: 10801 University Boulevard, Manassas, VA 20110-2209
Tel.: 703-365-2700
Fax: 703-365-2701
E-mail address: Product info: tech@atcc.org; Database help: help@atcc.org
Internet address: www.atcc.org
Products & services: Feeder layers, PCR mycoplasma detection kit, mycoplasma detection service, media, serum, cell lines, cell banks, hybridomas, plasmids

Atlas Clean Air, Ltd.
Address: Unit 1, Bentwood Road, Carrs Industrial Estate, Haslingden, Rossendale, Lancashire BB4 5HH, England
Tel.: 01706 831415
Fax: 01706 831410
Products & services: Laminar-flow hoods, safety cabinets

Baker Co., Inc.
Address: Old Sandford Airport Rd., P.O. Drawer E, Sanford, ME 04073
Tel.: 207-324-8773; 800-992-2537
Fax: 207-324-3869
Internet address: www.bakerco.com/hiband/
Products & services: Laminar-flow hoods, safety cabinets

Barnstead Thermolyne Corporation
Address: 2555 Kerper Boulevard, P.O. Box 797, Dubuque, Iowa 52004-0797
Tel.: 319-589-0538; 800-446-6060
Fax: 319-589-0530; 319-589-0516
E-mail address: mkt@barnsteadthermolyne.com
Internet address: www.barnsteadthermolyne.com
Products & services: Distillation, reverse osmosis, sterilizers, ovens, CO_2 incubators, CO_2 controllers, stirrers, hot plates, freezers, water purification, nitrogen freezers, pipettors, dispensers, repipet, recorders, heating tapes, heating mantles

Bayer Corp.
Address: Diagnostics Division, 511 Benedict Ave, Tarrytown, NY 10591
Tel.: 800-431-1970
Fax: 914-524-3978
Internet address: www.bayermhc.com/
Products & services: Mycoplasma elimination, cyprofloxacin, antibiotics, Miles Scientific, Lab-Tek, Lux, Thermanox coverslips

BD Biosciences (see Becton Dickinson)

Beckman Coulter Corporation
Address: P.O. Box 169015, Miami, FL 33116-9015
Tel.: 305-380-2530; 800-327-6531
Fax: 305-380-3699

Internet address: www.BeckmanCoulter.com
Products & services: Cell counters, Channelyzer attachment, cell sizing, flow cytometers, latex particles

Beckman Coulter Electronics, Ltd.
Address: Sales & Marketing, Northwell Drive, Luton, Beds LU3 3RH, England
Tel.: 01582 567000
Fax: 01582 490390
E-mail address: 106221.1426@COMPUSERVE.COM
Internet address: www.BeckmanCoulter.com
Products & services: Cell counters, Channelyzer attachment, cell sizing, flow cytometers, latex particles (standard), immunology reagents, DNA probes, immuno-histochemistry reagents

Beckman Coulter, Inc.
Address: 4300 N. Harbor Blvd., Box 3100, Fullerton, CA 92834-3100
Tel.: 714-871-4848; 800-233-4656
Fax: 714-773-8283; 800-643-4366
Internet address: www.beckman.com/;134.217.3.35/
Products & services: Centrifugal elutriator, centrifuges, densitometers, cell counters, cell sizing

Becton Dickinson Immunocytometry Division
Address: 2350 Qume Dr., San Jose, CA 95131-1807
Tel.: 408-432-9475; 800-952-3222
Fax: 408-954-2347
Internet address: www.bd.com
Products & services: FACS-Star, flow cytometers, cell separation, antibodies, filter sterilizer

Becton Dickinson and Co.
Address: 1 Becton Drive, Franklin Lakes, NJ 07417-1883
Tel.: Labware: 201-847-6800; 888-237-2762
       Microbiology systems, including difco: 201-847-6800
Fax: 201-847-2220; 201-847-4882
E-mail address: mail@bdl.com
Internet address: www.bd.com/labware; www.difco.com/difco/
Products & services: Automatic pipettes, flasks, multiwell plates, Petri dishes, tubes, Matrigel, growth factors, Fibronectin, laminin, collagen, filter well inserts, filters, flow cytometer, FACS, gas-permeable caps, permeable caps, organ culture grids, chamber slides, antibodies, filter plates, transepithelial electrical resistance (TEER) measurement, bronchial epithelium, Pharmingen, caco-2 assay, syringes, needles, Matrigel, Natrigel, Vitronectin, filter sterilizer, microbiological media, agar, bactopeptone, tryptose, Lactalbumin hydrolysate, nutrient broths

Becton Dickinson, U.K., Ltd.
Address: Between Towns Road, Cowley, Oxford OX4 3LY, England
Tel.: 01865 781615
Fax: 01865 781635
E-mail address: mail@cbpi.com
Internet address: www.bdl.bd.com
Products & services: Automatic pipettes, flasks, multiwell plates, Petri dishes, tubes, Matrigel, growth factors, Fibronectin, filter well inserts, filters, flow cytometer, FACS, gas-permeable caps, permeable caps, organ culture grids, chamber slides, antibodies, filter plates, transepithelial electrical resistance (TEER) measurement, bronchial epithelium, Pharmingen, caco-2 assay, syringes, needles,

Matrigel, Natrigel, collagen, laminin, Vitronectin, filter sterilizer, microbiological media, agar, bactopeptone, tryptose, Lactalbumin hydrolysate, nutrient broths

Bel-Art Products

Address: Pequannock, NJ 07440-1992

Tel.: 973-694-0500

Fax: 973-694-7199

E-mail address: Cservice@bel-art.com

Internet address: www.bel-art.com

Products & services: Siphon pipette washer, pipette cylinders, hods, pipettors, Pi-pump, pipette bulbs, flowmeters, disiccators, culture chambers, culture boxes, racks, bottles

Bellco Glass, Inc.

Address: 340 Edrudo Rd., Vineland, NJ 08360-3493

Tel.: 609-691-1075; 800-257-7043

Fax: 609-691-3247

E-mail address: sales@bellcoglass.com

Internet address: www.bellcoglass.com

Products & services: Pipette plugger, cloning rings, deionizers, magnetic stirrers, magnifying viewers, pipettors, pipette cans, flasks, incubators, roller racks, shakers, trypsinization flask, fermentors, bioreactors, borosilicate glass

Bennett & Co., Ltd.

Address: Elborough House, Banwell Rd., Locking, Weston Super Mare BS24 8PB, U.K.

Tel.: 0934 823823

Fax: 0934 823331

Products & services: Thermalog-S

Bibby Sterilin, Ltd.

Address: Tilling Drive, Stone, Staffordshire ST15 0SA, England

Tel.: 01785 812121

Fax: 01785 813748

E-mail address: bsl@bibby-sterilin.com

Internet address: www.bibby-sterilin.com

Products & services: Flasks, pipettes, dishes, universal containers, ESCO rubber, silicone tubing, silicone stoppers, bottle-cap liners, Silescol, butyl stoppers, micropipettes, cryovials, chamber slides, filter well inserts, filters

Bibby Sterilin (see also Dynalab Corp.)

Biofluids, Inc.

Address: 1146 Taft St., Rockville, MD 20850

Tel.: 301-424-4140; 800-972-5200

Fax: 301-424-3619

E-mail address: biofluids@erols.com

Internet address: www.biofluids.com

Products & services: Medium, selective medium, serum-free medium, FNC, collagen, Fibronectin, BSA coating, penicillin, streptomycin, antibiotics

Bio Medical Products Corp.

Address: 59 White Meadow Road, Rockaway, NJ 07866

Tel.: 201-627-6010

Fax: 201-627-7632

Products & services: Laboratory and hospital suppliers, disinfectants, Clorox, glutaraldehyde

Biomedical Technologies, Inc.

Address: 378 Page Street, Stoughton, MA 02072

Tel.: 781-344-9942

Fax: 781-341-1451

Products & services: Antibodies, matrix, collagen, laminin, fibronectin, poly-D-Lysine, growth factor

Bionique Testing Laboratories, Inc.

Address: Bloomingdale Rd., RR1, Box 2, Saranac Lake, NY 12983

Tel.: 518-891-2356

Fax: 518-891-5753

Internet address: www.bionique.com/

Products & services: culture chambers, in vitro toxicology

Biopool International, Inc.

Address: 6025 Nicolle St., Ventura CA 93003

Tel.: 805-654-0643; 800-944-4479

Fax: 805-654-0681

Products & services: Antibodies

Bioprocessing, Ltd.

Address: Consett No1, Industrial Estate, Medomsley Road, Consett Co, Durham DH8 6SZ, England

Tel.: 01207 581555

Fax: 01207 500944

E-mail address: biopro@biopro.co.uk

Products & services: Antibody purification, PROSEP, surface-modified beads

Bio-Rad Laboratories

Address: Life Science Group Div., 2000 Alfred Nobel Dr., Hercules, CA 94547

Tel.: 510-741-1000; 800-4BIORAD

Fax: 800-879-2289

Internet address: www.bio-rad.com

Products & services: Confocal microscope, DEAE dextran, electrophoresis, microcarriers, chromatography, image analysis, DNA transfer, gene transfection, plate readers

Bio-Rad Labs, Ltd.

Address: Bio-Rad House, Maylands Avenue, Hemel Hempstead, Herts HP2 7TD, England

Tel.: 0800 181134

Fax: 01442 259118

Internet address: www.bio-rad.com

Products & services: Confocal microscope, DEAE dextran, electrophoresis, microcarriers, chromatography, image analysis, DNA transfer, gene transfection, plate readers, microtitration software

BioReliance

Address: 14920 Broschart Rd., Rockville, MD 20850-3349

Tel.: 301-738-1000; 800-756-5658

Fax: 301-738-1036

E-mail address: info@bioreliance.com

Internet address: www.bioreliance.com

Products & services: Cell banking, cell culture production, cell line characterization, endotoxin testing and analysis, mycoplasma testing

BioReliance (UK)

Address: Stirling University Innovation Park, Hillfoots Road, Stirling FK9 4NF, Scotland

Tel.: 01786 451318

Fax: 01786 464764

Products & services: In vitro toxicology, quality assurance assays, cytotoxicity assays, virus screening

BioSupply Net

Internet address: www.biosupplynet.com/find.htm

Products & services: Internet listing of biological suppliers

Biotech Inst., Ltd.
Address: Biotech House, 75a High Street, Kimpton, Hertforshire SG4 8PU, England
Tel.: 01438 832555
Fax: 01438 833040
Internet address: www.biotinst.demon.co.uk/
Products & services: Confocal microscope, fermentors, electrophoresis

Biotech Trade & Service, GmbH
Address: Franz-Antoni-Str 22, D-6837, St. Leon-Rot 1, Germany
Tel.: 6227 51308
Fax: 6227 53694
Products & services: Serum, cell cultures, fibroblasts, melanocytes, keratinocytes, fermentors, serum-free medium

Biotechnology Companies
Internet address: www.infomatik.uni-rostock.de/; www.cato.com/biotech/bio-prod.html
Products & services: Internet list of major biotechnology companies

Bio-Tek Instruments, Inc.
Address: Highland Park, P.O. Box 998, Winooski, VT 05404-0998
Tel.: 802-655-4040; 800-451-5172
Fax: 802-655-7941
Internet address: www.biotek.com/
Products & services: Plate readers

BioWhittaker, Inc.
Address: 8830 Biggs Ford Road, P.O. Box 127, Walkersville, MD 21793-0127
Tel.: 301-898-7025; 800-638-8174
Fax: 301-845-8338
E-mail address: sales@biowhittaker.com
Internet address: www.biowhittaker.com
Products & services: BSS, Hanks', Earle's, medium, vitamins, growth factors, clonetics, selective media, serum-free media, skin, keratinocytes, endothelial cells, melanocytes, smooth muscle cells, neural progenitors, astrocytes, hepatocytes, prostate cells, mammary cells, female reproductive cells, renal cells, kidney epithelium, lung epithelium, penicillin, streptomycin, antibiotics

BioWhittaker U.K., Ltd.
Address: BioWhittaker House, 1 Ashville Way, Wokingham, Berkshire RG41 2PL, England
Tel.: 04411 89795234
Fax: 04411 89795231
E-mail address: sales@biowhittaker.com
Internet address: www.biowhittaker.com
Products & services: BSS, Hanks's, Earle's, medium, vitamins, growth factors, Clonetics, selective media, serum-free media, skin, keratinocytes, endothelial cells, melanocytes, smooth muscle cells, neural progenitors, astrocytes, hepatocytes, prostate cells, mammary cells, female reproductive cells, renal cells, kidney epithelium, lung epithelium, penicillin, streptomycin, antibiotics

BioWorld
Address: 4390 Tuller Rd., Dublin, OH 43017
Tel.: 614-792-8680; 800-860-9729
Fax: 614-792-8685
E-mail address: save@bio-world.com
Internet address: www.bio-world.com
Products & services: Antibiotics, penicillin G, streptomycin, gentamycin, hygromycin B, polyethylene glycol (PEG), HEPES, trypsin, vitamins, antibodies, fixatives, UV transilluminators, CCD cameras, photomicrography, image analysis, microscopes, colony counting, colony counters

Boehringer Mannheim (see Roche Diagnostics & Roche Molecular Biochemicals)

Boro Labs, Ltd.
Address: Aldermaston Berkshire RG7 4QU, England
Tel.: 01189 811731
Fax: 01189 811738
E-mail address: martin@borolabs.co.uk
Internet address: www.borolabs.co.uk
Products & services: CO$_2$ incubators, nitrogen freezers, Aire Liquide, temperature recorders, Dawson glassware washer, Sheldon data logger

Braun B Biotech, Inc.
Address: 999 Postal Rd, Allentown, PA 18103-9338
Tel.: 800-258-9000
Fax: 610-266-9319
Products & services: Fermentors, photographic developers, fixatives, and emulsions

British Oxygen Company, Ltd.
Address: The Priestly Centre, 10 Priestly Road, The Surrey Research Park, Guilford, Surrey GU2 5XY, England
Tel.: 01483 5798577
Fax: 0161 728 4309
Internet address: www.boc.com
Products & services: Gases, CO$_2$, vaccuum pumps

British Society for Cell Biology
Contact: Secretariat—Margaret Clements, BSCB Assistant, The Journal of Experimental Biology, Department of Zoology, Cambridge University, Downing Street, Cambridge, CB2 3EJ, England
Tel.: +44 (0)1223 311 788
Fax: + 44 (0)1223 353 980
E-mail address: zoo-jeb01@lists.cam.ac.uk

BS & S (Scotland), Ltd.
Address: 5/7 West Telferton, Portobello Industrial Estate, Edinburgh EH7 6UL, Scotland
Tel.: 0131 669 2282
Fax: 0131 657 4576
Products & services: Ecoscint

Buck Scientific, Inc.
Address: 58 Fort Point St., East Norwalk, CT 06855
Tel.: 203-853-9444; 800-562-5566
Fax: 203-853-0569
Products & services: Sterile packaging, cartridge paper, semipermeable nylon film

Burge Equipment
Address: 6 Langley Business Court, World's End, Beedon, Nr. Newbury, Berkshire RG20 8RY, England
Tel.: 01635 248171
Fax: 01635 248857
E-mail address: sales@burge.co.uk
Internet address: www.burge.co.uk/home.html
Products & services: Glassware washing machine

Calbiochem-Novabiochem Corp
Address: 10394 Pacific Ctr. Ct., San Diego, CA 92121

Tel.: 619-450-9600; 800-854-3417
Fax: 619-453-3552
Internet address: www.calbiochem.com
Products & services: Biochemicals, chemicals, cholera toxin, Dextran, immunochemicals, antibodies, hygromycin B, G418 sulfate, enzymes, glycobiology reagents, detergents, buffers

Calbiochem-Novabiochem (U.K.), Ltd.
Address: Padge Road, Beeston, Nottingham NG9 2JR, England
Tel.: 0115 9430840
Fax: 0115 9430951
E-mail address: customer.service@cruk.co.uk
Internet address: www.calbiochem.com
Products & services: Biochemicals, biologically active peptides, antibodies, proteins, immunochemicals, signal transduction, apoptosis, hygromycin B, G418 buffers, detergents, growth factors

Cambio, Ltd.
Address: 34 Newnham Rd., Cambridge CB3 9BY, England
Tel.: 01223 366 500
Fax: 01223 350 069
E-mail address: support@cambio.demon.co.uk
Internet address: www.cambio.co.uk
Products & services: Chromosome paints

Cambridge BioScience
Address: 24–25 Signet Court, Newmarket Road, Cambridge CB5 8LA, England
Tel.: 01223 316855
Fax: 01223 360732
E-mail address: cbio@cix.compulink.co.uk
Internet address: www.bioscience.co.uk
Products & services: Zymed, apoptosis, Oncogene Science, growth factors, cytokines, interleukin-2, ELISA, molecular probes, tetracycline, collagen, antibodies, cell cycle, DNA transfer, baculovirus, signal transduction, flow cytometry, oncogenes

Camlab, Limited
Address: Nuffield Road, Cambridge CB4 1TH, England
Tel.: 01223 424222
Fax: 01223 420856
E-mail address: info@camlab.co.uk
Internet address: www.camlab.co.uk
Products & services: Automatic pipette plugger, Duran, vials, containers, wash bottles, stirrers, sterilizing and drying ovens, plastics, glassware, ovens and incubators, pipettors, timers, washup and sterilization, ultrasonic baths

Carl Zeiss
Address: Microscope Division, One Zeiss Dr., Thornwood, NY 10594
Tel.: 914-747-1800; 800-233-2343
Fax: 914-681-7446
E-mail address: micro@zeiss.com
Internet address: www.zeiss.com/
Products & services: Microscopes, inverted microscopes, stereo microscopes, photomicroscopes, camera lucida, confocal microscopes, image analysis, fluorescence

Carl Zeiss Jena, GmBH
Address: Zeiss Gruppe/Mikroskopie, D-07740 Jena, Denmark
Tel.: 3641 64 16 16

Fax: 3641 64 31 44
E-mail address: mikro@zeiss.de
Internet address: www.zeiss.com/
Products & services: Microscopes, photomicroscopes, camera lucida, confocal microscopes, image analysis, fluorescence

Carl Zeiss, Ltd.
Address: P.O. Box 78, Woodfield Road, Wellwyn Garden City, Herts AL7 1LU, England
Tel.: 01707 871200
Fax: 01707 871287
Internet address: www.zeiss.co.uk/
Products & services: Microscopes, inverted microscopes, stereo microscopes, camera lucida, confocal microscopes, image analysis, fluorescence

Cellco, Incorporated
Address: 12321 Middlebrook Road, Germantown, MD 20874 USA
Tel.: 301-916-1000
Fax: 301-916-1010
Products & services: Bioreactors, scale-up, mass culture, perfusion, large-scale culture, hollow fiber

Cellex Biosciences, Inc.
Address: 8500 Evergreen Blvd, Minneapolis, MN 55433
Tel.: 612-786-0302
Fax: 612-786-0915
Products & services: Bioreactors, hollow fiber, Accusyst-S

Cellmark Diagnostics
Address: 20271 Goldenrod Lane, Suite 120, Germantown, MD 20876
Tel.: 301-428-4980; 800-872-5227
Fax: 301-428-4877
Products & services: DNA fingerprinting, DNA probes, Jeffreys probes

Cellmark Diagnostics U.K.
Address: Blacklands Way, Abingdon Business Park, Abingdon, Oxon OX14 1DY, England
Products & services: DNA fingerprinting, DNA probes

CELLON Sarl
Address: 204 route d'Arlon, L-8010 Strassen, Luxembourg
Tel.: 00 352 312 313
Fax: 00 352 311 052
Products & services: Scale-up, collagen, matrix, NASA, rotary cell culture, bioreactors, fermentors

CellPro, Inc.
Address: 22215 26th Ave., S.E., Bothell, WA 98021
Tel.: 425-485-7644; 800-221-2778
Fax: 425-489-8750
E-mail address: 425info@cellpro.com
Internet address: www.cellpro.com
Products & services: CEPRATE, cell separation, hemopoietic cells

Cell Systems Corp.
Address: 12815 NE 124th Street, Kirkland, WA 98034
Products & services: Specialized cell cultures

CELOX Laboratories
Address: 1311 Helmo Avenue, St. Paul, Minnesota 55128
Tel.: 612-730-1500
Fax: 612-730-8900
Products & services: BSS, Earle's, medium, serum-free medium, vitamins solution, serum replacements, cryopreservative

Charles River Laboratories
Address: 251 Ballardvale St., Wilmington, MA 01887
Tel.: 508-658-6000; 800-LABORATORY-RATS
Fax: 508-658-7132
Products & services: Opticell fermentor, bioreactor, laboratory animals

Charles River (U.K.), Ltd.
Address: Manson Rd., Margate, Kent CT9 4LT, England
Products & services: Opticell fermentor, bioreactor, laboratory animals

Chemap (see Alfa Leval)

Chiron Corp.
Address: 4560 Horton Street, Emeryville, CA 94608-2916
Tel.: 510-655-8730
Fax: 510-655-9910
E-mail address: corpcomm@cc.chiron.com
Internet address: www.chiron.com
Products & services: Pharmaceuticals, cytokines

Clandon Scientific, Ltd.
Address: Lysons Avenue Ahs Vale Aldershot Hants GU12 5QF, England
Tel.: 01252 514711
Products & services: Gonotec, osmometer

Clonetics Corporation (see also BioWhittaker)
Address: 8830 Biggs Ford Road, Walkersville, MD 21793
Tel.: 301-898-7025
Fax: 301-845-1008
E-mail address: cs@biowhittaker.com
Internet address: www.clonetics.com; www.biowhittaker.com
Products & services: Serum-free media, skin, keratinocytes, endothelial cells, melanocytes, smooth muscle cells, neural progenitors, astrocytes, hepatocytes, prostate cells, mammary cells, female reproductive cells, renal cells, kidney epithelium, lung epithelium

Cole–Parmer Instrument Co.
Address: 625 E. Bunker Ct., Vernon Hills, IL
Tel.: 847-549-7600; 800-323-4340
Fax: 847-549-7676
Internet address: www.coleparmer.com/
Products & services: Water baths, pumps, slide boxes, stirring hot plate

Collaborative Biomedical Products (Becton Dickinson)
Address: Labware Div., 2 Oak Park, Bedford, MA 01730
Tel.: 617-275-0004; 800-343-2035
Fax: 617-275-0043
Internet address: www.bd.com/labware
Products & services: Growth factors, matrix, collagen, laminin, Nu-Serum

Collagen Corporation
Address: 2500 Faber Place, Palo Alto, CA 94303
Tel.: 415-856-0200
Fax: 415-856-2238
Internet address: www.collagen.com/
Products & services: Collagen, Vitrogen 100

Comark, Limited
Address: Swallowfields, Welwyn Garden City, Herts AL7 1BR, England
Tel.: 01707 331 1051
Fax: 01707 331 1202
Internet address: www.co-mark.com/

Products & services: Electronic thermometers, recording thermometers

CompuCyte Corp.
Address: 12 Emily Street, Cambridge, MA 02139
Tel.: 617-492-1300
Fax: 617-492-1301
E-mail address: techsupport@compucyte.com

Contamination Control Products (see also Life Technologies, Roche Molecular Biochemicals, USB)
Address: 52 N. Main St., Bldg. C-9, Marlboro, NJ 07746
Tel.: 908-780-3211
Fax: 908-780-4855
Products & services: Laminar-flow hoods, safety cabinets

Controls & Automation, Ltd.
Address: Bury Mead Road, Hitchin, Herts SG5 1RT, England
Tel.: 01462 436161
Fax: 01462 451801
Products & services: Thermostats, proportional controllers

Coriell Cell Repository
Address: 401 Haddon Avenue, Camden, NJ 08103
Tel.: 609-757-4820
Fax: 609-964-0254
Internet address: locus.umdnj.edu/ccr
Products & services: Cell lines, cell bank

Corning Costar U.K.
Address: 1, The Valley Centre, Gordon Road, High Wycombe, Bucks HP13 6EQ, England
Tel.: 01494 684700
Fax: 01494 464891
Internet address: www.scienceproducts.corning.com
Products & services: Culture vessels, flasks, dishes, multiwell plates, glassware, pipettes, pipette aids, automatic dispensers, bottles, centrifuge tubes, filter well inserts, gas-permeable caps, permeable caps, micropipettes

Corning Costar Corp.
Address: One Alewife Center, Cambridge, MA 02140
Tel.: 617-868-6200; 800-492-1110
Fax: 617-868-6200
E-mail address: scienceproducts@corning.com
Internet address: www.scienceproducts.corning.com/
Products & services: Plastic culture vessels, flasks, dishes, multiwell plates, Pyrex glassware, pipettes, pipette aids, automatic dispensers, automatic pipettes, bottles, centrifuge tubes, filter well inserts, gas-permeable caps, permeable caps, micropipettes, filter sterilizer

Corning, Inc.
Address: Science Products Division, P.O. Box 5000, Corning, NY 14830
Tel.: 607-974-7740; 800-222-7740
Fax: 607-974-0345
Products & services: Bottles, centrifuge tubes, deionizers, Gilson Pipettman, pipettes, stills, Corex screw-top tubes, borosilicate glass

Corning Inc., Science Products Div.
Address: 45 Nagog Park, Acton, MA 01720
Tel.: 978-635-2200; 800-492-1110
Fax: 978-635-2476
E-mail address: cccwebmail@corning.com
Internet address: www.scienceproducts.corning.com

Products & services: Conductivity meters, electrodes, CO_2, pH meters, magnetic stirrers, temperature controllers, flasks

Corning Inc., Science Products Div.
Address: P.O. Box 5000, Corning NY, 14830
Tel.: 607-974-9000; 800-222-7740
Fax: 607-974-0345
E-mail address: labware@corning.com
Internet address: www.corninglabware.com
Products & services: Conductivity meters, electrodes, CO_2, pH meters, magnetic stirrers, temperature controllers, flasks

Corning Separations
Address: 45 Nagog Park, Acton, MA 01720
Tel.: 978-635-2200; 800-882-7711
Fax: 978-635-2990
E-mail address: separations@corning.com
Internet address: www.corningcostar.com
Products & services: Filters, sterile filtration, membranes

Covance Research Products Inc.
Address: P.O. Box 7200, Denver, PA 17517
Tel.: 717-336-4921; 800-345-4114
Fax: 717-336-5344
E-mail address: al.loch@covance.com
Internet address: www.covance.com
Products & services: Media, serum, Hazelton

Cryomed
Address: 51529 Birch St, New Baltimore, MI 48047
Tel.: 313-725-4614
Fax: 313-725-7501
Products & services: Liquid-nitrogen freezers, freezer racks, controlled-rate freezer, gloves

Cryoservice, Ltd.
Address: Blackpole Trading Estate East, Blackpole Road, Worcester WR3 85G, England
Tel.: 01905 754500
Fax: 01905 754060
Products & services: Gases, CO_2, liquid nitrogen

Cryotechnics
Address: Units B/B1 Albion Industrial Estate, 78 Albion Road, Edinburgh EH7 5QZ, Scotland
Tel.: 0131 652 2050
Fax: 0131 652 2299
E-mail address: sales@cryotechnics.co.uk
Internet address: www.cryotechnics.co.uk
Products & services: Liquid nitrogen freezers, cold rooms, cooled incubators, CO_2 incubators, low-temperature freezers, ultra deep freeze

Dage-MTI, Inc.
Address: 701 N. Roeske Ave, Michigan City, IN 46360
Tel.: 219-872-5514
Fax: 219-872-5559
E-mail address: dage@dagemti.com
Internet address: www.dagemti.com
Products & services: Time-lapse video, video cameras, image intensifiers, image processors, monitors, CCD cameras

Dako Corp.
Address: 6392 Via Real, Carpinteria, CA 93013
Tel.: 805-566-6655; 800-235-5763
Fax: 805-566-6688
E-mail address: techserv@dakousa.com

Internet address: www.dakousa.com/
Products & services: Antibodies, apoptosis reagents

Dako, Limited
Address: Marketing Dept., Denmark House, Angel Drove, Ecy., Cambridge CB2 4ET, England
Tel.: 01353 669911
Fax: 01353 668989
Internet address: www.dakoltd.co.uk
Products & services: Antibodies, apoptosis reagents

DAKO, A/S
Address: Produktionsvej 42, DK-2600 Glostrup, Denmark
Tel.: 44 85 95 00
Fax: 44 85 95 95
Internet address: www.dako.com
Products & services: Antibodies, apoptosis reagents

Day-Impex
Address: Station Works, Earls Colne, Colchester, Essex C06 2ER, England
Tel.: 01787 223232
Fax: 01787 224171
Products & services: Nitrogen freezers, plasticware

Decon Laboratories, Ltd.
Address: Conway Street, Hove, East Sussex BN3 3LY, England
Tel.: 01273 739241
Fax: 01273 722088
Internet address: www.deconlabs.com/
Products & services: Detergents, ultrasonic cleaning baths, marker pens, inks

DIFCO Laboratories (see Becton Dickinson and Co., Microbiology Systems)
Products & services: Microbiological medium, agar, bactopeptone, lactalbumin hydrolysate, tryptose phosphate broth, nutrient broths

DIFCO Laboratories (Europe) see Becton Dickinson, U.K., Ltd.

Distillers Co., Ltd.
Address: Cedar House, 39 London Rd., Reigate, Sussex RH2 9QE, England
Tel.: 0737 241133
Fax: 0737 241842
Products & services: Gases, CO_2

Don Whitley Scientific, Ltd.
Address: 14 Otley Road, Shipley, West Yorkshire, BD17 7SE, England
Tel.: 01274 595728
Fax: 01274 531197
Products & services: Colony counters, colony counting

Dow Corning Corp.
Address: P.O. Box 0994, Midland, MI 48686-0994
Tel.: 517-496-4000
Fax: 517-496-4586
E-mail address: usdccnn3@ibmmail.com
Products & services: Antifoam, deionizers, Methocel, methylcellulose, silicone tubing

DSMZ (Deutsche Sammlung von Mikroorganismen und Zellkulturen)
Address: Mascheroder Weg 1B, D-38124 Braunschweig, Germany
E-mail address: help@dsmz.de

Internet address: www.dsmz.de
Products & services: Cell bank, cell lines

Du Pont (U.K.), Ltd.
Address: Wedgwood Way, Stevenage, Hertfordshire SG1 4QN, England
Tel.: 01438 734000
Products & services: Sorvall centrifuges, scintillation fluid, serum substitutes, Serxtend, NEN, New England Nuclear, radioisotopes

Du Pont (see NEN Life Science Products)

Dynal, A.S.
Address: P.O. Box 158, Skoyen N-0212 Oslo, Norway
Tel.: 47 2206 1000
Fax: 47 2250 7015
E-mail address: dynal@dynal.no
Internet address: www.dynal.no
Products & services: Cell separation, magnetizable beads, immunoaffinity, cell separation, nylon mesh filters

Dynal, Inc.
Address: 5 Delaware Dr., Lake Success, NY 11042
Tel.: 516-326-3270; 800-638-9416
Fax: 516-326-3298
E-mail address: techsen@dynalusa.com
Internet address: www.dynal.no
Products & services: Cell separation, magnetizable beads, immunoaffinity, cell separation, nylon mesh filters

Dynal (U.K.), Ltd.
Address: 10 Thursby Road, Croft Business Park, Bromborough, Wirral L62 3PW, England
Tel.: 0151 346 1223
Internet address: www.dynal.no
Products & services: Cell separation, magnetizable beads, immunoaffinity, nylon mesh filters

Dynalab Corp.
Address: P.O. Box 112, Rochester, NY 14601-0112
Tel.: 716-334-2060; 800-828-6595
Fax: 716-334-9496

Dynex Technologies
Address: Daux Road, Billinghurst, West Sussex RH14 9SJ, England
Tel.: 01403 783381
Fax: 01403 784397
Products & services: Microtitration, plate readers, auto samplers, diluters, dispensers, mixers, plate washers, microtitration software

Dynex Technologies, Inc.
Address: 14340 Sullyfield Cir., Chantilly, VA 20151-1683
Tel.: 703-631-7800; 800-336-4543
Fax: 703-631-7816
E-mail address: info@dynextechnologies.com
Internet address: www.dynextechnologies.com
Products & services: Microtitration, plate readers, auto samplers, diluters, dispensers, mixers, plate washers, microtitration software

East Coast Biologics, Inc.
Address: P.O. Box 489, North Berwick, ME 03906-0489
Tel.: 207-676-7639
Fax: 207-384-4437
E-mail address: info@eastcoastbio.com
Internet address: www.eastcoastbio.com

Products & services: Antibodies

Eastman Kodak, Co.
Address: Scientific Imaging Systems Div, 343 State St, Rochester, NY 14652-4115
Tel.: 203-786-5600; 800-225-5352
Fax: 203-786-5694
Internet address: www.kodak.com
Products & services: Photographic developers, fixatives, and emulsions, hypoclearing agent, X-ray film

ECACC (European Collection of Cell Cultures)
Address: CAMR, Porton Down, Salisbury, Wilts SP4 0JG, England
Tel.: 01980 612512
Fax: 01980 611315
E-mail address: ecacc@camr.org.uk
Internet address: www.camr.org.uk
Products & services: Cell lines, characterization, validation, mycoplasma screening, cell line database, cell bank

EG & G Wallac
Address: 20 Vincent Avenue, Crownhill, Milton Keynes MK8 0AB, England
Tel.: 01908 265744
Fax: 01908 265956
Internet address: www.wallac.com
Products & services: Microtitration plate isotope counters, fluoroimager, luminometer, microtitration software

EG & G Wallac
Address: Analytical Systems Div., 9238 Gaither Rd., Gaithersburg, MD 20877
Tel.: 301-963-3200; 800-638-6692
Fax: 301-963-7780
E-mail address: info@wallacus.egginc.com
Internet address: www.wallac.com
Products & services: Microtitration plate isotope counters, fluoroimager, luminometer, microtitration software

Elga, Inc. (see also U.S. Filter)
Address: 430 Old Boston Road, Topsfield, MA 01983
Tel.: 508-887-6300; 800-354-2420
Fax: 508-887-6266
Products & services: Water purification equipment, reverse osmosis, ion exchange, filtration, photooxidation

Elga, Ltd.
Address: High Street, Lane End, High Wycombe Bucks HP14 3JH, England
Tel.: 01494 887 700
Fax: 0149 881007
E-mail address: elga@elga.demon.co.uk
Internet address: www.elga.co.uk/
Products & services: Water purification equipment, reverse osmosis, ion exchange, filtration, Photooxidation

Elkay Laboratory Products
Address: Unit 4, Marlborough Mews, Crockford Lane, Basingstoke, Hampshire RG24 8NA, England
Tel.: 01256 811118
Fax: 01256 811116
E-mail address: sales@elkay-uk.co.uk
Products & services: Tubes, vials, tips, tubing, micropipettes, pipettes, pipette tips, pastettes, Pasteur pipettes, marker pens, ripette, microtitration, plate sealers, media reservoirs,

biohazard bags, universal containers, transport tubes, disposable caps

Endotronics (see Cellex)

Envair, Ltd.
Address: York Ave, Haslingden, Rossendale, Lancashire BB4 4HX, England
Tel.: 01706 228416
Fax: 01706 831957
E-mail address: envair@envair.co.uk
Internet address: www.envair.co.uk
Products & services: Laminar-flow hoods, biohazard safety cabinets, CCTV monitoring

Eppendorf Scientific, Inc.
Address: 6524 Seybold Road, Madison, WI 53719
Tel.: 608-276-9855; 800-421-9988
Fax: 608-276-9866
E-mail address: custservice@eppendorf-usa.com
Internet address: www.eppendorf.com
Products & services: Micropipettes, tips, microcapillaries, microinjectors, micromanipulators, gridded cover slips

Eppendorf–Netheler–Hinz, GmbH
Address: Barkhausenweg 1, Hamburg 22339, Germany
Tel.: 49 40 538010
Fax: 49 50 53801556
E-mail address: eppendorf@eppendorf.com
Internet address: www.eppendorf.com
Products & services: Micropipettes, tips

ESA, Inc.
Address: 22 Alpha Road, Chelmsford, MA 01824-4171
Tel.: 001-978-250-7000
E-mail address: jmkent@esainc.co
Internet address: www.esainc.com
Products & services: Computer software, microtitration, Phoretix

European Cell Biology Organisation
Contact: Dennis Bray, Department of Zoology, Downing Street, Cambridge CB2 3EJ, UK
Tel.: +44 (0)1223 336602
Fax: +44 (0) 1223 336676
E-mail address: d.bray@zoo.cam.ac.uk
E-mail address (directory): zoo-jeb01@lists.cam.ac.uk

European Life Sciences Organisation
Contact: Ingeborg Fatscher
Tel.: +49 6224 929025
Fax: +49 6224 929026
E-mail address: fatscher@t-online.de

European Society for Animal Cell Culture Technology (ESACT)
Contact: Wolfgang Noe, Dept. of Biotechnical Production, Dr Karl Thomae GmbH, Birkendorfstrasse, 85, D-7950 Biberach an der Riss, Germany
Tel.: +49 735 154 4323
Fax: +49 735 154 7198
E-mail address: wolfgang.noe@bid.de

European Tissue Culture Society (ETCS)
Contact: Roland Graftstrom, Karolinska Institutet, Inst. Environmental Medicine, Box 210, S-171 77 Stockholm, Sweden
Tel.: 0046 8 301203
Fax: 0046 8 329402

E-mail address: roland.graftstrom@imm.ki.se
Internet address: www.uni-stuttgart.de/etcs/etcsmain.html

Falcon (see Becton Dickinson)

First Link (U.K.), Ltd.
Address: Unit 18H, Premier Estate, Leys Road, Brierley Hill, West Midlands DY5 3UP, England
Tel.: 01384 263862
Fax: 01384 480351
Products & services: PDGF, pituitary extracts, fibronectin, transferrin, medium, serum, albumin, growth factors, GMEMS, PDS, bovine pituitary extract, BPE, endothelial growth supplement, ECGS, horse serum, cytokines, interleukin-2

Fisher Scientific Corp.
Address: 2000 Park Ln, Pittsburgh, PA 15275
Tel.: 412-490-8300; 800-766-7000
Fax: 800-926-1166
Internet address: www.fisher1.com
Products & services: Chemicals, carboxymethylcellulose (CMC), centrifuges, cloning rings, CO_2 incubators, coverslips, crystal violet, diacetyl fluorescein, EDTA, Ehrlenmeyer flasks, electronic thermometers, freezers, Giemsa stain, glucose, glycerol, hemocytometers, dissecting instruments, magnetic stirrers, microcaps (Drummond), oxoid, pipettes, refractometer, sterilizing and drying oven, tape, vortex mixer, containers for emulsion (CytoMailer), temperature controllers, temperature recorders, stirring hot plate, iridectomy knives, Duran glass bottles, borosilicate glass

Fisher Scientific U.K.
Address: Bishop Meadow Road, Loughborough, Leicestershire LE11 5RG, England
Tel.: 01509 231166
Fax: 01509 231893
E-mail address: info@fisher.co.uk
Internet address: www.fisher.co.uk
Products & services: Laminar-flow hoods, safety cabinets, filters, microtitration, plate readers, plate fillers, chemicals, pumps, fermentors, CO_2 incubators, pipettors, Oxoid, autoclaves, carboxymethylcellulose, cloning rings, coverslips, crystal violet, diacetyl fluorescein, EDTA, Ehrlenmeyer flasks, Giemsa stain, glucose, glycerol, hemocytometers, dissecting instruments, magnetic stirrers, microcaps (Drummond), pipettes, refractometer, silica gel, silicone, sterilizing oven, tape, temperature controllers, stirring hot plate, iridectomy knives, Duran glass bottles, borosilicate glass bottles

Fluka Chemical Corp.
Address: 1001 W. St. Paul Ave, Milwaukee, WI 53233
Tel.: 414-273-3850; 800-358-5287
Fax: 414-273-4979
Internet address: www.sial.com
Products & services: Biochemicals, chemicals, EDTA, methylcellulose

Fluorochem, Ltd.
Address: Wesley Street, Old Glossop, Derbyshire SK13 9RY, England
Tel.: 01457 868921
Fax: 01457 869360
E-mail address: enquiries@fluorochem.co.uk

Products & services: Fluorochemicals, density medium

FMC BioProducts
Address: 191 Thomaston St., Rockland, ME 04841
Tel.: 800-341-1574; 207-594-3400
Fax: 207-594-3941; 800-362-5552
E-mail address: biotechserv@fmc.com
Internet address: www.bioproducts.com/
Products & services: DNA sequencing, DNA separation, protein separation, PCR, mutation detection

FMC BioProducts Europe
Address: Risingevej 1, DK-2665, Vallensbaek Strand, Denmark
Tel.: 42 73 11 22
Fax: 42 73 56 92
Internet address: www.bioproducts.com/
Products & services: DNA sequencing, DNA separation, protein separation, PCR, mutation detection

Forma Scientific, Inc.
Address: Millcreek Road, P.O. Box 649, Marietta, OH 45750
Tel.: 614-373-4763; 800-848-3080
Fax: 614-374-1817
E-mail address: DBERGENE: FORMA.COM
Internet address: www.forma.com
Products & services: Incubators, freezers, CO_2 controllers, CO_2 incubators, cryopreservation equipment, orbital shakers

Fuji Photo Film
Address: P.O. Box 015, Leamington Spa, Warwickshire CV 31 YA, England
Tel.: 01926 335537
Fax: 01926 887793
Internet address: www.fujifilm-europe.com/
Products & services: Photographic film, photographic developers, fixatives, emulsions, X-ray film

Fuji Medical Systems U.S.A., Inc.
Address: Bioimaging Group Division, P.O. Box 120035, Stamford, CT 06912
Tel.: 203-353-0300; 800-431-1850
Fax: 203-327-6485
E-mail address: ssg@fujimed.com
Internet address: www.fujimed.com
Products & services: Photographic film, photographic developers, fixatives, emulsions, X-ray film

Gen-Probe, Inc.
Address: 10210 Genetic Center Drive, San Diego, CA 92121
Tel.: 619-546-8000; 800-523-5001
Fax: 619-452-5848
Internet address: www.gen-probe.com/
Products & services: Mycoplasma detection

Genetic Research Instrumentation, Ltd.
Address: Gene House, Dunmow Rd., Felsted, Dunmow, Essex CM6 3LD, England
Tel.: 01371 821 081
Fax: 01371 820 131
Products & services: Hot-pack incubators, Labconco, Buchler, safety cabinets, water purification, controlled-rate freezer, cryogenic vials, fermentors, bioreactors, LEEC, refrigerated incubator, Statebourne Biostor nitrogen freezers, pipettors, micropipettes, pipette tips, Bellco roller rack, roller drum, spinner flasks, autoradiography film and cassettes, developer, fixer

Genzyme Diagnostics (see R&D Systems)

Genzyme Diagnostics (see R&D Systems, Europe)

Germfree Labs, Inc.
Address: Equipment Manufacturing Div., 7435 NW 41St. St., Miami, FL 33166
Tel.: 305-592-1780; 800-888-5357
Fax: 305-591-7280
E-mail address: info@germfree.com
Internet address: www.germfree.com
Products & services: Laminar-flow hoods, glove boxes, isolators, safety cabinets

Gibco BRL (see Life Technologies)

Gilson, Inc.
Address: 3000 W. Beltline Hwy., P.O. Box 620027, Middleton, WI 53562
Tel.: 608-836-1551; 800-445-7661
Fax: 608-831-4451
E-mail address: sales@gilson.com
Internet address: www.gilson.com/
Products & services: Micropipettes, automatic dispensers, peristaltic pumps, vacuum pumps, liquid handling

Gilson Medical Electronics (France)
Address: 72 rue Gambetta, B.P., 45, 95400 Villiers-le-Bel, France
Tel.: 33 1 34 29 50 50
Fax: 33 1 34 29 50 60
Products & services: Micropipettes, automatic dispensers, peristaltic pumps, vacuum pumps, liquid handling

Glaxo Wellcome, Inc.
Address: Five Moore Dr., Research Triangle Park, NC 27709
Tel.: 919-483-2574; 800-5-Glaxo5
Fax: 919-941-3275
Internet address: www.helix.com/
Products & services: Penicillin, streptomycin

Globepharm, Ltd.
Address: P.O. Box 89C, Sher, Kent, Surrey KT10 9ND, England
Tel.: 01372 465307
Fax: 01372 468818
Products & services: Serum

Grant Instruments (Cambridge), Ltd.
Address: Barrington, Cambridge, CB2 5QZ, England
Tel.: 01763 260811
Fax: 01763 262410
E-mail address: Marketing@grantinst.co.uk
Internet address: www.grant.co.uk
Products & services: Temperature recorders, recording thermometers, electronic thermometers, water baths

Greiner America Inc.
Address: P.O. Box 953279, Late Mary, FL 32795
Tel.: 407-333-2800; 800-884-4703
Fax: 407-333-3301
E-mail address: greineram@aol.com
Internet address: www.greineramerica.com/
Products & services: Flasks, dishes, plates

Greiner, GmbH
Address: Maybachstrasse 2, Postfach 1162, D-7443 Frickenhausen, Germany
Tel.: 7022 5010
Fax: 7022 501514

Products & services: Flasks, plates, dishes

Guest Medical, Ltd.
Address: Enterprise Way, Edenbridge, Kent TN8 6EW
Tel.: 01732 876466
Fax: 01732 876476
E-mail address: Guest Medical@BTInternet.com
Products & services: Biohazard spills kits, chlorine tablets, Aquasan-water bath bacteriostat, autoclavable disposal bags, glutaraldehyde

Haake Buchler
Address: 53 W Century Rd., Paramus, NJ 07652
Tel.: 201-265-7865
Fax: 201-265-1977
Internet address: www.haake-usa.com
Products & services: Electrophoresis, density gradient former, staining boxes

Harvard Apparatus Inc.
Address: 84 October Hill Rd., Holliston, MA 01746
Tel.: 508-893-8999; 800-272-2775
Fax: 508-429-5732
E-mail address: bioscience@harvardapparatus.com
Internet address: www.harvardapparatus.com
Products & services: Incubators, ovens, sterilizers

Hamamatsu Photonic Systems
Address: Div. of Hamamatsu Corp., 360 Foothill Rd., P.O. Box 6910, Bridgewater, NJ 08807
Tel.: 908-231-1116
Fax: 908-231-0852
Internet address: www.hamamatsu.com
Products & services: CCD cameras, closed-circuit TV, image analysis, videocameras, video recorders, time-lapse video

Hamamatsu Photonics, K.K.
Address: 325-6, Sunayama-cho, Hamamatsu City, 430 Japan
Tel.: (81) 53 452 2141
Fax: (81) 53 456 7889
Internet address: www.hamamatsu.com
Products & services: CCD cameras, closed-circuit TV, image analysis, videocameras, video recorders, time-lapse video

Hanna Instruments
Address: Eden Way, Pages Industrial Park, Leighton Buzzard, Beds LU7 8TZ, England
Tel.: 01525 850855
Fax: 01525 853668
Internet address: www.hannainst.com/
Products & services: Magnetic stirrers

Harlan Sera-Lab, Ltd.
Address: Hillcrest, Dodgeford Ln., Belton, Loughborough, Leicester LE12 9TE, England
Tel.: 01530 222123
Fax: 01530 224970
Products & services: Serum, antibodies

Harvard Apparatus, Inc.
Address: 22 Pleasant St., South Natick, MA 01760
Tel.: 508-655-7000, 800-272-2775
Fax: 508-655-6029
Products & services: Ovens and incubators, pressure cooker, bench-top autoclave, sterilizing oven

Hays Chemical Distribution, Ltd.
Address: 215 Tunnel Ave., East Greenwich, London SE10 0QE, England

Products & services: Chloros

Hazelton (see Covance)

Helena Laboratories
Address: P.O. Box 752, Beaumont, TX 77704-0752
Tel.: 409-842-3714; 800-231-5663
Fax: 409-842-9561
Internet address: www.helena.com/
Products & services: Densitometers

Henry Schein Rexodent
Address: 25/27 Merrick Rd., Southall, Middlesex UB2 4AU, England
Tel.: 0181 235 5050
Fax: 0181 235 5010
Products & services: Roccall

Heraeus Instruments
Address: Unit 9, Wates Way, Brentwood, Essex CM15 9TB, England
Tel.: 01277 231511
Fax: 01277 261856
Internet address: www.heraeus.de/index_e.htm
Products & services: Ovens, incubators, CO_2 incubators, centrifuges, Petriderm dishes, chamber slides, culture vessels, CO_2 incubators

Heraeus Instruments, GmbH
Address: Heraeusstr. 12-14, D-63450, Hanau, Germany
Tel.: 61 81 35 300
Fax: 61 81 35 59 44
Internet address: www.heraeus.de/index_e.htm
Products & services: Ovens, incubators, CO_2 incubators, centrifuges, Petriderm dishes, chamber slides, culture vessels

Heraeus Instruments, Inc.
Address: 111-A Corporate Blvd., South Plainfield, NJ 07080
Tel.: 908-754-0100, 0800-441-2554
Fax: 908-754-9494
Internet address: www.heraeus-instruments.com
Products & services: Ovens, incubators, CO_2 incubators, centrifuges, Petriperm dishes, chamber slides, culture vessels

Heto-Holten, A/S
Address: Gydevang 17–19, DK-3450 Allerod, Denmark
Tel.: 48 16 62 00
Fax: 48 16 62 97
Products & services: Laminar-flow hoods, class II safety cabinets

Heto-Holten (U.K.), Ltd.
Address: P.O. Box 31, Camberley, Surrey GU15 1TN, England
Tel.: 01276-63376
Fax: 01276 677 472
Products & services: Laminar-flow hoods, class II safety cabinets

High-Q, Inc.
Address: P.O. Box 440, Wilmette, IL 60091
Tel.: 800-474-2674; 847-853-9166
Fax: 800-474-2672; 847-788-7200
E-mail address: sales@high-q.com
Internet address: www.high-q.com
Products & services: Water purification, distillation

Hoechst U.K. (Ltd.)
Address: Chemical Division, Hoechst House, Salisbury Road, Hounslow, Middx, London TW4 6JH, England

Tel.: 01570 7712
Fax: 01236 6336
Products & services: Chemicals, Hoechst 33258, mycoplasma stain

Hotpack Corp.
Address: 10940 Dutton Road, Philadelphia, PA 19154
Tel.: 215-824-1700; 800-523-3608
Fax: 215-637-0519
E-mail address: hotpack@hotpack.com
Internet address: www.hotpack.com
Products & services: CO_2 incubators, autoclaves, laminar flow hoods, safety cabinets, freezers

HRP, Inc. (see Covance Research Products, Inc)

HyClone
Address: 1725 South HyClone Road, Logan, Utah 84321-6212
Tel.: 800-492-5663
Fax: 800-533-9450
Internet address: www.hyclone.com
Products & services: Medium, serum, serum-free medium

N.V. Hyclone Europe, S.A.
Address: Industriezone III, B-9320 Erembodegem-Aalst, Belgium
Tel.: 53 83 44 04
Fax: 53 83 76 38
Internet address: www.hyclone.com
Products & services: Medium, serum, serum-free medium

ICN Biochemicals, Inc.
Address: 3300 Hyland Ave., Costa Mesa, CA 92626
Tel.: 714-545-0100; 800-854-0530
Fax: 714-557-4872; 800-334-6999
Internet address: www.icnbiomed.com
Products & services: Medium, serum, automatic dispensers, antibodies, chicken serum, CO_2 incubators, collagen, detergents, disinfectants, glutamine, growth factors, HEPES, Ficoll-Hypaque density media, incubators, laminar-flow hoods, lymphocyte preparation media, PBS, pipettors, plate sealers, safety cabinets, serum-free media, trypsin, mycoplasma eradication, mycoplasma removal agent, MRA, Pluronic F68, penicillin, streptomycin, antibiotics, sodium metrizoate, Thermanox, Vitrogen

ICN Biomedicals, Ltd.
Address: 1 Elmwood, Chineham Business Park, Basingstoke, Hampshire RG24 8WG, England
Tel.: 0800 282 474
Fax: 0800 614 735
Internet address: www.icnbiomed.com
Products & services: Medium, serum, automatic dispensers, antibodies, chicken serum, chick plasma, CO_2 incubators, collagen, detergents, disinfectants, glutamine, growth factors, HEPES, Ficoll-Hypaque density media, incubators, laminar-flow hoods, lymphocyte preparation media, PBS, pipettors, plate sealers, safety cabinets, serum-free media, trypsin, mycoplasma eradication, mycoplasma removal agent, MRA, Pluronic F68, penicillin, streptomycin, antibiotics, sodium metrizoate, Thermanox, Vitrogen

Ilford, Ltd.
Address: Town Lane, Mobberley, Knutsford, Cheshire WA16 7HA, England

Tel.: 01565 684097
Fax: 01565 873035
Internet address: www.ilford.com
Products & services: Photographic emulsions, developers, fixatives, film

Ilford Photo Corp.
Address: W. 70 Century Road, Paramus, NJ 07653
Tel.: 201-265-6000
Internet address: www.acecam.com/crindex.html
Products & services: Photographic developers, fixatives, film, papers

Imaging Research Inc.
Address: Brock University, 500 Glenridge Avenue, St. Catherines, Ontario, L2S 3A1, Canada
Tel.: 905-688-2040
Fax: 9095-685-5861
Products & services: Image analysis, fluorescent probes

Imperial Laboratories
Address: West Portway, Andover, Hampshire SP10 3LF, England
Tel.: 01264 33 33 11
Fax: 01264 33 24 12
Products & services: Medium, serum, collagen, matrix, mycoplasma testing, mycoplasma removal agent, flasks, roller bottles, pipettes, stirrer flasks, biochemicals

Infors U.K., Ltd.
Address: Fortune House, 10 Bridgeman Terrace, Wigan, WN1 1SX, England
Tel.: 01942 825025
Fax: 01942 820412
E-mail address: tonyallman@inforsuk.prestel.co.uk
Internet address: www.infors.ch
Products & services: Shakers, incubators, incubator shakers, fermentors

Innovative Chemistry, Inc.
Address: P.O. Box 90, Marshfield, MA 02050
Tel.: 781-837-6709
Fax: 781-834-7325
Internet address: www.innovativechem.com
Products & services: Electrophoresis, isoenzyme analysis kit

Integra Biosciences, AG
Address: P.O. Box 74, CH-8304 Wallisellen, Switzerland
Tel.: 1 830 22 77
Fax: 1 830 78 52
E-mail address: INTEGRA_Biosciences@freeway.de
Internet address: www.integra-biosciences.com
Products & services: Roller racks, Roller bottles, Technomouse, perfused membrane culture system, magnetic stirrers, stirrer culture flasks, Bunsen burner, Maxisafe 2000, pipettors, dispensers, multipoint micropipettes, filter well inserts, filter flasks, safety cabinets, Cell-Spin, microtitration, fluid handling, scale-up, spinner flasks

Integra Biosciences, Inc.
Address: 44 Stedman Street, Lowell, MA 01851
Tel.: 800-866-8675; 508-934-9500
Fax: 508-934-9888
E-mail address: INTEGRA_Biosciences@freeway.de
Internet address: www.integra-biosciences.com
Products & services: Roller racks, roller bottles, Technomouse, perfused membrane culture system, magnetic stirrers, stirrer

culture flasks, Bunsen burner, Maxisafe 2000, pipettors, dispensers, multipoint micropipettes, filter well inserts, filter flasks, safety cabinets, Cell-Spin, microtitration, fluid handling, scale-up, spinner flasks

Integra BioSciences, Ltd.
Address: The Annexe, New Barnes Mill, Cottonmill Lane, St Albans Herts, AL1 2HB, England
Tel.: 01727 84 88 25; 01727 81 17 69
E-mail address: INTEGRA_Bioscience@freeway.de
Internet address: www.integra-biosciences.com
Products & services: Roller racks, roller bottles, Technomouse, perfused membrane culture system, magnetic stirrers, stirrer culture flasks, Bunsen burner, Maxisafe 2000, pipettors, dispensers, multipoint micropipettes, filter well inserts, filter flasks, safety cabinets, Cell-Spin, microtitration, fluid handling, scale-up, spinner flasks

Integen Company
Address: 2 Manhattanville Road, The Centre at Purchase, Purchase, NY 10577
Tel.: 914-694-1700
Fax: 914-694-1429
Internet address: www.intergenco.com
Products & services: Serum, trypsin, cytokine ELISA kits, cytokines, growth factors, bovine serum albumin, insulin, chymotrypsin, DNase, protease inhibitors, aprotinin, human serum, molecular biology products, probes, PCR, oncor

IEC/Labsystems
Address: 200 Second Ave., Needham Heights, MA 02194
Tel.: 781-449-8060; 800-843-1113
Fax: 781-444-6743; 781-455-9796
Products & services: Centrifuges, CO_2 controllers, CO_2 incubators, cytocentrifuge adaptors, Cytobuckets, liquid handling

Intracel, Limited
Address: Unit 4, Station Road, Shepreth, Royston, Herts SG8 6PZ, England
Tel.: 01763 262680
Fax: 01763 262676
Products & services: Specialized heated chambers for time-lapse studies

Irvine Scientific, Inc.
Address: 2511 Daimler St., Santa Ana, CA 94303
Tel.: 714-261-7800; 800-437-5706
Fax: 714-261-65022
E-mail address: nucleus@irvinesci.com
Internet address: www.irvinesci.com
Products & services: Medium, serum, serum-free medium

Invitrogen Corporation
Address: 1600 Faraday Avenue, Carlsbad, CA 92008
Tel.: 800-955-6288
Fax: 750-603-7201
E-mail address: tech_service@invitrogen.com
Internet address: www.invitrogen.com
Products & services: PCR, plasmids, enzymes, medium, transfection, DNA transfer, molecular biology, gene cloning, DNA cloning

Invitrogen BV
Address: 9704 CH Groningen, The Netherlands
Tel.: 0505 299 299

Fax: 0505 299 281
E-mail address: tech_service@invitrogen.nl
Internet address: www.invitrogen.com
Products & services: PCR, Plasmids, Enzymes, Medium, Transfection, DNA transfer, Molecular biology, Gene cloning, DNA cloning

Jackson ImmunoResearch Laboratories, Inc.
Address: 872 West Baltimore Pike, P.O. Box 9, West Grove, PA 19390
Tel.: 610-869-4024; 800-367-5296
Fax: 610-869-0171
E-mail address: cuserjaxn@aol.com
Internet address: www.jacksonimmuno.com
Products & services: Antibodies, FITC-conjugated donkey antisheep IgG

Japanese Cancer Cell Collection
Internet address: www.cellbank.nihs.go.jp/
Products & services: Cell bank, cell lines

JCRB (Japanese Collection of Research Bioresources)
Address: 1-1-43 Hoen-Zaka, Chuo-Ku, Osaka 540, Japan
E-mail address: cellbank@nih.go.jp
Internet address: www.cellbank.nihs.go.jp/
Products & services: Cell bank, cell lines

JTCS (Japanese Tissue Culture Association)
Address: Dr Masayoshi Namba, Dept. Cell Biology, Inst. Cellular and Molecular Biology, Okayama University Medical School, Okayama 700, Okayama, Japan
Tel.: 086 235 7393
Fax: 086 235 7400
E-mail address: mnamba@med.okayama_u.ac.jp
Products & services: Cell, tissue, organ culture, and biotechnology

Jencons (Scientific), Ltd.
Address: Cherrycourt Way Industrial Estate, Stanbridge Road, Leighton, Buzzard, Bedfordshire LU7 8BR, England
Tel.: 01525 372010
Fax: 01525 379547
E-mail address: Markoni@jencons.co.uk
Internet address: www.jencons.co.uk
Products & services: Automatic dispensers, CO_2 incubator, pipettors, safety cabinets, magnetic stirrers, sterilizing tape, stirrers, pipettors, micropipettes, liquid-nitrogen freezers, dewars, Taylor Wharton, neck plug controlled-rate freezer, biohazard bags

John Poulton
Address: 77-93 Tanner Street, Barking, Essex IG11 8RD, England
Tel.: 0181 594 4256
Fax: 0181 594 8419
Products & services: Pipettes, Volac, pipette plugger

Johnson & Johnson, Medical, Ltd.
Address: Coronation Road, Ascot, Berks SL5 9EY, England
Tel.: 01344 871 000
Fax: 01344 872 599
Products & services: Disinfectants, Precept, hypochlorite, swabs, gloves, Benchcote

Jouan, Inc.
Address: 170 Marcel Dr., Winchester, VA 22602
Tel.: 800-662-7477; 540-869-8623
Fax: 540-869-8626

E-mail address: info@jouaninc.com

Internet address: www.jouaninc.com

Products & services: Centrifuges, incubators, ovens, safety cabinets, water baths

Jouan, Ltd.

Address: Merlin Way, Quarry Hill Road, Iikeston, Derbys DE7 4RA, England

Tel.: 0115 944 7989

Fax: 0115 944 7080

Internet address: www.jouan.com

Products & services: Centrifuges, incubators, ovens, safety cabinets, water baths

Joyce-Loebl, Ltd.

Address: 390 Princes Way, Team Valley, Gateshead NA11 0T4, England

Tel.: 0191 491 0998

Fax: 0191 420 3030

Products & services: Densitometers, image analysis

JRH Biosciences

Address: 13804 West 107th St., P.O. Box 14848, Lenexa, KS 66215

Tel.: 913-469-5580; 800-255-6032

Fax: 913-469-5584

E-mail address: custsvc@jrhbio.com

Internet address: www.jrhbio.com

Products & services: Medium, serum, serum-free medium, vitamins, medium bags, bulk culture, Ventrex

J.T. Baker, Mallinckrodt-Baker Inc. Div.

Address: 222 Red School Ln., Phillipsburg, NJ 08865

Tel.: 908-859-2151; 800-582-2537

Fax: 908-859-9318

Products & services: Chemicals, reagents

Kabi Vitrum, Ltd.

Address: Diagnostica Dept., Riverside Way, Uxbridge, Middlesex 4B8 2YF, England

Tel.: 01895 51144

Products & services: Blood products, plasminogen

Labcaire Systems, Ltd.

Address: 15 Hither Green, Clevedon, Somerset, BS21 6XU, England

Tel.: 01275 340033

Fax: 01275 341313

Products & services: Class II safety cabinets, laminar-flow hoods

Lab Impex Research

Address: Lab Impex Res Ltd, Unit 5, Kingsway Business Park, Oldfield Road, Hampton, Middlesex TW12 2 HD, England

Tel.: 0181 296 1122

Fax: 0181 296 1123

Products & services: CO_2 controllers, disinfectants, refrigerators, freezers, freezer racks

Lab Safety Supply

Address: P.O. Box 1368, Janesville, WI 53547-1368

Tel.: 800-356-0783; 608-754-7160

Fax: 800-543-9910; 608-754-3937

Internet address: www.labsafety.com/

Products & services: Disinfectants, eyewash bottles, face masks, gloves, biohazard bags, labels, tags, sharps disposal, protective clothing, discard containers, sodium hypochlorite

Lab Vision Corporation

Address: 47770 Westinghouse Drive, Fremont, CA 94539

Tel.: 800-828-1628; 510-440-2820

Fax: 510-440-2826

E-mail address: Labvision@Labvision.com

Internet address: www.Labvision.com

Products & services: Antibodies, Heregulin, p21WAF1, MUC2, myelin, p16INK4a, BrdU, BUdR, Cyclin, EGFR, E-cadherin, fibronectin, gastric mucin, GFAP, cytokeratin, laminin receptor, nm23

Lab-Line Instruments, Inc. (see Barnstead Thermodyne)

Labmart (Cambridge), Ltd.

Address: 1 Pembroke Avenue, Waterbeach, Cambridge CB5 9QR, England

Tel.: 01223 441257

Fax: 01223 861990

Products & services: Autoclaves, cryobiological storage systems, liquid-nitrogen freezers, low-temperature freezers

Lancer U.K., Ltd.

Address: 1 Pembroke Avenue, Waterbeach, Cambridge CB5 9QR, England

Tel.: 01928 580783

Fax: 01928 580784

Products & services: Glassware washing machines

Lancer U.S.A., Inc.

Address: 705 State Road, Longwood, FL 32750

Tel.: 407-332-1855; 800-332-1855

Fax: 407-332-0040

Products & services: Glassware washing machines

Lawson Mardon Wheaton

Address: 1501 N. Tenth St., Millville, NJ 08332-2092

Tel.: 609-825-1100; 800-225-4124

Fax: 609-825-1368

E-mail address: chuck.carney@wheatonsci.com

Internet address: www.wheatonsci.com

Products & services: Glassware, plasticware, ampules, coverslips, dispensers, liquid handling, Ehrlenmeyer flasks, micropipettes, Gilson pipettman, roller racks, borosilicate glass

LEEC, Ltd.

Address: Private Road, No. 7 Colwick Industrial Estate, Nottingham NG4 2AJ, England

Tel.: 0115 9616222

Fax: 0115 9616680

Products & services: Incubators and ovens, CO_2 incubators, sterilizing ovens, bottle brush

Leica, Inc.

Address: 111 Deer Lake Rd., Deerfield, IL 60015

Tel.: 847-405-0123; 800-248-0123

Fax: 847-405-0147

Internet address: www.leica.com/

Products & services: Microscopes, cameras, inverted microscopes, closed-circuit TV, CCD, camera lucida, confocal microscopes, image analysis, micromanipulators

Leica Mikroscopie und Systeme, GmbH

Address: Ernst Leitz Strasse, P.O. Box 2040, @ 35530, Wetzlar-1, Germany

Tel.: 64 41 290
Fax: 64 41 29 25 99
Internet address: www.leica.com/
Products & services: Microscopes, cameras, inverted microscopes, closed-circuit TV, CCD, camera lucida, confocal microscopes, image analysis, micromanipulators

Leica U.K., Ltd.
Address: Davy Ave, Knowlhill, Milton Keynes, Bucks MK5 8LB, England
Tel.: 01908 246246
Fax: 01908 6099992
E-mail address: 106006, 1171@compuserve.com
Internet address: www.leica.com/
Products & services: Microscopes, cameras, inverted microscopes, closed-circuit TV, CCD, camera lucida, confocal microscopes, image analysis, micromanipulators

Life Sciences International
Address: Chadwich Road, Astmoor, Runcorn, Cheshire WA7 1PR, England
Tel.: 01928 566611
Fax: 01928 565845
Internet address: www.lifesciences.com
Products & services: Cytocentrifuge, electrophoresis, hemocytometers, microcaps (Drummond), pipette washer, pipette drier

Life Sciences International (U.K.), Ltd.
Address: Unit 5, The Ringway Centre, Edison Road, Basingstoke, Hampshire RG21 6ZZ, England
Tel.: 01256 817282
Fax: 01256 817292
E-mail address: Peter.brook@lifesciences.com
Internet address: www.lifesciences.com
Products & services: Liquid handling, micropipettes, Forma CO_2 incubators, microtitration, Denley, Finnpipette, Labsystems, IEC, Savant/E-C, low-temperature freezers

Life Technologies, Ltd. (Gibco BRL)
Address: P.O. Box 35, Trident House, Renfrew Road, Paisley PA3 4EF, Scotland
Tel.: 0141 814 6100
Fax: 0141 814 6268; 0800 243 485
Internet address: http://www.lifetech.com
Products & services: Medium, serum, serum-free medium, antibodies, agar, ampules, insect medium, chromosome painting, chick plasma, chick embryo extract, chicken serum, mycostatin, gentamicin, glutamine, growth factors, HEPES, kanamycin, microtitration plates, PBS, Pluronic F68, flasks, universal containers, trypsin, vitamins, yeast extract, Mycotect

Life Technologies, Inc. (Gibco BRL)
Address: 8400 Helgerman Ct., P.O. Box 6009, Gaithersburg, MD 20884
Tel.: 301-840-8000; 800-828-6686
Fax: 800-331-2286
Internet address: www.lifetech.com
Products & services: Medium, serum, antibiotics, antibodies, chick plasma, chick embryo extract, chick serum, fungicide, mycostatin, gentamicin, glutamine, growth factors, HEPES, kanamycin, microtitration plates, PBS, Pluronic F68, serum-

free media, flasks, universal containers, trypsin, vitamins, yeast extract, collagen, laminin, penicillin, streptomycin, antibiotics, Mycotect

LTE Scientific, Ltd.
Address: Greenbridge Lane, Greenfield, Oldham OL3 7EN, England
Tel.: 01457 876221
Fax: 01457 870131
E-mail address: mktg@lte.u-net.com
Products & services: Ovens, incubators, drying cabinets, autoclaves, cooled incubators, CO_2 incubators, bench-top autoclaves

Ludl Electronic Products
Address: 171 Brady Ave., Hawthorne, NY 10532
Tel.: Sales, 888-769-6111; Support, 914-769-6111
Fax: 914-769-4759
Products & services: Electronics components

MA Bioservices (see BioReliance)

Mallinckrodt Baker U.K.
Address: 17 Shenley Pavilions, Shenley Wood, Milton Keynes, Bucks MK5 6LB, England
Tel.: 01908 506000
Fax: 01908 503290
E-mail address: sales@mbuk.ccmail.compuserve.com
Products & services: Chemicals

Mallinckrodt-Baker, Inc.
Address: Mallinckrodt Division, 16305 Swingley Ridge Dr., Chesterfield, MO 63017
Tel.: 314-530-2000; 800-354-2050
Fax: 314-530-2328
Products & services: Chemicals

Markson LabSales, Inc.
Address: 5285 N.E. Elam Young Pkwy., State A-400, Hillsboro, OR 97124
Tel.: 503-648-0762; 800-528-5114
Fax: 503-648-8118
E-mail address: marksonls@eleport.com
Internet address: www.marksonls.com
Products & services: Chromatography reagents, filters, molecular-weight sieves, disinfectants

Matrix Technologies Corp.
Address: 44 Stedman St., Lowell, MA 01851
Tel.: 508-454-5690; 800-345-0206
Fax: 508-458-9174
Products & services: Matrix

MatTek Corp.
Address: 200 Homer Avenue, Ashland, MA 01721
Tel.: 508-881-6771
Fax: 508-879-1532
Products & services: Cytotoxicity assays, tissue replacements, skin equivalent, transplants, microwell dishes

MBL Medical & Biological Laboratories Co., Ltd.
Address: 5-10 Marunouchi 3 chome, Naka-ku, Nagoya 460-0002, Japan
Tel.: 52 971 2081
Fax: 52 971 2337
Internet address: www.mbl.co.jp
Products & services: Antibodies, apoptosis, receptors, signal transduction, cell proliferation, PCNA, BUdR

MDH, Limited
Address: Walworth Road, Andover, Hampshire, SP10 5AA, England
Tel.: 01264 362111
Fax: 01264 356452
E-mail address: 101376.1130@compuserve.com
Internet address: www.astec-microflow.co.uk
Products & services: Laminar-flow cabinets, safety cabinets, isolators, gas generators, clean rooms

Media-Cult, A/S
Address: Lerso Parkalle 42, Copenhagen DK-2100, Denmark
Tel.: 45 3916 6100
Fax: 45 3927 1533
E-mail address: Medi-cult@internet.dk
Products & services: RenCyte, serum-free medium, protein-free medium, CHO, BHK, fibroblasts, hybridomas

Medical Air Technology, Ltd.
Address: Wilton Street, Denton, Manchester M34 3LZ, England
Tel.: 0161 320 0652
Fax: 0161 335 0313
E-mail address: Sales@matltd.demon.co.uk
Products & services: Laminar-flow hoods, safety cabinets

Medical Research Apparatus International
Address: 696 12th Ave., N.E., Naples, FL 34120
Tel.: 941-455-2867
Fax: 941-455-2867
E-mail address: MRAINTL@aol.com
Products & services: Napco CO_2 incubators

Mettler-Toledo Inc.
Address: 1900 Polaris Parkway, Columbus, Ohio 43240
Tel.: 800 METTLER
Fax: 614-438-4900
E-mail address: us@mt-shop.com
Internet address: www.mt.com
Products & services: Pipettors, Dispensers, pH meters, Density meters, Balances

Mettler-Toledo Ltd.
Address: 64 Boston Road, Beaumont Leys, Leicester, LE4 2ZW
Tel.: 070000 MTSHOP
Fax: 0116 236 5500
E-mail address: uk@mt-shop.com
Internet address: www.mt-shop.com
Products & services: Pipettors, Dispensers, pH meters, Density meters, Balances

Merck, Inc.
Address: P.O. Box 2000, RY7-220, Rahway, NJ 07065
Tel.: 908-594-4600; 800-672-6372
Fax: 908-388-9778
Internet address: www.merck.com/
Products & services: Antifoam, biochemicals, carboxymethylcellulose (CMC), chemicals, crystal violet, dexamethasone, DMSO, EDTA, Giemsa stain, glucose, glycerol, hydrocortisone, mercaptoethanol, silica gel, stains, trypan blue, PEG

Merck, KGAA
Address: Frankfurter Strasse 250, Postfach 4119, D-6100 Darmstadt, Germany
Tel.: 49 61 51720

Fax: 49 61 5172 2000
Internet address: www.merck.com/
Products & services: Antifoam, biochemicals, carboxymethylcellulose (CMC), chemicals, crystal violet, dexamethasone, DMSO, EDTA, Giemsa stain, glucose, glycerol, hydrocortisone, mercaptoethanol, silica gel, stains, trypan blue, PEG

Merck, Ltd.
Address: Merck House, Poole, Dorset BH15 1TD, England
Tel.: 01202 669700
Fax: 01202 666536
Internet address: www.merck.com
Products & services: Antifoam, biochemicals, carboxymethylcellulose (CMC), chemicals, crystal violet, dexamethasone, DMSO, EDTA, Giemsa stain, glucose, glycerol, hydrocortisone, mercaptoethanol, silica gel, stains, trypan blue, PEG

Messer U.K., Ltd.
Address: Cedar House, 39 London Road, Reigate, Surrey RH2 9QE, England
Tel.: 01737 241133
Fax: 01737 241842
Products & services: Carbon dioxide, dry ice, gas mixtures, oxygen, nitrogen, argon, helium, acetylene, nitrous oxide, hydrogen, methane, calibration gas mixtures, regulators, gas-control equipment

Metachem Diagnostics, Ltd.
Address: 29 Forrest Rd., Piddington, Northampton, NN7 2DA, England
Tel.: 01604 870 370
Fax: 01604 870 194
Internet address: www.stemcell.com
Products & services: Hemopoietic progenitor culture, stem cell purification, cell separation, methylcellulose media, colony atlas, videos, dishes, chamber slides, spinner flasks, software, antibodies, cytokines, interleukin-2, StemCell Technologies

Microbiological Associates (see BioReliance)

Microgon, Inc.
Address: 23152 Verdugo Dr., Laguna Hills, CA 92653
Tel: 800-654-0111
Fax: 800-732-2929
Products & services: Filters, hollow-fiber perfusion, sterilizing filters, syringe-tip filters, DynaGard-ME

Miele Appliances, Inc.
Address: 22D Worlds Fair Dr., Somerset, NJ 08873
Tel.: 908-560-0899; 800-843-7231
Fax: 908-560-9649
E-mail address: moreinfo@mieleusa.com
Internet address: www.miele.de/E/VG_USA
Products & services: Glassware washing machines

Miele Co., Ltd.
Address: Abingdon, Oxon, OX14 1TW, England
Tel.: 01235 554455
Fax: 01235 554477
E-mail address: info@miele.de
Internet address: www.miele.de
Products & services: Glassware washing machines

Miele, Professional Products Div.
Address: 9 Independence Way, Princeton, NJ 08540

Tel.: 732-560-0899; 800-843-7231
Fax: 732-560-7469
E-mail address: products@mieleusa.com
Internet address: www.mieleusa.com
Products & services: Glassware washing machines

Miles, Inc. (see Bayer Corp.)

Miles, Ltd.
Address: Stoke Court, Stoke, Poges, Slough SL2 4LY,
England
Tel.: 012814 5151
Fax: 012814 3993
Products & services: Antifoam, cytocentrifuge, mycoplasma
eradication, immunostaining

Millipore Corp.
Address: 80 Ashby Rd., Bedford, MA 01730
Tel.: 617-275-9200; 800-MILLIPORE
Fax: 617-275-5550
Internet address: www.millipore.com/
Products & services: Filter sterilization, water purification,
filter holders, filter well inserts, filters, reverse osmosis,
molecular filtration, electrodeionization, carbon filtration,
deionization, ultrapure water, pumps, filter sterilizer,
Sterivex

Millipore (U.K.), Ltd.
Address: The Boulevard, Blackmoor Lane, Watford, Herts.,
WD1 8YW, England
Tel.: 01923 816375
Fax: 01923 818295/7
E-mail address: Olivia-wyatt@millipore.com
Internet address: www.millipore.com/
Products & services: Filter sterilization, water purification,
filter holders, filter well inserts, filters, reverse osmosis,
molecular filtration, electrodeionization, carbon filtration,
deionization, ultrapure water, pumps, filter sterilizer,
Sterivex

Miltenyi Biotec, Inc.
Address: 251 Auburn Ravine Road, Suite 208, Auburn, CA
95603
Tel.: 800-FOR-MACS; 530-888-8871
Fax: 530-888-8925
E-mail address: macs@miltenyibiotec.com
Internet address: www.miltenyibiotec.com
Products & services: Cell separation, magnetic sorting, MACS
magnet, antibodies, magnetic beads, CD34, AC133, stem
cell marker, nylon mesh filters

Miltenyi Biotec, Ltd.
Address: Almac House, Church Lane, Bisley, Surrey GU24
9DR, England
Tel.: 01483 799 800
Fax: 01483 799 811
Internet address: www.miltenyibiotec.com
Products & services: Cell separation, magnetic sorting, MACS
magnet, antibodies, magnetic beads, CD34, AC133, stem
cell marker, nylon mesh filters

Miltenyi Biotec, GmbH
Address: Friedrich-Ebert-Strasse 68, 51429 Bergisch
Gladbach, Germany
Tel.: 2204 8360 0
Fax: 2204 85197
E-mail address: macs@miltenyibiotec.de

Internet address: www.miltenyibiotec.com
Products & services: Cell separation, magnetic sorting, MACS
magnet, antibodies, magnetic beads, CD34, AC133, stem
cell marker, nylon mesh filters

Molecular Devices Corp.
Address: MaxLine Div., 1311 Orleans Dr., Sunnyvale, CA
94089
Tel.: 408-747-1700; 800-635-5577
Fax: 408-747-3602
E-mail address: info@moldev.com
Internet address: www.moleculardevices.com
Products & services: Plate readers, spectrophotometers,
fluorometers, microtitration software

Molecular Devices Ltd.
Address: 135 Wharfedale Road, Winnersh Triangle,
Winnersh, Wokingham, RG41 5RB, UK
Tel.: 0118 944 8000
Fax: 0118 944 8001
E-mail address: info@moldev.com
Internet address: www.moleculardevices.com
Products & services: Plate readers, Spectrophotometers,
Fluorometers, Microtitration software

Molecular Dynamics
Address: 928 E. Arques Ave., Sunnyvale, CA 94086
Tel.: 800-333-5703; 408-733-1222
Fax: 408-773-1343
Internet address: www.mdyn.com/
Products & services: Confocal microscopes, densitometers,
plate readers

MVE, Inc.
Address: 407 7th Street, N.W., P.O. Box 234, New Prague,
MN 56071-0234
Tel.: 612-758-4484; 888-MVE-CRYO
Fax: 612-758-8225
Internet address: www.mve-inc.com
Products & services: Liquid-nitrogen freezers

Nalge Nunc (Europe), Ltd.
Address: Foxwood Court, Rotherwas, Hereford HR2 6JQ,
England
Tel.: 01432 269333
Fax: 01432 351923
Products & services: Ampules, gas-permeable caps, flasks,
dishes, multiwell plates, chamber slides, filter well inserts,
cryopreservation vials, racks, cell factory, multilayered
culture vessels, filters, cell scrapers, filter bottom plates,
coverslips, pipettes, micropipette tips, culture tubes,
centrifuge tubes, microcarriers, filter sterilizer

Nalge Nunc International
Address: 2000 N. Aurora Rd., Naperville, IL 60563-1796
Tel.: 630-983-5700; 800-288-6862; 800-416-6862
Fax: 630-416-2556; 630-416-2519
Internet address: www.nalgenunc.com
Products & services: Ampules, filter well inserts, gas-
permeable caps, permeable caps, flasks, microcarriers,
conical tubes, cryotubes, cryovials, mutiwell plates,
microtitration plates, chamber slides, multisurface
propagator, filters, cell scrapers, coverslips, containers, filter
sterilizer

Nalge Nunc International
Address: P.O. Box 20365, Rochester, NY 14602-0365

Tel.: 716-586-8800; 800-625-4327
Fax: 800-NALGENE
Internet address: www.nalgenunc.com
Products & services: Filters, tubes, cryovials, sterile filtration

NapCO₂
Address: Merlin Way, Quarry Hill Road, Ilkeston, Derbyshire DE7 9BR, England
Tel.: 0115 944 7072
Fax: 0115 944 7080
E-mail address: napco@enterprise.net
Products & services: CO_2 incubators, glassware washers

Napco Scientific Co. (see also Medical Research Apparatus International)
Address: 20210 S.W. Teton, Tualatin, OR 97062-1000
Tel.: 503-692-4686; 800-547-2555
Fax: 503-692-6615
Products & services: CO_2 incubators, glassware washers

National Diagnostics, Inc.
Address: 305 Patton Dr., S.W. Atlanta, GA 30336-1817
Tel.: 404-699-2121; 800-526-3867
Fax: 404-699-2077
E-mail address: natldiagnos@mindspring.com
Internet address: www.nationaldiagnostics.com
Products & services: Scintillation fluid

NEN Life Science Products
Address: 549 Albany St., Boston, MA 02118
Tel.: 617-482-9595; 800-551-2121
Fax: 617-542-8468
E-mail address: techsupport@nenlifesci.com
Internet address: www.nenlifesci.com/
Products & services: Growth factors, NGF, bFGF, TGF-α, TGF-β, bioreactors, fermentors, centrifuges, molecular probes, cell separation, radioisotopes, SerXtend, image analysis, colony counters

NEN Life Science Products, U.K.
Address: P.O. Box 66, Hounslow TW5 9RT, England
Tel.: 0800 896046, 0800 891715
Fax: 0800 891714
Internet address: www.nenlifesci.com/
Products & services: Growth factors, NGF, bFGF, TGF-α, TGF-β, bioreactors, fermentors, centrifuges, molecular probes, cell separation, radioisotopes, SerXtend, image analysis, colony counters

New Brunswick Scientific Co., Inc.
Address: Box 4005, 44 Talmadge Road, Edison, NJ 08818-4005
Tel.: 908-287-1200
Fax: 908-287-4222
Internet address: www.nbsc.com/
Products & services: CO_2 incubators, fermentors, freezers, incubators, magnifying viewers, sterilizing oven, drying oven, roller racks, rocker platforms, gyratory shaker, shaking incubators, scale-up, bioreactors, roller culture

New Brunswick Scientific (U.K.), Ltd.
Address: Edison House, 163 Dixons Hill Road, North Mimms, Hatfield AL9 7JE, England
Tel.: 01707 75733
Fax: 01707 67859
Internet address: www.nbsc.com/

Products & services: CO_2 incubators, fermentors, freezers, incubators, magnifying viewers, sterilizing oven, drying oven, roller racks, rocker platforms, gyratory shaker, shaking incubators, scale-up, bioreactors, roller culture

New England Biolabs
Address: 32 Tozer Road, Beverly, MA 01915-5599
Tel.: 978-927-5054; 800-632-5227
Fax: 978-922-7085
E-mail address: bonventre@neb.com
Internet address: www.neb.com
Products & services: Molecular biology, vectors, signal transduction

New England Biolabs (U.K.), Ltd.
Address: Knowl Piece, Wilbury Way, Hitchin, Herts SG4 OTY, England
Tel.: 01462 420 616; 0800 318486
Fax: 01462 421 057; 0800 435682
E-mail address: info@uk.neb.com
Internet address: www.uk.neb.com
Products & services: Molecular biology, vectors, signal transduction

Nichols Institute Diagnostics Div.
Address: 33608 Ortega Highway, San Juan Capistrano, CA 92690-6130
Tel.: 714-728-4000; 800-NICHOLS
Fax: 714-728-4982

Nikon, Inc.
Address: 1300 Walt Whitman Road, Melville, NY 11747-3064
Tel.: 516-547-8500
Fax: 516-547-0306
Internet address: www.nikonusa.com
Products & services: Microscopes, cameras, inverted microscopes, camera lucida, colony marker, confocal microscopes, CCD, image analysis, micromanipulators

Nikon U.K., Ltd.
Address: Nikon House, 380 Richmond Road, Kingston upon Thames, Surrey KT2 5PR, England
Tel.: 0181 541 4440
Fax: 0181 541 4584
Internet address: www.nikon.co.uk
Products & services: Microscopes, cameras, inverted microscopes, camera lucida, colony marker, confocal microscopes, CCD, image analysis, micromanipulators

Novartis
Address: Wimblehirst Rd., Horsham, W. Sussex RH12 4AB, England
Tel.: 01403 272827
Internet address: www.novartis.com

NuAire, Inc.
Address: 2100 Fernbrook Ln., Plymouth, MN 55447
Tel.: 612-553-1270; 0800-328-3352
Fax: 612-553-0459
Internet address: www.nuaire.co.uk/
Products & services: Low-temperature freezers, laminar-flow hoods, safety cabinets, CO_2 incubators

Nuclepore Corp. (see also Corning Separations)
Address: Victoria House, 28-38 Desborough Street, High Wycombe, Bucks HP11 2NF, England
Tel.: 01494 471207
Fax: 01494 459540

Products & services: Filters, filtration equipment

Nunc (see Nalge Nunc International, Nalge (Europe) Ltd., and Life Technologies, U.K.)

Nycomed Pharma, AS

Address: P.O. Box 7220 Torshov, 0701 Oslo, Norway

Tel.: 47 23 18 50 50

Fax: 47 23 18 60 37

E-mail address: 6jh@nycomed.com

Products & services: Density medium, lymphocyte preparation medium, metrizamide, sodium metrizoate

Olympus

Address: 2-8 Honduras Street, London EC1Y 0TX, England

Tel.: 0171 250 4069

Fax: 0171 250 4677

E-mail address: SimonShelley@compuserve.com

Internet address: www.olympus, europa.com

Products & services: Microscopes, inverted microscopes, stereomicroscopes, photomicrography, camera lucida

Olympus America, Inc.

Address: Precision Instrument Div., Two Corporate Ctr. Dr., Melville, NY 11747

Tel.: 516-844-5000; 800-446-5967

Fax: 516-844-5112

E-mail address: olympus@performark.com

Internet address: www.olympus.com

Products & services: Microscopes, inverted microscopes, stereomicroscopes, photomirography, camera lucida

Omega Engineering, Inc.

Address: One Omega Drive, Box 4047, Stamford, CT 06907-4047

Tel.: 203-359-7700; 800-826-6342

E-mail address: info@omega.com

Internet address: www.omega.com

Products & services: Electronic thermometer, temperature indicator

Oncogene Research Products

Address: 84 Rogers Street, Cambridge, MA 02142

Tel.: 800-662-2616; 617-577-9333

Fax: 800-828-4871; 617-577-8015

E-mail address: customer.service@oncresprod.com

Internet address: www.apoptosis.com, www.neuroproducts.com

Products & services: ELISA kits, angiogenesis, VEGF, apoptosis, PCNA, IGF, LIF, TGF, TNF

Oncor, Inc. (see also Intergen, Appligene Oncor)

Address: Marketing Division, 209 Perry Parkway, Gaithersburg, MD 20877

Tel.: 510-249-1144

Fax: 510-249-1150

Internet address: www.oncor.com

Products & services: Chromosome paints

Optomax

Address: 9 Ash Street, P.O. Box 840, Hollis, NH 03049

Tel.: 603-465-3385

Fax: 603-465-2291

E-mail address: optomax@msn.com

Products & services: Colony counters

Orion Research, Inc.

Address: 500 Cummings Ctr., Beverly, MA 01915

Tel.: 508-922-4400; 800-225-1480

Fax: 508-927-3932

Internet address: www.orionres.com/

Oxford Instruments, Inc.

Address: 130A Baker Ave. Extension, Concord, MA 01742

Tel.: 508-369-9933; 800-447-4717

Fax: 508-369-6616

Products & services: Micropipettes

Oxoid (see Unipath, Fisher)

Paar Scientific, Ltd.

Address: 594 Kingston Rd., London SW20 8DN, England

Tel.: 0181 540 8553

Fax: 0181 543 8737

E-mail address: paar@psl.anton-paar.co.uk

Products & services: Density meter, viscometers, rheometers

Packard Instruments

Address: Brook House, 14 Station Road, Pangbourne, Berks, RG8 7AN, England

Tel.: 0118 984 4981

Fax: 01189 844059

E-mail address: pico-uk@packardinst.com

Internet address: www.packardinst.com

Products & services: Plate readers, scintillation counters, scintillation fluid, luminometers, vials, minivials, microtitration software

Packard Instrument Co., Inc.

Address: 800 Research Parkway, Meriden, CT 06450

Tel.: 203-238-2351; 800-323-1891

Fax: 203-639-2172

E-mail address: webmaster@packardinst.com

Internet address: www.packardinst.com

Products & services: Plate readers, scintillation counters, scintillation fluid, luminometers, vials, minivials, microtitration software

Paesel & Lorei GmbH & Co.

Address: Borsigallee 6 P.O. Box 630 347 D-6000 Frankfurt am Main 63 Federal Republic of Germany

Tel.: 069 42 20 95

Fax: 069 42 30 84

Products & services: Media, growth factors, antibodies

Pall Corporation

Address: Bio Support Div., 25 Harbor Park Dr., Port Washington, NY 11746

Tel.: 516-484-3600; 800-289-7255

Fax: 516-484-6129

E-mail address: custsvc@pall.com

Internet address: www.pall.com

Products & services: Filters, filter holders, filtration, pleated cartridge filter, pumps, vacuum pump, mycoplasma filtration, ultrafiltration, densitometers

Pall Gelman

Address: Laboratory Products Division, 50 Bearfoot Road, Northborough, MA 01532-1551

Tel.: 508-393-1800; 800-FILTRON

Fax: 508-393-1874

E-mail address: gelmanlab@pall.com

Internet address: www.pall.com/gelman

Products & services: Filters, filter holders, filtration, pleated cartridge filter, pumps, vacuum pump, mycoplasma filtration, ultrafiltration, densitometers

Pall Gelman Laboratory, Laboratory Products Div.

Address: 600 S. Wagner Rd., Ann Arbor, MI 48103

Tel.: 734-665-0651; 800-521-1520
Fax: 734-913-6114
E-mail address: gelmanlab@pall.com
Internet address: www.pall.com/gelman
Products & services: Molecular Biology reagents, media, cell separation, membranes, sterile filtration, ultrafiltration, pleated-cartridge filters

Pall Gelman Sciences
Address: Brackmills Business Park, Caswell Road, Northampton NN4 7EZ, England
Tel.: 01604 704704
Fax: 01604 704724
Internet address: www.pall.com/gelman
Products & services: Filters, filter holders, filtration, pleated cartridge filter, pumps, vacuum pump, mycoplasma filtration, ultrafiltration, densitometers

Pall Ultrafine Group
Address: Europa House, Havant Street, Portsmouth PO1 3PD, England
Tel.: 01705 303303
Fax: 01705 302506
E-mail address: ultrafine@pall.com
Internet address: www.pall.com
Products & services: Sterilizing filters, filter housings, mycoplasma filtration, pleated cartridge filters, tangential-flow filters, ultrafiltration, cell harvesters, hollow-fiber systems, protein purification

Pasteur Merieux Connaught
Address: Rt. 611, Box 187, Swiftwater, PA 18370
Tel.: 717-839-7187; 800-822-2463
Fax: 717-839-5415
Internet address: www.connaught.com
Products & services: Medium

Perceptive Instruments
Address: Blois Meadow Business Centre, Steeple Bumpstead, Haverhill, Suffolk CB9 7BN, England
Tel.: 01440 730773
Fax: 01440 730630
E-mail address: sales@perceptive.co.uk
Internet address: www.perceptive.co.uk
Products & services: Colony counters, counting clones

Pharmacia Biotech, Inc. (see also Amersham Pharmacia Biotech)
Address: 800 Centennial Ave., Box 1327, Piscataway, NJ 08855-1327
Tel.: 908-457-8000; 800-526-3593
Fax: 908-457-8130; 800-329-3593
Internet address: www.apbiotech.com
Products & services: Receptors, signal transduction, assays, electrophoresis, epidermal growth factor, fibronectin, Ficoll, Ficoll-Hypaque media, growth factors, microcarriers, chromotography media, DEAE dextran, marker beads, Percoll, phytohemagglutinin, peristaltic pumps, Ultraser-G

Pharmingen (see also Becton Dickinson Immunocylometry Div. and Becton Dickinson, U.K.)
Address: 10975 Torreyana Road, San Diego, CA 92121
Tel.: 619-812-8800
Fax: 619-812-8888
Internet address: www.pharmingen.com
Products & services: Signal transduction, cytokines, receptors

Philip Harris Scientific
Address: Novara House, Excelsior Road, Ashby Park, Ashby de la Zouch Leicestershire, LE65 1NG, United Kingdom
Tel.: 0845 604 0490
Fax: 01530 419300
E-mail address: branenq@scientific.philipharris.co.uk
Internet address: www.philipharris.co.uk/scientific
Products & services: Multiwell plates, Petri dishes, flasks, centrifuge tubes, Corning Costar, Greiner, roller racks, cloning rings, cloning cylinders, stirrer flasks, spinner flasks, Bellco, stainless-steel closures, stainless-steel caps

Phoretix International
Address: Cale Cross House, Newcastle upon Tyne NE1 6SU, England
Tel.: 0191 230 2121
Fax: 0191 230 2131
E-mail address: phoretix@phoretix.com
Internet address: www.phoretix.com
Products & services: Computer software, microtitration

Pierce Chemical Co.
Address: 3747 N. Meridan Rd., P.O. Box 117, Rockford, IL 61105
Tel.: 815-968-0747; 800-874-3723
Fax: 815-968-7316
E-mail address: ta@piercenet.com
Internet address: www.piercenet.com
Products & services: Chemicals, biochemicals, sample vials, protein assay, detergents, Pronectin, matrix-coated plates

Planer Select
Address: Windmill Road, Sunbury, Middlesex TW16 7HD, England
Tel.: 01932 779997
Fax: 01932 781151
Products & services: Liquid-nitrogen freezers, MVE, cell freezing, inventory control, controlled-rate freezers

Platon Instrumentation
Address: Jays Close, Viables Industrial Estate, Basingstoke, Hampshire RG22 4BS, England
Tel.: 01256 63345
Fax: 01256 460122
Products & services: Gas mixing, flowmeters, gases, flowstats, pressure regulators, flow regulators

Polaroid Corp.
Address: 575 Technology Square 9-P, Cambridge, MA 02139
Tel.: 617-577-2000; 800-225-1618
Fax: 617-386-6271
Internet address: www.polaroid.com/
Products & services: Instant cameras, closed-circuit TV, emulsion, Polaroid camera, microscope camera, photomicroscopy, instant film

Polaroid (U.K.), Ltd.
Address: Wheathampstead House, Codicote Road, Wheathampstead, Herts AL4 8SF, England
Tel.: 0800 010119
Fax: 01582 632001
Internet address: www.polaroid.com
Products & services: Instant cameras, closed-circuit TV, emulsion, Polaroid camera, microscope camera, photomicroscopy, instant film

PolyScience

Address: 6600 West Touhy Avenue, P.O. Box 48312, Niles, IL 60714

Tel.: 800-229-7569; 847-647-0611

Fax: 847-647-1155

E-mail address: polysci@polyscience.com

Internet address: www.polyscience.com/

Products & services: Chillers, water baths, circulators, analytical chemistry kits, Clorox, glutaraldehyde

Popper & Sons, Inc.

Address: 300 Denton Ave., New Hyde Park, NY 11040

Tel.: 516-248-0300

Fax: 516-747-1188

E-mail address: sales@popperandsons.com

Internet address: www.popperandsons.com

Products & services: Sterility indicators, Thermalog, needles

Portex, Ltd.

Address: The Reachfields, Hythe, Kent, England

Products & services: Semipermeable nylon film

Precision Scientific, Inc. (see also Jouan)

Address: 110-C Industrial Dr., Winchester, VA 22602

Tel.: 540-869-9892; 800-621-8820

Fax: 540-869-0130

Products & services: CO_2 incubators, incubators, sterilizing and drying oven, Vetra low-temperature freezers, Hi-Lo speed centrifuges, laminar-flow hoods, class II safety cabinets, vacuum concentration

Promega Corp.

Address: 2800 Woods Hollow Rd., Madison, WI 53711

Tel.: 608-274-4330; 800-356-9526

Fax: 608-277-2601

E-mail address: custserv@promega.com

Internet address: www.promega.com

Promega U.K.

Address: Delta House, Chilworth Research Centre, Southampton SO16 7NS, England

Tel.: 01703 760225; 0800 378994

Fax: 01703 767014; 0800 181037

E-mail address: ukmarketing@uk.promega.com

Internet address: www.euro.promega.com/uk

Products & services: Molecular biology reagents, gene transfection, DNA transfer, endothelial cells, growth factors, immortalization, cell proliferation assays, cytotoxicity assays, apoptosis assays

PromoCell, GmbH

Address: Handschuhsheimer Landstr. 12 69120 Heidelberg, Germany

Tel.: 6221 49049

Fax: 6221 484943

E-mail address: info@promocell.com

Internet address: www.promocell.com/

Products & services: Selective media, serum-free media, specialized cell cultures

Pronova Biopolymer, a.s.

Address: Gaustadalleen 21, N-0371, Oslo, Norway

Products & services: Alginate

Protein Polymer Technologies, Inc.

Address: 10655 Sorrento Valley Road, First Floor, San Diego, CA 92121

Tel.: 619-558-6064; 800-755-0407

Fax: 619-558-6477

E-mail address: aron_stern@ppti.com info@ppti.com

Internet address: www.ppti.com

Products & services: Pronectin, matrix, tissue adhesives, serum-free cell adhesion

Purite, Ltd.

Address: Bandet Way, Thame, Oxon OX9 3SJ, England

Tel.: 01844 217141

Fax: 01844 218098

E-mail address: mail@purite.com

Internet address: www.purite.com/

Products & services: Water purification

Queue Systems, Inc.

Address: 275 Aitken Road, Asheville, NC 28804

Products & services: Incubators, CO_2 incubators

R & D Systems

Address: 614 McKinley Place, NE, Minneapolis, MN 55413

Tel.: 800-328-2400; 612-379-2956

Fax: 612-379-6580

E-mail address: info@rndsystems.com

Internet address: www.rndsystems.com

Products & services: Antibodies, growth factors, cytokines, matrix, epidermal growth factor, fibronectin, ELISA assays, cytokine assays, microtitration, matrix metalloproteinases, Eitosanoid assay kits, neutrophins, microtitration software

R & D Systems Europe

Address: European Subsidiary of R&D Systems, Inc., 4-10 The Quadrant, Barton Ln., Abingdon, Oxon OX14 3YS, England

Tel.: 01235 529449

Fax: 01235 533420

Internet address: www.rndsystems.com

Products & services: Antibodies, growth factors, cytokines, matrix, epidermal growth factor, fibronectin, ELISA assays, cytokine assays, microtitration, matrix metalloproteinases, Eitosanoid assay kits, neutrophins, microtitration software

Radleys

Address: Shire Hill, Saffron Walden, Essex CB11 3AZ, England

Tel.: 01799 513320

Fax: 01799 513283

Products & services: Siphon pipette washer, pipette cylinders, hods, pipettors, Pi-pump, pipette bulbs, flowmeters, desiccators, culture chambers, culture boxes, racks, bottles, nitrile gloves

Rainin Instrument Co., Inc. (see also Anachem, Ltd.)

Address: Mack Road, P.O. Box 4026, Woburn, MA 01888-4026

Tel.: 617-935-3050; 800-472-4646

Fax: 617-938-1152

Products & services: Automatic pipettes

Raven Bio-Laboratories

Address: 8607 Park Dr., Omaha, NE 68106

Tel.: 402-556-6690; 800-728-5702

Fax: 402-556-4722

E-mail address: ravenbio@aol.com

Products & services: Slide boxes

Renco Corp.

Address: Manchester, MA

Tel.: 800-257-8284; 508-526-8494
Products & services: Nitrile gloves

Revco Scientific, Inc.
Address: 275 Aiken Road, Asheville, NC 28804
Tel.: 704-658-2711; 800-252-7100
Fax: 704-645-3368
E-mail address: sales@revco-sci.com
Internet address: www.revco-sci.com
Products & services: Ultra deep freezers

Richardsons of Leicester
Address: Evington Valley Road, Leicester LE5 5LJ, England
Tel.: 0116 273 6571
Products & services: Microscope slides (Berliner Glas KG)

Riken (Japan)
Address: 3-1-1 Koyadai, Tsukuba Science City, Ibaraki 305, Japan
Internet address: www.rtc.riken.go.jp/
Products & services: Cell lines, cell bank

Robbins Scientific
Address: Suite B, Greville Court, 1665 High Street, Knowle, West Midlands B93 0LL, England
Tel.: 01564 775525
Fax: 01564 775759
Products & services: Density media, lymphocyte preparation media

Roboz Surgical Instrument Co. Inc.
Address: 9210 Corporate Blvd., Ste. 220, Rockville, MD 20850
Tel.: 301-590-0055; 800-424-2984
Fax: 301-590-1290
E-mail address: mary@roboz.com
Internet address: www.roboz.com
Products & services: Sterilizers, syringes, canulas, instruments, sterilization indicator tape

Roche Diagnostics Ltd.
Address: Bell Lane Lewes West Sussex BN7 1LG, England
Tel.: 01273 480444; 0800 146399
Fax: 01273 480266; 0800 181 087
Internet address: biochem.boehringer-mannheim.com/
Products & services: Automatic pipettes, antibodies, biochemicals, ELISA, collagen, dispase, growth factors, micropipettes, medium, serum-free medium, vitamins, mycoplasma eradication, BM cyline, DAPI, apoptosis kits, DNA transfer, lipofection, DOTAP, DOSPER, Fugene, Nutridoma, purified trypsin, penicillin, streptomycin, antibiotics

Roche Diagnostics, GmbH
Address: Sandhofer Str 116, D68305 Mannheim 31, Germany
Tel.: 0621 759 8545
Fax: 0621 759 8509
Internet address: biochem.boehringer_mannheim.com/
Products & services: Automatic pipettes, antibodies, biochemicals, ELISA, collagen, dispase, growth factors, micropipettes, medium, serum-free medium, vitamins, mycoplasma eradication, BM cyline, DAPI, apoptosis kits, DNA transfer, lipofection, DOTAP, DOSPER, Fugene, Nutridoma, purified trypsin, penicillin, streptomycin, antibiotics

Roche Molecular Biochemicals
Address: 9115 Hague Rd., PO Box 50414, Indianapolis, IN 46250
Tel.: 800-428-5433, 800-262-1640
Fax: 800-428-2883
E-mail address: biochemts_us@bmc.boehringer-mannheim.com
Internet address: www.biochem.boehringer-mannheim.com
Products & services: Automatic pipettes, antibodies, biochemicals, ELISA, collagen, dispase, growth factors, micropipettes, medium, serum-free medium, vitamins, mycoplasma eradication, BM cyline, DAPI, apoptosis kits, DNA transfer, lipofection, DOTAP, DOSPER, Fugene, Nutridoma, purified trypsin, penicillin, streptomycin, antibiotics

Roche Laboratories (see also Roche Molecular Biochemicals)
Address: 340 Kingsland St., Nutley, NJ 07110
Tel.: 201-235-5000
Fax: 201-562-2739
Products & services: Heparin

RS Princeton Instruments
Address: 3660 Quakerbridge Road, Trenton, NJ 08619
Tel.: 609-587-9797
Fax: 609-587-1970
Products & services: Cameras and accessories

Rustrak Instruments
Address: The Hyde, Brighton BN2 4JU, England
Tel.: 01273 606271
Fax: 01273 609990
Products & services: Recorders, electronic thermometers

Rustrak Instruments
Address: 1900 S County Tr., East Greenwich, RI 02818
Tel.: 401-884-6800; 800-332-3202
Fax: 7401-884-4872
Products & services: Recorders, electronic thermometers

Safetech
Address: Enterprise House, Plassey Technological Park, Limerick, Ireland
Tel.: 353-61-338177
Products & services: Laminar-flow hoods, class II safety cabinets, cleanrooms

Safetec of America Inc.
Address: 1055 E. Delavan Ave., Buffalo, NY 14215
Tel.: 716-895-1822; 800-456-7077
Fax: 716-895-2969
Products & services: Laminar-flow hoods, class II safety cabinets, clean rooms

Sanyo Fisher Service Corp. (see also Sanyo Scientific)
Address: 1411 West 190th Street Suite 700, Gardena, CA 90248
Tel.: 310-769-5832
Fax: 310-243-6699
Internet address: www.broadband-guide.com/company/sanyofisher.html
Products & services: Incubators, freezers

Sanyo Gallenkamp, PLC
Address: Park House, Meridian East, Meridian Business Park, Leicester LE3 2UZ, England
Tel.: 0116 2630530

Fax: 0116 2630353
E-mail address: 100633.127@compuserv.com
Products & services: CO_2 incubators, freezers, MSE centrifuges

Sanyo Scientific
Address: 900 N. Arlington Heights Rd., Itasca, IL 60143
Tel.: 630-875-3530; 800-858-8442
Fax: 630-775-0044
Products & services: Incubators, freezers

Sartorius, AG
Address: P.O. Box 3243, Weender, Ondestrasse, 94-108, 3707 Goettingen, W. Germany
Tel.: 49 551 3080
Products & services: Balances, filters, disc filters, filter holders, filtration equipment, balances, chromatography media

Sartorius Corp.
Address: 131 Heartland Blvd., Edgewood, NY 11717
Tel.: 516-254-4249; 800-368-7178
Fax: 516-254-4253
E-mail address: 102233.432@compuserve.com
Internet address: www.sartorius.com/
Products & services: Balances, filters, disc filters, filter holders, filtration equipment, balances, chromatography media

Sartorius, Ltd.
Address: Blenheim Road, Longmead Industrial Estate, Epsom, Surry KT19 9QN, England
Tel.: 01372 737100
Fax: 013727 20799
Internet address: www.sartorius.com
Products & services: Balances, filters, disc filters, filter holders, filtration equipment, balances, chromatography media

Scanalytics, Incorporated
Address: 8550 Lee Highway, Suite 400, Fairfax, Virginia 22031-1515
Tel.: 703-208-2230
Fax: 703-208-1960
E-mail address: info@scanalytics.com
Internet address: www.scanalytics.com
Products & services: Imaging software, CCD cameras, image analysis

Schärfe System
Address: Kammerstrasse 22, D-72764 Reulingen, Germany
Tel.: 07121 38786-11
Fax: 07121 38786-99
E-mail address: mail@CASY-Technology.com
Internet address: www.CASY-Technology.com
Products & services: Cell counters, cell size analyzers, CASYton, counting fluid, CASYclean, detergent, CASYcups, sample vessels

Schleicher & Schuell
Address: VKF, P.O. Box 4, D-37582 Dassel, Germany
Tel.: 05561 791 0
Fax: 05561 791 533
E-mail address: filtration@s-und-s.de
Internet address: www.s-und-s.de
Products & services: Filter sterilization, blotting membranes, Centrex

Schott Corp.
Address: 3 Odell Plaza, Yonkers, NY 10701
Tel.: 914-968-8900; 800-633-4505

Fax: 914-968-8585
Internet address: www.schottglass.com
Products & services: Glassware, borosilicate glass, bottles, Duran glass bottles

Scientek
Address: 201–11151 Bridgeport Rd., Richmond, BC V6X 1T3, Canada
Tel.: 604-270-6119
Fax: 604-273-1262
Products & services: Glassware washing machine

Scientific Laboratory Supplies, Ltd.
Address: Unit 27, Nottingham South & Wilford Industrial Estate, Ruddington Lane, Wilford, Nottingham NG11 7EP, England
Tel.: 0115 982 1111
Fax: 0115 982 5275
Products & services: Cloning rings

Serotec
Address: 22 Bankside, Station Approach, Kidlington, Oxford OX5 1JE, England
Tel.: 01865 852700
Fax: 01865 373899
E-mail address: 100116.3413: compuserve.com
Internet address: www.serotec.co.uk
Products & services: Cytokines, antibodies, IL-6, OSM, KGF, VEGF, CAMS, anti-integrins, ELISA, epidermal growth factor, fibronectin, growth factors, media

Serotec Inc.
Address: 1017 Main Campus Dr., Ste. 2450, Raleigh, NC 27606
Tel.: 919-515-9980; 800-265-7376
E-mail address: serotec@serotec-inc.com
Internet address: www.serotec.co.uk
Products & services: Antibodies, cytokines, ELISA, growth factors, media

Serva Feinbiochemica, GmBH
Address: P.O. Box 105260, D-69042 Heidelberg, Germany
Products & services: Pluronic-F68, HEPES

Shamrock Scientific Specialty Systems, Inc.
Address: 34 Davis Dr., Bellwood, IL 60104
Tel.: 708-547-9005; 800-323-0249
Fax: 708-547-9021
E-mail address: slabels@AOL.com
Products & services: Sterile-indicating tape, pressure-sensitive tapes, labels, labeling tapes, tags, temperature indicator strips

Shandon Lipshaw
Address: 171 Industry Dr., Pittsburgh, PA 15275
Tel.: 412-788-1133; 800-547-7429
Fax: 412-788-1138
Internet address: www.shandon.com
Products & services: Centrifuge cytospin, CO_2 automatic change-over unit for cylinders, cytocentrifuge, electrophoresis, hemocytometers, microcaps (Drummond), pipette washer, pipette drier, cytospin

Sigma Chemical Co.
Address: P.O. Box 14508, St Louis, MO 63178
Tel.: 314-771-5750; 800-325-3010
Fax: 314-771-5757
E-mail address: sigma-techserv@sial.com

Internet address: www.sigma.sial.com
Products & services: Amino acids, antibodies, biochemicals, bovine serum albumin, chemicals, Colcemid, collagen, collagenase, dexamethasone, dextran, diacetyl fluorescein, disinfectants, DMSO, DNase, ethidium bromide, EDTA, formamide, glutamine, glutathione, growth factors, PEG, purified trypsin, penicillin, streptomycin, antibiotics, methotrexate, serum-free medium, HEPES, cytosine arabinoside, polyvinylpyrrolidine, phytohemaglutinin (PHA), Pluronic F68

Sigma-Aldrich Company, Ltd.
Address: Fancy Road, Poole, Dorset BH17 7NH, England
Tel.: 01202 733114; 0800 373 731
Fax: 08000 378 785
E-mail address: ukcustsv@vns.sial.com
Products & services: Antibodies, amino acids, biochemicals, bovine serum albumin, chemicals, Colcemid, collagen, collagenase, dexamethasone, dextran, diacetyl fluorescein, disinfectants, DMSO, DNase, ethidium bromide, EDTA, formamide, glutamine, glutathione, growth factors, HEPES, hyaluronidase, hydrocortisone, hypoxanthine, insulin, media, mercaptoethanol, nucleosides, penicillin, phytohemagglutinin, Pluronic F68, poly-D-lysine, polyvinyl pyrrolidone, Pronase, quinacrine dihydrochloride, sieves, α-thioglycerol, thymidine, heparin, Pluronic F68, PEG, purified trypsin, penicillin, streptomycin, antibiotics, methotrexate, serum-free medium, HEPES, cytosine arabinoside, polyvinylpyrrolidine, phytohemaglutinin (PHA)

Signal Instrument Co., Ltd.
Address: Standards House, 12 Doman Road, Camberley, Surrey GU15 3DF, England
Tel.: 01276 682 841
Fax: 01276 691302
Products & services: Gas mixers, gas blenders

S M L—Serum Medium Laborbedarf, Import–Export
Address: Silcherweg 27, D-88741 Lauphelm
Tel.: +49 7392 93362
Fax: +49 7392 93362
Products & services: Serum

Society for In Vitro Biology (SIVB)
Address: SIVB Business Office, 9315 Largo Drive West, Suite 255, Landover, MD 20774
Tel.: 301-324-5054
Fax: 301-324-5057
E-mail address: sivb@sivb.org
Internet address: www.sivb.org

Southern Biotechnology Associates, Inc.
Address: P.O. Box 26221, Birmingham, AL 35260
Tel.: 205-945-1774; 800-SBA-CALL
Fax: 205-945-8768
E-mail address: custserv@southernbiotech.com
Internet address: www.southernbiotech.com
Products & services: Antibodies, antimouse IgG3-phycoerythrein

Southern Watch & Clock Supplies, Ltd.
Address: Precista House, 48/56 High Street, Orpington, Kent, BR6 OJH, England
Tel.: 01689 824318; 01689 875206
Fax: 01689 870079

Products & services: Glass cement
Spectrum Medical, Inc.
Address: 23022 La Cadena Dr., Ste 100, Laguna Hills, CA 92653
Tel.: 714-581-3880; 800-634-3300
Fax: 714-855-6120
E-mail address: techsupport@spectrumlabs.com
Internet address: www.spectrumlabs.com
Products & services: Cellmax, hollow-fiber culture cartridges, perfusion

Staniar, J., & Co.
Address: 34 Stanley Road, Whitefield, Manchester, M45 8QX, England
Tel.: 0161 767 9026; 0161 767 9033
Fax: 0161 767 9049
Products & services: Gauze, nylon mesh, Nitex

StemCell Technologies, Inc.
Address: 808–777 West Broadway, Vancouver, BC, Canada, V5Z 4J7
Tel.: 604-877-0713; 800-667-0322
Fax: 604-877-0704; 800-567-2899
E-mail address: info@stemcell.com
Internet address: www.stemcell.com
Products & services: Hemopoietic progenitor culture, stem cell purification, cell separation, methylcellulose media, colony atlas, videos, dishes, chamber slides, spinner flasks, software, antibodies, cytokines

Steris Corporation
Address: Scientific Division, 2 Albany Court, Albany Park, Camberley, Surrey GU15 2XA, England
Tel.: 01276 683300
Fax: 01276 685662/3
Products & services: Autoclaves, glassware washing machines, stills, decontamination, freeze-drying, steam generators

Steris Corp.
Address: 5960 Heisley Rd., Mentor, OH 44060-1834
Tel.: 440-345-2600; 800-333-8838
Fax: 440-350-7081
Products & services: Autoclaves, glassware washing machines, stills, freeze-drying, steam generators

Stoelting Co.
Address: 620 Wheat Lane, Wood Dale, IL 60191
Tel.: 630-860-9700
Fax: 630-860-0775
E-mail address: physiology@stoeltingco.com
Internet address: www.stoeltingco.com
Products & services: Micromanipulators

Stratagene
Address: 11011 N. Torrey Pines Rd., La Jolla, CA 92037–1073
Tel.: 619-535-5400; 800-424-5444
Fax: 619-535-0034
E-mail address: tech_services@stratagene.com
Internet address: www.stratagene.com
Products & services: Products for molecular biology

Stratech Scientific, Ltd.
Address: 61-63 Dudley Street, Luton, Beds LU2 ONP, England
Tel.: 01582 481884

Fax: 01582 481895
Internet address: www.stratech.co.uk
Products & services: Antibodies, oncogenes, cell cycle, anticollagen, antimucin, anti-GFAP, Promega, Savant Instruments, concentrators, Jackson Immunoresearch, DNA extraction kits, Collaborative Research, growth factors, extracellular matrix, cytokines

Surgicon, Ltd.
Address: Wakefield Rd., Brighouse, W. Yorkshire HD6 1QL, England
Tel.: 01484 712 147
Fax: 01484 400 106
Products & services: Sterilization indicator tape (autoclave), gloves

TAAB Laboratories, Equipment, Ltd.
Address: 3 Minerva House, Calleva Park, Aldermaston, Berks, RG7 8NA, England
Tel.: 0118 9817775
Fax: 0118 9817881
E-mail address: sales@taab.co.uk
Products & services: Fixatives, glutaraldenhyde, buffers, sodium cacodylate, acetonitrile, immunochemistry, electron microscopy

Taylor-Wharton Cryogenic Products
Address: Postfach 14 70, D-25804 HUSLUM, Germany
Tel.: 49 48 41 985 0
Fax: 49 48 41 985 30
Internet address: www.taylor-wharton.com/cryohom.htm
Products & services: Liquid-nitrogen freezers, controlled-rate freezers for liquid N_2 freezing, dry shipper

Taylor-Wharton RDF Cryogenics
Address: P.O. Box 568, Theodore, AL 36590-0568
Tel.: 334-443-8680; 800-898-2657
Fax: 334-443-2250
E-mail address: twsales@taylor-wharton.com
Internet address: www.taylor-wharton.com/cryohom.htm
Products & services: Liquid-nitrogen freezers, controlled-rate freezer, dry shipper

TCS Biologicals, Ltd.
Address: Botolph, Claydon, Buckingham MK18 2LR, England
Tel.: 01296 714 071
Fax: 01296 715 753
E-mail address: sales@tcsgroup.co.uk
Products & services: Serum-free medium, matrix, Accutase, trypsinization, normal human cells, specialized media, serum, cell dispersal enzymes, antibodies, cell adhesion, apoptosis, signal transduction, cell cycle, antisense and hybridization probes, oligonucleotides, polyclonal antibody production, cytokine assays, neuroreceptors, toxins, flow cytometry, cell culture systems, Cultisper microcarriers, Upstate Biotechnology, Clonetics, Endogen, Biognostik, Caltag, IGEN

Techne (Cambridge), Limited
Address: Duxford, Cambridge, CB2 4PZ, England
Tel.: 01223 832401
Fax: 01223 836838
E-mail address: sales@techneuk.attmail.com

Internet address: www.techneuk.co.uk
Products & services: Magnetic stirrers, stirrer culture bottles, suspension culture, microcarrier culture bottles

Techne Inc.
Address: 743 Alexander Rd., Princeton, NJ 08540
Tel.: 609-452-9275; 800-225-9243
Fax: 609-987-8177
E-mail address: techneusa@worldnet.att.net
Internet address: www.techneuk.co.uk
Products & services: Stirrer culture vessels, suspension culture, microcarrier culture

Tekmar–Dohrmann
Address: 7143 E, Kemper Rd., Cincinnati, OH 45249
Tel.: 513-247-7000; 800-543-4461
Fax: 513-247-7050
E-mail address: sales@tekmar.com
Internet address: www.tekmar.com
Products & services: Nylon mesh, Nitex, media, stirrers

Thomas Scientific
Address: 99 High Hill Rd., P.O. Box 99, Swedesboro, NJ 08085
Tel.: 609-467-2000; 800-345-2100
Fax: 609-467-3087
Internet address: www.thomassci.com
Products & services: Lab equipment, silicone tubing, disinfectants, magnetic stirrers, iridectomy knives, Hamilton syringes, microsyringes

U.S. Filter (see also USF, Ltd.)
Address: 40–004 Cook Street, Palm Desert, CA 92211
Tel.: 760-340-0098
Fax: 760-341-9368
Internet address: www.usfilter.com
Products & services: Water purification, reverse osmosis, ion exchange, Permutit

Unipath Oxoid Division
Address: 800 Proctor Ave., Ogdensburg, NY 13669-2205
Tel.: 613-226-1318
Fax: 613-226-3728
Products & services: Microbiological medium, agar, PBS, tryptose phosphate broth

Universal Biologicals, Ltd.
Address: Fromehall Mill, Lodgemore Lane, Stroud, Gloucestershire, GL5 3EH, England
Tel.: 01453 753019
Fax: 01453 751448
Products & services: Growth factors, collagen

Upstate Biotechnology
Address: 1100 Winter Street, Suite 2300, Waltham, MA 02451
Tel.: 800-233-3991; 781-890-8845
Fax: 781-890-7738
E-mail address: info@upstatebiotech.com
Internet address: www.upstatebiotech.com
Products & services: Antibodies, growth factors, enzymes, assay kits, neurobiology, apoptosis, cell growth, cell proliferation assays

USB Corp.
Address: 26111 Miles Rd., Cleveland, OH 44128
Tel.: 216-765-5000; 800-321-9322

Fax: 800-535-0898; 216-464-5075
E-mail address: customerserv@usbweb.com
Internet address: www.usbweb.com
Products & services: Biochemicals, polyvinylpyrrolidine

USF, Ltd.
Address: Hareford Court, John Tate Road, Hertford SG13 7NW, England
Tel.: 01992 823300
Fax: 01992 501528
E-mail address: usf@usf.co.uk
Internet address: www.usfilter.com
Products & services: Water purification, reverse osmosis, ion exchange, Permutit

UVP, Inc.
Address: 2066 W 11th St., Upland, CA 91786
Tel.: 909-946-3197; 800-452-6788
Fax: 909-946-3597
E-mail address: uvp@uvp.com
Internet address: www.uvp.com
Products & services: Mercury vapor lamps, fluorescence, UV lamps, UV transilluminators

UVP, Ltd.
Address: Science Park, Milton Road, Cambridge CB4 4BN, England
Tel.: 0223 355722
E-mail address: uvp@uvp.com
Internet address: www.uvp.com
Products & services: Mercury vapor lamps, fluorescence, UV lamps, UV transilluminators

Valley Forge Instrument Co., Inc.
Address: 55 Buckwalter Rd., Phoenixville, PA 19460
Tel.: 610-933-1806
Products & services: Pressure cooker, bench-top autoclave

Vector Laboratories, Inc.
Address: 30 Ingold Road, Burlingame, CA 94010
Tel.: 650-697-3600
Fax: 650-697-0339
E-mail address: vector@vectorlabs.com
Internet address: www.vectorlabs.com
Products & services: Antibodies, ABC kit, enzyme immunoassays, hybridoma screening, fluorescence fade retardant, Vectastain, biotin/avidin, glucose oxidase, alkaline phosphatase, Vectashield, UV quench inhibitor, peroxidase, Vectabond, slide adhesive, lectins, fluorescence-quenching inhibitor, fluorescence bleaching retardant

Vector Laboratories, Ltd.
Address: 3 Accent Park, Blakewell Road, Orton Southgate, Peterborough PE2 6XS, England
Tel.: 01733 237999
Fax: 01733 237119
E-mail address: vector@vectorlabs.co.uk
Internet address: www.vectorlabs.com
Products & services: Antibodies, ABC kit, enzyme immunoassays, hybridoma screening, fluorescence fade retardant, Vectastain, biotin/avidin, glucose oxidase, alkaline phosphatase, Vectashield, UV quench inhibitor, peroxidase, Vectabond, slide adhesive, lectins, fluorescence-quenching inhibitor, fluorescence bleaching retardant

VWR Scientific Products
Address: 1310 Goshen Pkwy, W. Chester, PA 19380
Tel.: 610-431-1700; 800-932-5000
Fax: 610-436-1761
Internet address: www.vwrsp.com
Products & services: Lab supplies, Hamilton syringes, microsyringes, face masks

Vysis
Address: 3100 Woodcreek Dr., Downers Grove, IL 60515
Tel.: 708-271-7000
Fax: 708-271-7008
E-mail address: help@vysis.com
Internet address: www.vysis.com
Products & services: Molecular probes

Watson–Marlow/Bredel Pumps
Address: 220 Ballardvale St., Wilmington, MA 01887
Tel.: 978-658-6168; 800-282-8823
Fax: 978-658-0041
E-mail address: ussupport@watson-marlow.bredel.com
Internet address: www.watson-marlow.com
Products & services: Peristaltic pumps

Watson–Marlow, Ltd.
Address: Falmouth, Cornwall TR11 4RU, England
Tel.: 01326 370370
Fax: 01326 376009
Internet address: www.watson-marlow.com
Products & services: Peristaltic pumps

Wescor, Inc.
Address: 459, South Main Street, Logan, Utah 84321
Tel.: 801-752-6011
Fax: 801-752-4127
Products & services: Cytocentrifuge

Whatman Inc.
Address: 9 Bridewell Pl., P.O. Box 1197, Clifton, NJ 07014
Tel.: 973-773-5800; 800-631-7290
Fax: 973-773-6138
Products & services: Filter paper, filters, chromatography, Nuclepore

Whatman Polyfiltronics
Address: 136 Weymouth St., Rockland, MA 02370
Tel.: 781-878-1133; 800-434-7659
Fax: 781-878-0822
E-mail address: polyfil@polyfiltronics.com
Internet address: www.polyfiltronics.com
Products & services: Microarray chip technology, microtitration

Wheaton Science Products (see Lawson Mardon Wheaton)

Worthington Biochemical Corp.
Address: 730 Vassar Avenue, Lakewood, NJ 08701
Tel.: 732-942-1660
Fax: 732-942-9270
Internet address: www.worthington-biochem.com
Products & services: DNase, hyaluronidase, trypsin, purified trypsin

Zeiss (see Carl Zeiss)

Zinsser Analytic (U.K.), Ltd.
Address: Howarth Road, Stafferton Way, Maidenhead, Berks SL6 1AP, England
Tel.: 01628 773202
Fax: 01628 72199

Products & services: Automatic pipettes, vials, dispensers, liquids, peristaltic pumps, vacuum pumps

Zymed Laboratories, Inc.
Address: 458 Carlton Court, South San Francisco, CA 94080
Tel.: 800-874-4494; 650-871-4494
Fax: 415-871-4499
E-mail address: tech@zymed.com
Internet address: www.zymed.com
Products & services: Antibodies, Ki-67, cytokeratin, breast cancer

Glossary

[Modified after Schaeffer, 1990]

Adaptation. Induction or repression of synthesis of a macromolecule (usually a protein) in response to a stimulus; e.g., enzyme adaptation—an alteration in enzyme activity brought about by an inducer or repressor and involving an altered rate of enzyme synthesis or degradation.

Allograft. See *homograft*.

Amniocentesis. Prenatal sampling of the amniotic cavity.

Anchorage dependent. Requiring attachment to a solid substrate for survival or growth.

Anemometer. An instrument for measuring flow rate of air.

Aneuploid. Not an exact multiple of the haploid chromosome number. (See *haploid*.)

Aseptic. Free of microbial infection.

Autocrine. Receptor-mediated response of a cell to a factor produced by the same cell.

Autograft. A graft from one individual transplanted back to the same individual.

Autoradiography. Localization of radioisotopes in cells, tissue sections, and blots from electrophoresis preparations; achieved by exposure of a photographic emulsion placed in close proximity to the specimen.

Balanced salt solution. An isotonic solution of inorganic salts present in approximately the correct physiological concentrations; may also contain glucose, but is usually free of other organic nutrients.

Bioreactor. Culture vessel for large-scale production of cells, either anchored to a substrate or propagated in suspension.

Biostat. Culture vessel in which physical, physico-chemical, and physiological conditions, as well as cell concentration, are kept constant, usually by perfusion, monitoring, and feedback.

Carcinoma. A tumor derived from epithelium, usually endodermally or ectodermally derived cells.

Cell concentration. Number of cells per ml of medium.

Cell culture. Growth of cells dissociated from the parent tissue by spontaneous migration or mechanical or enzymatic dispersal.

Cell density. Number of cells per cm^2 of substrate.

Cell fusion. Formation of a single cell body by the fusion of two other cells, either spontaneously or, more often, by induced fusion with inactivated Sendai virus or polyethylene glycol.

Cell hybridization. See *hybrid cell*.

Cell line. A propagated culture after the first subculture.

Cell strain. A characterized cell line derived by selection or cloning.

Chemically defined. Made entirely from pure defined constituents (said of a medium); distinct from "serum free," in which other poorly characterized constituents may be used to replace serum.

Clone. A population of cells derived from one cell.

Commitment. Irreversible progression from a stem cell to a particular defined lineage endowing the cell with the potential to express a limited repertoire of properties.

Confluent. A monolayer of cells in which all cells are in contact with other cells all around their periphery, and no available substrate is left uncovered.

Constitutive. Expressed by a cell in the absence of external regulation.

Contact inhibition. Inhibition of plasma membrane ruffling and cell motility when cells are in complete contact with other adjacent cells, as in a confluent culture; often precedes, but is not necessarily causally related to, cessation of cell proliferation.

Continuous cell line or cell strain. Cell line or strain having the capacity for infinite survival. Previously known as "established" and often referred to as "immortal."

Cyclic growth. Growth from a low cell density to a high cell density with a regular subculture interval; regular repetition of the growth cycle for maintenance purposes.

Cytokine. A factor, released by cells, that will induce a receptor-mediated effect on the proliferation, differentiation, or inflammation of other cells; usually a short range paracrine, rather than systemic, effect.

Deadaptation. Reversible loss of a specific property due to the absence of the appropriate inducer (not always defined).

Dedifferentiation. Irreversible loss of the specialized properties that a cell would have expressed *in vivo*. As evidence accumulates that cultures dedifferentiate by a combination of the selection of undifferentiated stromal cells and deadaptation resulting from the absence of the appropriate inducers, the term is going out of favor. It is still correctly applied to mean the progressive loss of differentiated morphology in histological observations of, for example, tumor tissue.

Density limitation of growth. Mitotic inhibition correlated with an increase in cell density at confluence.

Diploid. Each chromosome represented as a pair, identical in the autosomes and female sex chromosomes and nonidentical in male sex chromosomes, and corresponding to the chromosome number and morphology of most somatic cells of the species from which the cells are derived.

Dome. A hemicystic or blister-like structure in a confluent epithelial monolayer implying ion transport across the monolayer and resulting in the accumulation of water below the monolayer.

Ectoderm. The outer germ layer of the embryo, giving rise to the epithelium of the skin.

Embryonic induction. The interaction (often reciprocal) of cells from two different germ layers, promoting differentiation.

Endocrine. Signaling factors, such as hormones, released by one tissue and having an effect on a distant tissue via the systemic vasculature.

Endoderm. The innermost germ layer of the embryo, giving rise to the epithelial component of organs such as the gut, liver, and lungs.

Endothelium. An epithelial-like cell layer lining spaces within mesodermally derived tissues, such as blood vessels, and derived from the mesoderm of the embryo.

Enzyme induction. An increase in synthesis of an enzyme produced by, for example, hormonal stimulation.

Epithelial. Cells derived from epithelium but often used more loosely to describe any cells of a polygonal shape with clear, sharp boundaries between them. More correctly, the latter should be referred to as epithelial-like or epithalioid.

Epithelium. A covering or lining of cells, as in the surface of the skin or lining of the gut, usually derived from the embryonic endoderm or ectoderm, but sometimes derived from mesoderm, as with kidney tubules and mesothelium lining body cavities.

Euploid. Exact multiple of the haploid chromosome set. The correct morphology characteristic of each chromosome pair in the species from which the cells are derived is not implicit in the definition, but is usually assumed to be the case; otherwise we should say "euploid, but with some chromosomal aberrations."

Explant. A fragment of tissue transplanted from its original site and maintained in an artificial medium.

FACS. See *fluorescence-activated cell sorter.*

Fermentor. Large-scale culture vessel, often used for cells in suspension; derived from same term applied to microbiological culture.

Fibroblast. A proliferating precursor cell of the mature differentiated fibrocyte.

Fibroblastic. Resembling fibroblasts (i.e., spindle shaped (bipolar) or stellate (multipolar)); usually arranged in parallel arrays at confluence if contact is inhibited. Often, the term is used indiscriminately for undifferentiated mesodermal cells, regardless of their relationship to the fibrocyte lineage; implies a migratory type or cell with processes exceeding the nuclear diameter by threefold or more. More correctly, fibroblast-like or fibroblastoid.

Ficoll-paque. Density medium made up of Ficoll combined with a radio-opaque iodinated substance, such as sodium metrizoate.

Finite cell line. A culture that has been propagated by subculture, but is capable of only a limited number of cell generations *in vitro* before dying out.

Flow cytometer. An instrument providing quantitative and qualitative analysis of individual cells in a population by scanning a single cell stream with a laser, or with multiple lasers of different wavelengths, and recording the light that is scattered or the fluorescence that is emitted.

Fluorescence-activated cell sorter (FACS). A cell separation device based on electromagnetic sorting of a single-cell suspension by means of the scattering of light or the fluorescent properties of individual cells revealed by a laser scanning a single cell stream. (See also *Flow cytometer.*)

Generation number. The number of population doublings (estimated from dilution at subculture) that a culture has undergone since explantation; necessarily contains an approximation of the number of generations in primary culture.

Generation time. The interval from one point in the cell division cycle to the same point in the cycle, one division later; distinct from population doubling time, which is derived from the total cell count of a population and therefore averages different generation times, including the effect of nongrowing cells.

Genotype. The total genetic characteristics of a cell.

Glycocalyx. Glycosylated peptides, proteins, and lipids, and glycosaminoglycans attached to the surface of the cell.

Growth curve. A semilogarithmic plot of the cell number on a logarithmic scale against time on a linear scale, for a proliferating cell culture; usually divided into the lag phase (the phase before growth is initiated), the log phase (the period of exponential growth), and the plateau (a stable cell count achieved when the culture stops growing at a high cell density).

Growth cycle. Growth interval from subculture to the top of the log phase, ready for a further subculture.

Growth factor. A factor, released by cells, that induces proliferation in other cells; mostly paracrine in effect, but may be released into the blood by platelets or endothelium.

Growth medium. The medium used to propagate a particular cell line; usually a basal medium with additives such as serum or growth factors.

Haploid. That chromosome number wherein each chromosome is represented once; in most higher animals, the number present in the gametes and half the number found in most somatic cells.

Heterokaryon. Cell containing two or more genetically different nuclei; usually derived by cell fusion.

Heteroploid. A culture in which the cells have chromosome numbers other than diploid and differing from each other.

Histotypic. A culture resembling a tissue-like morphology *in vivo*. Usually, a three-dimensional culture re-created from a dispersed cell culture that attempts to regain, by cell proliferation and multilayering or by reaggregation, a tissue-like structure. Organ cultures cannot be propagated, whereas histotypic cultures can.

Holding medium. Medium, usually without serum and growth factors, or with minimal serum, designed to maintain cells in a viable state without proliferation (e.g., for collecting biopsies or maintaining cells at a plateau with no further cell proliferation).

Homeothermic. Able to maintain a constant body temperature in spite of environmental fluctuations.

Homograft. (Allograft.) A graft derived from a genetically different donor of the same species as the recipient.

Homokaryon. Cell containing two or more genetically identical nuclei; usually a product of cell fusion.

Hybrid cell. Mononucleate cell that results from the fusion of two different cells, leading to the formation of a synkaryon. (See *synkaryon*.)

Ideogram. The arrangement of the chromosomes of a cell in order by size and morphology so that the karyotype may be studied and genetically analyzed.

Immortalization. The acquisition of an infinite life span. May be induced in finite cell lines by transfection with telomerase, oncogenes, or the large T-region of the SV40 genome, or by infection with SV40 (whole virus) or Epstein–Barr virus (EBV). Immortalization is not necessarily a malignant transformation, although it may be a component of malignant transformation.

Induction. An increase in effect produced by a given stimulus.

Infection. (Other than the commonplace definition.) Transfer of genomic DNA with a retroviral construct containing the DNA sequence under investigation, usually packaged with a promoter sequence and a reporter gene, such as β-galactosidase; the product of an infection may be detected by staining with a chromogenic substrate.

Isograft. (Syngraft.) A graft derived from a genetically identical or nearly identical donor of the same species as the recipient.

In ovo. In the egg—usually, the hen's egg.

In vitro. Literally, "in glass," but used conventionally to mean cultured outwith the host as cell cultures, organ cultures, or short-term organ bath preparations; also used to indicate biochemical and molecular reactions carried out in a test tube, but these reactions are better referred to as *cell free*.

In vivo. In the living plant or animal.

Karyotype. The distinctive chromosomal complement of a cell.

Laminar flow. The flow of a fluid that closely follows the shape of a streamlined surface without turbulence; said of hoods or cabinets characterized by a stable flow of air over the work area so as to minimize turbulence.

Laminar-flow hood or cabinet. A workstation with filtered air flowing in a laminar (nonturbulent) manner parallel to or perpendicular to the work surface, to maintain the sterility of the work; the parallel flow is called *horizontal* laminar flow, the perpendicular flow *vertical* laminar flow.

Leukemia. Malignant disease of the hemopoietic system, evident as circulating blast cells.

Lipofection. Transfection of DNA by fusion with lipid-encapsulated DNA.

Lymphoma. A solid tumor of lymphoid cells.

Log phase. See *growth curve.*

Macroautoradiography. Localisation of radioisotopes in whole body sections and blots from electrophoresis preparations, by exposure of a photographic emulsion placed in close proximity to the specimen, usually by placing the film with the blot in a cassette with an intensifier screen.

MACS. (Magnetic-activated cell sorting.) Sorting cells by the magnetic attraction of magnetizable antibody-coated ferritin beads that bind to specific cell surface antigens.

Malignant. Invasive or metastatic (i.e., colonizing other tissues); (said of a tumor). Usually progressive, leading to the destruction of host cells and, ultimately, death of the host.

Malignant transformation. The development of the ability to invade normal tissue without regulation in space or time; may also lead to metastatic growth (colonization of a distant site with subsequent unregulated invasive growth).

Manometer. A U-shaped tube containing liquid, the levels of which in each limb of the U reflect the pressure difference between the ends of the tube.

Medium. A mixture of inorganic salts and other nutrients capable of sustaining cell survival *in vitro* for 24 hours. *Growth medium:* That medium which is used in routine culture such that the cell number increases with time. *Maintenance medium:* A medium that will retain cell survival without cell proliferation (e.g., a low-serum or serum-free medium used with serum-dependent cells). The *plural* of medium is *media.*

Mesenchyme. Loose, often migratory embryonic tissue derived from the mesoderm, giving rise to connective tissue, cartilage, muscle, hemopoietic cells, etc., in the adult.

Mesoderm. A germ layer in the embryo arising between the ectoderm and endoderm and giving rise to mesenchyme, which, in turn, gives rise to connective tissue, etc. (See *mesenchyme.*)

Microautoradiography. Localization of radioisotopes in cells and tissue sections, by exposure of a photographic emulsion placed in close proximity to the specimen, usually by dipping it in the melted emulsion. After development the specimen may be viewed under a microscope.

Monoclonal. Derived from a single clone of cells.

Monoclonal antibody: Antibody produced by a clone of lymphoid cells either *in vitro* or *in vivo. In vitro,* the clone is usually derived from a hybrid of a sensitized spleen cell and a continuously growing myeloma cell.

Morphogenesis. The development of form and structure of an organism.

Myeloma. A tumor derived from myeloid cells; used in monoclonal antibody production when the myeloma cell can produce immunoglobulin.

Neoplastic. A new, unnecessary proliferation of cells giving rise to a tumor.

Neoplastic transformation. The conversion of a nontumorigenic cell into a tumorigenic cell.

Oncogene. A gene that, when transfected or infected into normal cells, induces malignant transformation; usually positively-acting genes coding for growth factors, receptors, signal transducers, or nuclear regulators.

Organ culture. The maintenance or growth of organ primordia or the whole or parts of an organ *in vitro* in a way that may allow differentiation and preservation of the architecture or function of the organ.

Organogenesis. The development of organs.

Organotypic. Histotypic culture involving more than one cell type to create a model of the cellular interactions characteristic of an organ *in vivo.* A reconstruction from dissociated cells or fragments of tissue is implied, as distinct from organ culture, in which the structural integrity of the explanted tissue is retained.

Osmolality. The concentration of osmotically active particles in an aqueous solution, expressed in osmols/kg.

Osmolarity. The concentration of osmotically active particles in an aqueous solution, expressed in osmoles/1.

Osmole. The amount of a substance containing 1 mole of osmotically active particles.

Paracrine. An effect of one cell on another, adjacent cell mediated by a soluble factor without involvement of the systemic vasculature.

Passage. The transfer or subculture of cells from one culture vessel to another; usually, but not necessarily, involves the subdivision of a proliferating cell population, enabling the propagation of a cell line or cell strain.

Passage number. The number of times a culture has been subcultured.

Pavement-like. Cells in a regular monolayer or polygonal cells. More correctly, epithelioid or epithelial-like.

Phenotype. The aggregate of all the expressed properties of a cell; the product of the interaction of the genotype with the regulatory environment.

Plateau. See *growth curve.*

Plating efficiency. The percentage of cells seeded at subculture that gives rise to colonies. If each colony can be said to be derived from one cell, plating efficiency is identical to cloning efficiency. Sometimes the plating efficiency is used loosely to describe the number of cells surviving after subculture, but this is better termed the *seeding efficiency.*

Ploidy. Relationship of chromosome number of a given type of cell to that found in normal somatic cells *in vivo.* See also *haploid, diploid, euploid, aneuploid,* and *heteroploid.*

Poikilothermic. Having a body temperature close to that of the environment and not regulated by metabolism.

Population density. The number of monolayer cells per unit area of substrate; for cells growing in suspension, the population density is identical to the cell concentration.

Population doubling time. The interval required for a cell population to double at the middle of the logarithmic phase of growth.

Primary culture. A culture started from cells, tissues, or organs taken directly from an organism, and before the first subculture.

Pseudodiploid. Numerically diploid chromosome number, but with chromosomal aberrations.

Quasidiploid. See *Pseudodiploid.*

Sarcoma. A tumor derived from mesodermally derived cells (e.g., connective tissue, muscle (*myosarcoma*), or bone (*osteosarcoma*)).

Saturation density. Maximum number of cells attainable per cm^2 (in a monolayer culture) or per ml (in a suspension culture) under specified conditions.

Seeding efficiency. The percentage of the inoculum that attaches to the substrate within a stated period of time (implying viability, or survival, but not necessarily proliferative capacity).

Somatic cell genetics. The study of cell genetics by the recombination and segregation of genes in somatic cells, usually by fusion.

Split ratio. The divisor of the dilution ratio of a cell culture at subculture (e.g., one flask divided into four, or 100 ml up to 400 ml, would be a split ratio of 4).

Subconfluent. Less than confluent; not all of the available substrate is covered.

Subculture. See *passage.*

Substrate. The matrix or solid underlay upon which a monolayer culture grows.

Superconfluent. Progressing beyond the state in which all the cells are attached to the substrate and multilayering occurs.

Suppressor gene. A gene that inhibits the transformed (malignant) phenotype, usually associated with dominant negative regulation of cell proliferation or cell migration; often, suppressor genes are mutated or deleted in transformed cells and cancer.

Suspension culture. A culture in which cells will multiply when suspended in growth medium.

Synkaryon. A hybrid cell that results from the fusion of the nuclei it carries.

Tetraploid. Twice the diploid (four times the haploid) number of chromosomes.

Tissue culture. Properly, the maintenance of fragments of tissue *in vitro,* but now commonly applied as a generic term denoting tissue explant culture, organ culture, and dispersed-cell culture, including the culture of propagated cell lines and cell strains.

Transdifferentiation. Cells from one lineage acquiring the ability to differentiate into cells of a different lineage.

Transfection. The transfer, by artificial means, of genetic material from one cell to another, when less than the whole nucleus of the donor cell is transferred. Transfection is usually achieved by transferring isolated chromosomes, DNA, or cloned genes.

Transformation. A permanent alteration of the cell phenotype, presumed to occur via an irreversible genetic change. May be spontaneous, as in the development of rapidly growing continuous cell lines from slow-growing early passage rodent cell lines, or may be induced by chemical or viral action. Usually produces cell lines that have an increased growth rate, an infinite life span, a lower serum requirement, and a higher plating efficiency and that are often (but not necessarily) tumorigenic.

Variant. A cell line expressing a stable phenotype that is different from the parental culture from which it was derived.

Viral transformation. A permanent phenotypic change induced by the genetic and heritable effects of a transforming virus.

Xenograft. Transplantation of tissue to a species different from that from which it was derived; often used to describe the implantation of human tumors in athymic (nude), immune-deprived, or immune-suppressed mice.

References

Aaronson, S. A., Bottaro, D. P., Miki, T., Ron, D., Finch, P. W., Fleming, T. P., Ahn, J., Taylor, W. G., & Rubin, J. S. (1991). Keratinocyte growth factor: A fibroblast growth factor family member with unusual target cell specificity. *Ann. NY Acad. Sci.* **638**:62–77.

Aaronson, S. A., & Todaro, G. J. (1968). Development of 3T3-like lines from Balb/c mouse embryo cultures: Transformation susceptibility to SV40. *J. Cell Physiol.* **72**:141–148.

Aaronson, S. A., Todaro, G. J., & Freeman, A. E. (1970). Human sarcoma cells in culture: Identification by colony-forming ability on monolayers of normal cells. *Exp. Cell. Res.* **61**:1–5

Abaza, N. A., Leighton, J., & Schultz, S. G. (1974). Effects of ouabain on the function and structure of a cell line (MDCK) derived from canine kidney; I: Light microscopic observations of monolayer growth. *In Vitro* **10**:172–183.

Abbott, N. J., Hughes, C. C., Revest, P. A., & Greenwood, J. (1992). Development and characterisation of a rat brain capillary endothelial culture: Towards an *in vitro* blood-brain barrier. *J. Cell Sci.* **103** (Pt 1):23–37.

Abercrombie, M., & Heaysman, J. E. M. (1954). Observations on the social behaviour of cells in tissue culture; II: "Monolayering" of fibroblasts. *Exp. Cell Res.* **6**:293–306.

Abney, E. R., Williams B. P., & Raff M. C. (1983). Tracing the development of oligodendrocytes from precursor cells using monoclonal antibodies, fluorescence activated cell sorting and cell culture. *Developmental Biol.* **100**:166–171.

Adams, D. O. (1979). Macrophages. In Jakoby, W. B., & Pastan, I. H. (eds.), *Methods of Enzymology: Vol. 57, Cell Culture.* New York, Academic Press, pp. 494–506.

Adams, R. L. P. (1980). In Work, T. S., & Burdon, R. H. (eds.): *Laboratory techniques in biochemistry and molecular biology: Cell Culture for Biochemists.* Amsterdam, Elsevier/North Holland Biomedical Press.

Adolphe, M. (1984). Multiplication and type II collagen production by rabbit articular chondrocytes cultivated in a defined medium. *Exp. Cell. Res.* **155**:527–536.

Adolphe, M., & Benya, P. D. (1992). Different types of cultured chondrocytes: The *in vitro* approach to the study of biochemical regulation. In Adolphe, M., ed., *Biological regulation of the chondrocytes.* Boca Raton, FL, CRC Press, pp. 105–139.

Advisory Committee on Dangerous Pathogens (1995a). Categorisation of Biological Agents According to Hazard and Categories of Containment. The Stationery Office, P. O. Box 276, London SW8 5DT, England.

Advisory Committee on Dangerous Pathogens (1995b). Protection against blood-borne infections in the workplace: HIV and hepatitis. The Stationery Office, P. O. Box 276, London SW8 5DT, England.

Agy, P. C., Shipley, G. D., & Ham, R. G. (1981). Protein-free medium for mouse neuroblastoma cells. *In Vitro* **17**:671–680.

Aitken, M. L., Villalon, M., Verdugo, P., & Nameroff, M. (1991). Enrichment of subpopulations of respiratory epithelial cells using flow cytometry. *Am. J. Resp. Cell. Mol. Biol.* **4**:174–178.

Alberts, B., Bray, D., Johnson, A., Lewis, J., Raff, J., Roberts, K., & Walter, P. (1997). *Essential cell biology.* New York, Garland.

Alberts, B., Bray, D., Lewis, J., Raff, M., Roberts, K., & Watson, J. D. (1994). *Molecular biology of the cell*, 3d ed. New York, Garland. (4th ed. 1997.)

Albrecht, A. M., Biedler, J. L., & Hutchison, D. J. (1972). Two different species of dihydrotolate reductase in mammalian cells differentially resistant to amethopterin and methasquin. *Cancer Res.* **32**:1539–1546.

Ali, S., Muller, C. R., & Epplen, J. T. (1986). DNA finger printing by oligonucleotide probes specific for simple repeats. *Human Genetics,* **74**:239–243.

Alley, M. C., Scudiero, D. A., Monks, A., Hursey, M. L., Czerwiniski, M. J., Fine, D. L., Abbot, B. J., Mayo, J. G., Shoemaker, R. H., & Boyd, M. R. (1988). Feasibility of drug screening with panels of human tumour cell lines using a microculture tetrazolium assay. *Cancer Res.* **48**:589–601.

Al-Rubeai, M., & Singh, R. P. (1998). Apoptosis in cell culture. *Curr. Opin. Biotechnol.* **9**:152–156.

Al-Rubeai, M., Welzenbach, K., Lloyd, D. R., & Emery, A. N. (1997). A rapid method for evaluation of cell number and viability by flow cytometry. *Cytotechnology* **24**:161–168.

Ames, B. N. (1980). Identifying environmental chemicals causing mutations and cancer. *Science* **204**:587–593.

Andersson, L. C., Nilsson, K., & Gahmberg, C. G. (1979a). K562—a human erythroleukemic cell line. *Int. J. Cancer* **23**:143–147.

Andersson, L. C., Jokinen, M., Klein, G., & Nilsson, K. (1979b). Presence of erythrocytic components in the K562 cell line. *Int. J. Cancer.* **24**:5–14.

Andersson, L. C., Jokinen, M., & Gahmberg, C. G. (1979c). Induction of erythroid differentiation in the human leukaemia cell line K562. *Nature* **278**:364–365.

Andreason, G. L., & Evans, G. A. (1988). Introduction and expression of DNA molecules in eukaryotic cells by electroporation. *Bio. Techniques* **6**:650–660.

Andreason, G. L., & Evans, G. A. (1989). Optimization of electroporation for transfection of mammalian cells. *Anal. Biochem.* **180**:269–275.

Andreoli, S. P., & McAteer, J. A. (1990). Reactive oxygen molecule-mediated injury in endothelial and renal tubular epithelial cells *in vitro. Kidney Int.* **38**:785–794.

Antoniades, H. N., Scher, C. D., & Stiles, C. D. (1979). Purification of human platelet-derived growth factor. *Proc. Natl. Acad. Sci. USA* **76**:1809.

Armati, P. J., & Bonner, J. (1990). A technique for promoting Schwann cell growth from fresh and frozen biopsy nerve utilizing D-valine medium. *In Vitro Cell Dev. Biol.* **26**:1116–1118.

Arrighi, F. E., & Hsu, T. C. (1974). Staining constitutive heterochromatin and Giemsa crossbands of mammalian chromosomes. In Yunis, J. (ed.), *Human chromosome methodology,* 2d ed. New York, Academic Press.

Arthursson, P., & Magnusson, C. (1990). Epithelial transport of drugs in cell culture II: Effect of extracellular calcium concentration on the parcellular transport of drugs of different lipophilicities across monolayers of intestinal epithelial (Caco-2) cells. *J. Pharmaceut. Sci.* **79**:595–600.

Ascenzioni F., Donini P., & Lipps H. J. (1997). Mammalian artificial chromosomes-vectors for somatic gene therapy. *Cancer Letters,* **118**:135–142.

Askanas, V., Bornemann, A., & Engel, W. K. (1990). Immunocytochemical localization of desmin at human neuromuscular junctions. *Neurology* **40**:949–953.

Au, A. M.-J., & Varon, S. (1979). Neural cell sequestration on immunoaffinity columns. *Exp. Cell Res.* **120**:269.

Auerbach, R., & Grobstein, C. (1958). Inductive interaction of embryonic tissues after dissociation and reaggregation. *Exp. Cell Res.* **15**:384–397.

Ausubel, F. M., Brent, R., Kingston, R. E., Moore, D. D., Seidman, J. G., Smith, J. A., & Struhl, K. (eds.) (1996). *Current protocols in molecular biology,* New York, John Wiley & Sons.

Babich, H., & Borenfreund, E. (1990). Neutral Red Uptake. In Doyle, A., Griffiths, J. B., & Newall, D. G. (eds.), *Cell and tissue culture: Laboratory procedures.* Chichester, U.K., Wiley, Module 4B:7.

Balin, A. K., Goodman, B. P., Rasmussen, H., & Cristofalo, V. J. (1976). The effect of oxygen tension on the growth and metabolism of WI-38 cells. *J. Cell Physiol.* **89**:235–250.

Balkovetz, D. F., & Lipschutz, J. H. (1999). Hepatocyte growth factor and the kidney: It is not just for the liver. *Int. Rev. Cytol.* **186**:225–260.

Ballard, P. L. (1979). Glucocorticoids and differentiation. *Glucocorticoid Horm. Action* **12**:439–517.

Ballard, P. L., & Tomkins, G. M. (1969). Dexamethasone and cell adhesion. *Nature* **244**:344–345.

Balmforth, A. J., Ball, S. G., Freshney, R. I., Graham, D. I., McNamee, B., & Vaughan, P. F. T. (1986). D-1 dopaminergic and beta-adrenergic stimulation of adenylate cyclase in a clone derived from the human astrocytoma cell line G-CCM. *J. Neurochem.* **47**:715–719.

Bansal, R., Stefansson K., & Pfeiffer, S. E. (1992). Proligodendroblast antigen (POA), a developmental antigen expressed by A007/O4-positive oligodendrocytes progenitors prior to the appearance of sulfatide and galactocerebroside. *J. Neurochem.* **58**:2221–2229.

Bard, D. R., Dickens, M. J., Smith, A. U., & Sarck, J. M. (1972). Isolation of living cells from mature mammalian bone. *Nature* **236**:314–315.

Barnes, D., & Sato, G. (1980). Methods for growth of cultured cells in serum-free medium. *Anal. Biochem.* **102**:255–270.

Barnes, W. D., Sirbasku, D. A., & Sato, G. H. (eds.). (1984a). *Cell culture methods for molecular and cell biology; Vol. 1: Methods for preparation of media, supplements, and substrata for serum-free animal cell culture.* New York, Alan R. Liss.

Barnes, W. D., Sirbasku, D. A., & Sato, G. H. (eds.). (1984b). *Cell culture methods for molecular and cell biology; Vol. 2: Methods for serum-free culture of cells of the endocrine system.* New York, Alan R. Liss.

Barnes, W. D., Sirbasku, D. A., & Sato, G. H. (eds.). (1984c). *Cell culture methods for molecular and cell biology; Vol. 3: Methods for serum-free culture of epithelial and fibroblastic cells.* New York, Alan R. Liss.

Barnes, W. D., Sirbasku, D. A., & Sato, G. H. (eds.). (1984d). *Cell culture methods for molecular and cell biology; Vol. 4: Methods for serum-free culture of neuronal and lymphoid cells.* New York, Alan R. Liss.

Barnett, S. C., Hutchins, A.-M., & Noble, M. (1993). Purification of olfactory nerve ensheathing cells of the olfactory bulb. *Dev. Biol.* **155**:337–350.

Bateman, A. E., Peckham, M. J., & Steel, G. G. (1979). Assays of drug sensitivity for cells from human tumours: *In vitro* and *in vivo* tests on a xenografted tumour. *Br. J. Cancer* **40**:81–88.

Battye, F. L., & Shortman, K. (1991). Flow cytometry and cell-separation procedures. *Curr. Opin. Immunol.* **3**:238–241.

Bazill, G. W., Haynes M., Garland J., & Dexter T. M. (1983). Characterisation and partial purification of a haemopoietic cell growth factor in WEHI-3 cell conditioned medium. *Biochem. J.* **210**:747–759.

Beddington, R. (1992). Transgenic mutagenesis in the mouse. *Trends Genetics* **8**:10.

Bedrin MS, Abolafia, C. M., & Thompson, J. F. (1997). Cytoskeletal association of epidermal growth factor receptor and associated signalling proteins is regulated by cell density in IEC-6 intestinal cells. *J. Cell. Physiol.* **172**:126–136.

Benda, P., Lightbody, J., Sato, G., Levine, L., & Sweet, W. (1968). Differentiated rat glial cell strain in tissue culture. *Science* **161**:370.

Benders, A. A. G. M., van Kuppevelt, T. H. M. S. M., Oosterhof, A., & Veerkamp, J. H. (1991). The biochemical and structural maturation of human skeletal muscle cells in culture: The effect of serum substitute, Ultroser. G. *Exp. Cell Res.* **195**:284–294.

Benya, P. D. (1981). Two dimensional CNBr peptide patterns of collagen types I, II and III. *Coll. Relat. Res.* **1**:17–26.

Benya, P. D., Padilla, S. R., & Nimmi, M. E. (1977). The progeny of rabbit articular chondrocytes synthesize collagen type I and III and I trimer, but not type II: Verification by cyanogen bromide peptide analysis. *Biochemistry* **16**:865–872.

Berdichevsky, F, Gilbert, C, Shearer, M., & Taylor-Papadimitriou, J. (1992). Collagen-induced rapid morphogenesis of human mammary epithelial cells: The role of the alpha 2 beta 1 integrin. *J. Cell Sci.* **102**:437–446.

Berenbaum, M. C. (1985). The expected effects of a combination of agents: The general solution. *J. Theor. Biol.* **114**:413–432.

Berger, S. L. (1979). Lymphocytes as resting cells. In Jakoby, W. B., & Pastan, I. H. (eds.), *Methods in enzymology; Vol. 57: Cell Culture.* New York, Academic Press, pp. 486–494.

Berky, J. J., & Sherrod, P. C. (eds.). (1977). *Short term in vitro testing for carcinogenesis, mutagenesis and toxicity.* Philadelphia, Franklin Institute Press.

Bernier, S. M., Desjardins, J., & Sullivan, A. K. (1990). Establishment of an osseous cell line from fetal rat calvaria using an immunocytolytic method of cell selection: Characterisation of the cell line and of derived clones. *J. Cell Physiol.* **145**:274–285.

Bernstein, A. (1975). Differentiation of clonal lines of teratocarcinoma cells: Formation of embryoid bodies *in vitro*. *Proc. Natl. Acad. Sci. USA* **72**:1441–1445.

Berry, M. N., & Friend, D. S. (1969). High yield preparation of isolated rat liver parenchymal cells: A biochemical and fine structural study. *J. Cell Biol.* **43**:506–520.

Bertheussen, K. (1993). Growth of cells in a new defined protein-free medium. *Cytotechnology* **11**, 219–231.

Bertoncello, I., Bradley, T. R., & Watt, S. M. (1991). An improved negative immunomagnetic selection strategy for the purification of primitive hemopoietic cells from normal bone marrow. *Exp. Hematol.* **19**:95–100.

Bettger, W. J., Boyce, S. T., Walthall, B. J., & Ham, R. G. (1981). Rapid clonal growth and serial passage of human diploid fibroblasts in a lipid-enriched synthetic medium supplemented with EGF, insulin and dexamethasone. *Proc. Natl. Acad. Sci. USA* **78**:5588–5592.

Bhargava, M., Joseph, A., Knesel, J., Halaban, R., Li, Y., Pang, S., Golberg, I., Setter, E., Donovan, M. A., Zarnegar, R., Faletto, D., & Rosen, E. M. (1992). Scatter factor and hepatocyte growth factor activities, properties, and mechanism. *Cell Growth Differentiation* **3**:11–20.

Bichko, V. V. (1998). Cationic liposomes. In Ravid, K., & Freshney, R. I. (eds.), *DNA transfer to cultured cells.* New York, Wiley-Liss, pp. 193–212.

Bickenbach, J. R., & Chism, E. (1998). Selection and extended growth of murine epidermal stem cells in culture. *Exp. Cell Res.* **244**:184–195.

Biedler, J. L. (1976). Chromosome abnormalities in human tumour cells in culture. In Fogh, J. (ed.), *Human Tumor Cells in vitro.* New York, Academic Press.

Biedler, J. L., Albrecht, A. M., Hutchinson, D. J., & Spengler, B. A. (1972). Drug response dihydrofolate reductase, and cytogenetics of amethopterin-resistant Chinese hamster cells *in vitro*. *Cancer Res.* **32**:151–161.

Biedler, J. L., & Spengler, B. A. (1976). A novel chromosomal abnormality in human neuroblastoma and anti-folate resistant Chinese hamster cell lines in culture. *J. Natl. Cancer Inst.* **57**:683–695.

Biggers, J. D., Gwatkin, R. B. C., & Heyner, S. (1961). Growth of embryonic avian and mammalian tibiae on a relatively simple chemically defined medium. *Exp. Cell Res.* **25**:41.

Bignami, A., Dahl, D., & Rueger, D. G. (1980). Glial fibrillary acidic (GFA) protein in normal neural cells and in pathological conditions. In Federoff, S., & Hertz, L. (eds.), *Advances in cellular neurobiology*, vol. 1. New York, Academic Press.

Biosafety in Microbiological and Biomedical Laboratories (1984) Division of Safety, BG31, ICO2, NIH, Bethesda, MD.

Birch, J. R., & Pirt, S. J. (1970). Improvements in a chemically-defined medium for the growth of mouse cells (strain LS) in suspension. *J. Cell Sci.* **7**:661–670.

Birch, J. R., & Pirt, S. J. (1971). The quantitative glucose and mineral nutrient requirements of mouse LS (suspension) cells in chemically-defined medium. *J. Cell Sci.* **8**:693–700.

Birnie, G. D., & Simons, P. J. (1967). The incorporation of ³H-thymidine and ³H-uridine into chick and mouse embryo cells cultured on stainless steel. *Exp. Cell Res.* **46**:355–366.

Bishop, J. M. (1991). Molecular themes in oncogenesis. *Cell* **64**:235–248.

Bissell, D. M., Arenson, D. M., Maher, J. J., & Roll, F. J. (1987). Support of cultured hepatocytes by a laminin-rich gel: Evidence for a functionally significant subendothelial matrix in normal liver. *J. Clin. Invest.* **79**:801–812.

Bjerkvig, R., Laerum, O. D., & Mella, O. (1986a). Glioma cell interactions with fetal rat brain aggregates *in vitro*, and with brain tissue *in vivo*. *Cancer Res.* **46**:4071–4079.

Bjerkvig, R., Steinsvag, S. K., & Laerum, O. D. (1986b). Reaggregation of fetal rat brain cells in a stationary culture system; I: Methodology and cell identification. *In Vitro* **22**:180–192.

Blaker, G. J., Birch, J. R., & Pirt, S. J. (1971). The glucose, insulin and glutamine requirements of suspension cultures of HeLa cells in a defined culture medium. *J. Cell Sci.* **9**:529–537.

Blanco, F. J., Geng, Y., & Lotz, M. (1995). Differentiation dependent effects of IL-1 and TGF-β on human articular chondrocyte proliferation are related to nitric oxide synthase expression. *J. Immunol.* **154**:4018–4026.

Blouin R., Grondin G., Beaudoin J., Arita Y., Daigle N., Talbot B. G., Lebel D., & Morisset J. (1997). Establishment and immunocharacterization of an immortalized pancreatic cell line derived from the H-2Kb-tsA58 transgenic mouse. *In Vitro Cell Dev. Biol.—Animal* **33**:717–726.

Bobrow, M., Madan, J., & Pearson, P. L. (1972). Staining of some specific regions on human chromosomes, particularly the secondary constriction of no. 9. *Nature* **238**:122–124.

Bochaton-Piallat, M. L., Gabbiani, F., Ropraz, P., & Gabbiani, G. (1992). Cultured aortic smooth muscle cells from newborn and adult rats show distinct cytoskeletal features. *Differentiation* **49**:175–185.

Bodnar, A. G., Ouellette, M., Frolkis, M., Holt, S. E., Chiu, C.-P., Morin, G. B., Harley, C. B., Shay, J. W., Lichsteiner, S., & Wright, W. E. (1998). Extension of life-span by introduction of telomerase into normal human cells. *Science* **279**:349–352.

Boggs, S. S., Gregg, R. G., Borenstein, N., & Smithies, O. (1986). Efficient transformation and frequent single-site, single-copy insertion of DNA can be obtained in mouse erythroleukemia cells transformed by electroporation. *Exp. Hematol.* **14**:988–994.

Bögler, O., Wren, D., Barnett, S. C., Land, H., & Noble, M. (1990). Cooperation between two growth factors promotes extended self-renewal and inhibits differentiation of oligodendrocyte-type-2 astrocyte (O-2A) progenitor cells. *Proc. Natl. Acad. Sci. USA* **87**:6368–6372.

Bolton, B. J., & Spurr, N. K. (1996). B-Lymphocytes. In Freshney, R. I., & Freshney, M. G., (eds.), *Culture of immortalized cells.* New York, Wiley-Liss, pp. 283–298.

Bonaventure J., Kadhom N., Cohen-Solal L., Ng K. H., Bourguignon J., Lasselin, C., & Freisinger P. (1994). Reexpression of cartilage-specific genes by dedifferentiated human articular chondrocytes cultured in alginate beads. *Exp. Cell Res.* **212**:97–104.

Booyse, F. M., Sedlak, B. J., & Rafelson, M. E. (1975). Culture of arterial endothelial cells: Characterization and growth of bovine aortic cells. *Thromb. Diathes. Ahemorrh.* **34**:825–839.

Borenfreund E., Babich H., & Martin-Alguacil N. (1990). Rapid chemosensitivity assay with human normal and tumor cells *in vitro*. *In Vitro Cell Dev. Biol.* **26**:1030–1034.

Bosco, D., Soriano, J. V., Chanson, M., Meda, P. (1994). Heterogeneity and contact-dependent regulation of amylase release by individual acinar cells. *J. Cell Physiol.* **160**:378–388.

Boshart, M., Weber, F., Jahn, G., Dorsch-Hasler, K., Fleckenstein, B., & Schaffner, W. (1985). A very strong enhancer is located upstream of an immediate early gene of human cytomegalovirus. *Cell* **41**:521–530.

Bottenstein J. E., & Sato, G. (1979). Growth of a rat neuroblastoma cell line in serum free supplemented medium. *Proc. Natl. Acad. Sci. USA* **76**:514–517.

Boukamp, P., Petrusevska, R. T., Breitkreutz, D., Hornung, J., & Markham, A. (1988). Normal keratinisation in a spontaneously immortalised, aneuploid human keratinocyte cell line. *J. Cell Biol.* **106**:761–771.

Bouzahzah, B., Nishikawa, Y., Simon, D., and Carr, B. I. (1995). Growth control and gene expression in a new hepatocellular carcinoma cell line, Hep40: Inhibitory actions of vitamin K. *J. Cell Physiol.* **165**:459–467.

Bowman, P. D., Betz, A. L., Ar, D., Wolinsky, J. S., Penney, J. B., Shivers, R. R., & Goldstein, G. (1981). Primary culture of capillary endothelium from rat brain. *In vitro* **17**:353–362.

Boxberger, H. J., Meyer, T. F., Grausam, M. C., Reich, K., Becker, H. D., & Sessler, M. J. (1997). Isolating and maintaining highly polarized primary epithelial cells from normal human duodenum for growth as spheroid-like vesicles. *In Vitro Cell Dev. Biol. Animal* **33**:536–545.

Boxman, D. L. A., Quax, P. H. A., Lowick, C. W. G. M., Papapoulos, S. E., Verheijen, J., & Ponec, M. (1995). Differential regulation of plasminogen activation in normal keratinocytes and SCC-4 cells by fibroblasts. *J. Invest. Dermatol.* **104**:374–378.

Boyce, S. T., & Ham, R. G. (1983). Calcium-regulated differentiation of normal human epidermal keratinocytes in chemically defined clonal culture and serum-free serial culture. *J. Invest. Dermatol.* **81**:33–40s.

Boyd, M., Cunningham, S. H., Brown, M. M., Mairs, R. J., & Wheldon, T. E. (1999). Noradrenaline transporter gene transfer for radiation cell kill by [131I]meta-iodobenzylguanidine. *Gene Ther.* (in press).

Boyd, M. R. (1989). Status of the NCI preclinical antitumor drug discovery screen. *Prin. Prac. Oncol.* **10**:1–12.

Boyum, A. (1968a). Isolation of leucocytes from human blood: A two-phase system for removal of red cells with methylcellulose as erythrocyte aggregative agent. *Scand. J. Clin. Lab. Invest.* (Suppl 97) **21**:9–29.

Boyum, A. (1968b). Isolation of leucocytes from human blood: Further observations—methylcellulose, dextran and Ficoll as erythrocyte aggregating agents. *Scand. J. Clin. Lab. Invest.* (Suppl 97) **31**:50.

Braa, S. S., & Triglia, D. (1991). Predicting ocular irritation using three-dimensional human fibroblast cultures. *Cosmetics Toiletries* **106**:55–58.

Braaten, J. T., Lee, M. J., Schewk, A., & Mintz, D. H. (1974). Removal of fibroblastoid cells from primary monolayer cultures of rat neonatal endocrine pancreas by sodium ethylmercurithiosalicylate. *Biochem. Biophys. Res. Comm.* **61**:476–482.

Bradford C. S., Sun, L., & Barnes, D. W. (1994a). Basic FGF stimulates proliferation and suppresses melanogenesis in cell cultures derived from early zebrafish embryos. *Mol. Mar. Biol. Biotech.* **3**:78–86.

Bradford, C. S., Sun, L., Collodi, P., & Barnes, D. W. (1994b). Cell cultures from zebrafish embryos and adult tissues. *J. Tiss. Cult. Meth.* **16**:99–107.

Bradford, M. (1976). A rapid and sensitive method for the quantitation of microgram quantities of protein using the principle of protein-dye binding. *Anal. Biochem.* **72**: 248–254.

Bradley, C., & Pitts, J. (1994). The use of genetic marking to assess the interaction of sensitive and multidrug resistant cells in mixed culture. *Br. J. Cancer* **70**:795–798.

Bradley, N. J., Bloom, H. J. G., Davies, A. J. S., & Swift, S. M. (1978). Growth of human gliomas in immune-deficient mice: A possible model for pre-clinical therapy studies. *Br. J. Cancer* **38**:263.

Bradley, T. R., Hodgson, G. S., & Rosendaal, M. (1978). The effect of oxygen tension on haemopoietic and fibroblast cell proliferation *in vitro*. *J. Cell Physiol.* **97**(Suppl 1):517–522.

Breder, J., Ruller, S., Ruller, E., Schlaak, M., & van der Bosch, J. (1996). Induction of cell death by cytokines in cell cycle-synchronous tumor cell populations restricted to G1 and G2. *Exp. Cell Res.* **223**:259–267.

Breen, G. A. M., & De Vellis, J. (1974). Regulation of glycerol phosphate dehydrogenase by hydrocortisone in dissociated rat cerebral cell cultures. *Dev. Biol.* **41**:255–266.

Breitkruetz, D., Stark, H.-J., Mirancea, N., Tomakidi, P., Steinbauer, H., Fusenig, N. E. (1997). Integrin and basement membrane normalization in mouse grafts of human keratinocytes: Implications for epidermal homeostasis. *Differentiation* **61**:195–209.

Breitman, T. R., Kene, B. R., & Hemmi, H. (1984). Studies of growth and differentiation of human myelomonocytic leukaemia cell lines in serum-free medium. In Barnes, D. W., Sirbasku, D. A., & Sato, G. H. (eds.), *Methods for serum-free culture of neuronal and lymphoid cells*. New York, Alan R. Liss, pp. 215–236.

Bretzel, R. G., Bonath, K., & Federlin, K. (1990). The evaluation of neutral density separation utilizing Ficoll-sodium diatrizoate and Nycodenz and centrifugal elutriation in the purification of bovine and canine islet preparations. *Hormone Metab. Res.* (Suppl) **25**:57–63.

Brewer, G. J. (1995). Serum-free B27/Neurobasal medium supports differentiated growth of neurons from the striatum, substantia nigra, septum, cerebral cortex, cerebellum, and dentate gyrus. *J. Neurosci. Res.* **42**:674–683.

Brindle, K. M. (1998). Investigating the performance of intensive mammalian cell bioreactor systems using magnetic resonance imaging and spectroscopy. *Biotech. Genet. Eng. Rev.* **15**:499–520.

British Standard BS5726 (1992). Microbiological Safety Cabinets, Parts 1–4. The Stationery Office, P. O. Box 276, London.

Brito Babapulle, V. (1981). Lateral asymmetry in human chromosomes 1, 3, 4, 15 and 16. *Cytogenet. Cell Genet.* **29**: 198–202.

Brockes, J. P., Fields, K. L., & Raff, M. C. (1979). Studies on cultured rat Schwann cells; I: Establishment of purified populations from cultures of peripheral nerve. *Brain Res.* **165**:105–118.

Brouty-Boyé, D., Kolonias, D., Savaraj, N., & Lampidis, T. J. (1992). Alpha-smooth muscle actin expression in cultured cardiac fibroblasts of newborn rat. *In Vitro Cell Dev. Biol.* **28A**:293–296.

Brower, M., Carney, D. N., Oie, H. K., Gazdar, A. F., & Minna, J. D. (1986). Growth of cell lines and clinical specimens of human nonsmall cell lung cancer in a serum-free defined medium. *Cancer Res.* **46**:798–806.

Brown, A. F., & Dunn, G. A. (1989). Microinterferometry of the movement of dry matter in fibroblasts. *J. Cell Sci.* **92**: 379–389.

Bruland, Ø., Fodstad, Ø., & Pihl, A. (1985). The use of multicellular spheroids in establishing human sarcoma cell lines *in vitro*. *Int. J. Cancer* **35**:793–798.

Brunk, C. F., Jones, K. C., & James, T. W. (1979). Assay for nanogram quantities of DNA in cellular homogenates. *Anal. Biochem.* **92**:497–500.

Brunton, V., Ozanne, B., Paraskeva, C., & Frame, M. (1997). A role for epidermal growth factor receptor, c-Src and focal adhesion kinase in an *in vitro* model for the progression of colon cancer. *Oncogene* **14**:283–293.

Bruynell, E. A., Debray, H., de Mets, M., Mareel, M. M., & Montreuil, J. (1990). Altered glycosylation in Madin-Darby canine kidney (MDCK) cells after transformation by murine sarcoma virus. *Clin. Expl. Metastasis* **8**:241–253.

Bryan, D., Sexton, C. J., Williams, D., Leigh, I. M., & McKay, I. (1995). Oral keratinocytes immortalized with the early region of human papillomavirus type 16 show elevated expression of interleukin 6, which acts as an autocrine growth factor for the derived T103C cell line. *Cell Growth & Diff.* **6**:1245–1250.

Bryan, T. M., & Reddel, R. R. (1997). Telomere dynamics and telomerase activity in *in vitro* immortalised human cells. *Eur. J. Cancer* **33**:767–773.

Bucana, C. D., Giavazzi, R., Nayar, R., O'Brian, C. A., Seid, C., Earnest, L. E., Fan, D. (1990). Retention of vital dyes correlates inversely with the multidrug-resistant phenotype of adriamycin-selected murine fibrosarcoma variants. *Exp. Cell Res.* **190**:69.

Buckingham, M. (1992). Making muscle in mammals. *Trends Genetics* **8**:144–149.

Buehring, G. C. (1972). Culture of human mammary epithelial cells: Keeping abreast of a new method. *J. Natl. Cancer Inst.* **49**:1433–1434.

Buick, R. N., Stanisic, T. H., Fry, S. E., Salmon, S. E., Trent, J. M., & Krosovich, P. (1979). Development of an agar-methyl cellulose clonogenic assay for cells of transitional cell carcinoma of the human bladder. *Cancer Res.* **39**: 5051–5056.

Buonassisi, V., Sato, G., & Cohen, A. I. (1962). Hormone-producing cultures of adrenal and pituitary tumor origin. *Proc. Natl. Acad. Sci. USA* **48**:1184–1190.

Burchell, J., Durbin, H., & Taylor-Papadimitriou, J. (1983). Complexity of expression of antigenic determinants recognised by monoclonal antibodies HMFG 1 and HMFG 2 in normal and malignant human mammary epithelial cells. *J. Immunol.* **131**:508–513.

Burchell, J., Gendler, S., Taylor-Papadimitriou, J., Girling, A., Lewis, A., Millis, R., & Lamport, D. (1987). Development and characterisation of breast cancer reactive monoclonal antibodies directed to the core protein of the human milk mucin. *Cancer Res.* **47**:5476–5482.

Burchell, J., & Taylor-Papadimitriou, J. (1989). Antibodies to human milk fat globule molecules. *Cancer Invest.* **17**:53–61.

Burgess, W. H., & Maciag, T. (1989). The heparin-binding fibroblast growth factor family of proteins. *Ann. Rev. Biochem.* **58**:575–606.

Burke, J. F., Price, T. N. C., & Mayne, L. V. (1996). Immortalization of human astrocytes. In Freshney, R. I., & Freshney, M. G. (eds.), *Culture of immortalized cells.* New York, Wiley-Liss, pp. 299–314.

Burwen, S. J., & Pitelka, D. R. (1980). Secretory function of lactating moose mammary epithelial cells cultured on collagen gels. *Exp. Cell Res.* **126**:249–262.

Buset, M., Winawer, S., & Friedman, E. (1987). Defining conditions to promote the attachment of adult human colonic epithelial cells. *In Vitro* **23**:403–412.

Butler, M., & Christie, A. (1994). Adaptation of mammalian cells to non-ammoniagenic media. *Cytotechnology* **15**:87–94.

Cahn, R. D., Cooh, H. G., & Cahn, M. B. (1967). In Wilt, F. H., & Wessells, N. K. (eds.), *Methods in developmental biology.* New York, Thomas Y. Crowell.

Campion, D. G. (1984). The muscle satellite cell—a review. *Int. Rev. Cytol.* **87**:225–251.

Cancela, M. L., Hu, B., & Price, P. A. (1997). Effect of cell density and growth factors on matrix GLA protein expression by normal rat kidney cells. *J. Cell Physiol.* **171**:125–134.

Caniggia, I., Tseu, I., Han, R. N., Smith, B. T., Tanswell, K., & Post, M. (1991). Spatial and temporal differences in fibroblast behavior in fetal rat lung. *Am. J. Physiol.* **261**:424–433.

Cao, D., Lin, G., Westphale, E. M., Beyer, E. C., & Steinberg, T. H. (1997). Mechanisms for the coordination of intercellular calcium signaling in insulin-secreting cells. *J. Cell Sci.* **110**:497–504.

Caputo, J. L. (1996). Safety procedures. In Freshney, R. I., & Freshney, M. G. (eds.), *Culture of immortalized cells.* New York, Wiley-Liss, pp. 25–51.

Carlsson, J., & Nederman, T. (1989). Tumour spheroid technology in cancer therapy research. *Eur. J. Cancer Clin. Oncol.* **25**:1127–1133.

Carmichael, J., DeGraff, W. G., Gazdar, A. F., Minna, D. J., & Mitchell, J. B. (1987a). Evaluation of a tetrazolium-based semiautomated colorimetric assay: Assessment of chemosensitivity testing. *Cancer Res.* **47**:936–942.

Carmichael, J., DeGraff, W. G., Gazdar, A. F., Minna, J. D., & Mitchell, J. B. (1987b). Evaluatioin of a tetrazolium-based semiautomated colorimetric assay: Assessment of radiosensitivity. *Cancer Res.* **47**:943–946.

Carney, D. N., Bunn, P. A., Gazdar, A. F., Pagan, J. A., & Minna, J. D. (1981). Selective growth in serum-free hormone-supplemented medium of tumor cells obtained by biopsy from patients with small cell carcinoma of lung. *Proc. Natl. Acad. Sci. USA* **78**:3185–3189.

Carpenter, G., & Cohen, S. (1977). Epidermal growth factor. In Acton, R. T., & Lynn, J. D. (eds.), *Cell culture and its application.* New York, Academic Press, pp. 83–105.

Carraway, K. L., Fregien, N., Carraway, K. L., III, & Carraway, C. A. (1992). Tumor sialomucin complexes as tumor antigens and modulators of cellular interactions and proliferation. *J. Cell Sci.* **103**:299–307.

Carrel, A. (1912). On the permanent life of tissues outside the organism. *J. Exp. Med.* **15**:516–528.

Cartwright, T., & Shah, G. P. (1994). Culture media. In Davis, J. M. (ed.), *Basic cell culture, a practical approach.* Oxford, U.K., IRL Press at Oxford University Press, pp. 58–91.

Caspersson, T., Farber, S., Foley, C. E., Kudynowski, J., Modest, E. J., Simonsson, E., Wagh, U., & Zech, L. (1968). Chemical differentiation along metaphase chromosomes. *Exp. Cell Res.* **49**:219–222.

Cataldo, L. M., Wang, Z., & Ravid, K. (1998). Electroporation of DNA into cultured cell lines. In Ravid, K., & Freshney, R. I. (eds.), *DNA transfer to cultures cells.* New York, Wiley-Liss, pp. 55–68.

Centers for Disease Control. (1988). Update: Universal precautions for prevention of transmission of human immunodeficiency virus, hepatitis B virus, and other blood-borne pathogens in healthcare settings. *MMWR* **37**:377–382, 387, 388.

Ceriani, R. L., Taylor-Papadimitriou, J., Peterson, J. A., & Brown, P. (1979). Characterization of cells cultured from early lactation milks. *In Vitro* **15**:356–362.

Chambard, M., Mauchamp J., & Chaband, O. (1987). Synthesis and apical and basolateral secretion of thyroglobulin by thyroid cell monolayers on permeable substrate: Modulation by thyrotropin. *J. Cell Physiol.* **133**:37–45.

Chambard, M., Vemer, B., Gabrion, J., Mauchamp, J., Bugeia, J. C., Pelassy, C., & Mercier, B. (1983). Polarization of thyroid cells in culture: Evidence for the basolateral localization of the iodide "pump" and of the thyroid-stimulating hormone receptor–adenyl cyclase complex. *J. Cell Biol.* **96**:1172–1177.

Chang, H., & Baserga, R. (1977). Time of replication of genes responsible for a temperature sensitive function in a cell cycle specific to mutant from a hamster cell line. *J. Cell Physiol.* **92**:333–343.

Chang, R. S. (1954). Continuous subcultivation of epithelial-like cells from normal human tissues. *Proc. Soc. Exp. Biol. Med.* **87**:440–443.

Chang, S. E., Keen, J., Lane, E. B., & Taylor-Papadimitriou, J. (1982). Establishment and characterisation of SV40-transformed human breast epithelial cell lines. *Cancer Res.* **42**:2040–2053.

Chang, S. E., & Taylor-Papadimitriou, J. (1983). Modulation of phenotype in cultures of human milk epithelial cells and its relation to the expression of a membrane antigen. *Cell Diff.* **12**:143–154.

Chaproniere, D. M., & McKeehan, W. L. (1986). Serial culture of adult human prostatic epithelial cells in serum-free medium containing low calcium and a new growth factor from bovine brain. *Cancer Res.* **46**:819–824.

Chen C.-S., Toda, K.-I., Maruguchi, Y., Matsuyoshi, N., Horriguchi, Y., & Imamura, S. (1997). Establishment and characterization of a novel *in vitro* angiogenesis model using a microvascular endothelial cell line, F-2C, cultured in chemically defined medium. *In Vitro Cell Dev. Biol.—Animal* **33**:796–802.

Chen, T. C., Curthoys, N. P., Lagenaur, C. F., & Puschett, J. B. (1989). Characterization of primary cell cultures derived from rat renal proximal tubules. *In Vitro Cell Dev. Biol.* **25**:714–722.

Chen, T. R. (1977). *In situ* detection of mycoplasm contamination in cell cultures by fluorescent Hoechst 33258 stain. *Exp. Cell Res.* **104**:255.

Cherry, R. S., & Papoutsakis, E. T. (1990). Understanding and controlling fluid-mechanical injury of animal cells in bioreactors. In Spier, R. E., & Griffiths, J. B., *Animal cell biotechnology*, Vol. 4, pp. 71–121.

Cho, J.-K., & Bikle, D. D. (1997). Decrease of Ca-ATPase activity in human keratinocytes during calcium-induced differentiation. *J. Cell Physiol.* **172**:146–154.

Chopra, D. P., Grignon, D. J., Joiakim, A., Mathieu, P. A., Mohamed, A., Sakr, W. A., Powell, I. J., & Sarkar, F. H. (1996). Differential growth factor responses of epithelial cell cultures derived from normal human prostate, benign prostatic hyperplasia and primary prostate carcinoma. *J. Cell Physiol.* **169**:269–280.

Christensen, B., Hansen, C., Kieler, J., & Schmidt, J. (1993). Identity of non-malignant human urothelial cell lines classified as transformation grade I (TGrI) and II (TGrII). *Anticancer Research* **13**:2187–2191.

Chu, G., Hayakawa, H., & Berg, P. (1987). Electroporation for the efficient transfection of mammalian cells with DNA. *Nucleic Acids Res.* **15**:1311–1326.

Ciccarone, V., Hawley-Nelson, P., Gebeyehu, G., Jessee, J. (1993). Cationic liposome-mediated transfection of eukaryotic cells high-efficiency nucleic-acid delivery with lipofectin®, Lipofectace™, and Lipofectamine™ reagents. *FASEB Journal* **7**:A1131

Cioni, C., Filoni, S., Aquila, C., Bernardini, S., & Bosco, L. (1986). Transdifferentiation of eye tissue in anuran amphibians: Analysis of the transdifferentiation capacity of the iris of *Xenopus laevis* larvae. *Differentiation* **32**:215–220.

Clark, J. M., & Pateman, J. A. (1978). Long-term culture of Chinese hamster Kupffer cell lines isolated by a primary cloning step. *Exp. Cell Res.* **112**:207–217.

Clarke, G. D., & Ryan, P. J. (1980). Tranquilizers can block mitogenesis in 3T3 cells and induce differentiation in Friend cells. *Nature* **287**:160–161.

Clayton, R. M., Bower, D. J., Clayton, P. R., Patek, C. E., Randall, F. E., Sime, C., Wainwright, N. R., & Zehir, A. (1980). Cell culture in the investigation of normal and abnormal differentiation of eye tissues. In Richards, R. J., & Rajan, K. T. (eds.), *Tissue Culture in Medical Research (II)*. Oxford, U.K., Pergamon Press.

Cobbold, P. H., & Rink, T. J. (1987). Fluorescence and bioluminescence measurement of cytoplasmic free Ca^{2+}. *Biochem. J.* **248**:313.

Cohen, C., Roguet, R., Cottin, M., Olive, C., Leclaire, J., & Rougier, A. (1997). The use of Episkin, a reconstructed epidermis, in the evaluation of the protective effect of sunscreens against chemical lipoperoxidation induced by UVA. *J. Invest. Dermatol.* **108**:77.

Cohen, S. (1962). Isolation of a mouse submaxillary gland protein accelerating incisor eruption and eyelid opening in the new-born animal. *J. Biol. Chem.* **237**:1555–1562.

Cole, J., Fox, M., Garner, R. C., McGregor, D. B., Thacker, J. (1990). Gene mutation assays in cultured mammalian cells. In Kirkland, D. J. (ed.), *Basic mutagenicity tests: UKEMS recommended procedures, Part 1 (revised)*. Cambridge, U.K., Cambridge University Press, pp. 87–114.

Cole, R. J., & Paul, J. (1966). The effects of erythropoietin on haem synthesis in mouse yolk sac and cultured foetal liver cells. *J. Embryol. Exp. Morphol.* **15**:245–260.

Cole, S. P. C. (1986). Rapid chemosensitivity testing of human lung tumour cells using the MTT assay. *Cancer Chemother. Pharmacol.* **17**:259–263.

Collins, S. J., Gallo, R. C., & Gallagher, R. E. (1977). Continuous growth and differentiation of human myeloid leukaemic cells in suspension culture. *Nature* **270**:347–349.

Collodi, P. (1998). DNA transfer to blastula-derived cultures from zebra fish. In Ravid, K., & Freshney, R. I. (eds.), *DNA transfer to cultured cells*. New York, Wiley-Liss, pp. 69–92.

Collodi, P., & Barnes, D. W. (1990). Mitogenic activity from trout embryos. *Proc. Natl. Acad. Sci. USA* **87**:3498–3502.

Collodi, P., Kamei, Y., Ernst, T., Miranda, C., Buhler, D. R., & Barnes, D. W. (1992a). Culture of cells from zebrafish (*Brachydanio rerio*) embryo and adult tissues. *Cell Biol. Tox.* **8**:43–61.

Committee for a Standardized Karyotype of *Rattus norvegicus*. (1973). Standard karyotype of the Norway rat, *Rattus norvegicus*. *Cytogenet. Cell Genet.* **12**:199–205.

Committee on Standardized Genetic Nomenclature for Mice. (1972). Standard karyotype of the mouse *Mus musculis*. *J. Hered.* **63**:69.

Compton, C. C., Warland, G., Nakagawa, H., Opitz, O. G., & Rustgi, A. K. (1998). Cellular characterization and successful transfection of serially subcultured normal human esophageal keratinocytes. *J. Cell Physiol.* **177**:274–281.

Coon, H. D. (1968). Clonal cultures of differentiated rat liver cells. *J. Cell Biol.* **39**:29a.

Coon, H. G., & Cahn, R. D. (1966). Differentiation in vitro: Effects of sephadex fractions of chick embryo extract. *Science* **153**:1116–1119.

Cooper, G. W. (1965). Induction of somite chondrogenesis by cartilage and notochord: A correlation between inductive activity and specific stages of cytodifferentiation. *Dev. Biol.* **12**:185–212.

Cooper, P. D., Burt, A. M., & Wilson, J. N. (1958). Critical effect of oxygen tension on rate of growth of animal cells in continuous suspended culture. *Nature* **182**:1508–1509.

Cooper, S., & Broxmeyer, H. (1994). Purification of murine granulocyte-macrophage progenitor cells (CFC-GM) using counterflow centrifugal elutriation. In Freshney, R. I., Pragnell, I. B., Freshney, M. G. (eds.), *Culture of haemopoietic cells*. New York, Wiley-Liss, pp. 223–234.

Cou, J. Y. (1978). Establishment of clonal human placental cells synthesizing human choriogonadotropin. *Proc. Natl. Acad. Sci. USA* **75**:1854–1858.

Courtenay, V. D., Selby, P. I., Smith, I. E., Mills, J., & Peckham, M. J. (1978). Growth of human tumor cell colonies from biopsies using two soft-agar techniques. *Br. J. Cancer* **38**:77–81.

Coutinho, L. H., Gilleece, M. H., de Wynter, E. A., Will, A., & Testa, N. G. (1992). Clonal and long-term cultures using human bone marrow. In Testa, N. G., & Molineaux, G. (eds.), *Haemopoiesis: A practical approach*. Oxford, U.K., IRL Press at Oxford University Press, pp. 75–106.

Crabb, I. W., Armes, L. G., Johnson, C. M., & McKeehan, W. L. (1986). Characterization of multiple forms of prostatropin (prostate epithelial growth factor) from bovine brain. *Biochem. Biophys. Res. Commun.* **136**:1155–1161.

Creasey, A. A., Smith, H. S., Hackett, A. I., Fukuyama, K., Epstein, W. L., & Madin, S. H. (1979). Biological properties of human melanoma cells in culture. *In Vitro* **15**: 342.

Croce, C. M. (1991). Genetic approaches to the study of the molecular basis of human cancer. *Cancer Res. (Suppl)* **51**: 5015s–5018s.

Cronauer, M. V., Eder, I. E., Hittmair, A., Sierek, G., Hobisch, A., Culig, Z., Thurnhui, M., Bartsch, G., & Klocker, H. (1997). A reliable system for the culture of human prostatis cells. *In Vitro Cell Dev. Biol.—Animal* **33**:742–744.

Crouch, E. C., Stone, K. R., Bloch, M., & McDivitt, R. W. (1987). Heterogeneity in the production of collagens and fibronectin by morphologically distinct clones of a human tumor cell line: Evidence for intratumoral diversity in matrix protein biosynthesis. *Cancer Res.* **47**(22):6086–6092.

Csoka, K., Nygren, P., Graf, W., Påhlman, L., Glimelius, B., & Larsson, R. (1995). Selective sensitivity of solid tumors to suramin in primary cultures of tumor cells from patients. *Int. J. Cancer* **63**:356–360.

Cuttitta, F., Carney, D. N., Mulshine, J., Moody, T. W., Fedorko, J., Fischler, A., & Minna, J. D. (1985). Bombesin-like peptides can function as autocrine growth factors in human small-cell lung cancer. *Nature* **316**:823.

Dairkee, S. H., Deng, G., Stampfer, M. R., Waldman, F. M., & Smith, H. S. (1995). Selective cell culture of primary breast carcinoma. *Cancer Res.* **55**:2516–2519.

Dairkee, S. H., Paulo, S. C., Traquina, P., Moore, D. H., Ljung, B. M., & Smith, H. S. (1997). Partial enzymatic degradation of stroma allows enrichment and expansion of primary breast tumor cells. *Cancer Res.* **57**:1590–1596.

Damon, D. H., Lobb, R. R., D'Amore, P. A., & Wagner, J. A. (1989). Heparin potentiates the action of acidic fibroblast growth factor by prolonging its biological half-life. *J. Cell Physiol.* **138**:221–226.

Dangles, V., Femenia, F., Laine, V., Berthelmy, M., LeRhun, D., Poupon, M. F., Levy, D., & Schwartz-Cornil, I. (1997). Two and three-dimensional cell structures govern epidermal growth factor survival function in human bladder carcinoma cell lines. *Cancer Res.* **57**:3360–3364.

Danielson, K. G., McEldrew, D., Alston, J. T., Roling, D. B., Damjanov, A., Damjanov, I., Daska, I., & Spinner, N. (1992). Human colon carcinoma cell lines from the primary tumor and a lymph node metastasis. *In Vitro Cell Dev. Biol.* **28A**:7–10.

Darcy, K. M., Shoemaker, S. F., Lee, P.-P. H., Vaughan, M. M., Black, J. D., & Ip M. M. (1995). Prolactin and epidermal growth factor regulation of the proliferation, morphogenesis, and functional differentiation of normal rat mammary epithelial cells in three dimensional primary culture. *J. Cell Physiol.* **163**:346–364.

Davies, B., Brown, P. D., East, N., Crimmin, M. J., & Balkwill, F. R. (1993). A synthetic matrix metalloproteinase inhibitor decreases tumour burden and prolongs survival of mice bearing ovarian carcinoma xenografts. *Cancer Res.* **53**, 2087–2091.

Davison, P. M., Bensch, R., Karasek, M. A. (1983). Isolation and long-term serial cultivation of endothelial cells from microvessels of the adult human dermis. *In Vitro* **19**:937–945.

Dawe, C. J., & Potter, M. (1957). Morphologic and biologic progression of a lymphoid neoplasm of the mouse *in vivo* and *in vitro*. *Am. J. Pathol.* **33**:603.

De Leij, L., Poppema, S., Nulend, I. K., Haar, A. T., Schwander, E., Ebbens, F., Postmus, P. E., & Hauw, The, T. (1985). Neuroendocrine differentiation antigen on human lung carcinoma and Kulchitski cells. *Cancer Res.* **45**: 2192–2200.

De Ridder, L. (1997). Autologous confrontation of brain tumor derived spheroids with human dermal spheroids. *Anticancer Res.* **17**:4119–4120.

De Ridder, L., & Calliauw, L. (1990). Invasion of human brain tumors *in vitro*: Relationship to clinical evolution. *J. Neurosurg.* **72**:589–593.

De Ridder, L., & Mareel, M. (1978). Morphology and ^{125}I-concentration of embryonic chick thyroids cultured in an atmosphere of oxygen. *Cell Biol. Int. Rep.* **2**:189–194.

De Silva, R., Moy, E. L., & Reddel, R. R. (1996). Immortalization of human bronchial epithelial cells. In Freshney, R. I., & Freshney, M. G. (eds.), *Culture of immortalized cells.* New York, Wiley-Liss, pp. 121–144.

De Silva, R., Whitaker, N. J., Rogan, E. M., & Reddel, R. R. (1994). HPV-16 E6 and E7 genes, like SV40 early region genes, are insufficient for immortalization of human mesothelial and bronchial epithelial cells. *Exp. Cell Res.* **213**:418–427.

De Vitry, F., Camier, M., Czernichow, P., Benda, P., Cohen, P., & Tixier-Vidal, A. (1974). Establishment of a clone of mouse hypothalamic neurosecretory cells synthesizing neurophysin and vasopressin. *Proc. Natl. Acad. Sci. USA* **71**:3575–3579.

De Vonne, T. L., & Mouray, H. (1978). Human α_2-macroglobulin and its antitrypsin and antithrombin activities in serum and plasma. *Clin. Chim. Acta* **90**:83–85.

De Vries, J. E., Dinjens, W. N. M., De Bruyne, G. K., Verspaget, H. W., van der Linden, E. P. M., de Bruine, A. P., Mareel, M. M., Bosman, F. T., & ten Kate, J. (1995). *In vivo* and *in vitro* invasion in relation to phenotypic characteristics of human colorectal carcinoma cells. *Brit. J. Cancer* **71**:271–277.

De Wynter, E., Allen, T., Coutinho, L., Flavell, D., Flavell, S. U., & Dexter, T. J. (1993). Localisation of granulocyte macrophage colony-stimulating factor in human long-term bone marrow cultures. *J. Cell Sci.* **106**:761–769.

Degos, L. (1997). Differentiation in acute promyelocytic leukemia: European experience. *J. Cell Physiol.* **173**:285–287.

Del Vecchio, P., & Smith, J. R. (1981). Expression of angiotensin converting enzyme activity in cultured pulmonary artery endothelial cells. *J. Cell Physiol.* **108**:337–345.

Del Vecchio, S., Stoppelli, M. P., Carriero, M. V., Fonti, R., Massa, O., Li, P. Y., Botti, G., Cerra, M., d'Aiuto, G., Espositi, G., & Salvatore, M. (1993). Human urokinase receptor concentration in malignant and benign breast tumours by *in vitro* quantitative autoradiography: Comparison with urokinase levels. *Cancer Res.* **53**:3198–3206.

De Larco, J. E., & Todaro, G. J. (1978). Epithelioid and fibroblastoid rat kidney cell clones: Epidermal growth factor receptors and the effect of mouse sarcoma virus transformation. *J. Cell Physiol.* **94**:335–342.

De Mars, R. (1957). The inhibition of glutamine of glutamyl transferase formation in cultures of human cells. *Biochim. Biophys. Acta* **27**:435–436.

Demers, G. W., Halbert, C. L., & Galloway, D. A. (1994). Elevated wild-type p53 protein levels in human epithelial cell lines immortalized by the human papillomavirus type 16 E7 gene. *Virology* **198**:169–174.

Department of Health and Social Security. (1986). *Good laboratory practice: The United Kingdom Compliance Programme.* London, HMSO.

Deshpande, A. K., Baig, M. A., Carleton, S., Wadgoankar, R., & Siddiqui, M. A. Q. (1993). Growth factors activate cardiogenic differentiation in avian mesodermal cells. *Mol. Cell Diff.* **1**:269–284.

Detrisac, C. J., Sens, M. A., Garvin, A. J., Spicer, S. S., & Sens, D. A. (1984). Tissue culture of human kidney epithelial cells of proximal tubule origin. *Kidney Int.* **25**:383–390.

Dexter, T. M., Allen, T. D., Scott, D., & Teich, N. M. (1979). Isolation and characterisation of a bipotential haematopoietic cell line. *Nature* **277**:417–474.

Dexter, T. J., Spooncer, E., Simmons, P., & Allen, T. D. (1984). Long-term marrow culture: An overview of technique and experience. In Wright, D. G., & Greenberger, J. S. (eds.), *Long-term bone marrow culture.* New York, Alan R. Liss, Kroc Foundation Series 18, pp. 57–96.

Dexter, T. M., Testa, N. G., Allen, T. D., Rutherford, S., & Scolnick, E. (1981). Molecular and cell biological aspects of erythropoiesis in long-term bone marrow cultures. *Blood* **58**:699–707.

Dickson, J. D., Flanigan, T. P., & Kemshead, J. T. (1983). Monoclonal antibodies reacting specifically with the cell surface of human astrocytes in culture. *Biochem. Soc. Trans.* **11**:208.

Dimitroulakos, J., Squire, J., Pawlin, G., & Yeger, H. (1994). NUB-7: A stable-1-type human neuroblastoma cell line inducible along N- and S-type cell lineages. *Cell Growth Diff.* **5**:373–384.

Dinnen, R., & Ebisuzaki, K. (1990). Mitosis may be an obligatory route to terminal differentiation in the Friend erythroleukemia cell. *Exp. Cell Res.* **191**149–152.

Dodson, M. V., Mathison, B. A., & Mathison, B. D. (1990). Effects of medium and substratum on ovine satellite cell attachment, proliferation and differentiation. *In Vitro Cell Diff. Dev.* **29**(1):59–66.

Doherty, P., Ashton, S. V., Moore, S. E., & Walsh, F. S. (1991). Morphoregulatory activities of NCAM and N-cadherin can be accounted for by G protein-dependent activation of L- and N-type neuronal Ca²⁺ channels. *Cell* **67**:21–33.

Dolbeare, F., & Selden, J. R. (1994). Immunochemical quantitation of bromodeoxyuridine: Application to cell-cycle kinetics. *Methods in Cell Biology* **41**:297–316.

Dorazio, J. A., Cole, B. C., & Steinstreilein, J. (1996). Mycoplasma-arthritidis mitogen up-regulates human NK cell activity. *Infection and Immunity* **64**(2):441–447.

Dotto, G. P., Parada, L. F., & Weinberg, R. A. (1985). Specific growth response of *ras*-transformed embryo fibroblasts to tumour promoters. *Nature* **318**:472–475.

Doucette, J. R. (1984). The glial cells in the nerve fibre layer of the rat olfactory bulb. *Anat. Record* **210**:285–391.

Doucette, J. R. (1990). Glial influences on axonal growth in the primary olfactory system. *Glia* **3**:433–449.

Douglas, J. L., & Quinlan, M. P. (1994). Efficient nuclear localization of the Ad5 E1A 12S protein is necessary for immortalization but not cotransformation of primary epithelial cells. *Cell Growth & Diff.* **5**:475–483.

Douglas, W. H. J., McAteer, J. A., Dell'Orco, R. T., & Phelps, D. (1980). Visualization of cellular aggregates cultured on a three-dimensional collagen sponge matrix. *In Vitro* **16**:306–312.

Dow, J., Lindsay, G., & Morrison, J. (1995). *Biochemistry: Molecules, cells and the body.* Workingham, U.K., Addison Wesley.

Doyle, A., & Bolton, B. J. (1994). The quality control of cell lines and the prevention, detection and cure of contamination. In Davis, J. M. (ed.), *Basic cell culture, a practical approach.* Oxford, U.K., IRL Press at Oxford University Press, pp. 242–271.

Doyle, A., Griffiths, J. B., Newall, D. G. (eds.). (1990–1999). Cell and Tissue Culture: Laboratory Procedures. Chichester, Wiley.

Doyle, A., Morris, C., & Mowles, J. M. (1990). Quality control. In Doyle, A., Hay, R., & Kirsop, B. E. (eds.), *Animal Cells, Living Resources for Biotechnology.* Cambridge, U.K., Cambridge University Press, pp. 81–100.

Drejer, J., Larsson, O. M., & Schousboe, A. (1983). Characterization of uptake and release processes for D- and L-aspartate in primary cultures of astrocytes and cerebellar granule cells. *Neurochem. Ref.* **8**:231–243.

Drexler, H. G., Gignac, S. M., Hu, Z.-B., Hopert, A., Fleckenstein, E., Viges, M., & Uphoff, C. C. (1994). Treatment of mycoplasma contamination in a large panel of cell cutlures. *In vitro Cell Dev. Biol.* **30A**:344–347.

Driever, W., & Rangini, Z. (1993). Characterization of a cell line derived from zebra-fish (Brachydanio rerio). *In Vitro Cell Dev. Biol. Anim.* **29A**:749–754.

Driever, W., Stemple, D., Schier, A., & Solnica-Krezel, L. (1994). Zebrafish: Genetic tools for studying vertebrate development. *Trends Genet.* **10**:152–159.

Duffy, M. J., Reilly, D., O'Sullivan, C., O'Higgins, N., Fennelly, J. J., & Andreasen, P. (1990). Urokinase-plasminogen activator, a new and independent prognostic marker in breast cancer. *Cancer Res.* **50**:6827–6829.

Duksin, D., Maoz, A., & Fuchs, S. (1975). Differential cytotoxic activity of anticollagen serum on rat osteoblasts and fibroblasts in tissue culture. *Cell* **5**:83–86.

Dulbecco, R., & Elkington, J. (1973). Conditions limiting multiplication of fibroblastic and epithelial cells in dense cultures. *Nature* **246**:197–199.

Dulbecco, R., & Freeman, G. (1959). Plaque formation by the polyoma virus. *Virology* **8**:396–397.

Dulbecco, R., & Vogt, M. (1954). Plaque formation and isolation of pure cell lines with poliomyelitis viruses. *J. Exp. Med.* **199**:167–182.

Duncan, E. L., De Silva, R., & Reddel, R. R. (1996). Immortalization of human mesothelial cells. In Freshney, R. I., & Freshney, M. G. (eds.), *Culture of immortalized cells.* New York, Wiley-Liss, pp. 239–258.

Dunham, L. J., & Stewart, H. L. (1953). A survey of transplantable and transmissible animal tumors. *J. Natl. Cancer Inst.* **13**:1299–1377.

Dunn, G. A., & Brown, A. F. (1990). Quantifying cellular shape using moment invariants. In Alt, W., & Hoffman, G. (eds.), *Biological motion: Lecture notes in biomathematics 89*. Berlin: Springer-Verlag, pp. 10–34.

Dunn, G. A., & Zicha, D. (1997). Using the DRIMAPS system of interference microscopy to study cell behavior. In Celis, J. E. (ed.), *Handbook of cell biology*, 2d ed. New York, Academic Press, pp. 44–53.

Dunn, M. E., Schilling, K., & Mugnaini, E. (1998). Development and fine structure of murine Purkinje cells in dissociated cerebellar cultures: Neuronal polarity. *Anat. Embryol.* **197**:9–29.

Eagle, H. (1955). The specific amino acid requirements of mammalian cells (stain L) in tissue culture. *J. Biol. Chem.* **214**:839.

Eagle, H. (1959). Amino acid metabolism in mammalian cell cultures. *Science* **130**:432.

Eagle, H. (1973). The effect of environmental pH on the growth of normal and malignant cells. *J. Cell Physiol.* **82**: 1–8.

Eagle, H., Foley, G. E., Koprowski, H., Lazarus, H., Levine, E. M., & Adams, R. A. (1970). Growth characteristics of virus-transformed cells. *J. Exp. Med.* **131**:863–879.

Earle, W. R., Schilling, E. L., Stark, T. H., Straus, N. P., Brown, M. F., & Shelton, E. (1943). Production of malignancy *in vitro*; IV: The mouse fibroblast cultures and changes seen in the living cells. *J. Natl. Cancer Inst.* **4**:165–212.

Easty, D. M., & Easty, G. C. (1974). Measurement of the ability of cells to infiltrate normal tissues *in vitro*. *Br. J. Cancer* **29**:36–49.

Ebendal, T., (1976). The relative roles of contact inhibition and contact guidance in orientation of axons extending on aligned collagen fibrils *in vitro*. *Exp. Cell Res.* **98**:159–169.

Echarti, C., & Maurer, H. R. (1989). Defined, serum-free culture conditions for the GM-microclonogenic assay using agar-capillaries. *Blut.* **59**:171–176.

Echarti, C., & Maurer, H. R. (1991). Lymphokine-activated killer cells: determination of their tumor cytolytic capacity by a clonogenic microassay using agar capillaries. *J. Immunol. Methods.* **143**:41–47.

Edelman, G. M. (1973). Nonenzymatic dissociations; B: Specific cell fractionation of chemically derivatized surfaces. In Kruse, P. F., Jr., & Patterson, M. K., Jr. (eds.), *Tissue culture methods and applications*. New York, Academic Press, pp. 29–36.

Edelson, J. D., Shannon, J. H., & Mason, R. J. (1988). Alkaline phosphatase: A marker of alveolar type II cell differentiation. *Am. Rev. Respir. Dis.* **138**:1268–1275.

Edington, K. G., Berry, I. J., O'Prey, M., Burns, J., Clark, L. J., Mitchell, R., Robertson, G., Soutar, D., Coggins, L. W., & Parkinson, E. K. *In vitro* analysis of multistage head and neck squamous cell carcinoma: Defective terminal maturation is an early and ubiquitous event. In Freshney, R. I., (ed.), *Culture of Tumor Cells*. New York, Wiley-Liss, (in preparation).

Edwards, P. A. W., Easty, D. M., & Foster, C. S. (1980). Selective culture of epithelioid cells from a human squamous carcinoma using a monoclonal antibody to kill fibroblasts. *Cell Biol. Int. Rep.* **4**:917–922.

Eisinger, M., Lee, J. S., Hefton, J. M., Darzykiewicz, A., Chiao, J. W., & Deharven, E. (1979). Human epidermal cell cultures—growth and differentiation in the absence of dermal components or medium supplements. *Proc. Natl. Acad. Sci. USA* **76**:5340.

Elliget, K. A., & Lechner, J. F. (1992). Normal human bronchial epithelial cell cultures. In Freshney, R. I. (ed.), *Culture of epithelial cells*. New York, Wiley-Liss, pp. 181–196.

Elvin, P., Wong, V., & Evans, C. W. (1985). A study of the adhesive, locomotory and invasive behaviour of Walker 256 carcinosarcoma cells. *Exp. Cell Biol.* **53**:9–18.

Emura, M., Katyal, S. L., Ochiai, A., Hirohashi, S., & Singh, G. (1997). *In vitro* reconstitution of human respiratory epithelium. *In Vitro Cell Dev. Biol.—Animal* **33**:602–605.

Eng, L. F., & Bigbee, J. W. (1979). Immunochemistry of nervous-system specific antigens. In Aprison, M. H. (ed.), *Advances in neurochemistry*. New York, Plenum Press, pp. 43–98.

Engel, L. W., Young, N. A., Tralka, T. S., Lippman, M. E., O'Brien, S. J., & Joyce, M. J. (1978). Establishment and characterization of three new continuous cell lines derived from breast carcinomas. *Cancer Res.* **38**:3352–3364.

Epstein, M. A., & Barr, Y. M. (1964). Cultivation *in vitro* of human lymphoblasts from Burkitt's malignant lymphoma. *Lancet* **1**:252.

Ermis, A., Muller, B., Hopf, T., Hopf, C., Remberger, K., Justen, H. P., Welter, C., & Hanselmann, R. (1998). Invasion of human cartilage by cultured multicellular spheroids of rheumatoid synovial cells—a novel *in vitro* model system for rheumatoid arthritis. *J. Rheumatol.* **25**:208–213.

Espmark, J. A., & Ahlqvist-Roth, L. (1978). Tissue typing of cells in cultures; I: Distinction between cell lines by the various patterns produced in mixed haemabsorption with selected multiparous sera. *J. Immunol. Methods* **24**: 141–153.

European Committee for Standardisation. (1998). Biotechnology performance criteria for microbiological safety cabinets [WI91]. CEN/TC233 Biotechnology, Central Secretariat, European Committee for Standardisation, Rue de Stassart 36, B-1050 Brussels, Belgium.

European Pharmacopoeia. (1980). *Biological tests*, 2d ed., Part 1, Vol. 2. Maisonneuve, S.A., France.

Evans, V. J., & Bryant, J. C. (1965). Advances in tissue culture at the National Cancer Institute in the United States of America. In Ramakrishnan, C. V. (ed.), *Tissue culture*. The Hague, W. Junk, pp. 145–167.

Evans, V. J., Bryant, J. C., Fioramonti, M. C., McQuilkin, W. T., Sanford, K. K., & Earle, W. R. (1956). Studies of nutrient media for tissue C cells *in vitro*; I: A protein-free chemically defined medium for cultivation of strain L cells. *Cancer Res.* **16**:77.

Fantini, J., Galons, J. P., Abadie, B., Canioni, P., Cozzone, P. J., Marvali, J., & Tirard, A. (1987). Growth in serum-free medium of human colonic adenocarcinoma cell lines on microcarriers: A two-step method allowing optimal cell spreading and growth. *In Vitro* **23**:641–646.

Farrant, J. (1980). General observations on cell preservation. In Ashwood-Smith, M. J., & Farrant, J. (eds.), *Low temperature preservation in medicine and biology*. London, Pitman Medical, 1980, pp. 1–18.

Federoff, S. (1975). In Evans, V. J., Perry, V. P., & Vincent, M. M. (eds.), *Manual of the Tissue Culture Association* 1:53–57.

Feinberg, A. P., & Vogelstein, B. (1983). A technique for radiolabeling DNA restriction fragments to high specific activity. *Anal. Biochem.* **132**:6–13.

Felgner, P., Gadek, T., Holm, M., Roman, R., Chan, H. W., Wenz, M., Northrop, J. P., Ringold G. M., and Danielsen, M. (1987). Lipofection: A highly efficient, lipid-mediated DNA-transfection procedure. *Proc. Natl. Acad. Sci. USA* **84**: 7413–7417.

Fell, H. B., & Robison, R. (1929). The growth, development and phophatase activity of embryonic avian femora and limb buds cultivated *in vitro*. *Biochem. J.* **23**:767–784.

Fergusson, R. J., Carmichael, J., Smyth, J. F. (1980). Human tumour xenografts growing in immunodeficient mice: A useful model for assessing chemotherapeutic agents in bronchial carcinoma. *Thorax* **41**:376–380.

Fernig, D. G., & Gallagher, J. T. (1994). Fibroblast growth factors and their receptors: An information network controlling tissue growth, morphogenesis and repair. *Progress in Growth Factor Research* **5**:353–377.

Finbow, M. E., & Pitts, J. D. (1981). Permeability of junctions between animal cells. *Exp. Cell Res.* **131**:1–13.

Fischer, G. A., & Sartorelli, A. C. (1964). Development, maintenance and assay of drug resistance. *Methods Med. Res.* **10**:247.

Fisher, H. W., Puck, T. T., & Sato, G. (1958). Molecular growth requirements of single mammalian cells: The action of fetuin in promoting cell attachment of glass. *Proc. Natl. Acad. Sci. USA* **44**:4–10.

Flier, J. S. (1995). The adipocyte storage depot or node on the energy information superhighway? *Cell* **80**:15–18.

Fogh, J. (1977). Absence of HeLa cell contamination in 169 cell lines derived from human tumors. *J. Natl. Cancer Inst.* **58**:209–214.

Fogh, J., & Trempe, G. (1975). In Fogh, J. (ed.), *Human tumor cells in vitro*. New York, Academic Press, pp. 115–159.

Folkman, I., Haudenschild, C. C., & Zetter, B. R. (1979). Long-term culture of capillary endothelial cells. *Proc. Natl. Acad. Sci. USA* **76**:5217–5221.

Folkman, I., & Moscona, A. (1978). Role of cell shape in growth control. *Nature* **273**:345–349.

Folkman, J. (1992). The role of angiogenesis in tumor growth. *Sem. Cancer Biol.* **3**:65–71.

Folkman, J., & D'Amore P. A. (1996). Blood vessel formation: what is its molecular basis? *Minireview. Cell* **87**: 1153–1155.

Folkman, J., & Haudenschild, C. (1980). Angiogenesis *in vitro*. *Nature* **288**:551–556.

Fontana, A., Hengarner, H., de Tribolet, N., & Weber, E. (1984). Glioblastoma cells release interleukin 1 and factors inhibiting interleukin 2-mediated effects. *J. Immunol.* **132**(4):1837–1844.

Food and Drug Administration. (1992). Federal Register *21: Code of Federal Regulations, Part 58*. Office of the *Federal Register* National Archives and Records, Washington, DC.

Foreman, J., & Pegg, D. E. (1979). Cell preservation in a programmed cooling machine: The effect of variations in supercooling. *Cryobiology* **16**:315–321.

Foster R., & Martin, G. S. (1992). A mutation in the catalytic domain of pp60v-src is responsible for the host- and temperature-dependent phenotype of the Rous sarcoma virus mutant tsLA33-1. *Virology* **187**:145–155.

Frame, M., Freshney, R. I., Shaw, R., & Graham, D. I. (1980). Markers of differentiation in glial cells. *Cell Biol. Int. Rep.* **4**:732.

Frame, M. C., Freshney, R. I., Vaughan, PO. F. T., Graham, D. I., & Shaw, R. (1984). Interrelationship between differentiation and malignancy-associated properties in glioma. *Br. J. Cancer*, **49**:269–280.

Franceschini, I. A., & Barnett, S. C. (1996). Low-affinity NGF-receptor and E-N-CAM expression define two types of olfactory nerve ensheathing cells that share a common lineage. *Dev. Biol.* **173**:327–343.

Franklin, J. M., Gilson, J. M., Franceschini, I. A., & Barnett, S. C. (1996). Schwann cell-like myelination following transplantation of an olfactory-nerve-ensheathing-cell line into areas of demyelination in the adult CNS. *Glia* **17**:217–224.

Frazier, J. M. (1992). *In vitro toxicity testing*. New York, Marcel Dekker.

Fredin, B. L., Seiffert, S. C., & Gelehrter, T. D. (1979). Dexamethasone-induced adhesion in hepatoma cells: The role of plasminogen activator. *Nature* **277**:312–313.

Freedman, V. H., & Shin, S. (1974). Cellular tumorigenicity in nude mice: Correlation with cell growth in semi-solid medium. *Cell* **3**:355–359.

Freshney, M. G. (1994). Colony-forming assays for CFC-GM, BFU-E, CFC-GEMM, and CFC-mix. In Freshney, R. I., Pragnell, I. B., & Freshney, M. G. (eds.), *Culture of hematopoietic cells*. New York, Wiley-Liss, pp. 265–268.

Freshney, R. I. (1972). Tumour cells disaggregated in collagenase. *Lancet* **2**:488–489.

Freshney, R. I. (1976). Separation of cultured cells by isopycnic centrifugation in metrizamide gradients. In Rickwood, D. (ed.), *Biological separations*. London and Washington, Information Retrieval, pp. 123–130.

Freshney, R. I. (1978). Use of tissue culture in predictive testing of drug sensitivity. *Cancer Topics* **1**:5–7.

Freshney, R. I. (1980). Culture of glioma of the brain. In Thomas, D. G. T., & Graham, D. I. (eds.), *Brain tumours: Scientific basic, clinical investigation and current therapy*. London, Butterworths, pp. 21–50.

Freshney, R. I. (1985). Induction of differentiation in neoplastic cells. *Anticancer Res.* **5**:111–130.

Freshney, R. I. (ed.). (1992). *Culture of epithelial cells*. New York, Wiley-Liss.

Freshney, R. I., Celik, F., & Morgan, D. (1982a). Analysis of cytotoxic and cytostatic effects. In Davis, W., Malvoni, C., & Tanneberger, St. (eds.), *The control of tumor growth and its biological base*, Fortschritte in der Onkologie, Band 10. Berlin, Akademie-Verlag, pp. 349–358.

Freshney, R. I., & Freshney, M. G. (eds.). (1996). *Culture of Immortalized Cells*. New York, Wiley-Liss.

Freshney, R. I., & Hart, E. (1982). Clonogenicity of human glia in suspension. *Br. J. Cancer* **46**:463.

Freshney, R. I., Hart, E., & Russell, J. M. (1982b). Isolation and purification of cell cultures from human tumours. In Reid, E., Cook, G. M. W., & Moore, D. J. (eds.), *Cancer cell organelles; methodological surveys* (*B*): *Biochemistry*, Vol. 2. Chicester, U.K., Horwood, pp. 97–110.

Freshney, R. I., Morgan, D., Hassanzadah, M., Shaw, R., & Frame, M. (1980a). Glucocorticoids, proliferation and the cell surface. In Richards, R. J., & Rajan, K. T. (eds.), *Tissue culture in medical research* (*II*). Oxford, U.K., Pergamon Press, pp. 125–132.

Freshney, R. I., Paul, J., & Kane, I. M. (1975). Assay of anticancer drugs in tissue culture: Conditions affecting their ability to incorporate ³H-leucine after drug treatment. *Br. J. Cancer* **31**:89–99.

Freshney, R. I., Pragnell, I. B., Freshney, M. G., (eds.). (1994). *Culture of Haemopoietic Cells*. New York, Wiley-Liss.

Freshney, R. I., Sherry, A., Hassanzadah, M., Freshney, M., Crilly, P., & Morgan, D. (1980b). Control of cell proliferation in human glioma by glucocorticoids. *Br. J. Cancer* **41**:857–866.

Freyer, J. P., & Sutherland, R. M. (1980). Selective dissociation and characterization of cells from different regions of multicell tumour spheroids. *Cancer Res.* **40**:3956–3965.

Friend, C., Patuleia, M. C., & Nelson, J. B. (1966). Antibiotic effect of tylosine on a mycoplasma contaminant in a tissue culture leukemia cell line. *Proc. Soc. Exp. Biol. Med.* **121**:1009.

Friend, C., Scher, W., Holland, J. G., & Sato, T. (1971). Hemoglobin synthesis in murine virus-induced leukemic cells *in vitro*; 2: Stimulation of erythroid differentiation by dimethyl sulfoxide. *Proc. Natl. Acad. Sci. USA* **68**:378–382.

Friend, K. K., Dorman, B. P., Kucherlapati, R. S., & Ruddle, F. H. (1976). Detection of interspecific translocations in mouse–human hybrids by alkaline Giemsa staining. *Exp. Cell Res.* **99**:31–36.

Fry, J., & Bridges, J. W. (1979). The effect of phenobarbitone on adult rat liver cells and primary cell lines. *Toxicol. Lett.* **4**:295–301.

Fuller, B. B., & Meyskens, F. L. (1981). Endocrine responsiveness in human melanocytes and melanoma cells in culture. *J. Natl. Cancer Inst.* **66**:799–802.

Furue, M., & Saito, S. (1997). Synergistic effect of hepatocyte growth factor and fibroblast growth factor-1 on the branching morphogenesis of rat submandibular gland epithelial cells. *Tiss. Cult. Res. Commun.* **16**:189–194.

Fusenig, N. E. (1994a). Epithelial–mesenchymal interactions regulate keratinocyte growth and differentiation *in vitro*. In Leigh, I. M., Lane, E. B., & Watt, F. M.. (eds.), *Keratinocyte handbook*, Vol. I. Cambridge, U.K., Cambridge University Press, pp. 71–94.

Fusenig, N. E. (1994b). Cell culture models: reliable tools in pharmacotoxicology? In Fusenig, N. E., & Graf, H. (eds.), *Cell culture in pharmaceutical research*. Berlin & Heidelberg, Springer-Verlag, pp. 1–7.

Fusenig, N. E., Limat, A., Stark, H.-J., & Breitkreutz, D. (1994). Modulation of the differentiated phenotype of keratinocytes of the hair follicle and from epidermis. *J. Dermatol. Sci.* **7**:142–151.

Gallagher, R., Collins, S., Trujillo, J., McCredie, K., Ahearn, M., Tsai, S., Metzgar, R., Aulakh, G., Ting, R., Ruscetti, F., & Gallo, R. (1979). Characterization of the continuous, differentiating myeloid cell line (HL-60) from a patient with promyelocytic leukemia. *Blood* **54**:713–733.

Gard, A. L., & Pfeiffer, S. E. (1990). Two proliferative stages of the oligodendrocyte lineage (A2B5+O4− and O4+GalC−) under different mitogenic control. *Neuron* **5**:615–625.

Garrido, T., Riese, H. H., Aracil, M., & Perez-Aranda, A. (1995). Endothelial cell differentiation into capillary-like structures in response to tumour cell conditioned medium: A modified chemotaxis chamber assay. *Br. J. Cancer* **71**:770–775.

Gartler, S. M. (1967). Genetic markers as tracers in cell culture. 2nd Decennial Review Conference on Cell, Tissue and Organ Culture. *NCI Monograph* 26, pp. 167–195.

Gartner, S., & Kaplan, H. S. (1980). Long-term culture of human bone marrow cells. *Proc. Natl. Acad. Sci. USA* **77**: 4756–4759.

Gaudernack, T., Leivestad, T., Ugelstad, J., & Thorsby, E. (1986). Isolation of pure functionally active CD8+ T cells. Positive selection with monoclonal antibodies directly conjugated to monosized magnetic microspheres. *J. Immunol. Meth.* **90**:179–187.

Gaush, C. R., Hard, W. L., & Smith, T. F. (1966). Characterization of an established line of canine kidney cells (MDCK). *Proc. Soc. Exp. Biol. Med.* **122**:931–933.

Gazdar, A. F., Carney, D. N., & Minna, J. D. (1983). The biology of nonsmall cell lung cancer. *Sem. Oncol.* **10**:3–19.

Gazdar, A. F., Carney, D. N., Russell, E. K., Sims, H. L., Baylin, S. B., Bunn, P. A., Guccion, J. G., & Minna, J. D. (1980). Establishment of continuous, clonable cultures of small cell carcinoma of the lung which have amine precursor uptake and decarboxylation properties. *Cancer Res.* **40**:3502–3507.

Gazdar, A. F., Zweig, M. H., Carney, D. N., Van Steirteghen, A. C., Baylin, S. B., & Minna, J. D. (1981). Levels of creatine kinase and its BB isoenzyme in lung cancer specimens and cultures. *Cancer Res.* **41**:2773–2777.

Germain, L., Rouabhia, M., Guignard, R., Carrier, L., Bouvard, V., & Auger, F. A. (1993). Improvement of human keratinocyte isolation and culture using thermolysin. *Burns* **19**:99–104.

Gey, G. O., Coffman, W. D., & Kubicek, M. T. (1952). Tissue culture studies of the proliferative capacity of cervical carcinoma and normal epithelium. *Cancer Res.* **12**:364–365.

Ghigo, D., Priotto, C., Migliorino, D., Geromin, D., Franchino, C., Todde, R., Costamagna, C., Pescarmona, G., & Bosia, A. (1998). Retinoic acid-induced differentiation in a human neuroblastoma cell line is associated with an increase in nitric oxide synthesis. *J Cell Physiol.* **174**:99–106.

Ghosh C., & Collodi, P. (1994). Culture of cells from zebrafish (*Brachydanio rerio*) blastula-stage embryos. *Cytotechnology* **14**:21–26.

Ghosh, D., Danielson, K. C., Alston, J. T., & Heyner, S. (1991). Functional differential of mouse uterine epithelial cells grown on collagen gels or reconstituted basement membranes. *In Vitro Cell Dev. Biol.* **27A**:713–719.

Giard, D. J., Aaronson, S. A., Todaro, G. J., Arnstein, P., Kersey, J. H., Dosik, K., & Parks, W. P. (1972). *In vitro* cultivation of human tumors: Establishment of cell lines derived from a series of solid tumors. *J. Natl. Cancer Inst.* **51**:1417.

Gignac, S. M., Steube, K., Schleithoff, L., Janssen, J. W., MacLeod, R. A., Quentmeier, H., & Drexler, H. G. (1993). Multiparameter approach in the identification of cross-contaminated leukemia cell lines. *Leukemia & Lymphoma* **10**(4–5):359–368.

Gilbert, S. F., & Migeon, B. R. (1975). D-Valine as a selective agent for normal human and rodent epithelial cells in culture. *Cell* **5**:11–17.

Gilbert, S. F., & Migeon, B. R. (1977). Renal enzymes in kidney cells selected by D-Valine medium. *J. Cell Physiol.* **92**:161–168.

Gilchrest, B. A., Albert, L. S., Karassik, R. L., & Yaar, M. (1985). Substrate influences human epidermal melanocyte attachment and spreading *in vitro*. *In Vitro* **21**:114.

Gillette, R. W. (1987). Alternatives to pristane priming for ascitic fluid and monoclonal antibody production. *J. Immunol. Meth.* **99**:21–23.

Gillis, S., & Watson, J. (1981). Interleukin-2 dependent culture of cytolytic T cell lines. *Immunol. Rev.* **54**:81–109.

Gimbrone, M. A., Jr., Cotran, R. S., & Folkman, J. (1974). Human vascular endothelial cells in culture, growth and DNA synthesis. *J. Cell Biol.* **60**:673–684.

Giovanella, B. C., Stehlin, J. S., & Williams, L. J. (1974). Heterotransplantation of human malignant tumors in "nude" mice; II: Malignant tumors induced by injection of cell cultures derived from human solid tumors. *J. Natl. Cancer Inst.* **52**:921.

Giron, J. A., Lange, M., & Baseman, J. B. (1996). Adherence, fibronectin-binding and induction of cytoskeleton reorganization in cultured human-cells by *Mycoplasma penetrans*. *Infection and Immunity* **64**:197–208.

Gjerset, R., Yu, A., & Haas, M. (1990). Establishment of continuous cultures of T-cell acute lymphoblastic leukemia cells at diagnosis. *Cancer Res.* **50**:10–14.

Glavin, G. B., Szabo, S., Johnson, B. R., Xing, P. L., Morales, R. E., Plebani, M., & Nagy L. (1996). Isolated rat gastric mucosal cells: Optimal conditions for cell harvesting, measures of viability and direct cytoprotection. *J. Pharmacol. Exp. Therapeut.* **276**:1174–1179.

Gluzman, Y. (1981). SV40-transformed simian cells support the replication of early SV40 mutants. *Cell* **23**:175–182.

Gobet, R., Raghunath, M., Altermatt, S., Meuli-Simmen, C., Benathan, M., Dietl, A., & Meuli, M. (1997). Efficacy of cultured epithelial autografts in pediatric burns and reconstructive surgery. *Surgery* **121**:546–661.

Goding, C. R., & Fisher, D. E. (1997). Meeting review—regulation of melanocyte differentiation and growth. *Cell Growth & Differentiation* **8**:935–940.

Goldberg, B. (1977). Collagen synthesis as a marker for cell type in mouse 3T3 lines. *Cell* **11**:169–172.

Goldman, B. I., & Wurzel, J. (1992). Effects of subcultivation and culture medium on differentiation of human fetal cardiac myocytes. *In Vitro Cell Dev. Biol.* **28A**:109–119.

Goldstein, S., Murano, S., Benes, H., Moerman, E. J., Jones, R. A., & Thweatt, R. (1989). Studies on the molecular genetic basis of replicative senscence in Werner syndrome and normal fibroblasts. *Exp. Gerontology* **24**(5–6):461–468.

Goldwasser, E. (1975). Erythropoietin and the differentiation of red blood cells. *Fed. Proc.* **34**:2285–2292.

Gomm, J. J., Coope, R. C., Browne, P. J., and Coombes, R. C. (1997). Separated human breast epithelial and myoepithelial cells have different growth factor requirements *in vitro* but can reconstitute normal breast lobuloalveolar structure. *J. Cell Physiol.* **171**:11–19.

Good, N. E., Winget, G. D., Winter, W., Connolly, T. N., Izawa, S., & Singh, R. M. M. (1966). Hydrogen ion buffers and biological research. *Biochemistry* **5**:467–477.

Goodwin, G., Shaper, J. H., Abezoff, M. D., Mendelsohn, G., & Baylin, S. B. (1983). Analysis of cell surface proteins delineates a differentiation pathway linking endocrine and nonendocrine human lung cancers. *Proc. Natl. Acad. Sci.* **80**:3807–3811.

Gospodarowicz, D. (1974). Localization of fibroblast growth factor and its effect alone and with hydrocortisone on 3T3 cell growth. *Nature* **249**:123–127.

Gospodarowicz, D., Delgado, D., & Vlodavsky, I. (1980). Permissive effect of the extracellular matrix on cell proliferation *in vitro*. *Proc. Natl. Acad. Sci. USA* **77**:4094–4098.

Gospodarowicz, D., Greenburg, G., Bialecki, H., & Zetter, B. R. (1978a). Factors involved in the modulation of cell proliferation *in vivo* and *in vitro*: The role of fibroblast and epidermal growth factors in the proliferative response of mammalian cells. *In Vitro* **14**:85–118.

Gospodarowicz, D., Greenburg, G., & Birdwell, C. R. (1978b). Determination of cell shape by the extracellular matrix and its correlation with the control of cellular growth. *Cancer Res.* **38**:4155–4171.

Gospodarowicz, D., & Moran, J. (1974). Growth factors in mammalian cell cultures. *Annu. Rev. Biochem.* **45**:531–558.

Gospodarowicz, D., Moran, J. S., & Braun, D. L. (1977). Control of proliferation of bovine vascular endothelial cells. *J. Cell Physiol.* **91**:377–386.

Gospodarowicz, D., Moran, J., Braun, D., & Birdwell, C. (1976). Clonal growth of bovine vascular endothelial cells: Fibroblast growth factor as a survival agent. *Proc. Natl. Acad. Sci. USA* **73**:4120–4124.

Grafstrom, R. C. (1990a). *In vitro* studies of aldehyde effects related to human respiratory carcinogenesis. *Mutation Res.* **238**:175–184.

Grafstrom, R. C. (1990b). Carcinogenesis studies in human epithelial tissues and cells *in vitro*: Emphasis on serum-free culture conditions and transformation studies. *Acta Physiologica Scandinavica* **140**:93–133.

Graham, F. L., Smiley, J., Russell, W. C., & Nairn, R. (1977). Characteristics of a human cell line transformed by DNA from human adenovirus type 5. *J. Gen. Virol.* **36**:59–72.

Graham, F. L., & Van der Eb, A. J. (1973). A new technique for the assay of infectivity of human adenovirus 5 DNA. *Virology* **52**:456–461.

Graham, G. J. Freshney, M. G., Donaldson, D., & Pragnell, I. B. (1992). Purification and biochemical characterisation of human and murine stem cell inhibitors. *Growth Factors* **7**:151–160.

Grampp, G. E., Sambanis, A., & Stephanopoulos, G. N. (1992). Use of regulated secretion in protein production from animal cells: An overview. *Adv. in Biochem. Eng.-Biotechnol.* **46**:35–62.

Grandolfo, M., d'Andrea, P., Paoletti, S., Martina, M., Silvestrini, G., Bonucci, E., & Vittur, F. (1993). Culture and differentiation of chondrocytes entrapped in alginate beads. *Calcif. Tiss. Int.* **52**:131–138.

Granner, D. K., Hayashi, S., Thompson, E. B., & Tomkins, G. M. (1968). Stimulation of tyrosine aminotransferase synthesis by dexamethasone phosphate in cell culture. *J. Mol. Biol.* **35**:291–301.

Graziadei, P. P. C., & Monti Graziadei, G. A (1979). Neurogenesis and neuron regeneration in the olfactory system of mammals; I: Morphological aspects of differentiation and structural organisation of the olfactory sensory neurons. *Journal of Neurocytology* **8**:1–18.

Graziadei, P. P. C., & Monti Graziadei, G. A. (1980). Neurogenesis and neuron regeneration in the olfactory system of mammals; III: Deafferentation and reinnervation of the olfactory bulb following section of the *fila olfactoria* in rat. *Journal of Neurocytology* **9**:145–162.

Green, A. E., Athreya, B., Lehr, H. B., & Coriell, L. L. (1967). Viability of cell cultures following extended preservation in liquid nitrogen. *Proc. Soc. Exp. Biol. Med.* **124**:1302–1307.

Green, H., & Kehinde, O. (1974). Sublines of mouse 3T3 cells that accumulate lipid. *Cell* **1**:113–116.

Green, H., Kehinde, O., & Thomas, J. (1979). Growth of cultured human epidermal cells into multiple epithelia suitable for grafting. *Proc. Natl. Acad. Sci. USA* **76**:5665–5668.

Green, H., & Thomas, J. (1978). Pattern formation by cultured human epidermal cells: Development of curved ridges resembling dermatoglyphs. *Science* **200**:1385–1388.

Greenberger, J. S. (1980). Self-renewaal of factor-dependent haemopoietic progenitor cell lines derived from long-term bone marrow cultures demonstrate significant mouse strain genotypic variation. *J. Supramol. Struct.* **13**:501–511.

Greider, C. W., & Blackburn, E. H. (1996). Telomeres, telomerase, and cancer. *Scientific American*, February 1996.

Griffiths, J. B. (1992). Scaling up of animal cell cultures. In Freshney, R. I. (ed.), *Animal cell culture, a practical approach*. Oxford, U.K., IRL Press at Oxford University Press, pp. 47–93.

Griffiths, J. B., & Pirt, G. J. (1967). The uptake of amino acids by mouse cells (Strain LS) during growth in batch culture and chemostat culture: The influence of cell growth rate. *Proc. R. Soc. Biol.* **168**:421–438.

Grizzle, W. E., & Polt, S. H. (1988). Guidelines to avoid personal contamination by infective agents in research laboratories that use human tissue. *J. Tiss. Cult. Meth.* **11**:191–200.

Gross, S. K., Lyerla, T. A., Williams, M. A., & McCluer, R. H. (1992). The accumulation and matabolism of glycosphingolipids in primary kidney cell cultures from beige mice. *Mol. Cell Biochem.* **118**:61–66.

Gugel, E. A., & Sanders, J. E. (1986). Needle-stick transmission of human colonic adenocarcinoma. *New Engl. J. Med.* **315**:1487.

Guguen-Guillouzo, C. (1992). Isolation and culture of animal and human hepatocytes. In Freshney, R. I. (ed.), *Culture of epithelial cells*. New York, Wiley-Liss, pp. 198–223.

Guguen-Guillouzo, C., Campion, J. P., Brissot, P., Glaise, D., Launois, B., Bourel, M., & Guillouzo A. (1982). High yield preparation of isolated human adult hepatocytes by enzymatic profusion of the liver. *Cell Biol. Int. Rep.* **6**:625–628.

Guguen-Guillouzo, C., Clement, B., Baffet, G., Beaument, C., Morel-Chany, E., Glaise, D., & Guillouzo, A. (1983). Maintenance and reversibility of active albumin secretion by adult rat hepatocytes co-cultured with another liver epithelial cell type. *Exp. Cell Res.* **143**:47–54.

Guguen-Guillouzo, C., & Guillouzo, A. (1986). Methods for preparation of adult and fetal hepatocytes. In Guguen-Guillouzo, C., & Guillouzo, A. (eds.), *Isolated and cultured hepatocytes*. Paris, Les Éditions INSERM, John Libbey Eurotext, pp. 1–12.

Guilbert, L. I., & Iscove, N. N. (1976). Partial replacement of serum by selenite, transferrin, albumin and lecithin in haemopoietic cell cultures. *Nature* **263**:594–595.

Guillouzo, A., Guguen-Guillouzo, C., & Bourel, M. (1981). Hepatocytes in culture: Expression of differentiated functions and their application to the study of metabolism. *Triangle (Sandoz J. Med. Sci.)* **20**:121–128.

Guillouzo, A. M. A. (1989). Méthodes *in vitro* en pharmacotoxiocologie. Les Éditions INSERM **170**:200.

Guiraud, J. M., Beuron, F., Sion, B., Brassier, G., Faivre, J., Thieulant, M. L., & Duval, J. (1991). Human prolactin producing pituitary adenomas in three dimensional culture. *In Vitro Cell Dev. Biol.* **27a**:188–190.

Gullino, P. M. (1985). Angiogenesis, tumor vascularization, and potential interference with tumor growth. In Mihich, E. (ed.), *Biological responses in cancer*. New York, Plenum.

Gullino, P. M., & Knazek, R. A. (1979). Tissue culture on artificial capillaries. In Jakoby, W. B., & Pastan, I. (eds.), *Methods in Enzymology, Vol. 58: Cell Culture*. New York, Academic Press, pp. 178–184.

Gumbiner, B. (1992). Epithelial morphogenesis. *Cell* **69**:385–387.

Gumbiner, B., & Simons, K. (1986). A functional assay for proteins involved in establishing an epithelial occluding barrier: Identification of a uvomorulin-like polypeptide. *J. Cell Biol.* **102**:457–468.

Gumbiner, B. M. (1995). Signal transduction of beta-catenin. *Curr. Opin. Cell Biol.* **7**:634–640.

Guner, M., Freshney, R. I., Morgan, D., Freshney, M. G., Thomas, D. G. T., & Graham, D. I. (1977). Effects of dexamethasone and betamethasone on *in vitro* cultures from human astrocytoma. *Br. J. Cancer* **35**:439–447.

Gupta, K., Ramakrishnan, S., Browne, P. V., Solovey, A., & Hebbel, R. P. (1997). A novel technique for culture of human dermal microvascular endothelial cells under either serum-free or serum-supplemented conditions: Isolation by panning and stimulation with vascular endothelial growth factor. *Exp. Cell Res.* **230**:244–251.

Gustafson, C. J., Eldh, J., & Kratz, G. (1998). Culture of human urothelial cells on a cell-free dermis for autotransplantation. *European Urology* **33**:503–506.

Hafny, B. E. L., Bourre, J.-M., & Roux, F. (1996). Synergistic stimulation of gamma-glutamyl transpeptidase and alkaline phosphatase activities by retinoic acid and astroglial factors in immortalised rat brain microvessel endothelial cells. *J. Cell. Physiol.* **167**:451–460.

Hallermeyer, K., & Hamprecht, B. (1984). Cellular heterogeneity in primary cultures of brain cells revealed by immunocytochemical localisation of glutamyl synthetase. *Brain Res.* **295**:1–11.

Halleux, C., & Schneider, Y.-J. (1994). Iron absorption by Caco-2 cells cultivated in serum-free medium as *in vitro* model of the human intestinal epithelial barrier. *J. Cell. Physiol.* **158**:17–28.

Ham, R. G. (1963). An improved nutrient solution for diploid Chinese hamster and human cell lines. *Exp. Cell Res.* **29**:515.

Ham, R. G. (1965). Clonal growth of mammalian cells in a chemically defined synthetic medium. *Proc. Natl. Acad. Sci. USA* **53**:288.

Ham, R. G. (1984). Growth of human fibroblasts in serum-free media. In Barnes, D. W., Sirbasku, D. A., & Sato, G. H. (eds.), *Cell culture methods for molecular and cell biology*, Vol. 3. New York, Alan R. Liss, pp. 249–264.

Ham, R. G., & McKeehan, W. L. (1978). Development of improved media and culture conditions for clonal growth of normal diploid cells. *In Vitro* **14**:11–22.

Ham, R. G., & McKeehan, W. L. (1979). Media and growth requirements. In Jakoby, W. B., & Pastan, I. H. (eds.), *Methods in enzymology; vol. 58: Cell culture.* New York, Academic Press, pp. 44–93.

Hamburger, A. W., & Salmon, S. E. (1977). Primary bioassay of human tumor stem cells. *Science* **197**:461–463.

Hamburger, A. W., Salmon, S. E., Kim, M. B., Trent, J. M., Soehnlen, B., Alberts, D. S., & Schmidt, H. J. (1978). Direct cloning of human ovarian carcinoma cells in agar. *Cancer Res.* **38**:3438–3444.

Hames, B. D., & Glover, D. M. (1991). *Oncogenes.* Oxford, U.K., IRL Press at Oxford University Press.

Hamilton, T. C., Young, R. C., McKoy, W. M., Grotzinger, K. R., Green, J. A., Chu, E. W., Whang-Peng, J., Rogan, A. M., Green, W. R., & Ozols, R. F. (1983). Characterization of a human ovarian carcinoma cell line (NIH:OVCAR-3) with androgen and estrogen receptors. *Cancer Res.* **43**:5379–5389.

Hamilton, W. G., & Ham, R. G. (1977). Clonal growth of Chinese hamster cell lines in protein-free media. *In Vitro* **13**:537–547.

Hammond, S. L., Ham, R. G., & Stampfer, M. R. (1984). Serum free growth of human mammary epithelial cells: Rapid clonal growth in defined medium and extended serial passage with pituitary extract. *Proc. Natl. Acad. Sci. USA* **81**:5435–5439.

Hanks, J. H., & Wallace, R. E. (1949). Relation of oxygen and temperature in the preservation of tissues by refrigeration. *Proc. Exp. Biol. Med.* **71**:196.

Hansson, B., Rönnbäck, L., Persson, L. I., Lowenthal, A., Noppe, M., Alling, C., & Karlsson, B. (1984). Cellular composition of primary cultures from cerebral cortex, striatum, hippocampus, brain-stem and cerebellum. *Brain Res.* **300**:9–18.

Hardman, P., Klement, B. J., & Spooner, B. S. (1990). Growth and morphogenesis of embryonic mouse organs on Biopore membrane. *In Vitro Cell Dev. Biol.* **26**:119–120.

Harrington, W. N., & Godman, G. C. (1980). A selective

inhibitor of cell proliferation from normal serum. *Proc. Natl. Acad. Sci. USA Biol. Sci.* **77**:423–427.

Harris, H., & Watkins, J. F. (1965). Hybrid cells from mouse and man: Artificial heterokaryons of mammalian cells from different species. *Nature* **205**:640–646.

Harris, L. W., & Griffiths, J. B. (1977). Relative effects of cooling and warming rates on mammalian cells during the freeze–thaw cycle. *Cryobiology* **14**:662–669.

Harrison, R. G. (1907). Observations on the living developing nerve fiber. *Proc. Soc. Exp. Biol. Med.* **4**:140–143.

Hart, I. R., & Fidler, I. J. (1978). An *in vitro* quantitative assay for tumor cell invasion. *Cancer Res.* **38**:3218–3224.

Hartley, R. S., & Yablonka-Reuveni, Z. (1990). Long-term maintenance of primary myogenic cultures on a reconstituted basement membrane. *In Vitro Cell Dev. Biol.* **26**:955–961.

Hass, R., Gunji, H., Hirano, M., Weichselbaum, R., & Kufe, D. (1993). Phorbol ester-induced monocytic differentiation is associated with G2 delay and down regulation of cdc25 expression. *Cell Growth Diff.* **4**:159–166.

Haudenschild, C. C., Zahniser, D., Folkman, J., & Klagsbrun, M. (1976). Human vascular endothelial cells in culture. *Exp. Cell Res.* **98**:175–183.

Hauschka, S. D., & Konigsberg, I. R. (1966). The influence of collagen on the development of muscle clones. *Proc. Natl. Acad. Sci. USA* **55**:119–126.

Häuselmann, H. J., Fernandes, R. J., Mok, S. S., Schmid, T. M., Block, J. A., Aydelotte, M. B., Kuettner, K. E., & Thonar, E. J. M. A. (1994). Phenotypic stability of bovine articular chondrocytes after long-term culture in alginate beads. *J. Cell Sci.* **107**:17–27.

Häuselmann, H. J., Masuda, K., Hunziker, E. B., Neidhart, M., Mok, S. S., Michel, B. A., & Thonar, E. J.-M. A. (1996). Adult human chondrocytes cultured in alginate form a matrix similar to native human cartilage. *Am. J. Physiol. (Cell Physiol.)* **40**:C742–C752.

Hawley-Nelson, P., Ciccarone, V., Gebevehu, G., Jessee, J. (1993). A new polycationic liposome reagent with enhanced activity for transfection. *FASEB Journal* **7**:A167.

Hay, E. D. (ed.). (1991). *Cell biology of extracellular matrix.* New York, Plenum Press.

Hay, R. J. (1979). Identification, separation and culture of mammalian tissue cells. In Reid, E. (ed.), *Methodological surveys in biochemistry; Vol. 8, Cell Populations.* London, Ellis Horwood, pp. 143–160.

Hay, R. J. (1991). Operator-induced contamination in cell culture systems. *Developments in Biological Standardization* **75**:193–204.

Hay, R. J. (2000). Cell line preservation and characterization. In Masters, J. R. W. (ed.). *Animal cell culture, a practical approach.* Oxford, U.K., IRL Press at Oxford University Press, pp. 95–148.

Hay, R. J., Caputo, J., & Macy, M. L. (1992). Establishing or verifying cell line identity. In *ATCC quality control methods for cell lines*, 2d ed. Rockville, MD, American Type Culture Collection, pp. 52–66.

Hay, R. J., & Cour, I. (1997). Testing for microbial contamination: Bacteria and fungi. In Doyle, A., Griffiths, J. B., & Newall, D. G. (eds.), *Cell and tissue culture: Laboratory procedures.* Chichester, U.K., Wiley, Module 7A:2.

Hay, R. J., & Strehler, B. L. (1967). The limited growth span of cell strains isolated from the chick embryo. *Exp. Gerontol.* **2**:123.

Hayashi, I., & Sato, G. H. (1976). Replacement of serum by hormones permits growth of cells in a defined medium. *Nature* **259**:132–134.

Hayflick, L. (1961). The establishment of a line (WISH) of human amnion cells in continuous cultivation. *Exp. Cell Res.* **23**:14–20.

Hayflick, L., & Moorhead, P. S. (1961). The serial cultivation of human diploid cell strains. *Exp. Cell Res.* **25**:585–621.

Haynes, L. W. (1999). *The neuron in tissue culture.* Chichester, U.K., John Wiley & Sons.

Heald, K. A., Hail, C. A., & Downing, R. (1991). Isolation of islets of Langerhans from the weanling pig. *Diabetes Res.* **17**:7–12.

Health and Safety Commission (1985). Approved Code of Practice "The Protection of Persons Against Ionising Radiation Arising from An Work Activity." HMSO, London.

Health and Safety Commission (1991). *Safe Working and the Prevention of Infection in Clinical Laboratories.* HMSO Publications, P. O. Box 276, London, SW8 5DT, England.

Health and Safety Commission (1991). *Safe Working and the Prevention of Infection in Clinical Laboratories—Model Rules for Staff and Visitors.* HMSO Publications, P. O. Box 276, London, SW8 5DT, England.

Health and Safety Commission (1992). *Genetically Modified Organisms (Contained Use) Regulations:* SI 1992/3217, HMSO, ISBN 0-11-025332-9.

Health and Safety Commission (1999). Carcinogens ACOP and Biological agents ACOP. *Control of Substances Hazardous to Health Regulations,* Approved Codes of Practice, L5, HSE Books, HMSO Publications Centre, P. O. Box 276, London SW8 5DT, England.

Health and Safety Commission (1999). *Control of Substances Hazardous to Health Regulations.* Statutory Instrument No. 437, HSE Books, Sudbury, U.K.

Health Services Advisory Committee (1992). *Safe Disposal of Clinical Waste.* HSE Books, Sudbury, U.K.

Heffelfinger, S. C., Hawkins, H. H., Barrish, J., Taylor, L., & Darlington, G. (1992). SK HEP-1: A human cell line of endothelial origin. *In Vitro Cell Dev. Biol.* **28A**:136–142.

Heldin, C. H., Westermark, B., & Wasteson, A. (1979). Platelet-derived growth factor: Purification and partial characterization. *Proc. Natl. Acad. Sci. USA* **76**:3722–3726.

Hemmati-Brivaniou, A., Kelly, O. G., & Melton, D. A. (1994). Follistatin, an antagonist of activin, is expressed in the Spemann organizer and displays direct neuralizing activity. *Cell* **77**:283–295.

Heyderman, E., Steele, K., & Ormerod, M. G. (1979). A new antigen on the epithelial membrane: Its immunoperoxidase localisation in normal and neoplastic tissue. *J. Clin. Pathol.* **32**:35–39.

Heyworth, C. M., & Spooncer, E. (1992). *In vitro* clonal assays for murine multipotential and lineage restricted myeloid progenitor cells. In Testa, N. G., & Molineux, G. (eds.), *Haemopoiesis: A practical approach.* Oxford, U.K., IRL Press at Oxford University Press, pp. 37–54.

Hicks, G. G., Chen, J., & Ruley, H. E. (1998). Production and use of retroviruses. In Ravid, K., & Freshney, R. I. (eds.). *DNA transfer to cultured cells.* New York, Wiley-Liss, pp. 1–26.

Higuchi, K. (1977). Cultivation of mammalian cell lines in serum-free chemically defined medium. *Methods Cell. Biol.* **14**:131.

Hill, B. T. (1983). An overview of clonogenic assays for human tumour biopsies. In Dendy, P. P., & Hill B. T. (eds.), *Human Tumour Drug Sensitivity Testing in vitro.* New York, Academic Press, pp. 91–102.

Hilwig, I., & Gropp, A. (1972). Staining of constitutive heterochromatin in mammalian chromosomes with a new fluorochrome. *Exp. Cell. Res.* **75**:122–126.

Hince, T. A., & Roscoe, J. P. (1980). Differences in pattern and level of plasminogen activator production between a cloned cell line from an ethylnitrosourea-induced glioma and one from normal adult rat brain. *J. Cell Physiol.* **104**:199–207.

Hino, H., Tateno, C., Sato, H., Yamasaki, C., Katayama, S., Kohashi, T., Aratani, A., Asahara, T., Dohi, K., & Yoshizato, K. (1999). A long-term culture of human hepatocytes which show a high growth potential and express their differentiated phenotypes. *Biochem. Biophys. Res. Comm.* **256**:184–189.

Hirai, Y., Takebe, K., Takashina, M., Kobayashi, S., & Takeichi, M. (1992). Epimorphin: A mesenchymal protein essential for epithelial morphogenesis. *Cell* **69**:471–481.

Hlubinova, K., Feldsamova, A., & Prachar, J. (1994). Evaluation of two methods for elimination of mycoplasma. *In Vitro Cell Dev. Biol.* **30A**:21–22.

Hoheisel, D., Nitz, T., Franke, H., Wegener, J., Hakvoort, A., Tilling, T., & Galla, H. J. (1998). Hydrocortisone reinforces the blood–brain properties in a serum free cell culture system. *Biochem. Biophys. Res. Comm.* **247**:312–315.

Holbrook, K. A., & Hennings, H. (1983). Phenotypic expression of epidermal cells *in vitro*: A review. *J. Invest. Dermatol.* **81**:11s–24s.

Hollenberg, M. D., & Cuatrecasas, P. (1973). Epidermal growth factor: Receptors in human fibroblasts and modulation of action by cholera toxin. *Proc. Natl. Acad. Sci. USA* **70**:2964–2968.

Holley, R. W., Armour, R., & Baldwin, J. H. (1978). Density-dependent regulation of growth of BSC-1 cells in cell culture: Growth inhibitors formed by the cells. *Proc. Natl. Acad. Sci. USA* **75**:1864–1866.

Holt, S. E., Shay, J. W., & Wright, W. E. (1996). Refining the telomere–telomerase hypothesis of ageing and cancer. *Nature Biotechnology* **14**:836–839.

Hopert, A., Uphoff, C. C., Wirth, M., Hauser, H., & Drexler, H. G. (1993). Mycoplasma detection by PCR analysis. *In Vitro Cell Dev. Biol.* **29A**:819–821.

Hopps, H., Bernheim, B. C., Nisalak, A., Tjio, J. H., & Smadel, J. E. (1963). Biologic characteristics of a continuous cell line derived from the African green monkey. *J. Immunol.* **91**:416–424.

Horibata, K., & Harris, A. W. (1970). Mouse myelomas and lymphomas in culture. *Exp. Cell Res.* **60**:61–77.

Horita, A., & Weber, L. J. (1964). Skin penetrating property of drugs dissolved in dimethylsulfoxide (DMSO) and other vehicles. *Life Sci.* **3**:1389–1395.

Hornsby, P. J., Yang, L., Lala, D. S., Cheng, C. Y. & Salmons, B. (1992). A modified procedure for replica plating of mammalian cells allowing selection of clones based on gene expression. *Biotechniques* **12**:244–251.

Horster, M. (1979). Primary culture of mammalian nephron epithelia: Requirements for cell outgrowth and proliferation from defined explanted nephron segments. *Pflugers Arch.* **382**:209–215.

Hotamisligil, G. S., Arner, P., Caro, J. F., Atkinson, R. L., & Speigelman, B. M. (1995). Increased adipose tissue expression of tumor necrosis factor-α in human obesity and insulin resistance. *J. Clin. Invest.* **95**:2409–2415.

Howard, B. V., Macarak, E. J., Gunson, D., & Kefalides, N. A. (1976). Characterization of the collagen synthesized by endothelial cells in culture. *Proc. Natl. Acad. Sci. USA* **73**:2361–2364.

Howard, M., Kessler, S., Chused, T., & Paul, W. E. (1981). Long term culture of normal mouse B lymphocytes. *Proc. Natl. Acad. Sci. USA* **78**:5788–5792.

Howie Report. (1978). *Code of practice for prevention of infection in clinical laboratories and post-mortem rooms.* London, H.M. Stationery Office.

Hoyer, L. W., de los Santos, R. P., & Hoyer, J. R. (1973). Antihemophilic factor antigen: Localization in endothelial cells by immunofluorescence microscopy. *J. Clin. Invest.* **52**:2737–2744.

Hsu, T. C., & Benirschke, K. (1967). *Atlas of Mammalian Chromosomes*, Vols. 1–4. New York, Springer.

Hu, G.-F., Riordan, J. F., & Vallee, B. L. (1997). A putative angiogenin receptor in angiogenin-responsive human endothelial cells. *Proc. Natl. Acad. Sci. USA* **94**:204–209.

Hughes, S. E. (1996). Functional characterization of the spontaneously transformed human umbilical vein endothelial cell line ECV304: Use in an *in vitro* model of angiogenesis. *Exp. Cell Res.* **225**:171–185.

Hull, R. N., Cherry, W. R., & Weaver, G. W. (1976). The origin and characteristics of a pig kidney cell strain, LLC-PKI. *In Vitro* **12**:670–677.

Human Cytogenetic Nomenclature: *See* International System for Human Cytogenetic Nomenclature.

Hume, D. A., & Weidemann, M. J. (1980). *Mitogenic lymphocyte transformation.* Amsterdam, Elsevier/North Holland Biomedical Press.

Huschtscha, L. I., & Holliday, R. (1983). Limited and unlimited growth of SV40-transformed cells from human diploid MRC-5 fibroblasts. *J. Cell Sci.* **63**:77–99.

Hynes, R. O. (1973). Alteration of cell-surface proteins by viral transformation and by proteolysis. *Proc. Natl. Acad. Sci. USA* **70**:3170–3174.

Hynes, R. O. (1974). Role of cell surface alterations in cell transformation: The importance of proteases and cell surface proteins. *Cell* **1**:147–156.

Hynes, R. O. (1992). Integrins: Versatility, modulation, and signaling in cell adhesion. *Cell* **69**:11–25.

Hyvonen, T., Alakuijala, L., Andersson, L., Khomutov, A. R., Khomutov, R. M. & Eloranta, T. O. (1988). 1-Amino-oxy-3-aminopropane reversibly prevents the proliferation of cultured baby hamster kidney cells by interfering with polyamine synthesis. *J. Biol. Chem.* **263**:1138–1144.

Ikeno, M., Grimes B., Okazaki T., Nakano M., Saitoh K., Hoshino, H., McGill, N. I., Cooke, H., & Masumoto, H. (1998). Construction of YAC-based mammalian artificial chromosomes. *Nature Biotechnology.* **16**(5):431–439.

Ilsie, A. W., & Puck, T. T. (1971). Morphological transformation of Chinese hamster cells by dibutyryl adenosine cycline 3′:5′-monophosphate and testosterone. *Proc. Natl. Acad. Sci. USA* **2**:358–361.

Inamatsu, M., Matsuzaki, T., Iwanari, H., & Yoshizato, K. (1998). Establishment of rat dermal papilla cell lines that sustain the potency to induce hair follicles from afollicular skin. *J. Invest. Dermatol.* **111**:767–775.

Inokuchi, S., Handa, H., Imai, T., Makuuchi, H., Kidokoro, M., Tohya, H., Aizawa, S., Shimamura, K., Ueyama, Y., Mitomi, T., & Sawada, Y. (1995). Immortalisation of human oesophageal epithelial cells by a recombinant SV40 adenovirus vector. *Br. J. of Cancer* **71**:819–825.

Inoué, S., & Spring, K. R. (1997). *Video microscopy*, 2d ed. New York and London, Plenum Press.

International System for Human Cytogenetic Nomenclature (1978). *Report of the Standing Committee on Human Cytogenetic Nomenclature.* Washington, DC, National Foundation of the March of Dimes.

Ireland, G. W., Dopping-Hepenstal, P. J., Jordan, P. W., & O'Neill, C. H. (1989). Limitation of substratum size alters cytoskeletal organization and behaviour of Swiss 3T3 fibroblasts. *Cell Biol. Int. Rep.* **13**:781–790.

Iscove, N., & Melchers, F. (1978). Complete replacement of serum by albumin, transferrin and soybean lipid in cultures of lipopolysaccharide-reactive B lymphocytes. *J. Exp. Med.* **147**:923–933.

Iscove, N. N., Guilbert, L. W., & Weyman, C. (1980). Complete replacement of serum in primary cultures of erythropoitin-dependent red cell precursors (CFU-E) by albumin, transferrin, iron, unsaturated fatty acid, lecithin and cholesterol. *Exp. Cell Res.* **126**:121–126.

Ishida, Y., Levin, J., Baker, G., Stenberg, P., Yamada, Y., Sasaki, H., & Inoué, T. (1993). Biological and biochemical characteristics of murine megakaryoblastic cell line L8057. *Exp. Hematol.* **21**:289–298.

Itagaki, A., & Kimura, G. (1974). TES and HEPES buffers in mammalian cell cultures and viral studies: Problems of carbon dioxide requirements. *Exp. Cell Res.* **83**:351–360.

Iyer, V. R., Eisen, M. B., Ross, D. T., Schuler, G., Moore, T., Lee, J. C. F., Trent, J. M., Staudt, L. M., Hudson, J., Jr., Boguski, M. S., Lashkari, D., Shalon, D., Botstein, D., & Brown, P. O. (1999). The Transcriptional Program in the Response of Human Fibroblasts to Serum. *Science*, **283**: 83–87.

Izutsu, K. T., Fatherazi, S., Belton, C. M., Oda, D., Cartwright, F. D., & Kenny, G. E. (1996). Mycoplasma orale infection affects K and Cl currents in the HSG salivary gland cell line. *In vitro Cell Dev. Biol.—Animal* **32**:361–365.

Jacobs, J. P. (1970). Characteristics of a human diploid cell designated MRC-5. *Nature* **227**:168–170.

Jacobs, J. P., Garrett, A. J., & Meron, R. (1979). Characteristics of a serially propagated human diploid cell designated MRC-9. *J. Biol. Stand.* **7**:113–122.

Jaffe, E. A., Nachman, R. L., Becker, G. C., & Ninick, C. R. (1973). Culture of human endothelial cells derived from umbilical veins. *J. Clin. Invest.* **52**:2745–2756.

Jain, D., Ramasubramamanyan, K., Gould, S., Lenny, A., Candelore, M., Tota, M., Strader, C., Alves, K., Cuca, C., Tung, J. S., Hunt, G., Junker, B., Buckland, B. C., & Silberklang, M. (1991). In Speir, R. E., Griffiths, J. B., & Meignier, B. (eds.), *Production of biologicals from animal cells in culture.* Oxford, U.K., Butterworth–Heinemann, pp. 345–351.

Jain, R. K. Schlenger, K., Hockel, M., & Yuan, F. (1997). Quantitative angiogenesis assays: Progress and problems. *Nature Medicine* **3**:1203–1208.

Jeffreys, A. J., Wilson, V., & Thein, S. L. (1985). Individual specific "fingerprints" of human DNA. *Nature* **316**:76–79.

Jeng, Y.-J., Watson, C. S., & Thomas, M. L. (1994). Identification of vitamin D-stimulated phosphatase in IEC-6 cells, a rat small intestine crypt cell line. *Exp. Cell Res.* **212**:338–343.

Jenkins, N. (ed.). (1992). *Growth factors, a practical approach.* Oxford, U.K., IRL Press at Oxford University Press.

Jessell, T. M., & Melton, D. A. (1992). Diffusible factors in vertebrate embryonic induction. *Cell* **68**:257–270.

Jinard, F., Sergent-Engelen, T., Trouet, A., Remacle, C., & Schneider, Y.-J. (1997). Compartment coculture of procine arterial endothelial and smooth muscle cells on a microporus membrane. *In Vitro Cell Dev. Biol.—Animal* **33**: 92–103.

Johnson, G. D. (1989). Immunofluorescence. In Catty, D. (ed.), *Antibodies; Vol. II: A practical approach.* Oxford, U.K., IRL Press at Oxford University Press, pp. 179–200.

Jones, E. L., & Gregory, J. (1989). Immunoperoxidase methods. In Catty, D. (ed.), *Antibodies; Vol. II: A practical approach.* Oxford, U.K., IRL Press at Oxford University Press, pp. 155–177.

Joukov, V., Kaipainen, A., Jeltsch, M., Pajusola, K., Olofsson B., Kumar, V., Eriksson, U., & Alitalo, K. (1997). Vascular endothelial growth factors VEGF-B and VEGF-C. *J. Cell Physiol.* **173**:211–215.

Jozan, S., Roche, H., Cheutin, F., Carton, M., & Salles, B. (1992). New human ovarian cell line OVCCR1/sf in serum-free medium. *In Vitro Cell Dev. Biol.* **28A**:687–689.

Kaartinen, L., Nettesheim, P., Adler, K. B., & Randell, S. H. (1993). Rat tracheal epithelial cell differentiation *in vitro*. *In Vitro Cell Dev. Biol.* **29A**:481–492.

Kahn, P., & Shin, S.-L. (1979). Cellular tumorigenicity in nude mice: Test of association among loss of cell-surface fibronectin, anchorage independence, and tumor-forming ability. *J. Cell Biol.* **82**:1.

Kaltenbach, J. P., Kaltenbach, M. H., & Lyons, W. B. (1958). Nigrosin as a dye for differentiating live and dead ascites cells. *Exp. Cell Res.* **15**:112–117.

Kaminska, B., Kaczmarek, L., & Grzelakowska-Sztabert, B. (1990). The regulation of G0–S transition in mouse T lymphocytes by polyamines. *Exp. Cell Res.* **191**:239–245.

Kane, D. A., Warga, R. M., & Kimmel, C. B. (1992). Mitotic domains in the early embryo of the zebrafish. *Nature* **360**: 735–737.

Kao, F. T., Chasin, L., & Puck, T. T. (1969). Genetics of somatic mammalian cells; X: Complementation analysis of glycine-requiring auxotrophs. *Proc. Natl. Acad. Sci. USA* **64**:1284–1291.

Kao, F. T., & Puck, T. T. (1968). Genetics of somatic mammalian cells; VII: Induction and isolation of nutritional mutants in Chinese hamster cells. *Proc. Nalt. Acad. Sci. USA* **60**:1275–1281.

Kao, W.-Y., & Prockop, D. I. (1977). Proline analogue removes fibroblasts from cultured mixed cell populations. *Nature* **266**:63–64.

Karenberg, J. R., & Freelander, E. F. (1974). Giemsa technique for the detection of sister chromatid exchanges. *Chromosoma* **48**:355–360.

Kawa, S., Kimura, S., Hakomori, S., & Igarashi, Y. (1997). Inhibition of chemotactic motility and trans-endothelial migration of human neutrophils by sphingosine 1-phosphate. *FEBS Letters.* **420**:196–200.

Kédinger, M., Simon-Assmann, P., Alexandre, E., & Haffen, K. (1987). Importance of a fibroblastic support for *in vitro* differentiation of intestinal endodermal cells and for their response to glucocorticoids. *Cell Diff.* **20**:171–182.

Keen, M. J. & Rapson, N. T. (1995). Development of a serum-free culture medium for the large scale production of recombinant protein from a Chinese hamster ovary cell line. *Cytotechnology* **17**:153–163.

Keles, G. E., Berger, M. S., Lim, R., & Zaheer, A. (1992). Expression of glial fibrillary acidic protein in human medulloblastoma cells treated with recombinant glia maturation factor-beta. *Oncology Research* **4**:431–437.

Kelley, D. S., Becker, J. E., & Potter, V. R. (1978). Effect of insulin, dexamethasone, and glucagon on the amino acid transport ability of four rat hepatoma cell lines and rat hepatocytes in culture. *Cancer Res.* **38**:4591–4601.

Kempson, S. A., McAteer, J. A., Al-Mahrouq, H. A., Dousa, T. P., Dougherty, G. S., & Evan, A. P. (1989). Proximal tubule characteristics of cultured human renal cortex epithelium. *J. Lab. Clin. Med.* **113**:285–296.

Kenworthy, P., Dowrick, P., Baillie-Johnson, H., McCann, B., Tsubouchi, H., Arakaki, N., Daikuhara, Y., & Warn, R. M. (1992). The presence of scatter factor in patients with metastatic spread to the pleura. *Br. J. Cancer* **66**:243–247.

Kern, P. A., Knedler, A., & Eckel, R. H. (1983). Isolation and culture of microvascular endothelium from human adipose tissue. *J. Clin. Invest.* **71**:1822–1829.

Khan, M. Z., Spandidos, D. A., Kerr, D. J., McNicol, A. M., Lang, J. C., de Ridder, L., & Freshney, R. I. (1991). Oncogene transfection of mink lung cells: effect on growth characteristics *in vitro* and *in vivo*. *Anticancer Res.* **11**:1343–1348.

Kibbey, M. C., Royce, L. S., Dym, M., Baum, B. J., & Kleinman, H. K. (1992). Glandular-like morphogenesis of the human submandibular tumour cell line A253 on basement membrane components. *Exp. Cell Res.* **198**:343–351.

Kimhi, Y. H., Palfrey, C., & Spector, I. (1976). Maturation of neuroblastoma cells in the presence of dimethyl sulphoxide. *Proc. Natl. Acad. Sci. USA* **73**:462–466.

Kinard, F., De Clercq, L., Billen, B., Amory, B., Hoet, J.-J., & Remacle, C. (1990). Culture of endocrine pancreatic cells in protein-free chemically defined media. *In Vitro Cell Dev. Biol.* **26**:1004–1010.

Kingsbury, A., Gallo, V., Woodhams, P. L., & Balazs, R. (1985). Survival, morphology and adhesion properties of cerebellar interneurons cultured in chemically defined and serum-supplemented medium. *Dev. Brain Res.* **17**:17–25.

Kinsella, J. L., Grant, D. S., Weeks, B. S., & Kleinman, H. K. (1992). Protein kinase C regulates endothelial cell tube formation on basement membrane matrix, Matrigel. *Exp. Cell Res.* **199**:56–62.

Kirkland, S. C., & Bailey, I. G. (1986). Establishment and characterisation of six human colorectal adenocarcinoma cell lines. *Br. J. Cancer* **53**:779–785.

Klagsbrun, M., & Baird, A. (1991). A dual receptor system is required for basic fibroblast growth factor activity. *Cell* **67**:229–231.

Klein, B., Pastink, A., Odijk, H., Westerveld, A., & van der Eb, A. J. (1990). Transformation and immortalization of diploid xeroderma pigmentosum fibroblasts. *Exp. Cell Res.* **191**:256–262.

Kleinman, H. K., McGoodwin, E. B., Rennard, S. I., & Martin, G. R. (1979). Preparation of collagen substrates for cell attachment: Effect of collagen concentration and phosphate buffer. *Anal. Biochem.* **94**:308–312.

Klevjer-Anderson, P., & Buehring, G. C. (1980). Effect of hormones on growth rates of malignant and nonmalignant human mammary epithelia in cell culture. *In Vitro* **16**:491–501.

Klingel, S., Rothe, G., Kellermann, W., & Valet, G. (1994). Flow cytometric determination of cysteine and serine proteinase activities in living cells with rhodamine 110 substrates. *Methods in Cell Biology* **41**:449–459.

Klöppinger, M., Fertig, G., Fraune, E., & Miltenburger, H. G. (1991). High cell density perfusion culture of insect cells for production of baculovirus and recombinant protein. In Speir, R. E., Griffiths, J. B., Meignier, B. (eds.): *Production of biologicals from animal cells in culture.* Oxford, U.K., Butterworth–Heinemann, pp. 470–474.

Knazek, R. A., Gullino, P., Kohler, P. O., & Dedrick, R. (1972). Cell culture on artificial capillaries: An approach to tissue growth *in vitro. Science* **178**:65–67.

Knazek, R. A., Kohler, P. O., & Gullino, P. M. (1974). Hormone production by cells grown *in vitro* on artificial capillaries. *Exp. Cell Res.* **84**:251.

Knedler, A., & Ham, R. G. (1987). Optimized medium for clonal growth of human microvascular endothelial cells with minimal serum. *In Vitro* **23**(7):481–491.

Kneuchel, R., & Masters, J. R. W. (1999). Bladder Cancer. In Masters, J. R. W., & Palsson, B. (eds.), *Human Cell Culture,* Vol. I, Kluwer, Dordrecht, pp. 213–230.

Knott, C. L., Kuus-Reichel, K., Liu, R., & Wolfert, R. L. (1997). Development of antibodies for diagnostic assays. In Price, C., & Newman, D. (eds.), *Principles and Practice of Immunoassay,* 2d ed. New York, Stockton Press, pp. 36–64.

Knowles, B. B., Howe, C. C., & Aden, D. P. (1980). Human hepatocellular carcinoma cell lines secrete the major plasma proteins and hepatitis B surface antigen. *Science* **209**:497–499.

Kohler, G., & Milstein, C. (1975). Continuous cultures of fused cells secreting antibody of predefined specificity. *Nature* **256**:495–497.

Kohlhepp, E. A., Condon, M. E., & Hamburger, A. W. (1987). Recombinant human interferon-α enhancement of reti-

noic acid induced differentiation of HL-60 cells. *Exp. Hematol.* **15**:414–418.

Kondo, S., Kooshesh, F., & Sauder, D. N. (1997). Penetration of keratinocyte-derived cytokines into basement membrane. *J. Cell Physiol.* **171**:190–195.

Kono, T. (1997). Nuclear transfer and reprogramming. *Rev. Reprod.* **2**:74–80.

Koren, H. S., Handwerger, B. S., & Wunderlich, J. R. (1975). Identification of macrophage-like characteristics in a murine tumor cell line. *J. Immunol.* **114**:894–897.

Korenberg, J. R., Chen, X. N., Adams, M. D., & Venter, J. C. (1995). Toward a cDNA map of the human genome. *Genomics* **29**:364–370.

Koschier, F. J., Roth, R. N., Wallace, K. A., Curren, R. D., & Harbell, J. W. (1997). A comparison of three-dimensional human skin models to evaluate the dermal irritation of selected petroleum products. *In Vitro Toxicology* **10**:391–405.

Kralovanszky, J., Harrington, F., Greenwell, A. (1990). Isolation of viable intestinal epithelial cells and their use for *in vitro* toxicity studies. *In Vivo* **4**:201–204.

Kreisberg, J. L., Sachs, G., Pretlow, T. G. E., & McGuire, R. A. (1977). Separation of proximal tubule cells from suspensions of rat kidney cell by free-flow electrophoresis. *J. Cell. Physiol.* **93**:169–172.

Kruse, P. F., Jr., Keen, L. N., & Whittle, W. L. (1970). Some distinctive characteristics of high density perfusion cultures of diverse cell types. *In Vitro* **6**:75–78.

Kucherlapati, R., & Skoultchi, A. (1984). Introduction of purified genes into mammalian cells. *CRC Critical Reviews in Biochemistry* **16**:349–379.

Kujoth, G. C., & Fahl, W. E. (1997). c-sis/platelet-derived growth factor-B promoter requirements for induction during the 12-O-tetradecanoylphorbol-13-acetate-mediated megakaryoblastic differentiation of K562 human erythroleukemia cells. *Cell Growth Diff.* **8**:963–977.

Kuriharcuch, W., & Green, H. (1978). Adipose conversion of 3T3 cells depends on a serum factor. *Proc. Natl. Acad. Sci. USA* **75**:6107–6110.

Kurtz, J. W., & Wells, W. W. (1979). Automated fluorometric analysis of DNA, protein, and enzyme activities: Application of methods in cell culture. *Anal. Biochem.* **94**:166.

Labarca, C., & Paigen, K. (1980). A simple, rapid, and sensitive DNA assay procedure. *Anal. Biochem.* **102**:344–352.

Laferte S., & Loh, L. C. (1992). Characterization of a family of structurally related glycoproteins expressing beta 1-6-branched asparagine-linked oligosaccharides in human colon carcinoma cells. *Biochem. J.* **283**:192–201.

Lag, M., Becher, R., Samuelsen, J. T., Wiger, R., Refsnes, M., Huitfeldt, H. S., & Schwarze, P. E. (1996). Expression of CYP2B1 in freshly isolated and proliferating cultures of epithelial rat lung cells. *Exp. Lung Res.* **22**:627–649.

Lamb, R. F., Hennigan, R. F., Katsanakis, K. D., Turnbull, K., MacKenzie, E. D., Birnie, G. D., & Ozanne, B. W. (1997). AP-1-mediated invasion requires increased expression of the hyaluronan receptor, CD44. *Mol. Cell Biol.* **17**:963–976.

Lan, S., Smith, H. S., & Stampfer, M. R. (1981). Clonal growth of normal and malignant human breast epithelia. *J. Surg. Oncol.* **18**:317–322.

Lane, E. B. (1982). Monoclonal antibodies provide specific intramolecular markers for the study of tonofilament organisation. *J. Cell Biol.* **92**:665–673.

Lang, M. S., Hovenkamp, E., Savelkoul, H. F. J., Knegt, P., & van Ewijk, W. (1995). Immunotherapy with monoclonal antibodies directed against the immunosuppressive domain of p15E inhibits tumour growth. *Clin Exp Immunol.* **102**:468–475.

Lange, W., Brugger, W., Rosenthal, F. M., Kanz, L., & Lindemann, A. (1991). The role of cytokines in oncology. *Int. J. Cell Clon.* **9**:252–273.

Lasfargues, E. Y. (1973). Human mammary tumors. In Kruse, P., & Patterson, M. K. (eds.), *Tissue culture methods and applications*. New York, Academic Press, pp. 45–50.

Lasnitzki, I. (1992). Organ culture. In Freshney, R. I. (ed.), *Animal cell culture, a practical approach*. Oxford, U.K., IRL Press at Oxford University Press, pp. 213–261.

Latt, S. A. (1973). Microfluorometric detection of DNA replication in human metaphase chromosomes. *Proc. Natl. Acad. Sci. USA* **70**:3395–3399.

Latt, S. A. (1981). Sister chromatid exchange formation. *Ann. Rev. Genet.* **15**:11–55.

Laug, W. E., Tokes, Z. A., Benedict, W. F., & Sorgente, N. (1980). Anchorage independent growth and plasminogen activator production by bovine endothelial cells. *J. Cell Biol.* **84**:281–293.

Lavappa, K. S. (1978). Survey of ATCC stocks of human cell lines for HeLa contamination. *In Vitro* **14**(5):469–475.

Law, L. W., Dunn, T. B., Boyle, P. J., & Miller, J. H. (1949). Observations on the effect of a folic acid antagonist on transplantable lymphoid leukemia in mice. *J. Natl. Cancer Inst.* **10**:179–192.

Le Poole, I. C., van den Berg, F. M., van den Wijngaard, R. M., Galloway, D. A., van Amstel, P. J., Buffing, A. A., Smiths, H. L., Westerhof, W., Das, P. K. (1997). Generation of a human melanocyte cell line by introduction of HPV16 E6 and E7 genes. *In Vitro Cell Dev. Biol.—Animal* **33**:42–49.

Le Roith, D., Raizada, M. K. (eds.). (1989). *Molecular and cellular biology of insulin-like growth factors and their receptors*. New York, Plenum.

Leake, R. E., Freshney, R. I., & Munir, I. (1987). Steroid responses *in vivo* and *in vitro*. In Green, B., & Leake, R. E., (eds.), *Steroid hormones, a practical approach*, Oxford, U.K., IRL Press at Oxford University Press, pp. 205–218.

Lebeau, M. M., & Rowley, J. D. (1984). Heritable fraile sites in cancer. *Nature* **308**:607–608.

Lechardeur, D., Schwartz, B., Paulin, D., & Scherman, D. (1995). Induction of blood–brain barrier differentiation in a rat brain-derived endothelial cell line. *Exp. Cell Res.* **220**:161–170.

Lechner, J. F., Haugen, A., Autrup, H., McClendon, I. A., Trump, B. F., & Harris, C. C. (1981). Clonal growth of epithelial cells from normal adult human bronchus. *Cancer Res.* **41**:2294–2304.

Lechner, J. F., & LaVeck, M. A. (1985). A serum free method for culturing normal human bronchial epithelial cells at clonal density. *J. Tissue Cult. Methods* **9**:43–48.

Leder, A., & Leder, P. (1975). Butyric acid, a potent inducer of erythroid differentiation in cultured erythroleukemic cells. *Cell* **5**:319–322.

Lee, T. H., Baik, M. G., Im, W. B., Lee, C. S., Han, Y. M., Kim, S. J., Lee, K. K., & Choi, Y. J. (1996). Effects of EHS matrix on expression of transgenes in HC11 cells. *In Vitro Cell Dev. Biol.—Animal* **32**:454–456.

Leibo, S. P., & Mazur, P. (1971). The role of cooling rates in low-temperature preservation. *Cryobiology* **8**:447–452.

Leibovitz, A. (1963). The growth and maintenance of tissue cell cultures in free gas exchange with the atmosphere. *Am. J. Hyg.* **78**:173–183.

Leigh, I. M., & Watt, F. M. (1994). *Keratinocyte methods*. Cambridge, U.K., Cambridge University Press.

Leighton, J. (1991). Radial histophysiologic gradient culture chamber rationale and preparation. *In Vitro Cell Dev. Biol.* **27A**:786–790.

Leighton, J., Mark, R., & Rush, G. (1968). Patterns of three-dimensional growth in collagen coated cellulose sponge: Carcinomas and embryonic tissues. *Cancer Res.* **28**:286–296.

Lemare, F., Steimberg, N., Le Griel, C., Demignot, S., and Adolphe, M. (1998). Dedifferentiated chondrocytes cultured in alginate beads: Restoration of the differentiated phenotype and of the metabolic response to Interleukin-1β. *J. Cell Physiol.* **176**:303–313.

Lesuffleur, T., Barbat, A., Dussaulx, E., & Zweibaum, A. (1990). Growth adaptation to methotrexate of ht-29 human colon-carcinoma cells is associated with their ability to differentiate into columnar absorptive and mucus-secreting cells. *Cancer Res.* **50**:6334–6343.

Lever, J. (1986). Expression of differentiated functions in kidney epithelial cell lines. *Min. Elec. Metab.* **12**:14–19.

Levi-Montalcini, R. C. P. (1979). The nerve-growth factor. *Sci. Am.* **240**:68.

Levin, D. B., Wilson, K., Valadares de Amorim, G., Webber, J., Kenny, P., & Kusser, W. (1995). Detection of p53 mutations in benign and dysplastic nevi. *Cancer Res.* **55**:4278–4282.

Levine, E. M., & Becker, B. G. (1977). Biochemical methods for detecting mycoplasma contamination. In McGarrity, G. T., Murphy, D. G., & Nichols, W. W. (eds.), *Mycoplasma infection of cell cultures*. New York, Plenum Press, pp. 87–104.

Ley, K. D., & Tobey, R. A. (1970). Regulation of initiation of DNA synthesis in Chinese hamster cells; II: Induction of DNA synthesis and cell division by isoleucine and glutamine in G_1-arrested cells in suspension culture. *J. Cell Biol.* **47**:453–459.

Li, A. P., Roque, M. M., Beck, D. J., & Kaminski, D. L. (1992). Isolation and culturing of hepatocytes from human livers. J. Tissue Cult. Methods **14**:139–146.

Li, Y., Field, P. M., & Raisman, G. (1997). Repair of adult rat corticospinal tract by transplants of olfactory ensheathing cells. *Science* **277**:2000–2002.

Lieberman, D., & Sachs, L. (1978). Nuclear control of neurite induction in neuroblastoma cells. *Exp. Cell Res.* **113**:383–390.

Lillie, I. H., MacCallum, D. K., & Jepsen, A. (1980). Fine structure of subcultivated stratified squamous epithelium grown on collagen rafts. *Exp. Cell Res.* **125**:153–165.

Limat, A., Breitkreutz, D., Thiekötter, G., Klein, E. C., Braathen, L. R., Hunziker, T., & Fusenig, N. E. (1995). Formation of a regular neo-epidermis by cultured human outer root sheath cells grafted on nude mice. *Transplantation* **59**:1032–1038.

Limat, A., Hunziker, T., Boillat, C., Bayreuther, K., & Noser, F. (1989). Post-mitotic human dermal fibroblasts efficiently support the growth of human follicular keratinocytes. *J. Invest. Dermatol.* **92**:758–762.

Limat, A., Mauri, D., & Hunziker, T. (1996). Successful treatment of chronic leg ulcers with epidermal equivalents generated from cultured autologous outer root sheath cells. *J. Invest. Dermatol.* **107**:128–135.

Lin, C. C., & Uchida, I. A. (1973). Fluorescent banding of chromosomes (Q-bands). In Kruse, P. F., & Patterson, M. K. (eds.), *Tissue Culture Methods and Applications.* New York, Academic Press, pp. 778–781.

Lin, M. A., Latt, S. A., & Davidson, R. L. (1974). Identification of human and mouse chromosomes in human-mouse hybrids by centromere fluorescence. *Exp. Cell Res.* **87**:429–433.

Linser, P., & Moscona, A. A. (1980). Induction of glutamine synthetase in embryonic neural retina-localization in Muller fibers and dependence on cell interaction. *Proc. Natl. Acad. Sci. USA* **76**:6476–6481.

Liotta, L. (1987). The role of cellular proteases and their inhibitors in invasion and metastasis: Introductionary overview. *Cancer Metastasis Rev.* **9**:285–287.

Lippincott-Schwartz, J., Glickman, J., Donaldson, J. G., Robbins, J., Kreis, T. E., Seamon, K. B., Sheetz, M. P., Klausner, R. D. (1991). Forskolin inhibits and reverses the effects of Brefeldin A on Golgi morphology by a cAMP-independent mechanism. *J. Cell Biol.* **112**:567.

Littlefield, J. W. (1964a). Selection of hybrids from matings of fibroblasts *in vitro* and their presumed recombinants. *Science* **145**:709–710.

Littlefield, J. W. (1964b). Three degrees of guanylic acid pyrophosphorylase deficiency in mouse fibroblasts. *Nature* **203**:1142–1144.

Litwin, J. (1973). Titanium disks. In Kruse, P. F., & Patterson, M. K. (eds.), *Tissue culture methods and applications.* New York, Academic Press, pp. 383–387.

Liu, L., Delbe, J., Blat, C., Zapf, J., & Harel, L. (1992). Insulin like growth factor binding protein (IGFBP-3), an inhibitor of serum growth factors other than IGF-I and -II. *J. Cell Physiol.* **153**:15–21.

Liu, M., Xu, J., Souza, P., Tanswell, B., Tanswell, A. K., & Post, M. (1995). The effect of mechanical strain on fetal rat lung cell proliferation: Comparison of two- and three-dimensional culture systems. *In Vitro Cell Dev. Biol.—Animal* **31**:858–866.

Lodish, H., Baltimore, D., Berk, A., Zipursky, S. L., Matsudaira, P., & Darnell, J. (1995). *Molecular cell biology.* Scientific American Books. New York, Freeman. (New ed. Nov. 1999)

Lopez-Casillas, F., Wrana, J. L., & Massague, J. (1993). Betaglycan presents ligand to the TGF-β signalling receptor. *Cell* **73**:1435–1444.

Lotan, R., & Lotan, D. (1980). Simulation of melanogenesis in a human melanoma cell line by retinoids. *Cancer Res.* **40**:33–45.

Lounis, H., Provencher, D., Godbout, C., Fink, D., Milot, M.-J., & Mes-Masson, A.-M. (1994). Primary cultures of normal and tumoral ovarian epithelium: A powerful tool for basic molecular studies. *Exp. Cell Res.* **215**:303–309.

Lovelock, J. E., & Bishop, M. W. H. (1959). Prevention of freezing damage to living cells by dimethyl sulphoxide. *Nature* **183**:1394–1395.

Luikart, S. D., Maniglia, C. A., Furcht, L. T., McCarthy, J. B., & Oegama, T. R. (1990). A heparan sulphate-containing fraction of bone marrow stroma induces maturation of HL-60 cell *in vitro. Cancer Res.* **50**:3781–3785.

Lundqvist, M., Mark, J., Funa, K., Heldin, N. E., Morstyn, G., Weddell, B., Layton, J., & Oberg, K. (1991). Characterisation of a cell line (LCC-18) from a cultured human neuroendocrine-differentiated colonic carcinoma. *Eur. J. Cancer* **12**:1662–1668.

Lutz, M. P., Gaedicke, G., & Hartmann, W. (1992). Large-scale cell separation by centrifugal elutriation. *Anal. Biochem.* **200**:376–380.

Maas-Szabowski, N., & Fusenig, N. E. (1996). Interleukin-1-induced growth factor expression in postmitotic and resting fibroblasts. *J. Invest. Dermatol.* **107**:849–855.

Maas-Szabowski, N., Fusenig, N. E., & Shimotoyodome, A. (1999). Keratinocyte growth regulation in fibroblast cocultures via a double paracrine mechanism. *J. Cell Sci.* **112**:1843–1853.

MacDonald, C. M., Freshney, R. I., Hart, E., & Graham, D. I. (1985). Selective control of human glioma cell proliferation by specific cell interaction. *Exp. Cell Biol.* **53**:130–137.

Macé, K., Gonzalez, F. J., McConnell, I. R., Garner, R. C., Avanti, O., Harris, C. C., & Pfeifer, A. M. A. (1994). Activation of promutagens in a human bronchial epithelial cell line stably expressing human cytochrome P450 1A2. *Mol. Carcinogen.* **11**:65–73.

MacGregor, G., & Caskey, C. (1989). Construction of plasmids that express *E. coli* beta-galactosidase in mammalian cells. *Nucleic Acids Res.* **17**:2363–2365.

Maciag, T., Cerondolo, J., Ilsley, S., Kelley, P. R., & Forand, R. (1979). Endothelial cell growth factor from bovine hypothalamus—identification and partial characterization. *Proc. Natl. Acad. Sci. USA* **76**:5674–5678.

Macieira-Coelho, A. (1973). Cell cycle analysis; A: Mammalian cells. In Kruse, P. F., & Patterson, M. K. (eds.), *Tissue culture methods and applications.* New York, Academic Press, pp. 412–422.

Macklis, J. D., Sidman, R. L., & Shine, H. D. (1985). Cross-linked collagen surface for cell culture that is stable, uniform, and optically superior to conventional surfaces. *In Vitro* **21**:189–194.

MacLeod, R. A. F., Dirks, W. G., Kaufmann, M., Matsuo, Y., Milch, H., & Drexler, H. G. (1999). Widespread intraspecies cross-contamination of human tumor cell lines arising at source. *Int. J. Cancer* (In Press).

Macpherson, I. (1973). Soft agar technique. In Kruse, P. F., & Patterson, M. K. (eds.), *Tissue culture methods and applications.* New York, Academic Press, pp. 276–280.

Macpherson, I., & Bryden, A. (1971). Mitomycin C treated cells as feeders, *Exp. Cell Res.* **69**:240–241.

Macpherson, I., & Montagnier, L. (1964). Agar suspension culture for the selective assay of cells transformed by polyoma virus. *Virology* **23**:291–294.

Macpherson, I., & Stoker, M. (1962). Polyoma transformation of hamster cell clones—an investigation of genetic factors affecting cell competence. *Virology* **16**:147.

Macville, M., Veldman, T., Padilla-Nash, H., Wangsa, D., O'Brien, P., Schrock, E., & Ried, T. (1997). Spectral karyotyping, a 24-colour FISH technique for the identification of chromosomal rearrangements. *Histochemistry* **108**:299–305.

Macy, M. (1978). Identification of cell line species by isoenzyme analysis. *Manual Am. Tissue Cult. Assoc.* **4**:833–836.

Mahato, R. I., Kawabata, K., Nomura, T., Takakura, Y., & Hashida, M. (1995a). Physicochemical and pharmacokinetic characteristics of plasmid DNA/cationic liposome complexes. *Journal of Pharmaceutical Sciences* **84**:1267–1271.

Mahato, R. I., Kawabata, K., Takakura, Y., & Hashida, M. (1995b). *In vivo* disposition characteristics of plasmid DNA complexed with cationic liposomes. *Journal of Drug Targeting* **3**:149–157.

Mahdavi, V., & Hynes, R. O. (1979). Proteolytic enzymes in normal and transformed cells. *Biochim. Biophys. Acta* **583**:167–178.

Mairs, R. J., & Wheldon, T. E. (1996). Experimental tumour therapy with targeted radionuclides in multicellular tumour spheroids. In Hagen, U., Jung, H., & Streffer, C. (eds.), *Radiation Research, 1895–1995* (Proceedings of the 10th International Congress of Radiation Research, Wurzburg, Germany).

Maltese, W. A., & Volpe, I. J. (1979). Induction of an oligodendroglial enzyme in C-6 glioma cells maintained at high density or in serum-free medium. *J. Cell Physiol.* **101**:459–470.

Management of Health and Safety at Work Regulations. (1992). Health and Safety Executive, Broad Lane, Sheffield S3 7HQ, England.

Maniatis, T., Hardison, R. C., Lacy, E., Lauer, J., O'Connell, C., Quon, D., Sim, G. K., & Efstradiadis, A. (1978). The isolation of structural genes from libraries of eukaryotic DNA. *Cell* **15**:687–701.

Maramorosch, K. (1976). *Invertebrate tissue culture.* New York, Academic Press.

Marchionni, M. A., Goodearl, A. D., Chen, M. S., Bermingham-McDonogh, O., Kirk, C., Hendricks, M., Danehy, F., Misumi, D., Sudhalter, J., Kobayashi, K., Wroblewski, D., Lynch, C., Baldassare, M., Hiles, I., Davis, J. B., Hsuan, J. J., Totty, N. F., Otsu, M., McBurney, R. N., Waterfield, M. D., Stroobant, P., & Gwynne, D. (1993). Glial growth factors are alternatively spliced erbB2 ligands expressed in the nervous system. *Nature* **362**:312–318.

Marcus, M., Lavi, U., Nattenberg, A., Ruttem, S., & Markowitz, O. (1980). Selective killing of mycoplasmas from contaminated cells in cell cultures. *Nature* **285**:659–660.

Mardh, P. H. (1975). Elimination of mycoplasmas from cell cultures with sodium polyanethol sulphonate. *Nature* **254**:515–516.

Mareel, M., Kint, J., & Meyvisch, C. (1979). Methods of study of the invasion of malignant C3H-mouse fibroblasts into embryonic chick heart *in vitro*. *Virchows Arch. B Cell Pathol.* **30**:95–111.

Mareel, M. M., Bruynell, E., & Storme, G. (1980). Attachment of mouse fibrosarcoma cells to precultured fragments of embryonic chick heart. *Virchows Arch. B Cell Pathol.* **34**:85–97.

Mark, J. (1971). Chromosomal characteristics of neurogenic tumours in adults. *Hereditas* **68**:61–100.

Marks, P. A., Richon, V. M., Kiyokawa, H., & Rifkind, R. A. (1994). Inducing differentiation of transformed cells with hybrid polar compounds: a cell cycle-dependent process. *Proc. Nat. Acad. Sci. USA* **91**.10251 10254

Markus, G., Takita, H., Camiolo, S. M., Corsanti, J., Evers, J. L., & Hobika, J. H. (1980). Content and characterization of plasminogen activators in human lung tumours and normal lung tissue. *Cancer Res.* **40**:841–848.

Marshall, C. J. (1991). Tumor suppressor genes. *Cell* **64**:313–326.

Marte, B. M., Meyer, T., Stabel, S., Standke, G. J. R., Jaken, S., Fabbro D., & Hynes, N. E. (1994). Protein kinase C and mammary cell differentiation: Involvement of protein kinase C alpha in the induction of beta-casein expression. *Cell Growth Diff.* **5**:239–247.

Martin, G. R. (1975). Teratocarcinomas as a model system for the study of embryogenesis and neoplasia. *Cell* **5**:229–243.

Martin, G. R., (1978). Advantages and limitations of teratocarcinoma stem cells as models of development. In Johnson, M. H. (ed.), *Development in mammals*, Vol. 3. Amsterdam, North-Holland Publishing, p. 225.

Martin, G. R., & Evans, M. J. (1974). The morphology and growth of a pluripotent teratocarcinoma cell line and its derivatives in tissue culture. *Cell* **2**:163–172.

Martinsen, A., Skjåk-Bræk, G., Smidsrød, O. (1989). Alginate as immobilization material; 1: Correlation between chemical and physical properties of alginate gel beads. *Biotechnol. Bioeng.* **33**:79–89.

Maskell, R., & Green, M. (1995). Applications of the comet assay technique. *Int. Micr. Lab.*, **6**:2–5.

Massague, J., Cheifetz, S., Laiho, M., Ralph, D. A., Weis, F. M., & Zentella, A. (1992). Transforming growth factor-beta. *Cancer Surveys* **12**:81–103.

Masson, E. A., Atkin, S. L., & White, M. C. (1993). D-valine selective medium does not inhibit human fibroblast growth *in vitro*. *In Vitro Cell Dev. Biol. Animal* **29A**:912–913.

Masters, J. R., Bedford, P., Kearney, A., Povey, S., & Franks, L. M. (1988). Bladder cancer cell line cross-contamination: Identification using a locus-specific minisatellite probe. *British Journal of Cancer* **57**(3):284–286.

Masters, J. R. W. & Palsson, B., (eds.) (1999). Human Cell Culture, Vol. I, Kluwer, Dordrecht.

Masui, T., Lechner, J. F., Yoakum, G. H., Willey, J. C., & Harris, C. C. (1986b). Growth and differentiation of normal and transformed human bronchial epithelial cells. *J. Cell Physiol. (Suppl)* **4**:73–81.

Masui, T., Wakefield, L. M., Lechner, J. F., LaVeck, M. A., Sporn, M. B., & Harris, C. C. (1986a). Type beta transforming growth factor is the primary differentiation-inducing serum factor for normal human bronchial epithelial cells. *Proc. Natl. Acad. Sci. USA* **83**:2438–2442.

Mather, J. (1979). Testicular cells in defined medium. In Jakoby, W. B., & Pastan, I. H. (eds.), *Methods in enzymology*, *Vol. 57: Cell culture*. New York, Academic Press, p. 103.

Mather, J. P. (1998). Making informed choices: Medium, serum, and serum-free medium; how to choose the appropriate medium and culture system for the model you wish to create. *Meth. Cell Biol.* **57**:19–30.

Mather, J. P., & Sato, G. H. (1979a). The growth of mouse melanoma cells in hormone supplemented, serum-free medium. *Exp. Cell Res.* **120**:191.

Mather, J.P., & Sato, G. H. (1979b). The use of hormone supplemented serum free media in primary cultures. *Exp. Cell Res.* **124**:215.

Matsui, A., Zsebo, K., & Hogan, B. L. M. (1992). Derivation of pluripotential embryonic stem cells from murine primordial germ cells in culture. *Cell* **70**:841–847.

Mayne, L. V., Price, T. N. C., Moorwood, K., & Burke, J. F. (1996). Development of immortal human fibroblast cell lines. In Freshney, R. I., & Freshney, M. G. (eds.), *Culture of immortalized cells*. New York, Wiley-Liss, pp. 77–93.

Mayne, L. V., Priestly, A., James, M. R., & Burke, J. F. (1986). Efficient immortalisation and morphological transformation of human fibroblasts with SV40 DNA linked to a dominant marker. *Exp. Cell Res.* **162**:530–538.

McAteer, J. A., Kempson, S. A., & Evan, A. P. (1991). Culture of human renal cortex epithelial cells. *J. Tissue Cult. Methods* **13**:143–148.

McCall, E., Povey, J., & Dumonde, D. C. (1981). The culture of vascular endothelial cells on microporous membranes. *Thromb. Res.* **24**:417–431.

McCormack, S. A., Viar, M. J., Tague, L., & Johnston, L. R. (1996). Altered distribution of the nuclear receptor rar (*β*) accompanies proliferation and differentiation changes caused by retinoic acid in Caco-2 cells. *In Vitro Cell Dev. Biol.—Animal* **32**:53–61.

McCormick C., & Freshney, R. I. (2000). Activity of growth factors in the IL-6 group in the differentiation of human lung adenocarcinoma. *Brit. J. Cancer* **82**:881–890.

McCormick, C., Freshney, R. I., & Speirs, V. (1995). Activity of interferon alpha, interleukin 6 and insulin in the regulation of differentiation in A549 alveolar carcinoma cells. *Brit. J. Cancer* **71**:232–239.

McGarrity, G. J. (1982). Detection of mycoplasmic infection of cell cultures. In Maramorosch, K. (ed.), *Advances in cell culture*, Vol. 2. New York, Academic Press, pp. 99–131.

McGowan, J. A. (1986). Hepatocyte proliferation in culture. In Guillouzo, A., & Guguen-Guillouzo, C. (eds.), *Isolated and cultured hepatocytes*. Paris, Les Editions Inserm, John Libbey Eurotext, pp. 13–38.

McGregor, D. B., Edwards, I, Riach, C. J., Cattenach, P., Martin, R., Mitchell, A. & Caspary, W. J. (1988). Studies of an S9 based metabolic activation system used in the mouse lymphoma L51768Y cell mutation assay. *Mutagenesis* **3**:485–490.

McIlwrath, A., Vasey, P., Ross, G., Brown, R. (1994). Cell cycle arrests and radiosensitity of human tumour cell lines: Dependence on wild-type p53 for radiosensitivity. *Cancer Research* **54**:3718–3722.

McKay, I., & Taylor-Papadimitriou, J. (1981). Junctional communication pattern of cells cultured from human milk. *Exp. Cell Res.* **134**:465–470.

McKeehan, W. L. (1977). The effect of temperature during trypsin treatment on viability and multiplication potential of single normal human and chicken fibroblasts. *Cell Biol. Int. Rep.* **1**:335–343.

McKeehan, W. L., Adams, P. S., & Rosser, M. P. (1982). Modified nutrient medium MCDB 151 (WJAC401), defined growth factors, cholera toxin, pituitary factors, and horse serum support epithelial cell and suppress fibroblast proliferation in primary cultures of rat ventral prostate cells. *In Vitro* **18**:87–91.

McKeehan, W. L., Adams, P. S., & Rosser, M. P. (1984). Direct mitogenic effects of insulin, epidermal growth factor, cholera toxin, unknown pituitary factors and possibly prolactin, but not androgen, on normal rat prostate epithelial cells in serum-free primary cell culture. *Cancer Res.* **44**:1998–2010.

McKeehan, W. L., & Ham, R. G. (1976a). Stimulation of clonal growth of normal fibroblasts with substrata coated with basic polymers. *J. Cell Biol.* **71**:727–734.

McKeehan, W. L., & Ham, R. G. (1976b). Methods for reducing the serum requirement of growth *in vitro* of non-transformed diploid fibroblasts. *Dev. Biol. Standard.* **37**: 97–98.

McKeehan, W. L., Hamilton, W. G., & Ham, R. G. (1976). Selenium is an essential trace nutrient for growth of WI-38 diploid human fibroblasts. *Proc. Natl. Acad. Sci. USA* **73**:2023–2027.

McKeehan, W. L., & McKeehan, K. A. (1979). Oxocarboxylic acids, pyridine nucleotide-linked oxidoreductases and serum factors in regulation of cell proliferation. *J. Cell Physiol.* **101**:9–16.

McKeehan, W. L., McKeehan, K. A., Hammond, S. L., & Ham, R. G. (1977). Improved medium for clonal growth of human diploid cells at low concentrations of serum protein. *In Vitro* **13**:399–416.

McLean, J. S., Frame, M. C., Freshney, R. I., Vaughan, P. F. T., & Mackie, A. E. (1986). Phenotypic modification of human glioma and non-small cell lung carcinoma by glucocorticoids and other agents. *Anticancer Res.* **6**:1101–1106.

Mege, R. M., Matsuzaki, F., Gallin, W. J., Goldberg, J. I., Cummingham, B. A., & Edelman, G. M. (1989). Construction of epithelioid sheets by transfection of mouse sarcoma cells with cDNAs for chicken cell adhesion molecules. *Proc. Natl. Acad. Sci. USA* **85**:7274–7278.

Melera, P. W., Wolgemuth, D., Biedler, J. L., & Hession, C. (1980). Antifolate-resistant Chinese hamster cells: Evidence from independently derived sublines for the overproduction of two dihydrofolate reductases encoded by different mRNAs. *J. Biol. Chem.* **255**:319–322.

Ment, L. R., Stewart, W. B., Scaramuzzino, D., & Madri, J. A. (1997). An *in vitro* three-dimensional coculture model of cerebral microvascular angiogenesis and differentiation. *In Vitro Cell Dev. Biol.—Animal* **33**:684–691.

Messing, E. M., Fahey, I. L., deKernion, I. B., Bhuta, S. M., & Bubbers, I. E. (1982). Serum-free medium for the *in vitro* growth of normal and malignant urinary bladder epithelial cells. *Cancer Res.* **42**:2392–2397.

Metcalf, D. (1990). The colony stimulating factors. Cancer **65**:2185–2195.

Meyer, W., Latouche, G. N., Daniel, H. M., Thanos, M., Mitchell, T. G., Yarrow, D., Schonian, G., & Sorrell, T. C. (1997). Identification of pathogenic yeasts of the imperfect genus *Candida* by polymerase chain reaction fingerprinting. *Electrophoresis* **18**:1548–1549.

Meyskens, F. L., & Fuller, B. B. (1980). Characterization of the effects of different retinoids on the growth and differentiation of a human melanoma cell line and selected subclones. *Cancer Res.* **40**:2194–2196.

Michalopoulos, G., & Pitot, H. C. (1975). Primary culture of parenchymal liver cells on collagen membranes: Morphological and biochemical observations. *Exp. Cell Res.* **94**:70–78.

Michler-Stuke, A., & Bottenstein, J. (1982). Proliferation of glial-derived cells in defined media. *J. Neurosci. Res.* **7**:215–228.

Midgley, C. A., Craig, A. L., Hite, J. P., & Hupp, T. R. (1998). Baculovirus expression and the study of the regulation of the tumor suppressor protein p53. In Ravid, K., & Freshney, R. I. (eds.), *DNA transfer to cultured cells.* New York, Wiley-Liss, pp. 27–54.

Mikulits, W., Dolznig, H., Edelmann, H., Sauer, T., Deiner, E. M., Ballou, L., Beug, H., & Mullner, E. W. (1997). Dynamics of cell cycle regulators: Artefact-free analysis by recultivation of cells synchronized by centrifugal elutriation. *DNA & Cell Biology* **16**:849–859.

Miller, G. G., Walker, G. W. R., & Giblack, R. E. (1972). A rapid method to determine the mammalian cell cycle. *Exp. Cell Res.* **72**:533–538.

Miller, R. G., & Phillips, R. A. (1969). Separation of cells by velocity sedimentation. *J. Cell Physiol.* **73**:191–201.

Mills, K. J., Volberg, T. M., Nervi, C., Grippo, J. F., Dawson, M. I., & Jetten, A. M. (1996). Regulation of retinoid-induced differentiation in embryonal carcinoma PCC4, aza1R cells: Effects of retinoid-receptor selective ligands. *Cell Growth & Differentiation* **7**:327–337.

Milo, G. E., Ackerman, G. A., & Noyes, I. (1980). Growth and ultrastructural characterization of proliferating human keratinocytes *in vitro* without added extrinsic factors. *In Vitro* **16**:20–30.

Minna, I. D., Carney, D. N., Cuttitta, F., & Gazdar, A. F. (1983). The biology of lung cancer. In Chabner, B. (ed.), *Rational basis for chemotherapy.* New York, Alan R. Liss.

Mirskey, R., Dubois, C., Morgan, L., & Jessen, K. R. (1990). O4 and A007 sufatide antibodies bind to embryonic Schwann cells prior to the appearance of galactocerebroside: Regulation by axon-Schwann cell signals and cyclic AMP. *Development* **109**:105–116.

Mitaka, T., Norioka, K.-I., & Mochizuki, Y. (1993). Redifferentiation of proliferated rat hepatocytes cultures in L15 medium supplemented with EGF and DMSO. *In Vitro Cell Dev. Biol.* **29A**:714–722.

Mitaka, T., Sattler, C. A., Sattler, G. L., Sargent, L. M., & Pitot, H. C. (1991). Multiple cell cycles occur in rat hepatocytes cultured in the presence of nicotinamide and epidermal growth factor. *Hepatology* **13**:21–30.

Mitelman, F., (ed.) (1995). ISCN. An international system for human cytogenetic nomenclature, Basel, S. Karger.

Miura, Y., Akimoto, T., Kanazawa, H., & Yagi, K. (1986). Synthesis and secretion of protein by hepatocytes entrapped within calcium alginate. *Artif. Organs* **10**:460–465.

Moll, R., Franke, W. W., & Schiller, D. L. (1982). The catalog of human cytokeratins: Patterns of expression in normal epithelia, tumours and cultured cells. **Cell 31**:11–24.

Montagnier, L. (1968). Correlation entre la transformation des cellule BHK21 et leur resistance aux polysaccharides acides en milieu gélifié. *CR Acad. Sci. D* **267**:921–924.

Montesano, R., Matsumonto, K., Nakamura, T., & Orci, L. (1991). Identification of a fibroblast-derived epithelial morphogen as hepatocyte growth factor. *Cell* **67**:901–908.

Montesano, R., Soriano, J. V., Pepper, M. S., & Orci, L. (1997). Induction of epithelial branching tubulogenesis *in vitro. J. Cell Physiol.* **173**:152–161.

Moore, A. E., Sabachewsky, L., & Toolan, H. W. (1955). *Cancer Res.* **15**:598.

Moore, G. E., Gerner, R. E., & Franklin, H. A. (1967). Culture of normal human leukocytes. *J. Am. Med. Assoc.* **199**:519–524.

Moreno, R. F. (1990). Enhanced conditions for DNA fingerprinting with biotinylated M13 bacteriophage. *J. Forensic Science* **35**:831–837.

Morgan, J. E., Moore, S. E., Walsh, F. S., & Partridge, T. A. (1992). Formation of skeletal muscle *in vivo* from the mouse C2 cell line. *J. Cell Sci.* **102**:779–787.

Morgan, J. G., Morton, H. J., & Parker, R. C. (1950). Nutrition of animal cells in tissue culture; I: Initial studies on a synthetic medium. *Proc. Soc. Exp. Biol. Med.* **73**:1.

Morton, H. J. (1970). A survey of commercially available tissue culture media. *In Vitro* **6**:89–108.

Moscona, A. A. (1952). Cell suspension from organ rudiments of chick embryos. *Exp. Cell Res.* **3**:535.

Moscona, A. A., & Piddington, R. (1966). Stimulation by hydrocortisone of premature changes in the developmental pattern of glutamine synthetase in embryonic retina. *Biochim. Biophys. Acta* **121**:409–411.

Mosmann, T. (1983). Rapid colorimetric assay for cellular growth and survival: Application to proliferation and cytotoxicity assays. *J. Immunol. Methods* **65**:55–63.

Moulton, D. G. (1974). Dynamics of cell populations in the olfactory epithelium. *Ann. NY Acad. Sci.* **237**:52–61.

Mowles, J. (1988). The use of ciprofloxacin for the elimination of mycoplasma from naturally infected cell lines. *Cytotechnology* **1**:355–358.

Muirhead, E. E., Rightsel, W. A., Pitcock, J. A., & Inagami, T. (1990). Isolation and culture of juxtaglomerular and renomedullary interstitial cells. *Methods Enzymol.* **191**:152–167.

Muller, U., Wang, D., Denda, S., Meneses, J. J., Pedersen, R. A., & Reichardt, L. F. (1997). Integrin alpha 8 beta 1 is critically important for epithelial–mesenchymal interactions during kidney morphogenesis. *Cell* **88**:603–613.

Mullin, J. M., Marano, C. W., Laughlin, K. V., Nuciglio, M., Stevenson, B. R., & Soler, A. P. (1997). Different size limitations for increased transepithelial paracellular solute flux across phorbol ester and tumour necrosis factor treated epithelial cell sheets. *J. Cell Physiol.* **171**:226–233.

Munthe-Kaas, A. C., & Seglen, P. O. (1974). The use of metrizamide as a gradient medium for isopycnic separation of rat liver cells. *FEBS Lett.* **43**:252–256.

Murakami, H. (1984). Serum-free cultivation of plasmacytomas and hybridomas. In Barnes, D. W., Sirbasku, D. A., & Sato, G. H. (eds.), *Methods for serum-free culture of neuronal and lymphoid cells.* New York, Alan R. Liss, pp. 197–206.

Murakami, H., & Masui, H. (1980). Hormonal control of human colon carcinoma cell growth in serum-free medium. *Proc. Natl. Acad. Sci. USA* **77**:3464–3468.

Murao, S., Gemmell, M. A., & Callaghan, M. F. (1983). Control of macrophage cell differentiation in human promyelocytic HL-60 leukemia cells by 1,25-dihydroxyvitamin D_3 and phorbol-12-myristate-13-acetate. *Cancer Res.* **43**:4989–4996.

Murphy, D. S., Hoare, S. F., Going, J. J., Mallon, E. E., George, W. D., Kaye, S. B., Brown, R., Black, D. M., & Keith, W. N. (1995). Characterization of extensive genetic alterations in ductal carcinoma *in situ* by fluorescence *in situ* hybridization and molecular analysis. *J. Natl. Cancer Inst.* **87**:1694–1704.

Murphy, S. J., Watt, D. J., & Jones, G. E. (1992). An evaluation of cell separation techniques in a model mixed cell population. *J. Cell Sci.* **102**:789–798.

Naeyaert, J. M., Eller, M., Gordon, P. R., Park, H.-Y., & Gilchrest, B. A. (1991). Pigment content of cultured human melanocytes does not correlate with tyrosinase message level. *Br. J. Dermatol.* **125**:297–303.

Nagaoka, S., Tansawa, H., & Suzuki, J. (1990). Cell proliferation of hydrogels. *In Vitro Cell Dev. Biol.* **26**:51–61.

Nardone, R. M., Todd, G., Gonzalez, P., & Gaffney, E. V. (1965). Nucleoside incorporation into strain L cells: Inhibition by pleuropneumonia-like organisms. *Science* **149**:1100–1101.

National Bioethics Advisory Commission (NBAC). (1999). *Research involving human biological materials: Ethical issues and policy guidance,* August 1999 (available on www.bioethics.gov/pubs.html).

National Research Council. (1989). *Biosafety in the laboratory: Prudent practices for the handling and disposal of infectious materials.* Washington, DC, National Academy Press.

National Sanitation Foundation. (1983). *Standard 49: Class II (laminar flow) biohazard cabinetry.* Ann Arbor, Michigan.

Nelson-Rees, W. A., Daniels, D., Flandermeyer, R. R. (1981). Cross-contamination of cells in culture. *Science* **212**:446–452.

Nelson-Rees, W., & Flandermeyer, R. R. (1977). Inter- and intraspecies contamination of human breast tumor cell lines HBC and BrCa5 and other cell cultures. *Science* **195**:1343–1344.

Neugut, A. I., & Weinstein, I. B. (1979). Use of agarose in the determination of anchorage-independent growth. *In Vitro* **15**:351.

Neumann, J. R., Morency, C. A., Russian, K. O. (1987). A novel rapid assay for chloramphenicol acetyltransferase gene-expression. *Biotechniques,* **5**:444.

Newbold, R. F., & Cuthbert, A. P. (1998). Mapping human senescence genes using microcell-mediated chromosome transfer. In Ravid, K., & Freshney, R. I. (eds.), *DNA transfer to cultured cells.* New York, Wiley-Liss, pp. 237–264.

Neyfakh, A. A. (1987). Use of fluorescent dyes as molecular probes for the study of multidrug resistance. *Exp. Cell Res.* **174**:168.

Nichols, W. W., Murphy, D. G., & Christofalo, V. J. (1977). Characterization of a new diploid human cell strain, IMR-90. *Science* **196**:60–63.

Nicola, N. A. (1987). Hemopoietic growth factors and their interactions with specific receptors. *J. Cell Physiol. (Suppl.)* **5**:9–14.

Nicolson, G. L. (1976). Trans-membrane control of the receptors on normal and tumor cells; II: Surface changes associated with transformation and malignancy. *Biochim. Biophys. Acta* **458**:1–72.

Nicosia, R. F., & Leighton, J. (1981). Angiogenesis *in vitro*: Light microscopic, radioautographic and ultrastructural studies of rat aorta in histophysiological gradient culture. *In Vitro* **17**:204.

Nicosia, R. F., & Ottinetti, A. (1990). Modulation of microvascular growth and morphogenesis by reconstituted basement membrane gel in three-dimensional cultures of rat aorta: A comparative study of angiogenesis in Matrigel, collagen, fibrin, and plasma clot. *In Vitro Cell Dev. Biol.* **26**:119–128.

Nicosia, R. F., Tchao, R., & Leighton, J. (1983). Angiogenesis-dependent tumor spread in reinforced fibrin clot culture. *Cancer Res.* **43**:2159–2166.

Nielsen, V. (1989). Vibration patterns in tissue culture vessels. *Nunc Bulletin 2* (May 1986; rev March 1989). Roskilde, Denmark, A/S Nunc.

Nilos, R. M., & Makarski, J. S. (1978). Control of melanogenesis in mouse melanoma cells of varying metastatic potential. *J. Natl. Cancer Inst.* **61**:523–526.

Nims, R. W., Shoemaker, A. P., Bauternschub, M. A., Rec, L. J., & Harbell, J. W. (1998). Sensitivity of isoenzyme analysis for the detection of interspecies cell line cross-contamination. *In Vitro Cell Dev. Biol—Animal* **34**:35–39.

Noble, M., & Barnett, S. C. (1996). Production and growth of conditionally immortal primary glial cell cultures and cell lines. In Freshney, R. I., & Freshney, M. G. (eds.), *Culture of immortalized cells.* New York, Wiley-Liss, pp. 331–366.

Noble, M., & Murrey, K. (1984). Purified astrocytes promote the *in vitro* division of a bipotential glial progenitor cell. *EMBO Journal* **3**:2243–2247.

Norwood, T. H., Zeigler, C. J., & Martin, G. M. (1976). Dimethyl sulphoxide enhances polyethylene glycol-mediated somatic cell fusion. *Somatic Cell Genet.* **2**:263–270.

O'Brien, S. J., Shannon, J. E., & Gail, M. H. (1980). Molecular approach to the identification and individualization of human and animal cells in culture: Isozyme and allozyme genetic signatures. *In Vitro* **16**:119–135.

Oda, D., & Watson, E. (1990). Human oral epithelial cell culture; I: Improved conditions for reproducible culture in serum-free medium. *In Vitro Cell Dev. Biol.* **26**:589–595.

O'Farrell, P. H. (1975). High resolution two-dimensional electrophoresis of proteins. *J. Biol. Chem.* **250**:4007–4021.

Office of Nuclear Regulatory Research *see U.S. Nuclear Regulatory Commission.*

O'Hare, M. J., Ellison, M. L., & Neville, A. M. (1978). Tissue culture in endocrine research: Perspectives, pitfalls, and potentials. *Curr. Top. Exp. Endocrinol.* **3**:1–56.

Ohmichi, H., Koshimizu, U., Matsumoto, K., & Nakamura, T. (1998). Hepatocyte growth factor (HGF) acts as a mesenchyme-derived morphogenic factor during fetal lung development. *Development* 125:1315–1324.

Ohno, T., Saijo-Kurita, K., Miyamoto-Eimori, N., Kurose, T., Aoki, Y., & Yosimura, S. (1991). A simple method for *in situ* freezing of anchorage-dependent cells including rat liver parenchymal cells. *Cytotechnology* 5:273–277.

Oie, H. K., Russell, E. K., Carney, D. N., & Gazdar, A. F. (1996). Cell culture methods for the establishment of lung cancer cell lines. *J. Cell Biochem.*, No. S24, 24–31.

Olie, R. A., Looijenga, L. H. J., Dekker, M. C., de Jong, F. H., van Dissel-Emiliani, F. M. F., de Rooij, D. G., van der Holt, B., & Oosterhuis, J. W. (1995). Heterogeneity in the *in vitro* survival and proliferation of human seminoma cells. *Brit. J. Cancer* 71:13–17.

Olsson, I., & Ologsson, T. (1981). Induction of differentiation in a human promyelocytic leukemic cell line (HL-60). *Exp. Cell Res.* 131:225–230.

Orellana, S. A., Neff, C. D., Sweeney, W. E. & Avner, E. D. (1996). Novel Madin Darby canine kidney cell clones exhibit unique phenotypes in response to morphogens. *In Vitro Cell Dev. Biol.—Animal* 32:329–339.

Orly, J., Sato, G., & Erickson, G. F. (1980). Serum suppresses the expression of hormonally induced function in cultured granulosa cells. *Cell* 20:817–827.

Osborne, C. K., Hamilton, B., Tisus, G., & Livingston, R. B. (1980). Epidermal growth factor stimulation of human breast cancer cells in culture. *Cancer Res.* 40:2361–2366.

Osborne, H. B., Bakke, A. C., & Yu, J. (1982). Effect of dexamethasone on HMBA-induced Friend cell erythrodifferentiation. *Cancer Res.* 42:513–518.

Osborne, R., Durkin, T., Shannon, H., Dornan, E., & Hughes, C. (1999). Performance of open-fronted microbiological safety cabinets: the value of operator protection tests during routine servicing. *J. Appl. Microbiol.* 86:962–970.

Ostertag, W., & Pragnell, I. B. (1978). Changes in genome composition of the Friend virus complex in erythroleukaemia cells during the course of differentiation induced by DMSO. *Proc. Natl. Acad. Sci. USA* 75:3278–3282.

Ostertag, W., & Pragnell, I. B. (1981). Differentiation and viral involvement in differentiation of transformed mouse and rat erythroid cells. *Curr. Topics. Microbiol. Immunol.* 94/95:143–208.

Otterlei, M., Østgaard, K., Skjåk-Bræk, G., Smidsrød, O., Soon Shiong, P., & Espevik, T. (1991). Induction of cytokine production from human monocytes stimulated with alginate. *J. Immunother.* 10:286–291.

Otto, E., Zalewski, C., Kaloss, M., del Giudice, R. A., Gardella, R., and McGarrity, G. J. (1996). Quantitative detection of cell culture Mycoplasmas by a one step polymerase chain reaction method. *Meth. Cell Sci.* 18:261–268.

Owens, R. B., Smith, H. S., & Hackett, A. J. (1974). Epithelial cell culture from normal glandular tissue of mice. Mouse epithelial cultures enriched by selective trypsinisation. *J. Natl. Cancer Inst.* 53:261–269.

Paddenberg, R., Wulf, S., Weber, A., Heimann, P., Beck, L., & Mannherz, H. G. (1996). Internucleosomal DNA fragmentation in cultured cells under conditions reported to induce apoptosis may be caused by mycoplasma endonucleases. *Europ. J. Cell Biol.* 71:105–119.

Pantel, K., Dickmanns, A., Zippelius, A., Klein, C., Shi, J., Hoechtlen-Vollmar, W., Schlimok, G., Weckermann, D., Oberneder, R., Fanning, E., & Rietmüller, G. (1995). Establishment of micrometastatic cell lines: A novel source of tumor cell vaccines. *J. Natl. Cancer Inst.* 87:1162–1168.

Paraskeva, C., Buckle, B. G., Sheer, D., & Wigley, C. B. (1984). The isolation and characterisation of colorectal epithelial cell lines at different stages in malignant transformation from familial polyposis coli patients. *Int. J. Cancer* 34:49–56.

Paraskeva, C., Buckle, B. G., & Thorpe, P. E. (1985). Selective killing of contaminating human fibroblasts in epithelial cultures derived from colorectal tumors using an anti-Thy-1 antibody-ricin conjugate. *Br. J. Cancer* 51:131–134.

Paraskeva, C., & Williams, A. C. (1992). The colon. In Freshney, R. I. (ed.), *Culture of epithelial cells.* New York, Wiley-Liss, pp. 82–105.

Pardee, A. B., Cherington, P. V., & Medrano, E. E. (1984). On deciding which factors regulate cell growth. In Barnes, D. W., Sirbasku, D. A., & Sato, G. H. (eds.), *Methods for serum-free culture of epithelial and fibroblastic cells.* New York, Alan R. Liss, pp. 157–166.

Paris Conference (1971, Suppl 1975). Standardization in human cytogenetics. *Cytogenet. Cell Genet.* 15:201–238.

Park, H. Y., Murphy M., & Gilchrest, B. A. (1999). Increasing PKC-β is the rate limiting step in cAMP-induced human melanogenesis. *J. Invest. Dermatol.* (in press).

Park, H. Y., Russakovsky, V., Ohno, S., & Gilchrest, B. A. (1993). The β isoform of protein kinase C stimulates melanogenesis by activating tyrosinase in pigment cells. *J. Biol. Chem.* 268:11742–11749.

Park, J. G., & Gazdar, A. F. (1996). Biology of colorectal and gastric cancer cell lines. *J. Cell Biochem.*, No. S24:131–141.

Parker, R. C., Castor, L. N., & McCulloch, E. A. (1957). Altered cell strains in continuous culture. Special publications. *NY Acad. Sci.* 5:303–313.

Parkinson, E. K., & Yeudall, W. A. (1992). The epidermis. In Freshney, R. I. (ed.), *Culture of epithelial cells.* New York, Wiley-Liss, pp. 59–80.

Parks, W. M., Gingrich, R. D. Dahle, C. E., & Hoak, J. C. (1985). Identification and characterization of an endothelial, cell-specific antigen with a monoclonal antibody. *Blood* 66:816–823.

Patel, K., Moore, S. E., Dickinson, G., Rossell, R. J., Beverley, P. C., Kemshead, J. T. & Walsh, F. S. (1989). Neural cell adhesion molecule (NCAM) is the antigen recognised by monoclonal antibodies of similar specificity in small-cell lung carcinoma and neuroblastoma. *Int. J. Cancer* 44:573–578.

Paul, J. (1975). *Cell and tissue culture.* Edinburgh, Churchill Livingstone, pp. 172–184.

Paul, J., Conkie, D., & Freshney, R. I. (1969). Erythropoietic cell population changes during the hepatic phase of erythropoiesis in the foetal mouse. *Cell Tissue Kinet.* 2:283–294.

Pavelic, K., Antonic, M., Pavelic, L., Pavelic, J., Pavelic, Z., & Spaventi, S. (1992). Human lung cancers growing on extracellular matrix: Expression of oncogenes and growth factors. *Anticancer Research* 12, 2191–2196.

Peat, N., Gendler, S. J., Lalani, N., Duhig, T., & Taylor-Papadimitriou, J. (1992). Tissue-specific expression of a human polymorphic epithelial mucin (MUC1) in transgenic mice. *Cancer Res.* **52**:1954–1960.

Peehl, D. M., & Ham, R. G. (1980). Clonal growth of human keratinocytes with small amounts of dialysed serum. *In Vitro* **16**:526–540.

Pegolo, G., Askanas, V., & Engel, W. K. (1990). Expression of muscle-specific isozymes of phosphorylase and creatine kinase in human muscle fibers cultured aneurally in serum-free, hormonally/chemically enriched medium. *Int. J. Dev. Neurosci.* **8**:299–308.

Peppelenbosch, M. P., Tertoolen, L. G. J., DeLaat, S. W., & Zivkovic, D. (1995). Ionic responses to epidermal growth factor in zebrafish cells. *Exp. Cell Res.* **218**:183–188.

Pereira, M. E. A., & Kabat, E. A. (1979). A versatile immunoadsorbent capable of binding lectins of various specificities and its use for the separation of cell populations. *J. Cell Biol.* **82**:185–194.

Pereira-Smith, O., & Smith, J. (1988). Genetic analysis of indefinite division in human cells: Identification of four complementation groups. *Proc. Natl. Acad. Sci. USA* **85**: 6042–6046.

Perl, A.-K., Wilgenbus, P., Dahl, U., Semb, H., & Christofori, G. (1998). A causal role for E-cadherin in the transition from adenoma to carcinoma. *Nature* **392**:190–193.

Perry, P., & Wolf, S. (1974). New Giemsa method for the differential staining of sister chromatids. *Nature* **251**: 156–158.

Pertoft, H., & Laurent, T. C. (1982). Sedimentation of cells in colloidal silica (Percoll). In Pretlow, T. G., & Pretlow, T. P. (eds.), *Cell separation, methods and selected applications*, Vol. 1. New York, Academic Press, pp. 115–152.

Peters, D. M., Dowd, N., Brandt, C., & Compton, T. (1996). Human papilloma virus E6/E7 genes can expand the lifespan of human corneal fibroblasts. *In Vitro Cell Dev. Biol. —Animal* **32**:279–284.

Petersen, D. F., Anderson, E. C., & Tobey, R. A. (1968). Mitotic cells as a source of synchronized cultures. In Prescott, D. M. (ed.), *Methods in cell physiology*. New York, Academic Press, pp. 347–370.

Petit, B., Masuda, K., d'Souza, A., Otten, L., Pietryla, D., Hartmann, D. J., Morris, N. P., Uebelhart, D., Schmid, T. M., & Thonar, E. J.-M. A. (1996). Characterization of crosslinked collagens synthesized by mature articular chondrocytes cultures in alginate beads: Comparison of two distinct matrix compartments. *Exp. Cell Res.* **225**: 151–161.

Pfeffer, L. M., & Eisenkraft, B. L. (1991). The antiproliferative and antitumour effects of human alpha interferon on cultured renal carcinomas correlate with the expression of a kidney-associated differentiation antigen. *Interferons Cytokines* **17**:30–31.

Phillips, P., Kumar, P., Kumar, S., & Waghe, M. (1979). Isolation and characterization of endothelial cells from adult rat brain white matter. *J. Anat.* **129**:261–272.

Phillips, P. D., & Cristofalo, V. J. (1988). Classification system based on the functional equivalency of mitogens that regulate WI-38 cell proliferation. *Exp. Cell Res.* **175**: 396–403.

Pignata, S., Maggini, L., Zarrilli, R., Rea, A., & Acquaviva, A. M. (1994). The enterocyte-like differentiation of the

Caco-2 tumour cell line strongly correlates with responsiveness to cAMP and activation of kinase A pathway. *Cell Growth & Differentiation* **5**:967–973.

Pignatelli, M., & Bodmer, W. F. (1988). Genetics and biochemistry of collagen binding-triggered glandular differentiation in a human colon carcinoma cell line. *Proc. Natl. Acad. Sci. USA* **85**:5561–5565.

Pipia, G. G., & Long, M. W. (1997). Human hematopoietic progenitor cell isolation based on galactose-specific cell surface binding. *Nature Biotechnology* **15**:1007–1011.

Pitot, H., Periano, C., Morse, P., & Potter, V. R. (1964). Hepatomas in tissue culture compared with adapting liver *in vitro*. *Natl. Cancer Inst. Monogr.* **13**:229–245.

Pizzonia, J. H., Gesek, F. A., Kennedy, S. M., Coutermarach, B. A., Bacskal, B. J., & Friedman, P. A. (1991). Immunomagnetic separation, primary culture, and characterisation of cortical thick ascending limb plus distal convoluted tubule cells from mouse kidney. *In Vitro Cell Dev. Biol.* **27A**:409–416.

Planas-Silva, M. D., & Weinberg, R. A. (1997). The restriction point and control of cell proliferation. *Curr. Opin. Cell Biol.* **9**:768–772.

Planz, B., Wang, Q., Kirley, S. D., Lin, C. W., & McDougal, W. S. (1998). Androgen responsiveness of stromal cells of the human prostate: Regulation of cell proliferation and keratinocyte growth factor by androgen. *J. Urol.* **160**: 1850–1855.

Platsoucas, C. D., Good, R. A., & Gupta, S. (1979). Separation of human lymphocyte-T subpopulations (T-mu, T-gamma) by density gradient electrophoresis. *Proc. Natl. Acad. Sci. USA* **76**:1972.

Plumb, J. A., Milroy, R., & Kaye, S. B. (1989). Effects of the pH dependence of 3-(4,5-dimethylthiazol-2-yl)-2,5-diphenyltetra-zolium bromide-formazan absorption on chemosensitivity determined by a novel tetrazolium-based assay. *Cancer Res.* **49**:4435–4440.

Pollack, M. S., Heagney, S. D., Livingston, P. O., & Fogh, J. (1981). HLA-A, B, C & DR alloantigen expression on forty-six cultured human tumor cell lines. *J. Natl. Cancer Inst.* **66**:1003–1012.

Pollard, J. W., & Walker, J. M. (eds.). (1990). *Animal cell culture: Methods in molecular biology*, 5. Clifton, NJ, Humana Press, pp. 83–97.

Pollock, G. S., Franceschini, I. A., Graham, G., & Barnett, S. C. (1999). Neuregulin is a mitogen and survival factor for olfactory bulb ensheathing cells and a related isoform is produced by astrocytes. *European J. Neurosci.* **11**:769–780.

Pollock, M. F., & Kenny, G. E. (1963). Mammalian cell cultures contaminated with pleuro-pneumonia-like organisms; III: Elimination of pleuro-pneumonia-like organisms with specific antiserum. *Proc. Soc. Exp. Biol. Med.* **112**: 176–181.

Pontecorvo, G. (1975). Production of mammalian somatic cell hybrids by means of polyethylene glycol treatment. *Somat. Cell Genet.* **1**:397–400.

Pontén, J. (1975). Neoplastic human glia cells in culture. In: Fogh, J. (ed.), Human Tumor Cells *in Vitro*. New York, Plenum, pp. 175–185.

Pontén, J., & Westermark, B. (1980). Cell generation and aging of nontransformed glial cells from adult humans.

In Fedorof, S., & Hertz, L. (eds.), *Advances in cellular neurobiology*, Vol. 1. New York, Academic Press, pp. 209–227.

Poot, M., Hoehn, H., Kubbies, M., Grossmann, A., Chen, Y., & Rabinovitch, P. S. (1994). Cell-cycle analysis using continuous bromodeoxyuridine labeling and Hoechst 33358-ethidium bromide bivariate flow cytometry. *Methods in Cell Biology* **41**:327–340.

Post, M., Floros, J., & Smith, B. T. (1984). Inhibition of lung maturation by monoclonal antibodies against fibroblast-pneumocyte factor. *Nature* **308**:284–286.

Potter, H., Weir, L., & Leder, P. (1984). Enhancer-dependent expression of human k immunoglobulin genes introduced into mouse pre-B lymphocytes by electroporation. *Proc. Natl. Acad. Sci. USA* **81**:7161–7165.

Powers, D. A. (1989). Fish as model systems. *Science* **246**: 352–357.

Pragnell, I. B., Wright, E. G., Lorimore, S. A., Adam, J., Rosendaal, M., deLamarter, J. F., Freshney, M., Eckmann, L., Sproul, A., & Wilkie, N. (1988). The effect of stem cell proliferation regulators demonstrated with an *in vitro* assay. *Blood* **72**:196–201.

Prasad, K. N., & Edwards-Prasad, J., Ramanujam, S., & Sakamoto, A. (1980). Vitamin E increases the growth inhibitory and differentiating effects of tumour therapeutic agents on neuroblastoma and glioma cells in culture. *Proc. Soc. Exp. Biol. Med.* **164**:158–163.

Preissmann, A., Wiesmann, R., Buchholz, R., Werner, R. G., & Noe, W. (1997). Investigations on oxygen limitations of adherent cells growing on macroporous microcarriers. *Cytotechnology* **24**:121–134.

Pretlow, T. G., Delmoro, C. M., Dilley, G. G., Spadafora, C. G., & Pretlow, T. P. (1991). Transplantation of human prostatic carcinoma into nude mice in Matrigel. *Cancer Res.* **51**:3814–3817.

Pretlow, T. G., & Pretlow, T. P. (1989). Cell separation by gradient centrifugation methods. *Methods Enzymol.* **171**: 462–482.

Prince, G. A., Jenson, A. B., Billups, L. C., & Notkins, A. L. (1978). Infection of human pancreatic beta cell cultures with mumps virus. *Nature* **271**:158–161.

Provision and use of work equipment. (1992). Health and Safety Executive, Broad Lane, Sheffield S3 7HQ, England.

Pruckler, J. M., & Ades, E. W. (1995). Detection by polymerase chain-reaction of all common mycoplasma in a cell-culture facility. *Pathobiology* **63**:9–11.

Puck, T. T., Cieciura, S. J., & Robinson, A. (1958). Genetics of somatic mammalian cells; III: Long term cultivation of euploid cells from human and animal subjects. *J. Exp. Med.* **108**:945–956.

Puck, T. T., & Marcus, P. I. (1955). A rapid method for viable cell titration and clone production with HeLa cells in tissue culture: The use of X-irradiated cells to supply conditioning factors. *Proc. Natl. Acad. Sci. USA* **41**:432–437.

Punchard, N., Watson, D., Thomson, R., & Shaw, M. (1996). Production of immortal human umbilical vein endothelial cells. In Freshney, R. I., & Freshney, M. G. (eds.), *Culture of immortalized cells*. New York, Wiley-Liss, pp. 203–238.

Quarles, J. M., Morris, N. G., & Leibovitz, A. (1980). Carcinoembryonic antigen production by human colorectal adenocarcinoma cells in matrix-perfusion culture. *In Vitro* **16**:113–118.

Quax, P. H., Frisdal, E., Pedersen, N., Bonavaud, S., Thibert, P., Martelly, I., Verheijen, J. H., Blasi, F., & Barlovatz-Meimon, G. (1992). Modulation of activities and RNA level of the components of the plasminogen activation system during fusion of human myogenic satellite cells *in vitro*. *Dev. Biol.* **151**:166–175.

Quintanilla, M., Brown, K., Ramsden, M., & Balmain, A. (1986). Carcinogen specific mutation and amplification of Ha-ras during mouse skin carcinogenesis. *Nature* **322**: 78–79.

Quon, M. J. (1998). Transfection of rat adipose cells by electroporation. In Ravid, K., & Freshney, R. I. (eds.), *DNA transfer to cultured cells*. New York, Wiley-Liss, pp. 93–110.

Rabito, C. A., Tchao, R., Valentich, J., & Leighton, J. (1980). Effect of cell substratum interaction of hemicyst formation by MDCK cells. *In Vitro* **16**:461–468.

Raff, M. C. (1990). Glial cell diversification in the rat optic nerve. *Science* **243**:1450–1455.

Raff, M. C., Abney, E., Brockes, J. P., & Hornby-Smith, A. (1978). Schwann cell growth factors. *Cell* **15**:813–822.

Raff, M. C., Fields, K. L., Hakomori, S. L., Minsky, R., Pruss, R. M., & Winter, J. (1979). Cell-type-specific markers for distinguishing and studying neurons and the major classes of glial cells in culture. *Brain Res.* **174**:283–309.

Raisman, G. (1985). Specialized neuroglial arrangement may explain the capacity of vomeronasal axons to reinnerveate central neurons. *Neuroscience* **14**:237–254.

Rak, J., Mitsuhashi, Y., Erdos, V., Huang, S.-N., Filmus, J., & Kerbel, R. S. (1995). Massive programmed cell death in intestinal epithelial cells induced by three-dimensional growth conditions: Suppression by mutant c-H-ras oncogene expression. *J. Cell Biol.* **131**:1587–1598.

Ramaekers, F. C. S., Puts, J. J. G., Kant, A., Moesker, O., Jap, P. H. K., & Vooijs, G. P. (1982). Use of antibodies to intermediate filaments in the characterization of human tumors. *Cold Spring Harbor Symp. Quant. Biol.* **46**:331–339.

Ramon-Cueto, A., Plant, G. W., Avila, J., & Bunge, M. B. (1998). Long-distance axonal regeneration in the transected adult rat spinal cord is promoted by olfactory ensheathing glia transplants. *J. Neurosci.* **18**:3803–3815.

Ranscht, B., Clapshaw, P. A., Price, J., Noble, M., & Seifert, W. (1982). Development of oligodendrocytes and Schwann cells studied with monoclonal antibody against galactocerebroside. *Proc. Natl. Acad. Sci. USA* **79**:2709–2713.

Rathjen, P. D., Lake, J., Whyatt, L. M., Bettess, M. D., & Rathjen, J. (1998). Properties and uses of embryonic stem cells: Prospects for application to human biology and gene therapy. *Reproduction, Fertility, & Development* **10**:31–47.

Rattner, A., Sabido, O., Massoubre, C., Rascle, F., & Frey, J. (1997). Characterization of human osteoblastic cells: Influence of the culture conditions. *In Vitro Cell Dev. Biol. —Animal* **33**:757–762.

Ravid, K., Doi, T., Beeler, D. L., Kuter, D. J., & Rosenberg, R. D. (1991). Transcriptional regulation of the rat platelet factor 4 gene: Interaction between an enhancer/silencer domain and the GATA site. *Mol. Cell. Biol.* **11**: 6116–6127.

Ravid, K., & Freshney, R. I. (eds.). (1998). *DNA transfer to cultured cells*. New York, Wiley-Liss.

Ray, T. (1989). Antibodies in HLA serology. In Catty, D., (ed.), *Antibodies, Vol. II, a Practical Approach*, Oxford, U.K., IRL Press at Oxford University Press, pp. 31–75.

Raz, A. (1982). B16 melanoma cell variants: Irreversible inhibition of growth and induction of morphologic differentiation by anthracycline antibiotics. *J. Natl. Cancer Inst.* **68**:629–638.

Reaven, G. M. (1995). The fourth musketeer—from Alexandre Dumas to Claude Bernard. *Diabetologia* **38**:3–13.

Reddel, R., Ke, Y., Gerwin, B. I., McMenamin, M. G., Lechner, J. F., Su, R. T., Brash, D. E., Park, J. B., Rhim, J. S., & Harris, C. C. (1988). Transformation of human bronchial epithelial cells by infection with SV40 or adenovirus-12 SV40 hybrid virus or transfection via strontium phosphate coprecipitation with a plasmid containing SV40 early region genes. *Cancer Res.* **48**:1904–1909.

Reel, J. R., & Kenney, F. T. (1968). "Superinduction" of tyrosine transaminase in hepatoma cell cultures: Differential inhibition of synthesis and turnover by actinomycin D. *Proc. Natl. Acad. Sci. USA* **61**:200–206.

Reeves, M. E. (1992). A metastatic tumour cell line has greatly reduced levels of a specific homotypic cell adhesion molecule activity. *Cancer Res.* **52**:1546–1552.

Reid, L. M. (1990). Stem cell biology, hormone/matrix synergies and liver differentiation. *Curr. Opin. Cell. Biol.* **2**: 121–130.

Reitzer, L. J., Wice, B. M., & Kennel, D. (1979). Evidence that glutamine, not sugar, is the major energy source for cultured HeLa cells. *J. Biol. Chem.* **254**:2669–2677.

Repesh, L. A. (1989). A new *in vitro* assay for quantitating tumor cell invasion. *Invas. Metast.* **9**:192–208.

Rheinwald, J. G., & Beckett, M. A. (1981). Tumorigenic keratinocyte lines requiring anchorage and fibroblast support cultured from human squamous cell carcinomas. *Cancer Res.* **41**:1657–1663.

Rheinwald, J. G., & Green, H. (1975). Serial cultivation of strains of human epidermal keratinocytes: The formation of keratinizing colonies from single cells. *Cell* **6**:331–344.

Rheinwald, J. G., & Green, H. (1977). Epidermal growth factor and the multiplication of cultured human keratinocytes. *Nature* **265**:421–424.

Richard, O., Duittoz, A. H., & Hevor, T. K. (1998). Early, middle, and late stages of neural cells from ovine embryo in primary cultures. *Neuroscience Research* **31**:61–68.

Richler, C., & Yaffe, D. (1970). The *in vitro* cultivation and differentiation capacities of myogenic cell lines. *Dev. Biol.* **23**:1–22.

Richmond, A., Lawson, D. H., Nixon, D. W. & Chawla, R. K. (1985). Characterization of autostimulatory and transforming growth factors from human melanoma cells. *Cancer Res.* **45**:6390–6394.

Rickwood, D., & Birnie, G. D. (1975). Metrizamide, a new density gradient medium. *FEBS Lett.* **50**:102–110.

Ried, T., Liyanage, M., du Manoir, S., Heselmeyer, K., Auer, G., Macville, M., & Schrock, E. (1997). Tumor cytogenetics revisited: Comparative genomic hybridization and spectral karyotyping. *J. Mol. Med.* **75**:801–814.

Rindler, M. J., Chuman, L. M., Shaffer, L., & Saier, M. H., Jr. (1979). Retention of differentiated properties in an established dog kidney epithelial cell line (MDCK). *J. Cell Biol.* **81**:635–648.

Robertson, K. M., & Robertson, C. N. (1995). Isolation and growth of human primary prostate epithelial cultures. *Methods in Cell Science* **17**:177–185.

Robins, D. M., Ripley, S., Henderson, A. S., & Axel, R. (1981). Transforming DNA integrates into the host chromosome. *Cell* **23**:29–39.

Rodbell, M. (1964). Metabolism of isolated fat cells: Effects of hormones on glucose metabolism and lipolysis. *J. Biol. Chem.* **239**:375–380.

Rofstad, E. K. (1994). Orthotopic human melanoma xenograft model systems for studies of tumor angiogenesis, pathophysiology, treatment sensitivity and metastatic pattern. *Br. J. Cancer* **70**:804–812.

Rogers, A. W. (1979). *Techniques of autoradiography*, 3d ed. Amsterdam, Elsevier/North-Holland Biomedical Press.

Roguet, R., Cohen, C., & Rougier, A. (1994). A reconstituted human epidermis to assess cutaneous irritation, photoirritation and photoprotection *in vitro*. *In Vitro Skin Toxicology* **10**:141–149.

Rojkind, M., Gatmaitan, Z., Mackensen, S., Giambrone, M. A., Ponce, P., & Reid, L. M. (1980). Connective tissue biomatrix: Its isolation and utilization for long term cultures of normal rat hepatocytes. *J. Cell Biol.* **87**:255–263.

Rooney, D. E., & Czepulkowski, B. H. (eds.). (1986). *Human cytogenetics, a practical approach*. Oxford, U.K., IRL Press at Oxford University Press.

Rooney, S. A., Young, S. L., & Mendelson, C. R. (1995). Molecular and cellular processing of lung surfactant. *FASEB Journal* **8**:957–967.

Rosenfeld, M. A., Yoshimura, K., Trapnell, B. C., Yoneyama, K., Rosenthal, E. R., Dalemenas, W., Fukayama, M., Bargon, J., Stier, L. E., Stratford-Perricaudet, L., Perricaudet, M., Guggino, W. B., Pavirani, A., Lecocq, J.-P., & Crystal, R. G. (1992). *In vivo* transfer of the human cystic fibrosis transmembrane conductance regulator gene to the airway epithelium. *Cell* **68**:143–155.

Rosenman, S. J., & Gallatin, W. M. (1991). Cell surface glycoconjugates in intercellular and cell-substratum interactions. *Sem. Cancer Biol.* **2**:357–366.

Rossi, G. B., & Friend, C. (1967). Erythrocytic maturation of (Friend) virus-induced leukemic cells in spleen clones. *Proc. Natl. Acad. Sci. USA* **58**:1373–1380.

Rothfels, K. H., & Siminovitch, L. (1958). An air drying technique for flattening chromosomes in mammalian cells growth *in vitro*. *Stain Technol.* **33**:73–77.

Rotman, B., & Papermaster, B. W. (1966). Membrane properties of living mammalian cells as studied by enzymatic hydrolysis of fluorogenic esters. *Proc. Natl. Acad. Sci. USA* **55**:134–141.

Royal College of Physicians of London. (1990). *Guidelines on the practice of ethics committees in medical research involving human subjects*, 2d ed. ISBN 0 900596 902.

Rudland, P. S. (1992). Use of peanut lectin and rat mammary stem cell lines to identify a cellular differentiation pathway for the alveolar cell in the rat mammary gland. *J. Cell Physiol.* **153**:157–168.

Ruff, M. R., & Pert, C. B. (1984). Small cell carcinoma of the lung: Macrophage-specific antigens suggest hematopoietic stem cell origin. *Science* **225**:1034–1036.

Rules and guidance for pharmaceutical manufacturers and distributors. (1997). The Stationery Office, Ltd., London, U.K.

Rundlett, S. E., Gordon, D. A., & Miesfeld, R. L. (1992). Characterisation of a panel of rat ventral prostate epithelial cell lines immortalized in the presence or absence of androgens. *Exp. Cell Res.* **203**:214–221.

Ruoff, N. M., & Hay, R. J. (1979). Metabolic and temporal studies on pancreatic exocrine cells in culture. *Cell Tissue Res.* **204**:243–252.

Russo, A. A., Tong, L., Lee, J. O., Jeffrey, P. D., & Pavletich, N. P. (1998). Structural basis for inhibition of the cyclin-dependent kinase Cdk6 by the tumour suppressor p16INK4a. *Nature* **395**:237–243.

Rutzky, L. P., Tomita, J. T., Calenoff, M. A., & Kahan, B. D. (1979). Human colon adenocarcinoma cells; III: *In vitro* organoid expression and carcino-embryonic antigen kinetics in hollow fiber culture. *J. Natl. Cancer Inst.* **63**:893–902.

Rygaard, K., Moller, C., Bock, E., & Spang-Thomsen, M. (1992). Expression of cadherin and NCAM in human small cell lung cancer cell lines and xenografts. *Br. J. Cancer* **65**:573–577.

Sabatini, L. M., Allen-Hoffmann, B. L., Warner, T. F., Azen, E. A. (1991). Serial cultivation of epithelial cells from human and macaque salivary glands. *In vitro Cell Dev. Biol.* **27A**:939–948.

Sachs, L. (1978). Control of normal cell differentiation and the phenotypic reversion of malignancy in myeloid leukaemia. *Nature* **274**:535–539.

Safe working and the prevention of infection in clinical laboratories. (1991). Health and Safety Commission, HMSO Publication, P. O. Box 276, London, SW8 5DT, England.

Sager, R. (1992). Tumor suppressor genes in the cell cycle. *Curr. Opin. Cell Biol.* **4**:155–160.

Saier, M. H. (1984). Hormonally defined, serum free medium for a proximal tubular kidney epithelial cell line, LLC-PKI. In Barnes, W. D. (ed.), *Methods for serum free culture of epithelial and fibroblastic cells.* New York, Alan R. Liss, pp. 25–31.

Sambrook, J., Fritsch, E. F., & Maniatis, T. (1989). *Molecular cloning: A laboratory manual,* 2d ed. Cold Spring Harbor, NY, Cold Spring Harbor Laboratory Press, 3 vols.

Sandberg, A. A. (1982). Chromosomal changes in human cancers: Specificity and heterogeneity. In Owens, A. H., Coffey, D. S., & Baylin, S. B. (eds.), *Tumour cell heterogeneity.* New York, Academic Press, pp. 367–397.

Sanes, J. R., Rubenstein, J. L., Nicolas, J. F. (1986). Use of recombinant retrovirus to study post-implantation cell lineage in mouse embryos. *EMBO J.* **5**:3133–3142.

Sanford, K. K., Earle, W. R., Evans, V. J., Waltz, H. K., & Shannon, I. E. (1951). The measurement of proliferation in tissue cultures by enumeration of cell nuclei. *J. Natl. Cancer Inst.* **11**:773.

Sanford, K. K., Earle, W. R., & Likely, G. D. (1948). The growth *in vitro* of single isolated tissue cells. *J. Natl. Cancer Inst.* **9**:229.

Sasaki, M., Honda, T., Yamada, H., Wake, N., & Barrett, J. C. (1996). Evidence for multiple pathways to cellular senescence. *Cancer Res.* **54**:6090–6093.

Sato, G. H., & Yasumura, Y. (1966). Retention of differentiated function in dispersed cell culture. *Trans. NY Acad. Sci.* **28**:1063–1079.

Sattler, G. A., Michalopoulos, G., Sattler, G. L., & Pitot, H. C. (1978). Ultrastructure of adult rat hepatocytes cultured on floating collagen membranes. *Cancer Res.* **38**: 1539–1549.

Saunders, N. A., Bernacki, S. H., Vollberg, T. M., & Jetten, A. M. (1993). Regulation of transglutaminase type-I expression in squamous differentiating rabbit tracheal epithelial cells and human epidermal keratinocytes—effects of retinoic acid and phorbol esters. *Mol. Endocrinol.* **7**: 387–398.

Scanlon, E. F., Hawkins, R. A., Fox, W. W., & Smith, W. S. (1965). Fatal homotransplanted melanoma (a case report). *Cancer* **18**:782–789.

Schaeffer, W. I. (1990). Terminology associated with cell, tissue and organ culture, molecular biology and molecular genetics. *In Vitro Cell Dev. Biol.* **26**:97–101.

Schaffer, K., Herrmuth, H., Mueller, J., Coy, D. H., Wong, H. C., Walsh, J. H., Classen, M., Schusdziarra, V., & Schepp, W. (1997). Bombesin-like peptides stimulate somatostatin release from rat fundic D cells in primary culture. *Am. J. Physiol.* **273**(3 Pt 1):G686–G695.

Scher, W., Holland, J. G., & Friend, C. (1971). Hemoglobin synthesis in murine virus-induced leukemic cells *in vitro*; I: Partial purification and identification of hemoglobins. *Blood* **37**:428–437.

Schimmelpfeng, L., Langenberg, U., & Peters, I. M. (1968). Macrophages overcome mycoplasma infections of cells *in vitro*. *Nature* **285**:661.

Schlechte, W., Brattain, M., & Boyd, D. (1990). Invasion of extracellular matrix by cultured colon cancer cells: Dependence on urokinase receptor display. *Cancer Comm.* **2**: 173–179.

Schlessinger, J., Lax, I., & Lemmon, M. (1995). Regulation of growth factor activation by proteoglycans: What is the role of the low affinity receptors? *Cell* **83**:357–360.

Schmidt, R., Reichert, U., Michel, S., Shrott, B., & Boullier, M. (1985). Plasma membrane transglutaminase and cornified envelope competence in cultured human keratinocytes. *FEBS Lett.* **186**:204.

Schneider, E. L., & Stanbridge, E. I. (1975). A simple biochemical technique for the detection of mycoplasma contamination of cultured cells. *Methods Cell Biol.* **10**:278–290.

Schoenlein, P. V., Shen, D.-W., Barrett, J. T., Pastan, I. T., & Gottesman, M. M. (1992). Double minute chromosomes carrying the human multidrug resistance 1 and 2 genes are generated from the dimerization of submicroscopic circular DNAs in colchicine-selected KB carcinoma cells. *Mol. Biol. Cell* **3**:507–520.

Schor, S. L. (1994). Cytokine control of cell motility modulation and mediation by the extracellular matrix. *Prog. Growth Factor Res.* **5**:223–248.

Schousboe, A., Thorbek, P., Hertz, L., & Krogsgaard-Larsen, P. (1979). Effects of GABA analogues of restricted conformation on GABA transport in astrocytes and brain cortex slices and on GABA receptor binding. *J. Neurochem.* **33**:181–189.

Schulman, H. M. (1968). The fractionation of rabbit reticulocytes in dextran density gradients. *Biochim. Biophys. Acta* **148**:251–255.

Schwartz Albiez, R., Heidtmann, H.-H., Wolf, D., Schirr-macher, V., & Moldenhauer, G. (1991). Three types of human lung tumour cell lines can be distinguished according to surface expression of endogenous urokinase and their capacity to bind exogenous urokinase. *Br. J. Cancer* **65**:51–57.

Schwartz, S. M. (1978). Selection and characterization of bovine aortic endothelial cells. *In Vitro* **14**:966–980.

Scotto, K. W., Biedler, I. L., & Melera, P. W. (1986). Amplification and expression of genes associated with multi-drug resistance in mammalian cells. *Science* **232**:751–755.

Seeds, N. W. (1971). Biochemical differentiation in reaggregating brain cell culture. *Proc. Natl. Acad. Sci. USA* **68**: 1858–1861.

Seglen, P. O. (1975). Preparation of isolated rat liver cells. *Methods Cell Biol.* **13**:29–83.

Seifert, W., & Müller, H. W. (1984). Neuron–glia interaction in mammalian brain: Preparation and quantitative bioassay of a neurotropic factor (NTF) from primary astrocytes. In Barnes, D. W., Sirbasku, D. A., & Sato, G. H. (eds.), *Methods for serum-free culture of neuronal and lymphoid cells.* New York, Alan R. Liss, pp. 67–78.

Seigel, G. A. (1996). Establishment of an E1a-immortalised retinal cell culture. *In Vitro Cell Dev. Biol.—Animal* **32**:66–68.

Selby, P. J., Thomas, M. J., Monaghan, P., Sloane, J., & Peckham, M. J. (1980). Human tumour xenografts established and serially transplanted in mice immunologically deprived by thymectomy, cytosine arabinoside and whole-body irradiation. *Br. J. Cancer* **41**:52.

Selden, R. F., Howie, K. B., Rowe, M. E., Goodman, H. M., & Moore, D. (1986). Human growth hormone as a reporter gene in regulation studies employing transient gene expression. *Mol. Cell Biol.* **6**:3173–3179.

Shall, S. (1973). Sedimentation in sucrose and Ficoll gradients of cells grown in suspension culture. In Kruse, P. F., Patterson, M. K. (eds.), *Tissue culture methods and applications.* New York, Academic Press, pp. 198–204.

Shall, S., & McClelland, A. J. (1971). Synchronization of mouse fibroblast LS cells grown in suspension culture. *Nature New Biol.* **229**:59–61.

Sharpe, P. T. (1988). *Methods of cell separation.* Amsterdam, Elsevier.

Shay, J. W., & Wright, W. E. (1989). Quantitation of the frequency of immortalization of normal human diploid fibroblasts by SV40 large T antigen. *Exp. Cell Res.* **184**: 109–118.

Shay, J. W., Wright, W. E., Brasiskyte, D., & der Haegen, A. (1993). E6 of human papillomavirus type 16 can overcome the M1 stage of immortalization in human mammary epithelial cells but not in human fibroblasts. *Oncogene* **8**:1407–1413.

Shepel, L. A., Morrissey, L. W., Hsu, L. C., & Gould, M. N. (1994). Bivariate flow karyotyping, sorting, and peak assignment of all rat chromosomes. *Genomics* **19**:75–85.

Shipley, G. D., & Ham, R. G. (1983). Multiplication of Swiss 3T3 cells in a serum-free medium. *Exp. Cell Res.* **146**:249–260.

Silvestri, F. F., Banavale, S. D., Hulette, B. C., Civin, C. I., & Preisler, H. D. (1991). Isolation and characterization of

the CD34$^+$ hematopoietic progenitor cells from the peripheral blood of patients with chronic myeloid leukemia. *Int. J. Cell Cloning* **9**:474–490.

Simon-Assmann, P., Kédinger, M., & Haffen, K. (1986). Immunocytochemical localization of extracellular matrix proteins in relation to rat intestinal morphogenesis. *Differentiation* **32**:59–66.

Simonian, M. H., White, M. L., & Foggia, D. A. (1987). Clonal growth and culture life span of bovine adrenocortical cells in a serum-free medium. *In Vitro* **23**(4):247–256.

Sinha, M. K., Buchanan, C., Raineri-Maldonado, C., Khazanie, P., Atkinson, S., DiMarchi, R., & Caro, J. F. (1990). IGF-II receptors and IGF-II-stimulated glucose transport in human fat cells. *Am. J. Physiol.* **258**:E534–E542.

Skehan, P., Storeng, R., Scudiero, D., Monks, A., McMahon, J., Vistica, D., Warren, J. T., Bokesch, H., Kenney, S., Boyd, M. R. (1990). New colorimetric cytotoxicity assay for anticancer-drug screening. *J. Natl. Cancer Inst.* **82**:1107–1112.

Skobe, M., & Fusenig, N. E. (1998). Tumorigenic conversion of immortal human keratinocytes through stromal cell activation. *Proc. Natl. Acad. Sci. USA* **95**:1–6.

Smith, A. D., Datta, S. P., Smith, G. H., Campbell, P. N., Bentley, R., McKenzie, H. A. (eds.). (1997). *Oxford Dictionary of Biochemistry and Molecular Biology*, Oxford, Oxford University Press.

Smith, A. G., Heath, J. K., Donaldson, D. D., Wong, G. G., Moreau, J., Stahl, M., & Rodgers, D. (1988). Inhibition of pluripotential stem cell differentiation by purified polypeptides. *Nature* **336**:688–690.

Smith, H. S., Lan, S., Ceriani, R., Hackett, A. J., & Stampfer, M. R. (1981). Clonal proliferation of cultured non-malignant and malignant human breast epithelia. *Cancer Res.* **41**:4637–4643.

Smith, H. S., Owens, R. B., Hiller, A. J., Nelson-Rees, W. A., & Johnston, J. O. (1976). The biology of human cells in tissue culture; I: Characterization of cells derived from osteogenic sarcomas. *Int. J. Cancer* **17**:219–234.

Smith, M. D., Summers, M. D., & Frazer, M. J. (1983). Production of human beta interferon in insect cells infected with a baculovirus expression vector. *Mol. Cell Biol.* **3**: 2156–2165.

Smith, S., & de Lange, T. (1997). TRF1, a mammalian telomeric protein. *Trends in Genetics* **113**:21–26.

Smith, S. M., & Schroedl, N. A. (1992). Heme containing compounds replace chick embryo extract and enhance differentiation in avian muscle cell culture. *In Vitro Cell Dev. Biol.* **28A**:387–390.

Smith, W. L., & Garcia-Perez, A. (1985). Immunodissection: Use of monoclonal antibodies to isolate specific types of renal cells. *Am. J. Physiol.* **248**:F1–F7.

Smola, H., Stark, H.-J., Thiekötter, G., Mirancea, N., Krieg, T., & Fusenig, N. E. (1998). Dynamics of basement membrane formation by keratinocyte–fibroblast interactions in organotypic skin culture. *Exp. Cell Res.* **239**:399–410.

Smola, H., Thiekötter, G., & Fusenig, N. E. (1993). Mutual induction of growth factor gene expression by epidermal–dermal cell interaction. *J. Cell Biol.* **122**:417–429.

Smola, H., Thiekötter, G., Stark, H.-J., Breitkreutz, D., Hafner, G., Krieg, T., & Fusenig, N. E. *Expression of vimentin in normal human keratinocytes associated with adhesional stress* (*in preparation*).

Soder, A. I., Going, J. J., Kaye, S. B., & Keith, W. N. (1998). Tumour specific regulation of telomerase RNA gene expression visualized by *in situ* hybridization, *Oncogene* **16**: 979–983.

Soder, A. I., Hoare, S. F., Muir, S., Going, J. J., Parkinson, E. K., & Keith, W. N. (1997). Amplification, increased dosage and *in situ* expression of the telomerase RNA gene in human cancer. *Oncogene* **14**:1013–1021.

Sommer, I., & Schachner, M. (1981). Cells that are O4-antigen-positive and O1-antigen-negative differentiate into O1 antigen-positive oligodendrocytes. *Neuroscience Letters* **29**:183–188.

Soon-Shiong, P., Feldman, E., Nelson, R., Komtebedde, J., Smidsrød, O., Skjak-Bræk, G., Espevik, T., Heintz, R., & Lee, M. (1992). Successful reversal of spontaneous diabetes in dogs by intraperitoneal microencapsulated islets. *Transplantation* **54**:769–774.

Sordillo, L. M., Oliver, S. P., & Akers, R. M. (1988). Culture of bovine mammary epithelial cells in D-valine modified medium: Selective removal of contaminating fibroblasts. *Cell Biol. Int. Rep.* **12**:355–364.

Soriano, V., Pepper, M. S., Nakamura, T., Orci, L., & Montesano, R. (1995). Hepatocyte growth factor stimulates extensive development of branching duct-like structures by cloned mammary gland epithelial cells. *J. Cell Sci.* **108**: 413–430.

Sorour, O., Raafat, M., El-Bolkainy, N., & Mohamad, R. (1975). Infiltrative potentiality of brain tumors in organ culture. *J. Neurosurg.* **43**:742–749.

Soule, H. D., Maloney, T. M., Wolman, S. R., Teterson, W. D., Brenz, R., McGrath, C. M., Russo, J., Pauley, R. J., Jones, R. F., & Brooks, S. C. (1990). Isolation and characterization of a spontaneously immortalized human breast epithelial cell line, MCF-10. *Cancer Res.* **50**:6075–6086.

Soule, H. D., Vasquez, J., Long, A., Albert, S., & Brennan, M. (1973). A human cell line from a pleural effusion derived from a breast carcinoma. *J. Natl. Cancer Inst.* **51**:1409–1416.

Southam, C. M. (1958). Homotransplantation of human cell lines. *Bull. NY Acad. Med.* **34**:416–423.

Southern, P. J., & Berg, P. (1982). Transformation of mammalian cells to antibiotic resistance with a bacterial gene under control of the SV40 early region promoter. *J. Mol. App. Genet.* **1**:327–341.

Spandidos, D. A., & Wilkie, N. M. (1984a). Malignant transformation of early passage rodent cells by a single mutated human oncogene. *Nature* **310**:469–475.

Spandidos, D. A., & Wilkie, N. M. (1984b). Expression of exogenous DNA in mammalian cells. In Hames, B. D., & Higgins, S. J. (eds.), *In vitro transcription and translation—a practical approach.* Oxford, U.K., IRL Press, pp. 1–48.

Speir, R., & Griffiths, J. B. (1985–1990). *Animal cell biotechnology.* London, Academic Press, 4 vols.

Speir, R. E., Griffiths, J. B., & Meignier, B. (eds.). (1991). *Production of biologicals from animal cells in culture.* Oxford, U.K., Butterworth–Heinemann.

Speirs, V., Green, A. R., & White, M. C. (1996). Collagenase III: A superior enzyme for complete disaggregation and improved viability of normal and malignant human breast tissue. *In Vitro Cell Dev. Biol.—Animal* **32**:72–74.

Speirs, V., Ray, K. P., & Freshney, R. I. (1991). Paracrine control of differentiation in the alveolar carcinoma, A549, by human foetal lung fibroblasts. *Br. J. Cancer* **64**: 693–699.

Spinelli, W., Sonnenfeld, K. H., & Ishii, N. (1982). Effects of phorbol ester tumor promoters and nerve growth factor on neurite outgrowth in cultured human neuroblastoma cells. *Cancer Res.* **42**:5067–5073.

Splinter, T. A. W., Beudeker, M., & Beek, A. V. (1978). Changes in cell density induced by isopaque. *Exp. Cell Res.* **111**:245–251.

Spooncer, E., Eliason, J., & Dexter, T. M. (1992). Long-term mouse bone marrow cultures. In Testa, N. G., & Molineux, G. (eds.), *Haemopoiesis: a practical approach.* Oxford, U.K., IRL Press at Oxford University Press, pp. 55–74.

Spremulli, E. N., & Dexter, D. L. (1984). Polar solvents: A novel class of antineoplastic agents. *J. Clin. Oncol.* **2**:227–241.

Sredni, B., Sieckmann, D. G., Kumagai, S. H., Green, I., & Paul, W. E. (1981). Long term culture and cloning of non-transformed human B-lymphocytes. *J. Exp. Med.* **154**: 1500–1516.

Stacey, G. N., Bolton, B. J., & Doyle, A. (1993). Multilocus DNA fingerprinting used for definitive isolation of HeLa contamination in cell lines and determination of genetic diversity amongst HeLa cell clones. *In Vitro Cell Dev. Biol.* **29A**:123A.

Stacey, G. N., Bolton, B. J., Morgan, D., Clark, S. A., & Doyle, A. (1992). Multilocus DNA fingerprint analysis of cellbanks: Stability studies and culture identification in human B-lymphoblastoid and mammalian cell lines. *Cytotechnology* **8**:13–20.

Stacey, G. N., Masters, J. R. M., Hay, R. J., Drexler, H. G., MacLeod, R. A. F., Freshney, R. I. (2000) Cell contamination leads to inaccurate data: we must take action now. *Nature,* **403**:456.

Stacey, G. N., Hoelzl, H., Stehenson, J. R., & Doyle, A. (1997). Authentication of animal cell cultures by direct visualisation of repetitive DNA, aldolase gene PCR and isoenzyme analysis. *Biologicals* **25**:75–85.

Stampfer, M., Halcones, R. G., & Hackett, A. J. (1980). Growth of normal human mammary cells in culture. *In Vitro* **16**:415–425.

Stanbridge, E. J., & Doersen, C.-J. (1978). Some effects that mycoplasmas have upon their injected host. In McGarrity, G. J., Murphy, D. G., & Nichols, W. W. (eds.), *Mycoplasma infection of cell cultures.* New York, Plenum Press, pp. 119–134.

Stanley, M. A., & Parkinson, E. (1979). Growth requirements of human cervical epithelial cells in culture. *Int. J. Cancer* **24**:407–414.

Stanners, C. P., Eliceri, G. L., & Green, H. (1971). Two types of ribosome in mouse–hamster hybrid cells. *Nature New Biol.* **230**:52–54.

Stanton, B. A., Biemesderfer, D., Wade, J. B., & Giebisch, G. (1981). Structural and functional study of the rat distal nephron: Effects of potassium adaptation and depletion. *Kidney Int.* **19**:36–48.

Stark, H.-J., Baur, M., Breitkreutz, D., Mirancea, N., & Fusenig, N. E. Organotypic keratinocyte cocultures in defined medium with regular epidermal morphogenesis and differentiation. *J. Inv. Derm.* **112**:681–691.

States, B., Foreman, J., Lee, J., & Segal, S. (1986). Characteristics of cultured human renal cortical epithelia. *Biochem. Med. Metab. Biol.* **36**:151–161.

Steel, G. G. (1979). Terminology in the description of drug-radiation interactions. *Int. J. Radiat. Oncol. Biol. Phys.* **5**: 1145–1150.

Steele, M. P., Levine, R. A., Joyce-Brady, M., & Brody, J. S. (1992). A rat alveolar type II cell line developed by adenovirus 12SE1A gene transfer. *Am. J. Resp. Cell Mol. Biol.* **6**:50–56.

Stein, H. G., & Yanishevsky, R. (1979). Autoradiography. In Jakoby, W. B., & Pastan, I. H. (eds.), *Methods in enzymology; vol. 57: Cell culture*. New York, Academic Press, pp. 279–292.

Steinberg, M. L. (1996). Immortalization of human epidermal keratinocytes by SV40. In Freshney, R. I., & Freshney, M. G. (eds.), *Culture of immortalized cells*. New York, Wiley-Liss, pp. 95–120.

Stewart, C. E. H., James, P. L., Fant, M. E., & Rotwein, P. (1996). Overexpression of insulin-like growth factor-II induces accelerated myoblast differentiation. *J. Cell. Physiol.* **169**:23–32.

Stoker, M., O'Neill, C., Berryman, S., & Waxman, B. (1968). Anchorage and growth regulation in normal and virus transformed cells. *Int. J. Cancer* **3**:683–693.

Stoker, M., Perryman, M., & Eeles, R. (1982). Clonal analysis of morphological phenotype in cultured mammary epithelial cells from human milk. *Proc. R. Soc. Lond., Ser. B* **215**:231–240.

Stoker, M. G. P. (1973). Role of diffusion boundary layer in contact inhibition of growth. *Nature* **246**:200–203.

Stoker, M. G. P., & Rubin, H. (1967). Density dependent inhibition of cell growth in culture. *Nature* **215**:171–172.

Stoner, G. D., Katoh, Y., Foidart, J.-M., Trump, B. F., Steinert, P., & Harris, C. C. (1981). Cultured human bronchial epithelial cells: Blood group antigens, keratin, collagens and fibronectin. *In Vitro* **17**:577–587.

Strange, R., Li, F., Fris, R. R., Reichmann, E., Haenni, B., & Burri, P. H. (1991). Mammary epithelial differentiation *in vitro*: Minimum requirements for a functional response to hormonal stimulation. *Cell Growth & Differentiation* **2**: 549–559.

Strauss, W. M. (1998). Transfection of mammalian cells with yeast artificial chromosomes. In Ravid, K., Freshney, R. I. (eds.), *DNA transfer to cultured cells*. New York, Wiley-Liss, pp. 213–236.

Strickland, S., & Beers, W. H. (1976). Studies on the role of plasminogen activator in ovulation: *In vitro* response of granulosa cells to gonadotropins, cyclic nucleotides, and prostaglandins. *J. Biol. Chem.* **251**:5694–5702.

Stryer, L. (1995). Biochemistry, 4th ed. New York, W. H. Freeman, p. 505.

Stubblefield, E. (1968). Synchronization methods for mammalian cell cultures. In Prescott, D. M. (ed.), *Methods in cell physiology*. New York, Academic Press, pp. 25–43.

Styles, J. A. (1977). A method for detecting carcinogenic organic chemicals using mammalian cells in culture. *Br. J. Cancer* **36**:558.

Su, H. Y., Bos, T. J., Monteclaro, F. S., & Vogt, P. K. (1991). Jun inhibits myogenic differentiation. *Oncogene* **6**:1759–1766.

Subramanian, M., Madden, J. A., & Harder, D. R. (1991). A method for the isolation of cells from arteries of various sizes. *J. Tiss. Cult Meth.* **13**:13–20.

Subramanian, S. V., Fitzgerald, M. L., & Bernfield, M. (1997). Regulated shedding of syndecan-1 and -4 ectodomains by thrombin and growth factor receptor activation. *J. Biol. Chem.* **272**:14713–14720.

Suggs, J. E., Madden, M. C., Friedman, M., & Edgell, C.-J. S. (1986). Prostacyclin expression by a continuous human cell line derived from vascular endothelium. *Blood 1986* **4**:825–829.

Sun, L., Bradford, C. S., & Barnes, D. W. (1995a). Feeder cell cultures for zebrafish embryonal cells *in vitro. Mol. Mar. Biol. Biotech.* **4**:43–50.

Sun, L., Bradford, C. S., Ghosh, C., Collodi, P., & Barnes, D. W. (1995b). ES-like cell cultures derived from early zebrafish embryos. *Mol. Mar. Biol. Biotech.* **4**:193–199.

Sundqvist, K., Liu, Y., Arvidson, K., Ormstad, K., Nilsson, L., Toftgård, R., & Grafström, R. C. (1991). Growth regulation of serum-free cultures of epithelial cells from normal human buccal mucosa. *In Vitro Cell Dev. Biol.* **27A**: 562–568.

Sutherland, R. M. (1988). Cell and micro environment interactions in tumour microregions: The multicell spheroid model. *Science* **240**:117–184.

Swope, V. B., Supp, A. P., Cornelius, J. R., Babcock, G. F., & Boyce, S. T. (1997). Regulation of pigmentation in cultured skin substitutes by cytometric sorting of melanocytes and keratinocytes. *J. Invest. Dermatol.* **109**:289–295.

Sykes, J. A., Whitescarver, J., Briggs, L., & Anson, J. H. (1970). Separation of tumor cells from fibroblasts with use of discontinuous density gradients. *J. Natl. Cancer Inst.* **44**: 855–864.

Tagawa, M., Yokosuka, O., Imazeki, F., Ohto, M., & Omata, M. (1996). Gene expression and active virus replication in the liver after injection of duck hepatitis B virus DNA into the peripheral vein of ducklings. *J. Hepatol.* **24**:328–334.

Takahashi, K., & Okada, T. S. (1970). Analysis of the effect of "conditioned medium" upon the cell culture at low density. *Dev. Growth Diff.* **12**:65–77.

Takahashi, K., Suzuki, K., Kawahara, S., & Ono, T. (1991). Effects of lactogenic hormones on morphological development and growth of human breast epithelial cells cultivated in collagen gels. *Japanese J. Cancer Res.* **82**:553.

Takeda, K., Minowada, J., & Bloch, A. (1982). Kinetics of appearance of differentiation-associated characteristics in ML-1, a line of human myeloblastic leukaemia cells, after treatment with TPA, DMSO, or Ara-C. *Cancer Res.* **42**: 5152–5158.

Tarella, C., Ferrero, D., Gallo, E., Luyca Pagliardi, G., & Ruscetti, F. W. (1982). Induction of differentiation of HL-60 cells by dimethylsulphoxide: Evidence for a stochastic model not linked to the cell division cycle. *Cancer Res.* **42**:445–449.

Tashjian, A. H., Jr. (1979). Clonal strains of hormone-producing pituitary cells. In Jakoby, W. B., & Pastan, I.

H. (eds.), *Methods in enzymology; Vol. 57: Cell culture*. New York, Academic Press, pp. 527–535.

Tashjian, A. H., Yasamura, Y., Levine, L., Sato, G. H., & Parker, M. (1968). Establishment of clonal strains of rat pituitary tumor cells that secrete growth hormone. *Endocrinology* **82**:342–352.

Taub, M. L., Yang, S. I., & Wang, Y. (1989). Primary rabbit proximal tubule cell cultures maintain differentiated functions when cultured in a hormonally defined serum-free medium. *In Vitro Cell Dev. Biol.* **25**:770–775.

Taylor, J. H. (1958). Sister chromatid exchanges in tritium labeled chromosomes. *Genetics* **43**:515–529.

Taylor-Papadimitriou, J., Purkiss, P., & Fentiman, I. S. (1980). Choleratoxin and analogues of cyclic AMP stimulate the growth of cultured human epithelial cells. *J. Cell Physiol.* **102**:317–322.

Taylor-Papadimitriou, J., Shearer, M., & Stoker, M. G. P. (1977). Growth requirement of human mammary epithelial cells in culture. *Int. J. Cancer* **20**:903–908.

Tedder, R. S., Zuckerman, M. A., Goldstone, A. H., Hawkins, A. E., Fielding, A., Briggs, E. M., Irwin, D., Blair, S., Gorman, A. M., Patterson, K. G., Linch, D. C., Heptonstall, J., & Brink, N. S. (1995). Hepatitis B transmission from contaminated cryopreservation tank. *Lancet*, **346**:137–140.

Temin, H. M. (1966). Studies on carcinogenesis by avian sarcoma viruses; III: The differential effect of serum and polyanions on multiplication of uninfected and converted cells. *J. Natl. Cancer Inst.* **37**:167–175.

Teofili, L., Rutella, S., Pierelli, L., Ortu la Barbera, E., di Mario, A., Menichella, G., Rumi, C., & Leone, G. (1996). Separation of chemotherapy plus G-CSF-mobilized peripheral blood mononuclear cells by counterflow centrifugal elutriation: *In vitro* characterization of two different CD34[+] cell populations. *Bone Marrow Transplantation* **18**:421–425.

Terasaki, T., Kameya, T., Nakajima, T., Tsumuraya, M., Shimosato, Y., Kato, K., Ichinose, H., Nagatsu, T., & Hasegawa, T. (1984). Interconversion of biological characteristics of small cell lung cancer cells depending on the culture conditions. *Gann* **75**:1689–1699.

Testa, N. G., & Molineux, G. (eds.). (1993). *Haemopoiesis*. Oxford, U.K., Oxford University Press.

Thacker, J., Webb, M. J. T., & Debenham, P. G. (1988). Fingerprinting cell lines: Use of human hypervariable DNA probes to characterise mammalian cell cultures. *Som. Cell and Mol. Gen.* **14**:519–525.

Thomas, D. G. T., Darling, J. L., Paul, E. A., Mott, T. C., Godlee, J. N., Tobias, J. S., Capra, L. G., Collins, C. D., Mooney, C., Bozek, T., Finn, G. P., Arigbabu, S. O., Bullard, D. E., Shannon, N., & Freshney, R. I. (1985). Assay of anti-cancer drugs in tissue culture: Relationship of relapse free interval (FRI) and *in vitro* chemosensitivity in patients with malignant cerebral glioma. *Br. J. Cancer* **51**:525–532.

Thomas, S., Gray, E., & Robinson, C. J. (1997). Response of HUVEC and EAhy926 and fibroblast growth factors. *In Vitro Cell Dev. Biol.—Animal* **33**:492–494.

Thompson, L. H., & Baker, R. M. (1973). Isolation of mutants of cultured mammalian cells. In Prescott, D. (ed.),

Methods in cell biology, Vol. 6. New York, Academic Press, pp. 209–281.

Thomson, A. A., Foster, B. A., & Cunha, G. R. (1997). Analysis of growth factor and receptor mRNA levels during development of the rat seminal vesicle and prostate. *Development* **124**:2431–2439.

Thomson, A. W. (ed.). (1991). *Cytokine handbook*. London, Academic Press.

Thomson, J. A., Itskovitz-Eldor, J., Shapiro, S. S., Waknitz, M. A., & Swiergiel, J. J. (1998). Embryonic stem cell lines derived from human blastocysts. *Science* **282**:1145–1147.

Thornton, S. C., Mueller, S. N., & Levine, E. M. (1983). Human endothelial cells: Use of heparin in cloning and long-term serial cultivation. *Science* **222**:623–625.

Thorsell, A., Blomqvist, A. G., & Heilig, M. (1996). Cationic lipid-mediated delivery and expression of prepro-neuropeptide Y cDNA after intraventricular administration in rat: Feasibility and limitations. *Regulatory Peptides* **61**:205–211.

Till, J. E., & McCulloch, E. A. (1961). A direct measurement of the radiation sensitivity of normal mouse bone marrow cells. *Radiation Res.* **14**:213–222.

Tobey, R. A., Anderson, E. C., & Petersen, D. F. (1967). Effect of thymidine on duration of G1 in chinese hamster cells. *J. Cell Biol.* **35**:53–67.

Todaro, G. J., & DeLarco, I. E. (1978). Growth factors produced by sarcoma virus-transformed cells. *Cancer Res.* **38**:4147–4154.

Todaro, G. J., Green, H. (1963). Quantitative studies of the growth of mouse embryo cells in culture and their development into establed lines. *J. Cell Biol.* **17**:299–313.

Toji, L. H., Lenchitz, T. C., Kwiatkowski, V. A., Sarama, J. A., & Mulivor, R. A. (1998). Validation of mycoplasma testing by PCR. *In Vitro Cell Dev. Biol.—Animal* **34**:356–358.

Tomakidi, P., Fusenig, N. E., Kohl, A., & Komposch, G. (1997). Histomorphological and biochemical differentiation capacity in organotypic co-cultures of primary gingival cells. *J. Periodont Res.* **32**:388–400.

Toneguzzo, F., Keating, A., Glynn, S., & McDonald, K. (1988). Electrical field mediated gene transfer: Characterization of DNA transfer and patterns of integration in lymphoid cells. *Nucleic Acids Res.* **16**:5515–5532.

Topley, P., Jenkins, D. C., Jessup, E. A., & Stables, J. N. (1993). Effect of reconstituted basement membrane components on the growth of a panel of human tumour cell lines in nude mice. *Br. J. Cancer* **67**:953–958.

Torday, J. S., & Kourembanas, S. (1990). Fetal rat lung fibroblasts produce a TGF-β homologue that blocks type II cell maturation. *Dev. Biol.* **13**:35–41.

Tozer, B. T., & Pirt, S. J. (1964). Suspension culture of mammalian cells and macromolecular growth promoting fractions of calf serum. *Nature* **201**:375–378.

Traganos, F., Darzynkiewicz, Z., Sharpless, T., & Melamed, M. R. (1977). Nucleic acid content and cell cycle distribution of five human bladder cell lines analyzed by flow cytofluorometry. *Int. J. Cancer* **20**:30–36.

Trapp, B. D., Honegger, P., Richelson, E., & Webster, H. de F. (1981). Morphological differentiation of mechanically dissociated fetal rat brain in aggregating cell cultures. *Brain Res.* **160**:235–252.

Trickett, A. E., Ford, D. J., Lam-Po Tang, P. R. L., & Vowels, M. R. (1990). Comparison of magnetic particles for immunomagnetic bone marrow purging using an acute lymphoblastic leukaemia model. *Transpl. Proc.* **22**:2177–2178.

Triglia, D., Braa, S. S., Yonan, C., & Naughton, G. K. (1991). Cytotoxicity testing using neutral red and MTT assays on a three-dimensional human skin substrate. *Toxic. In Vitro* **5**:573–578.

Trotter, J., & Schachner, M. (1988). Cells positive for the O4 surface antigen isolated by cell sorting are able to differentiate into oligodendrocytes and type-2 astrocytes. *Dev. Brain Res.* **46**:115–122.

Trowell, O. A. (1959). The culture of mature organs in a synthetic medium. *Exp. Cell Res.* **16**:118–147.

Troyer, D. A., & Kreisberg, J. I. (1990). Isolation and study of glomerular cells. *Methods Enzymol.* **191**:141–152.

Tsao, M. C., Walthall, B. I., & Ham, R. G. (1982). Clonal growth of normal human epidermal keratinocytes in a defined medium. *J. Cell Physiol.* **110**:219–229.

Tsao, S.-W., Mok, S. C., Fey, E. G., Fletcher, J. A., Wan, T. S. K., Chew, E.-C., Muto, M. G., Knapp, R. C., & Berkowitz, R. S. (1995). Characterisation of human ovarian surface epithelial cells immortalized by human papilloma viral oncogenes (HPV-E6E7 ORFs). *Exp. Cell Res.* **218**:499–507.

Tsuruo, T., Hamilton, T. C., Louis, K. G., Behrens, B. C., Young, R. C., & Ozols, R. F. (1986). Collateral susceptibility of adriamycin-, melphalan-, and cisplatin-resistant human ovarian tumor cells to bleomycin. *Jpn. J. Cancer Res.* **77**:941–945.

Tumilowicz, J. J., Nichols, W. W., Cholon, J. J., & Greene, A. E. (1970). Definition of a continuous human cell line derived from neuroblastoma. *Cancer Res.* **30**:2110–2118.

Turner, R. W. A., Siminovitch, L., McCulloch, E. A., & Till, J. E. (1967). Density gradient centrifugation of hemopoietic colony-forming cells. *J. Cell Physiol.* **69**:73–81.

Tuszynski, M. H., Roberts, J., Senut, M. C., U, H. S., & Gage, F. H. (1996). Gene therapy in the adult primate brain: Intraparenchymal grafts of cells genetically modified to produce nerve growth factor prevent cholinergic neuronal degeneration. *Gene Therapy* **3**:305–314.

Tveit, K. M., & Pihl, A. (1981). Do cells lines *in vitro* reflect the properties of the tumours of origin? A study of lines derived from human melanoma xenografts. *Br. J. Cancer* **44**:775–786.

Twentyman, P. R. (1980). Response to chemotherapy of EMT6 spheroids as measured by growth delay and cell survival. *Eur. J. Cancer* **42**:297–304.

Uchida, I. A., & Lin, C. C. (1974). Quinacrine fluorescent patterns. In Yunis, J. (ed.), *Human chromosome methodology*, 2d ed. New York, Academic Press, pp. 47–58.

UKCCCR. (1999). *Guidelines for the use of human cell lines in cancer research.* UKCCCR, P. O. Box 123, Lincoln's Inn Fields, London WC2A 3PX, England.

United States Pharmacopeia. (1985). *Sterility tests*, 21st revision. United States Pharmacopeial Convention, Inc., pp. 1156–1160.

Unkless, I., Dano, K., Kellerman, G., & Reich, E. (1974). Fibrinolysis associated with oncogenic transformation: Partial purification and characterization of cell factor, a plasminogen activator. *J. Biol. Chem.* **249**:4295–4305.

Ure, J. M., Fiering, S., & Smith, A. G. (1992). A rapid and efficient method for freezing and recovering clones of embryonic stem cells. *Trends Genet.* **8**:6.

U.S. Department of Health and Human Services (1993). *Biosafety in microbiological and biomedical laboratories*, 3d ed. Publication (CDC) 93-8395, Centers for Disease Control, US Govt. Printing Office, Washington, DC.

U.S. Nuclear Regulatory Commission (1997). Draft regulatory guide DG-0006. Guide for the preparation of applications for commercial nuclear pharmacy Licenses. Office of Nuclear Regulatory Research, U.S. Nuclear Regulatory Commission, Washington, DC 20555.

Vachon, P. H., Perreault, N., Magny, P., & Beallieu, J.-F. (1996). Uncoordinated transient mosaic patterns of intestinal hydrolase expression in differentiating human enterocytes. *J. Cell Physiol.* **166**:198–207.

Vago, C. (ed.). (1971). *Invertebrate tissue culture*, Vol. 1. New York, Academic Press.

Vago, C. (ed.). (1972). *Invertebrate tissue culture*, Vol. 2. New York, Academic Press.

Vaheri, A., Ruoslahti, E., Westermark, B., & Pontén, J. (1976). A common cell-type specific surface antigen in cultured human glial cells and fibroblasts: Loss in malignant cells. *J. Exp. Med.* **143**:64–72.

Van Diggelen, O., Shin, S., & Phillips, D. (1977). Reduction in cellular tumorigenicity after mycoplasma infection and elimination of mycoplasma from infected cultures by passage in nude mice. *Cancer Res.* **37**:2680–2687.

Van Helden, P. D., Wiid, I. J., Albrecht, C. F., Theron, E., Thornley, A. L., & Hoal-van Helden, E. G. (1988). Cross-contamination of human esophageal squamous carcinoma cell lines detected by DNA fingerprint analysis. *Cancer Res.* **48**:5660–5662.

Van Roozendahl, C. E. P., van Ooijen, B., Klijn, J. G. M., Claasen, C., Eggermont, A. M. M., Henzen-Logmans, S. C., & Foekens, J. A. (1992). Stromal influences on breast cancer cell growth. *Br. J. Cancer* **65**:77–81.

Varga Weisz, P. D., & Barnes, D. W. (1993). Characterization of human plasma growth inhibitory activity on serum-free mouse embryo cells. *In vitro Cell Dev. Biol.* **29A**:512–516.

Varner, H. H., Hewitt, A. T., & Martin, G. R. (1984). Isolation of chondronectin. In Barnes, D. W., Sirbasku, D. A., & Sato, G. H. (eds.), *Cell culture methods for molecular and cell biology*, Vol. 1. New York, Alan R. Liss, pp. 239–244.

Varon, S., & Manthorpe, M. (1980). Separation of neurons and glial cells by affinity methods. In Fedoroff, S., & Hertz, L. (eds.), *Advances in cellular neurobiology*, Vol. 1. New York, Academic Press, pp. 405–442.

Vaughan, A., & Milner, A. (1989). Fluorescence activated cell sorting. In Catty, D. (ed.), *Antibodies; Volume II: A practical approach.* Oxford, U.K., IRL Press at Oxford University Press, pp. 201–222.

Vaziri, H., & Benchimol, S. (1998). Reconstitution of telomerase activity in normal human cells leads to elongation of telomeres and extended replicative life-span. *Current Biology* **8**:279–282.

Velcich, A., Palumbo, L., Jarry, A., Laboisse, C., Racevskis, J., & Augenlicht, L. (1995). Patterns of expression of lineage-specific markers during the *in vitro*-induced differentiation of HT29 colon carcinoma cells. *Cell Growth & Diff.* **6**:749–757.

Venitt, S. (1984). *Mutagenicity testing, a practical approach.* Oxford, U.K., IRL Press.

Verbruggen, G., Veys, E. M., Wieme, N., Malfait, A. M., Gijselbrecht, L., Nimmeegers, J., Almquist, K. F., & Broddelez, C. (1990). The synthesis and immobilization of cartilage-specific proteoglycan by human chondrocytes in different concentratons of agarose. *Clin. Exp. Rheumatol.* **8**:371–378.

Vierick, J. L., McNamara, P., & Dodson, M. V. (1996). Proliferation and differentiation of progeny of ovine unilocular fat cells (adipofibroblasts). *In Vitro Cell Dev. Biol.— Animal* **32**:564–572.

Vilamitjana-Amedee, J., Bareile, R., Rouais, F., Caplan, A. I., & Harmand, M. F. (1993). Human bone marrow stromal cells express an osteoblastic phenotype in culture. *In Vitro Cell Dev. Biol.* **29A**:699–707.

Visser, J. W., & De Vries, P. (1990). Identification and purification of murine hematopoietic stem cells by flow cytometry. *Methods Cell Biol.* **33**:451–468.

Vistica, D. T., Skehan, P., Scudiero, D., Monks, A., Pittman, A., & Boyd, M. R. (1991). Tetrazolium-based assays for cellular viability: A critical examination of selected parameters affecting formazan production. *Cancer Res.* **51**: 2515–2520.

Vlodavsky, I., Lui, G. M., & Gospodarowicz, D. (1980). Morphological appearance, growth behavior and migratory activity of human tumor cells maintained on extracellular matrix versus plastic. *Cell* **19**:607–617.

Vogel, F. R., & Powell, M. F. (1995). A compendium of vaccine adjuvants and excipients. In Powell, M. F., & Newman, M. (eds.), *Vaccine design: The subunit and adjuvant approach.* New York, Plenum Publishing, pp. 141–228.

Von der Mark, K. (1986). Differentiation, modulation and dedifferentiation of chondrocytes. *Rheumatology* **10**:272–315.

Von Hoff, D. D., Clark, G. M., Weis, G. R., Marshall, M. H., Buchok, J. B., Knight, W. A., & Lemaistre, C. F. (1986). Use of *in vitro* dose response effects to select antineoplastics for high dose or regional administration regimens. *J. Clin. Oncol.* **4**:18–27.

Vonen, B., Bertheussen, K., Giaever, A. K., Florholmen, J., & Burhol, P. G. (1992). Effect of a new synthetic serum replacement on insulin and somatostatin secretion from isolated rat pancreatic islets in long term culture. *J. Tiss. Cult. Meth.* **14**:45–50.

Voyta, J. C., Via, D. P., Butterfield, C. W., & Zetter, B. R. (1984). Identification and isolation of endothelial cells based on their increased uptake of acetylated–low density lipoprotein. *J. Cell Biol.* **99**:2034–2040.

Vries, J. E., Benthem, M., & Rumke, P. (1973). Separation of viable from nonviable tumor cells by flotation on a Ficoll-triosil mixture. *Transplantation* **5**:409–410.

Wada, T., Dacy, K. M., Guan, X.-P., & Ip, M. M. (1994). Phorbol 12-myristate 13 acetate stimulates proliferation and ductal morphogenesis and inhibits functional differentiation of normal rat mammary epithelial cells in primary culture. *J. Cell Physiol.* **158**:97–109.

Waleh, N. S., Brody, M. D., Knapp, M. A., Mendonca, H. L., Lord, E. M., Koch, C. J., Laderoute, K. R., & Sutherland, R. M. (1995). Mapping of the vascular endothelial growth factor-producing hypoxic cells in multicellular tumour spheroids using a hypoxia-specific marker. *Cancer Res.* **55**: 6222–6226.

Walston, J., Silver, K., Bogardus, C., Knowler, W. C., Celi, F. S., Austin, S., Manning, B., Strosberg, A. D., Stern, M. P., Raben, N., Sorkin J. D., Roth, J., & Shuldiner, A. R. (1995). Time of onset of non-insulin-dependent diabetes mellitus and genetic variation in the beta 3-adrenergic-receptor gene. *N. Engl. J. Med.* **333**:343–347.

Walter, H. (1975). Partition of cells in two-polymer aqueous phases: A method for separating cells and for obtaining information on their surface properties. In Prescott, D. M. (ed.), *Methods in cell biology.* New York, Academic Press, pp. 25–50.

Walter, H. (1977). Partition of cells in two-polymer aqueous phases: A surface affinity method for cell separation. In Catsimpoolas, N. (ed.), *Methods of cell separation.* New York, Plenum Press, pp. 307–354.

Wang, H. C., & Fedoroff, S. (1972). Banding in human chromosomes treated with trypsin. *Nature New Biol.* **235**:52–53.

Wang, H. C., & Fedoroff, S. (1973). Karyology of cells in culture: Trypsin technique to reveal G-bands. In Kruse, P. F., & Patterson, M. J. (eds.), *Tissue culture methods and applications.* New York, Academic Press, pp. 782–787.

Wang, R. I. (1976). Effect of room fluorescent light on the deterioration of tissue culture medium. *In Vitro* **12**:19–22.

Warburton, P. E., & Kipling, D. (1997). Providing a little stability. *Nature,* **386**:553–555.

Ward, J. P., & King, J. R. (1997). Mathematical modelling of avascular tumour growth. *IMA J. Math. App. Med. Biol.* **14**: 39–69.

Warnock, M. (1985). *A question of life: The Warnock report on human fertilisation and embryology.* Oxford, U.K., Basil Blackwell.

Watanabe, T., Kondo, K., & Oishi, M. (1991). Induction of *in vitro* differentiation of mouse erythroleukemia cells by genistein, an inhibitor of tyrosine kinases. *Cancer Res.* **51**: 764–768.

Watt, F. (1991). Annual Meeting of European Tissue Culture Society, Kraków, Poland.

Watt, J. L., & Stephen, G. S. (1986). Lymphocyte culture for chromosome analysis. In Rooney, D. E., & Czepulkowski, B. H. (eds.), *Human cytogenetics, a practical approach.* Oxford, U.K., IRL Press at Oxford University Press, pp. 39–56.

Waymouth, C. (1959). Rapid proliferation of sublines of NCTC clone 929 (Strain L) mouse cells in a simple chemically defined medium (MB752/1). *J. Natl. Cancer Inst.* **22**: 1003.

Waymouth, C. (1970). Osmolality of mammalian blood and of media for culture of mammalian cells. *In Vitro* **6**:109–127.

Waymouth, C. (1974). To disaggregate or not to disaggregate: Injury and cell disaggregation, transient or permanent? *In Vitro* **10**:97–111.

Waymouth, C. (1979). Autoclavable medium AM 77B. *J. Cell Physiol.* **100**:548–550.

Waymouth, C. (1984). Preparation and use of serum-free culture media. In Barnes, W. D., Sirbasku, D. A., & Sato, G. H. (eds.), *Cell culture methods for molecular and cell biology*; *Vol. 1: Methods for preparation of media, supplements, and substrata for serum-free animal cell culture*. New York, Alan R. Liss, pp. 23–68.

Weibel, E. R., & Palade, G. E. (1964). New cytoplasmic components in arterial endothelia. *J. Cell Biol.* **23**:101–102.

Weichselbaum, R., Epstein, I., & Little, J. B. (1976). A technique for developing established cell lines from human osteosarcomas. *In Vitro* **12**:833–836.

Weinberg, R. A. (ed.). (1989). *Oncogenes and the molecular origins of cancer*. Cold Spring Harbor, NY, Cold Spring Harbor Laboratory Press.

Weiss, M. C., & Green, H. (1967). Human–mouse hybrid cell lines containing partial complements of human chromosomes and functioning human genes. *Proc. Natl. Acad. Sci. USA* **58**:1104–1111.

Wells, D. L., Lipper, S. L., Hilliard, J. K., Stewart, J. A., Holmes, G. P., Herrmann, K. L., Kiley, M. P., & Schonberger, L. B. (1989). *Herpesvirus simiae* contamination of primary rhesus monkey kidney cell cultures: CDC recommendations to minimize risks to laboratory personnel. *Diagn. Microbiol. Infect. Dis.* **12**:333–335.

Wessells, N. K. (1977). *Tissue interactions and development*. Menlo Park, CA, W. A. Benjamin.

Wessels, D., Titus, M., & Soll, D. R. (1996). A Dictyostelium myosin I plays a crucial role in regulating the frequency of pseudopods formed on the substratum. *Cell Motility and the Cytoskeleton* **33**:64–79.

Westerfield, M. (1993). *The zebrafish book: A guide for the laboratory use of zebrafish* (Brachydanio rerio). Eugene, OR, University of Oregon Press.

Westermark, B. (1974). The deficient density-dependent growth control of human malignant glioma cells and virus-transformed glialike cells in culture. *Int. J. Cancer* **12**:438–451.

Westermark, B. (1978). Growth control in miniclones of human glial cells. *Exp. Cell Res.* **111**:295–299.

Westermark, B., Pontén, J., & Hugosson, R. (1973). Determinants for the establishment of permanent tissue culture lines from human gliomas. *Acta Pathol. Microbiol. Scand. A* **81**:791–805.

Westneat, D. F., Noon, W. A., Reeve, H. K., & Aquadro, C. F. (1988). Improved hybridisation conditions for DNA fingerprints probed with M13. *Nucleic Acids Research* **16**, 4161.

Whitehead, R. H., & Jospeh, J. L. (1994). Derivation of conditionally immortalized cell lines containing the min mutation from the normal colonic mucosa and other tissues of an "immortomouse"/min hybrid. *Epith. Cell Biol.* **3**: 119–125.

Whitlock, C. A., Robertson, D., & Witte, O. N. (1984). Murine B cell lymphopoiesis in long term culture. *J. Immunol. Methods* **67**:353–369.

Whur, P., Magudia, M., Boston, I., Lockwood, J., & Williams, D. C. (1980). Plasminogen activator in cultured Lewis lung carcinoma cells measured by chromogenic substrate assay. *Br. J. Cancer* **42**:305–312.

Wienberg, J., & Stanyon, R. (1997). Comparative painting of mammalian chromosomes. *Curr. Opin. Genet. & Dev.* **7**: 784–791.

Wilkenheiser, K. A., Vorbroker, D. K., Rice, W. R., Clark, J. C., Bachurski, C. J., Oie, H. K., & Whitsett, J. E. (1991). Production of immortalized distal respiratory epithelial cell lines from surfactant protein C/simian virus 40 large tumor antigen transgenic mice. *Proc. Natl. Acad. Sci. USA* **90**:11029–11033.

Wilkins, L. Gilchrest, B. A., Szabo, G., Weinstein, R., & Maciag, T. (1985). The stimulation of normal human melanocyte proliferation *in vitro* by melanocyte growth factor from bovine brain. *J. Cell Physiol.* **122**:350.

Willard, H. F. (1998). Centromeres: the missing link in the development of human artificial homosomes. *Curr. Opin. Genet. & Dev.* **8**:219–225.

Willey, J. C., Moser, C. F., Jr., Lechner, J. F., & Harris, C. C. (1984). Differential effects of 12-0-tetradecanoylphorbol-13-acetate on cultured normal and neoplastic human bronchial epithelial cells. *Cancer Res.* **44**:5124–5126.

Williams, B. P., Abney, E. R., & Raff, M. C. (1985). Macroglial cell development in embryonic rat brain: Studies using monoclonal antibodies, fluorescence-activated cell sorting and cell culture. *Dev. Biol.* **112**:126–134.

Williams, G. M., & Gunn, J. M. (1974). Long-term cell culture of adult rat liver epithelial cells. *Exp. Cell Res.* **89**: 139–142.

Wilson, A. P., Dent, M., Pejovic, T., Hubbold, L., & Rodford, H. (1996). Characterisation of seven human ovarian tumour cell lines. *Br. J. Cancer* **74**:722–727.

Wilson, P. D., Dillingham, M. A., Breckon, R., & Anderson, R. J. (1985). Defined human renal tubular epithelia in culture: Growth, characterization, and hormonal response. *Am. J. Physiol.* **248**:F436–F443.

Wilson, P. D., Schrier, R. W., Breckon, R. D., & Gabow, P. A. (1986). A new method for studying human polycystic kidney disease epithelia in culture. *Kidney Int.* **30**:371–378.

Winterton, A. (1989). *Review of the guidance on the research use of fetuses and fetal material*. London, Her Majesty's Stationery Office, Cm 762.

Witkowski, J. A. (1990). The inherited character of cancer—an historial survey. *Cancer Cells* **2**:229–257.

Wolf, D. P., Meng, L., Ely, J. J., & Stouffer, R. L. (1998). Recent progress in mammalian cloning. *J. Assist. Reprod. Genet.* **15**:235–239.

Wolff, E. T., & Haffen, K. (1952). Sur une méthode de culture d'organes embryonnaires *in vitro*. *Tex. Rep. Biol. Med.* **10**:463–472.

Wolswijk, G., & Noble, M. (1989). Identification of an adult-specific glial progenitor cell. *Development* **105**:387–400.

Wright, K. A., Nadire, K. B., Busto, P., Tubo, R., McPherson, J. M., & Wentworth, B. M. (1998). Alternative delivery of keratinocytes using a polyurethane membrane and the implications for its use in the treatment of full-thickness burn injury. *Burns* **24**:7–17.

Wu, Y. J., Parker, L. M., Binder, N. E., Beckett, M. A., Sinard, J. H., Griffiths, C. T., & Rheinwald, J. G. (1982). The mesothelial keratins: A new family of cytoskeletal proteins identified in cultured mesothelial cells and nonkeratinizing epithelia. *Cell* **31**:693–703.

Wuarin, L., Verity, M. A., & Sidell, N. (1991). Effects of interferon-gamma and its interaction with retinoic acid on human neuroblastoma differentiation. *Int. J. Cancer* **48**:136–141.

Wurster-Hill, D., Cannizzaro, L. A., Pettengill, O. S., Sorenson, G. D., Cate, C. C., & Maurer, L. H. (1984). Cytogenetics of small cell carcinoma of the lung. *Cancer Genet. Cytogenet.* **13**:303–330.

Wyllie, F. S., Bond, J. A., Dawson, T., White, D., Davies, R., & Wynford-Thomas, D. (1992). A phenotypically and karyotypically stable human thyroid epithelial line conditionally immortalized by SV40 large T antigen. *Cancer Res.* **52**:2938–2945.

Wysocki, L. J., & Sata, V. L. (1978). "Panning" for lymphocytes: A method for cell selection. *Proc. Natl. Acad. Sci. USA* **75**:2844–2848.

Yamada, K. M., & Geiger, B. (1997). Molecular interactions in cell adhesion complexes. *Curr. Opin. Cell Biol.* **9**:76–85.

Yamada, T., Placzek, M., Tanaka, H., Dodd J., & Jessell, T. M. (1991). Control of cell pattern in the developing nervous system: Polarizing activity of the floor plate and notochord. *Cell* **64**:635–647.

Yamaguchi, N., Yamamura, Y., Koyama, K., Ohtsuji, E., Imanishi, J., & Ashihara, T. (1990). Characterization of new pancreatic cancer cell lines which propagate in a protein-free chemically defined medium. *Cancer Res.* **50**: 7008–7014.

Yan, G., Fukabori, Y., Nikolaropoulost, S., Wang, F., & McKeehan, W. L. (1992). Heparin binding keratinocyte growth factor is a candidate stromal to epithelial cell andromedin. *Mol. Endocrinol.* **6**:2123–2128.

Yanai, N., Suzuki, M., & Obinata, M. (1991). Hepatocyte cell lines established from transgenic mice harboring temperature-sensitive simian virus 40 large T-antigen gene. *Exp. Cell Res.* **197**:50–56.

Yasumura, Y., Tashijian, A. H., & Sato, G. (1966). Establishment of four functional clonal strains of animal cells in culture. *Science* **154**:1186–1189.

Yen, A., Coles, M., & Varvayanis, S. (1993). 1,25-dihydroxy vitamin D3 and 12-0-tetradecanoyl phorbol-13-acetate synergistically induce monocytic cell differentiation: FOS and RB expression. *J. Cell Physiol.* **156**:198–203.

Yeoh, G. C. T., Hilliard, C., Fletcher, S., & Douglas, A. (1990). Gene expression in clonally derived cell lines produced by *in vitro* transformation of rat fetal hepatocytes: Isolation of cell lines which retain liver-specific markers. *Cancer Res.* **50**:75–93.

Yerganian, G., & Leonard, M. J. (1961). Maintenance of normal *in situ* chromosomal features in long-term tissue cultures. *Science* **133**:1600–1601.

Yevdokimova, N., & Freshney, R. I. (1997). Activation of paracrine growth factors by heparan sulphate induced by glucocorticoid in A549 lung carcinoma cells. *Brit. J. Cancer* **76**:261–289.

Yoshida, M., & Beppu, T. (1990). In Doyle, A., Griffiths, J. B., & Newell, D. G., *Cell and tissue culture: Laboratory procedures.* Chichester, U.K., John Wiley & Sons, Module 4E2.

Yoshioka, M., Nakajima, Y., Ito, T., Mikami, O., Tanaka, S., Miyazaki, S., & Motoi, Y. (1997). Primary culture and expression of cytokine mRNAs by lipopolysaccharide in bovine Kupffer cells. *Vet. Immunol. Immunopathol.* **58**:155–163.

Yuhas, J. M., Li, A. P., Martinez, A. O., & Ladman, A. J. (1977). A simplified method for production and growth of multicellular tumour spheroids (MTS). *Cancer Res.* **37**: 3639–3643.

Yuspa, S. H., Koehler, B., Kulesz-Martin, M., & Hennings, H. (1981). Clonal growth of mouse epidermal cells in medium with reduced calcium concentration. *J. Invest. Dermatol.* **76**:144–146.

Yusufi, A. N. K., Szczepanska-Konkel, M., Kempson, S. A., McAteer, J. A., & Dousa, T. P. (1986). Inhibition of human renal epithelial Na$^+$/Pi cotransport by phosphonoformic acid. *Biochem. Biophys. Res. Commun.* **139**:679–686.

Zaroff, L., Sato, G. H., & Mills, S. E. (1961). Single-cell platings from freshly isolated mammalian tissue. *Exp. Cell Res.* **23**:565–575.

Zeltinger, J., & Holbrook, K. A. (1997). A model system for long-term serum-free suspension organ culture of human fetal tissues: Experiments on digits and skin from multiple body regions. *Cell & Tissue Research* **290**:51–60.

Zetter, B. R. (1981). The endothelial cells of large and small blood vessels. *Diabetes* **30(suppl 2)**:24–28.

Zhang, Y., Proenca, R., Maffei, M., Barone, M., Leopold, L., & Friedman, J. M. (1994). Positional cloning of the mouse obese gene and its human homologue. *Nature* **372**:425–432.

Zhu, S. Y., Cunningham, M. L., Gray, T. E., & Nettesheim, P. (1991). Cytotoxicity, genotoxicity and transforming activity of 4-(methylnitrosamino)-1-(3-pyridyl)-1-butanone (NNK) in rat trachea epithelial cells. *Mutation Res.* **261**(4): 249–259.

Zimmermann, U., & Vienken, J. (1982). Electric field-induced cell-to-cell fusion. *J. Mem. Biol.* **67**:165–182.

Zwain, I. H., Morris, P. L., & Cheng, C. Y. (1991). Identification of an inhibitory factor from a Sertoli clonal cell line (TM4) that modulates adult rat Leydig cell steroidogenesis. *Mol. Cell Endocrinol.* **80**:115–126.

General Textbooks for Further Reading

Adolphe, M., & Barlovatz-Meimon, G. (1985). *Culture de cellules animales; methodologies, applications.* Paris, Editions INSERM. *A collection of technique-oriented chapters on basic and advanced aspects of tissue culture.*

Alberts, B., Bray, D., Lewis, J., Raff, M., Roberts, K., & Watson, J. D. (1997). The Molecular Biology of the Cell, 4th edition. Garland, New York.

Barnes, D. W., Sirbasku, D. A., & Sato, G. H. (eds.). (1984). *Cell culture methods for molecular and cell biology.* 4 vols. New York, Alan R. Liss.

Butler, M. (ed.). (1991). *Mammalian cell biotechnology, a practical approach.* Oxford, U.K., IRL Press at Oxford University Press. *A useful introduction to basic biotechnology.*

Davis, J. M. (1994). *Basic cell culture, a practical approach.* Oxford, U.K., IRL Press at Oxford University Press.

Dealtry, G. B., & Rickwood, D. (1992). *Cell biology LabFax.* Oxford, U.K., Bios Scientific Publishers. *A useful collection of data on microscopy, cell structure, oncogenes, growth factors, inhibitors, and radioisotopes in biology.*

Dixon, R. A. (1985). *Plant cell culture, a practical approach.* Oxford, IRL Press.

Doyle, A., Griffiths, J. B., & Newell, D. G. (eds.). (1993). *Cell and tissue culture: Laboratory procedures.* Chichester, U.K., John Wiley & Sons. *A loose-leaf compendium of general and specialized techniques with regular updates. Very expensive, but a very good source for a wide variety of techniques.*

Doyle, A., Hay, R., & Kirsop, B. E. (eds.). (1990). *Living resources for biotechnology.* Cambridge, U.K., Cambridge University Press. *Useful information on databases and quality control.*

Freshney, R. I. (ed.). (1992). *Culture of epithelial cells.* New York, Wiley-Liss. *Invited chapters on specialized culture of epithelium; technique oriented.*

Freshney, R. I. (1999). Freshney's *Culture of animal cells, a multimedia guide.* New York, Wiley-Liss.

Freshney, R. I., & Freshney, M. G. (1996). *Culture of immortalized cells.* New York, Wiley-Liss.

Freshney, R. I., Pragnell, I. B., & Freshney, M. G. (eds.). (1994). *Culture of haemopoietic cells.* New York, Wiley-Liss

(in press). *Second in the series "Culture of Specialized Cells." Invited chapters on specialized techniques.*

Haynes, L. W. (ed.). *The neuron in tissue culture.* Chichester, U.K., John Wiley & Sons.

Leigh, I. M., Lane, E. B., & Watt, F. M. (eds.). (1994). *The keratinocyte handbook.* Cambridge, U.K., Cambridge University Press.

Leigh, I. M., & Watt, F. M. (eds.). (1994). *Keratinocyte methods.* Cambridge, U.K., Cambridge University Press.

Lodish, H., Baltimore, D., Berk, A., Zipursky, S. L., Matsudaira, P., & Darnell, J. (1999). Molecular Cell Biology. Scientific American Books, Freeman, New York.

Masters, J. R. W., (ed.), *Animal cell culture, a practical approach,* 2d ed. Oxford, U.K., IRL Press (in press).

Masters, J. R. W. (ed.). (1991). *Human cancer in primary culture.* London, Kluwer. *Product of a European Tissue Culture Society workshop.*

Masters, J. R. W., & Palsson, B. (1999). (eds.). *Human cell culture.* Dordrecht, the Netherlands, Kluwer.

Pollack, R. (ed.). (1981). *Reading in mammalian cell culture,* 2nd ed. Cold Spring Harbor, NY, Cold Spring Harbor Laboratory. *A very good compilation of key papers in the field. Used as a tutorial, for general interest, and for teaching.*

Ravid, K., & Freshney, R. I. (1998). *DNA transfer to cultured cells.* New York, Wiley-Liss.

Reinert, J., & Yeoman, M. M. (1982). *Plant cell and tissue culture: a laboratory manual.* Berlin, Heidelberg, & New York, Springer-Verlag.

Sato, G., Pardee, A. B., & Sirbasku, D. A. (1982). *Growth of cells in hormonally defined media.* Cold Spring Harbor Conferences on Cell Proliferation, Vol. 9. Cold Spring Harbor, NY, Cold Spring Harbor Laboratory. *A good review of serum-free culture.*

Shahar, A., de Vellis, J., Vernadakis, A., & Haber, B. (1989). *A dissection and tissue culture manual of the nervous system.* New York, Wiley-Liss. *Useful short protocols; well illustrated.*

Spier, R. E., & Griffiths, J. B. (eds.). (1985). *Animal cell biotechnology,* 4 vols. New York, Academic Press. *Invited chapters covering a wide range of biotechnological applications of cell culture.*

Useful Journals

Technique-Oriented Tissue Culture
Cytotechnology
In Vitro Cell and Development Biology
Methods in Cell Science
Tissue Culture Research Communications (Japanese)

Cell Biology
Cell
Cell Biology, International Reports
Cellular Biology
Cell Growth & Differentiation
Current Opinion in Cell Biology
European Journal of Cell Biology

Experimental Cell Biology
Experimental Cell Research
Journal of Cell Biology
Journal of Cellular Physiology
Journal of Cell Science
Nature Biotechnology

Cancer
British Journal of Cancer
Cancer Research
European Journal of Cancer and Clinical Oncology
International Journal of Cancer
Journal of the National Cancer Institute

Index